AF802886

Math for Teachers

An Exploratory Approach
Second Edition

Robert G. Stein
with Laura Wallace

Kendall Hunt
publishing company

Cover photo courtesy of Fotosearch.

Kendall Hunt
publishing company

www.kendallhunt.com
Send all inquiries to:
4050 Westmark Drive
Dubuque, IA 52004-1840

ISBN 978-0-7575-8106-9

Printed in the United States of America
10 9 8 7 6 5 4 3 2

To Roni, Joey, Justine, Lucy, and the next generation

Contents

Preface

This new edition reflects years of experience with the previous version.

- Material on geometry has been rearranged to flow more smoothly.
- Chapters on statistics and probability have been added.
- New examples, problems, and explorations have been added.
- Additional historical references are provided.
- Pictures, diagrams, and uses of manipulatives have been enhanced to provide more concrete foundations for abstract ideas and to model how the material can be taught in schools..

What has not changed is the goal of giving those who teach K-8 mathematics the opportunity to build a strong foundation for teaching mathematics and to experience the joy and satisfaction that can bring. Mathematical exploration, the heart of this book, is enhanced in this new edition with the inclusion of many new explorations.

Why the emphasis on exploration?

1. What we discover for ourselves is particularly vivid and meaningful.
2. Explorations offer an alternative to the false choice between conceptual understanding and computational skill. They allow students to experience the joy of mathematical investigation, even as they build the practical skills necessary to use mathematics in daily life.
3. Exploration is a paradigm for teaching as well as for learning. In a classroom based on exploration, the open-ended nature of the activity keeps everyone involved, thus freeing the teacher to work with students who need help. Without some such provision such for individual differences, "Leave no child behind" slips easily into "Let no child get ahead."

Criticism of the mathematics curriculum as "a mile wide and an inch deep" is particularly apt in the area of arithmetic. The hope that calculators would obviate the need to master arithmetic and thus free up time for higher-level thinking was used to excuse a lack of attention to arithmetic in schools. However, students often accepted uncritically results of electronic calculation, thereby failing to exercise the higher thinking skills they were supposed to develop. Today it is clear that the original premise was false. Students need to master both basic skills and underlying concepts in order to use mathematics effectively, even with calculators. For example, to check an electronic calculation of a square root, they need to understand what square roots are conceptually, and they need to be able to estimate reasonably, which, of course, involves arithmetic skills. Therefore, we make arithmetic the primary focus of our first eight chapters, recognizing it as a rich and fascinating area of mathematics in its own right, and certainly much more than a collection of procedures that can be dispensed with by using a calculator.

We highlight analogies between algebra and arithmetic as a way to shed light on both subjects. That way, algebra grows naturally out of arithmetic and at the same time reinforces understanding of arithmetic. It also helps teachers see how the all the mathematics they teach, including geometry, probability, and statistics, is related to the mathematics taught at other levels of the curriculum.

Because the subject is immense and open ended, this book inevitably includes more material than can be covered in a year. We hope that in future years you revisit the mathematics in this book and explore further. That will enrich your teaching and add interest for your students.

It is a pleasure to acknowledge the contributions of others to this project. Among the many students and colleagues who contributed to this new edition, Laura and I are especially grateful to Joyce Ahlgren, Davida Fischman, Richard Hobson, Shawn McMurran, and Larry Stevenson. Jon Hughes Fuller at Bent Tree Press and Jessica Canavo at Kendall/Hunt strongly supported the project, as did Peter Williams, chair of the CSUSB math department and the department staff, especially Nancy Pennington, Paula Cramer, and Leeann Hartson. Perhaps most critically of all, this project benefits from Elena Strzheletzka's careful proofreading and Jen Wreisner's skillful production. To all these people we are deeply grateful. We could not have done this alone.

We are pleased to present this new edition and hope that you, who read it, will pass on to your students the joy and fascination of this wonderful but challenging subject.

How to Use This Book

When you tell family and friends about this class, you will often hear comments like, "Oh, I was never good at math." Maybe you feel that way yourself. But this book is different from other math books. Therefore some ideas on how to use this book are in order.

You already know some of the mathematics in this book.
Even if your own experiences with mathematics were dreadful, you will find a great deal you already know. Here you will revisit the material from a different point of view. You will find many opportunities deepen your knowledge and to fill in gaps along the way.

Teach material from this book to others. Helping classmates will help you learn the mathematics yourself. Teaching the material to a class of your own will help you even more.

Explorations are the heart of the book
Do as many explorations as you can. Write out your work completely and carefully, on full-sized paper, as if you were doing it for a school assignment. Since the explorations are open-ended, they typically include some questions that are difficult or at least require considerable thought. Therefore, do not feel that you must finish each exploration before moving on or starting another one. Save your work to revisit in the future.

Use the Answers and Hints at the back of the book. This section is there to help you check your work and to give you hints when you feel stuck or unsure. Do not hesitate to use it.

Make this book part of your personal collection of ideas and approaches for teaching mathematics. This is a math book, not a "methods" book, but it is written for teachers. The focus is mathematics, but many issues of teaching and the curriculum are included. Ideas are presented in ways that you can easily adapt for your own teaching. Assemble a collection of ideas, perhaps in folders or in a large ring binder, sorted by topic (e.g. Adding fractions). Put parts of this book in the folders, and supplement them with ideas from other sources, such as the internet, other books, professional journals, and conferences you attend. Over the years, this collection will evolve into a treasured set of teaching resources.

And if someone makes a negative comment about mathematics, we hope you respond with a gentle invitation to teach them some of the ideas you learn here.

Bob Stein
bstein@csusb.edu

Laura Wallace
Wallace@csusb.edu

Chapter One

The Exploratory Approach

The teacher who would educate the mathematical imagination of pupils must begin by cultivating his own.
–Mary E. Boole, *Preparation of the Child for Science*, Oxford, 1904.

This chapter introduces exploration as an approach to teaching and learning mathematics. It sets the tone for this entire book. Even if you are familiar with many of the ideas you meet here, you may nevertheless gain new insights into them as you go along.

1.1 Box Puzzles: Exploring Basic Properties of Addition

Adding across the rows and down the columns of a little box like figure 1.1.1 leads to some interesting discoveries.

2	8	
7	5	

Figure 1.1.1

Add 2 and 8 across the top row, and put 10 in the upper right-hand corner. Next, add 7 and 5 in the middle row and put the sum, 12, in the square just below the 10, as in figure 1.1.2. (Numbers being added are called **addends**. The result of adding is called the **sum**.)

2	8	10
7	5	12

Figure 1.1.2

Now add the original numbers down the columns, leaving only the lower right-hand corner empty, as in figure 1.1.3.

2	8	10
7	5	12
9	13	

Figure 1.1.3

How should this box be filled in? You might add the 10 and the 12 down the right-hand column and put 22 in the box. If you add the 9 and 13 in the bottom row, you also get 22, so the puzzle will be completed as in figure 1.1.4.

2	8	10
7	5	12
9	13	22

Figure 1.1.4

You might think filling in the box ends the exercise, but we take a different point of view. Instead of the box puzzle itself, think of the questions that it can lead to. For example, we got the same number, 22, by adding down the right hand column as we got adding across the bottom row. Is this a numerical coincidence? One way to approach this question is to repeat the experiment with different numbers, starting with figure 1.1.5.

3	13	
10	12	

Figure 1.1.5

Adding across the rows and down the columns yields figure 1.1.6.

3	13	16
10	12	22
13	25	

Figure 1.1.6

Now, adding down the right-hand column yields 38, as does adding across the bottom row. There is indeed something to investigate!

Initially, an exploration is like experimental science. Gather evidence by experimenting with a variety of box puzzles. Each of our first two examples began with two even numbers and two odd numbers. Is that significant? One way to find out is to try a puzzle with, say, all odd or all even numbers. Then try other variations to find out when the sum down the right-hand column is the same as the sum across the bottom row.

- What if one or more of the original four numbers is zero?
- What if very large numbers are involved?
- What if some or all of the original four numbers are fractions or decimals?
- What if some or all are negative numbers?
- What if addition is replaced throughout by subtraction or multiplication?
- What if the format is altered, for example, as in figure 1.1.7?

5	10	13	
1	17	20	

3	11	2	
15	4	7	
8	6	10	

Figure 1.1.7

Explore these questions, making up examples and completing them. If your questions lead you into areas of mathematics which you do not know or are rusty at, skip them. We will review those areas in detail as we come to them. **Read further only after you have actually worked a variety of puzzles, exploring the above questions as much as possible on your own.**

Are you convinced yet that more than coincidence is involved? Did you notice that for addition box puzzles the number in the lower right hand corner is the sum of the numbers you began with? In our first puzzle (figures 1.1.1-1.1.4), for example, 22 is the sum of the original four numbers, 2, 8, 7, and 5. Adding first down the columns, we found 22 as (2+8)+(7+5) or 10+12, and adding first across the rows we found 22 as (2+7)+(8+5) or 9+13. Similar reasoning applies to other examples. At this point, if you know some algebra, you might consider a more general example, calling the four initial numbers a, b, c, and d, as in figure 1.1.8.

a	b	
c	d	

Figure 1.1.8

Adding across rows and down columns now yields figure 1.1.9.

a	b	$a+b$
c	d	$c+d$
$a+c$	$b+d$	

Figure 1.1.9

To complete the puzzle, add across the bottom row to get $(a+c)+(b+d)$, or add down the right-hand column to get $(a+b)+(c+d)$. These numbers are the same no matter what numbers a, b, c, and d are, because neither grouping nor order has any bearing on the final result of addition. What at first looked like coincidence is always true, a consequence of two basic properties of addition:[1]

1. Grouping does not affect a sum. This is the **associative property of addition**, symbolized for any three numbers p, q, and r, as

$$(p+q)+r=p+(q+r).$$

In words, "For any numbers p, q, and r, adding r to the sum of p and q is the same as adding the sum of q and r to p."

2. Order does not affect a sum. This is the **commutative property of addition**, symbolized for any two numbers r and s as

$$r+s=s+r.$$

In words, "For any numbers r and s, the sum of r and s is the same as the sum of s and r."

[1] The associative and commutative properties were not explicitly named until the 19th century, when mathematicians began to reexamine the basic assumptions underlying arithmetic. Use of the plus and minus signs to denote addition and subtraction originated in Germany during the late 15th century. The plus may have originated as a short form of the Latin word "et," meaning "and."

It is easy to confuse these two properties. The commutative property involves the order in which the numbers are written, whereas the associative property deals only with grouping. The distinction becomes clear in English. If the English language were commutative, "race horse" would mean the same thing as "horse race" and "house cat" would mean the same as "cat house." If the English language were associative, Ella Wheeler Wilcox would not have been upset when her poem, "My Soul Is a Lighthouse Keeper" appeared in print as "My Soul Is a Light Housekeeper."

The associative and commutative properties of addition are indeed the keys to understanding box puzzles, as far as mathematicians are concerned. Kids, however, are not likely to be convinced by an appeal to abstract properties. Another explanation, which has more intuitive appeal to kids, comes from thinking of the numbers in the puzzles as numbers of objects. For example, think of figure 1.1.10 as a pictorial version of figure 1.1.1.

2	8	
7	5	

Figure 1.1.1 (again)

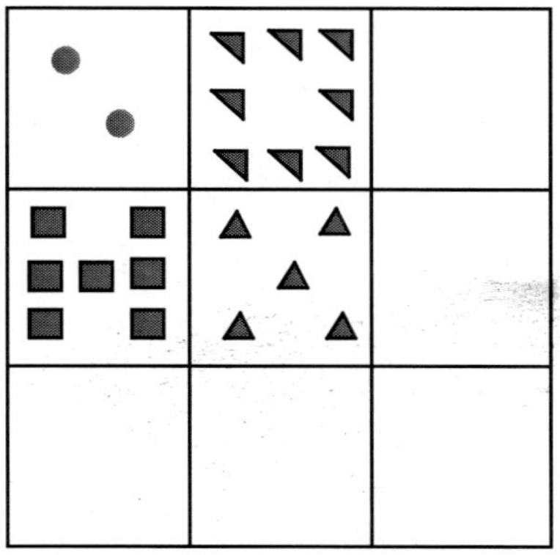

Figure 1.1.10

With this interpretation, adding across the top row may be pictured as putting all the objects from the top row into the upper right hand corner. Similarly, adding across the middle row puts the seven squares and five triangles into the middle square in the right hand column, as in figure 1.1.11.

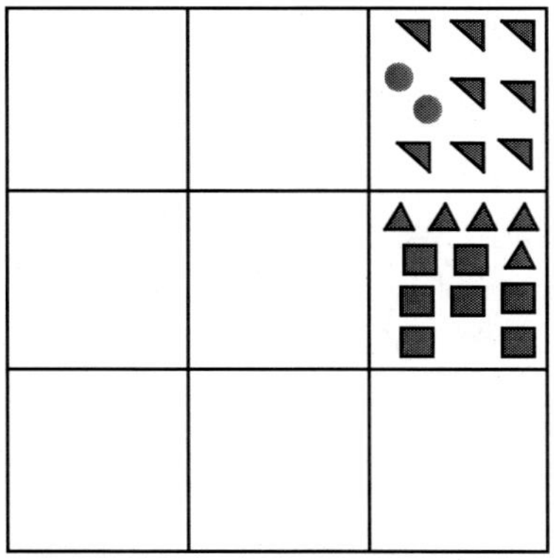

Figure 1.1.11

Next, adding down the right-hand column puts all of the objects, which were originally in the four upper left-hand squares, into the lower right-hand box. That is why the number we get in the lower right hand corner is the sum of the four numbers we began with. If you add first down the columns and then along the bottom row, you still end up with all the original objects in the lower right-hand corner. This reasoning holds regardless of how many objects are in the upper left boxes; that is why you always get the same sum along the bottom row and down the right-hand column.

Using this exploration in a classroom
Which of the two explanations above is better?

Advantages of the second explanation:

1. It is down-to-earth and concrete, where the first explanation is abstract. Kids find it relatively easy to understand, especially if they actually put objects down in boxes, move them around, and count them, as suggested in figures 1.1.10 and 1.1.11.
2. This explanation works equally well with puzzles, which, like those in figure 1.1.7, have various numbers of rows and columns.

The second explanation has the disadvantage of being restricted to box puzzles with whole number entries, where the operation used is addition. Only in such cases can addition be interpreted in terms of putting together collections of objects.

Advantages of the first explanation:

1. It is concise.
2. It is not restricted to whole numbers but works equally well with fractions, negative numbers, etc.
3. It shows clearly the role of the associative and commutative properties.
4. We will soon see that multiplication, too, is associative and commutative, so if you replace all the additions in a box puzzle with multiplications, the box puzzle will still work.

The concrete nature of the second explanation makes it the easier one for beginners, but ultimately each explanation has its unique advantages.

Just what to explain, as well as how and when, depends on the students. For very young children, an explanation in terms of the associative and commutative properties, especially when it involves the use of letters to stand for numbers, is meaningless.

Perhaps the most important thing about explanations is not to rush into them. Exploring box puzzles provides excellent practice. Why cut it short? Students stop exploring as soon as the mystery is cleared up. Robert Wirtz, who used box puzzles extensively in the *Math Workshop for Children*, suggested using them as follows in grades 3 to 6:

1. Introduce the puzzles somewhat as we did here.
2. Ask each student to make up a few box puzzles that work.
3. Ask if the box puzzles always work, and challenge the students to find one that doesn't.

The resulting student activity may continue at home or in subsequent classes. The classroom need not be silent; allow students to consult with neighbors. Move about the room, looking at student work, observing their experiments. Students with little confidence in their mathematical skills limit their experiments to puzzles with small, whole numbers. More confident students move on quickly to examples they find challenging. These observations give you valuable information about the students without diagnostic testing, which many students find stressful. At the same time, students get plenty of practice in basic skills at levels they each select. You are spared the nuisance of reading and correcting student papers, since the only ones you read at all are examples which kids say do not work. Even these you need not grade; just find the errors, and ask the kids to check their work in the appropriate row or column ("Is 17+45 really 52?") and then see if the puzzle works. This puts you in a role of helper rather than judge.

Reflecting back on this exploration
The foregoing exploration models sensible priorities in mathematics teaching. Even in an age of calculators and computers, everyone needs basic computation skills, and these in turn require practice. But there is an important distinction between useful practice and senseless drill. Well-chosen practice can increase both understanding and skill without killing curiosity, whereas poorly chosen practice, all too common in the form of drill exercises, can kill motivation without increasing comprehension. How can you tell the difference?

In sensible practice:

a. Students have a voice in deciding which problems to work on and even make up their own problems.
b. The questions have built-in checks to help students evaluate their own work without consulting an external authority such as a teacher or an answer sheet.
c. Practice is not an end in itself but instead is built into a larger question that has intrinsic appeal.

In drill exercises, the situation is reversed:

a. Students play no role in deciding what they work on.
b. The questions have no built-in checks. The teacher or book becomes an authority figure, and mathematics is seen as an authoritarian subject.
c. The skill being drilled is the goal; the student does not see the work in a broader context.

The role of exploration
Explorations offer a paradigm for how to deal with the wide range of abilities and attitudes toward mathematics often found in one class. Teachers often feel forced to choose between boring those who find the material easy and leaving behind those who are struggling. Explorations provide a third option. They begin in a straightforward manner; everyone who does the basic portions of an exploration can be considered to have succeeded at a basic level. However, those who find those basic goals easy need not sit around bored, because the explorations offer plenty of additional challenge. In this book, begin as many explorations as you can, and carry them as far as your time and interest allow.

Exercises 1.1

1. You often use the associative and commutative properties in mental arithmetic. For example, to add (87+71)+13, if you first use the commutative property to write this as (71+87)+13 and then use the associative property to write that as 71+(87+13), you get 71+100 or 171.

   Which of the following involve the associative property and which the commutative?

   a. 37+65=65+37 b. (25+48)+12=25+(48+12)

   c. 36+(50+24)=36+(24+50) d. 11+(19+7)=(19+7)+11

2. A bidder at an art auction cabled a wealthy client for advice. The reply, NO PRICE TOO HIGH, was cabled without punctuation, which changed the meaning. What was missing? This example shows again that English is not associative. Another example: A child misunderstood the question, "Is your dog house broken?" Explain, and try to think of other such examples.

3. Why do these exercises not begin with several box puzzles for you to practice on, with answers at the back of the book?

4. "As you struggle through life this may help you fight longer: Anything that doesn't kill you outright only makes you stronger."

   Piet Hein, Grooks[2]

   "If you want to help someone learn, double the rate at which he makes mistakes."

   Attributed to Henry Ford

   Comment on the view that schools should be places where people can safely make mistakes and learn from them.

5. If your class has been exploring box puzzles, how can you decide when they are ready to move on to another topic?

1.2 A Closer Look at Addition

Fundamental to the process of learning is each learner's personal reinvention of each concept. To foster this learning of basic number concepts, provide plenty of opportunities for young children to use numbers in varied settings. Some favorites are games like dominos, with their pictures of small numbers, and games that involve hopping along a path. Board games that use dice to determine the moves are particularly rich in learning opportunities:

1. Dice, like dominos, help beginners to picture small numbers.

[2] Piet Hein, noted Danish mathematician and scientist, is best known for his invention of the board game hex. "Grooks" were popular in Denmark as spontaneous expressions resistance to German occupation in World War 2. Later Hein published a collection of them in a book called, naturally, *Grooks*.

2. Adding the numbers on a pair of dice is good practice with basic sums.
3. Hopping an object on a board while counting fosters kinesthetic learning.

Many more sophisticated manipulatives are available today. Two of these, Cuisenaire rods and base ten blocks, are introduced briefly below as offering ways to model addition.

Cuisenaire rods are colored rectangular blocks with lengths from 1 to 10 centimeters (cm). If these rods are new to you, take time to examine them, compare them, and arrange them in patterns before setting out to use them in mathematics[3]. Try building a staircase using one rod of each color. Note that there is a different color rod for each length from 1 unit to 10 units long. Figure 1.2.1 shows the first five steps of the staircase. When you complete the staircase, draw a scale model on graph paper and color it.

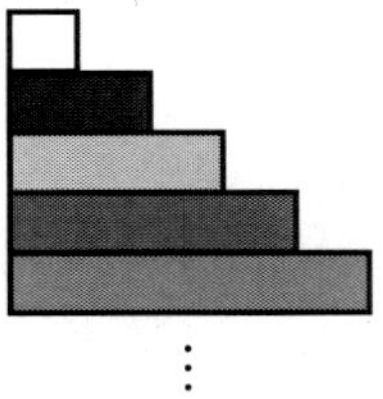

The 1-rod is white.
The 2-rod is red.
The 3-rod is light green.
The 4-rod is purple.
The 5-rod is yellow.
⋮

Figure 1.2.1

Rods laid end to end are called a **train** of rods.

Figure 1.2.2 uses trains to show 2+3=3+2, which illustrates the commutative property of addition. Both trains have length 5.

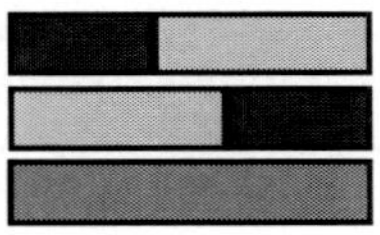

Figure 1.2.2

Similarly, figure 1.2.3 shows with rods that 2+(3+4)=(2+3)+4.

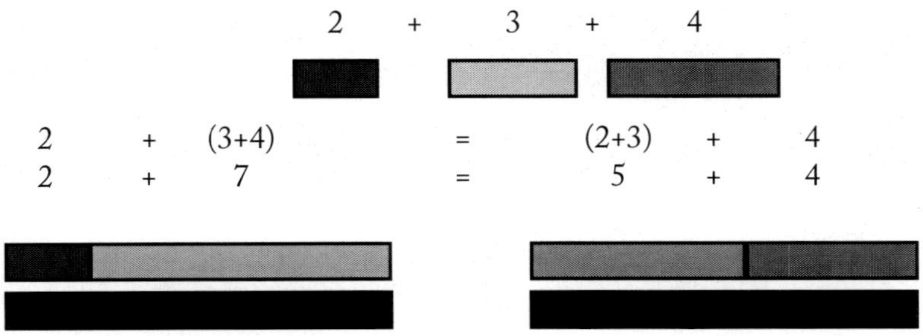

Figure 1.2.3

The 2+(3+4) and (2+3)+4 trains have the same length, which illustrates the associative property of addition.

[3] Young students need considerable time to become familiar with the rods. If they are given time, they will play with them like blocks. This play, however, involves quite a bit of learning and builds the familiarity necessary to use the rods as a tool for learning mathematics.

Base-ten blocks

We can hold mental pictures of numbers only if they are relatively small. To count a large pile of identical 236 little cubes (also called **units**), for example, is both tedious and subject to error, but if they are grouped as two square arrays of 100 cubes each, three linear arrays of 10 cubes each, and six extra cubes, you can easily see that there are 236 cubes in all. This is the basis of the decimal system of numeration, which **base-ten blocks** (figure 1.2.4) model concretely. Here we focus on how base ten numeration influences procedures for addition. A deeper, more detailed treatment of numeration is in chapters 5 and 8.

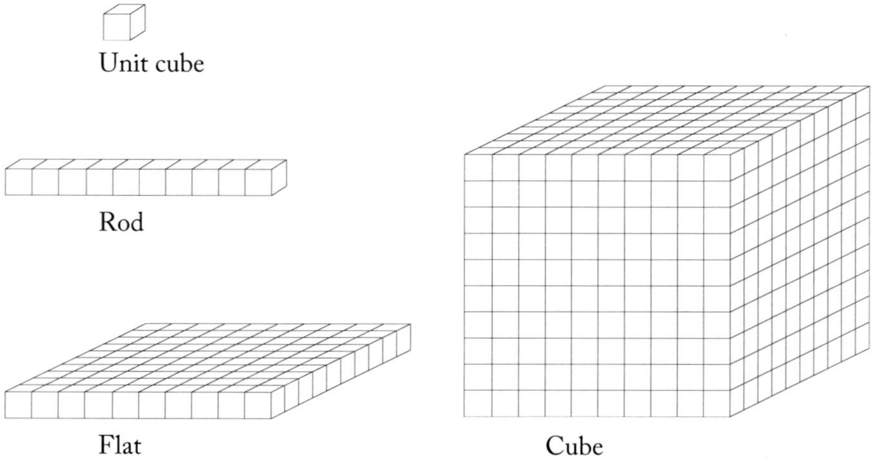

Figure 1.2.4

As with Cuisenaire rods, allow some time to get acquainted with base-ten blocks before trying to do mathematics with them. Note that 10 units equal 1 rod, 10 rods equal 1 flat, and 10 flats equal 1 cube. Then, when at last you begin to do mathematics, focus first on:

1. Addition questions which do not need carrying
2. Exchanges without addition, such as those below.

1. Gina has 27 units. If she exchanges twenty of them for two rods, she will have two rods and seven units.

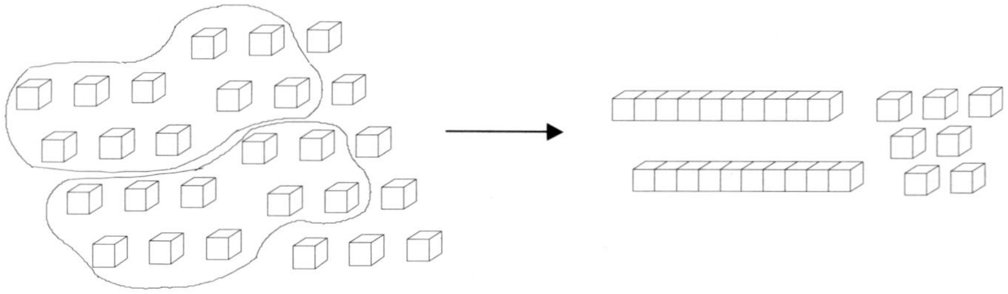

2. If Colby has 12 rods, he can exchange ten of them for a flat, giving him 1 flat and two rods.

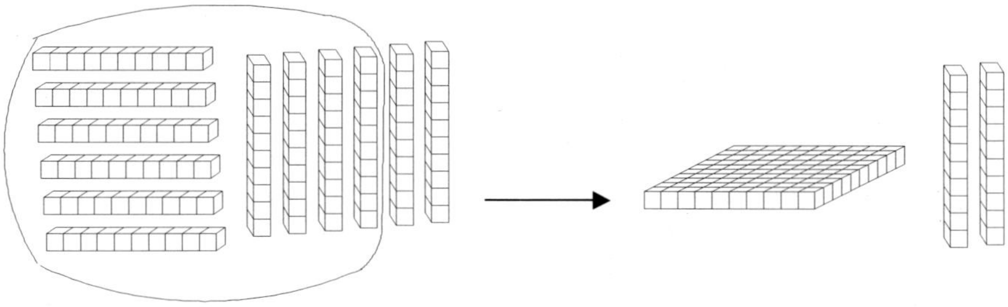

3. Maria has 2,532 units. Since ten units make a rod, she can exchange 2,530 units for 253 rods.

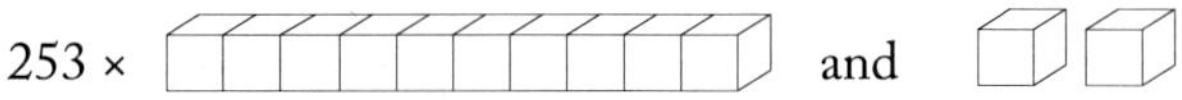

Now, since ten rods make a flat, Maria can exchange 250 rods for 25 flats.

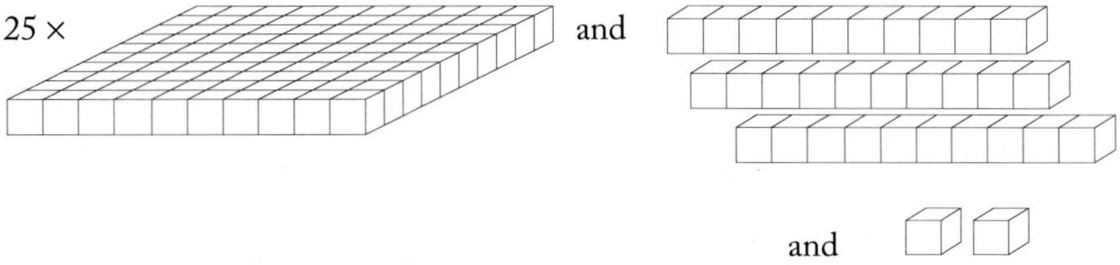

Finally, Maria can exchange 20 of her flats for 2 cubes, giving her a total of 2 cubes, 5 flats, 3 rods and 2 units.

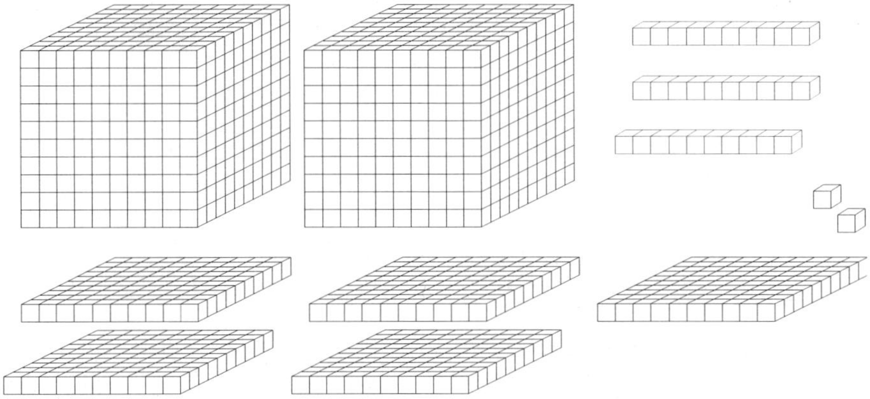

When students can represent numbers in several different ways with base-ten blocks, they are ready to model additions that involve carrying.

In the United States today, addition is usually taught by the following procedure: (It helps to write the numbers above each other, so that their units digits, tens digits, and any other digits of the addends lie in vertical columns of like type.)

1. Add the units digits. If the total of these digits is greater than 9, write the units digit below the units digit of the addends, and "carry" the excess multiple of ten, writing it above the tens digits (if any) of the addends.

2. Repeat step one for the tens digits, adding in anything that was carried from step 1.
3. Repeat step 2 for digits successively further to the left, until there are no more digits to add.

Example 1 shows how to model this procedure.

Example 1 Model 38+26 with base-ten blocks and show how this relates to the "standard" algorithm for addition (adding the units column, then the tens, and so on working right to left).

Solution:

Base-Ten Blocks	Shorthand for Picture	Algorithm
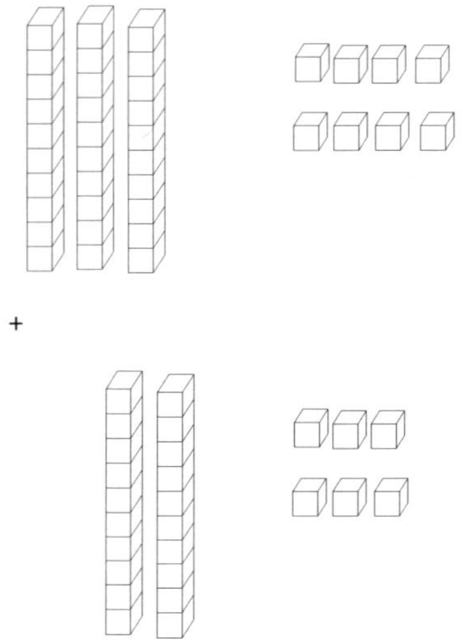	3 tens + 8 ones plus 2 tens + 6 ones	tens ones 3 8 + 2 6
Combine like pieces 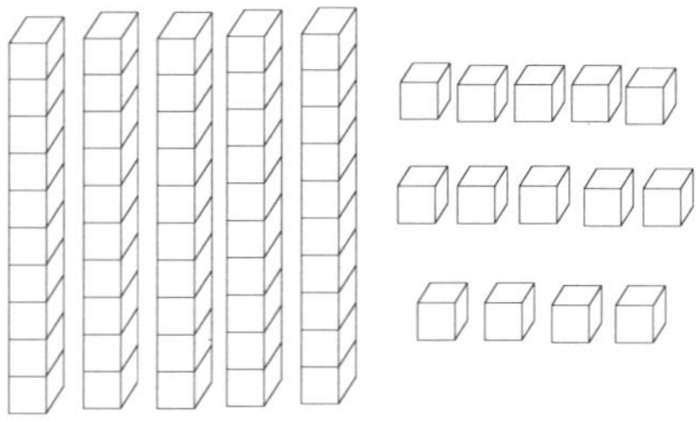	5 tens + 14 ones	tens ones 3 8 + 2 6 5 14
Trade 10 ones for 1 ten 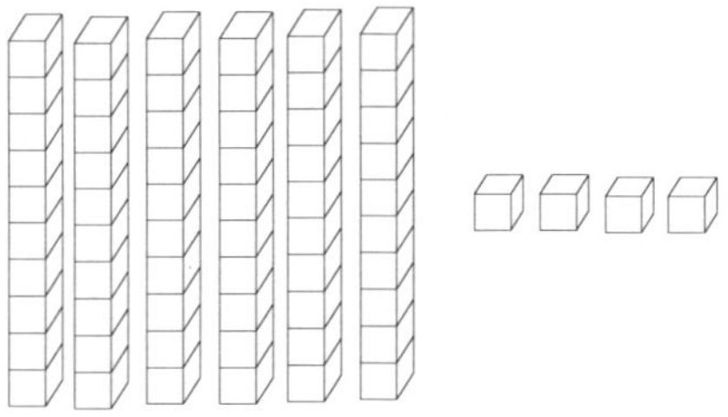	5 tens + 10 ones + 4 ones equals 5 tens + 1 ten + 4 ones equals 6 tens + 4 ones equals 64	tens ones 1 3 8 + 2 6 6 4

This **algorithm** (a systematic, step-by-step procedure)[4] is so widely used that teachers are sometimes surprised to find their students using any other procedure to add. However, there are other correct algorithms for addition, and it helps to be aware of them. Here is a brief sampler:

1. Add by columns, but work from left to right:

   $$\begin{array}{r} 38 \\ +\underline{26} \\ 5 \\ \underline{14} \\ 64 \end{array}$$

   First add the tens digits to get 5, then add the units to get 14. Carrying the ten units as another ten yields 64.

   Many Europeans learn this "left to right" algorithm for addition, and they appear to perform and understand addition as well as those who learn it from right to left.

2. Use a box puzzle, as in section 1.1, to add the units and tens separately

30	8	38
20	6	26
50	14	64

3. Add the tens from the second number, then the ones:

   38+26 = 38+(20+6) = (38+20)+6 = 58+6 = 64

4. Make the first number a multiple of ten using the second number:

   38+26 = 38+(2+24) = (38+2)+24 = 40+24 = 64

5. Add the closest multiple of 10 of the second number to the first number, then compensate:

   38+30 = 68 (This is 4 too many, since 30=26+4, so subtract 4)

   68 - 4 = 64

Adding 26 by adding 30-4 is a bit like paying for a 26 cent item with 30 cents and getting 4 cents change. This method is the basis for the trick described below in question 5.

Regardless of which algorithms you teach or use, you need to be aware of other algorithms for addition, if only to be able to figure out whether your students' work is correct.

[4] The word "algorithm" (also sometimes written "algorism") is believed to come from the Latin translation of an Arabic text, which begins with the words, "Dixit Algorismi," which means "Algorismi says." The writer of the Arabic text was Muhammad ibn Musa Al-Khowarizmi (c.780-850) a scholar at the House of Wisdom (see section 5.4), where he wrote the *Kitab al-jam'wal tafriq bi hisab al-Hind (Book on Addition and Subtraction after the Method of the Indians)*, which brought decimal numeration to the Muslim world and later to the West, where it is sometimes inaccurately called Arabic numeration. Al-Khowarizmi also wrote on algebra and geography.

Some educators, noting that algorithms are often used by rote with little understanding, have suggested that they not be taught in schools, and that instead children invent their own approaches to computation such as adding and multiplying. This is an extreme position, which leaves teachers the heavy burden of trying to evaluate the hodgepodge of individual ideas that a class can produce. The point of an algorithm is to reduce or eliminate the need for creative thought, thus saving time and effort for other questions. Basic calculation should not preoccupy students or teachers as a problem-solving activity indefinitely; at some point, after whatever initial exploration is appropriate, algorithms should be introduced, explained, and used.

Exercises 1.2

1. In the text, pictures of Cuisenaire rods show that 2+3=3+2 and that (2+3)+4=2+(3+4). It is stated that these examples illustrate the commutative and associative principles for addition. The word "illustrate" was carefully chosen. Why would it be incorrect to state that these examples **prove** those properties?

2. Fill in the "exchange" chart below. Some of it has already been done for you.

Start with this number of units	After exchanging units for rods		After exchanging rods for flats			After exchanging flats for cubes			
	Rods	Units	Flats	Rods	Units	Cubes	Flats	Rods	Units
2,532	253	2	25	3	2	2	5	3	2
503	50	3	5	0	3	0	5	0	3
1,786									
						9	0	8	7
			42	3	1				
	786	0							
		6	12					5	
11									

3. Use base-ten blocks to model the standard algorithm for 326+187. Draw pictures of the blocks, and explain each step. Also give a written translation using expanded form or place value columns.

4. Show how to use two alternate algorithms (not the standard one) to add 326 and 187.

5. This trick is a source of both entertainment and practice. It also reinforces an interesting basic lesson about addition. Ask two students to write four-digit numbers, such as the last four digits of their phone numbers, on the board, one below the other. They might look like this:

   7,556 (first student's number)
   5,120 (second student's number)

   Now, appearing causal, write a number yourself.

   7,556 (first student's number)
   5,120 (second student's number)
   4,879 (your number)

   Ask for a third four-digit number from the class and, finally, add another four-digit number yourself. You may now have something like this on the blackboard.

   7,556 (first student's number)
   5,120 (second student's number)
   4,879 (a number you write)
   3,115 (third student's number)
   6,884 (a number you write)

Now challenge the class to add up these five numbers. While they work on that, write the sum, 27,554, on a piece of paper, then put it where nobody can see it, and walk away from it. When the class has finished the addition, have a student unfold your paper and read the sum. Challenge the class to figure out how you did it.

a. How did you do it? Hint: The numbers you wrote, apparently casually, were in fact carefully chosen. Find out how they were chosen by adding each of your numbers to the student number immediately above it. A quick way to add 9,999 is to add 10,000 and subtract 1.
b. Can you explain why this works?
c. How would you modify the trick if a student contributes the number 9,999?
d. How could you modify the trick for a class which can add only two-digit numbers?

6. This old favorite is a challenging, if unwelcome message:

```
  SEND
  MORE
------
MONEY
```

Each letter stands for a digit in a correct addition. Different letters stand for different digits, and repeated letters all stand for the same digit. (For example, if one M is 7, so is the other M, and other letters, such as E, cannot be 7.) What is the addition?

1.3 Multiplication and an Interesting Application

For whole numbers, multiplication is just repeated addition of the same number. For example, 3+3+3+3+3 is a sum obtained by repeated addition of threes. This is called 5 times 3 or five threes. Finding this sum is called **multiplying** 3 by 5. Numbers that are multiplied are called **factors**, and a number obtained by multiplying is called a **product.** In our example, 3 and 5 are factors, and 15 is the product. In elementary schools the symbol × is usually used for multiplication (3×5), but in more advanced, work a centered dot is often used, perhaps to avoid confusing × with the letter x.[5] Write the dot high enough not to look like a decimal point (3·5).

You can picture multiplication of whole numbers with objects in a rectangular array. For example, figure 1.3.1, an array with three rows of five dots each, shows 5+5+5, or 5 multiplied by 3.

Figure 1.3.1

You can also view this as five columns of three dots each, or 3+3+3+3+3, or 3 multiplied by 5. You may think of the array either as five threes or as three fives, as you wish. Either way, the number of dots is 15. More generally, if m and n are whole numbers greater than 0, we may picture the product $m \times n$ as a rectangular array, either with m rows and n columns or with n rows and m columns, so that $m \times n = n \times m$. In these easily pictured cases you can see that multiplication, like addition, is commutative. We assume this to be the case for all numbers.

Multiplication is also associative. Figure 1.3.2 shows this for the case, 2×(3×4)=(2×3)×4. To understand this figure, first look at the 2×3 array **ABCD**, which forms the left side of the figure. To count the balls in the figure,

[5] The tilted "St Andrew's cross" to denote multiplication was introduced in England shortly before 1620. The earliest use of the dot for multiplication was by Gottfried Leibniz in 1698. Leibniz (1646-1716) was a philosopher, lawyer. While on diplomatic assignment in Paris, he met other scholars and became interested in mathematics and science. He invented binary arithmetic, invented much of the calculus independently of Isaac Newton, and designed the first calculating machine that could add, subtract, multiply, and divide.

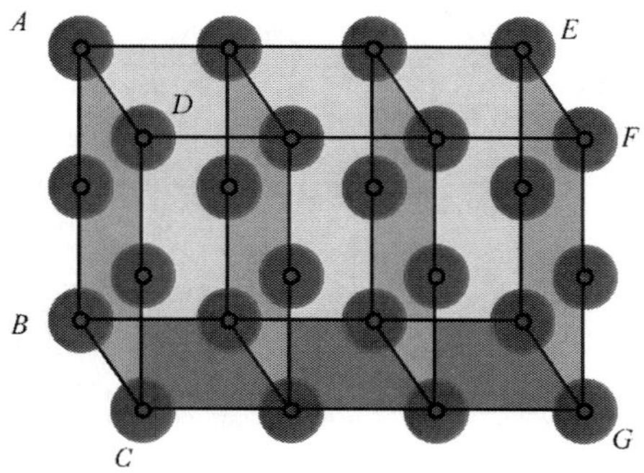

Figure 1.3.2

you can think of four such arrays or (2×3)×4. But you can also count the balls in this figure by starting with the 3×4 array **DCGF**. There are two such of these, so there are 2×(3×4) balls in all. Motivated by pictures such as this, we define multiplication, even for negative numbers, fractions, etc., so that it is always commutative and associative. The box below summarizes the commutative and associative properties as well as special properties of 1 and 0.

Commutative property of multiplication: For any numbers *a* and *b*,
$ab=ba$.

Associative property of multiplication: For any numbers *a*, *b*, and *c*,
$(ab)c=a(bc)$.

Two special numbers for multiplication:

Multiplication by 1 leaves any number alone.
In symbols, for any number *a*, $1\times a=a\times 1=a$.

Any number multiplied by 0 is 0.
In symbols, for any number *a*, $0\times a=a\times 0=0$.

It often helps to picture multiplication in terms of area. Area is usually measured in terms of unit squares of a specified size. If the squares are one foot on edge, the area is measured in square feet, but if the squares are one kilometer on edge, the area is measured in square kilometers[6]. Figure 1.3.3 shows how, if the length and width are whole numbers, the region inside the rectangle can be viewed as a rectangular array of unit squares.

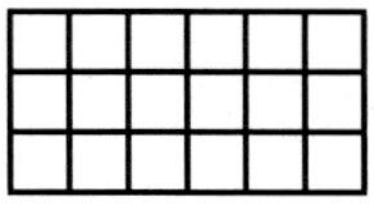

Figure 1.3.3

This leads to a general definition.

Definition: The area of a rectangle is the product of its length and width.

[6] Names of traditional area units sometimes hide their definitions. A section is a square mile, an acre is $\frac{1}{640}$ of a section, and a hectare is 10,000 square meters.

The mathematician Karl F. Gauss, is said to have found a clever use of the ideas behind box puzzles when he was only ten years old.[7] His teacher had him add up all the whole numbers from 1 through 100. He thought it would occupy Karl for a while, but the boy did the problem entirely in his head and got the correct answer almost instantly. The story may be apocryphal, but it is believable once you see the idea. Before reading on, take a minute or two to try to think up creative ways to add the whole numbers from 1 to 100.

You can get Gauss' idea from a simpler problem: Add the whole numbers from 1 to 7. You could do that in your head, but instead, look at the box puzzle in figure 1.3.4. The top row contains the numbers to be added, and so does the second row, only in the reverse order.

1	2	3	4	5	6	7	*
7	6	5	4	3	2	1	*

Figure 1.3.4

The completed box puzzle will have the sum 1+2+3+ . . . +7 in each starred box. (Three dots in mathematics mean that the pattern continues). You could find that sum by adding across a horizontal row; that is the direct, but naïve approach. Instead, add first down the columns and then across the bottom row. The sum is the same in every column. Since there are seven columns, the number that goes in the lower right-hand corner is 7×8. Then the sum we want is half of 56, or 28.

Gauss' idea has a geometric interpretation.

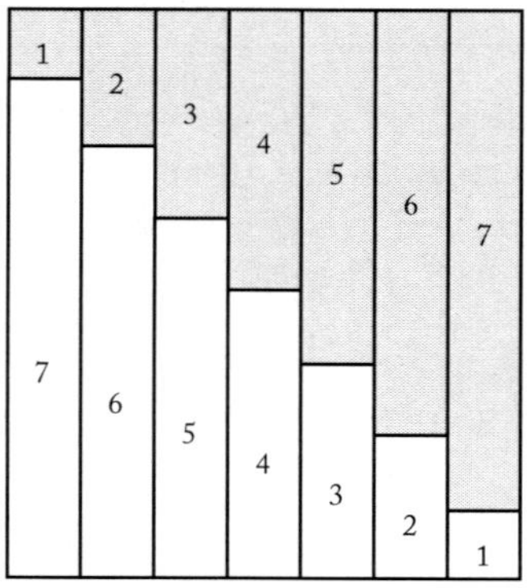

Figure 1.3.5

[7] Karl Friedrich Gauss (1777-1855) was born in Braunschweig, Germany, to a poor stonemason and a loving but uneducated mother. A local duke financed his education through the University of Göttingen, and Gauss later became the director of the observatory there, a post he held for the rest of his life. He made numerous and deep contributions in mathematics and physics. The "Gaussian distribution" in statistics and the unit of magnetic force, the Gauss, are named for him. He avidly read newspapers, magazines, and novels, as well as serious literature, often in the original Danish, English, French, Greek, Italian, Latin, Russian, Spanish, or Swedish. He made major contributions to virtually all areas of the astronomy, mathematics, and physics of his day. Adding the numbers from 1 to 100 is based on a story a colleague put into a memorial volume after Gauss died.

View each entry in the puzzle as the area of a rectangle one unit wide. Then each star denotes the area of a stair-step pattern in figure 1.3.5. This can, of course, be found by adding up the areas of the vertical strips, but the elegant approach is to observe that the two stair-step patterns have equal areas, so each has half the area of the rectangle formed by the two of them together. The rectangle is 8 units tall and 7 wide, so its area is 56, and the area of each stair-step pattern is half that, or 28.

Gauss' clever idea saves no work in adding 1+2+3+ . . . + 7, but it helps a lot on longer problems like 1+2+3+ . . . +100. Imagine a long box puzzle with the whole numbers from 1 to 100 written in ascending order along the top row and in descending order along the bottom row. Think of this puzzle as having been put on a long scroll of paper and then folded up as in figure 1.3.6.

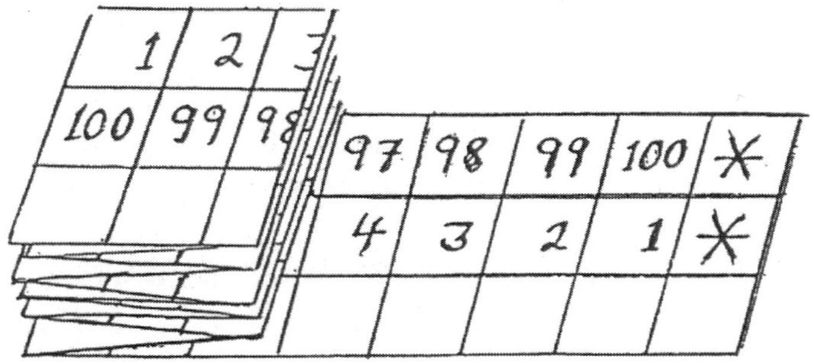

Figure 1.3.6

Once again, the number we seek will eventually go in each starred box, but now it would be tedious to add across either of the long rows. Adding down the columns is much easier; each sum is 101. Since there are 100 columns, the number which goes in the lower right-hand box is 100×101=10,100. Each starred box gets half this number, or 5,050, so 1+2+3+ . . . +100=5,050.

Example 1. Find the sum of the whole numbers from 1 to 1,000,000.
Solution: A box puzzle a million squares long staggers the imagination, but here are the beginning and end (the middle is torn out). Once again, all the

1	2		99,999	1,000,000	*
1,000,000	99,999		2	1	*

column sums are the same, in this case 1,000,001. There are a million columns, so the box in the lower right hand corner will contain 1,000,001×1,000,000 or 1,000,001,000,000. Each starred box gets half of that, or 500,000,500,000, and that is the sum 1+2+. . . +1,000,000. With Gauss' method the sum of the first million numbers is as easy to find as the sum of the first 100.

Example 2. Add up 11+13+15+ . . . +29 (odd numbers between 10 and 30).
Solution: As before, make a box puzzle with the numbers to be added written along the top row and, in reverse order, on the next row down.

11	13	15	17	19	21	23	25	27	29	
29	27	25	23	21	19	17	15	13	11	

Adding across the row might be tedious, but adding vertically you get 40 in each bottom cell, so the cell in the lower right is 10×40 or 400. Half of that is the desired sum: 11+13+15+ . . . +29=200.

Gauss' trick works whenever writing the numbers forwards and backwards in a box puzzle makes all the sums in the columns the same. That always happens when the numbers to be added increase by a fixed number. A list of numbers with that property is called an **arithmetic progression** or **arithmetic sequence.** (Are there other cases?) Adding up numbers in an arithmetic sequence is a standard topic in high school math, where most students memorize a formula for the sum, never understanding the elegant idea behind it. Why include this topic here? There are several reasons:

1. It is well within the reach of 5th and 6th graders, and they enjoy it.
2. It involves the practice of basic skills in an interesting context.
3. It illustrates beautifully the power of the ideas which underlie box puzzles, and indeed much of mathematics. Using only the most basic properties of numbers, you can add up all the numbers from 1 to a trillion, even though The United States has not existed as a nation for a trillion seconds. What power!

Exercises 1.3

1. What multiplications correspond to these pictures?

a. 

b. 

c. 

d. 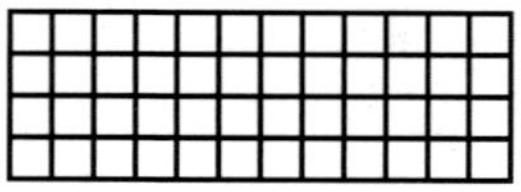

e. 

f. 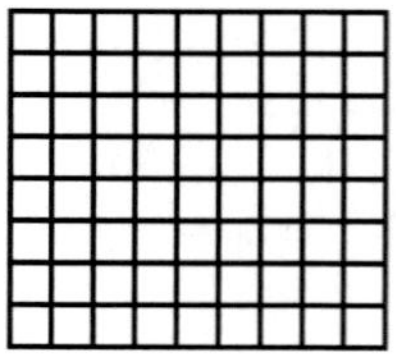

2. Draw arrays to picture: a. 3×7 b. 2×1 c. 5×8

3. Fill in these box puzzles, which illustrate multiplication as repeated addition.

a.

5	5	
5	5	

b.

3	3	3	
3	3	3	

c.

8	8	8	
8	8	8	
8	8	8	

4. a. How many square inches are there in a square foot? (There are 12 inches in a foot.)
   b. How many square feet are there in a square yard? (There are three feet in a yard.)
   c. How many square inches are there in a square mile?

5. SHARP SHARON: Choose your favorite big league baseball team. For the coming season, I'll add the number of runs they score in each game, and you multiply the same numbers. For example, in the first three games, if they score 3, 7, and 2 runs, I'll have 3+7+2=12, and you'll have 3×7×2=42. At the end of the season, I'll bet you my sum is higher than your product.

   GULLIBLE GULLIVER: Sure!
   Explain why Sharon is very likely to win the bet.

6. Verify by adding that $1+2+3+\cdots+7=28$.

7. Many adults see figure 1.3.5 as a very clear and concrete illustration of Gauss' trick, but youngsters will not understand this picture as readily. For them, even a picture is an abstraction, one step removed from the "reality" that it is supposed to show. Devise an even more concrete representation of the idea in figure 1.3.5 using physical Cuisenaire rods.

8. Find the sum $2+4+6+8+\cdots+14$ by the following three different methods, and check that all three ways yield the same sum.
   a. Use a box puzzle somewhat like figure 1.3.4. (Pay close attention to how **many** numbers are to be added.)
   b. Notice that the numbers to be added are just double the corresponding numbers illustrated in figure 1.3.4, so their sum ought to be twice what was found in that case.
   c. As a final check, compute the sum by direct, if slightly tedious, addition.

9. a. Draw a picture similar to figure 1.3.5 to show 11+13+15+ · · · +29.
   b. In the sum 11+13+15+ · · · +29, how many numbers are to be added?
   c. Compute the sum 11+13+15+ · · · +29.

10. A man walked 100 yards on January 1, 200 yards on January 2, 300 yards on January 3, and he kept on increasing his daily stroll by 100 yards per day for the rest of the year, which was not a leap year.
    a. How many yards did he walk on the last day of the year?
    b. What was the total length of these daily walks for the year?

11. Make up a word problem which, like problem 10, involves finding the sum of an arithmetic progression.

12. Add the whole numbers between 1 and 100 which end in either 2 or 8; that is, add 2+8+12+18+ · · · +92+98. Note: The numbers to be added in this case do not form an arithmetic progression, because they increase alternately by 6 and 4 rather than by a constant. However, Gauss' idea can be used here anyway. Here are three ways to do this:
    a. Separate the sum into two parts, one consisting of numbers which end in 2 and the other consisting of numbers which end in 8. (What do you observe about 2+12+22+ · · · +92 and 8+18+28+ · · · +98?)
    b. Even though the increases in the progression 2+8+12+18+ . . . +92+98 are not constant, write them into a box puzzle forwards and backwards and find the sum. You can still use Gauss' method. Explain!
    c. Use the associative property of addition to write the sum as (2+8) + (12+18) + … + (92+98) and then add the resulting arithmetic progression.

13. Add the whole numbers between 0 and 1,000 which end in 1, 3, 7, or 9. There are several ways to approach this problem, much as in question 12. Try to find at least two ways to do this, then do it more than one way.

1.4 Distributivity: Box Puzzles with Borders

Figure 1.4.1 shows an interesting variation on box puzzles.

	5	2	
3	*	*	
7	*	*	

Figure 1.4.1

The numbers along the upper and left-hand borders are keys for filling in the boxes. To fill in each starred square, multiply the border numbers for its row and column, as in figure 1.4.2.

	5	2	
3	15	6	
7	35	14	

Figure 1.4.2

For example, the 14 in the center box is 7×2, since 7 and 2 are the border numbers for the center box. Now fill in the remaining boxes by adding, as in earlier box puzzles, to get figure 1.4.3.

	5	2	?
3	15	6	21
7	35	14	49
?	50	20	70

Figure 1.4.3

Figure 1.4.3 has question marks at the bottom of the left-hand border and the right of the top border. To replace them, add along the top border and down the left-hand border, as in figure 1.4.4.

	5 →	2 →	7
3 ↓	15	6	21
7 ↓	35	14	49
10	50	20	70

Figure 1.4.4

The addition part of this puzzle is no surprise in view of section 1.1, but the connection between the numbers on the borders and those in the boxes is new. Note, for example, that the border numbers may be multiplied to fill in all nine boxes, and not only the four we did that way. Does this lead to new ideas or discoveries? What questions should you or your students ask at this point? (If you have trouble thinking of good questions, look at those in section 1.1.) Experiment with these box puzzles with borders for a while before reading further. Your experiments should include, at a minimum, making up several box puzzles with borders and trying to find out if and why they work.

To understand these puzzles, think of the products as areas. Figure 1.4.5 shows a large rectangle partitioned into four smaller rectangles. The lengths and widths of the smaller rectangles are the border numbers on the top and the left-hand sides of the large rectangle.

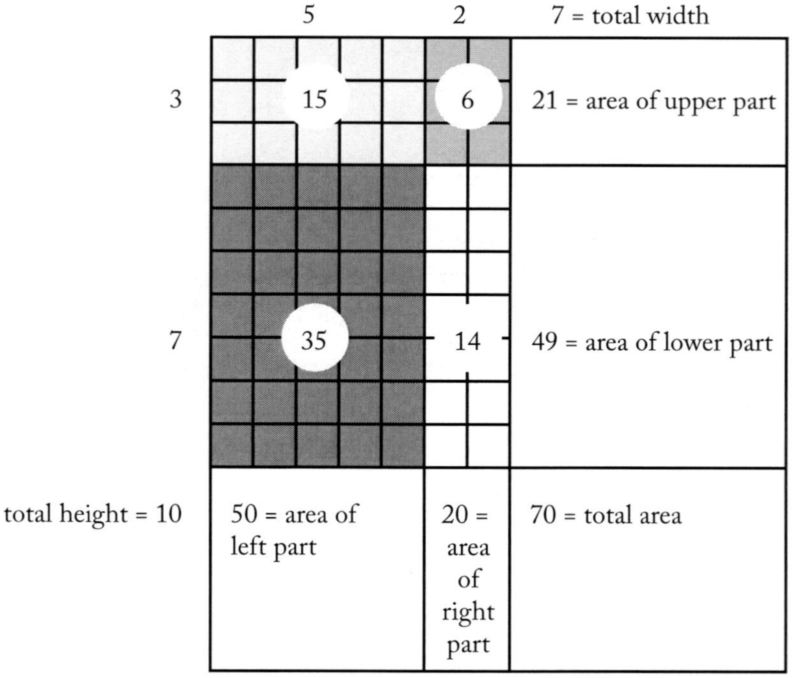

Figure 1.4.5

To find the area of the big rectangle in figure 1.4.5, you can add up the areas of the four smaller rectangles, each of which is found by multiplying its height by its width, or you can simply multiply the total width by the total height. Figure 1.4.5 is just figure 1.4.4 drawn to scale.

These puzzles illustrate the **distributive** property, which relates addition to multiplication:

> **Distributive Property: For any numbers a, b, and c,**
> $$a(b+c)=ab+ac.$$
> **Reminder, when no operation is indicated, multiplication is understood.**

Sometimes this property is written as (b+c)a=ba+ca, but order does not affect products, so this is really no different. Figure 1.4.6, which is the upper part of Figure 1.4.5, shows the link between areas of rectangles and the distributive property with a=3, b=5, and c=2.

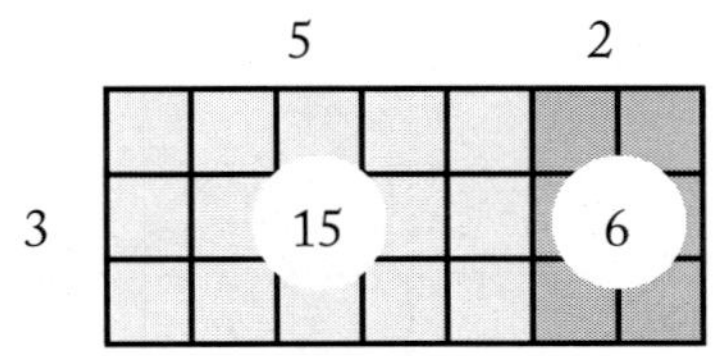

3(5 +2)= 3(5) + 3(2)

3(7) = 15 + 6

21 = 21

Figure 1.4.6

A word about parentheses: An English phrase such as "Three times five plus two" can reasonably be interpreted in two ways. On the one hand, it might mean 3×5 (which is 15) plus 2, which is 17. On the other hand, it might mean 3 times the sum 5+2, which is 3×7 or 21. In mathematics, clarity comes first, so we use the following convention.

> **CONVENTION ON ORDER OF OPERATIONS**
> **Unless otherwise indicated, expressions involving both addition and multiplication are to be evaluated by first doing the multiplications and then doing the additions. If there are several multiplications or several additions, they are to be carried out in order from left to right. To indicate a different order of operations, use parentheses.**

Then, for example, 3·5+2 means 15+2=17 and not 21. To make the 3 multiply both the 5 and the 2, write 3(5+2). That way the addition is done before the multiplication. The parentheses force us to view whatever they include as a single quantity. Other symbols, such as { } and [ ] are often used the same way.

Connection to Long Multiplication

Multiplications that are too long to do mentally are done digit by digit. The process, called **long multiplication**, is not difficult, but many people know it only by rote. As an example, consider 351×82. Write this as

$$\begin{array}{r} 351 \\ \underline{\times\ 82} \end{array}$$

and then say (possibly under your breath) something like this:

> "2×1=2, so write 2; now 2×5=10, so put 0 to the left of the 2 and remember the 1; 2×3=6, which, with the remembered 1 makes 7, which goes to the left of the 0. Now 8×1=8, which goes below the 0 of 702; 8×5=40, so put 0 to the left of the 8 and remember the 4; 8×3=24, which, with the 4 remembered, is 28. Finally, add (figure 1.4.7) to get 351×82=28,782."

What is really going on here? To see how the process works, compare figures 1.4.7 and 1.4.8. Using box puzzles in schools could help students understand long multiplication.

$$\begin{array}{r} 351 \\ \underline{\times\ 82} \\ 702 \\ \underline{28\ 08} \\ 28{,}782 \end{array}$$

Figure 1.4.7

	300	50	1	351
2	600	100	2	702
80	24,000	4,000	80	28,080
82				28,782

Figure 1.4.8

Example 1. Make a box puzzle to illustrate 325×107.

Solution:

	300	20	5	
7				
100				

Long multiplication may also be modeled with base-10 blocks, as in question 11 below.

Alternate Methods of Multiplication

As you can see, box puzzles provide a method for multiplication. This method involves more writing and is therefore slower than the usual long multiplication, but it is easy to understand. As a teacher, you may find it useful to be familiar with alternative methods, whether or not you use them very much with your students. You may teach students who learned to multiply by one of those methods, possibly in a different country, and you will want to understand what they are doing. At their heart, all these methods rely on the distributive property, but they differ in other details.

"Friendly Tens"

This method is convenient for doing relatively small multiplications mentally. For example, to multiply 31 by 24, think of 31 as 10+10+10+1. It is easy to multiply by 10 (just put a zero at the end of the number), so

$$\begin{aligned}24\times31&=24(10+10+10+1)\\&=240+240+240+24=720+24=724\end{aligned}$$

Lattice Multiplication

This method was widely used in Europe during the renaissance. To find 56×47, make a 2×2 array (since each factor has 2 digits), with each box cut diagonally as in figure 1.4.9 (parts a,b,c), then write the factors at the top and right as in figure 1.4.9b.

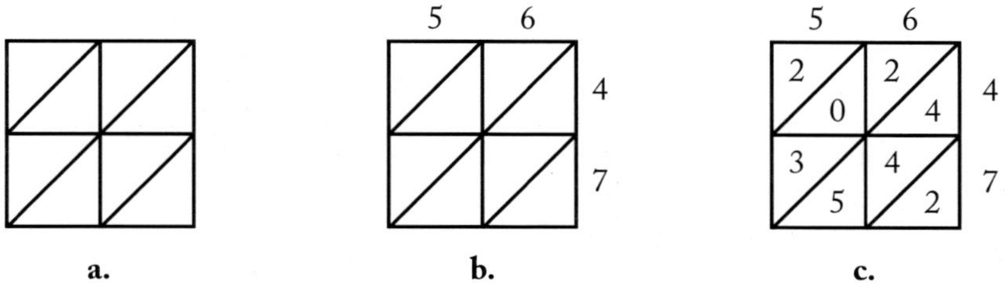

Figure 1.4.9

Next, multiply as you would in a box puzzle, but write units digits below the diagonals and the tens digits above them, as in figure 1.4.9c. Finally, add along the diagonals, starting with the 2 in the lower right-hand corner. The next diagonal has 4+4+5, or 13, from which the 3 is written, and the 1 is carried into the next diagonal addition (2+0+3) to get 6 in all. Finally, the 2 at the upper left has nothing to add to it. The product 56×47=2,632 is bold in figure 1.4.9d.

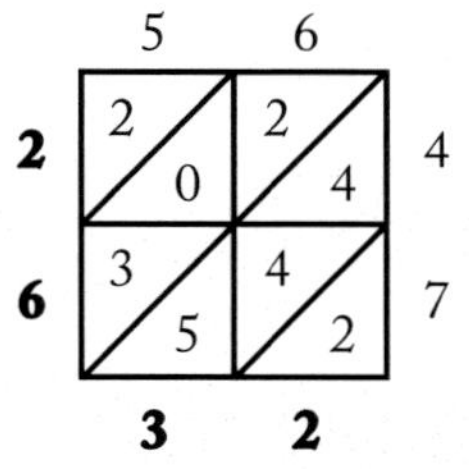

Figure 1.4.9d

Successive Doubling

This method is described in an Egyptian papyrus from around 1650 B.C.E. Again consider 56×47. Regard the problem as one of putting together a large number of 47s, and begin by doubling to find:

1×47=47
2×47=94
4×47=188
8×47=376
16×47=752
32×47=1,504

Stop here, since the next step would be 64×47, which is more 47s than wanted. By trial and error, 56=32+16+8, so add the corresponding multiples: 56×47= 1,504+752+376=2,632. This doubling method is slow, but it might help students who are having trouble learning basic multiplication facts.

Exercises 1.4

1. a. Fill in the bottom row of figure 1.4.8.

   b. The partial products 28,080 and 702, which correspond to 80×351 and 2×351, respectively, appear in figures 1.4.7 and 1.4.8. On the other hand, the partial products 24,600, 4,100, and 82, which appear in the bottom row of figure 1.4.8, do not appear at all in figure 1.4.7. Why not?

2. Make a careful sketch, using graph paper if possible, to illustrate the following examples of the distributive property in terms of lengths, widths, and areas of rectangles.

   a. 3(4+5)=3×4+3×5 b. 7×8+3×8=10×8

3. Pierre bought three cookies for 40¢ each, and then Alicia bought five more at the same price. How much did their aunt, who paid for the cookies, pay in all? Show that there are two ways to figure this out, and that these ways are related by the distributive property.

4. Write a question, which, like question 3, uses the distributive property.

5. The distributive property can help with mental multiplication. For example, to find 17×7, think of (10+7)×7 and simply add 10×7 to 7×7. In a similar way, explain how to multiply in your head:

   a. 6×18 b. 9×14

6. Make up box puzzles to illustrate several multiplications with factors of two or more digits. Then picture the multiplications in terms of areas, as in figure 1.4.5.

7. a. Fill in the box puzzle from example 1 in the text to complete the multiplication.

   b. Make a box puzzle to picture and compute 54×109.

   c. Make a box puzzle to picture and compute 781×1,729.

8. Use the lattice method to illustrate the multiplications in question 7.

9. Is the following work correct? Explain.

$$\begin{array}{r} 65 \\ \times\ 23 \\ \hline 115 \\ 138 \\ \hline 1{,}495 \end{array}$$

10. Here is a way to make lattice multiplication easy using Napier's rods.[8]

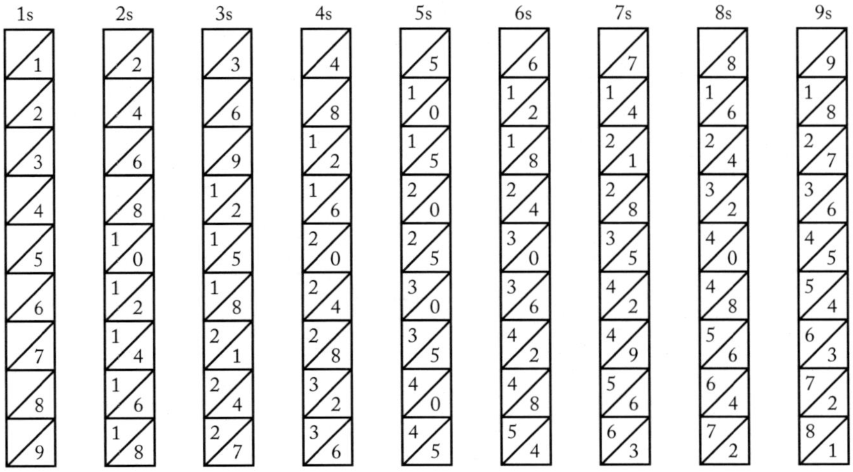

These are one-digit multiplication tables written as for lattice multiplication on rods, often made of wood. Cardboard works well for classroom use.

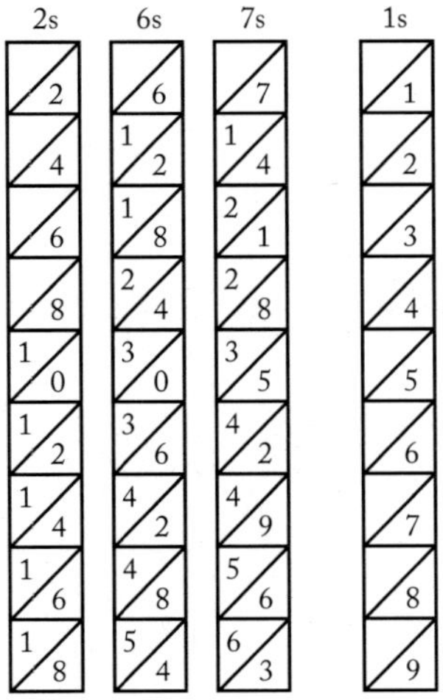

534 = 267×2

1,602 = 267×6

$$\begin{array}{r} 267 \\ \times 26 \\ \hline 1{,}602 \\ 5\,34 \\ \hline 6{,}942 \end{array}$$

Use Napier's rods to illustrate 56×47, and compare the result to the example of lattice multiplication in the text above.

[8] Napier's rods were invented by John Napier (1550-1617). For more on Napier, see section 17.5.

11. a. Explain how this picture, with base-ten blocks, models a multiplication example.

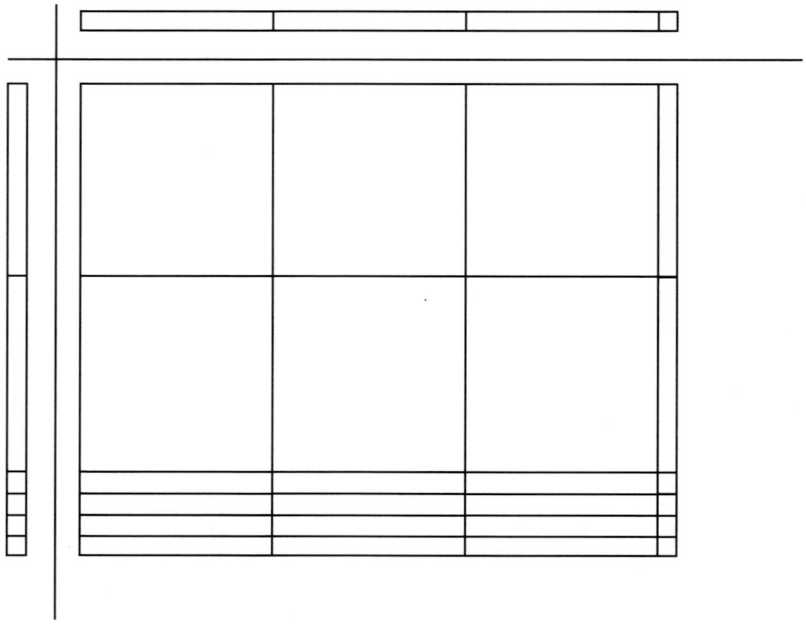

b. Show 47×23 with base-ten blocks.

c. What limitations would you encounter, if you tried to model multiplication with three-digit factors by means of base-ten blocks?

12. Algebra students are often taught to use FOIL to multiply binomials. FOIL is an acronym for First, Outer, Inner, Last. To use it, for example, on $(x+2)(y+3)$, see the picture:

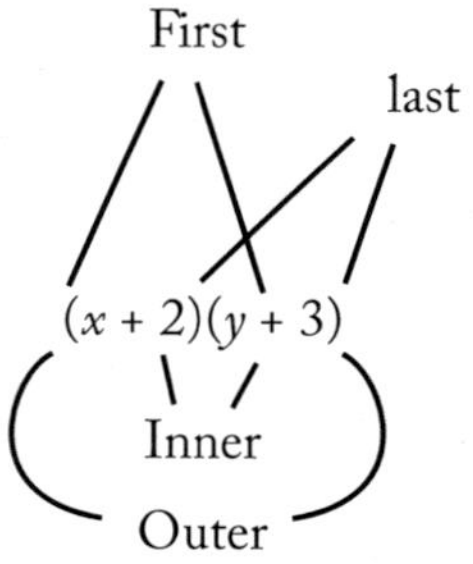

Make a box puzzle to multiply $(x+2)(y+3)$ and use it to show that FOIL simply states the distributive property in this case.

13. **Another Exploration** Circle any number in the square array below. Then circle a second number, not in the same row or column as the first number. (For example, if the first number circled is 7, then the 6, 8, 9, and 10 may not be circled as the second number, since they are in the same horizontal row as 6, and 2, 12, 17, and 22 may not be circled, since they are in the same column as 7.) Now circle a third number, in a different

row and column from each of the first two numbers, then a fourth number, in a new row and column, and finally circle a fifth number, in a different row and column from any of the first four numbers.

1	2	3	4	5
6	7	8	9	10
11	12	13	14	15
16	17	18	19	20
21	22	23	24	25

a. Add your five circled numbers.

b. Repeat this process of circling five numbers and adding, varying the numbers you circle. What do you notice?

c. Explain your observation, picturing numbers with nickels and pennies, like this:

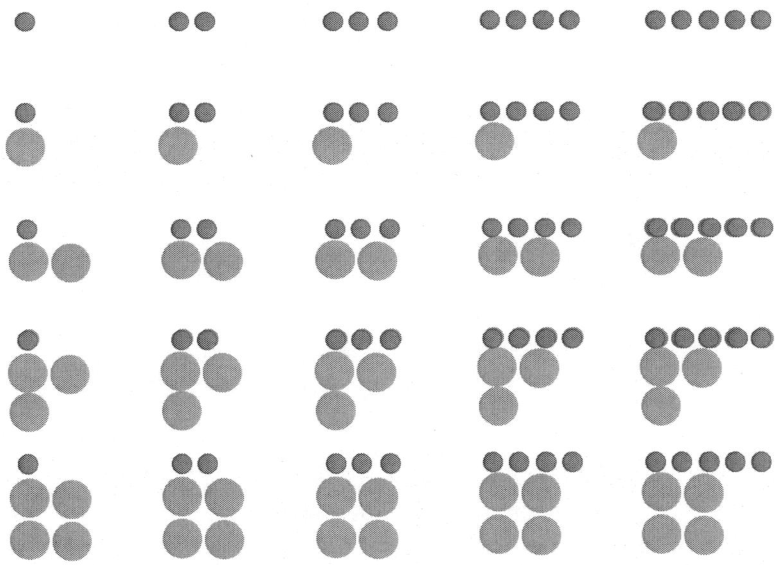

d. Use the commutative, associative, and possibly the distributive properties, to explain the result of part b a second way

e. State a generalization of the results above and check it for a square array of the numbers from 1 to 36.

Vocabulary for Chapter 1

Addend (1.1)
Algorithm (1.2)
Area (1.3)
Arithmetic progression (1.3)
Associative property (1.1)
Commutative property (1.1)
Distributive property (1.4)
Factor (1.3)
Lattice multiplication (1.4)
Product (1.3)
Sum (1.1)

Self-Quiz for Chapter 1

1. Complete this box puzzle

27	93	
49	74	

2. Draw an array to picture 5×8.

3. Show, by means of a box puzzle, how to add up all the whole numbers from 1 to 500, and actually calculate the sum.

4. a. Complete this box puzzle.

	3	7	
8			
6			

   b. Illustrate the box puzzle in part a in terms of areas.

5. Write a box puzzle that illustrates 36×84, and use it to calculate that product.

6. Show how to calculate 36×84 both by lattice multiplication and by repeated doubling.

Chapter Two
Introduction to Negative Numbers

Since people first began to count, the concept of number has been repeatedly extended and refined. One such extension is to numbers "below zero." Numbers below zero are known as **negative** numbers, a name with unfortunate connotations. (A negative attitude, for example, is generally considered bad.) The medieval church, which associated negative numbers with debt and the sin of usury, regarded those numbers as sinful. This effectively banned negative numbers until the Renaissance, and even then they were written in the devil's color, red, a tradition which endures in modern bookkeeping.[1]

2.1 The Number Line

When you use a ruler or a thermometer, you probably take for granted the correspondence of numbers with points on a line. Basic as it is, this correspondence has far-reaching consequences. Imagine an inchworm inching along the number line in figure 2.1.1. Starting at 0, its first "step" takes it to 1, its second to 2, and so on. Continuing, the inchworm determines a point on the line for each number in the endless list of **whole numbers**, 0, 1, 2, 3, 4, 5, ...

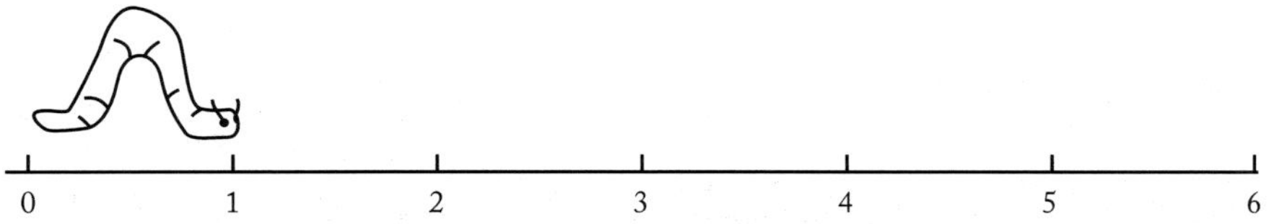

Figure 2.1.1

What if the inchworm moves in the opposite direction from 0? Now its first move lands not at the point we have called 1 but at a new point, which we label -1 (read "negative one"). Continuing in that direction, it reaches -2, -3, -4, -5, ..., as shown in figure 2.1.2.

[1] Negative numbers were used in China some 2,000 years ago, and they were also used in India for several centuries before they were accepted in Europe. The Alexandrian Greek mathematician Diophantus, thought to have lived around 250 C.E., used negative numbers quite extensively, but the level of his work with algebra and number theory, and the use of negative numbers, were not attained again in Europe for more than a millennium.

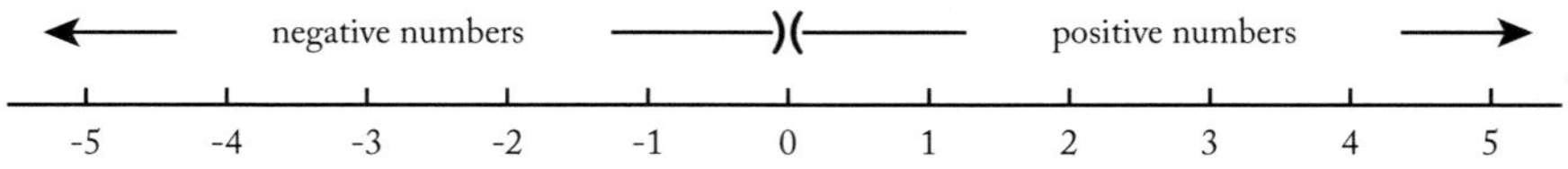

Figure 2.1.2

Numbers on the same side of 0 as 1 are called **positive**, while those on the opposite side are called **negative**; 0 itself is neither positive nor negative.

In figure 2.1.2, you can see that 0 is midway between 1 and -1, between 2 and -2, etc. In general, if n is a number, $-n$ is the number symmetrical to n on the other side of 0; we say $-n$ is the **opposite** of n.[2] This expression is even applied to negative numbers; for example, the opposite of -3, which is written -(-3), means the opposite of the opposite of 3, or 3 itself.

This chapter is concerned only with the numbers that can be obtained by starting with 0, adding 1 or -1, and repeating these additions. These numbers are called **integers**, a name related to our word "integrity" (the characteristic of a person who acts and speaks as a whole person, not internally divided). Later chapters treat positive and negative fractions.

It is not always easy to illustrate negative numbers in practical settings familiar to elementary school children. Indeed, beyond debt and subzero temperatures such examples may be hard to find. Do not worry too much about this. Numbers are tools of thought, intrinsically worth studying and mastering. It is easy to think of situations where fractions make no sense, but fractions are worth studying nonetheless. The same is true of negative numbers; in time you will see many uses for them.

Exercises 2.1

1. In English we often describe situations that involve negative numbers without actually using the word "negative." For example, you wouldn't say, "I won negative $200 in Las Vegas." Instead you would say "I lost $200 in Las Vegas." Restate each of the following sentences in more normal, standard English.
   a. Our profit last month was negative.
   b. The meeting starts in -5 minutes.
   c. The airplane is climbing at -500 feet per minute.
   d. In a recession the economy has a negative growth rate.
   e. The rocket is approaching Earth at -3 miles per second.
   f. The north poles of two magnets attract each other negatively.
   g. The quarterback gained -7 yards.
   h. Welfare is a negative tax.

[2] The opposite of a number is sometimes called the **negative** of the number, but that term can be confusing. If the number n is itself negative, its "negative" is positive. More theoretical treatments call $-n$ the **additive inverse** of n, but that is clumsy terminology for elementary school children. Therefore, we prefer to call $-n$ the **opposite** of n.

2. It was mentioned above that -(-3)=3, since -(-3) is the opposite of -3.
   a. By the same reasoning, what is -(-(-3))?
   b. Extend the result from part a to -(-(-(-3))).
   c. Extend the result from part b to -(-(-(-(-3)))).
   d. Find and state a general rule to tell from the number of minus signs whether expressions like those in parts a–c are positive or negative.

2.2 Adding Negative and Positive Numbers

A Linear Model for Adding Integers

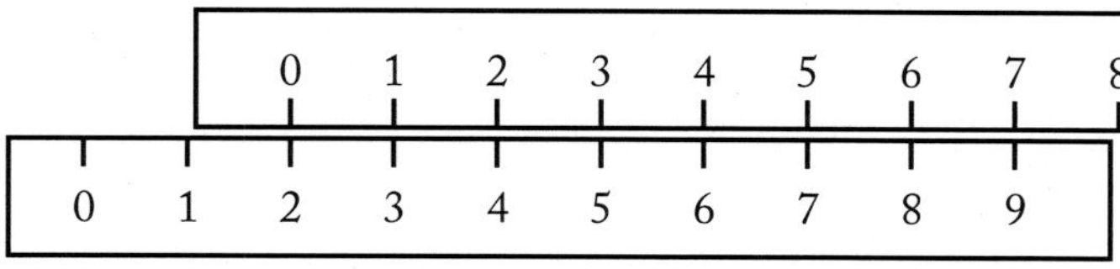

Figure 2.2.1

Figure 2.2.1 shows another way to model addition. Can you use it to read 2+5? How about 2+4? What other sums can you read from the figure? Can you extend the number lines in figure 2.2.1 to make it show 2+10? How about 2+(-1), or 2+(-2), or even 2+(-3)? Make a pair of number lines like those in Figure 2.2.1 and experiment with them, trying to answer the questions stated and other questions as well. For example, can you put the number lines in a position that illustrates 3+5? How about 3+(-2)? After you have experimented, read on.

The above questions lead to a slide rule for adding. Here is how to make such a device simply, inexpensively, with no fancy equipment (with mathematics teaching devices, as with cooking, homemade is best).

1. Fold a piece of lined paper perpendicular to the lines as in figure 2.2.2.

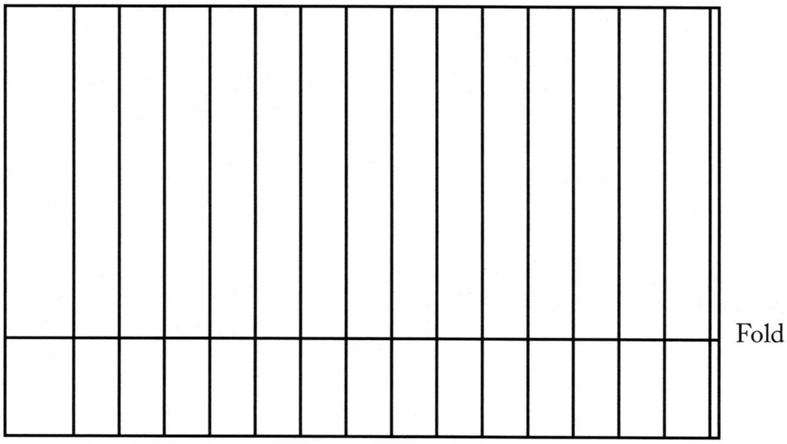

Figure 2.2.2

2. Number the lines (not the spaces) on the back of the fold as shown in figure 2.2.3.

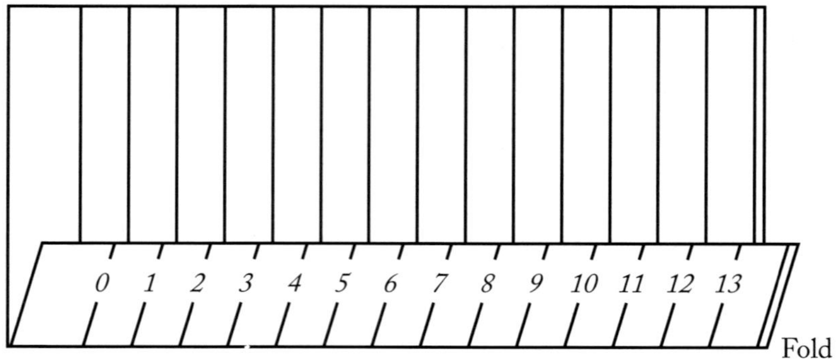

Figure 2.2.3

3. Take a second sheet of paper, and number its lines just like the first sheet, putting the numbers far enough from the edge that they can be seen when the second sheet is placed in the fold, as in figure 2.2.4.

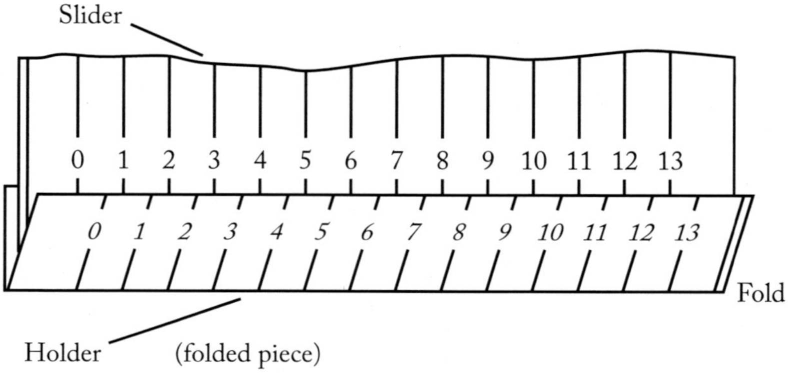

Figure 2.2.4

To use the slide rule, put the slider's 0 opposite one of the addends (numbers being added) on the holder. Find the second addend on the slider, and the sum is the corresponding number on the holder. For example, to add 2+5 move the slider so 0 on the slider is opposite the 2 of the holder (lower scale). Now 5 on the slider is opposite 7 on the carrier, showing 2+5=7.

Once you see how this works with positive numbers, turn the fold backward and number the other side of the slider and holder to get the slide rule in figure 2.2.5, which works for negative numbers as well as positive.

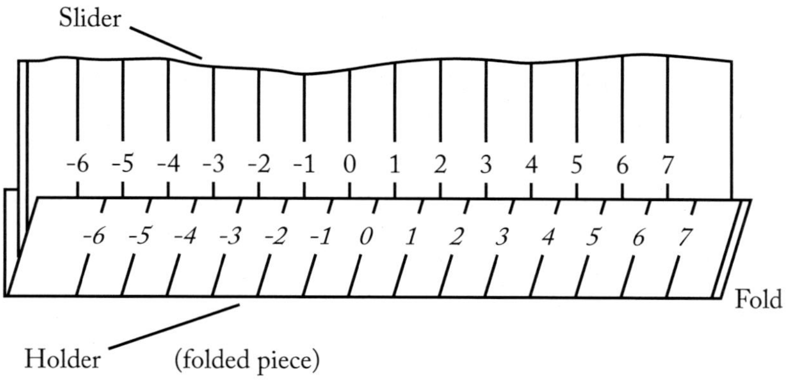

Figure 2.2.5

Experiment with this slide rule. Compute several sums, including a few with both numbers positive, some with one positive and one negative, and some with both numbers negative.

The slide rule is based on adding directed distances, which is not the same concept of addition as putting together sets of objects. Adding 2 and 5, for example, may be pictured by marking off a distance of 2 on one number line, then marking off a distance of 5, starting where the first distance ends, as in figure 2.2.6. Arrows allow us to picture these directed distances. We use a solid triangle to mark the starting point and a hollow triangle to mark the ending point.

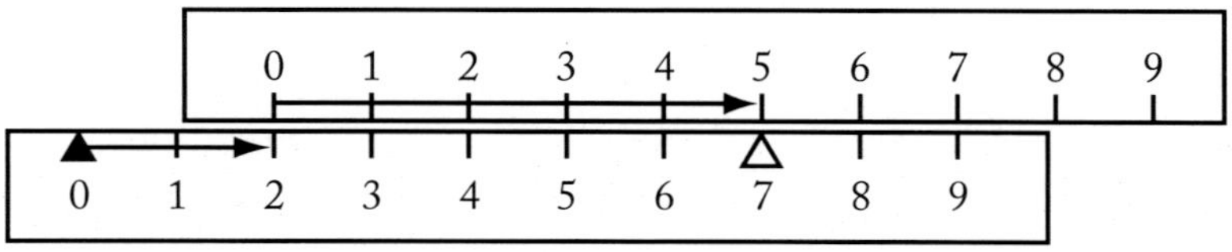

Figure 2.2.6

To use this idea for negative numbers as well as positive, picture negative numbers as arrows pointing in the opposite direction from the positive ones. Figure 2.2.7 shows how to add -3 and 7 by putting an arrow of 7 to the right (toward the positive numbers, because 7 is positive) at the tip of an arrow of -3 (toward the negative numbers, because -3 is negative). It shows that -3+7=4.

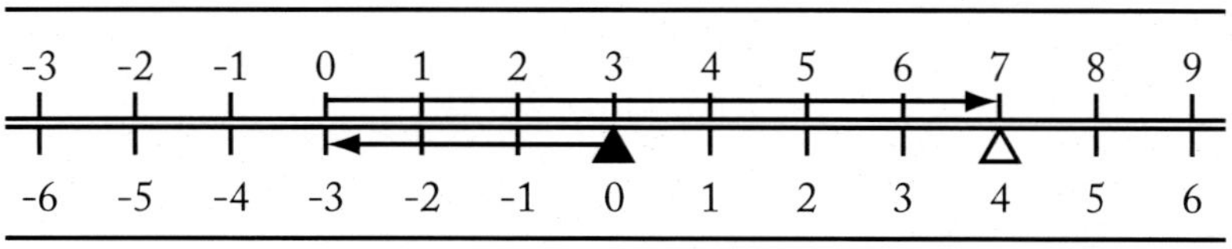

Figure 2.2.7

Quantities which can be thought of as directed distances are known technically as **vectors**. Figures 2.2.6 and 2.2.7 illustrate vector addition on the number line. If you have met vectors in another context, such as a physics course, it may help you to realize that addition of positive and negative numbers is vector addition, but that terminology will not be emphasized in this book.

A Set Model for Adding Integers: Algebra Tiles

The most elementary concept of addition comes from putting together sets of objects. Putting together a pile of n objects with m objects in another pile makes a single pile with $m+n$ objects (assuming n and m are whole numbers). To extend this idea to add negative and positive numbers, we use objects with two colors such as algebra tiles or two-color chips.

Using the unit squares in a set of algebra tiles, designate one color (gray) to be negative and one color (black) to be positive. Define a **zero pair** to be two opposite colored tiles paired together.

So  is a way to picture 1+(-1), which is, of course, 0.

This can be extended; 2+(-2) may be represented as two zero pairs, 3+(-3) as three zero pairs, and, in general, $n+(-n)$ as n zero pairs. This leads to the following general statement.

Any number plus its opposite is 0. In symbols, for any number n,

$$n+(-n)=(-n+n)=0.$$

Example 1. 5+(-3) can be translated as "Add (or combine) 5 positives and 3 negatives" and pictured as follows:

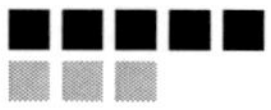

Three times you can pair up one positive black tile with one negative gray tile to make zero pairs. Remove each zero pair, and two positive tiles remain, so 5+(-3)=2.

Example 2. (-4)+(-6) can be translated as "Add 4 negatives and 6 negatives" and pictured as

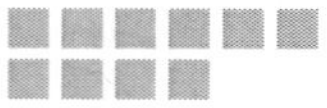

Here there are no zero pairs, so just count the negative tiles. Since there are 10 negative tiles, (-4)+(-6)=-10.

Example 3. 4+(-7) can be pictured as:

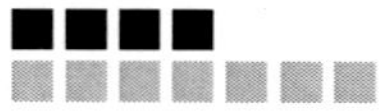

After removing the zero pairs, 3 negative tiles remain, so 4+(-7)=-3.

In practice, you need to be able to add positive and negative numbers without depending on a mechanical device or counting on a number line. Here is one way to do that; it uses some imagination, but in time you will find it easy. Imagine two rival parties, the **Positives** and the **Negatives**, in an election. Votes for Positives offset votes for Negatives one for one. Thus, if there are 2 votes for Negatives and 5 for Positives, the 2 Negative votes cancel out 2 of the Positive votes, leaving 3 Positives. Figure 2.2.8 models -2+5=3 as vector addition on the number line and shows two of the Positives offset the two Negatives to get back to 0.

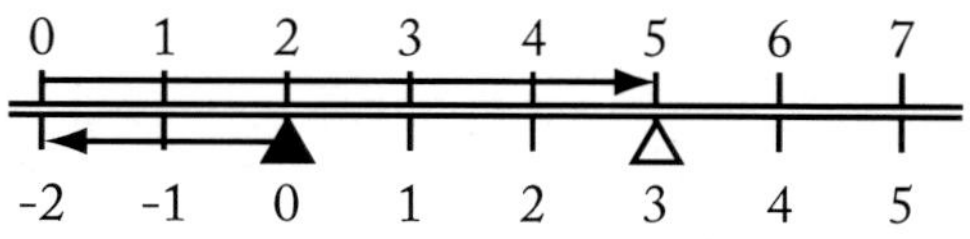

Figure 2.2.8

Again, if there are 50 Positive votes and 75 Negatives, the Negatives win by 25, which means that 50+(-75)=-25.

If the votes are all Positive or all Negative, they reinforce each other instead of cancelling. For example, 7 Negative votes plus 9 more Negatives are just 16 Negatives, showing that -7+(-9)=-16.

Exercises 2.2

1. What additions are pictured here?

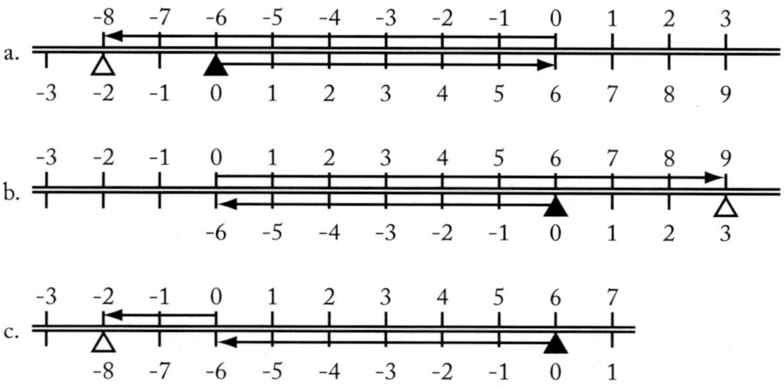

2. Picture as in question 1:
   a. -5+2 b. -5+(-7) c. 10+(-4)

3. a. Show pictorially (using either a number line or algebra tiles) that:
   -4+8=8+(-4) and -11+6=6+(-11)
   b. What property of addition have you illustrated in parts a and b?

4. Draw pictures of algebra tiles, explaining the following:
   a. -9+7 b. 6+(-2) c. -5+(-3)

5. a. Here is a way to check the addition of a positive and a negative number illustrated in figure 2.2.7. Write -3+7 as -3+(3+4) and then use the associative property: -3+(3+4)=(-3+3)+4. But any number, added to its opposite, yields 0, so this is 0+4 or 4.
   b. Use figure 2.2.8, and reasoning from part a to explain why 5+(-2)=3.

6. Would your slide rule help you to add fractions? Mark where $1\frac{1}{2}$ and $2\frac{1}{2}$ belong on the number lines in the slide rule, and, then use the slide rule to show that $1\frac{1}{2} + 2\frac{1}{2} = 4$.

7. Repeat question 6 with:

   a. $-1\frac{1}{2}$ and $2\frac{1}{2}$ b. $1\frac{1}{2}$ and $-2\frac{1}{2}$

   c. $-1\frac{1}{2}$ and $-2\frac{1}{2}$

Exploration: The Adder (questions 8-13)

8. Figure 2.2.9 shows a simple device for adding, consisting of three parallel rows of numbered dots. Experiment to see if you can figure out how to use it.

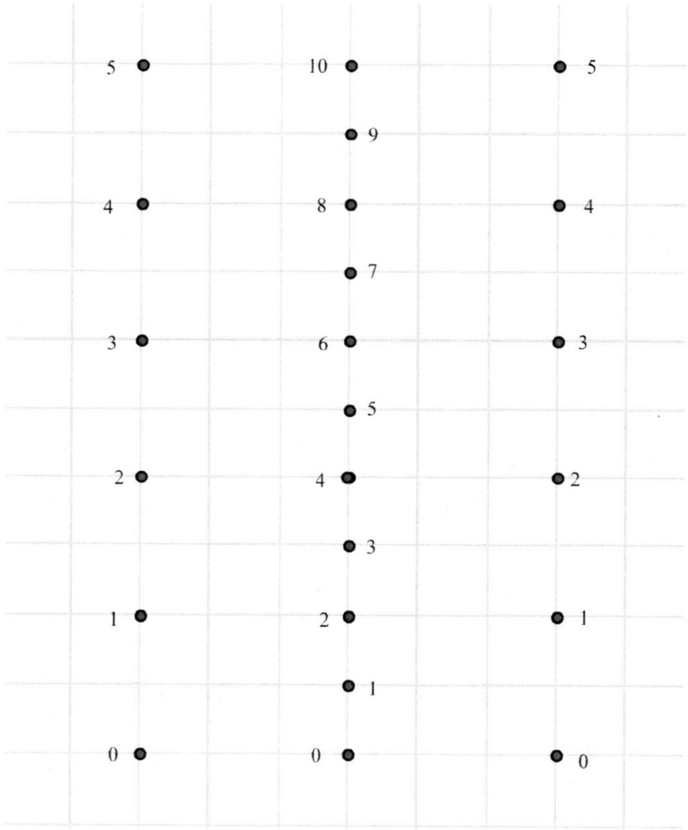

Figure 2.2.9

9. To see how the adder works, consider a specific case, 3+2. Lining up the dots numbered 3 and 2 on opposite sides makes the shape shown in figure 2.2.10. This shape is a right **trapezoid**, a name related to the word "trapeze." A **trapezoid** is a four-sided figure with two of the sides parallel. A **right trapezoid** is a trapezoid with one side perpendicular to a pair of parallel sides.

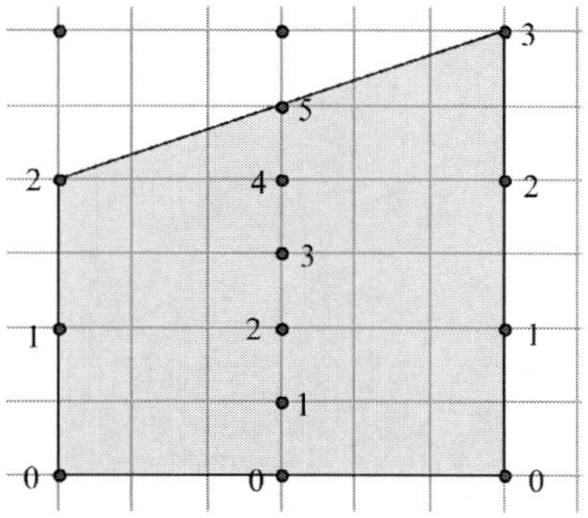

Figure 2.2.10

Two copies of this trapezoid fit together to form a rectangle (figure 2.2.11).

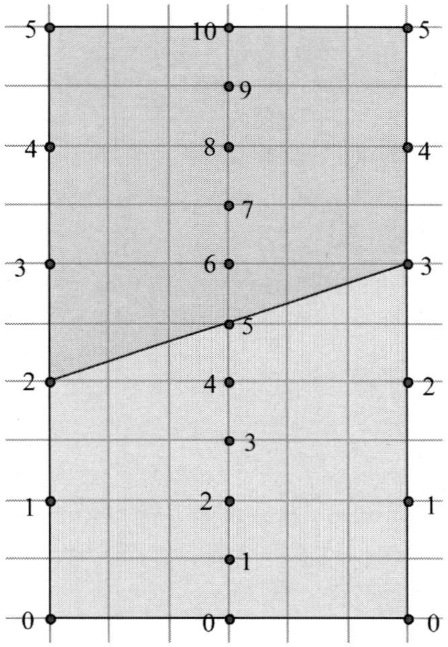

Figure 2.2.11

The rectangle is 5 units from top to bottom, and this is the sum of the numbers we are adding. The middle column is the same height, but it is measured in units only half as big, and in terms of the smaller units it is 10 units tall. The sloping line cuts the middle column of numbers half-way up, which is to say at 5, so 2+3=5.

Make pictures like figures 2.2.10 and 2.2.11 to illustrate 3+7.

10. Like any discovery, the adder should lead you to new questions. As you did with box puzzles, think of questions that grow out of the adder and list at least three good ones to explore.

11. (Skip this if you are not confident with the fractions.)

    a. Write in and label dots for $\frac{1}{2}$, $\frac{1}{4}$, and $\frac{3}{4}$ in all columns of the adder.

b. Use the adder to add $\frac{1}{2}+\frac{1}{2}$.

c. Use the adder to add $\frac{1}{2}+\frac{1}{4}$.

12. a. Make a copy of the adder, using graph paper for accurate spacing, much like the one started here but extending from -10 to 10. (The spacing is crucial. The picture on the next page uses 4 squares between columns. Be sure to have as many squares between the center and left columns as between the center and right.)

    b. Experiment with positive and negative numbers on the adder.

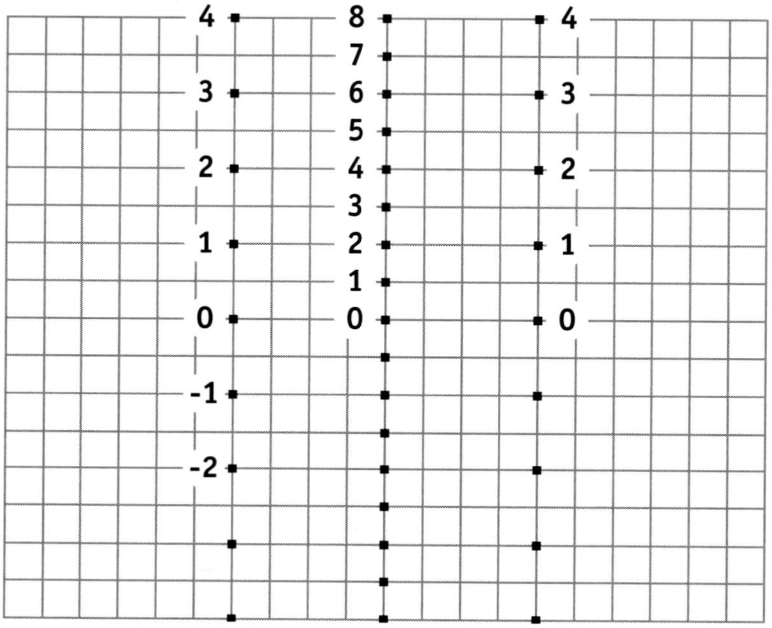

13. Two 3rd graders set out to make a big adder by taping pieces of graph paper together, making a long dot pattern, and numbering it. They numbered up to 200, which took a long time, and eventually they gave up the project, because it was awkward to use and inaccurate. (The teacher wisely let them take it as far as they wanted, and they learned from the experience.) What are the problems in making a huge adder practical, and what could children learn by trying to make one?

14. Add (in your head if you can):

a. -12 +(-27)	b. 91+(-92)	c. -48+1
d. -73+2	e. -194+195	f. -72+71
g. -931+0	h. -80+80	i. 3+(-61)
j. -190+100	k. -1+(-100)	l. -40+(-43)

15. Do box puzzles work with negative numbers?
    Try these to find out:

a.

-9	13	
7	-6	

b.

20	-11	
-31	18	

c.

-2	0	16	
35	-4	5	
10	-9	-70	

16. Redo question 15 after reversing all the signs of the given numbers (i.e., put minus signs before the positive numbers and take the minus signs off of the negative numbers). How does this affect all the other numbers in the puzzle? Explain in a sentence or two.

17. Add all the integers from -10 to 10, using the method from section 1.3.

18. a. Add the integers from -25 to 35, inclusive, by the method of Section 1.3.
    b. Add $26+27+28+\cdots+35$ by the method of Section 1.3.
    c. Explain the relationship between your answers for parts a and b.

2.3 Multiplying with Negative Numbers

Multiplying one positive and one negative number

Multiplication with positive integers is repeated addition. Can multiplication with negative numbers also be thought of as repeated addition? A few cases offer hope. For example, you can think of $3(-4)$ as $(-4)+(-4)+(-4)$, which is -12. You can check it with the distributive property. For example, $3\times(10+(-4))$ can be found by first adding 10 to -4 to get 6, then multiplying the result by 3 to get 18. On the other hand, by the distributive property,

$$3\times(10+(-4))=3\times10+3\times(-4)$$
$$=30+3\times(-4).$$

Then $18=30+3\times(-4)$, which will be true only if $3\times(-4)=-12$.

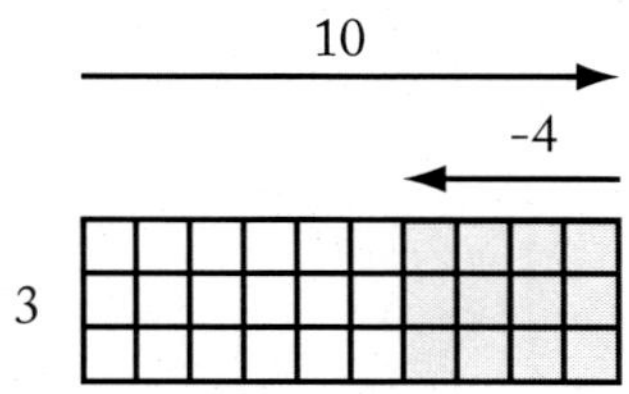

Figure 2.3.1

Figure 2.3.1 illustrates this. The clear squares are $3\times6=3\times(10+(-4))$, and 3×10 is the area of the entire rectangle. To cut the big rectangle down to 3×6, the 12 shaded squares must be removed, so it makes sense that $3\times(-4)=-12$.

You can also picture multiplication of a positive times a negative with algebra tiles. For example, 2×(-3) translates as "add 2 groups of 3 negatives"

We can see that there are 6 negatives, so 2×(-3)=-6.
This example illustrates that (**positive**)×(**negative**)=**negative**.

The box puzzles with borders from section 1.4 are based on the distributive property, so they offer a way to practice what we are doing with negative numbers and check that the results are reasonable. Fill them in; in each case use what you know about adding positive and negative numbers, and interpret products of a positive and negative number as a repeated addition.

Example 1. Complete this box puzzle.

	8	-5	
3			
4			

Solution: There is no doubt about how to fill in the boxes at the left, which involve only positive numbers. Fill in those boxes first.

	8	-5	
3	24		
4	32		
7	56		

Next, add across the top to put a 3 at the upper right, and complete the right hand column by multiplying.

	8	-5	3
3	24		9
4	32		12
7	56		21

Now you can see what to put in the middle column in order to make the horizontal additions correct. For example, putting -15 in the top row leaves us with $24+(-15)=9$, which is correct, thus confirming that $3(-5)=-15$.

	8	-5	3
3	24	-15	9
4	32	-20	12
7	56	-35	21

Finally, note that the addition in the middle column, $-15+(-20)=-35$, is correct.

Box puzzles like this one should give you confidence to use the distributive property when negative numbers are involved. Also, you can now see a rule:

When you multiply a positive number by a negative number, the product is negative. Specifically, if m and n are positive numbers, $(-m)\times n=-(m\times n)$.

The effect of this rule is that when you multiply a positive number by a negative number, you can start out by ignoring the minus sign and then just attach it to the product when you are done. For example, to multiply 5 by –7, temporarily ignore the minus sign and multiply 5 by 7 to get 35, then adjoin the minus sign to the product to get –35.

When the negative factor is –1, the above rule boils down to $(-1)\times n=-(1\times n)$. Since $1\times n=n$, this means $(-1)\times n=-n$. Question 4 below explores this further.

To model this with algebra tiles, translate $(-2)\times 3$ as, "Remove 2 groups of 3 positives," since it is equal to $-(2\times 3)$. Here it is important to start with zero pairs so that you have something to remove. You can start with any number of zero pairs as long as you have enough to remove 2 groups of 3 positives. Let's suppose we start with 8 zero pairs:

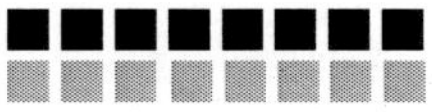

From this (0), remove 2 groups of 3 positives or 6 positives to be left with:

Notice the two zero pairs still left. Remove these since they won't change the net effect of the product. The final result is 6 negatives:

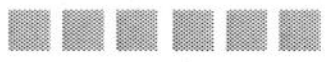

So $(-2)\times 3=-6$.
In this case, the ideal number of zero pairs to start with is 6.
This example shows that **(negative)**×**(positive)**=**negative**.

Multiplying two negative numbers

Example 2. What is (-4)(-5)? Repeated addition makes no sense here, but the box puzzles (i.e., the distributive property) still show the way.

Solution: Save the upper left-hand corner, which is (-4)(-5), for last, to be figured out indirectly.

	-5	5	
-4			
4			

Fill in the products with at most one negative factor; they involve nothing new.

	-5	5	
-4		-20	
4	-20	20	

Add across row and down columns.

	-5	5	0
-4		-20	
4	-20	20	0
0		0	

Since 0 × anything = 0,

	-5	5	0
-4		-20	0
4	-20	20	0
0	0	0	0

In the upper left square, (-4)(-5), whatever it is, if added to -20, yields 0. But then (-4)(-5) must be 20.

Multiplication with negative numbers is defined to extend the distributive property to negative numbers. Therefore,

> **The product of two negative numbers is always positive. Specifically, if m and n are positive numbers then $(-m)\times(-n)=m\times n$.**

This, too, can be illustrated with algebra tiles, translating (-2)×(-3) as "Remove 2 groups of 3 negatives." Again start with zero pairs. (This time we use 6 zero pairs):

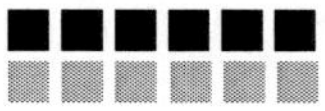

Now remove 2 groups of 3 negatives or 6 negatives in all, to be left with 6 positives:

Then (-2)×(–3)=6 and, in general, **(negative)**×**(negative)=positive**.

Exercises 2.3

1. Which of these products are positive? Which are negative?
   a. (-5)(-13) b. (-71)(40)(-11) c. (-3)(-3)(-3)(-3)(-3)
   d. 1(-2)(3)(-4)(5)(-6)(7)(-8)(9)(-10)(11)(-12)

2. Complete these box puzzles and make sure they check.

a.

	-4	-6	
3			
8			

b.

	2	9	
-1			
-2			

c.

	-11	5	
-8			
-2			

d.

	15	-24	
17			
-13			

3. a. Write three box puzzles with borders involving multiplication of negative numbers.
   b. Complete the box puzzles you made up in part a.

4. **Multiplication by -1**

   If you multiply any number except 0 by -1, you get the opposite of that number. For example, if you multiply 5 by -1, you get -5, and if you multiply -2 by -1, you get 2.

   Compute and compare (parts a-c):

   a. -(3+5) and -3+(-5) b. -(7+11) and -7+(-11)

   c. -(6+4) and -6+(-4).

   d. Based on your answers from parts a, b, and c, complete this general statement: -(s+t)=

   e. Test your answer from part d on several examples, where either or both of s and t is negative.

   f. Explain your answer from part d by interpreting -(s+t) as (-1)(s+t) and applying the distributive property.

5. Fill in the blanks in the table below to make a multiplication table for positive and negative numbers. The numbers to multiply are in boldface type. A few other numbers are put in as checks on your work. This is a good way to review the table and see the patterns in it from a fresh perspective. For example, the 2s table, which is usually seen as 2, 4, 6, . . increasing by 2, is now extended in the other direction as well, diminishing by 2 each time even below 0.

									7								
									6								
									5								
									4		12						
					-9				**3**								
									2								
-8	**-7**	**-6**	**-5**	**-4**	**-3**	**-2**	**-1**	**0**	**1**	**2**	**3**	**4**	**-5**	**-6**	**7**	**-8**	**9**
									0								
									-1						-7		
								0	**-2**								
									-3								
									-4								
									-5								
									-6								
									-7								

6. The two rules, **(-*m*)×*n*=-(*m*×*n*)** and **(-*m*)×(-*n*)=*m*×*n*** are presented above assuming that the numbers m and n are positive. However, that restriction is not necessary. Make up examples in which one or both of the numbers m and n is negative (so that $-m$ and/or $-n$ is then positive), and show that in those cases the rules still apply.

7. Use algebra tiles to picture the following multiplications, explaining each step as you would to a class.

   a. 3(-5) b. (-3)(4) c. (-2)(-4)

8. Many algebra books prove the rules for multiplying with positive and negative numbers this way:

   First, note that $m\times 0=0$, since anything times 0 is 0. But if you write 0 as $-n+n$, this becomes. $m\times(-n+n)=0$. Therefore, by the distributive property, $m\times(-n)+m\times n=0$. Then $m\times(-n)$ must be the opposite of $m\times n$. Would school children find this convincing? Explain your reasoning about this.

Vocabulary for Chapter 2

Additive inverse (2.1)
Integer (2.1)
Negative (2.1)
Opposite (2.1)
Positive (2.1)
Trapezoid (2.2)
Vector (2.2)

Self-Quiz for Chapter 2

1. Show how to use the addition slide rule to find:

   a. 3+8 b. -6+3

2. Calculate and explain why your answers are correct:

   a. -32+55 b. 121+(-150)

   c. -48+(-67)

3. Complete this box puzzle

-27	93	
49	-74	

4. a. Explain how to find -7×4 in terms of repeated addition.

   b. Explain why -3×(-2) is positive.

5. Complete this box puzzle.

	-4	-9	
12			
-7			

Chapter Three
Figurate Numbers

Since ancient times, people have arrayed pebbles, dots, or coins as geometric figures, such as triangles, squares, or cubes.[1] The resulting **figurate numbers** are an endless source of patterns, which turn up throughout mathematics. We use them as a source of explorations, which are accessible to anyone who can add and multiply whole numbers. Like adults, young students sometimes find the patterns so interesting that they hardly notice the calculation practice built in.

Concrete to Abstract

Abstractions, including ideas like truth and beauty, as well as mathematical concepts, grow out of experience. (The idea of three, for example, is an abstraction based on specific examples, such as three fingers, three people, or three coins.) Teachers must provide concrete examples as the foundation on which their students can build abstract ideas. What is abstract for one person may be concrete to another. For example, you may see this picture:

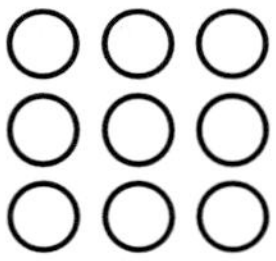

as a way to represent a square number, but if you find pictures abstract, you may grasp the idea more easily if you make the array with buttons, washers, or coins. This is natural; we learn with our senses. Physical objects bring touch and a kinesthetic element to learning. At later stages, drawing or coloring pictures may help the transition from concrete to abstract. In time, pictures themselves are concrete, and there is less need for physical objects. An architect, for example, can visualize a building from a set of plans, but other people may need a physical model or artist's rendition. People vary widely in their comfort with abstraction, so teachers must present ideas as concretely as possible, involving as many senses as possible. Often, this calls for manipulatives, but they need not be fancy or expensive. Manipulatives you make yourself may be preferable to high priced commercial kits, because you understand well what you make yourself. Manipulatives students make themselves, like the slide rules in section 2.2, are best of all, because the students, too, learn from the creation process.

[1] The Pythagoreans were a mystical cult that formed around Pythagoras (c.572-497 B.C.E.) in Crotona, a Greek town in what is now Italy. They had strict codes of conduct and believed that "All is number." They studied figurate numbers and the relationship of the lengths of vibrating strings to the notes they produced. They certainly knew of what we now call the Pythagorean theorem (see Ch.12), but it is not known whether Pythagoras or his followers discovered the theorem or if he had learned it on his earlier travels in Egypt and Mesopotamia.

3.1 Figurate Numbers in the Plane

What do the pictures in figure 3.1.1 have in common?

Photographs by Robert Whitehead

Figure 3.1.1 Triangular numbers at CSUSB

Both show triangular numbers. Figure 3.1.2 shows more triangular numbers.

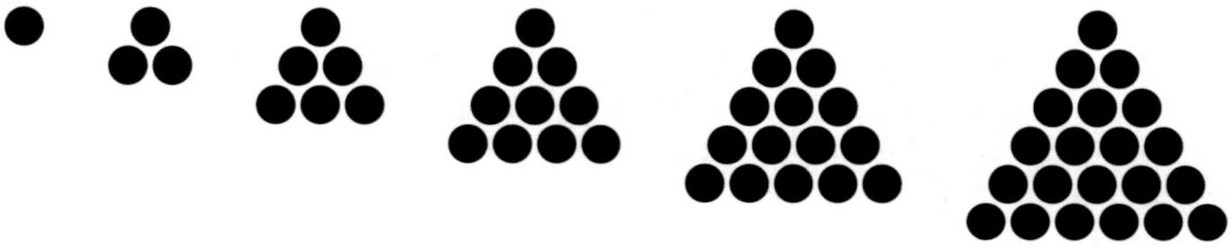

Figure 3.1.2

Counting the dots in figure 3.1.2 shows the first triangular numbers are 1, 3, 6, 10, 15, and 21. There are, of course, larger triangular numbers. Square numbers, together with triangular numbers, turn up in countless curious patterns and relationships. You will encounter some of these patterns in Exercises 3.1.

Exercises 3.1: Exploring Figurate Number Patterns

1. a. Fill in the missing numbers to fit the patterns.

$$1 = 1$$
$$1 + 2 = \_$$
$$1 + 2 + 3 = \_$$
$$1 + 2 + \_ + 4 = \_$$
$$1 + \_ + 3 + 4 + 5 = \_$$
$$1 + \_ + 3 + 4 + \_ + \_ = \_$$

   b. What do you notice about the sums on the right? Identify them by name.

   c. How is the pattern in these sums related to figure 3.1.2?

2. a. Fill in the blanks and look for patterns.
   Half of 1×2= ____
   Half of 2×3= ____
   Half of 3×4= ____
   Half of 4×5= ____
   Half of 5×6= ____
   b. Identify by name the family of numbers you found in part a.

3. This picture explains why half of 4×5 is a triangular number.

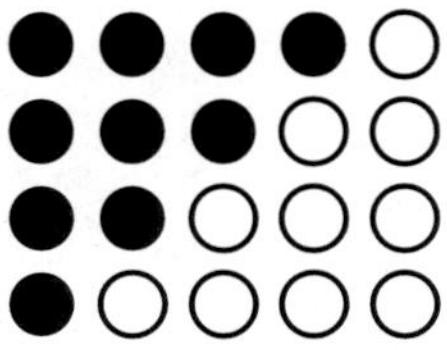

   a. Make a similar picture to show that half of 3×4 is a triangular number.
   b. Make a similar picture to show that half of 5×6 is a triangular number.
   c. Explain why half of 100×101 is a triangular number. (This number was encountered in section 1.3, in adding up the numbers from 1 to 100. Compare the picture above with figure 1.3.5.)
   d. Is the sum 1+2+ . . . +n always a triangular number? Explain.

Another family of figurate numbers is the **square numbers**. Figure 3.1.3 shows the first five square numbers.

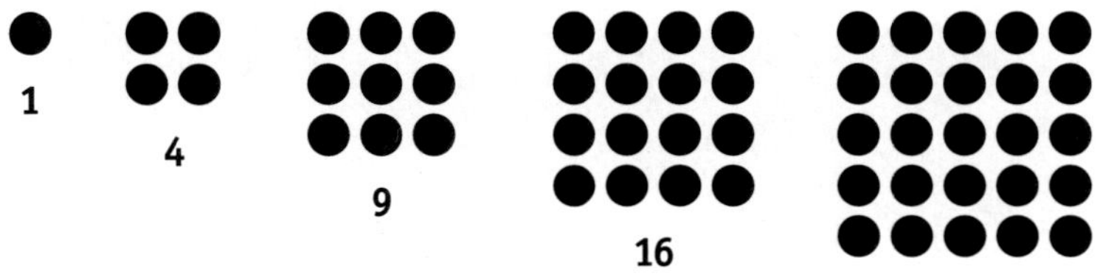

Figure 3.1.3

Of course, one dot is not much of an array, but 1 fits the numerical pattern of square numbers. What are the next few square numbers? Can you see why a number multiplied by itself is said to be **squared?**

4. a. Look for patterns similar to those in question 1 above.

$$
\begin{aligned}
1 &= 1\\
1+3 &= \_\\
1+3+5 &= \_\\
1+3+\_+7 &= \_\\
1+\_+5+7+\_ &= \_\\
\_+\_+5+7+\_+\_ &= \_
\end{aligned}
$$

b. Try to explain these patterns, using figure 3.1.4.

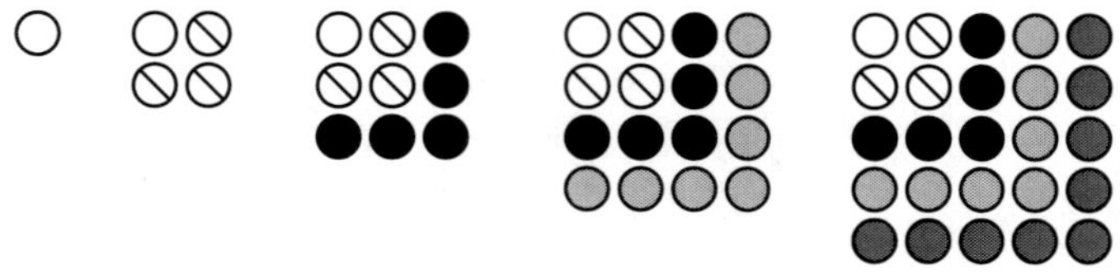

Figure 3.1.4

5. Try adding each triangular number to the one just preceding, like this (In the case of 1, there is no smaller triangular number to add.)

1 = 1
3 + 1 = __
6 + 3 = __
10 + 6 = __
15 + 10 = __
21 + 15 = __

The sums are numbers you should recognize. What are they?

6. Extend question 5 for at least three more lines. Do the patterns continue? Write a sentence which sums up the overall pattern.

7. Using figure 3.1.5, explain the patterns from questions 5 and 6.

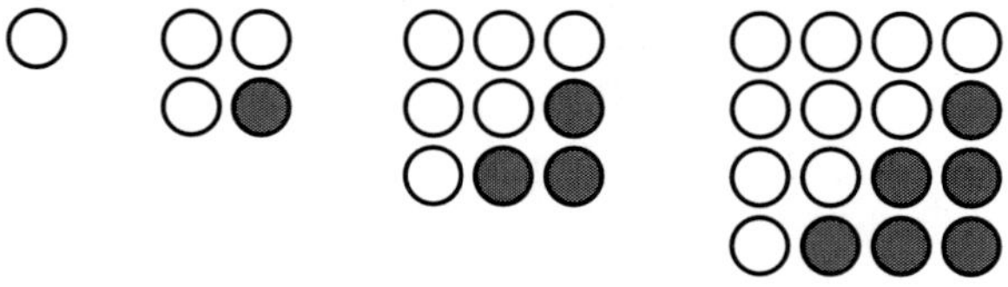

Figure 3.1.5

Make similar drawings (or arrays of colored objects) for the next two cases.

8. a. Multiply triangular numbers by 8, then add 1:

   1×8+1 = __

   3×8+1 = __

   6×8+1 = __

   10×8+1 = __

   15×8+1 = __

   21×8+1 = __

   b. Describe the numbers obtained this way precisely.

   c. Extend this exercise for at least three more lines. Then do the arithmetic. Does the pattern continue?

9. Figure 3.1.6 may be used to explain the patterns in question 8. Note how each picture shows 8 copies of a triangular number, along with one extra dot in the center, and note the overall shape of each array. Make a similar illustration of 10×8+1, either with colored objects or in a picture.

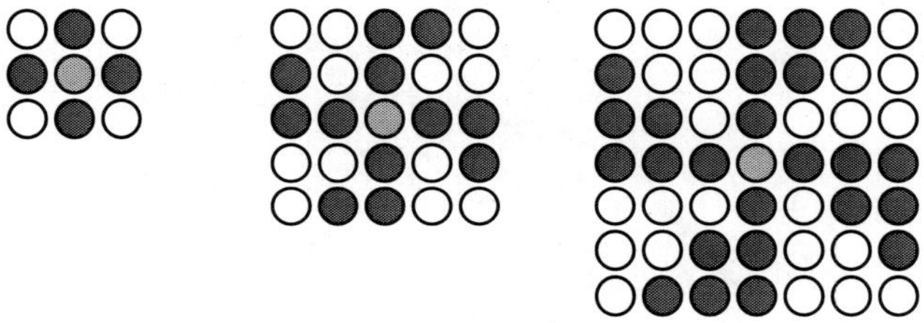

Figure 3.1.6

10. Counting 0 (which is 0×0) as a square, the first few squares are:

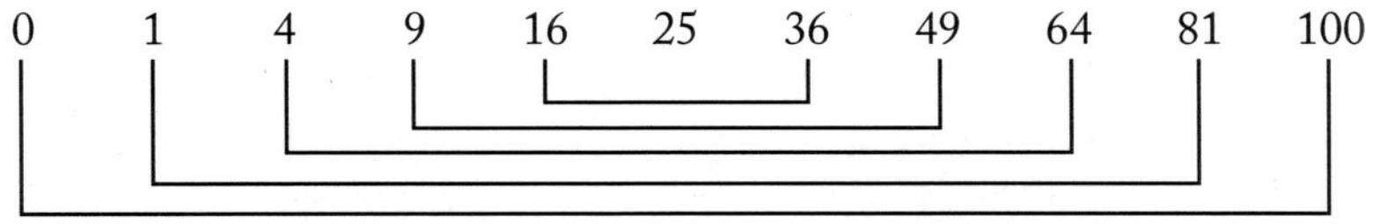

Here the lines bring out symmetry in the units digits.

   a. Compute the next ten squares and find out whether the pattern continues.

   b. Are there any square numbers which end in 2, 3, 7, or 8? Try to justify your answer by considering how the units digit of a number being squared determines the units digit of the resulting square number.

11. a. List all the triangular numbers up to 190, beginning your list with 0 in analogy with question 10.

   b. Describe a pattern in the units digits of the triangular numbers similar to that found in question 10.

12. Compute:
    a. $0\times1\times2\times3+1$ b. $1\times2\times3\times4+1$
    c. $2\times3\times4\times5+1$ d. $3\times4\times5\times6+1$
    e. Predict $4\times5\times6\times7+1$, then compute it to check your prediction.

13. Compute:
    a. $6\times3+3\times1=$ b. $6\times6+3\times3=$
    c. $6\times10+3\times6=$ d. $6\times15+3\times10=$
    e. How does this sequence of calculations involve figurate numbers? (Hints: What kind of numbers are being multiplied by 6 and 3 each time? What kind of numbers are the results of these calculations?) State the general pattern accurately in words.
    f. Study figure 3.1.7, which pictures case b above, then make similar pictures or physical arrangements of colored objects to illustrate cases a and c.

Figure 3.1.7

    g. If you view figure 3.1.7 as a "triangular number of triangular numbers," you can see why 3 and 6, which are consecutive triangular numbers, play an important role in examples a-d above. Make up a similar sequence of examples with a different pair of consecutive triangular numbers, such as 1 and 3. Then do the arithmetic to see if the calculations again yield triangular numbers. Finally, draw a picture similar to figure 3.1.7.

14. Compute:
    a. $14\times1+2\times3+1=$
    b. $14\times3+2\times6+1=$
    c. $14\times6+2\times10+1=$
    d. Name the sequence of numbers which are multiplied by 14 and 2 in parts a, b, and c.
    e. The results of the computations are certain triangular numbers. Describe which triangular numbers they are.
    f. What triangular number would you expect to get if you extend the list of examples in parts a-c one more line? Do the arithmetic to check your prediction.

15. At a meeting, everyone present shakes hands just once with everyone else. How many handshakes does this involve if the number of people present is:

    a. 2 b. 3 c. 4 d. 5

    e. Identify the sequence of numbers found as answers in parts a-d.

    f. When the 1st person arrives, no handshakes occur (there is nobody else to shake hands with). When the 2nd person arrives, there is one handshake. The 3rd person to arrive shakes hands with each of the two already there, the 4th shakes hands with each of the three already there, and so on. If you were the 101st person to arrive, how many hands would you shake?

    g. If there are 101 members at the meeting, how many handshakes would be required in all?

    h. Section 1.3 showed a way to add 1+2+3+... +100. Compare that method with the method used in problem 3 above.

16. (A pattern discovered by Terry Trotter) Take the three consecutive triangular numbers 3, 6, and 10. Multiply the first and last (3×10), then add 6 to get 36, which is the square of the second of the three consecutive triangular numbers. Test the pattern with 6, 10, 15 and again with 10, 15, and 21.

3.2 Figurate Numbers in Three Dimensions

Figurate numbers in 3-D are especially interesting. You will meet these numbers here in a set of exploratory exercises, as well as in later chapters. Visualization in three dimensions can be difficult, so physical models help. We begin with **tetrahedral numbers**, whose name indicates a three-sided pyramid or **tetrahedron**. (The prefix *tetra,* Greek for four, refers to the base as a fourth side.)

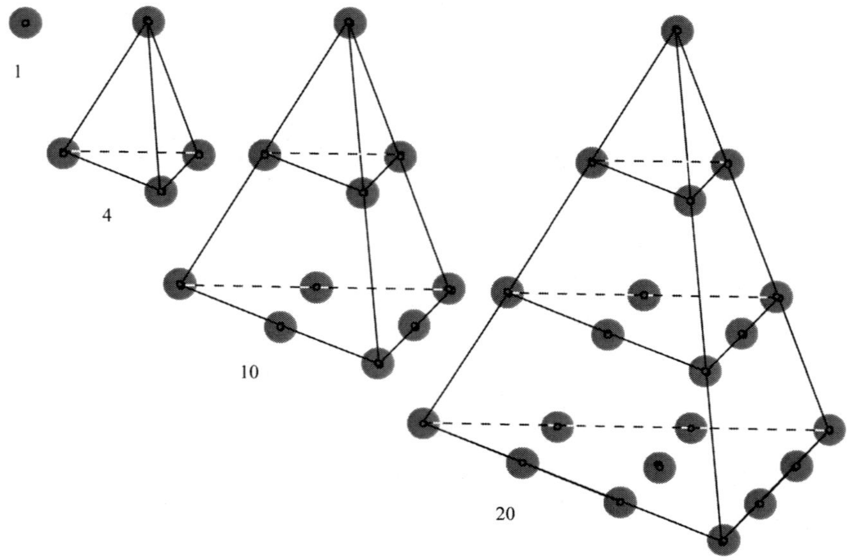

Figure 3.2.1 The first four tetrahedral numbers

The three-dimensional analog of a square is called a **cube**. The three smallest cubes are shown in figure 3.2.2.

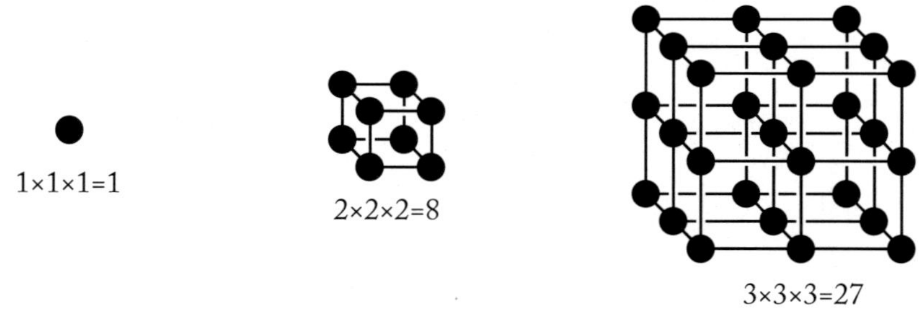

Figure 3.2.2

A good way to get to know cubes and tetrahedral numbers is to make models of them with gumdrops and toothpicks. In a classroom, this works well in small groups. Wash hands before beginning; nibbling, authorized or not, may happen. Allow time for cleanup.

Exercises 3.2

1. Get some gumdrops and toothpicks, and make models of the tetrahedral numbers and cubes pictured in figures 3.2.1 and 3.2.2. Then build the next two larger tetrahedral numbers, which are 20 and 35. You might also want to build the next larger cube, but that is a longer project, since it needs 4×4×4=64 gumdrops.

2. Find the patterns in these additions:

$$\begin{aligned} 1 &= \\ 1+3 &= \\ 1+3+6 &= \\ 1+3+6+10 &= \\ 1+3+6+10+15 &= \end{aligned}$$

   a. Name the family of numbers being added on the left-hand sides.
   b. Carry out the additions, and name the family of numbers which appears on the right-hand sides, which you have also met recently.
   c. Try to explain these patterns geometrically, by looking at gumdrop models or pictures like figure 3.2.1.

3. The first ten cubes (counting 0×0×0=0 as a cube) are:

   0, 1, 8, 27, 64, 125, 216, 343, 512, 729, 1,000.

   a. In each of these cubes, underline the units digit. Find a pattern in these units digits analogous to that found for square numbers in problem 10 of Exercises 3.1.
   b. Extend the list of cubes up to 20×20×20, and check to see whether or not the pattern discovered in part a continues.

4. When Ramanujan, a mystic who regarded numbers as his friends, was ill, his friend G. H. Hardy visited him.[2] To cheer him up by talking about numbers, Hardy said he had just ridden in cab number 1,729, a number which he found uninteresting. Ramanujan replied that, to the contrary, 1,729 is a very interesting number, since it is the smallest number which can be expressed in two different ways as the sum of two cubes. What are those ways?

---

2 Srinivasa Ramanujan (1887-1920), a self-taught mathematician from Madras, India, sent a list of his mathematical discoveries to Cambridge University, where Hardy was professor of mathematics. Hardy found the list astonishing; some of Ramanujan's results were previously unknown and remarkably deep. Ramanujan gave no proofs for his discoveries, which were based on his unique intuition about numbers. He went to Cambridge and spent years there in productive collaboration with Hardy, returning to India when he was already gravely ill with tuberculosis.

5. a. Carry out the following additions:

$$1=$$
$$3+5=$$
$$7+9+11=$$
$$13+15+17+19=$$
$$21+23+25+27+29=$$

   b. Name the family of numbers which appear as sums in part a.

   c. Use the pattern to write the next line down and predict the sum.

   d. Check your prediction from part c by carrying out the addition.

6. a. Multiply:

$$1\times2\times3=$$
$$2\times3\times4=$$
$$3\times4\times5=$$
$$4\times5\times6=$$
$$5\times6\times7=$$

   b. Multiply tetrahedral numbers by six:

$$1\times6=$$
$$4\times6=$$
$$10\times6=$$
$$20\times6=$$
$$35\times6=$$

   c. Compare your answers in parts a and b.

   d. Make a prediction about $6\times7\times8$ and six times a certain tetrahedral number, then do the arithmetic to check it.

7. The first eight tetrahedral numbers are: 1, 4, 10, 20, 35, 56, 84, 120. Select any three in a row from this list, multiply the middle one by 4 and add the other two, as in these examples:

$$1+4\times4+10=$$
$$4+10\times4+20=$$
$$10+20\times4+35=$$
$$20+35\times4+56=$$
$$35+56\times4+84=$$

   a. Carry out the calculations.

   b. Identify the family of numbers which you get as sums in part a.

   c. Extend the pattern two more lines, then check to see if you got the sums you expected.

Exploration: Tetrahedral numbers and squares in square grids (problems 8-11)

8. How many squares can be found in this picture?

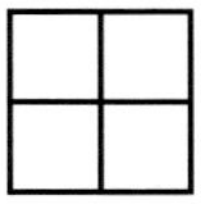

There are four small squares and one large 2×2 square, for a total of five

a. How many squares are there in this picture?

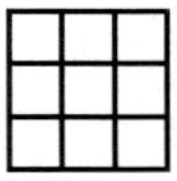

Here are two systematic ways to count them.

Classify the squares by size: How many 1×1 squares? ______

How many 2×2 squares? ______

How many 3×3 squares? ______

Total number of squares ______

b. Count how many squares have their upper right-hand corners at each point in the square grid. For example, vertex A

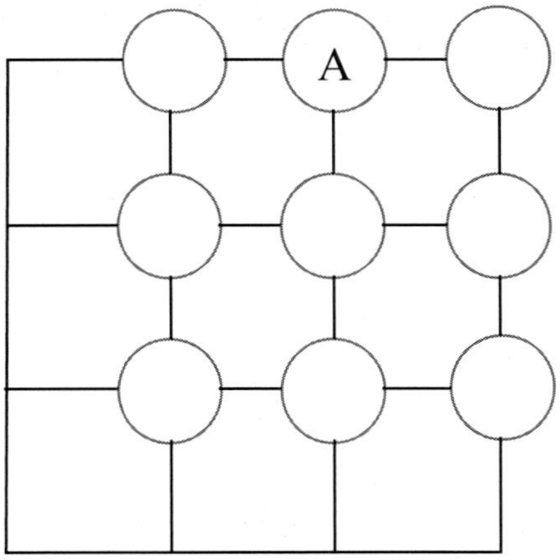

pictured is the upper right-hand corner of two squares, as shown below,

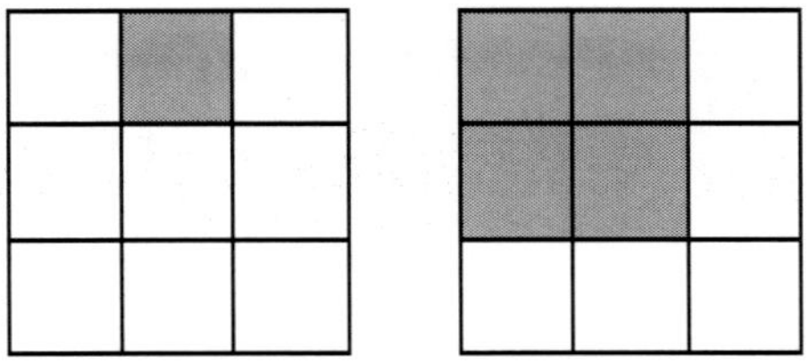

so write 2 in the circle that matches A. Fill in all the circles that way.

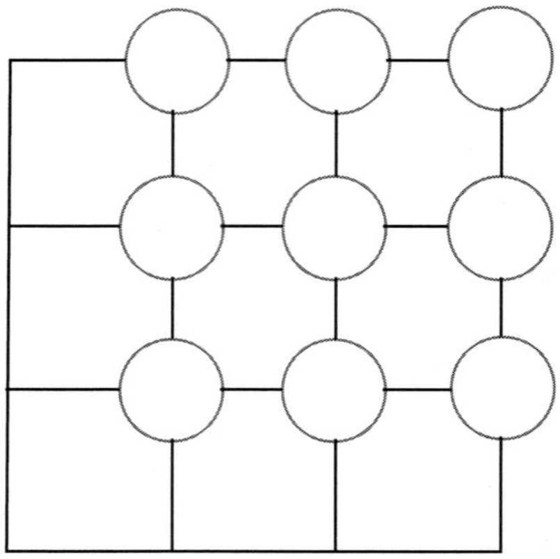

c. Make a model with toothpicks and gumdrops, with the bottom level a 3×3 square, the second level a 2×2 square, and the top level a 1×1 square, using different colors of gumdrops to match the first three squares pictured below. Stack the squares so that each gumdrop is directly above one of the same color.

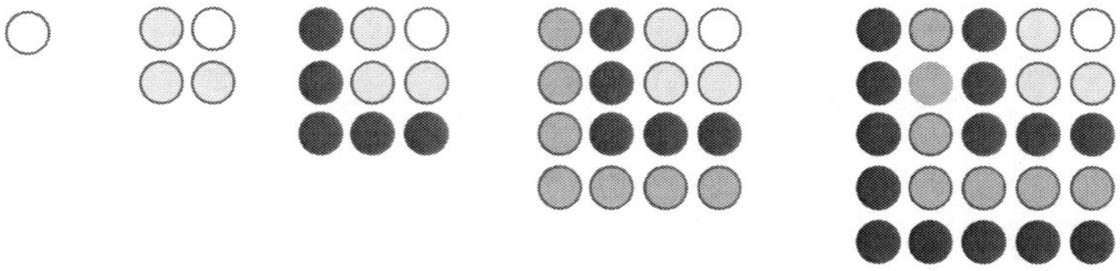

d. Following parts b and c above, explain why the number of squares in a ruled 4×4 square is the sum of the first four squares, 1+4+9+16.

9. a. Add to check that the following statements are true.

$$1=1$$
$$1+4=1+4$$
$$1+4+9=4+10$$
$$1+4+9+16=10+20$$
$$1+4+9+16+25=20+35$$

b. Identify the family of numbers added on the left-hand sides.

c. Identify the family of numbers added on the right-hand sides.

d. Extend the pattern another line and check to see whether or not the resulting equation is true.

10. a. Which line from question 9 is pictured here?

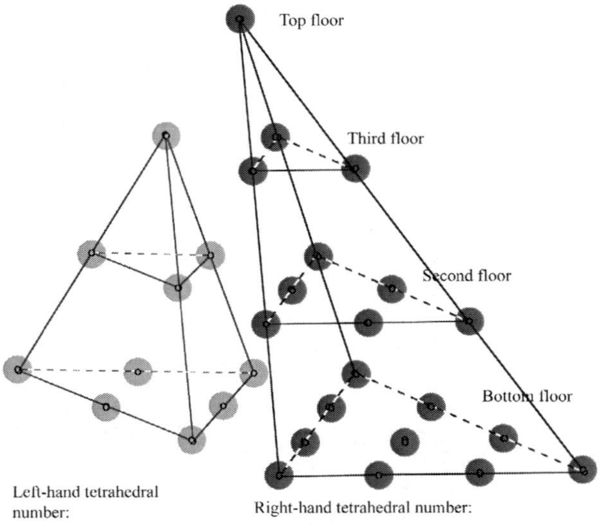

b. What family of numbers are represented by the different "floors"?

c. What family of numbers are represented by the balls of just one shade of gray on each level? (Compare with problems 5-7 of exercises 3.1.)

d. What numbers are represented by the totality of balls (all levels) of just one shade of gray?

e. Do the other lines in question 9 also fit this geometric interpretation? Explain for the bottom line: 1+4+9+16+25=20+35.

11. a. Compute:

$$1 =$$
$$4 =$$
$$1 + 9 =$$
$$4 + 16 =$$
$$1 + 9 + 25 =$$
$$4 + 16 + 36 =$$

b. Identify precisely the family of numbers added on the left.

c. Identify the family of numbers you get as sums.

d. Extend the pattern two more lines, then check it by doing the arithmetic.

12. a. Compute:

$$8 \times 0 + 1 =$$
$$8 \times 1 + 2 =$$
$$8 \times 4 + 3 =$$
$$8 \times 10 + 4 =$$
$$8 \times 20 + 5 =$$

b. What family of numbers is multiplied by 8 here?

c. Describe as accurately as possible the set of numbers resulting from part a.

d. Extend the patterns another line or two, then do the arithmetic to see if the results remain accurate.

13. a. Compute:

$$\begin{aligned} 4(1+4) &= \\ 4(4+10) &= \\ 4(10+20) &= \\ 4(20+35) &= \end{aligned}$$

b. Describe accurately what is being added in the parentheses.

c. Extend the pattern another line or two and check it.

14. Take any three consecutive tetrahedral numbers. Multiply the middle one by six, then add the other two, like this:

$$\begin{aligned} 0+6\times 1+4 &= \\ 1+6\times 4+10 &= \\ 4+6\times 10+20 &= \\ 10+6\times 20+35 &= \end{aligned}$$

a. Do the arithmetic and describe clearly the numbers which result.

b. Extend the pattern another line or two, then check the arithmetic.

15. a. Calculate:

$$\begin{aligned} 1\cdot 1 &= \\ 1\cdot 2+2\cdot 1 &= \\ 1\cdot 3+2\cdot 2+3\cdot 1 &= \\ 1\cdot 4+2\cdot 3+3\cdot 2+4\cdot 1 &= \\ 1\cdot 5+2\cdot 4+3\cdot 3+4\cdot 2+5\cdot 1 &= \end{aligned}$$

b. What family of numbers are the answers?

c. Extend the pattern two more lines and check it.

16. The array of numbers below was discovered by Roberto Soto in 2006.

$$\begin{aligned} 1 &= 1 \\ 8 &= 2+6 \\ 27 &= 3+12+12 \\ 64 &= 4+18+24+18 \\ 125 &= 5+24+36+36+24 \\ 216 &= 6+30+48+54+48+30 \end{aligned}$$

a. What family of numbers appear on the left-hand sides of these equations?

b. What family of numbers appears just to the right hand side of the equals signs?

c. Describe the numbers in the vertical column beginning with 6.

d. Describe the numbers in the vertical column beginning with 12.

e. Describe the numbers in the vertical column beginning with 18.

f. Extend the pattern for two more lines.

g. Do the arithmetic to check whether or not the extended patterns claimed from part f are indeed correct.

17. a. Multiply triangular numbers starting with 3 by successive integers, then add, to calculate:

$$1 \cdot 3 =$$
$$1 \cdot 3 + 2 \cdot 6 =$$
$$1 \cdot 3 + 2 \cdot 6 + 3 \cdot 10 =$$
$$1 \cdot 3 + 2 \cdot 6 + 3 \cdot 10 + 4 \cdot 15 =$$

b. What kind of figurate numbers are the sums from part a?.

c. Sharpen your answer from part b to describe precisely the numbers found as sums in part a.

d. Extend the pattern from part a two more lines. Then do the arithmetic in the new lines and see if you get what you would predict from your answer to part c.

Vocabulary for Chapter 3

Cube (3.2)
Square number (3.1)
Tetrahedral number (3.2)
Tetrahedron (3.2)
Triangular number (3.1)

Self-Quiz for Chapter 3

1. Picture the sum 1+2+3+4+5+6+7+8+9 as a triangular number.

2. Picture the sum 1+3+5+7+9+11+13 as a square number.

3. Make a model to show that 8 is a cube.

4. Make a model to show that 10 is a tetrahedral number.

5. Show pictorially that the sum of two consecutive triangular numbers is a square number.

6. What digits cannot possibly be the last (units) digit of a square number?

Chapter Four

Multiples, Factors, And Primes

This chapter is about the structure of the system of whole numbers as it relates to multiplication. Students still memorize multiplication tables, as did Mesopotamian students 4,000 years ago, though the need for doing so today is sometimes questioned. We will discuss that issue, but our main goal is deeper understanding of this part of mathematics, which is fundamental for higher mathematics and interesting in its own right.

4.1 Multiples and the Marvelous Multiplication Table

If you multiply a whole number n by other whole numbers you get **multiples** of n. For example, 10, 20, 30, and 40 are multiples of 10, since you can get them by multiplying 10, respectively, by 1, 2, 3, and 4. Figure 4.1.1a models multiples of 3 as rectangular areas using Cuisenaire rods, while figure 4.1.1b models 9 as a multiple of 3 by showing that a train of 3 rods has the same length as a 9 rod.

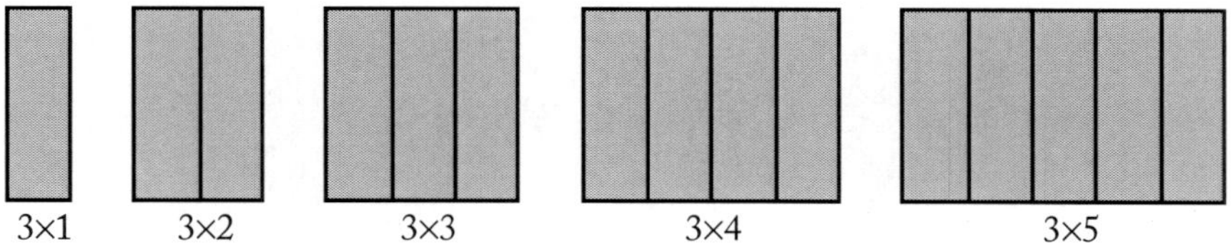

Figure 4.1.1a

Figure 4.1.1b

You use multiples of one-digit numbers in long multiplication and division. You also use multiples to find common denominators when you add fractions. For example, to add $\frac{3}{10}+\frac{5}{6}$, the common denominator must be a multiple of both 10 and 6. One obvious multiple of both 6 and 10 is 6x10, but there are others. To find them, you could list multiples of 10 and multiples of 6 and look for numbers that appear in both lists.

Multiples of 10: 10, 20, 30, 40, 50, 60, 70, 80, 90, 100, 120, . . .
Multiples of 6: 6, 12, 18, 24, 30, 36, 42, 48, 54, 60, 66, 72, . . .

These lists continue without end, but already you can see two numbers, 30 and 60, which appear in both lists. Numbers common to the two lists, like 30 and 60, are said to be **common multiples** of 6 and 10. You could use 30, 60, or any other common multiple of 6 and 10 as common denominator to add $\frac{3}{10} + \frac{5}{6}$. Procedures for finding least common multiples, useful in adding fractions, are discussed in section 5.2.

Multiples are so useful that generations of students memorized the table in figure 4.1.2.

10	10	20	30	40	50	60	70	80	90	100
9	9	18	27	36	45	54	63	72	81	90
8	8	16	24	32	40	48	56	64	72	80
7	7	14	21	28	35	42	49	56	63	70
6	6	12	18	24	30	36	42	48	54	60
5	5	10	15	20	25	30	35	40	45	50
4	4	8	12	16	20	24	28	32	36	40
3	3	6	9	12	15	18	21	24	27	30
2	2	4	6	8	10	12	14	16	18	20
1	1	2	3	4	5	6	7	8	9	10
	1	**2**	**3**	**4**	**5**	**6**	**7**	**8**	**9**	**10**

Figure 4.1.2

You can picture the table in terms of arrays in figure 4.1.3.

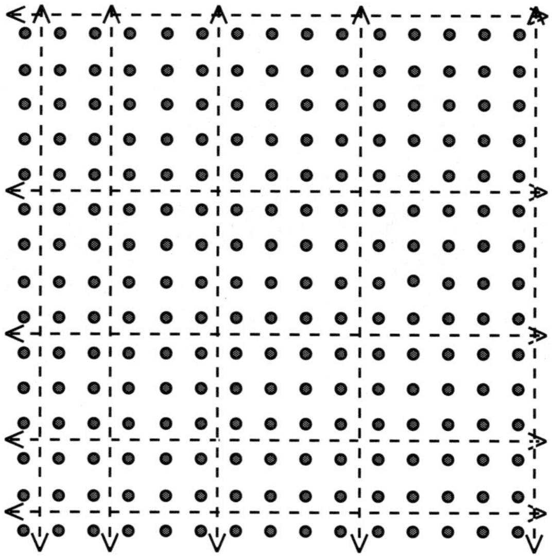

Figure 4.1.3

The table extends endlessly to multiples of larger numbers, but knowing one-digit multiplication facts, together with long multiplication, allows you to multiply numbers of any size.[1] Therefore, grade schoolers traditionally learn multiplication facts up to 10 times 10. Unfortunately, the beauty of the multiplication table is often overlooked in the rush to memorize it. Even more serious, the push to have students learn facts by rote can bring out authoritarian teaching styles that may lead students to hate mathematics. Is it really important for students to memorize the multiplication table? Or is that, as some claim, a waste of time in an age of electronic calculation?

Rote learning alone is inadequate in today's world, but the multiplication table, and all basic arithmetic, remain fundamental, even in an electronic age. Students should learn multiplication facts, but the process need not be unpleasant. Here are some ways to help students understand and appreciate the multiplication table as they learn it.

1. **Introduce multiplication as repeated addition** (as in chapter 1). Reinforce this with activities like addition box puzzles with all entries the same (see Exercises 1.3) and questions about, say, the value of a number of nickels.
2. **Use activities that involve counting by 2s, 3s, and other numbers.** One such activity: Count aloud quickly in unison, shouting multiples of a certain number and saying the other numbers more and more softly until they are silent. e.g for 3, up to 30:

First time:	1,2,**3**,4,5,**6**,7,8,**9**,. . . (emphasize multiples of 3)
Second time:	1,2,**3**, 4, 5, **6**,7,8,**9**,. . .(very soft except multiples of 3)
Third time-	1, 2, **3**,4, 5, **6** , 7, 8, **9**, . . (whisper except multiples of 3)
Fourth time:	_,_,**3**,_,_,**6**,_,_,**9**, . . .(silent, but shout multiples of 3)

3. **Use songs to help students learn multiplication tables.** Music and poetry can make memorization easy and fun.
4. **Study multiples from the point of view of patterns.** Some of these patterns are explored in the exercises below.
5. **Give students multiplication tables with some of the numbers left blank, and have them use the patterns to fill the blanks.** Repeat this periodically, leaving out more and more entries.
6. **Post a multiplication table up to 10×10 prominently in the classroom** for students to refer to in doing mathematics. Explore patterns in the table. Tell the class that they are expected to learn the table.
   a. Warn them that you will cover up the multiples of 1 in the posted table. This is no loss, since the 1s table is so easy.
   b. Warn that you will cover the 2s, and help the students memorize them. Point out that they can just add the

[1] During the Renaissance, our current decimal system of writing numbers replaced Roman numerals in Europe, and people began to learn arithmetic for commerce. In 1820 August Crelle, founder of an early mathematical research journal, published a multiplication table up to 1,000x1,000.

number to itself. Cover up the 2s only when the class has them pretty well learned.

c. Show how easy the 10s are and how to use them to find the 9s. (Take a a number from 10 times itself to get 9 times it.). After warning and practice, cover up the tens.

d. The 5s have an easy pattern to learn, and anyhow they can be found by cutting the 10s in half. After some practice, cover up the 5s.

e. Point out a pattern in the 9s; the units digits decrease as the tens digits increase, so that their total is 9. After ample warning and practice, cover up the 9s.

f. Point out that doubling the 2s yields 4s, as does halving the 8s. Then, after ample warning and practice, cover up the 4s. Those who need help can still look at the 8s and take half to get the 4s.

g. By now there are only a small number of combinations left to learn, and they, too, have patterns.

7. **Use multiplication practice activities available on line.**

8. **Have students make sets of Napier's rods (page 29) and use them.** Making rods reinforces learning the multiplication table, and using rods reinforces the connection between the table and long multiplication.

After all this, some students may still have trouble learning some multiplication facts. Do not make them feel like failures for this, but encourage them to continue their efforts.

Exercises 4.1

Exploration: Patterns in multiples (questions 1-8). Ideally, do questions 1-3 on plastic transparencies and save them for use in later exercises.

1. a. Circle all multiples of 2, up to 30.

1	2	3	4	5	6	7	8	9	10
11	12	13	14	15	16	17	18	19	20
21	22	23	24	25	26	27	28	29	30
31	32	33	34	35	36	37	38	39	40
41	42	43	44	45	46	47	48	49	50
51	52	53	54	55	56	57	58	59	60
61	62	63	64	65	66	67	68	69	70
71	72	73	74	75	76	77	78	79	80
81	82	83	84	85	86	87	88	89	90
91	92	93	94	95	96	97	98	99	100

Figure 4.1.4 Multiples of 2

b. Find a pattern in the circles and extend it to multiples of 2 up to 100.

2. a. Circle all multiples of 3, up to 30.

1	2	3	4	5	6	7	8	9	10
11	12	13	14	15	16	17	18	19	20
21	22	23	24	25	26	27	28	29	30
31	32	33	34	35	36	37	38	39	40
41	42	43	44	45	46	47	48	49	50
51	52	53	54	55	56	57	58	59	60
61	62	63	64	65	66	67	68	69	70
71	72	73	74	75	76	77	78	79	80
81	82	83	84	85	86	87	88	89	90
91	92	93	94	95	96	97	98	99	100

Figure 4.1.5 Multiples of 3

b. Find a pattern in the circles from part a. Once you are sure the pattern is correct, extend it to circle all the multiples of 3 up to 100.

3. As in questions 1 and 2 above, circle the multiples indicated:

1	2	3	4	5	6	7	8	9	10
11	12	13	14	15	16	17	18	19	20
21	22	23	24	25	26	27	28	29	30
31	32	33	34	35	36	37	38	39	40
41	42	43	44	45	46	47	48	49	50
51	52	53	54	55	56	57	58	59	60
61	62	63	64	65	66	67	68	69	70
71	72	73	74	75	76	77	78	79	80
81	82	83	84	85	86	87	88	89	90
91	92	93	94	95	96	97	98	99	100

Figure 4.1.6 Multiples of 4

1	2	3	4	5	6	7	8	9	10
11	12	13	14	15	16	17	18	19	20
21	22	23	24	25	26	27	28	29	30
31	32	33	34	35	36	37	38	39	40
41	42	43	44	45	46	47	48	49	50
51	52	53	54	55	56	57	58	59	60
61	62	63	64	65	66	67	68	69	70
71	72	73	74	75	76	77	78	79	80
81	82	83	84	85	86	87	88	89	90
91	92	93	94	95	96	97	98	99	100

Figure 4.1.7 Multiples of 5

1	2	3	4	5	**6**	7	8	9	10
11	**12**	13	14	15	16	17	**18**	19	20
21	22	23	**24**	25	26	27	28	29	**30**
31	32	33	34	35	36	37	38	39	40
41	42	43	44	45	46	47	48	49	50
51	52	53	54	55	56	57	58	59	60
61	62	63	64	65	66	67	68	69	70
71	72	73	74	75	76	77	78	79	80
81	82	83	84	85	86	87	88	89	90
91	92	93	94	95	96	97	98	99	100

Figure 4.1.8 Multiples of 6

1	2	3	4	5	6	**7**	8	9	10
11	12	13	**14**	15	16	17	18	19	20
21	22	23	24	25	26	27	**28**	29	30
31	32	33	34	35	36	37	38	39	40
41	42	43	44	45	46	47	48	49	50
51	52	53	54	55	56	57	58	59	60
61	62	63	64	65	66	67	68	69	70
71	72	73	74	75	76	77	78	79	80
81	82	83	84	85	86	87	88	89	90
91	92	93	94	95	96	97	98	99	100

Figure 4.1.9 Multiples of 7

1	2	3	4	5	6	7	**8**	9	10
11	12	13	14	15	**16**	17	18	19	20
21	22	23	**24**	25	26	27	28	29	30
31	32	33	34	35	36	37	38	39	40
41	42	43	44	45	46	47	48	49	50
51	52	53	54	55	56	57	58	59	60
61	62	63	64	65	66	67	68	69	70
71	72	73	74	75	76	77	78	79	80
81	82	83	84	85	86	87	88	89	90
91	92	93	94	95	96	97	98	99	100

Figure 4.1.10 Multiples of 8

1	2	3	4	5	6	7	8	**9**	10
11	12	13	14	15	16	17	**18**	19	20
21	22	23	24	25	26	**27**	28	29	30
31	32	33	34	35	36	37	38	39	40
41	42	43	44	45	46	47	48	49	50
51	52	53	54	55	56	57	58	59	60
61	62	63	64	65	66	67	68	69	70
71	72	73	74	75	76	77	78	79	80
81	82	83	84	85	86	87	88	89	90
91	92	93	94	95	96	97	98	99	100

Figure 4.1.11 Multiples of 9

1	2	3	4	5	6	7	8	9	**10**
11	12	13	14	15	16	17	18	19	**20**
21	22	23	24	25	26	27	28	29	**30**
31	32	33	34	35	36	37	38	39	40
41	42	43	44	45	46	47	48	49	50
51	52	53	54	55	56	57	58	59	60
61	62	63	64	65	66	67	68	69	70
71	72	73	74	75	76	77	78	79	80
81	82	83	84	85	86	87	88	89	90
91	92	93	94	95	96	97	98	99	100

Figure 4.1.12 Multiples of 10

4. Comparing the circle patterns for multiples of 2 and of 6, you can see that the multiples of 6 are some, but not all, of the multiples of 2. Find another number whose multiples include all the multiples of 6.

5. a. Find several other numbers, all of whose multiples are included among the multiples of 2.
   b. If you were to circle all multiples of 11, 12, 13, 14, and 15, which, if any, of these would be included among the multiples of 2? Experiment briefly to check your answer.
   c. What distinguishes numbers whose multiples are all multiples of 2?

6. What distinguishes the numbers which are common multiples of:
   a. 2 and 3
   b. 2 and 5
   c. 3 and 5
   d. 2, 3, and 5 (all three)
   e. 4 and 5
   f. 2 and 4
   g. 4 and 6
   h. 2, 3, 4, and 6 (all four)

7. The multiples of 9 (9, 18, 27, 36, 45, 54, 63, 72, 81, 90) have a pattern in their digits, which may help people learn them. Describe that pattern in your own words.

8. Find or make up an activity, which, like questions 1-3 above, could be used to help children learn about multiples visually and kinesthetically.

Exploration. Differences of squares (problems 9-13)

The pattern explored here will be encountered over and over again in more advanced work.

10	20	30	40	50	60	70	80	90	**100**
9	18	27	36	45	54	63	72	**81**	90
8	16	24	32	40	48	56	**64**	72	80
7	14	21	28	35	42	**49**	56	63	70
6	12	18	24	30	**36**	42	48	54	60
5	10	15	20	**25**	30	35	40	45	50
4	8	12	**16**	20	24	28	32	36	40
3	6	**9**	12	15	18	21	24	27	30
2	**4**	6	8	10	12	14	16	18	20
1	2	3	4	5	6	7	8	9	10

9. a. Describe where in the multiplication table above the sequence of square numbers is found.

   b. Starting at 49 in the table, if you move one entry to the right and one down, you are at 48. Try this with the other square numbers in the line you found in part a. Describe what you observe in a sentence or two.

   c. Use an area model of multiplication and a picture to show $6\times8=7\times7-1$.

   d. State an algebraic generalization relating $(a-1)\times(a+1)$ to $a\times a$.

   e. Use the pattern from part d and the fact that $30\times30=900$, to predict the value of 29×31.

   f. Check your prediction from part e by doing the arithmetic.

10. a. Starting at 49 in the multiplication table above, if you move two entries to the right and two down, you are at 45. Try this starting with several other square numbers. Describe what you observe in a sentence or two.

    b. Use an area model of multiplication and a picture to show $5\times9=7\times7-4$.

    c. Use an area model of multiplication and a picture to show $7\times11=9\times9-4$.

    d. State an algebraic generalization relating $(a-2)\times(a+2)$ to $a\times a$.

    e. Use the fact that $70\times70=4{,}900$, together with the pattern from part f, to find 68×72.

    f. Check your answer from part g by doing the arithmetic.

11. a. Starting at 49 in the multiplication table above, if you move three entries to the right and three down, you are at 40. Try this starting with several other square numbers. Describe what you observe in a sentence or two.

  b. Use an area model of multiplication and a picture to show $4\times10=7\times7-9$.

  c. Knowing that $20\times20=400$, use the pattern from part a to predict the value of 17×23. Then check your prediction by multiplying.

  d. Knowing that $16\times16=256$, use the pattern from part a to predict the value of 13×19. Then check your prediction by multiplying.

  e. Knowing that $19\times19=361$, use the pattern from part a to predict the value of 16×22. Then check your prediction by multiplying.

  f. State an algebraic generalization relating $(a-3)\times(a+3)$ to $a\times a$.

12. a. Based on the patterns from questions 9-11, guess an algebraic relation between $(a-4)\times(a+4)$ and $a\times a$.

  b. Show how the pattern from part a applies to 26×34.

  c. Use the pattern from part a to calculate 26×34.

  d. Show how the pattern from part a applies to 46×54.

  e. Use the pattern from part a to calculate 46×54.

13. a. Put the patterns from questions 9-12 together into one more general pattern relating $(a-b)\times(a+b)$ to $a\times a$.[2]

  b. Make up and do three examples to test your pattern from part a.

14. a. Adding along each diagonal indicated in the multiplication table shown, the first three sums are: 1, 2+2=4, and 3+4+3=10. Find the next seven sums.

10	20	30	40	50	60	70	80	90	100
9	18	27	36	45	54	63	72	81	90
8	16	24	32	40	48	56	64	72	80
7	14	21	28	35	42	49	56	63	70
6	12	18	24	30	36	42	48	54	60
5	10	15	20	25	30	35	40	45	50
4	8	12	16	20	24	28	32	36	40
3	6	9	12	15	18	21	24	27	30
2	4	6	8	10	12	14	16	18	20
1	2	3	4	5	6	7	8	9	10

[2] This pattern was known in ancient times as the "regula Nichomachi" after Nichomachus of Gerasa, who included this and many other properties of figurate numbers in his *Introduction to Arithmetic*, written late in the first century.

b. Identify by name the family of numbers found in part a.

c. Use this picture of a tetrahedral number, with the tetrahedron perched on an edge, to explain the discovery from parts a and b.

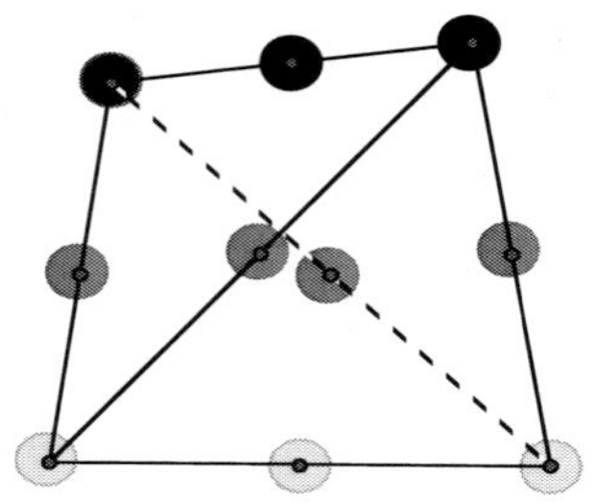

d. Use toothpicks and gumdrops, as in section 3.2, to model the tetrahedral number 20, showing it as $1\times4+2\times3+3\times2+4\times1$.

Exploration. Rectangles in a ruled square (problems 15-21)

15. How many rectangles are there in this picture?

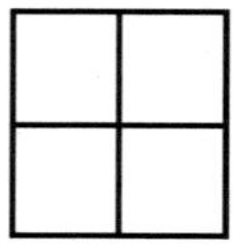

Squares are rectangles, too, so there are actually several rectangles, some overlapping, as shown here:

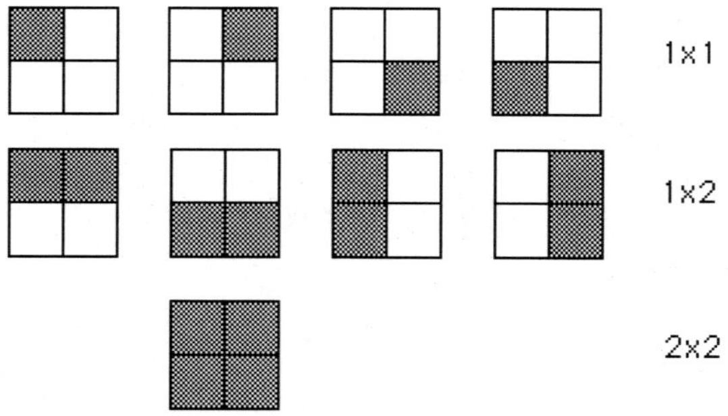

16. How many rectangles are there in this picture?

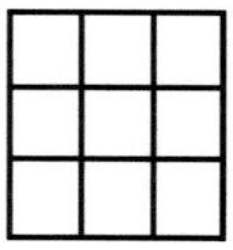

Use the copies below (enough are provided) to shade in first all 1×1 rectangles, then all 1×2 rectangles, all 1×3, all 2×2, all 2×3, and all 3×3. When you have finished, count and complete the table below:

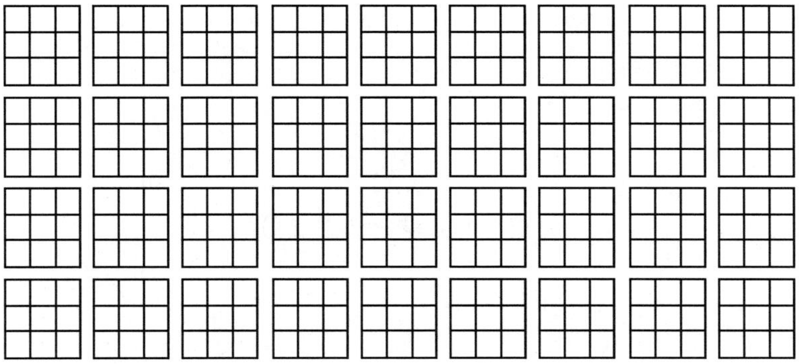

Based on your pictures, count and fill in the blanks:

a. The number of rectangles with shortest side 3 is _____

b. The number of rectangles with shortest side 2 is _____

c. The number of rectangles with shortest side 1 is ____

d. From what family of numbers are your answers to parts a, b, and c?

e. Add your answers from parts a, b, and c to find the total number of rectangles in the ruled square.

17. How many rectangles are there in this 4×4 ruled grid?

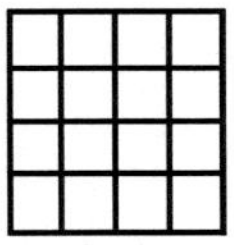

With whatever sketches you want, find how many rectangles are:

a. 1×1

b. 1×2

c. 1×3

d. 1×4

e. 2×2

f. 2×3

g. 2×4

h. 3×3

i. 3×4

j. 4×4

k. Add the answers from parts a-d to find the number of rectangles with shortest side 1.

l. Add the answers from parts e-g to find the number of rectangles with shortest side 2.

m. Add the answers from parts h and i to find the number of rectangles with shortest side 3.

n. Add the answers from parts j-m to find the total number of rectangles in the 4×4 grid.

18. A remarkable pattern can be found in the following sums:

$$\begin{aligned} 1 &= \\ 1+8 &= \\ 1+8+27 &= \\ 1+8+27+64 &= \\ 1+8+27+64+125 &= \end{aligned}$$

a. Identify the family of numbers being added on the left-hand sides.

b. Carry out the additions.

c. Identify the sums you get on the right-hand side. (They are square numbers, but they are not **all** of the square numbers. Which ones are they?)

d. Extend the table one more line, predict what the sum should be according to your pattern, and then check your prediction by carrying out the addition.

19. There is a connection between the numbers involved in problem 18 and those obtained in problems 15, 16, and 17.

a. Find this connection, and describe it in detail.

b. Use the result from part a to predict the number of rectangles in this 6×6 ruled square.

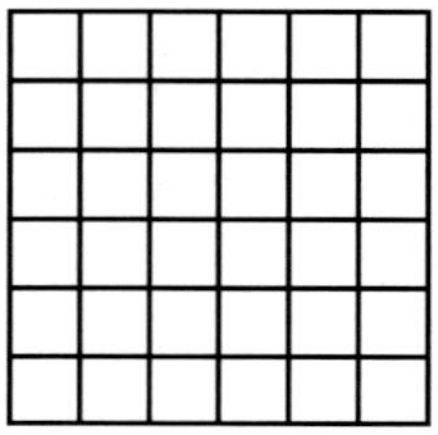

c. Check your prediction from part a by counting the number of rectangles in this 6×6 ruled square.

20. In problems 15-19 the rectangles in a ruled square are classified by width and length, then counted by asking how many rectangles have each possible length and width. Another strategy for counting the rectangles in a ruled square is to count the number of such rectangles with their upper right-hand corners at a given point. For example, how many rectangles have their upper right-hand corner at vertex A in the picture below?

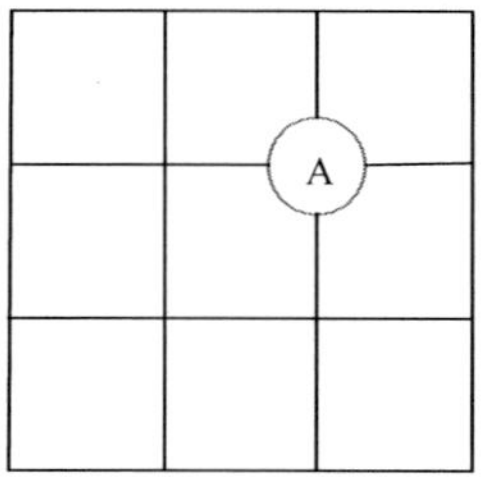

A little experimenting turns up four such rectangles.

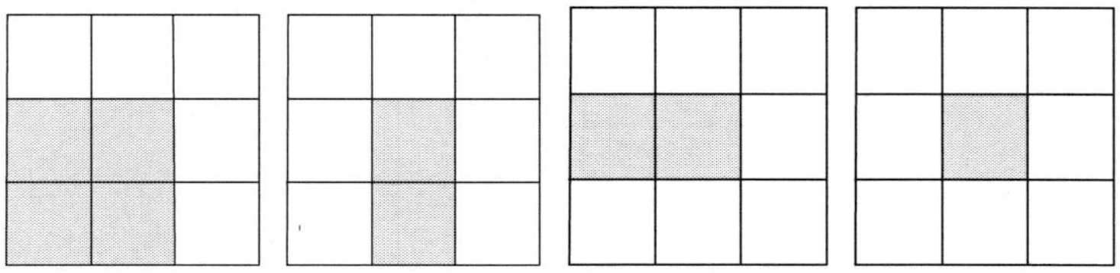

Therefore, 4 is written in place of A, like this.

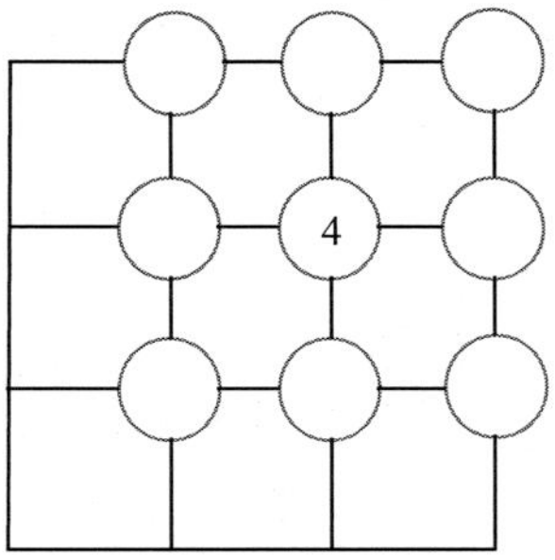

a. Find out how many rectangles have their upper right-hand vertices at each circled vertex, and write that number in the circle.

b. Describe the pattern of numbers in the circles after completing part a.

c. Use the distributive property to explain why the total number of rectangles in a 3×3 ruled square is the square of (1+2+3).

d. Follow the steps from parts a-c to explain why the number of rectangles in a 4×4 ruled square is the square of (1+2+3+4).

21. Problems 15-20 suggest that the sum of the first n cubes is the same as the square of the sum of the first n positive integers-i.e., that $1^3+2^3+...+n^3 = (1+2+3+...+n)^2$. This result was known in ancient India and actually proved by Al Karaji[3] (died, 1019), who observed that in figure 4.1.3 the number of dots on each side is 1+2+3+4+5, so there are $(1+2+3+4+5)^2$ dots in the picture.

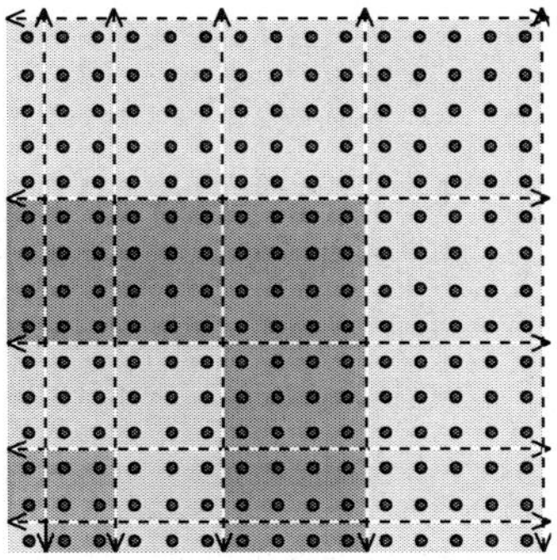

He also observed that the number of dots in each L-shape is a cube. For example the dots in 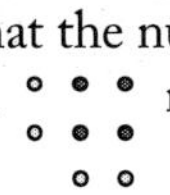 may be reassembled as the cube 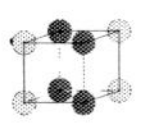

[3] Abu Bakr al-Karaji (died 1019) lived in Baghdad and wrote extensively on mathematics and engineering. His proof of this result about sums of cubes is in his algebra book, *al-Fakhri* (*The Marveloous*)

a. Explain why the numbers of dots in the next larger L-shape, 3+6+9+6+3, is the cube number 3•3•3. Use the cube below and Al-Karaji's array in your explanation.

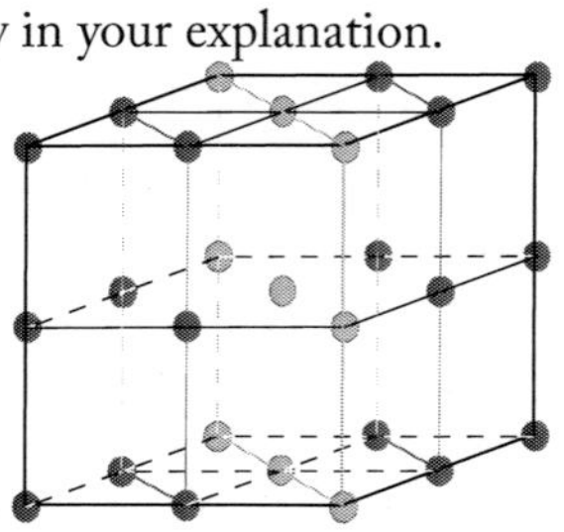

b. Color the 4•4•4 cube shown below analogously to the cube in part a, to show that 1•4+2•4+3•4+4•4+3•4+2•4+1•4=4•4•4.

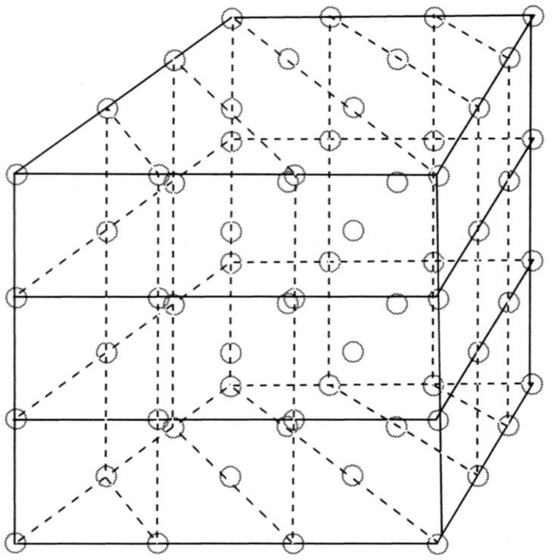

22. a. In the multiplication table add all the numbers that lie in a square with the 1 in the lower left hand corner.

8	8	16	24	32	40	48	56	64
7	7	14	21	28	35	42	49	56
6	6	12	18	24	30	36	42	48
5	5	10	15	20	25	30	35	40
4	4	8	12	16	20	24	28	32
3	3	6	9	12	15	18	21	24
2	2	4	6	8	10	12	14	16
1	1	2	3	4	5	6	7	8
	1	**2**	**3**	**4**	**5**	**6**	**7**	**8**

The first four of those squares are

1

2	4
1	2

3	6	9
2	4	6
1	2	3

4	8	12	16
3	6	9	12
2	4	6	8
1	2	3	4

b. Identify the sums found in part a in terms of figurate numbers.

c. Check your statement from part b on the next two cases.

23. (From a colleague, Joe Chavez) Choose two sets of numbers A={2, 4, 5, 9} and B={3, 7, 11, 21}, and fill in a grid with the products of one factor from set A and one from set B in random order.

84	14	55	189
35	99	42	44
22	27	28	15
12	105	63	6

a. Player 1 selects one number from A and one from B and puts an X on their product in the grid. Then Player 2 repeats this with a new pair of numbers. Play continues in this way.
Whoever gets three Xs or Os in a row, as in tic tac toe, wins.

b. Make a square array that would allow you to play the same game with the sets C={3, 4, 5, 13} and D={2, 5, 6, 7} instead of sets A and B above.

c. Repeat the game, making up your own sets A and B.

4.2 Factors and Primes

Now we will turn around ideas from the previous section. Instead of looking at multiples of a number, we will start with a whole number and ask of which numbers it is a multiple. For example, if we start with 25, we find it is a multiple of 1, of 5, and of 25. We use the word **factor** when we take this point of view, saying that 1, 5, and 25 are factors of 25. Again, if we start with 30, we find it is a multiple of 1, 2, 3, 5, 6, 10, 15, and 30, so we say these are the factors of 30. The word **divisor** is often used interchangeably with the word **factor**. The idea behind that word is that a divisor of a number is one that divides into it with zero remainder. For example, if you divide 30 by 6, you have zero remainder, so 6 is a divisor of 30. The words **divisor** and **factor** really have the same meaning, but for now we use **factor**, since we have not yet discussed division and remainders.

The only whole number with just one factor is 1. Any other whole number is a multiple of itself and of 1 and therefore has at least two factors.

A whole number with just two factors is called a **prime number** or, for short, a **prime.** The numbers 2, 3, and 7 are examples of primes.

A whole number with more than two factors is called a **composite number**. Examples of composite numbers are 4, 6, and 9.

When you express a number as a product, you **factor** it. Any number can be factored as 1 times itself, but this is usually not of interest. This is called **trivial factorization** and is sometimes not counted as factoring at all, even though 1 and the number itself are factors of any number. A prime number, such as 7, can only be factored trivially, or, as is sometimes said, cannot be factored. On the other hand, composite numbers can be factored, possibly in several ways. For example, 30 can be factored as 2×15, 3×10, or 5×6, whereas 15 can only be factored non trivially as 3×5.

Example 1. Which, if any, of 30, 31, 32, and 33 is (are) prime?
Solution: You can factor 30 as 2×15, so it is not prime. Any even number greater than 2, such as 32, is a multiple of 2 and so is composite. Now, 33 is not even, but still it can be factored as 3×11, so 33 is not prime. On the other hand, no matter how hard you try, you cannot factor 31 (except as 1×31), so 31 is a prime.

A number is said to be **completely factored** if it is factored as a product of primes. If 18 is factored as 3×6 or as 2×9, it is not completely factored, since 6 and 9 are not primes. However, the factorization of 18 as 2×3×3 is complete, since all the factors are primes. To factor 18 completely, you could start by factoring it as 3×6, then factor the 6 as 2×3, or you could start with 2×9 and then factor the 9 as 3×3. These processes are illustrated in figure 4.2.1 by means of **factor trees**. To see 18 completely factored, express it as a product of the numbers at the ends of the branches.

Figure 4.2.1

The phrase **factored into primes** is often used for **completely factored**, and **complete factorization** or **prime factorization** refer to the result of this process.

The factoring process shown in figure 4.2.1 applies generally. Start with any composite number and factor it. If any of the factors are composite, factor them in turn. Continue this process until all of the numbers at the ends of the branches are primes. Interestingly, **the prime factorizations of a composite number differ only in the order in which the factors are written.** (For example, for 18 you might have found 3×2×3 or 3×3×2 instead of 2×3×3, but you could not have come up with anything else.) Since multiplication is commutative, these factorizations are essentially the same, a situation which turns out to be true for any number.

Uniqueness of Prime Factorization

> **If a number is factored completely, the set of primes involved will always be the same, though the order in which they appear may be different.**

As another illustration, if you factor 84 into primes, you might get 2×2×3×7, 2×7×3×2, or some other arrangement of two factors of 2, one factor of 3, and one factor of 7, but you always end up with two 2s, one 3, and one 7.

We will illustrate this uniqueness of factorization in the exercises below, and we will use it implicitly by speaking of **the** prime factorization of a number. Just because a composite number can be factored does not mean it is easy to do so. Multiplying whole numbers is cut and dried, but factoring involves trial and error. This is easy for small numbers, but factoring very large numbers is tedious, even with a computer, and it may not be easy to tell ahead whether or not it can be done at all. The number may, after all, be a prime.

Example 2. Is 197 a prime number?
Solution: If 197 is not a prime, then it must have a factor between 2 and 197. Do we need to try all the numbers from 2 to 197? No. For example, 197 surely is not a multiple of 196. In fact, you can see that no number larger than half of 197 could possibly help factor 197. But we don't even have to go up that far, because if half of 197 were a factor, 197 would be 2 times that number, so 2 would be a factor, and we would find that out when we work up the list of possible factors, starting with 2. Similarly, if 3 is not a factor, then there is no factor which is one-third of 197, and we need not check for factors that large. As you can check, it happens that neither 2 nor 3 is a factor of 197. Next, you might think we should try 4, but there is no need to do so. We saw in section 4.1 that every multiple of 4 is also a multiple of 2. Since 197 is not a multiple of 2, it cannot be a multiple of 4. The same reasoning applies to 6, which would be a factor only if both 2 and 3 were factors. Therefore, **we need try only prime numbers.** It is easy to see that 5 is not a factor. (To be a multiple of 5, 197 would have to end in a 0 or a 5.) The next primes to try are 7, 11, and 13, but none of them turn out to be factors. Do we need to try higher primes? No. For example, if 17 were a factor of 197, then 197 would be 17 times a number smaller than 17, since 17×17 is greater than 197. But then some smaller prime would be a factor of 197. The same reasoning applies to any prime whose square is greater than 197. Therefore, search no further for factors; 197 is prime.

Testing a number to see if it is prime

1. **You need check only primes as possible factors.**
2. **Work your way up the list of primes starting with 2, looking for factors. You need not try any prime whose square is larger than the number you want to factor.**

It would be nice to have a quick way to determine whether or not a given number is prime, but no such test is known. There are divisibility tests to show quickly whether or not a number is a multiple of a particular number. These tests can cut down the work in trying to factor large numbers, but there are no easy divisibility tests for most of the higher primes.

Divisibility Tests

A number is a multiple of 2 if and only if its last digit is even.
You can see this is so by looking at figure 4.1.2.

A number is a multiple of 3 if and only if its digits add up to a multiple of 3.

Illustrations:

a. 52 **is not** a multiple of 3, and 5+2=7, which **is not** a multiple of 3.
b. 489 **is** a multiple of 3, and 4+8+9=21 **is** a multiple of 3.

A number is a multiple of 4 if and only if its last two digits are a multiple of 4. You can see this by extending figure 4.1.5. The last two digits match those in figure 4.1.5 itself. The first two rows after 100 are

101	102	103	(104)	105	106	107	(108)	109	110
111	(112)	113	114	115	(116)	117	118	119	(120)

where the circled numbers end in 4, 8, 12, 16, 20. The multiples of 4 from 100 to 200 thus end in the same two digits as the multiples of 4 from 0 to 100. The same is true from 200 to 300, 300 to 400, and so on for all higher numbers.

Illustrations:

a. 31,724 **is** a multiple of 4, since 24 **is** a multiple of 4.

b. 4,398 **is not** a multiple of 4, since 98 **is not** a multiple of 4.

A number is a multiple of 5 if and only if its last digit is 0 or 5. You can see this from figure 4.1.6.

A number is a multiple of 9 if and only if its digits add up to a multiple of 9. This is explored further in Exercises 6.3.

Illustrations:

a. 837 **is** a multiple of 9, since 8+3+7=18, which **is** a multiple of 9.

b. 94 **is not** a multiple of 9, since 9+4=13 **is not** a multiple of 9.

A number is a multiple of 11 if and only if you get a multiple of 11 (possibly negative or 0) by alternately adding and subtracting the digits.

Illustrations:

a. 475 **is not** a multiple of 11, since 4-7+5 = 2, which **is not** a multiple of 11.

b. 23,470,216 **is** a multiple of 11, since 2-3+4-7+0-2+1-6=-11, which **is** a negative multiple of 11.

There are no simple tests for divisibility by most other primes.

Example 3. Plato, like many ancient Greeks, had a mystical view of numbers. In *The Laws* he stated that an ideal city has just 5,040 lots. (He did not relate this number to the size of the population!) Find a way to factor 5,040, which shows why the number might appeal to a mystic.
Solution: Start by factoring 5,040 as 504×10. Now factor each of the factors, 504 and 10, continuing until 5,040 is completely factored as 2×2×2×2×3×3×5×7. With some recombining, this leads to 5,040=1×2×3×4×5×6×7, which must have been significant to Plato.

Example 4 (Difference of squares). Factor 8,099 completely.
Solution: The basic divisibility tests show that 2, 3, 5, and 11 are not factors. It turns out that 7 is a factor, but even after factoring out 7, it appears a difficult task remains. However, this example is included to illustrate a trick that sometimes helps in such cases. Note that 8,099 is just one less than 8,100, which is a perfect square (90×90). That means you can use, in reverse, the multiplication trick from question 9 of exercises 4.1 to factor 8,099 as 89×91. Now 89 is prime, but 91=7×13, so 8,099=7×13×89.

Questions 9-13 of exercises 4.1 give us a result which is even more general:

Any difference of squares $a \times a - b \times b$ can be factored as $(a-b)(a+b)$.
Symbolically, $a \times a - b \times b = (a-b)(a+b)$

This amounts to using the result of exercises 4.1, question 13, in reverse, as the following picture shows.

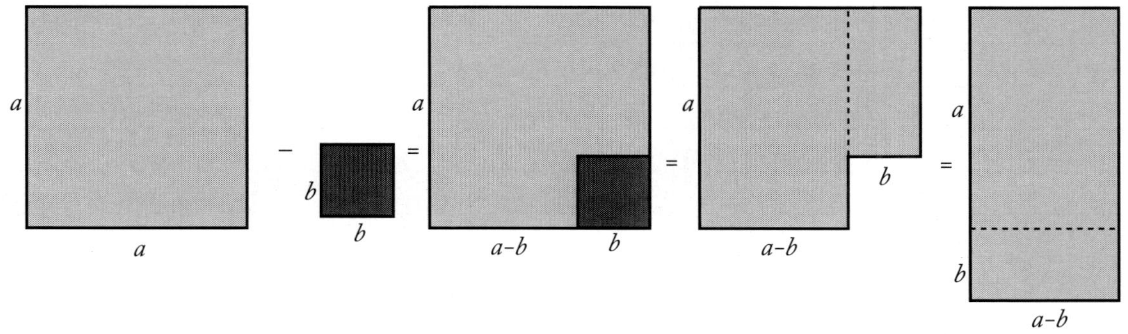

This observation is most useful in factoring numbers like 713, where trial and error is tedious, because no small numbers are factors. Since 713=729-16, which is 27×27-4×4, it can be factored as (27-4)(27+4)=23×31. It helps to know the square numbers.

Exercises 4.2

1. Explain the difference between a number x being a **factor** of a second number y and x being a **multiple** of a second number y.

2. Use Cuisenaire rods to show that:
   a. 4 is a factor of 8
   b. 2 is a factor of 6
   c. 9 is a multiple of 3
   d. 9 is not a multiple of 4
   e. 3 is not a factor of 10

3. a. Use square tiles to form all the rectangular arrays possible with the number of tiles named at the left in the table below. Then use the information to complete the rest of the table.

   b Look for relationships between the number of rectangles that can be formed and the number of factors of *n*.

Number of tiles *n*	**Length and width of possible rectangles**	**How many rectangles?**	**Factors of *n***	**How many factors?**	**Prime?**
1					
2					
3					
4					
5					
6					
7					
8	1×8, 2×4, 4×2, 8×1	4	1, 2, 4, 8	4	No
9					
10					
11					
12					
13					
14					
15					
16					
17					
18					
19					
20					
21					
22					
23					
24					
25					

Teaching note: This activity can be used to introduce factoring. As teachers, we may use our knowledge of factoring to fill in the table quickly. For example, knowing that 8=4×2=8×1, we can see that 8 squares may be arranged as a rectangle 4 units long and 2 wide or as a rectangle 8 units long and 1 wide. Children may not see this so fast. For them, forming a rectangle with a given number of squares and finding the dimensions of that rectangle (factoring the number) involves trial and error, moving unit squares until a rectangle is formed.

4. a. Factor 60 in as many ways as you can using just two factors.
   b. Starting with each of the different factorizations you found in part a, factor 60 into primes.
   c. Compare the various prime factorizations of 60 found in part b. Do they indeed differ at most in the order in which the prime factors appear?

5. a. Factor 168 in as many ways as you can using just two factors.
   b. Starting with each of the different factorizations you found in part a, factor 168 into primes.
   c. Compare the various prime factorizations of 168 found in part b. Do they indeed differ at most in the order in which the prime factors appear?

6. Which, if any, of these numbers are primes? Find prime factorizations of those which are composite.

a. 23	b. 25	c. 33	d. 47	e. 51	f. 81
g. 87	h. 89	i. 91	j. 93	k. 97	l. 99
m. 117	n. 119	o. 121	p. 143	q. 149	r. 297

7. If you think of the multiplication table (Figure 4.1.2) as extending without end, some numbers will appear in it more often than others. For example, 5 appears only twice, as the products 1×5 and 5×1.
   a. How many times does 6 appear?
   b. How many times does 8 appear?
   c. How many times does 12 appear?
   d. Describe the set of numbers which appear exactly twice in the multiplication table.
   e. Describe the set of numbers which appear exactly three times in the multiplication table.
   f. Describe the set of numbers which appear an odd number of times in the multiplication table.

8. Use divisibility tests to determine **without actually dividing,** whether or not:
   a. 5 is a factor of 2,385,880
   b. 3 is a factor of 889,922,335,544,117,766
   c. 4 is a factor of 56,800,278
   d. 9 is a factor of 415,298,301
   e. 11 is a factor of 10,203,103,020

9. **Exploration: the Sieve of Eratosthenes**[4]
The sieve of Eratosthenes is a way to find all the primes less than a given number. To illustrate, we use it to find all primes less than 100. Begin with the table used in exploring multiples and follow the steps (parts a-h).

1	2	3	4	5	6	7	8	9	10
11	12	13	14	15	16	17	18	19	20
21	22	23	24	25	26	27	28	29	30
31	32	33	34	35	36	37	38	39	40
41	42	43	44	45	46	47	48	49	50
51	52	53	54	55	56	57	58	59	60
61	62	63	64	65	66	67	68	69	70
71	72	73	74	75	76	77	78	79	80
81	82	83	84	85	86	87	88	89	90
91	92	93	94	95	96	97	98	99	100

a. The smallest prime is 2. Since each higher multiple of 2 is composite (cannot possibly be a prime), cross them all out. The primes less than 100 are among the numbers **not** crossed out.

b. The next smallest prime is 3, since it is not a multiple of any smaller prime. Using a different color pencil, cross out every multiple of 3 except for 3 itself.

c. The next smallest prime is 5. Using a third color, cross out all higher multiples of 5 i.e., 10, 15, 20, ...

d. The next smallest prime is 7. Using a fourth color, eliminate all higher multiples of 7.

e. At this point, every prime less than 100 is among the numbers which have not been crossed out. Are all those numbers primes? No, 1 is not a prime. But are there any composite numbers less than 100 which are not yet crossed out? To find out, observe:

   The first number crossed out as a multiple of **2** is **4**.

   The first number crossed out as a multiple of 3 is 6, but 6 was already crossed out. The first multiple of **3** which was not already crossed out is **9**.

   The first number crossed out as a multiple of 5 is 10, but 10 was already crossed out. The first multiple of **5** which was not crossed out before is **25**.

   The first number crossed out as a multiple of 7 is 14, but 14 was already crossed out. The first multiple of **7** which was not crossed out before is **49**.

[4] Eratosthenes of Cyrene (about 274-194 B.C.E.) taught in Alexandria after about 240 B.C.E., where he was librarian. His main interests were poetry and geography, but he is also remembered in mathematics for his "sieve" for determining all the prime numbers up to any given size. His nickname was "Beta." Problem 9 of exercises 18.3 shows Eratosthenes' elegant calculation of the size of the Earth.

Ignoring numbers already crossed out:

The first number crossed out by **2** is **4**

The first number crossed out by **3** is **9**

The first number crossed out by **5** is **25**

The first number crossed out by **7** is **49**

Do you see the pattern? If the pattern continues, what would be the first number to be crossed out by 11?

f. Try to explain the pattern in part e. (Hint: If a number is a multiple of more than one prime, which one eliminates that number first?)

g. Make a list of all the prime numbers less than 100.

h. If you use the sieve of Eratosthenes to determine all the primes less than 200, what is the largest prime whose multiples you have to cross out?

10. (This problem extends the technique mentioned in example 4.)

a. Factor 6,399 into primes by first noting that 6,399=6,400-1.

b. Factor 2,491 into primes, by first noting that 2,491=2,500-9=50×50-3×3.

c. Factor 9,991 into primes.

11. A common error. A student factored 149 incorrectly as 2×3×3×7. Here is his attempt at a factor tree for this problem.

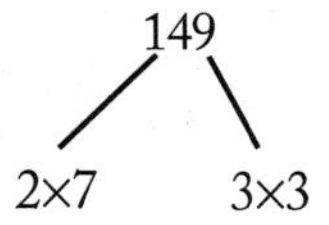

a. What did this student do wrong?

b. How would you help this student?

12. If you used transparencies to do questions 1-3 of exercises 4.1, how could you use those transparencies to do the sieve of Eratosthenes?

13. Show that the triangular number 210 can be expressed as a product either of two consecutive numbers or of three consecutive numbers.

14. Use differences of squares to factor:

a. 1,599	b. 1,591	c. 1,519
d. 2,499	e. 891	f. 2,419
g. 3,599	h. 3,591	i. 3,519

15. $A4273B$ is a six-digit number in which A and B are digits, and the number is a multiple of 72. Find the value of A and the value of B. (Used, with permission, from George Lenchner: *Math Olympiad Contest Problems for Elementary and Middle Schools,* MOESM, 1997.)

4.3 Using Prime Factorization

With numbers factored into primes, it is easy to find the answers to certain questions. For example, is 196 a multiple of 7? You could answer this by listing multiples of 7, or you could divide 196 by 7 to see if there is a remainder, but if you happen to know that the prime factorization of 196 is $2{\cdot}2{\cdot}7{\cdot}7$, you can see at a glance that 7 is a factor of 196, or, what amounts to the same thing, that 196 is a multiple of 7. You can also see that $7{\cdot}28=196$, since $28=2{\cdot}2{\cdot}7$. The same reasoning applies to larger numbers.

Example 1. Is 18,791 a multiple of 437?
Solution: To answer this you might list multiples of 437 and see if 18,791 is in the list. Alternatively, you could divide 18,791 by 437 and see if there is a non-zero remainder. However, if you happen to have these numbers completely factored, you can answer the question at a glance. It turns out that $437=19{\cdot}23$, while $18{,}791=19{\cdot}23{\cdot}43$. By the associative property of multiplication, $18{,}791=(19{\cdot}23){\cdot}43=437{\cdot}43$. Thus 18,791 is indeed a multiple of 437.

Sometimes we can use our knowledge of prime factors to answer this kind of question without factoring completely. Example 2 is a case in point.

Example 2. Is 15 a factor of 4,389?
Solution: Recall that every multiple of 15 is also a multiple of 5, and any multiple of 5 ends in 5 or 0. But 4,389 does not end in 5 or 0, so it is not a multiple of 5, and therefore not of 15 either. Then 15 is not a factor of 4,389.

Example 3. Is 168 a common multiple of 28, 12, and 42?
Solution: Factor 168 into primes: $168=2{\cdot}2{\cdot}2{\cdot}3{\cdot}7$. Can you make 28, 12, and 42 from the prime factors of 168? Yes, since

$$\begin{aligned} &28=2{\cdot}2{\cdot}7\\ &12=2{\cdot}2{\cdot}3\\ \text{and}\quad &42=2{\cdot}3{\cdot}7. \end{aligned}$$

Then $168 = 6{\cdot}\mathbf{28} = 14{\cdot}\mathbf{12} = 4{\cdot}\mathbf{42}$.
Therefore 168 is indeed a common multiple of 28, 12, and 42. Of course, you could learn that without using prime factorization, but in time we will come to cases for which this method is particularly helpful.

To find all the factors (prime or not) of a number can be difficult if you have to start from scratch, but it is easy if you know the number's prime factorization.

Example 4. Find all the factors of 45.
Solution: First factor into primes; $45=3{\cdot}3{\cdot}5$. A factor of 45 has as its own factors at most a single 5 and two 3s. Specifically, the factors of 45 are:

1 (1 is a factor of any number)
3
5
$9=3\cdot3$
$15=3\cdot5$
$45=3\cdot3\cdot5$ (any number is a factor of itself)

The work is systematically presented in the following table.

	1	3	9
1	1	3	9
5	5	15	45

Example 5. Find all the factors of 102.
Solution: You could solve this problem just by trying all the whole numbers up to 102, but prime factorization helps you do the job more efficiently. Factor $102=2\cdot3\cdot17$, and then begin by finding those factors which do not use more than one prime. They are 1, 2, 3, and 17. Next, those factors which use no more than the primes 2 and 3 can be tabulated.

	1	2
1	1	2
3	3	6

The factors of 102 that are not multiples of 17 appear in the body of this table. Now make a new table, with the factors 1, 2, 3, 6 along one edge and 1 and 17 along the other.

	1	2	3	6
1	1	2	3	6
17	17	34	51	102

All the factors of 102 are listed in the body of this table.

The next example applies this same method to a number with quite a few factors. Without a systematic procedure, it would be easy to miss factors in such cases, and even if you listed them all, you would be nagged by the idea that perhaps you had overlooked one or more factors.

Example 6. Find all the factors of 1,800.
Solution: Factor 1,800 into primes. $1{,}800=2\cdot2\cdot2\cdot3\cdot3\cdot5\cdot5$

The factors involving no more than one prime are:
With 2s 1, 2, 4, and 8
With 3s 1, 3, and 9
With 5s 1, 5, and 25

First tabulate all factors which are not multiples of 5.

	1	2	4	8
1	1	2	4	8
3	3	6	12	24
9	9	18	36	72

Now make a new table, with the entries from the body of the above table on one edge (they are factors which do not involve the prime 5) and 1, 5, and 25 along the other edge.

	1	2	4	8	3	6	12	24	9	18	36	72
1	1	2	4	8	3	6	12	24	9	18	36	72
5	5	10	20	40	15	30	60	120	45	90	180	360
25	25	50	100	200	75	150	300	600	225	450	900	1,800

The factors of 1,800 are all the numbers in the body of this table. Had there been a fourth prime factor involved, we would have made a third table, and so on for larger numbers of primes. It is worth taking some time to convince yourself that this method really does list all the factors of 1,800.

Exercises 4.3

1. Is 237 a multiple of 6? Explain your reasoning.

2. Is 9,634 a multiple of 35? Explain your reasoning.

3. Factored into primes, 8=2·2·2, and the factors of 8 (prime or not) are 1, 2, 4, and 8. To list them, start with 1 and then use 2 as a factor one, two, or three times. No number can be a factor of 8 if it has any prime factor other than 2 or if it has more than three factors of 2. In a similar way, list all the factors of:
   a. 32
   b. 27

4. The prime factorization of 196 is 2·2·7·7. Use this to list all the factors of 196, prime or not. (Don't forget 1.)

5. List all the factors of:
   a. 72
   b. 847=7·11·11

6. The prime factorization of 369 is 3·3·41. Use that to decide which of the following, if any, are factors of 369.
   a. 1 b. 2 c. 3 d. 6 e. 9
   f. 36 g. 41 h. 43 i. 82 j. 123

7. The prime factorization of 168 is 2·2·2·3·7. Use that to decide which of the following, if any, are factors of 168.
   a. 6 b. 7 c. 9 d. 12 e. 16
   f. 21 g. 22 h. 40 i. 49

8. A number is called a **common factor** of two other numbers if it is a factor of both of them. For example, 2 is a common factor of 8 and 12.
   a. Factor 72 and 90 into primes and use the result to list all common factors of 72 and 90. (Don't forget 1).
   b. Use prime factorization to find all common factors of 56 and 84.

9. As in example 2, determine without actually dividing, whether or not:
   a. 15 is a factor of 135,725
   b. 45 is a factor of 612,345
   c. 36 is a factor of 1,877,400
   d. 55 is a factor of 4,489,950,105

10. If you multiply out $1 \cdot 2 \cdot 3 \cdot 4 \cdot 5 \cdot 6 \cdot 7 \cdot 8 \cdot 9 \cdot 10 \cdot 11 \cdot 12 \cdot 13 \cdot 14 \cdot 15$, how many zeros will there be at the end of the product?

4.4 The Fascination of Primes

I particularly loved prime numbers, the fact that they were indivisible, could not be broken down, were inalienably themselves. (I had no such confidence in myself, for I felt I was being divided, alienated, broken down, more every week.). Primes were the building blocks of all other numbers, and there must be, I felt, some meaning to them. Why did primes come when they did? Was there any pattern, any logic to their distribution? Was there any limit to them, or did they go on forever? I spent innumerable hours factoring, searching for primes, memorizing them. . . .

I made a grid, ten by ten, of the first hundred numbers with the primes blacked in, but I could see no pattern, no logic to their distribution. I made larger tables, twenty squared, thirty squared, but still could discern no obvious pattern. And yet I was convinced that there must be one.

Oliver Sacks. ***Uncle Tungsten. Memories of a Chemical Boyhood****, p26*

Questions about prime numbers, some of them remarkably difficult to answer, have long been a source of fascination and intrigue.

How many primes are there?
Even the simple question of whether or not a given number is prime can be baffling. True, you might quickly recognize that a number is **not** prime, if, say, it is even or a multiple of 5. In general, however, for large numbers, the only ways to test numbers for primeness are based on some sort of trial and error. Primes get scarcer as we search larger and larger numbers, because each number, to be a prime, must pass the tests of not being a multiple of any smaller primes. Higher numbers have to pass more such tests and are therefore less likely to qualify as primes. You might wonder, then, whether or not the primes get so scarce among very large numbers that they eventually die out entirely. In short, is there a largest prime? The largest number known to be a prime at this writing has some

12.9 million digits, but Euclid, in his *Elements*[5], proved that there is no largest prime, because the number of primes is infinite.

Here is the main idea. Suppose someone thought that 2 and 3 were the only primes. Their product, 6, is a multiple of both these primes; if you add 1 to that product, you get 7, which is not a multiple of either 2 or 3. This is a new prime number. Again, if a person thought that 2, 3, and 5 were the only primes, you could multiply them together to get a multiple of all of them, then add 1 to get a number, 31, which is not a multiple of any of them and is another prime number. No matter how many primes a person thought there were, you could multiply them all together (or at least imagine it) and then add 1 to get a number which is not a multiple of any of those primes. This new number might turn out to be a prime, as did 7 and 31, but prime or not, its prime factors are none of the primes you multiplied together, so in any case there is at least one prime other than those you multiplied together. Evidently, then, you could not have multiplied all of the primes together in the first place, so the number of primes must be infinite.

This is an example of **indirect reasoning**, which is best understood in terms of an analogy. Imagine yourself deep in a tunnel, which you want to leave. If one way leads out, while the other dead ends, how can you tell which is which? If you try one direction, you might be lucky and get out, or you might find it is a dead end. In either case your question is answered, since reaching a dead end tells you which way leads out. The number of primes is either infinite or finite. These two possibilities correspond to the directions in the tunnel. We started in one direction, assuming the number of primes is finite, and we came to a contradiction, a logical dead end. Therefore, the other way is the correct one; the number of primes must be infinite.

People who meet indirect reasoning for the first time sometimes feel swindled by it, because it seems contrived, and indeed it is. It is a device for finding that a tunnel dead-ends; without it, we could wander for a long time without realizing that we would eventually reach a dead end.

Twin primes

Since 2 is the only even prime, 2 and 3 are the only consecutive numbers which are both prime. However, there are many pairs of consecutive odd numbers which are both primes, such as 3 and 5, 11 and 13, 17 and 19, or 71 and 73, or 140,737,488,353,699 and 140,737,488,353,701. Such pairs are called **twin primes**. Nobody knows whether or not there are infinitely many pairs of twin primes. As of August, 2004, the largest known pair of twin primes have 51,090 digits each. To find the latest records, look up "twin primes" on the web.

Primes in arithmetic progression

The only trio of consecutive odd numbers which are all primes is 3, 5, 7, since every third odd number is a multiple of 3. However, there are longer lists of primes which form arithmetic progressions. For example, if you begin with 5 and go up in steady jumps of 6, you get 5, 11, 17, 23, and 29, which are five primes in arithmetic progression. However, the search for longer arithmetic progressions

---

[5] Little is known of Euclid's life. Even the dates of his birth and death are uncertain. He was probably Athenian, and he taught mathematics at the university in Alexandria. His *Elements*, written around 300 B.C.E., summarized the geometry and number theory known at the time, and in logical organization and rigor it was a model of exposition unequalled for over 2,000 years. Evidence of the amazing influence of this work over the centuries is the fact that it has appeared in numerous languages and over 1,000 printed versions, second only to the *Bible*.

containing only prime numbers rapidly becomes difficult. In 1995 an arithmetic progression of seven primes was discovered, and that was bettered in 1997, when a list of eight primes in arithmetic progression was found. In March, 1998 the record was shattered yet again when an arithmetic progression with ten consecutive prime number terms was discovered.[6] In 2004 it was proved that there are sequences of primes in arithmetic progression of any finite length, and an example of such a sequence 23 primes long was found.

Goldbach's conjecture

Another baffling question about primes was raised in 1742 by a Swiss mathematics teacher named Christian Goldbach. He tried to see what numbers he could make by adding pairs of prime numbers. He found that he could make every **even** number he tried, except for 2. For example, 12 = 5+7. For 4 and 6 he had to use the same prime twice (4=2+2 and 6=3+3), but in any case, he conjectured (guessed) that every even number greater than 2 can be expressed as the sum of two primes. Goldbach's conjecture has been checked all the way up to 400,000,000,000,000,000, and no exceptions have been found. It is generally believed that his conjecture is true, but nobody has proved it, and prospects for a proof in the near future seem dim. In March, 2000 the Farber publishing company, in connection with their publication of a novel, *Uncle Petros and Goldbach's Conjecture,* offered a prize of $1,000,000 for the first proof of Goldbach's conjecture. The prize was not claimed by the April, 2002 deadline.

There are many other puzzles and unsolved questions involving prime numbers. A few are given in the exercises below, and a few others are saved until we introduce additional mathematical tools, such as exponents.

Exercises 4.4

1. Find at least one prime number between:

   a. 1 and 4 b. 4 and 9 c. 9 and 16
   d. 16 and 25 e. 25 and 36 f. 36 and 49
   g. 49 and 64 h. 64 and 100

   It is widely believed that there is at least one prime number between any pair of consecutive squares To date, however, nobody has been able to prove this.

2. Check Goldbach's conjecture for even numbers up to 50. When it is possible to express an even number as a sum of two primes in more than one way (for example, 20 is 3+17 or 7+13), do so in as many ways as possible.

3. a. Find three prime numbers whose squares add up to the prime 83.
   b. Find another prime less than 200 which is the sum of the squares of three primes. (It is widely believed that there are infinitely many prime numbers which are the

[6] See : http://www.ltkz.demon.co.uk/ar2/10primes.htm

sums of the squares of three primes, but this has never been proved.)

4. An indication that primes are rarer among larger numbers is that there are longer and longer strings of consecutive composite numbers. Here is one way to find such strings. To find a string of at least three consecutive composite numbers, start with 2·3·4, or 24.

   Since 24 is a multiple of 2, so is 24+2 or 26.

   Since 24 is a multiple of 3, so is 24+3 or 27.

   Since 24 is a multiple of 4, so is 24+4 or 28.

   Therefore, the three consecutive numbers, 26, 27, and 28 are all composite. (In this case 24 and 25 are also composite, so it happens that the string of consecutive composite numbers is even longer, but that is not guaranteed by this reasoning.) By this method we can find a string of at least 49 consecutive composite numbers. First, multiply all the numbers from 2 to 50. Adding 2 to this product yields a multiple of 2, adding 3 to the product yields a multiple of 3, etc. How would you use this idea to find a string of at least 100 consecutive composite numbers?

5. What numbers can you make by adding three primes-not necessarily different from each other?

   6=2+2+2 7=2+2+3 8=2+3+3

   a. Make every number from 9 up to 40 in this way.
   b. It is believed that every whole number greater than 6 can be expressed as a sum of three primes, but this has never been proved. If Goldbach's conjecture were known to be true, you could use it to prove that every whole number greater than 6 can be expressed in at least one way as a sum of three primes. Describe this reasoning clearly.

Vocabulary for Chapter 4

Common multiple (4.1)
Complete factorization (4.2)
Composite number (4.2)
Divisor (4.2)
Divisibility test (4.2)
Factor or divisor (4.2)
Factor tree (4.2)
Indirect reasoning (4.4)
Multiple (4.1)
Prime factorization (4.2)
Prime number (4.2)
Sieve of Eratosthenes (4.2)
Trivial factorization (4.2)
Twin primes (4.4)

Self-Quiz for Chapter 4

1. List five multiples of 13.

2. List three common multiples of 12 and 15.

3. Circle the primes: 77 87 97 107 117

4. Factor completely: a. 1,000 b. 196 c. 143

5. Use divisibility tests to decide without actually dividing, whether or not:
   a. 5 is a factor of 3,265
   b. 3 is a factor of 987,654,321
   c. 4 is a factor of 97,653,724
   d. 9 is a factor of 125,863,452
   e. 11 is a factor of 22,334,455

6. Use divisibility tests to decide without actually dividing, whether or not:
   a. 15 is a factor of 12,345
   b. 45 is a factor of 12,345
   c. 36 is a factor of 335,512
   d. 55 is a factor of 56,655,665

7. Use the fact that $21 \times 21 = 441$ to factor 437.

8. List all the factors (prime or not) of:
   a. 81 b. 90 c. 37

Chapter Five
Exponents

Just as repeated addition of the same number leads to multiplication, repeated multiplication by the same number leads to another process, called **raising to a power** or **exponentiation**. This new process plays an important role in mathematics. Here you will meet exponentiation and explore some of its properties.

5.1 Exponents and Repeated Multiplication

In Chapter 3 you met the square numbers, 1, 4, 9, 16, 25, . . . , which result from multiplying two identical factors, and the cube numbers, which result from multiplying three identical factors. Rather than write out repeated factors, we often use a more compact notation, writing squares as 1^2, 2^2, 3^2, . . . and cubes as 1^3, 2^3, 3^3, ... The raised (superscript) numbers are called **exponents**.[1] They indicate how many factors of the given number are to be multiplied.

1 The word "exponent" was introduced by Michel Stiefel in his 1544 book *Cos.* Stiefel is best known today as a preacher and friend of Martin Luther. Stiefel predicted the end of the world on October 18, 1533. He had to be rescued from angry parishioners who, believing the prediction, had abandoned their worldly possessions. Exponents were first written in the modern way by René Descartes in 1637. Descartes (1596-1650) was born near Tours, France, and attended the Jesuit school at La Flèche where Mersenne (see note, section 5.3) taught. There he studied the mathematical works of Christoph Clavius (see note, section 10.4). Descartes convinced school authorities to let him stay in bed until 11 a.m. every day, saying he did his best thinking at that time. He maintained that habit throughout his life. After years of travel in the army of the Dutch Duke of Orange, he settled in Holland, where he spent 20 years investigating mathematics, philosophy, and science. His *Discourse on the Method of Rightly Conducting the Reasoning and Seeking Truth in the Sciences*, published in 1637, contained the beginnings of plane analytic geometry in an appendix, which also had his new notation for exponents. In 1649 Queen Christina of Sweden invited Descartes to Stockholm to teach her mathematics and philosophy. She took seriously his advice to doubt everything including the Lutheran state religion, of which she was the highest official. Christina, an early riser, wanted lessons very early. Descartes accommodated her but died from pneumonia caught on cold pre-dawn walks to the palace, or so history books say. But a recently discovered letter from Christina's personal physician describes Descartes' symptoms as characteristic not of pneumonia but of arsenic poisoning. It has been speculated that an Augustinian monk feared Descartes' skepticism would spoil a scheme to convert Christina to Catholicism. After Descartes died, Christina converted, abdicated, and lived out her days in sunny Italy.

number or 1st power	square or 2nd power	cube or 3rd power
$1=1^1$	$1=1\cdot 1=1^2$	$1=1\cdot 1\cdot 1=1^3$
$2=2^1$	$4=2\cdot 2=2^2$	$8=2\cdot 2\cdot 2=2^3$
$3=3^1$	$9=3\cdot 3=3^2$	$27=3\cdot 3\cdot 3=3^3$
$4=4^1$	$16=4\cdot 4=4^2$	$64=4\cdot 4\cdot 4=4^3$
$5=5^1$	$25=5\cdot 5=5^2$	$125=5\cdot 5\cdot 5=5^3$
$6=6^1$	$36=6\cdot 6=6^2$	$216=6\cdot 6\cdot 6=6^3$
$7=7^1$	$49=7\cdot 7=7^2$	$343=7\cdot 7\cdot 7=7^3$
$8=8^1$	$64=8\cdot 8=8^2$	$512=8\cdot 8\cdot 8=8^3$
$9=9^1$	$81=9\cdot 9=9^2$	$729=9\cdot 9\cdot 9=9^3$
$10=10^1$	$100=10\cdot 10=10^2$	$1{,}000=10\cdot 10\cdot 10=10^3$

We call the second and third powers the square and cube, even when we do not picture them geometrically. For example, even though you might not picture a square with sides of length -4, you would still speak of $(-4)^2$ as the square of -4. Exponents larger than 3 work the same way. For example, 7^5 means 7×7×7×7×7 and is read "**seven to the fifth power**" or, for short, "seven to the fifth."

Example 1: Use the fact that $7^2=49$ to compute 70^2.
Solution: $70^2=(7\times10)^2=(7\times10)(7\times10)$. By the associative and commutative properties of multiplication, this may be written as (7×7)(10×10), which is 49×100 or 4,900. This example shows that to square any multiple of ten, which may be written $b\times10$, just multiply the square of b by 100. Thus, for example, $80^2=8^2\times10^2=6{,}400$, which is just the square of 8 (i.e., 64) followed by two zeros. Problems 15-17 below exploit this fact for mental multiplications.

An exponent applies only to the number by which it is written, unless grouping symbols indicate otherwise. For example, $1+2^3$ is 1+8=9, whereas in $(1+2)^3$ the parentheses tell us to first add 1+2, then apply the exponent to the sum to get $3^3=27$. Lacking parentheses, a spoken phrase like "one plus two cubed" is inherently ambiguous. To avoid confusion, it is necessary to resort to something like "the cube of the sum of 1 and 2" for $(1+2)^3$ or "1 plus the cube of 2" for $1+2^3$. Though labored, these phrases are clear, and in mathematics clarity is paramount. Therefore, we extend the convention on order of operations (section 1.3) to include exponents.

Convention On Order Of Operations

Unless parentheses indicate otherwise, evaluate expressions involving addition, multiplication, and powers in this order:

1. **Calculate powers,**
2. **Multiply,**
3. **Add.**

If there are several exponentiations, multiplications or additions, do them in order from left to right.

Exercises 5.1

1. Evaluate:
   a. 3^4 and 4^3
   b. 5^2 and 2^5
   c. By parts a and b, in general is $p^q = q^p$?

2. a. Use area to picture $(3+5)^2$.
   b. Use area to picture 3^2+5^2?
   c. Use your pictures from parts a and b to show why $(3+5)^2 \neq 3^2+5^2$.

3. Folding a sheet of paper doubles its thickness.
   a. If a folded paper is folded in half again, how many sheets thick is it?
   b. If a twice-folded paper is folded in half a third time, how many sheets thick is it?
   c. How many sheets thick would the paper be after five such folds?
   d. Based on parts a-c, explain in a sentence or two why it is a safe bet that nobody can fold an ordinary piece of paper in half nine times (never unfolding it) with bare hands alone?

4. In parts a-d compute as indicated.
   a. $1^3+5^3+3^3=$ b. $3^3+7^3+0^3=$
   c. $3^3+7^3+1^3=$ d. $4^3+0^3+7^3=$
   e. Write a sentence describing something remarkable about the computations in parts a-c and their results. (These numbers, along with 0 and 1, are the only numbers with this property.)

5. a. In 1879 one Georges Doster noticed that,

$$3^2+4^2=5^2$$

$$10^2+11^2+12^2=13^2+14^2$$

$$21^2+22^2+23^2+24^2=25^2+26^2+27^2$$

   Do the arithmetic to check these equations. A calculator may help.
   b. Look at the numbers that are squared at the beginning of each line. Identify them as precisely as possible in terms of figurate numbers.
   c. The numbers whose squares are just to the left of the equals signs (4, 12, 24, . . .) are interesting. You may not recognize them as a family, but you would if you divided each one by 4. Describe them.
   d. Extend the pattern one more line and see if the resulting equation is true.

6. A blacksmith offers to shoe a horse and charge for the job as follows: One cent for the first nail, two cents for the second, four cents for the third, eight cents for the fourth, and so on, doubling the price for each successive nail. The blacksmith intends to put six nails in each of the four hooves.
   a. What will the twenty-fourth nail cost?
   b. Find a quick way (no tedious addition) to find the total cost of all 24 nails.

7. Versions of this puzzle have been traced back to ancient Egypt:

   As I was going to St. Ives
   I met a man with seven wives.
   Each wife had seven sacks,
   Each sack had seven cats,
   Each cat had seven kits.
   Kits, cats, sacks, and wives,
   How many were going to St. Ives?

   The answer, of course, is supposed to be one, the speaker. If all those the speaker met, including kits, cats, and sacks, were also going to St. Ives, what would the answer be?

8. Some numbers, which are not squares, can be written as a sum of two squares. For example, 5=1+4, and 8=4+4. Still other numbers are neither squares nor sums of two squares, but they can be written as a sum of three squares. The smallest example, 3, can be written 1+1+1 but cannot be written as a sum of fewer than three squares. Another number that is a sum of three squares and no fewer is 11=9+1+1.
   a. Can every number be written as a sum of no more than three squares? Show that this is not the case, by finding a number less than ten which can be written as a sum of four squares, but no fewer.
   b. Write each integer from 1 to 40 as either a square or a sum of at most four squares. (Every positive integer, no matter how large, can be expressed as a sum of at most four squares. There are similar results for cubes and higher powers, but this area of mathematics has many unsolved problems.)

9. a. Compute: $1^3-1=$
   $2^3-2=$
   $3^3-3=$
   $4^3-4=$
   $5^3-5=$
   b. Each number found in part a, starting with 2^3-2, is a multiple of 6. What family of numbers appears here as the other factors?

Exploration: Sums of Consecutive Positive Integers (questions 10-13)
Some numbers can be expressed as sums of two or more consecutive positive integers. For example, 5=2+3 and 6=1+2+3. Some numbers can be done in more than one way (e.g., 9=4+5=2+3+4), but others, such as 2, cannot be done at all.

10. By experimenting, find which numbers from 0 to 40 can be expressed as a sum of consecutive positive integers. For those numbers that can be expressed that way, list all the ways to do it. (This is a relatively time-consuming task, but it makes a good group activity for grades 3 and up.) Record your work in a chart.

11. a. In your chart from question 10, circle in red all the sums of just two numbers. Then circle in green all the sums of just three numbers, then in blue all the sums of 4 numbers, and in black all the sums of five numbers.
    b. Based on your work from part a, fill in the blanks in the following statements:

       Starting with 3, every second number can be made with 2 addends.

       Starting with ___, every ______ number can be made with 3 addends.

       Starting with ___, every ______ number can be made with 4 addends.

       Starting with ___, every ______ number can be made with 5 addends.
    c. Fill in the blanks in this statement, which generalizes part b:

       Starting with the nth triangular number, every __th number can be made with __ addends.

12. a. List all the numbers below 50 which cannot be written as a sum of two or more consecutive positive integers.
    b. Name (identify) the set of numbers which cannot be expressed as sums of two or more consecutive positive integers.

13. Find and state a rule for how many ways (possibly 0) a number can be expressed as a sum of two or more consecutive positive integers? (This is difficult; feel free to consult the hint in the back.)

14. In Exercises 4.1 you may have seen a remarkable pattern: If you add the whole numbers from 1 to any n and square the sum, you get the same number as if you added the cubes of those numbers. For example, you could add 1+2+3 to get 6, and $6^2=36$, which is what you get when you add the cubes of the original numbers: $1^3+2^3+2^3=1+8+27=36$. This holds not just for the consecutive numbers from 1 to n but for any array of numbers that includes 1 in a rectangular corner of the multiplication table.

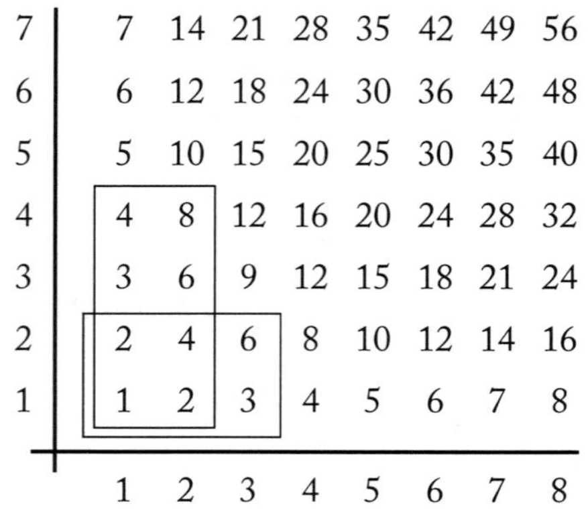

a. For the numbers in this array:

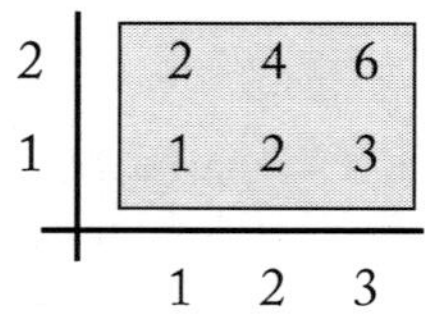

First compute the cubes of the entries and add them:$1^3+2^3+2^3+4^3+3^3+6^3$, then add 1+2+3+2+4+6 and square the sum. Show that you get the same number both ways.

b. Repeat the experiment with the numbers in this array:

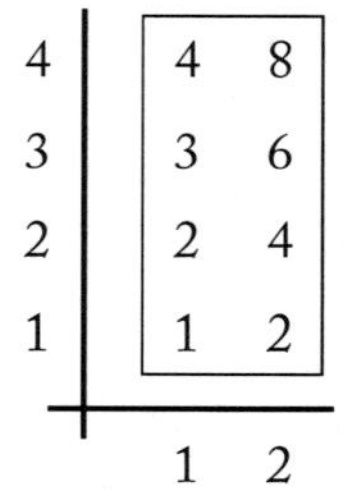

c. Check this pattern in two more cases, using a calculator if you like.

Exploration: Differences of squares and mental multiplication (problems 15-17). These problems use the difference of squares pattern for mental multiplication. Review problems 9-13 from Exercises 4.1 before doing these.

15. Calculate 60^2 mentally as in Example 1 of this section, and then use the difference of squares pattern $(a-b)(a+b)=a^2-b^2$ to calculate mentally:

a. 59×61 b. 62×58 c. 57×63
d. 66×54 e. 51×69

16. Calculate 90^2 mentally as in Example 1 above, and then use the difference of squares pattern $(a-b)(a+b)=a^2-b^2$ to calculate mentally:

a. 92×88 b. 85×95 c. 84×96
d. 98×82 e. 79×101

17. Calculate mentally:

    a. 23×37 b. 11×29 c. 56×44

18. (By Larry Stevenson) Find a three digit number which, if doubled yields a square, but if multiplied by 11 yields a cube.

Exploration: Pascal's triangle (questions 19-23)

19. The array called **Pascal's triangle**[2] appears in many parts of mathematics and contains many wonderful patterns, some of which are explored here. Before going on, see what patterns you can discover for yourself by just looking at the triangle for a few minutes.

```
                  1
               1     1
            1     2     1
         1     3     3     1
      1     4     6     4     1
   1     5    10    10     5     1
1     6    15    20    15     6     1
1  7    21    35    35    21     7     1
```

a. Add:

```
                         1 =
                     1 + 1 =
                 1 + 2 + 1 =
             1 + 3 + 3 + 1 =
         1 + 4 + 6 + 4 + 1 =
    1 + 5 + 10 + 10 + 5 + 1 =
1 + 6 + 15 + 20 + 15 + 6 + 1 =
1 + 7 + 21 + 35 + 35 + 21 + 7 + 1 =
```

[2] Blaise Pascal (1623-1662) was born in Clermont, France. His mother died when he was four, after which he and his three sisters were raised by their single father, Étienne Pascal. Étienne, a tax official and a noted mathematician in his own right, moved the family to Paris and home schooled the children. He recognized his son's gifts and began to take him to scholarly meetings organized by Marin Mersenne (see note, section 5.3) At age 16 Pascal presented to this group some deep new results in geometry. A few years later he invented a mechanical calculator, of which 55 copies were built. Always in poor health, in his late twenties he turned increasingly to the study of theology. His *Traité du triangle arithmetique (Treatise of the Arithmetic Triangle)*, written in 1654, was not published until after his death, but it led to his name being attached to the triangle. The triangle had been discovered in many contexts and cultures in prior centuries, but Pascal's treatise, which appears to be based largely on work presented earlier by Mersenne, was the first systematic study of the triangle to be published.

b. Add and subtract as indicated:

$$
\begin{array}{c}
1 \\
1 - 1 = \\
1 - 2 + 1 = \\
1 - 3 + 3 - 1 = \\
1 - 4 + 6 - 4 + 1 = \\
1 - 5 + 10 - 10 + 5 - 1 = \\
1 - 6 + 15 - 20 + 15 - 6 + 1 = \\
1 - 7 + 21 - 35 + 35 - 21 + 7 - 1 =
\end{array}
$$

c. Find and circle the sequence of triangular numbers in Pascal's triangle.

d. Find and circle the sequence of tetrahedral numbers in Pascal's triangle.

e. Try adding two consecutive entries in any row. Where does their sum appear? Use this to extend the triangle two more rows.

f. Here is a map of one way streets. Starting from A, how many routes lead to each lettered intersection? What do you notice?

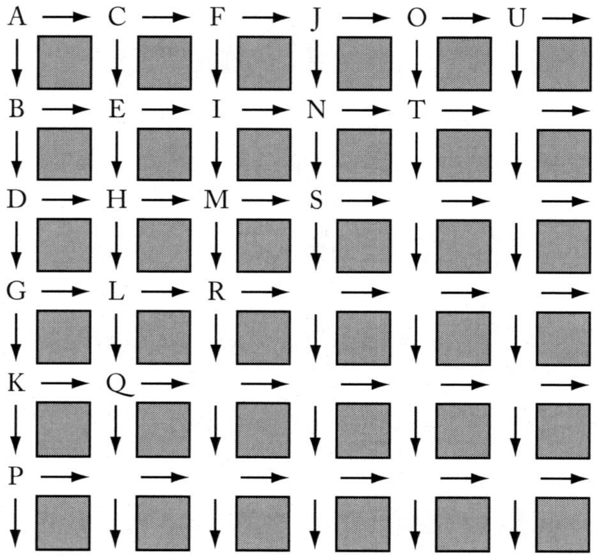

In questions 20-23 you can use positive integers and 0 as often as you like, but no other numbers. For this exercise, consider the same addends written in a different order as different. An example: How many different ways can you make 3 as a sum of three numbers? List them.

Answer:	3+0+0	1+2+0	
	2+0+1	0+0+3	
	2+1+0	0+1+2	
	1+0+2	0+2+1	
	1+1+1	0+3+0	There are ten ways in all.

20. In the same spirit, find out how many ways you can make:
    a. 1 as a sum of 2 numbers ______
    b. 2 as a sum of 2 numbers ______
    c. 3 as a sum of 2 numbers ______
    d. 4 as a sum of 2 numbers ______
    e. (Generalize) n as a sum of 2 numbers ______

21. In the same spirit, find out how many ways you can make:
    a. 1 as a sum of 3 numbers ______
    b. 2 as a sum of 3 numbers ______
    c. 3 as a sum of 3 numbers ______
    d. 4 as a sum of 3 numbers ______
    e. (Generalize) n as a sum of 3 numbers ______

22. In the same spirit, find out how many ways you can make:
    a. 1 as a sum of 4 numbers ______
    b. 2 as a sum of 4 numbers ______
    c. 3 as a sum of 4 numbers ______
    d. 4 as a sum of 4 numbers ______
    e. (Generalize) n as a sum of 4 numbers ______

23. Arrange your answers to questions 20-22 in vertical columns next to each other. Where have you seen this pattern of numbers before?

Exploration: Fibonacci numbers (questions 24-27)

The numbers 1, 1, 2, 3, 5, 8, 13, 21, 34, 55, 89, 144, 233, ... are called the **Fibonacci**[3] **numbers**. After the initial pair of ones, each number in the list is the sum of the two before. These exercises explore some of their properties.

24. Add the squares of two consecutive Fibonacci numbers:
    a. $1^2+1^2=$
    b. $1^2+2^2=$
    c. $2^2+3^2=$
    d. $3^2+5^2=$
    e. $5^2+8^2=$
    f. $8^2+13^2=$
    g. Look for a pattern in your answers to parts a-f, and use it to predict 13^2+21^2 and 21^2+34^2.
    h. Check your predictions from part g by doing the arithmetic.
    i. Describe the pattern from parts a-h in carefully chosen words.

[3] The Fibonacci numbers appear prominently in *The DaVinci Code*, by Dan Brown. For details about Fibonacci himself, see note, section 5.4.

25. The following lines suggest another pattern among the Fibonacci numbers.

$$1^2 + 1^2 = 1 \times 2$$
$$1^2 + 1^2 + 2^2 = 2 \times 3$$
$$1^2 + 1^2 + 2^2 + 3^2 = 3 \times 5$$
$$1^2 + 1^2 + 2^2 + 3^2 + 5^2 = 5 \times 8$$

  a. Extend the pattern two more lines.
  b. Do the arithmetic to check your answer from part a.
  c. Describe the pattern from parts a and b in carefully chosen words.

26. Here is another curious pattern involving Fibonacci numbers.
  a. Do the arithmetic to check that:

$$1 \times 3 = 2^2 - 1$$
$$2 \times 5 = 3^2 + 1$$
$$3 \times 8 = 5^2 - 1$$
$$5 \times 13 = 8^2 + 1$$
$$8 \times 21 = 13^2 - 1$$

  b. Extend the pattern two more lines, then check those cases by doing the arithmetic.
  c. Describe the pattern from parts a and b in carefully chosen words.

27. a. Add along the slanted lines in Pascal's triangle to find a pattern.

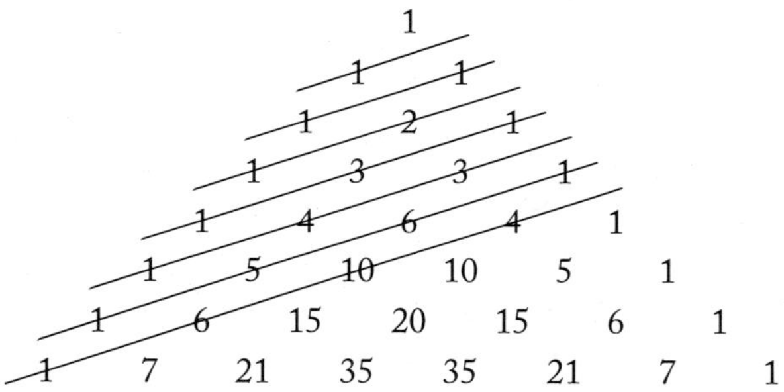

  b. Extend the triangle at least three more lines to see if the pattern continues.
  c. Identify by name the family of numbers obtained as sums.

Explorations starting with "trains" of Cuisenaire rods (questions 28-29)

For these explorations, consider trains as different, even if they differ only in the order of the rods.

28. You can see there is only one possible train of length 1, and there are two possible trains of length 2.

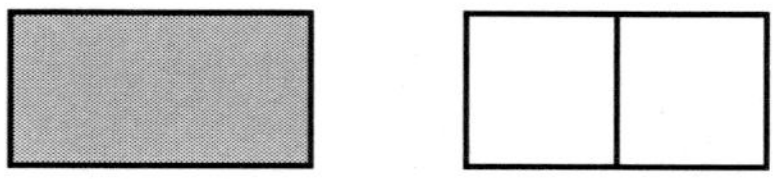

a. There are four trains of length 3. They are 3, 21, 12, and 111. How many trains have length 4?
b. How many trains have length 5?
c. In general, for any given length n, how many trains have length n?

29. Answer question 28 if you are restricted to using only 1 rod and 2 rods, but you can use as many as you need of each kind.
    a. How many trains have length 1?
    b. How many trains have length 2?
    c. How many trains have length 3?
    d. How many trains have length 4?
    e. How many trains have length 5?
    f. Look for a pattern in the answers to parts a-e. What numbers do you notice?

5.2 Least Common Multiples

Primes and factoring are crucial for understanding how to add fractions. That understanding is often missed in elementary schools, where often rote manipulation is all too accepted instead. Students who do not really understand adding fractions in arithmetic will find the same topic even more difficult in algebra; others find adding fractions in algebra easy, having learned it correctly in arithmetic. The connection between primes, factoring, and fractions is that a common denominator must be a common multiple of the given denominators. The smallest of these common multiples, known as the **least common multiple (LCM)**, is the one which keeps the work as simple as possible. Students usually find common denominators by trial and error, but knowledge of the LCM is important for more difficult examples.

There are three basic methods for finding LCMs, which we will first name and then illustrate:

1. Mental trial and error
2. List the multiples of all the numbers involved, then scan the lists of multiples for numbers found in all lists. Use the smallest of these common multiples.
3. Factor each of the numbers, and then systematically create the LCM from the factorizations. (This is explained below.)

The first method works well on small numbers but is not helpful in more difficult cases. The second method is very simple in principle, but for large numbers it, too, is tedious. The third method is the most complicated in principle but is actually the most convenient for larger numbers. It also brings out the underlying concepts most clearly. The following examples illustrate these methods.

Example 1. Find the LCM of 4, 12, and 36.
Solution: Since 36 is a multiple of both 4 and 12, 36 is the LCM.

Example 2. Find the LCM of 30 and 42.
Solutions: **First method: mental trial and error**

Try this briefly if you like, then read on.

Second method: lists

Multiples of 30: 30, 60, 90, 120, 150, 180, 210, 240, 270, . . .

Multiples of 42: 42, 84, 126, 168, 210, 252, 294, . . .

The smallest number common to both lists, 210, is the LCM of 30 and 42.

Third method: systematic procedure

Factor 30 into primes: $30=2\cdot3\cdot5$

Factor 42 into primes: $42=2\cdot3\cdot7$

Build the LCM from these factorizations, using enough factors of each prime to be able to find the original numbers as factors. To find 30 as a factor (since $30=2\cdot3\cdot5$) we need: a 2, a 3, and a 5. To find 42 as a factor (since $42=2\cdot3\cdot7$) we need: a 2 (which we already have from 30), a 3 (which we already have from 30), and a 7.

The work can be organized in a table, which has one row for each of the given numbers and one for the LCM, and which has one column for each prime involved. The last row uses each prime the maximum number of times it appears in any cell above.

Factors of 30	**2**	**3**	**5**	
Factors of 42	**2**	**3**		**7**
Factors of LCM	**2**	**3**	**5**	**7**

Then the LCM is $2\cdot3\cdot5\cdot7$, which is 210. You can see in $2\cdot3\cdot5\cdot7$ all the factors needed to make 30 or to make 42.

The LCM, in effect, combines the factor trees of 30 and 42.

To find the LCM of several numbers

> **1. Factor each number into primes.**
> **2. Build the LCM from the primes found in step 1, using each prime the greatest number of times it occurs in any of the given numbers.**

Example 3. Find the LCM of 8, 30, and 54.
Solution: Guessing and lists are impractical here, so use the third method.

Step 1: Factor into primes: $8=2^3$
$30=2\cdot3\cdot5$
$54=2\cdot3^3$

Step 2: Build the LCM using each prime factor found in step one the greatest number of times it occurs in any of the given numbers.

Use 3 factors of 2 to get a multiple of 8,

Use 1 factor of 5,

Use 3 factors of 3.

Then the LCM is $2^3 \cdot 3^3 \cdot 5 = 1{,}080$. As a check note that

$$\mathbf{2^3} \cdot 3^3 \cdot 5 = \mathbf{8} \cdot 3^3 \cdot 5$$
$$= \mathbf{2} \cdot 2 \cdot 2 \cdot \mathbf{3} \cdot 3 \cdot 3 \cdot \mathbf{5} = \mathbf{30} \cdot 2 \cdot 2 \cdot 3 \cdot 3$$
$$= 2 \cdot 2 \cdot \mathbf{2} \cdot \mathbf{3} \cdot \mathbf{3} \cdot \mathbf{3} \cdot 5 = 2 \cdot 2 \cdot \mathbf{54} \cdot 5.$$

With the numbers factored into primes, the work can be organized in a table.

Factors of 8	2^3		
Factors of 30	2	3	5
Factors of 54	2	3^3	
Factors of LCM	2^3	3^3	5

Here, in each column, the highest power is circled. The circled entries are re-entered in the bottom row. Their product, $2^3 \cdot 3^3 \cdot 5 = 1{,}080$, is the LCM.

Example 4. Find the LCM of $2^5 \cdot 7^2 \cdot 11$, $2^4 \cdot 5 \cdot 7^3 \cdot 11^2$, and $5^7 \cdot 7^4 \cdot 31$.
Solution: The primes involved are 2, 5, 7, 11, and 31.
The highest power of 2 in any of the given numbers is 2^5.
The highest power of 5 in any of the given numbers is 5^7
The highest power of 7 in any of the given numbers is 7^4.
The highest power of 11 in any of the given numbers is 11^2.
The highest power of 31 in any of the given numbers is 31.
Then the LCM of the given numbers is $2^5 \cdot 5^7 \cdot 7^4 \cdot 11^2 \cdot 31$.
This example is designed to illustrate the process, which is straightforward, even when large numbers are involved, if they are already completely factored. Of course, the LCM, too, is presented in factored form, but in later applications we shall see cases where that is actually an advantage. You are asked to make the table for this example in problem 4 below.

An alternate approach to finding the LCM of several numbers is given in problem 6 below. Example 5 illustrates the fact that finding the LCM in algebra is virtually the same process as in arithmetic. If you are unsure of your algebra, skip that example for now. We will return to this in chapter 13.

Example 5. (The algebra connection) Find the LCM of $8x^2y$ and $10xy^3$.
Solution: Factor the numerical coefficients completely, and treat any letters involved like prime numbers, with different letters acting like different primes. By the method above, the LCM of 2^3x^2y and $2 \cdot 5xy^3$ is then $2^3 \cdot 5x^2y^3$, or $40x^2y^3$. The table below summarizes the work.

Factors of $8x^2y$	2^3		x^2	y
Factors of $10xy^3$	2	5	x	y^3
Factors of LCM	2^3	5	x^2	y^3

Here is a slightly different kind of question that is easy to answer with the LCM.

Example 6. Find the smallest number that is a multiple of 4 and 6 and is one more than a multiple of 5.
Solution: Since the number we want is a multiple of 4 and 6, it is a multiple of the LCM of 4 and 6, which is 12. Now look for a multiple of 12 which is one more than a multiple of 5. The list of multiples of 12 is: 12, 24, 36, 48, ... We need go no further, because 36 is one more than a multiple of 5.

Exercises 5.2

1. a. Complete this table for finding the LCM of $5^2 \cdot 7$ and $3 \cdot 7^2$.

Factors of $5^2 \cdot 7$			
Factors of $3 \cdot 7^2$			
Factors of LCM			

   b. What is the LCM of $5^2 \cdot 7$ and $3 \cdot 7^2$?

2. To find the LCM of 12, 15, and 20:
   a. Factor 12 into primes.
   b. Factor 15 into primes.
   c. Factor 20 into primes.
   d. Fill in this table as in examples 3 and 4 of the text.

Factors of 12			
Factors of 15			
Factors of 20			
Factors of LCM			

   e. Circle the entry in each column with the greatest number of factors
   f. In the bottom row enter, in factored form, the LCM of 12, 15, and 20.
   g. Multiply out your answer from part f.

3. Repeat the procedure from problem 2 to find the LCM of 9, 12, and 42.

4. Make the table for example 4.

5. Find the LCM of:

   a. $3^5 \cdot 5^6$ and $3^8 \cdot 5^3$ b. 2^8 and 2^{10} c. $5^2 \cdot 23$ and 23^3

   d. $2^{10} \cdot 3^2 \cdot 13^3$ and $5 \cdot 7$ e. $2 \cdot 3^2 \cdot 11^3$ and $2^4 \cdot 9$

6. Euler, in his *Einleitung zur Rechenkunst (Introduction to the Art of Reckoning)*[4] gave another way to find the LCM of several numbers, illustrated here for 6, 8, and 16. First find the LCM of two of the numbers, say 6 and 8. This LCM is 24. Now find the LCM of 24 and the third number, 16. This LCM, 48, is the LCM of 6, 8, and 16.

   a. Use this method to find the LCM of 8, 30, and 54.

   b. Check your LCM against that found in example 3 in the text above.

7. Use any method you like to find the LCM of:

   a. 5, 6, 8 b. 9, 10, 12 c. 16, 18, 20

   d. 15, 20, 24, 25 e. $4p^2$, $6pq$, and $10q^3$

   f. $2xyz$, $6xy^2z$, and $4xy^3$

8. Hot dogs come in packs of 10, while hot dog buns come in packages of 12. What is the smallest number of each kind of package you would need in order to have as many hot dogs as buns?

9. a. One warning light flashes every 25 seconds, and a second one flashes every 45 seconds. If they both flash at the stroke of midnight, when will they they next flash simultaneously?

   b. A third light flashes every 30 seconds. If it, too, flashes at the stroke of midnight, when will all three lights next flash simultaneously?

10. a. What is the smallest amount of money that could buy a number of 42-cent stamps or a number of 27-cent stamps and get no change in either case?

    b. Repeat the question for 44-cent stamps and 28-cent stamps

11. **LCM Bingo**

    Make up a fairly large set of numbers, and find the LCM of all pairs of these numbers. Make up cards of these LCMs, and deal them out to the players. At random, select pairs of numbers from the chosen set and call them out. Any player

[4] Euler wrote this two-volume book in 1738-1740 for those in charge of schools in Russia, where he was working at the time. See note on Euler in section 6.4.

whose card has the LCM of those numbers may cover it on his/her card. The winner is the first player to cover a predetermined number of spots on his/her card (or who covers spots in a predetermined pattern, like the five in a row used in ordinary Bingo).

12. What is the smallest number that is a multiple of 6, 8, and 9 and is three more than a multiple of 5?

5.3 Perfect Numbers

Six is a perfect number in itself, and not because God created all things in six days; rather the inverse is true; God created all things in six days because this number is perfect. And it would remain perfect even if the work of six days did not exist.

Saint Augustine

Many ancient mystics, such as St. Augustine, studied perfect numbers. In his 12th century book, *Healing of Souls,* Rabbi Josef ben Jehuda Ankin recommends their study. Today, however, perfect numbers are known primarily for the fascinating mathematical problems related to them.

A number is said to be **perfect** if it is the sum of all its factors other than itself. Thus 6 is perfect, since its factors, not counting 6 itself, are 1, 2, and 3, which add up to 6. In contrast, 8 is not a perfect number, since its factors (again, not counting 8 itself) are 1, 2, and 4, and these do not add up to 8.

Perfect numbers are rare. The eight smallest are

6
28
496
8,128
33,550,336
8,589,869,056
137,438,691,328
2,305,843,008,139,952,128

All these were listed by the French amateur mathematician Mersenne in his *Cogitata (Thoughts)* in 1644.[5] He also listed what he claimed to be the next three higher perfect numbers, but two of them have since been shown not to be perfect.

All the 46 perfect numbers known to date conform to the elegant formula $2^{n-1}(2^n - 1)$, which was known to Euclid, though of course he did not state it in modern symbolism. Curiously, this formula yields perfect numbers only

[5] Marin Mersenne (1588-1648) was a Minim friar who was also an excellent mathematician, musician, and scientist. He was in regular contact with most of the leading European thinkers of his time, including Descartes, Fermat, and Pascal, and he regularly organized meetings at which these people discussed their ideas.

for certain values of n. Specifically, it yields perfect numbers when n is 2, 3, 5, or 7, but not when n is 4, 6, 8, 9, or 10. From this you might guess that the formula yields perfect numbers whenever n is prime and not when n is not prime. However, things are not quite that simple; if you try n=11 in the formula, you do not get a perfect number. It turns out that the formula yields perfect numbers whenever 2^n-1 is a prime, which is never the case when n is composite and only sometimes happens when n is prime. Naturally, this has led to a search for values of n for which 2^n-1 is prime. The list of these "Mersenne primes" has been extended considerably in recent years with the help of ever more powerful computers, but at this writing there are still only 46 of them known, of which the largest is $2^{43,112,609}-1$.[6] This number is unimaginably large. To write it out in full would take about 12.9 million digits!

Some 2,000 years after Euclid, Euler (see note, section 6.4) proved that all even perfect numbers conform to Euclid's formula, but nobody knows whether or not there are any odd perfect numbers. Any odd perfect number, **if** there are any, has at least 300 digits and is divisible by at least eight different prime numbers, two of which must be greater than 1,000,000,007.

Exercises 5.3

1. a. Evaluate $2^{n-1}(2^n-1)$ for n=2, 3, 4, 5, and 6.
   b. List all the factors of each answer from part a.
   c. Using part b, find the sum of the factors of $2^{n-1}(2^n-1)$ for n=2, 3, 4, 5, and 6. Use this to show that $2^{n-1}(2^n-1)$ is perfect for n=2, 3, and 5 but not for n=4 or 6.

2. The following sums of odd cubes look rather like some of the sums in Chapter 3, but the relation they reveal is even more beautiful and surprising than those discovered there.
   a. Compute:
      $1^3+3^3=$
      $1^3+3^3+5^3+7^3=$
      $1^3+3^3+5^3+7^3+9^3+11^3+13^3+15^3=$
   b. Describe clearly and accurately the general pattern suggested by part a. Your pattern must take into account not only the fact that the cubes to be added are odd, but also how many are added and what the sum is.

3. A number is called **multiply perfect** if the sum of its factors other than itself is a multiple of the number. (For perfect numbers, this multiple is 1). Other than perfect numbers themselves, many multiply perfect numbers are known, the two smallest being 120 and 672. Verify that each of these is indeed multiply perfect.

[6] A brief history of the discoveries of the first 36 known perfect numbers is given in Besuska and Kenney, "*Even Perfect Numbers Update*," ***The Mathematics Teacher***, volume 90, #8 (November, 1997), pages 629-633. In 1996, the Great Internet Prime Search (GIMPS) was founded, linking the computers of many thousands of volunteers in a coordinated search. GIMPS has resulted in the discovery of new Mersenne primes at a rate of about one per year. The GIMPS home page is http://www.mersenne.org.

4. a. The pair of numbers 1,210 and 1,184 has a property that is related to perfect numbers; the sum of the factors of each of these numbers (except for the number itself) is the other number. Verify this by listing all the factors of each of these numbers except for the number itself and then adding them up. These numbers are an example of an **amicable pair**. Only one such pair (in part b below) was known in ancient times.[7]

   b. A Bible scholar has suggested that Jacob's present of 220 goats to Esau (Genesis 32:14) is especially significant, since 220 is one member of a pair of amicable numbers. Find the other member of this pair of numbers, and verify that they do in fact form an amicable pair—that is, show that the factors of each, not counting the number itself, add up to the other.

5. The following diagram, due to Don Goldberg, of Occidental College, illustrates the fact that 496 is a perfect number. Explain, and make a similar picture showing that 28 is a perfect number.

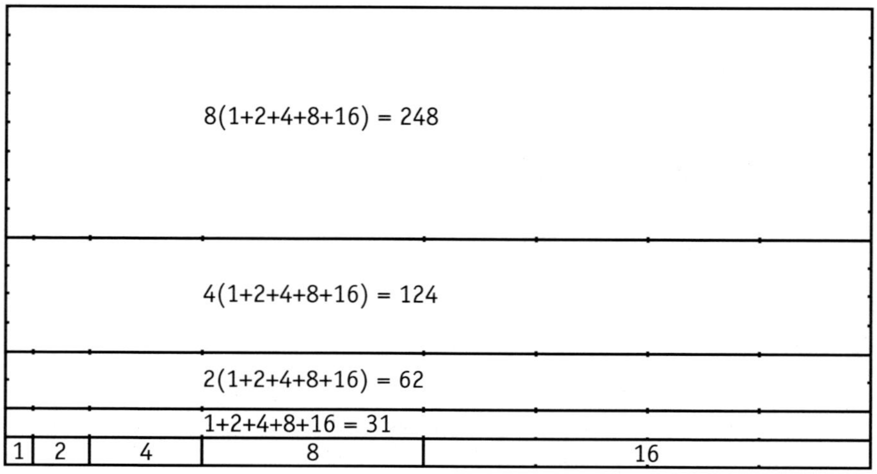

5.4 Numeration Systems and Powers

What is in this box?

LEMONADE

The box is not wet. There is no lemonade in it, just printed marks which, together, represent lemonade as a symbol. Likewise, 4 is just a symbol. It stands for an idea, the number 4. In daily life, we often blur the distinction between symbols and what they represent, but here it is important. **Numbers** are abstract ideas, whereas

[7] The mathematician al Banna (1256-1321) of Marrakech found a second amicable pair, later rediscovered by Fermat, and Descartes found a third pair, 9,363,584 and 9,437,056, in 1638. Over a century later, Euler came up with 57 more amicable pairs, most of which were very large. Curiously, all of these scholars missed the pair 1,210 and 1,184, which was discovered in 1866 by B. N. I. Paganini, then age 16.

numerals are written symbols for numbers. Some other numerals which stand for the same number as 4 are: 2^2, 6-2, •••• (ancient Mayan), IV (Roman), and δ (Greek).

Our modern decimal system of numeration is believed to have originated in India between 300 B.C.E. and 800 C.E. The details are not known today, but the *Aryabhatiya*, a 6th century book of verses summarizing mathematics of the time by Aryabhata, contains the maxim, "From place to place each is ten times the preceding," which eloquently captures the idea. Over a century later, a book by Brahmagupta specifically used zero and negative numbers. Later, the Arab caliph al-Mamun (809-833) established a "House of Wisdom" in Baghdad, where translating books from other cultures into Arabic was a major activity. There the mathematician and astronomer al-Khwarizmi wrote *Concerning the Hindu Art of Reckoning,* which introduced Hindu numeration to the Arab world. Over 300 years later, Leonardo of Pisa learned Hindu numeration from Arabs, and in 1202 he published a book, the *Liber Abaci,* to explain Hindu numeration to Europeans.[8] At the time, Europeans still used Roman numerals, which are so clumsy that the few people who could do arithmetic without an abacus or counting board were sometimes suspected of supernatural powers.

Decimal numeration has three important features that are conspicuously lacking in Roman numeration:

- A set of ten symbols, called **digits**, are used for all numbers, no matter how large or small.
- The meaning of a digit depends on where it is written. For example, the 8s in 83 and 48 have different meanings. A child who does not understand this thinks that 301 and 103 are closer to each other than 301 and 299. Locating a page by a number written in decimal numeration is like looking up a word in a dictionary. The far left digit, like the first letter of the word, gives us the first clue. Successive digits (letters) to the right narrow the search until the page (word) is found. This is the decimal pattern. How is the meaning of each digit determined? By multiplying it by an appropriate power of 10. The far right digit is multiplied by 1 and is called the **units digit**, the next digit to the left is multiplied by 10 and called the **tens digit**, and so on, as in figure 5.4.1.

[8] Leonardo Pisano (ca. 1170-1250) was given the nickname, Fibonacci, which means "Son of Bonaccio" by a 19th century editor of his work. (Leonardo's father was named Guilielmi; Bonaccio may have been another ancestor, such as a grandfather.) Born in Pisa, he learned Hindu notation in Bugia, in what is today Algeria, where his father was sent with Pisan traders. Later he travelled throughout the Eastern Mediterranean and along the way learned a great deal of Islamic mathematics. Other Europeans had used decimal numeration. (One, Gerbert D'Aurillac, who later became Pope, had done so over 200 years earlier, but he had only very crude methods for calculating with the new numerals.) The *Liber Abaci* did much to popularize decimal numeration. Fibonacci is widely known today for the so-called Fibonacci numbers, 1, 1, 2, 3, 5, 8, ... (See Exercises 5.1).

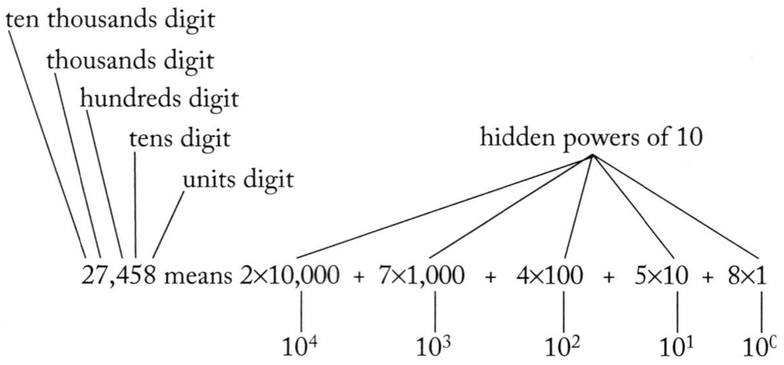

Figure 5.4.1

- There is a digit 0, which can serve as a **place holder.**[9] In 301, for example, the 0 occupies the position of the tens digit, pushing the 3 into the hundreds column. Contrast it with the 0s in 007, which do not influence the position of the 7 and could be left out without changing the meaning.

Place values are named in groups of three. The pattern of ones, tens, and hundreds is repeated in the next three place values thousands, ten thousands, and hundred thousands. It is repeated again in the next three places: millions, ten millions, and hundred millions, and in subsequent groups of three places, as shown in figure 4.5.2. You can find the names of even larger numbers at http://www.kokogiak.com/megapenny/nineteen.asp.

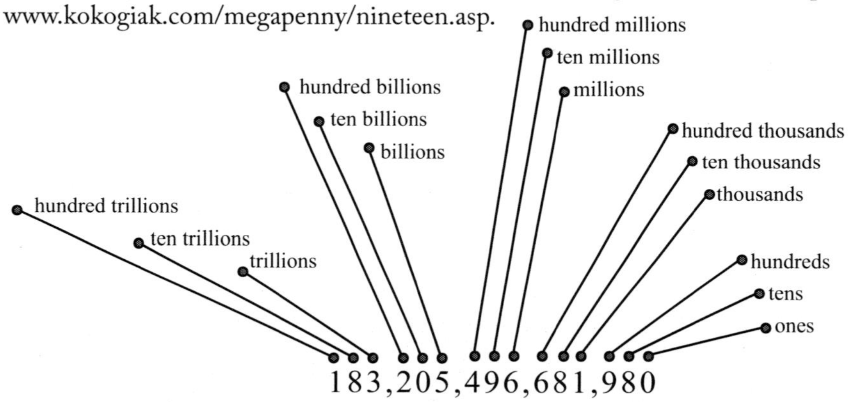

One hundred eighty three trillion, two hundred five billion, four hundred ninety six million, six hundred eighty one thousand, nine hundred and eighty
Figure 5.4.2

There is, literally, more to decimal numeration than meets the eye. Helping children learn this system is a major goal of the primary grades, since carrying, borrowing, and many other key concepts in arithmetic depend on understanding decimal numeration.

Figure 5.4.1 shows 27,458 as 2×10,000+7×1,000+4×100+5×10+8×1. To express a number this way is to write it in **expanded notation.** The multipliers to the right of the multiplication signs are called **place values**, and for us the important observation is that in base ten the place values are all powers of ten. This leads us, as in figure

9 Nobody knows when or where zero was first invented. The earliest documented use of zero is in an inscription from India dating from 876, centuries after the other nine digits were invented. At roughly the same time in history, the Maya of Yucatan used a numeration system based on both five and twenty, which included a symbol for zero.

5.4.1, to identify 1 with 10^0, a notion which may strike you as strange and which deserves a careful look for teaching purposes. Exponents, after all, were introduced as a notation for repeated multiplication, but there is no need to multiply the 8 in the units place of figure 5.4.1 by anything at all. Students often ask why, if 10^0 is to be defined at all, it is not defined as 0, since there is no need to multiply 8 by anything in figure 5.4.1. The answer is simple: One is certainly the place value of 8, and it fits Aryabhata's maxim, "from place to place, each is ten times the preceding." On the other hand, 10^0 fits the pattern of the other powers in figure 5.4.1. Students who want to define 10^0 as 0 think that multiplying by 0 ("multiplying by nothing") is the same as not multiplying, but that is not true. Multiplying by 1 is the same as not multiplying at all, since multiplying by 1 does not change any number. No number beside 1 has this property. In particular, multiplying by 0 changes all numbers to 0 rather than leaving them unchanged. With this in mind, we define

The 0th power of any number n is 1 unless n=0. We do not define 0^0.

Rounding

Sometimes, for example if we are estimating, we may be more interested in a convenient approximation to a number than in the precise number itself. For example, the dates 300 and 800 given above in connection with the history of decimal numeration are approximations for dates that are not known precisely. Even though these numbers are approximations, they are useful. Again, if one car costs $12,385 and another costs $15,890, the main point is that one is a bit over $12,000, while the other is nearly $16,000. We say in this case that, rounding $12,385 to the nearest thousand yields 12,000, and rounding 15,890 to the nearest thousand yields 16,000. More generally, to **round** a number n is to find a number ending in a certain number of zeros that is as close as possible to n. If the number must end in two zeros, we are rounding "to the nearest hundred," if to three zeros, we are rounding "to the nearest thousand," and so on. Rounding the car prices above to the nearest hundred dollars yields, respectively, $12.400 and $15,900.

Example 1. Round the number N=183,205,496,681,980 in figure 5.4.2

a. To the nearest trillion

b. To the nearest thousand

Solution: a. On a number line seen from great distance, so that that only the trillions can be seen, the number in question looks like this:

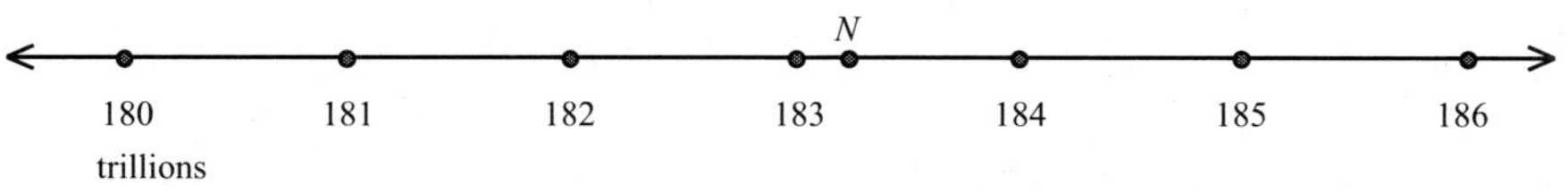

N is closer to 183 trillion than to any other whole number of trillions, so rounding N to the nearest trillion yields 183,000,000,000,000.

b. To round N to the nearest thousand, we want the number that ends in three zeros and is as close to N as possible. Just replacing the final three digits, 980, by zeros yields 183,205,496,681,000, which is less than N by 980. But adding only 16 to N yields 183,205,496,682,000. Therefore, rounding N to the nearest thousand yields 183,205,496,682,000.

Example 2. Round 350 to the nearest hundred.

Solution: Since 350 is between 300 and 400, you would expect to round to one of those numbers. Trouble is, 350 is precisely as far from (or close to) 300 as 400.

The alternatives of rounding down to 300 or up to 400 are equal in merit. In such cases, it may be best to avoid rounding entirely. If you must round, choose either 300 or 400, but state clearly how the rounding was done.

Bases of numeration other than ten

The choice of ten as the base of our numeration is not a mathematical necessity. The reasons for it are historical and no doubt come from counting on fingers, which is so natural that it is found all over the world. In Southern Africa, the Sesotho word for 6 translates as "crossing over," since, the person counting, having used all the fingers on one hand, crosses over to the other hand. Natural as base ten is, numeration systems other than base ten, known as **non-decimal** systems, also have been used. Europeans, presumably barefoot, used their toes as well as their fingers and so based their early counting on twenty. This appears today in French, where the word for eighty is, literally, "four twenties," and in old English, where the **score** ("Four score and seven years ago. . .") is twenty. Some South American tribes base their traditional counting on five, in effect using the fingers of just one hand, a tradition we also use in tallying like this

𝍸 𝍸 𝍷𝍷

Non-decimal numeration also has modern uses. For example, computers do internal arithmetic in base two, eight, or sixteen. Our purpose in examining non-decimal systems is to rethink place value and give you an idea of what children face in trying to learn the decimal system.

Introduction to Base Five

People who use the fingers of only one hand count like this: one, two, three, four, hand, hand and one, hand and two, hand and three, hand and four, two hands, two hands and one, two hands and two, and so on. Here the word "hand" represents what we call five, so, for example, two hands and one ("twandy one?") is two fives and one, which we call eleven. Of course, these people would represent numbers with only five digits, and if we use our familiar 0, 1, 2, 3, and 4 for them, their system would be summed up in figure 5.4.3. (The word "digit," used for counting, is also used for fingers and toes, and it is easy to see why.)

Tally	Base Five Numeral	Spoken in Base Five	Base Ten
𝍷	1	one	1
𝍷𝍷	2	two	2
𝍷𝍷𝍷	3	three	3
𝍷𝍷𝍷𝍷	4	four	4
𝍸	10	hand	5
𝍸 𝍷	11	hand and one	6
𝍸 𝍷𝍷	12	hand and two	7
𝍸 𝍷𝍷𝍷	13	hand and three	8
𝍸 𝍷𝍷𝍷𝍷	14	hand and four	9
𝍸 𝍸	20	two hands	10
𝍸 𝍸 𝍷	21	two hands one	11
𝍸 𝍸 𝍷𝍷	22	two hands two	12
𝍸 𝍸 𝍷𝍷𝍷	23	two hands three	13
𝍸 𝍸 𝍷𝍷𝍷𝍷	24	two hands four	14
𝍸 𝍸 𝍸	30	three hands	15

Figure 5.4.3

When using nondecimal numeration, take special care to read numbers digit by digit. For example, read 11 in base five as "one one base five" and **not** as "eleven." The number eleven is written 21 in base five and would be read as "two one base five."

Counting higher, we soon reach 44 (four hands and four or 24 in base ten), which is the largest number that can be written with two digits in base five, since 4 is the largest digit. The next higher number, written 100 in base five, is what we, in base ten, write as 25. You can use pennies, nickels, and quarters to represent ones, fives, and twenty fives. Evidently, base five numeration is like base ten, but the only digits are 0, 1, 2, 3, and 4, and the place values are powers of five. To paraphrase Aryabhata, from place to place, each is five times the preceding. We will picture higher place values in base five by extending the tally system to a picture of twenty five:

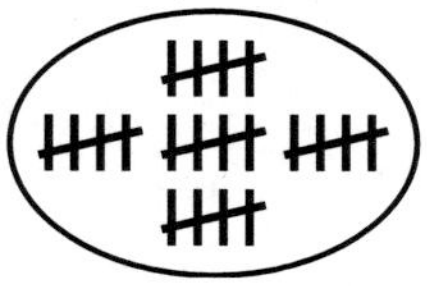

Twenty-five, or a "hand of hands"

We would circle five of these to picture the next place value, 125 (a "hand of hands of hands") and so on. Now we can represent the base five numeral 234 this way:

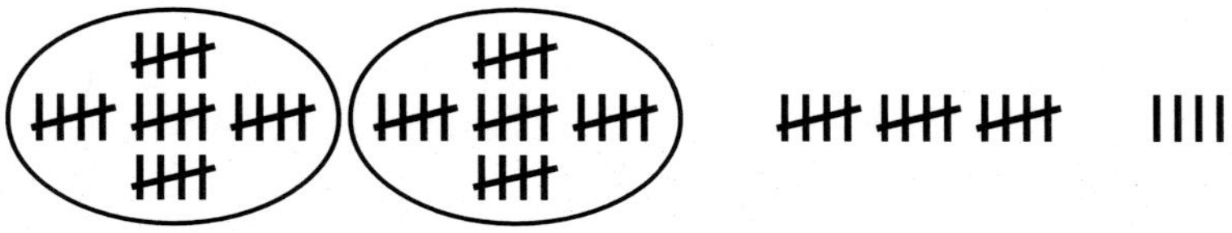

Example 3. If 2,104 represents a number in base five, find the base ten representation for the same number.
Solution: First think of the columns, which determine the place values.

5^3	5^2	5^1	5^0
125	25	5	1
2	1	0	4

Now interpret the base five numeral 2,104 in base ten as

$$2\times125+1\times25+0\times5+4\times1=250+25+4=279.$$

Computation in base five

Computations in other bases are like those in base ten, except that if there is any carrying or borrowing, it involves the base (fives in base five) instead of tens. Example 4 illustrates the idea.

Example 4. Add in base five, 134+123.
Solution: Picture the addition as in figure 5.4.4.

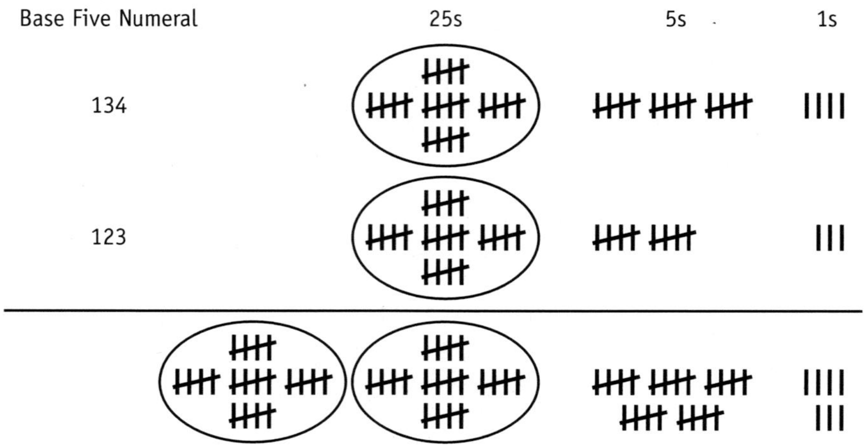

Figure 5.4.4

Carrying out the addition by columns produces seven in the 1s place and five in the 5s place, so we must carry, since in base five, the largest digit is 4. First, carry five from the 1s to the 5s column, leaving 2 in the 1s place (figure 5.4.5).

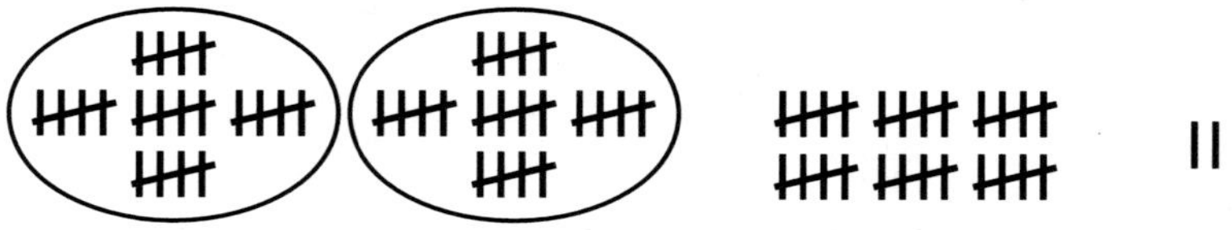

Figure 5.4.5

Now, there are six fives in the 5s place, so carry five of them over to the 25s place, leaving (figure 5.4.6)

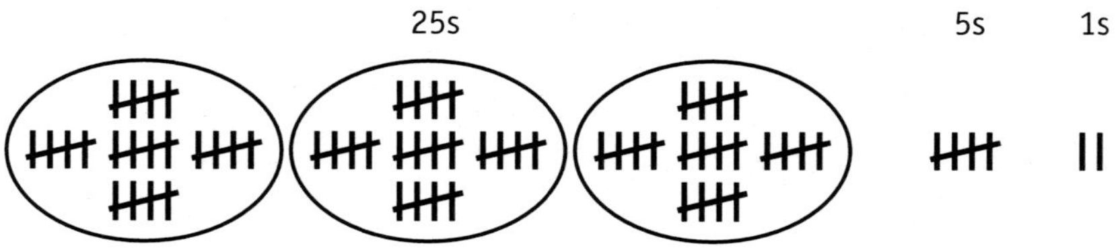

Figure 5.4.6

This represents 312 in base five, which in base ten is written as 82. A youngster growing up with base five would learn to add by saying something like this:

$$\begin{array}{r} 134 \\ +\underline{123} \end{array}$$

In the 1s column, 3+4=seven (written 12 in base five), so write 2 and carry one to the 5s column:

$$\begin{array}{r} 1 \\ 134 \\ +\underline{123} \\ 2 \end{array}$$

In the 5s column, 1+3+2=11, so write a 1 and carry one to the 25s column.

$$\begin{array}{r} 11 \\ 134 \\ +\underline{123} \\ 12 \end{array}$$

Now, add the 25s column to get 312 for the sum. You can check this by doing the addition in base ten.

The other operations, too, are much the same in base five as in base ten. Of course, all borrowing and carrying is done with fives, and the basic combinations, shown in figure 5.4.6, look a bit strange at first. Base five users need to learn relatively few addition and multiplication facts (figure 5.4.7), but they usually need more digits to express numbers than we do in base ten.

+	**0**	**1**	**2**	**3**	**4**
4	4	10	11	12	13
3	3	4	10	11	12
2	2	3	4	10	11
1	1	2	3	4	10
0	0	1	2	3	4

×	**0**	**1**	**2**	**3**	**4**
4	0	4	13	22	31
3	0	3	11	14	22
2	0	2	4	11	13
1	0	1	2	3	4
0	0	0	0	0	0

Figure 5.4.7

Example 3 illustrates the use of these tables.

Example 3. Multiply 34 by 22 in base five.
Solution: Proceed as in base ten, using the tables in figure 5.4.7.

$$\begin{array}{r} 34 \\ \times\underline{22} \end{array}$$

Since 2×4=13 (by the table), write the 3 and carry the one.

$$\begin{array}{r} 34 \\ \times\underline{22} \\ 3 \end{array}$$

Now, 2×3=11 (by the table), and adding the carried one yields

$$\begin{array}{r} 34 \\ \times\underline{22} \\ 123 \end{array}$$

Now repeat the process, moving the partial product left by one place:

$$\begin{array}{r} 34 \\ \times\underline{22} \\ 123 \\ 123 \end{array}$$

Finally, add the partial products, remembering this is base five.

$$\begin{array}{r} 34 \\ \times\underline{22} \\ 123 \\ \underline{123} \\ 1403 \end{array}$$

To check this, convert both factors and the product to base ten.

You can see by now that the only changes in going from base ten to base five are in the basic combinations (figure 5.4.6) and carrying. In all other ways, the algorithms for base ten carry over to other number bases.

Example 4. Express 2,212 (base three) in base ten.
Solution: The base three place values are powers of three: 1, 3, 9, 27, 81, 243, 729, ... Thus, 2,212 in base three means 2×27+2×9+1×3+2×1=77 in base ten, as shown in figure 5.4.8.

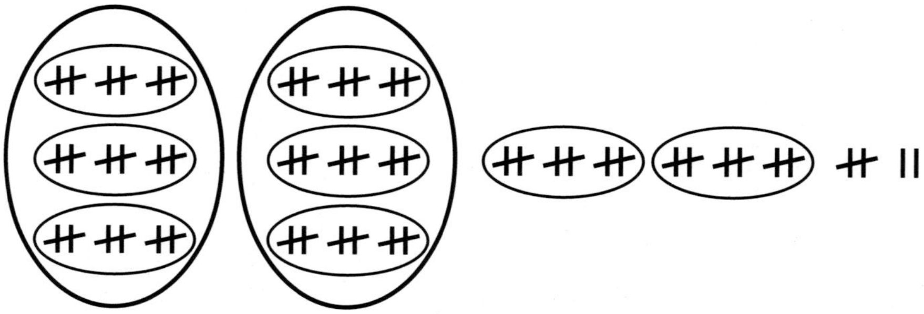

Figure 5.4.8

To translate from any base whatever into base ten, write down the place values, which are powers of the given base, multiply them by the appropriate digits, and add up the results.

Exercises 5.4

1. Round to the nearest hundred:

   a. 2,145 b. 31,990 c. 1,001 d. 650

2. Extend the table in figure 5.4.3 all the way through twenty-seven.

For questions 3-17, you may want to refer to figure 5.4.7.

3. What base five numerals are represented by these tally pictures?

a. 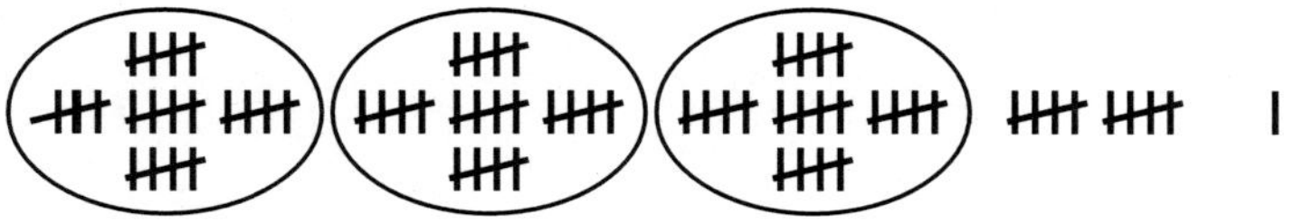

b. 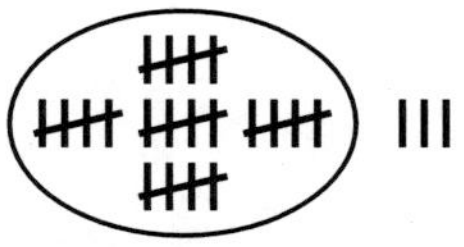

c. 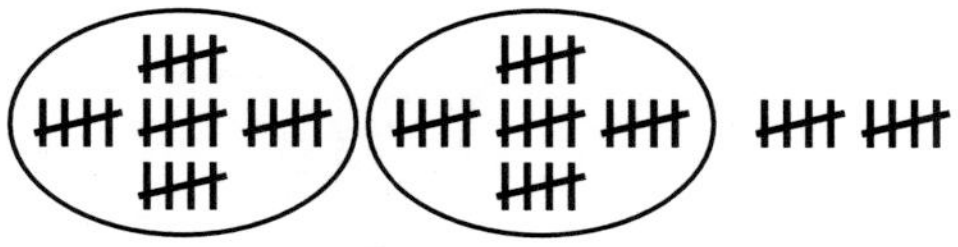

4. Draw tally pictures to represent these base five numerals.

   a. 314 b. 103 c. 420

5. a. What is the largest number that can be written in base five with three digits?

   b. How would that number be written in base ten?

6. This number line was labeled with base five numerals, but some were erased. Can you fill it in?

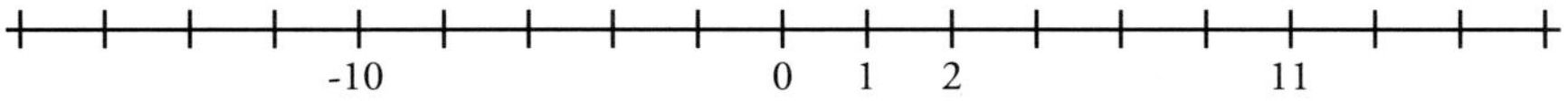

7. What base five additions do these pictures represent?

a.

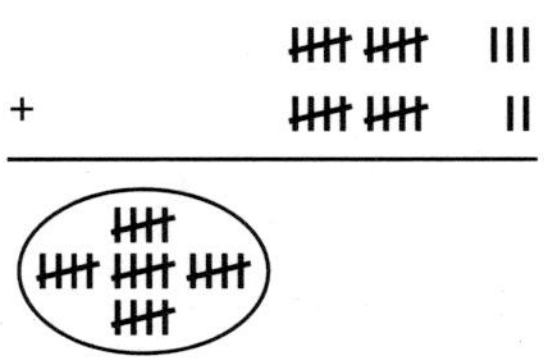

b.

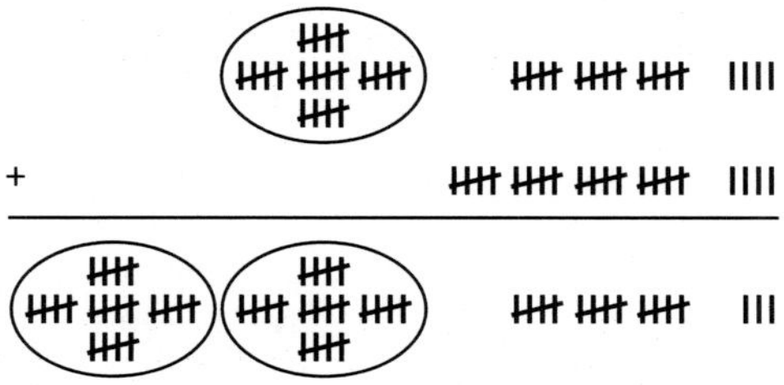

8. Add in base five, using tally pictures to check your work.

   a. 41+102 b. 34+43

   c. 33+42 d. 24+314

9. All the numbers in this base five addition puzzle were correctly written in base five, but some have since been erased. Can you fill in the missing parts?

	31	112
41		211

10. Complete this base five box puzzle with borders.

	2	4	
3			
4			

11. Fill in the missing parts of this base five box puzzle with borders.

	4		
	0		
2			31

12. Pictured are numbers in base three. Write the base three numeral for each number pictured, and then write the same number in base ten.

a.
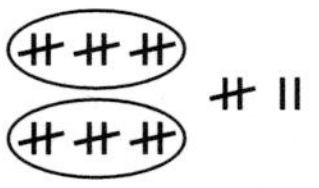

b.
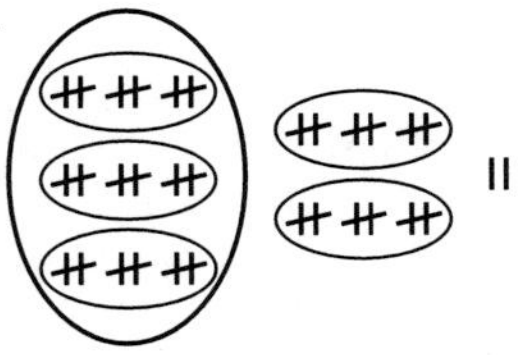

c.
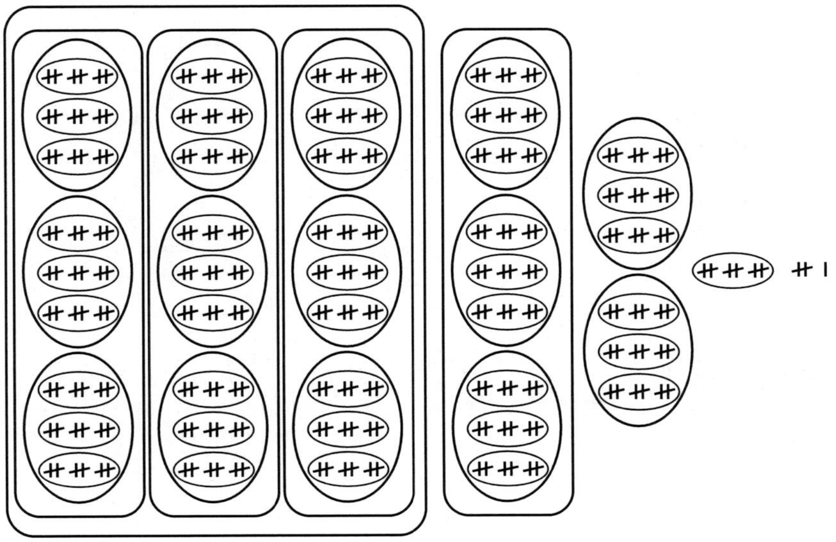

13. Draw tally pictures for these base three numbers.
    a. 1,222 b. 2,020 c. 2,202

14. These numbers are written in base three. Express each one in base ten.
    a. 21,201 b. 112,012 c. 200,212

15. a. What base four numeral is pictured here?

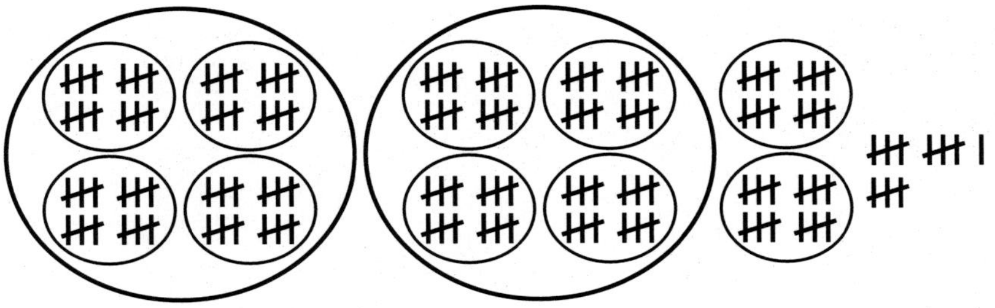

    b. Express that same number in base ten.

16. Make pictures like that in problem 15 for these base four numerals.
    a. 301 b. 120 c. 1,023

5.5 Fun with Base Two

There are 10 kinds of mathematicians — those who use base two and those who don't.
– Author unknown.

Base two, or *binary* arithmetic was discussed extensively by Leibniz[10] in his *"Explication de l'arithmétique binaire, qui se sert des seuls caractères 0 et 1, avec des remarques sur son utilité, et sur ce qu'elle donne des anciennes figures Chinoises de Fohy," (An Explanation of Binary Arithmetic Using only the Characters 0 and 1, with Remarks about its Utility and the Meaning it Gives to the Ancient Chinese Figures of Fuxi).* Leibniz identified 0 with nothing and 1 with God, and he cited the simplicity of binary calculation. Today binary arithmetic is at the core of computers, with 0 and 1 corresponding to "off" and "on" states of an electrical switch.

A Guessing Trick

Here is an old but enjoyable trick. Write or post the array of numbers shown in figure 5.5.1 for all to see, and ask someone to think of a number from 1 through 15 without telling it to you. The number appears in one or more columns in the array. Now ask, "Is your number in the left most column? Is it in the next column over? It is in the next column over (the one next to the right)? Is it in the column at the right?" From the answers, you can "guess" the number. Of course you could search the array of numbers carefully, but you do it instantly, without looking. Read on to see how.

8	4	2	1
9	5	3	3
10	6	6	5
11	7	7	7
12	12	10	9
13	13	11	11
14	14	14	13
15	15	15	15

Figure 5.5.1

Despite tongue-in-cheek references to your photographic memory or other mystical powers, kids will try to figure out what you are doing, and part of the fun is helping them do so. After baffling them for a while, give them a big hint by changing your questions and answers something like this:

You: Sarah, have you a number?

Sarah: Yes.

You: Is it in the column with 8 at the top?

Sarah: Yes.

You: OK, I'll remember the 8. Is it in the column with 4 at the top?

Sarah: No.

You: OK, then I won't remember the 4. So far, I just have 8. Is it in the column with 2 at the top?

Sarah: Yes.

[10] See note on Leibniz, section 1.3.

You: OK, I'll remember the 2 and the 8. That's 10 so far. Is it in the column with 1 at the top?

Sarah: Yes.

You: OK, I'll remember the 1, along with the 10 I had before. Your number is 11.

Eventually the kids figure out that you are using the numbers 8, 4, 2, and 1, which is all you have memorized, and you are adding those where they say "Yes." At first, you hid this by referring to the "left-hand column," etc., without mentioning the numbers at the tops. If you work from left to right, you can do the mental addition almost as fast as the students say "Yes" or "No."

As kids begin to catch on to the trick, bring them up before the class to play it on the others. A classroom session with this trick usually ends up with the kids playing the trick on each other (It is amazing how long they will eagerly do this) and eventually copying figure 5.5.1 so they can play the trick on family and friends after school.

Knowing how to play the trick is not the same as knowing why it works, but it helps, and it gives enjoyable practice in mental addition. Eventually, however, the kids will want to know why the trick works. The secret lies in the array of numbers, but, by the title of this section if nothing else, you probably suspect it is related to base two numeration. Indeed it is. Can you find the connection? It is explained after a second trick is considered.

A Sorting Trick

Make a deck of cards like those in figure 5.5.2 from 3×5 cards, spacing the holes and notches carefully so that when the cards are stacked with their cut corners at the upper left, you can poke a straight object, such as a toothpick, right through the deck at each hole.

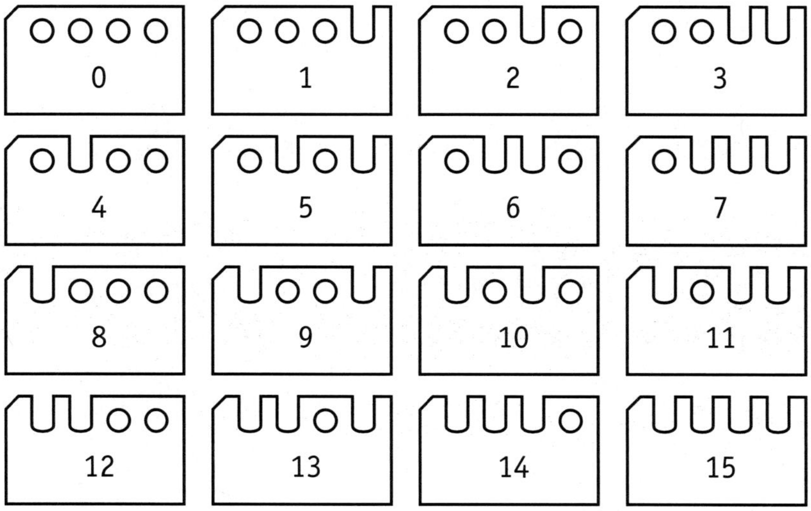

Figure 5.5.2

Shuffle the cards, then stack them with the trimmed corners at your upper right, so the numbers on the cards face away from you and can be read by your audience or class. Now, put a toothpick or other straight object (a straightened paper clip will do) through the left most hole in the deck (as you see it) and lift.

The cards with holes in the left most position will be picked up, while those with notches in that position remain behind. You may have to jiggle the cards a bit as you pick them up to make sure no cards with notches are picked up by friction. Put the cards you picked up on the side of the deck toward the audience. Repeat this with each of the other holes, each time moving one place farther to the right as you see it. When you are done, the cards will be in order!

Why does this trick work? For starters, you can see why the 0 card passes every other card, since each other card has at least one notch and therefore stays behind at some stage when the 0 card moves to the front. Similarly, the 15 card, with all notches and no holes, is passed by all other cards, which all have at least one hole, and therefore 15 ends up at the back of the deck. Likewise, the 1 card is passed by the 0 card at the first step; after that, it passes every other card, since all other cards have notches in at least one of the remaining places. Similarly, the 14 card stays behind all cards except for the 15 card. You can continue in this manner and eventually explain the trick.

The two tricks are closely related, the first trick depends on the sequence of yes and no answers that goes with each number, and the second trick depends on the pattern of holes and notches for each card. How are these related? Consider an example: Somebody thinking of 9 in the first trick would have answered "Yes" (my number is in the left-hand column), "No, No, Yes." This is related to the card for 9 in the second trick, as figure 5.5.3 shows.

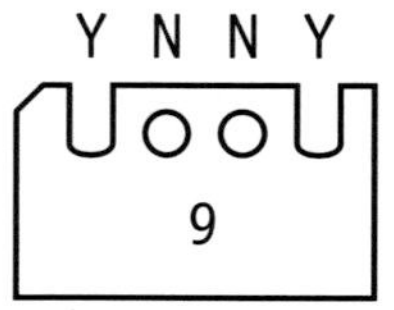

Figure 5.5.3

Evidently, notches correspond to "Yes" and holes to "No." This illustrates something in mathematics which is not rare but always interesting: a single pattern turning up in settings which at first seem unrelated. What is the pattern in this case? Base two numeration! The base two numeral for nine is 1,001. Check a few other cases. The base two numeral for 13 is 1,101. Can you predict the card for 13 from that? Try, then check your answer against figure 5.5.2.

The connection to base two shows how the first trick works and how figure 5.5.1 was made up. The array was made up by heading the columns with 8, 4, 2, and 1, and then writing other numbers under whichever top numbers add up to them. For example, 13 is written under 8, 4, and 1, because 8+4+1=13. The first trick can be extended to higher numbers this way. Start a new column, headed by 16, and put 17 beneath 16 and 1, 18 beneath 16 and 2, 19 beneath 16, 2, and 1, etc., as in figure 5.5.4.

16	8	4	2	1
17	9	5	3	3
18	10	6	6	5
19	11	7	7	7
	12	12	10	9
	13	13	11	11
	14	14	14	13
	15	15	15	15
			18	17
			19	19

Figure 5.5.4

"Russian Peasant" Multiplication

Here is a fine chance for some ham acting with a script something like this: Russian peasants need no multiplication tables. To multiply, they add numbers to themselves and cut numbers in half. Take an example, 75×231. Write the numbers on one line with space between them, this way:

$$75 \quad \times \quad 231$$

Now, cut 75 in half and add 231 to itself to get a new line like this

$$37 \quad \times \quad 462$$

If someone objects that half of 75 is really 37.5, not 37, say that Russian peasants don't need fractions either! Continue this way until you reach 1 by halving. Write the work in columns like this:

75	×	231
37		462
18		924
9		1,848
4		3,696
2		7,392
1		14,784

Now, whenever the number in the first column is even, cross out the entire row, and add what is left in the doubling column, like this:

	75	×	231
	37		462
	~~18~~		~~924~~
	9		1,848
	~~4~~		~~3,696~~
	~~2~~		~~7,392~~
	1		14,784
Total not crossed out			17,235

When students check the multiplication, they are amazed and wonder if they wasted their time with tables and long multiplication.

"Russian peasant multiplication" is actually a variant of the Egyptian method explained in a papyrus written by Ahmes, a scribe, around 1650 B.C.E. What is new is the idea of matching the doubling, which the Egyptians used, with halving and crossing out any rows where halving produced even numbers. The reason this works is explored in exercises 6.4.

Exercises 5.5

1. Look for patterns in figure 5.5.4, and extend the table up to 31.

2. Play the first trick on someone, and, eventually, help your victim discover how it works.

3. Check Russian peasant multiplication for several cases, until you feel confident you can use it correctly.

4. In Russian peasant multiplication, does it matter which number is cut in half and which is doubled? Try several examples both ways.

5. Section 5.3 introduced perfect numbers, of which the four smallest are 6, 28, 496, and 8,128.
   a. Express each of these in base two.
   b. Find a pattern in your answers to part a and describe it.
   c. Explain the pattern you found in part a, using the formula for perfect numbers given in section 5.3.

Exploration: AP-free sequences (questions 6-8)
Some terminology is needed for questions 6-8. Recall from section 1.3 that an arithmetic progression (AP) is a sequence of three or more numbers increasing or decreasing by a constant amount. Examples:

a.	3, 4, 5, 6, 7	increasing by 1
b.	100, 105, 110,	increasing by 5
c.	10, 9, 8, 7, 6	decreasing by 1 (or increasing by -1)
d.	1, 1.3, 1.6, 1.9	increasing by 0.3
e.	5, 8, 9, 11	not an AP
f.	0, 3, 7, 12	not an AP

If you had the numbers from example e to use or not use, as you please, you could form an AP with 5, 8, and 11. You could not do that with the numbers in example f, so it is said to be AP-free. Similarly, 4, 7, 8, 11, 13 is AP-free, but 3, 9, 10, 13, 14, 17 is not. (Can you find the hidden AP?)

6. Find any APs hidden in these examples. If there are none, state that it is "AP-free."
   a. 2, 4, 5, 8, 9, 13, 14
   b. 2, 3, 5, 6, 10
   c. 1, 5, 7, 8, 12
   d. 5, 6, 9, 10 12, 14, 17, 19
   e. 21, 22, 25, 30, 31, 38

7. Now you are ready to explore a curious sequence of numbers. It is formed by starting with 0 and admitting each successive whole number to the list, provided that the list itself remains AP-free. Thus, the second number is 1, since a sequence of two numbers is too short to be an AP, but having admitted 0 and 1, we may not admit 2, since 0, 1, 2 would be an AP. The next candidate for admission, 3, is accepted, since 0, 1, 3 is AP-free. Find the 4th, 5th, 6th, 7th and 8th numbers in the sequence.

8. The sequence generated in question 7 may seem irregular, but it is based on a remarkable pattern. To see it, follow these steps:
   a. Express each of the eight numbers in the base three.
   b. Look for a pattern in your answers to part a. (Hint: Interpret each number as if it were written in base two.)
   c. Use the result of part b to predict the next two numbers in the sequence.
   d. Check your predictions to see that the sequence you are generating remains AP-free.

Vocabulary for Chapter 5

Amicable pairs (5.3)
Base of numeration (5.4)
Binary (5.5)
Consecutive (5.1)
Digit (5.4)
Expanded notation (5.4)
Exponent (5.1)
Fibonacci numbers (5.1)
Least common multiple (5.2)
Numeral (5.4)
Pascal's triangle (5.1)
Perfect number (5.3)
Place value (5.4)
Powers (5.1)
Rounding (5.4)

Self-Quiz for Chapter 5

1. Evaluate:

   a. 2^7 b. 5^3 c. 3^2+7^2 d. $(3+7)^2$

2. Find the least common multiple of 5, 6, 8, and 10.

3. The eighth line of Pascal's triangle is

   1 8 28 56 70 56 28 8 1

   Use this to write the ninth line.

4. Explain the difference between a number and a numeral.

5. A number expressed in base five is 3,124. What base ten numeral represents the same number?

6. Write the base five numeral for one hundred sixty eight.

7. Write the base three numeral for sixty seven.

8. What number is written as 100,101 in base two?

9. Do these computations in base five, then check the work by changing over to base ten.

   a. $234+1{,}234$ b. 23×431

Chapter Six
Subtraction and Division

Subtraction and division are understood better in relation to addition and multiplication than as totally new processes. Having treated addition and multiplication, we now turn to subtraction and division.

6.1 What Is Subtraction?

Here are three common ways to think of subtraction.

Take away School children think of subtraction as "take away." To them, a subtraction question like 8–5 asks how many will be left if five objects are taken away from eight. This interpretation is easy to represent concretely and is, in short, a natural for early grades.

Add the opposite Take away becomes murky when negative numbers are involved. What, for example, does –7–(–9) mean? When students first meet subtraction with negative numbers, often in late elementary school or junior high, they also meet a new definition of subtraction. They are told that to subtract a number *b* means to add its opposite, –*b*. For example, in –7–(–9), they are told that to subtract –9 from –7 means to add the opposite of –9, or 9, so that –7–(–9)= –7+9=2. This definition, with no basis in prior experience, seems to come out of the blue, and examples to justify it with bank accounts, subzero temperatures, etc., often come across as contrived.

Missing addend Hope for alleviating confusion about subtraction comes from the **missing addend**. This view of subtraction asks what must be added to one number to get another. For example, 8-5 is the number which must be added to 5 to get 8. If we represent the numbers 5 and 8 as vectors that begin at 0, then the missing addend may be viewed as a vector which begins at 5 and ends at 8 (emphasized in figure 6.1.1a). This vector is three units long and points to the right, indicating that 8-5=3. A negative number would be shown as a vector pointing to the left. Similarly, -5-(-7) is the number which must be added to -7 to get -5, pictured as a vector (emphasized in figure 6.1.1b) beginning at -7 and ending at -5. This vector is 2 units long and points to the right, so -5-(-7)=2.

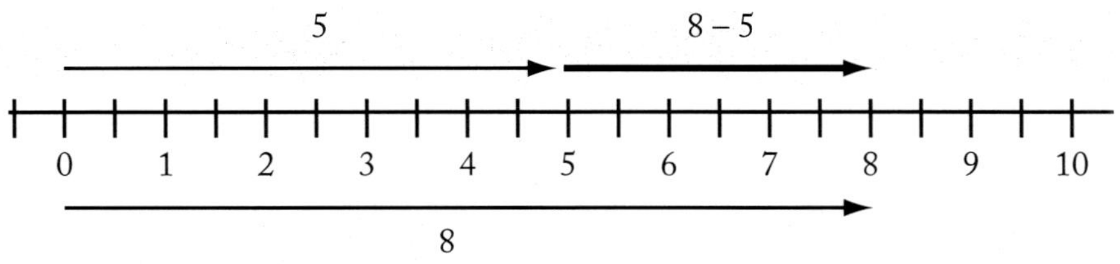

Figure 6.1.1a

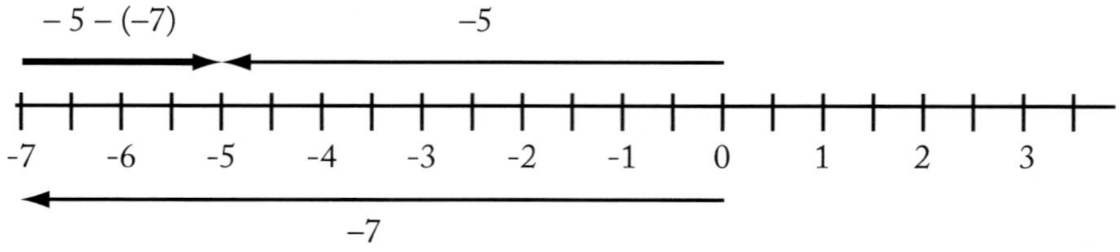

Figure 6.1.1b

Although the three views of subtraction are equivalent, we will define subtraction in terms of the missing addend.

> **Definition: The difference of *y* from *x*, written *x*-*y*, is the number which may be added to *y* to get *x*.**

Note: Subtraction is not commutative: x-y is not the same as y-x, and English can mislead students into subtracting in the wrong order. In English, "Subtract 3 from 7," has the numbers 7 and 3 in the opposite order from 7-3.

Unlike take away, the missing addend definition applies easily to negative numbers. Examples like 2-(-3)=5 puzzle students who are locked into the "take away" notion of subtraction, and do not understand that the subtraction asks what must be added to –3 to get 2. This subtraction is revisited in Example 1 below.

Extending "take away" to negative numbers If you prefer the "take away" view of subtraction, here is a way to extend it to subtractions with negative numbers. The idea is based on adding a suitable number of zero pairs.

Example 1. Carry out the subtraction 2-(-3) as a take away problem.
Solution: To take away -3, first add three zero pairs in form 3+(-3).

$$\begin{aligned} 2&=2+0 \\ &=2+(3+(-3)) \\ &=(2+3)+(-3) \\ &=5+(-3) \end{aligned}$$

Now take away -3, leaving 5.

To model this with algebra tiles, start with 2 positives: 

We must take away 3 negatives.
Since there are no negatives
to take away, add 3 zero pairs: 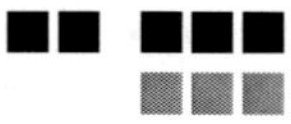

Take away 3 negatives,
and 5 positives remain. 

Example 2. Subtract 2 from –6.
First Solution: Take away We must find –6–2. Note, once again, that the -6 and 2 appear in the reverse order from the verbal statement of the problem. We can restate the problem: From six negatives, take away two positives.

Start with 6 negative tiles 

Since there are no positives
to take away, add 2 zero pairs

Take away 2 positives leaving 8 negatives
So $-6 - 2 = -8$.

Second solution: Missing addend What must be added to 2 to get –6? Starting at 2 on the number line, moving to –6 involves moving to the left, which means adding a negative number. Moving 2 spaces to the left gets you down to 0, and then you still have to go 6 mores spaces to the left. In all, you had to hop 8 spaces to the left, so you added –8. Thus –6-2=-8. This is pictured below as a vector 8 units long pointing to the left.

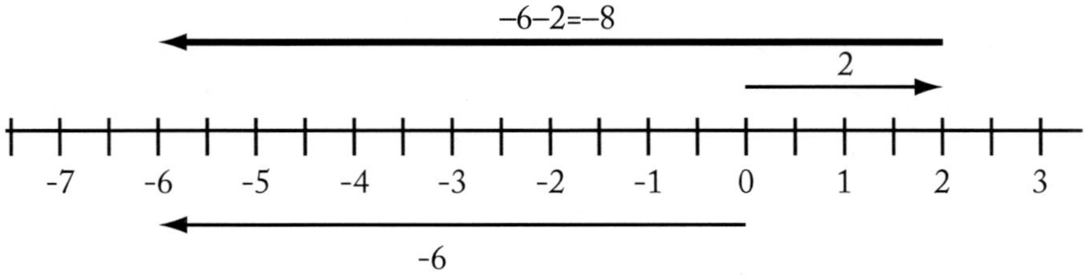

Third solution: Add the opposite To subtract 2 from –6, add –2.
Then $-6-2=-6+(-2)=-8$.

Meanings of the minus sign The minus sign has two meanings, related but conceptually distinct. When used immediately in front of a number, the minus sign indicates the opposite number. When used between two numbers, the minus sign indicates the operation of subtraction. To keep the two meanings of the minus sign clear, elementary mathematics books sometimes raise the minus sign when it is used to indicate the opposite of a number, so that $^{-}4$ would indicate negative 4. Calculators make this distinction by means of two keys, one marked (-) or *chg sgn* and the other marked -. The first key indicates the opposite of a number, the second subtraction. If you press the wrong key, the calculator will indicate an error.

The meaning of the minus sign is always clear from the context, if you follow this simple convention:

> **Never write two signs for operations (+, -, ×, ÷) without something else in between. If necessary, use grouping signs such as parentheses to avoid consecutive operation signs.**

If you write –5–3, the first minus sign appears immediately before the 5 and is not between two numbers, so it indicates the opposite of 5, whereas the second minus sign appears between two numbers and indicates subtraction. On the other hand, –5+–3 is not simply poor style, it is meaningless. A calculator will not know what to do with this and will indicate an error. The error message will be avoided by putting parentheses around the –3 to get –5+(–3).

Example 3. Subtract –2 from –6 — i.e., calculate $-6-(-2)$.

First solution: Missing addend What must be added to −2 to get −6?
If you start at −2, you must move 4 units to the left (in the negative direction) to get to −6, so $-6-(-2)=-4$.

Second solution: Take away
Start with 6 negative tiles: 

We have enough negative tiles to take away 2 negatives, so we will not need to add any zero pairs.

Take away 2 negatives leaving 4 negatives: 
Then $-6-(-2)=-4$.

Third solution: Add the opposite $-6-(-2)=-6+2=-4$

Algebra tile pictures can be used to show why take away is the same as addition of the opposite. For instance, look at example 1: $2-(-3)$

After putting in the zero pairs and taking away 3 negative tiles, look at what remains:

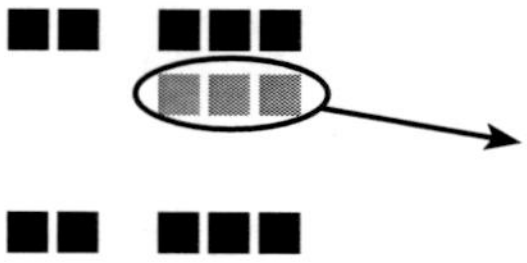

In addition to the original 2 positive tiles there are now 3 more positive tiles. Subtraction of −3 has been replaced by addition of 3, so $2-(-3)=2+(-(-3))=2+3=5$.

Similarly, in example 2: $-6-2$

After putting in the zero pairs and taking away 2 positive tiles, look at what remains.
Taking away of two positive tiles has been replaced by addition of two negative tiles.

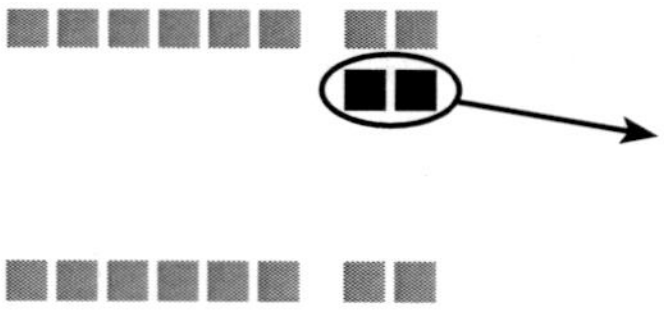

Another way of seeing that take away is the same as addition of the opposite is shown below.

To show that $-6-(-2)=-6+2$, start with 6 negatives:

For $-6-(-2)$:

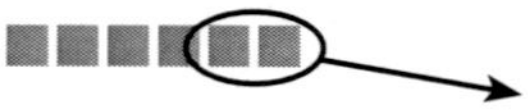

Take away 2 negatives.

For $-6+2$:

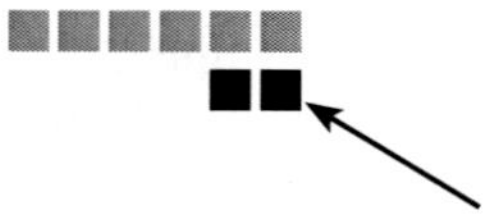

Add 2 positives.
Remove zero pairs.

Either way, you end up with 4 negatives:

Evidently, taking away negative tiles has the same net effect as adding positive tiles.

What views of subtraction to teach

The three views of subtraction are all correct and all needed, but adding the opposite makes sense only to students who already understand addition with positive and negative numbers. Students who have not yet learned to add positive and negative numbers will be best served if they encounter both the take away and the missing addend views of subtraction from the beginning.

Exercises 6.1

1. The following sequence of subtractions begins with four examples which can be done by the usual take away approach.
   a. Extend the list of examples as far as 8–(–5), looking for the pattern in the sequence of answers.

      8–4

      8–3

      8–2

      8–1

      8–0

      8–(–1)

   b. Picture each of these problems as in figure 6.1.1.

2. Complete the table below.

Subtraction a-b	Answer to subtraction	Equivalent addition a+(-b)	Result of adding the opposite
14-9	5	14+(-9)	5
2-9			
-12-13			
-4-(-9)			
7-(-4)			
-1-(-3)			

   b. For each subtraction question in the chart above, picture the answer as a vector on the number line. In each case, the vector should start at the number being subtracted and go to the number being subtracted from. (e.g., for 14-9, the arrow should go from 9 to 14.)

   c. In each case shown in the table, verify that the answer to the subtraction question and the result of adding the opposite are the same.

   d. Use algebra tiles to illustrate how to do each subtraction question from the table above as a take away problem.

3. Josh completed these addition box puzzles, but gremlins erased them. Can you reconstruct them?

a.

		13
12		25
	22	

b.

	5	
		14
11		27

c.

		21
	13	
18		50

d.

	22	57
	13	41

4. What concept of subtraction best describes the activity in question 3? In a sentence or two, explain your answer.

5. Make up several (at least three) deleted box puzzle examples like those in question 3, using negative entries as well as positive ones.

6. In each deleted box puzzle in question 3, four entries are given at the start, and the other five are to be found by subtraction and/or addition. In making up such examples, can you simply put numbers into any four arbitrarily selected boxes? If so, show why. If not, which sets of four boxes could you put numbers into so that the puzzle could be completed in just one way? What would happen if you began with fewer than four boxes filled in, or with more than four? (Note: This question is deliberately left wide open for exploration. It can generate practice in adding and subtracting, but its main focus is on questions which are ideal for small group work.)

7. The adder of exercises 2.2 can also be used for subtraction.
   a. Show how this can be done.
   b. Check to see that the adder subtracts correctly, even when negative numbers are involved.
   c. What concept of subtraction is used here?

8. Maps give the elevation of Badwater, the lowest point in Death Valley, CA, as –282 feet, which means 282 feet below mean sea level (msl). The highest point in California, the top of Mount Whitney, which can be seen from Death Valley, is 14,495 feet above msl.
   a. How much higher is the top of Mount Whitney than Badwater?
   b. What subtraction does this illustrate?

9. Repeat problem 8 with the extreme elevations of China's Xinjiang Province, which range from the Turpan Depression, 505 feet below mean sea level (msl), to K2 (Mount Godwin Austen) 28,250 feet above msl.

10. A football team with 5 yards to go took a 17-yard loss on a play.
    a. How far did they then have to go?
    b. What subtraction does this illustrate?

11. One wintry night in an Alaskan town the temperature fell to –47° F. By noon the next day, however, it had risen to a balmy –27° F.
    a. How much did the temperature rise in that time?
    b. What subtraction does this illustrate?

12. Make up a problem like those in questions 8–11, which illustrates a subtraction involving negative numbers.

13. It is mentioned above that subtraction is not commutative. Is subtraction associative? Explain your answer in a sentence or two.

14. Compute and compare these pairs of problems:
    a. 7–(4–1) and 7–4+1
    b. 5–(9–2) and 5–9+2
    c. –12–(8–3) and –12–8+3
    d. 15–(–4–5) and 15–(–4)+5
    e. Sum up your results from parts a-d by completing this statement: For any number x, y, and z, x-(y-z)= ______.

15. A traveler in the desert has accidentally spilled his water and now has only an empty 8 quart gourd. A second traveler, who has full gourds of 3 and 5 quart capacities agrees to share his water equally with the first. The gourds are irregular, and there is no way to tell how much each one contains by looking at it or weighing it. How can you, by pouring water back and forth between the gourds, measure out exactly four quarts? (This is based on a problem in *Kraitchik's Mathematical Recreations*.)

6.2 Methods of Subtraction

The way students subtract reflects the concept of subtraction they use. The algorithm for subtraction usually taught in elementary school is based, as you might expect, on a take away view of subtraction. Let's look carefully at how elementary school students learn to subtract whole numbers. In all but the simplest cases, subtractions of whole numbers are done column-wise. For example, to subtract 62 from 97, subtract the 2 from the 7 to get 5 and the 6 from the 9 to get 3, so 97–62=35. The rationale for column-wise subtraction lies in expanded notation. Write 62 as 60+2, and 97 as 90+7, and the process is

$$\begin{array}{rcrcr} 97 & = & 90 + 7 & = & 90 + 7 \\ \underline{-62} & = & \underline{-(60 + 2)} & = & \underline{-60 - 2} \\ 35 & = & 30 + 5 & = & 30 + 5 \end{array}$$

In effect, the original problem has been broken into two simpler ones: Subtract 60 from 90 and subtract 2 from 7. The results of those two problems, 30 and 5, are then put together again as 35.

If we try this method with a problem like 52-37, the work would look like this

$$\begin{array}{rcr} 52 & = & 50 + 2 \\ \underline{-37} & = & -\underline{(30 + 7)} \\ & & 20 + (-5) \end{array}$$

The answer is correct, but to avoid a negative units digit we first rewrite 52 as 40+12 instead of 50+2. Then the work looks like this:

$$\begin{array}{rcr} 52 & = & 40 + 12 \\ \underline{-37} & = & -\underline{(30 + 7)} \\ 15 & & 10 + 5 = 15 \end{array}$$

Rewriting 50+2 as 40+12 is called **regrouping** or **borrowing**.[1] It is not a difficult process, but it does confuse some beginners, especially where, as in the next example, it must be done more than once.

Example 1. Subtract 98 from 207.
Solution: (as typically taught in elementary schools)

1. Write the number to be subtracted directly beneath the number being subtracted from, so that the units digits are directly above one another. (Then other corresponding digits, such as the tens or hundreds, will also line up in vertical columns.) Put a horizontal line beneath the bottom number.

$$\begin{array}{r} 2\,0\,7 \\ \underline{-\,9\,8} \end{array}$$

2. Next take away vertically in each column, if possible, starting with the units and moving toward the left. However, if take away would lead to a negative number in some column, first "borrow" from the column immediately to the left. In this case, we must borrow from the tens in order to subtract in the units column, but the tens digit is 0, meaning there are no tens available to borrow. Therefore, we must first get some tens by borrowing from the hundreds:

[1] Liping Ma, in *Knowing and Teaching Elementary Mathematics,* points out that the term "borrowing" may mislead children into thinking that what is borrowed is owed or should be repaid.

Begin with

	hundreds	tens	ones
	2	0	7
minus		9	8

Rewrite this as

	hundreds	tens	ones
	1	10	7
minus		9	8

Now borrow one of the tens to rewrite this again as

	hundreds	tens	ones
	1	9	17
minus		9	8

Now taking away in each column yields 109. We can picture this with base ten blocks as follows.

Start with 207:

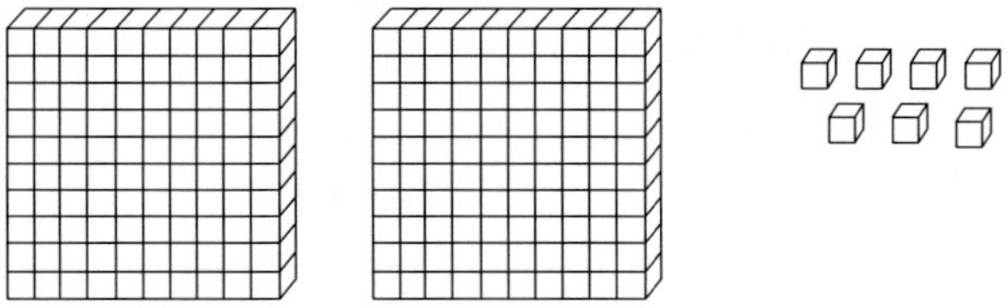

Trade 1 hundred for 10 tens:

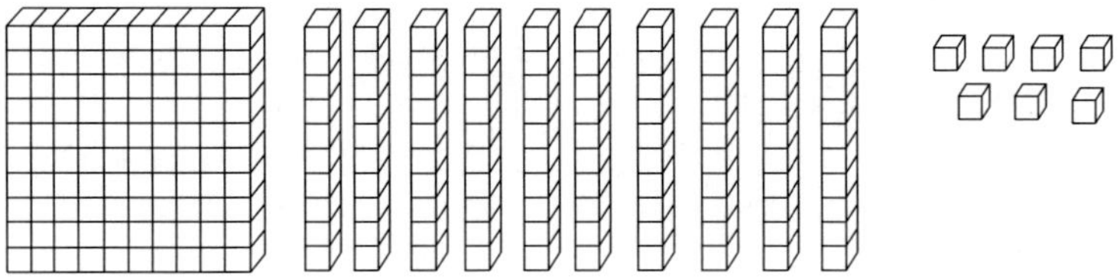

Trade 1 ten for 10 ones:

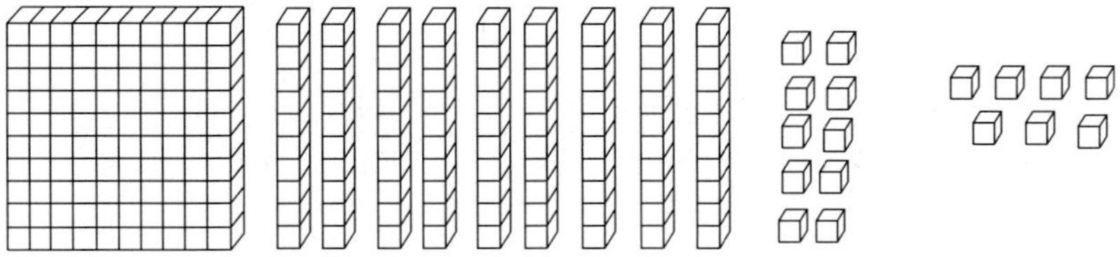

Take away 8 ones from 17 ones:

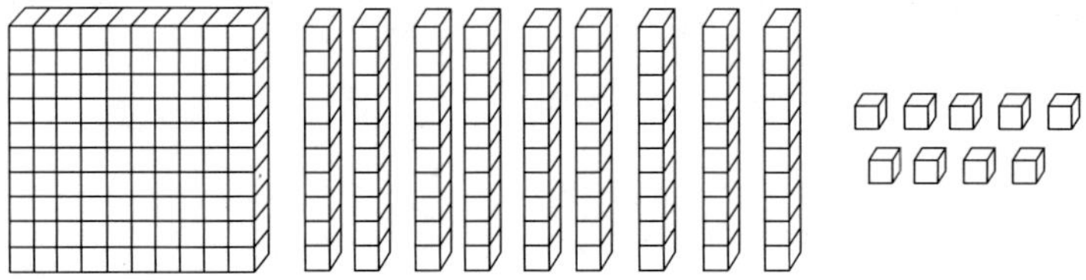

Take away 9 tens from 9 tens:

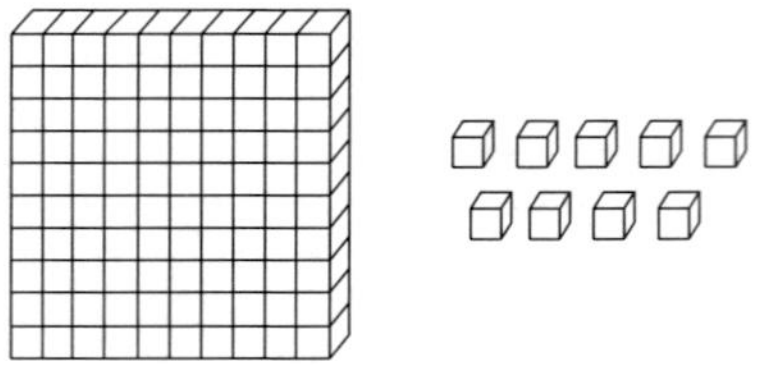

This leaves 1 hundred and 9 ones or 109.

Money provides another concrete way to represent this process, with 207 as two dollars and seven cents. Then the first borrowing step amounts to changing a dollar for ten dimes, and the second amounts to changing one of the dimes for ten pennies.

The algorithm illustrated above is easy to visualize, it accurately reflects the take away concept of subtraction, and it works well for numbers with many digits. However, some students have trouble mastering this process, especially where repeated borrowing is required, as in the previous example. What should you, as teacher, do to help a student who has trouble with the usual subtraction algorithm? The answer to this question must, of course, vary from student to student, but a few general remarks are nonetheless in order.

1. **Extra practice alone can be counter-productive.** Before practicing further, the student needs to understand how to avoid the mistakes that are causing problems.
2. **Try to express the process in terms which are as concrete as possible**. Use base blocks, money, or other manipulatives that allow students to learn with the eye and hand as well as with pencil and paper.
3. **Don't let a difficulty become a roadblock**. Limit damage to the student's self-confidence by finding areas of mathematics in which he or she can continue to progress while you work on this problem area. To do this, you may have to be flexible about the order in which topics are covered. In this book, for example, the ideas in the chapters prior to this one involve little or no subtraction, as do some topics which appear later, such as much of the work on fractions. On the other hand, postpone long division with this student, since it uses subtraction extensively.
4. **Look for alternate approaches**. In the case of subtraction, the missing addend approach, which avoids borrowing altogether, can help students who have trouble with borrowing.

Example 1 (reconsidered). Subtract 98 from 207.
Solution based on missing addend: Start with 98 and count up to 207.

One way to do this is:	Add	2	to get to	100
	Add	100	to get to	200
	Add	7	to get to	207
	Total added	109		

Then 207-98=109.

Like solution 1, this is easy to see in terms of money; indeed, it is a good way to make change. For subtractions involving many digits, the process of counting up may be slower than the usual way of taking away and borrowing, but today, with calculators widely used, speed is not a reason to select one process over another. More than ever before, the only meaningful criterion is whether it makes sense to students.

Exercises 6.2

1. Explain the process of subtraction, both with and without borrowing, using physical materials such as base blocks to illustrate the process. First, use the materials to explain to yourself, then use them to help another person (possibly, but not necessarily, a youngster who is in grade 2 or 3) understand. Start with very simple examples and work up to more complex ones, which, like example 1 above, involve repeated borrowing.

2. Explain subtraction to someone using the missing addend technique. Illustrate the idea with money, having the person count out the change on a purchase of $.79 (tax included) if given:

   a. A $1 bill b. A $5 bill c. A $10 bill

3. Explain how to use the missing addend (counting up) approach to compute.

   a. 200-197

   b. 300-197

   c. 312-197

   d. 1,312-197

4. a. Learn, if you do not know it, the game of blackjack (also called 21), leaving out the betting that often goes with it. Remark: To avoid offending the sensibilities of people who may react negatively to the use of a game linked with gambling, rename the game and change the target number from 21 to something else (for example, 22).

   b. Describe how you could use this game in the classroom to give extensive practice in basic addition and subtraction in a small-group setting.

5. Describe, in a paragraph, how the game of Monopoly contributes to development of mathematical skills. In particular, show where the game helps students practice:

   a. Basic addition facts

   b. Counting

   c. Addition

   d. Subtraction

   e. Multiplication

   What other mathematical ideas are used in Monopoly?

6. Make a list of at least three other games or group activities which include extensive practice with basic mathematical ideas and are suitable for students in grade schools. Describe what mathematical ideas are encountered in each activity and what basic skills are needed before a person can enjoy playing the game.

7. **(Suggested by Larry Stevenson)**
This question concerns the square numbers, augmented by 0: 0, 1, 4, 9, 16, 25, 36, 49, 64, 81, 100, 121, 144 ... What whole numbers can you make by subtracting one of these from another? Some numbers, such as 9 may be done on more than one way (9-0 and 25-16). Others, such as 2, cannot be done at all. Curiously, the number of ways a whole number can be expressed as a difference of squares is related to the number of ways it can be expressed as a product of two odd factors or two even factors (not one even and one odd). For example, 40 can be expressed as a product of two even factors as 4×10 and 2×20 (we do not count 8×5 or 40×1, since they use one odd factor and one even). Also, 40 can be written as 49-9 and as 121-81.
   a. Explore this equation by filling in the table below.

Whole number $n>1$	**Ways to write** n **as a difference of squares**	**Ways to express** n **as a product of two factors**	**Number of ways to express** n **as a product of two factors, both even or both odd**
2			
3	4-1		
4	4-0		
5	9-4		
6			
7	16-9		
8	9-1		
9	9-0, 25-16		
10			
11			
12			
13			
14			
15	16-1, 64-49		
16			
17			
18			
19			
20			
21			

   b. Find a pattern in the table and describe it.

8. **Exploration: Difference Patterns**
In the arrays below, you get each entry by subtraction in the line above. In each case, subtract the number above and left from the one above and to the right. For example, if you had
48 73, you would put 25 in the next row down, like this
25.
A few numbers are filled in for your use in checking your work. If you work carefully, you will find some interesting patterns.

a. Squares

0^2 1^2 2^2 3^2 4^2 5^2 6^2 7^2 8^2 9^2 10^2

___ ___ 4 ___ 16 25 ___ ___ ___ ___ ___

___ ___ ___ 7 9 ___ ___ ___ ___ ___

___ ___ ___ 2 ___ ___ ___ ___ ___

b. Cubes

0^3 1^3 2^3 3^3 4^3 5^3 6^3 7^3 8^3 9^3 10^3

___ ___ 8 ___ 64 125 ___ ___ ___ ___ ___

___ ___ ___ 37 61 ___ ___ ___ ___ ___

___ ___ ___ 24 ___ ___ ___ ___ ___

___ ___ ___ ___ ___ ___ ___ ___

c. 4th powers. (First try to predict what you will get, using parts a and b, then go ahead with the arithmetic to check your predictions.)

0^4 1^4 2^4 3^4 4^4 5^4 6^4 7^4 8^4 9^4 10^4

___ ___ 16 ___ 256 625 ___ ___ ___ 6561 ___

___ ___ ___ 175 369 ___ ___ ___ ___ ___

___ ___ ___ 194 ___ ___ ___ ___ ___

___ ___ ___ ___ ___ ___ ___ ___

___ ___ ___ ___ ___ ___ ___

d. Extend the tables from parts a, b, and c to powers of negative numbers as far as -10. Do the patterns continue to hold in the extended tables?

e. Find a pattern in the results from parts a, b, and c above that would allow you to predict what you would find if you made a difference table for fifth powers. To verify the prediction directly would be a lot of work. Instead, check your overall pattern in a very simple case, first powers.

9. Look back at question 8 from the point of view of a teacher.
   a. What lower level skills are involved?
   b. Would question 8 be appropriate for students who have just begun to work with exponents or whose subtraction skills are weak? Could question 8 be modified to make it less demanding, yet still reach the same goals?
   c. What would be gained and what would be lost if students did all the calculations in question 8 on a calculator or computer?

6.3 What Is Division?

As with subtraction, there are three common views of division.

Repeated subtraction If 8 children share 24 cookies equally, how many will each get? To solve this division problem, you could dole the cookies out, much like dealing cards, until they are all gone. Dealing one cookie to each child removes 8 cookies from those left to share; dealing out a second cookie to each removes 8 more. The question, then, is how many times you can take 8 from 24, with the understanding that you cannot go below zero. From that point of view, a division problem is really **repeated subtraction**.

Missing factor If you give n cookies to each of 8 children, you give out $8n$ cookies. If you give out all 24 cookies, the question of how many cookies you can give to each child comes down to finding a number n for which $8n=24$. This is a multiplication question in reverse. Given the product, 24, and one factor, 8, the question is to find the **missing factor**, which, in this context, is called the **quotient**. The known factor (the number you divide by) is called the **divisor**, and the number being divided (24 in this case) is called the **dividend**.

Multiplication by the reciprocal A third view of division, especially helpful for dividing by a fraction, is that **division by any number is really multiplication by the reciprocal**. The **reciprocal** of a number n is whatever number can be multiplied by n to get 1. For example, the reciprocal of 3 is $\frac{1}{3}$; and the reciprocal of $\frac{2}{7}$ is $\frac{7}{2}$. This will be considered in detail in chapter 7, on fractions.

Elementary students encounter this view of division when they divide by fractions ("invert and multiply"), and advanced textbooks often simply define division as multiplication by the reciprocal. That is, they define division by $\frac{2}{7}$ as multiplication by $\frac{7}{2}$.

Which view of division is best? Ultimately, all three ways contribute to understanding. Since multiplication is introduced as repeated addition, it is natural to introduce division as repeated subtraction, especially since, in simple examples, it is easy to model concretely. However, like take away in subtraction, repeated subtraction is practical only in relatively simple examples. For example, to divide -2/3 by 5 by repeated subtraction would, at best, stretch the idea. Some divisions, such as $(-100)\div(-5)$ are easy to grasp in terms of a missing factor, so

[2] Johann Heinrich Rahn introduced the sign ÷ for division in an algebra book published in Zürich in 1659. Some earlier writers had used the same sign to denote subtraction. Rahn's use of the sign did not meet with favor in his native Switzerland but was soon adopted in England.

that idea, too, should be introduced early.[2] Multiplication by the reciprocal is also a useful way to think of division, and it is invaluable where fractions or solving equations is involved, but it makes no sense to students who have not yet learned enough about fractions. That view of division, therefore, should be introduced only to students with enough background in fractions to understand it. With that in mind, we will wait to consider division as multiplication by the reciprocal until chapter 7. For now, we define division as follows:

Definition: For any numbers a, b, and c, where $b \neq 0$, $a \div b = c$ means $a = b \times c$.

Division by 0 is avoided entirely. To see why, consider the problem of dividing 5 by 0, which amounts to finding a number n for which $5 = n \times 0$. But $n \times 0 = 0$, not 5. Since there is no such number n, there is no reasonable way to divide 5 by 0. Thus we avoid division by 0 and regard it as meaningless or undefined.

3 views of subtraction	**3 views of division**
Take away	Repeated subtraction
Missing addend	Missing factor
Add the opposite	Multiply by the reciprocal

Classroom models of division

Many classroom examples of division involve partitioning a set of objects into equal groups. For example, if you want five children to share 17 cookies equally, you could deal out the cookies, one to each child, much as you would deal cards, going around three times. You could keep the other two cookies as a **remainder**, or you could break them and give each child an extra $\frac{2}{5}$ of a cookie, so each child gets $3\frac{2}{5}$ cookies in all. In the first case, we would say the quotient is 3 and the remainder 2. In the second case, we would say the quotient is $3\frac{2}{5}$.

(This is discussed further in chapter 7, on fractions.) Which of these approaches to use depends on the context. In cases where partial objects make no sense (e.g., with light bulbs or ping pong balls) use a whole number quotient and leave extra objects as a remainder.

Division based on partitioning a set of objects into a given number of equal groups is called **partitive** division. Contrast that with the following problem. Suppose that you work in a factory, packing ball point pens from a large bin into boxes of a dozen each. The bin has 1,850 pens in it. How many boxes of a dozen can you pack with the pens in this bin? This is not a question of partitioning the pens equally; you cannot solve it by dealing out the pens. Instead, you need to pack (or imagine packing) pens a dozen at a time until the supply is too low to fill another box. Now the quotient is the number of boxes you can pack, and the remainder is the number of pens left over (possibly 0 but in any case no more than 11). This example illustrates **measurement** or **quotative** division; it involves measuring out a given number of objects into number of groups, which is to be determined.

The area of a rectangle is, by definition, the product of its length and width. Using this in reverse is what might be called an **area model** of division. Specifically, if you know the area of a rectangle and the length of one side, finding the length of the other side amounts to dividing the area by the known length. For example, if the area is 12 and the length of one side is 5, the width of the rectangle is $2\frac{2}{5}$. However, if you are working with unbreakable tiles that are each one square foot, you cannot use fractional lengths. Then the best you can do is a 5×2 rectangle with 2 squares left over. We say that 12÷5=2 with a remainder of 2. This is pictured below.

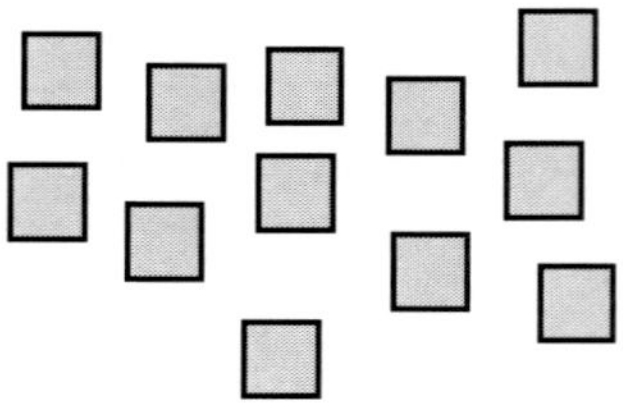

Twelve tiles

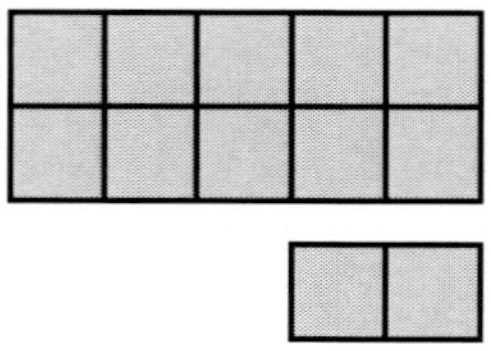

Rectangle 5 units long is 2 units wide, leaves 2 extra tiles

Another example: 25÷7.

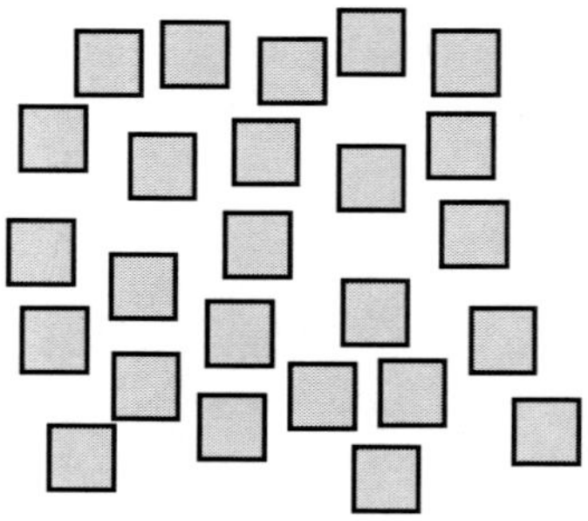

Twenty five tiles

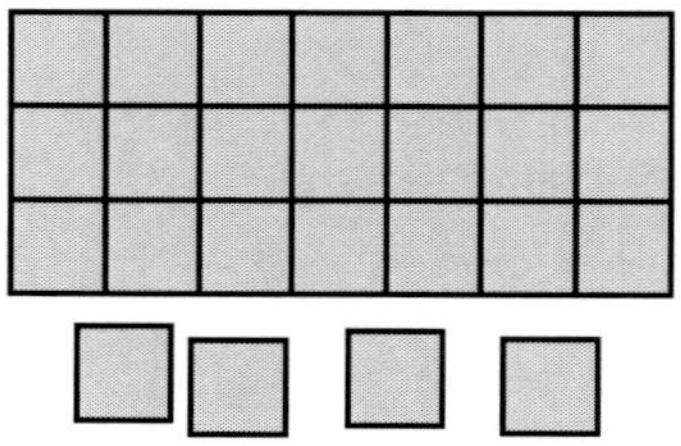

Rectangle 7 units long is 3 units wide, leaves 4 extra tiles

Then 25÷7=3 with remainder 4. The remainder is $\frac{4}{7}$ of a row of tiles. If the tiles may be cut, the quotient (width) is $3\frac{4}{7}$ and the remainder 0.

The distinctions between the models of division presented above depend more on the context in which the mathematics takes place than on the mathematics itself. As a teacher, expose your students to a variety of these contexts. Only then will they understand division from many points of view and be able to apply it in a wide variety of situations.

A further note on order of operations

For purposes of order of operations, we consider subtractions to be additions of opposites and divisions to be multiplications by reciprocals. For example, to evaluate 3×8÷2×6, work from left to right to get 24÷2×6=12×6=72. This is standard in mathematics, but occasionally you will encounter a computing language or a calculator, which uses a non-standard convention. For example, some calculators or computer languages first execute indicated multiplications (left to right), and only then do they go back and perform any indicated divisions, again working from left to right.

Division with negative numbers

The examples above involve positive numbers exclusively. To understand division with negative numbers, begin with the definition of division: $p \div q = r$ means $qr = p$, and experiment. For example, -36÷9 is a number, n for which, $9n = -36$. Clearly

n is negative, since if n were positive, $9n$ would be positive as well. A little further thought will convince you that n must be -4. Similar experiments (as in questions 1, 3, and 4 below) will show you that a quotient $p \div q$ is positive if p and q have like signs and negative if they have opposite signs.

Exercises 6.3

1. Sally did several box puzzles, but her little sister erased most of them. Can you help? One is totally erased. In that one, make up clues for someone else to do the puzzle as you have done in the other six. What concept of division is involved in these examples?

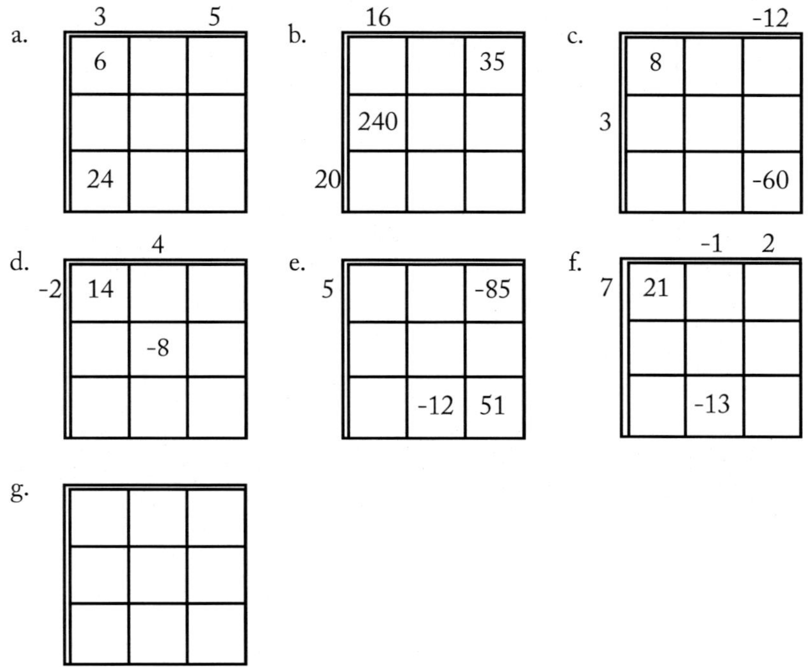

2. a. Picture $18 \div 7$ by trying to use 18 tiles to make a rectangle with length 7.

   b. Write a problem in which $18 \div 7$ can be found by dealing out objects one at a time.

3. In parts a-c, compute and check by multiplication:

   a. $(-24) \div (-8)$ b. $(-24) \div 8$ c. $24 \div (-8)$

   d. Model each division in parts a–b using algebra tiles.

4. Based on problem 3, decide whether $x \div y$ is positive or negative if:

   a. x and y are both positive
   b. x and y are both negative
   c. x is positive and y is negative
   d. x is negative and y is positive

5. In parts a-f calculate, using the rules for order of operations found in the text.
   a. 7-2(3-8)-5(6)
   b. 7-2[3-8(-5)6]
   c. $4+3(-5)^2$
   d. $(4+3)(-5)^2$
   e. $[4+3(-5)]^2$
   f. $[(4+3)(-5)]^2$
   g. 5×8÷10×2
   h. What would a calculator show in part g if it first did all multiplications from left to right and then did all divisions?

6. In the following table:
   a. Circle each multiple of 3.
   b. Put a triangle around each number which is 1 more than a multiple of 3.
   c. Put a square around each number which is 2 more than a multiple of 3.

1	2	3	4	5	6	7	8	9	10
11	12	13	14	15	16	17	18	19	20
21	22	23	24	25	26	27	28	29	30
31	32	33	34	35	36	37	38	39	40
41	42	43	44	45	46	47	48	49	50
51	52	53	54	55	56	57	58	59	60
61	62	63	64	65	66	67	68	69	70
71	72	73	74	75	76	77	78	79	80
81	82	83	84	85	86	87	88	89	90
91	92	93	94	95	96	97	98	99	100

   d. Look at the visual patterns created in the table above by your answers to parts a, b, and c. Do these patterns continue if the table is extended? Extend the table two more rows and then check to find out.

7. (Question 6, continued)
   a. If you add two numbers with triangles around them in the table above, what shape do you get? Try this on several pairs.
   b. If you add a pair of numbers with squares around them in the table above, what shape do you get? Try this on several such pairs.
   c. If you add a number with a triangle around it to a number with a square around it, what shape do you get? Try this on several such pairs.

d. Extend the results of parts a, b, and c to complete this addition table.

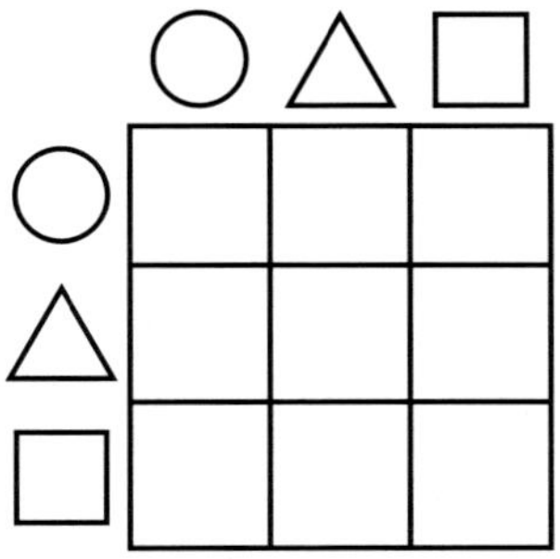

8. Find the remainders in each case (ignore the quotients).

   a. $1^2 \div 4$ $3^2 \div 4$ $5^2 \div 4$ $7^2 \div 4$ $9^2 \div 4$ $11^2 \div 4$

   b. $2^2 \div 4$ $4^2 \div 4$ $6^2 \div 4$ $8^2 \div 4$ $10^2 \div 4$ $12^2 \div 4$

   c. Make a general statement about remainders that expresses the results of parts a and b above.

9. Among odd prime numbers (those under 50 are: 3, 5, 7, 11, 13, 17, 19, 23, 29, 31, 37, 41, 43, 47), some can be expressed as a sum of two squares (e.g., $5=1^2+2^2$ and $29=5^2+2^2$), while others, such as 7 and 11, cannot be expressed as the sum of two squares.

   a. Find and list all the odd primes less than 50 which **can** be expressed as a sum of two squares. Then make a second list of all the odd primes under 50 which **cannot** be expressed as a sum of two squares.

   b. Divide each odd prime under 50 by 4 and look only at the remainder. How do the remainders relate to the lists you made in part a? Write your answer in a clear sentence or two.

   c. Are there any cases in part a when you wrote a prime as the sum of two **odd** squares? How about two **even** squares? Could the sum of two even numbers be odd? How about the sum of two odd squares?

   d. Combine the results of part c with those of question 8 to give a partial explanation of your observation in part b.

   e. Find and explain the relation of this question to question 9 of exercises 3.1.

Exploration: Digital Roots and the Magic of 9 (questions 10–16) The **digital root** of a number is, by definition, its remainder after division by 9. For example, the digital root of 29 is 2, since if you divide 29 by 9, the remainder is 2. Curiously, you can find the digital root of a number by finding the digital root of the sum of its digits. For example, with 29, you could just add 2+9=11; then you can even play the game again. The digital root of 11 is the same as the digital root of the sum of its digits, 1+1=2.

As another example, we find the digital root of 679. Divide by 9 to find $679=9\times75+4$, so the digital root of 679 is 4. You could also find this by adding digits: 6+7+9=22. Now add those digits: 2+2=4.

If the digit addition method leads to a final outcome of 9, the method must be modified, as you can see from an example, 45. The digital root of 45 is 0, since dividing 45 by 9 leaves 0 remainder, yet adding the digits yields 4+5=9. Therefore, **if adding the digits leads to 9, the digital root is 0.**

10. Find the digital roots of these numbers by adding digits. Then check your answers by dividing by 9.

    a. 57 b. 487 c. 3,289 d. 4,237

11. Sometimes the process of finding digital roots is called **casting out nines,**[3] because any time 9 is encountered as a digit or a sum of digits, it can be cast out or discarded. For example, to find the digital root of 341,598, you could of course divide by 9 to get the remainder. Alternatively, you could use the digit addition method. Easiest of all, however, is to "cast out" the 9, cast out the 4 and 5 (they add to 9), and cast out the 1 and 8 (1+8=9), leaving 3 as the digital root of 341,598. Use casting out nines to find the digital roots of

    a. 1,926 b. 2,899

    c. 556,274 d. 4,673,006

12. This problem shows why the digital root of a number is the same as the digital root of the sum of the digits. As an example, consider 385.

    Rewrite this in expanded form 3×100+8×10+5

    This in turn is 3(99+1)+8(9+1)+5 or $\underline{3\times 99+8\times 9}$+**3+8+5**.

    Here the boldfaced part is indeed the sum of the digits of 385.

    a. Use the distributive property to show that the underlined part is a multiple of 9.

    b. Mimic this explanation with 4,682.

13. Casting out nines is sometimes used to check addition. Consider this example:

problem	**digital roots**
4,802	5
+1,469	+2
6,271	7

The digital root of 6,271 is 7; that is also the sum of the digital roots of 4,802 and 1,469, so the addition checks. Had we come out with 6,261 (forgetting to carry), its digital root, 6, is not 5+2, indicating an error.

a. Cast out nines to check these additions.

22	149	3,904
45	203	4,470
20	765	2,529
87	1,017	10,903

b. Show an addition error which casting out nines will not catch.

[3] Casting out nines was used in many European mathematics texts of the 16th century.

14. Casting out nines can be modified for use as a subtraction check, by subtracting digital roots instead of adding them. For example, to check 3,158–495=2,663, proceed this way

problem	**digital roots**
3,158	8
– 495	–0
2,663	8

Since 8–0=8, which is also the digital root of 2,663, the subtraction checks. If subtracting digital roots leads you to a negative number, add 9 to it, as in this example.

problem	**digital roots**
2,036	2
– 404	–8
1,632	–6

This seems not to check, since the digital root of 1,632 is not –6. However, add 9 to it, which yields 3, and that is the digital root of 1,632, so the subtraction checks. (In cases with many digits, you may have to add 9 more than once to get up to a number between 0 and 9.)

Use casting out nines to check the following subtractions.

a. 39,812–17,734=22,178 b. 36,142–14,095=22,147

15. To use casting out nines to check multiplication, multiply digital roots rather than adding or subtracting them. For example, to check whether 437×502=219,374, multiply the digital roots of 439 and 502, cast out nines in the result, and compare that to the digital root of 219,374. Then cast out nines to find any and all incorrect multiplications below.

a. 237×529=125,273

b. 589×4,897=2,874,333

c. 4,998×1,274=6,365,452

d. 3,654×7,802=28,507,508

16. Can casting out nines be used to check division? Yes. Every division that ends without a remainder can be checked by multiplying. For example, 112÷8=14 can be checked by multiplying 8×14 and seeing if you come out with 112. To check this by casting out nines, see question 15. To check a division with remainder, you need to add in the remainder at the end. For example, to check 114÷8=14 with remainder 2, first note that this is equivalent to saying that 8×14+2=114. Use casting out nines to check

a. 179÷11=16 remainder 3 b. 295÷14=22 remainder 1

17. (By Larry Stevenson) Find the largest four digit multiple of 45 whose digits add up to:

a. 27 b. 24. (This is a trick question; think!)

18. (By Larry Stevenson) Find all possible pairs of digits X and Y for which the seven-digit number 8,271,XY6 is a multiple of 72.

6.4 Long Division

Long division[4] is often a source of frustration for students and teachers, and there have even been calls to drop it from the curriculum altogether. However, if properly taught, long division can give students understanding that carries over from arithmetic to algebra. Here we will examine long division, why students find it hard, and some ideas for teaching this technique effectively.

To find, say, 995 ÷ 24 by long division, you probably learned to write $24\overline{)995}$ and then say something like this: 24 times 4 is 96, so write 4 above the second 9, 96 below 99, and subtract:

$$\begin{array}{r}
4\\
24\overline{)995}\\
\underline{96}\\
3
\end{array}$$

Now bring down the 5 and write it next to the 3.

$$\begin{array}{r}
4\\
24\overline{)995}\\
\underline{96}\\
35
\end{array}$$

Next, since 24 goes once into 35, put a 1 next to the 4 at the top, and write 24 beneath the 35 like this

$$\begin{array}{r}
41\\
24\overline{)995}\\
\underline{96}\\
35\\
24
\end{array}$$

Finally, subtract the 24 from the 35 to get 11, which is a remainder, since it is smaller than 24.

$$\begin{array}{rl}
41 & \\
24\overline{)995} & \\
\underline{96} & \\
35 & \\
\underline{24} & \\
11 & \text{remainder}
\end{array}$$

To check the work, multiply 24 by 41 to get 984, and then add the remainder, 11, to get 995.

Long division has many opportunities for error. It requires accuracy in subtraction and multiplication, as well as care in writing numbers in just the right places. Worst of all, it may seem incomprehensible. Why divide 24 into 99? Why bring down the 5?

[4] Long division, as it is known in the U.S. today, was introduced by Henry Briggs (1561-1630). Briggs is best known for his work on logarithms (see section 17.5), in the course of which he pioneered many new computational techniques. Long division is one of several ways used around the world to arrange the writing in division problems.

To understand long division, go back to repeated subtraction. In the example above, the question is how many times we can subtract 24 before using up a supply of 995 objects. The work shows we would have to subtract 24 a total of 41 times, a tedious prospect. Long division speeds up the repeated subtraction and makes it more efficient, in effect, dealing out more than one object at a time. Since it is easy to multiply by ten and 10×24=240, subtracting ten 24s would still leave most of the 995. In fact, subtracting twenty 24s would only subtract 480, and we can even double that to subtract forty 24s or 960. That leaves 995-960 or 35 still to give out. We might arrange the work something like this:

```
      41
  24)995
     960  40 twenty-fours subtracted
      35     remaining
      24   1 more twenty-four subtracted
      11     remaining
          41 twenty-fours subtracted in all
```

This looks like the long division above, but now you can see that subtracting 96 from 99 and bringing down 5 is just a way to subtract 960 from 995. Overall, you can understand what is going on as repeated subtraction.

Here is a concrete way to model long division with base ten blocks.

Example 1. Illustrate the division 437÷3 with base ten blocks.
Solution: Imagine three people who want to share 437 objects equally. Picture 437 with base ten blocks like this

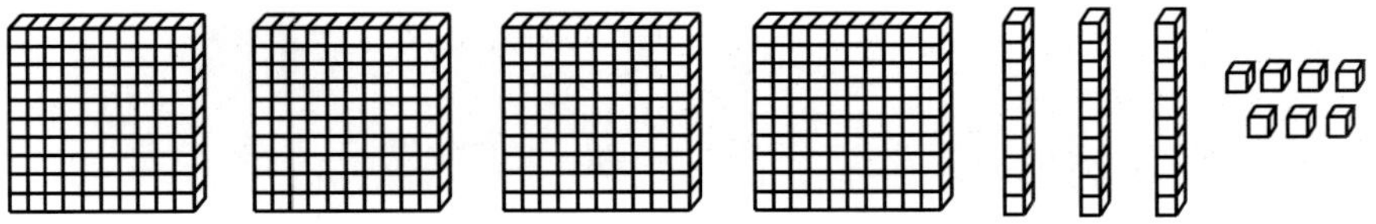

1. Give out a flat of 100 to each of the three people. That leaves 137 objects available to distribute.

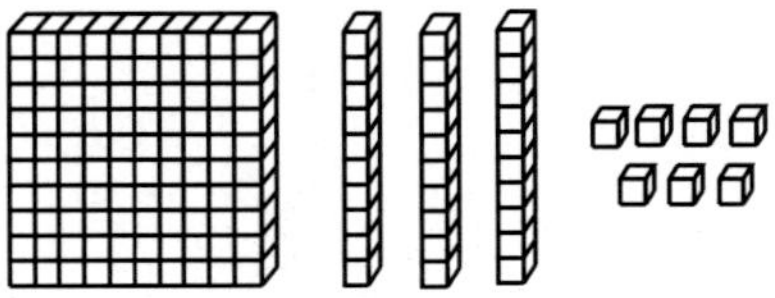

2. Trade in the remaining flat for ten rods.

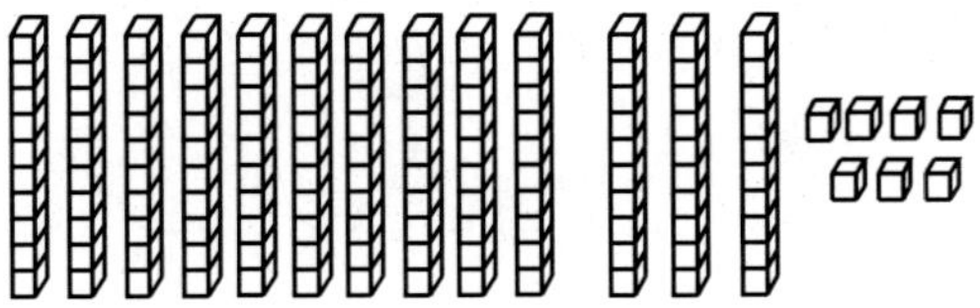

3. Give four ten rods to each, leaving one ten rod and seven, or 17 objects available to distribute.

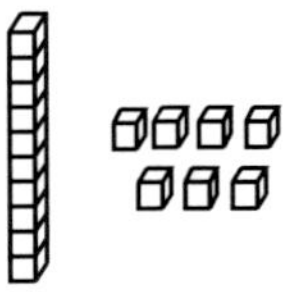

4. Trade the ten rod for ten ones.

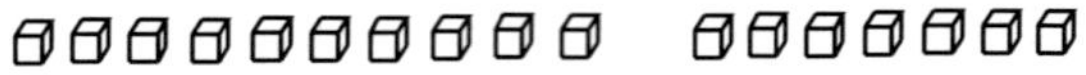

5. Give five ones to each, leaving two as remainder. Each person has 145 objects, and two remain.

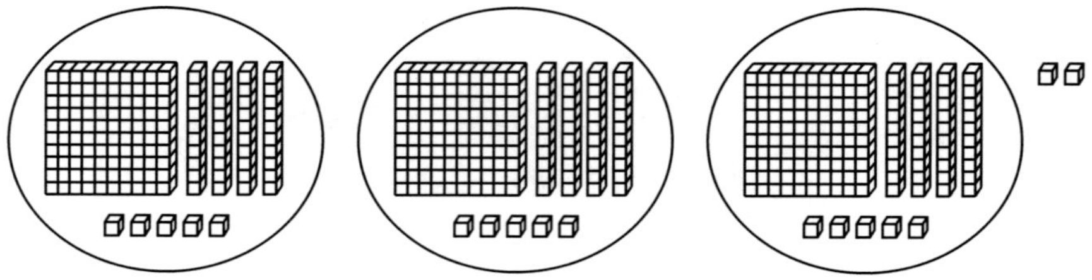

All the 437 objects we began with are accounted for, since 3×145+2=437.

The partitioning method was practical in this case, because the divisor or number of groups was small. If the divisor is large, another model may be more helpful. The next example uses the area model with a two-digit divisor.

Example 2. Find the quotient and remainder of 769÷24.
Solution: Build a rectangle using 7 hundreds, 6 tens, and 9 ones with the length of one side as 24. Our goal is to find the length of the other side (the quotient). Start with:

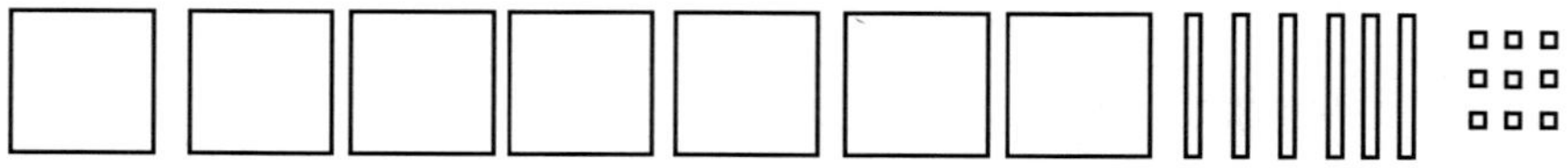

To form the rectangle, lay down 2 hundreds and 4 tens vertically. This corresponds to 10 groups of 24.

Then trade 1 hundred for 10 tens, and lay down 2 more hundreds and 4 more tens. So far this gives 20 groups of 24:

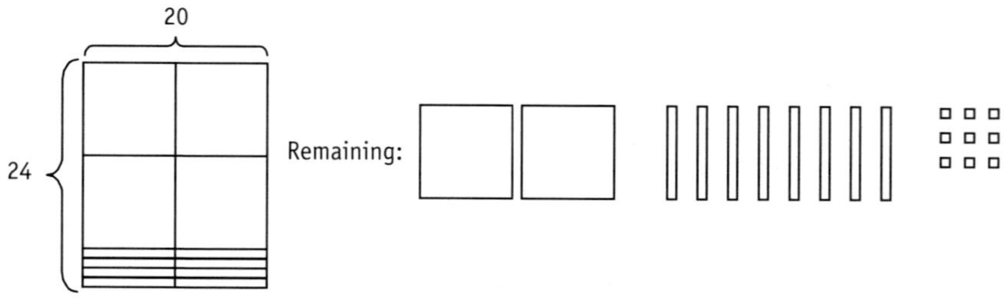

Continue building the rectangle with the remaining 2 hundreds and 4 more tens.

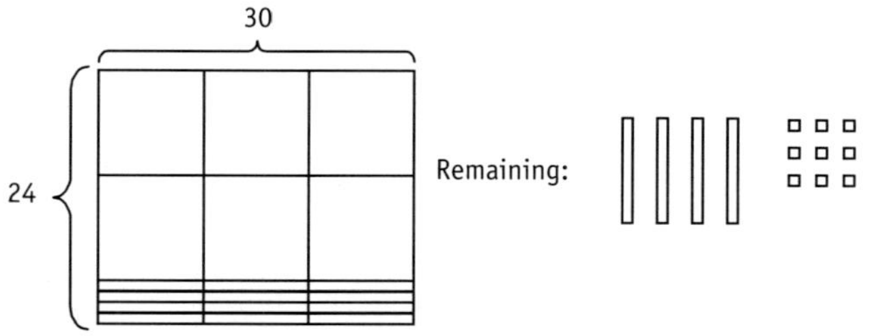

Extend the rectangle with 2 tens and 4 ones. This gives 1 more group of 24:

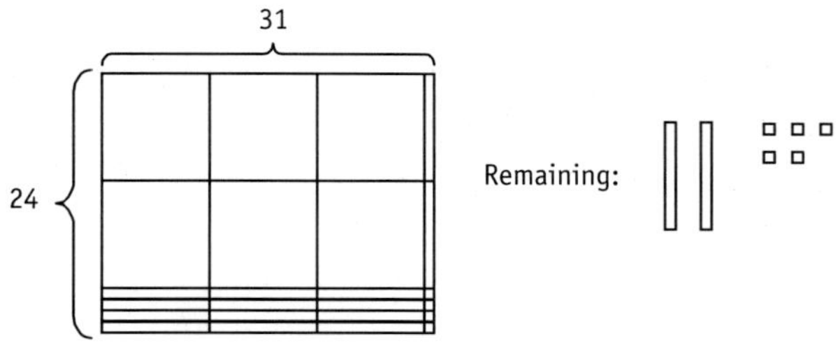

Use 2 tens and 4 ones once more giving a 24×32 rectangle with one left over.

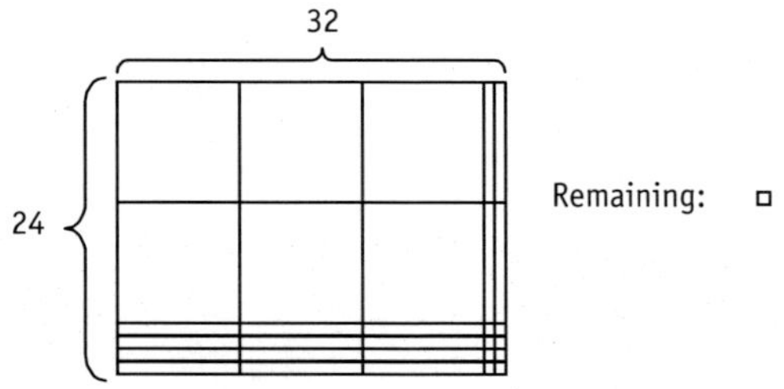

Then the quotient is 32, and the remainder is 1.

Each vertical strip of two hundreds and four tens represents 240, and each time we put down such a strip, we subtracted 240 from the objects remaining. Evidently, forming a rectangle in this way is a visual model of repeated subtraction:

769			
-240 →	2 hundreds and 4 tens		
529			
-240 →	2 hundreds and 4 tens		
289			
-240 →	2 hundreds and 4 tens		
49			
- 24 →		2 tens and 4 ones	
25			
- 24 →		2 tens and 4 ones	
1 →			remainder 1
	Total: 6 hundreds,	16 tens, and 8 ones,	remainder 1.
	or 7 hundreds,	6 tens, and 8 ones,	remainder 1.

Tips for teaching long division

1. Before you teach long division, be sure that the kids can:
   • Multiply
   • Subtract
   • Understand division as repeated subtraction.
2. Work through several examples, as we did above, using concrete representations and showing how to streamline the repeated subtraction.
3. When at last you introduce long division, keep the problems simple at first, possibly using only single digit divisors, and do not assign many long divisions all at once.
4. Use squared paper to help students organize their work in an orderly and readable manner and avoid errors brought about by sloppiness.
5. Have them check their results by multiplying.
6. Warn your students not to expect instant success. With care and persistence, they can learn long division.

Exercises 6.4

1. Mimic the discussion in the text to show how long division streamlines the repeated subtraction process for
   a. 563÷11 b. 2,749÷52

2. If a school trip for 106 students uses school busses, and each bus seats 34 students, how many busses will be needed?

3. A rectangular plot of land 98 meters long has an area of 4,802 square meters. How wide is this plot of land?

4. a. Model 624÷5 using base ten blocks.
   b. Model 743÷35 using an area representation like example 2 above.

5. **A trick (and a source of division practice)**
   a. Ask your students to each choose a 3-digit number (perhaps the first three digits of a phone number), then copy it next to itself to get a 6-digit number. For example, someone who chose 382 now has 382,382.
   b. Have your students divide this six-digit number by 7, and assure them that there will be no remainder, if they divide correctly.
   c. Have your students divide their answers (quotients) from

step b by 11. Again they will not get a remainder unless they make a mistake.

d. Finally, have your students divide the quotient from step c by 13, and repeat the promise about remainders. If they divide correctly, they will be amazed, not only at the fact that they get no remainder, but also at the quotient itself. Try it yourself and see what you get.

e. Can you explain the trick?

6. In parts a-d, divide 7 into the given number:

a. 646,464 b. 383,838 c. 252,525 d. 797,979

e. What do you notice about the remainders in each of these divisions?

f. Divide each quotient from parts a-d by 3.

g. What do you notice about the remainders in each of these divisions?

h. Divide each quotient from part f by 13.

i. What do you notice about the remainders in each of these divisions?

j. Divide each quotient from part h by 37.

k. What do you notice about your quotients in part j?

l. Make a general statement based on the pattern in these examples.

m. To explain this, divide 797,979 by 79, then divide 252,525 by 25. What do you notice about the quotients?

n. Factor 10,101 into primes. Then show how this explains parts a-j above.

7. In parts a-j, check that:

a. 2^2-2 is a multiple of 2

b. 2^3-2 is a multiple of 3

c. 2^4-2 is not a multiple of 4

d. 2^5-2 is a multiple of 5

e. 2^6-2 is not a multiple of 6

f. 2^7-2 is a multiple of 7

g. 2^8-2 is not a multiple of 8

h. 2^9-2 is not a multiple of 9

i. $2^{10}-2$ is not a multiple of 10

j. $2^{11}-2$ is a multiple of 11

k. Ancient Chinese mathematicians thought, as you may, too, that 2^n-2 is a multiple of n only if n is prime. But, for some composite numbers n, 2^n-2 is a multiple of n. The smallest such n is 341. What are its prime factors?

8. Fermat (see section 12.4) observed that

   $2^{2^0}+1=3$ (Remember, $2^0=1$)

   $2^{2^1}+1=5$

   $2^{2^2}+1=17$

   $2^{2^3}+1=257$

   $2^{2^4}+1=65{,}537$

   are all primes, and he stated that $2^{2^n}+1$ is prime for all whole numbers n. Later, Euler[5] observed that Fermat's guess does not hold for $n=5$, since $2^{2^5}+1=4{,}294{,}967{,}297$ is a multiple of the prime 641. Verify Euler's claim, using a calculator. (Euler, who had no calculating machines, was as skillful in arithmetic as he was in higher mathematics. In this case, he did not use long division, but instead thought up a clever shortcut).

   Numbers of the form $2^{2^n}+1$ are now called **Fermat numbers**, and those listed above are the only Fermat numbers known to be primes. Indeed, the question now is whether there are any higher Fermat primes. That question took on added interest when Gauss (see note, section 1.3) showed at the age of 19 that a regular polygon with a prime number of sides can be constructed by classical Greek methods if and only if the prime is a Fermat prime. This discovery so delighted Gauss that he changed his study from linguistics to math and science.

9. This puzzle is from Maurice Kraitchik's *Mathematical Recreations.* Each X covers a digit in this long division, but all covered digits are not the same. Can you determine the missing digits?

```
          X X 8 X X
    X X)X X X X X X X X
        X X X
              X X
              X X
                X X X
                X X X
                    1
```

[5] Leonhard Euler (1707-1783) was born near Basel, Switzerland, but spent most of his life in Berlin and St. Petersburg. Mathematics came so easily to him that he often worked on mathematics while playing with some of his 13 children. He produced a staggering quantity of high quality work. His 873 books and articles make up about a third of all the mathematics, theoretical physics, and engineering mechanics published between 1725 and 1800. In addition, he wrote about 5,000 letters, some very long and technical, to leading figures of his time. He produced hundreds of published articles, most involving numerous equations and complicated formulas, after he had become totally blind! As a youngster he memorized many long poems as well as the squares, cubes, fourth, fifth, and sixth powers of the first 100 prime numbers. He remembered both the numbers and the poems all his life.

10. If you are given a number in base ten, you can express it in another number base by repeated division by the base. For example, to express 37 in base five:

    37÷5 has quotient 7 and remainder 2

    7÷ 5 has quotient 1 and remainder 2.

    1÷5 has quotient 0 and remainder 1.

    Stop dividing by the base after you obtain a 0 quotient. The remainders, in reverse order, are the base five numeral for the original number. Here 37 is 122 in base five.

    a. Use this method to express 684 in base five, then check the result by changing it back into base ten as in section 5.4.
    b. Use this method to express 172 in base three. (Divide repeatedly by 3 this time), then check your result by changing it back to base ten.
    c. Use this method to express 75 in base two.

11. In section 5.5 Russian peasant multiplication was illustrated with

	75 ×	231	
	37	462	
	~~18~~	~~924~~	
	9	1,848	
	~~4~~	~~3,696~~	
	~~2~~	~~7.392~~	
	1	14,784	
Total not crossed out		17,235	=75×231.

    To see why this method works, we need some preliminary observations.

    a. How is the **parity** (oddness or evenness) of a number reflected in its base two expression?
    b. How is the parity of a number indicated by the remainder on division by 2?
    c. In base ten, to multiply a whole number by ten, write a 0 at the right of the number. For example 357×10=3,570. In base two, what is the effect of writing a 0 to the right of a whole number?
    d. Begin with 1 and work up the list of numbers that are halved above: To retrace the list, double at each step. Then, if the number you need to reach is odd, add 1.

	Number	Odd or even	Binary 1s digit	Binary form
	1	odd	**1**	1
Double	**2**	even	**0**	10
Double	**4**	even	**0**	100
Double	8	even	0	1000
Add one	**9**	odd	**1**	1001
Double	**18**	even	**0**	10010
Double	36	even	0	100100
Add 1	**37**	odd	**1**	100101
Double	74	even	0	1001010
Add 1	**75**	odd	**1**	1001011

e. The bold numbers in the first column are the numbers in the halving column from multiplying 75×231. The bold numbers in the binary 1s digit column are, reading from top down, the binary form of 75. You can see that the 0 digits correspond to the **even** numbers in the halving numbers at left in bold type. This is the reason for crossing out lines in which the halving number is even in Russian Peasant multiplication.

6.5 Greatest Common Factors

What is the largest number which is a factor of both 8 and 44? This is not a difficult question. Just list the factors of 8: 1, 2, 4, 8, and see which of them is also a factor of 44. You quickly find that 1, 2, and 4 are factors of both 8 and 44, so they are called **common factors** of 8 and 44. The greatest of these, 4, is called the **greatest common factor** (GCF) of 8 and 44. The GCF is useful, for example, in reducing fractions to lowest terms (see chapter 7). If the numbers involved are not too large, the tabular approach to finding least common multiples given in chapter 5 is also helpful for finding the GCF.

Example 1. Find the greatest common factor of 30 and 42.
Solution 1: Factor 30 into primes: $30=2\cdot3\cdot5$
Factor 42 into primes: $42=2\cdot3\cdot7$
The two numbers both have a factor of 2 and a factor of 3 in common. Then both 2 and 3 are common factors of 30 and 42, and the greatest common factor of these numbers is $2\cdot3$ or 6.

It is often helpful to organize the work in a table like this:

Factors of 60	2^2	3	5	
Factors of 42	2	3		7
Factors of GCF	2	3		

This may remind you of the tables used for least common multiples, and indeed the GCF and LCM are similar. But in making the table for the LCM, the greatest power of each prime was circled, and for the GCF we circle the smallest power of each prime (possibly an empty box) and use those smallest powers to calculate the GCF. An empty box indicates no factors of the prime in that column are used. An empty box is equivalent to a box with a 1 in it. If the last row of the table consists entirely of empty boxes, the desired GCF is 1.

Another systematic way to find the GCF of two numbers was given by Euclid in his *Elements* and is now known as **Euclid's algorithm**.[6] Euclid's algorithm does not require factoring, but it does use repeated division.

Euclid's algorithm is based on the simple observation that if you add multiples of any number, you get a multiple of that same number. If you add one multiple of 5 to another, you will always get a multiple of 5, since $5\times3+5\times11=5(3+11)=5(14)$, for example. This follows directly from the distributive property. The same goes

[6] See note on Euclid in section 4.4.

for subtracting multiples of the same number, though we usually restrict the term "multiple" to positive numbers. To see how this helps us find the GCF of 8 and 44, for example, note that since 8 and 44 are both multiples of the GCF, so are

$$44 - 8 = 36$$
$$44 - 16 = 28$$
$$44 - 24 = 20$$
$$44 - 32 = 12$$
$$44 - 40 = 4$$

Then 4 is a multiple of the GCF of 8 and 44, and you can easily guess that 4 is, in fact, the GCF. To shorten the work, replace repeated subtraction with division; 4 is the remainder if you divide 44 by 8. Applying this idea repeatedly is the basis of Euclid's algorithm.

Example 2. Find the GCF of 36 and 280 by Euclid's algorithm.
Solution: Step 1. Divide 280 by 36 to find 280=36×7+28. This shows the GCF of 36 and 280 is also a factor of 28. You could now check factors of 28, but instead just repeat the process, since the number we seek is also the GCF of 28 and 36.
Step 2. Divide 36 by 28 to find 36=28×1+8. Therefore the GCF of 36 and 28 is also the GCF of 8 and 28.
Step 3. Divide 28 by 8 to find that 28=8×3+4, so the GCF of 8 and 28 is also the GCF of 4 and 8. Since 8 is a multiple of 4, 4 is the GCF of 36 and 280.

Here, then, is **Euclid's algorithm. To find the GCF of two positive integers *x* and *y* (we use *x* for the smaller of the two):**

1. Divide y by x and call the remainder r_1.
   If $r_1=0$, x is the GCF of x and y. Otherwise go to step 2.
2. Divide x by r_1 and call the remainder r_2.
   If $r_2=0$, r_1 is the GCF of x and y. Otherwise go to step 3.
3. Divide r_1 by r_2 and call the remainder r_3.
   If $r_3=0$, r_2 is the GCF of x and y. Otherwise continue.
4. Continue to divide each remainder into the previous divisor until you get a 0 remainder. The last non-0 remainder is the GCF of x and y.

Example 3. Find the GCF of 135 and 560 by Euclid's algorithm.
Solution:

1. Divide 560 by 135. The remainder is 560–540=20.
2. Divide 135 by 20. The remainder is 135–120=15.
3. Divide 20 by 15. The remainder is 20–15=5.
4. Divide 15 by 5. The remainder is 0.
   The last nonzero remainder is 5. Therefore, 5 is the GCF of 135 and 560.

Schematically,

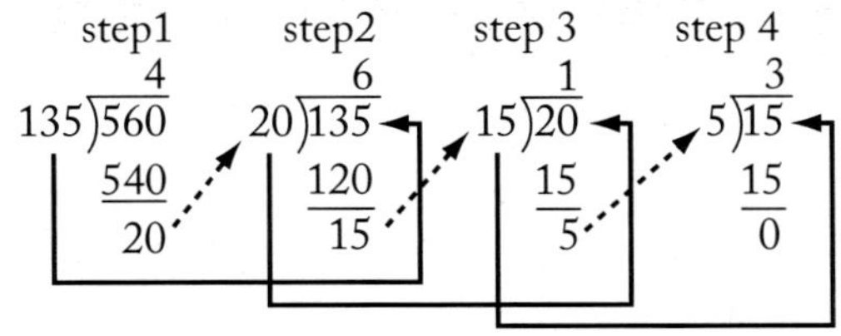

Exercises 6.5

1. Find the GCF of:
   a. 150 and 468 b. 260 and 836 c. 580 and 1,194

2. a. Fill in the following table:

numbers a, b	GCF of a, b	LCM of a, b	product ab
4, 6			
6, 10			
5, 8			
4, 8			
7, 6			
12, 20			

   b. Look for a pattern in part a, then state what a rule relating the product to the GCF and the LCM.

3. **Exploration:**
   a. In each cell put the GCF of the numbers at the borders. A few are filled in as checks.

	1	2	3	4	5	6	7	8	9	10	11
1											
2			1	2							
3						3					
4											
5											
6											
7											
8											
9											
10								2		10	
11											

The above table has some interesting patterns.

   b. The first entry in each row and each column is 1. Name the family of numbers whose rows and columns have 1 as their **second** entry.

c. Name the family of numbers whose rows and columns have **only 1 and one other number**.

d. Name the family of numbers which have rows and columns **involving 1 and exactly two other numbers**.

Vocabulary for Chapter 6

Borrowing or regrouping (6.2)
Common factor (6.5)
Difference (6.1)
Digital root (6.3)
Dividend (6.3)
Divisor (6.3)
Euclid's algorithm (6.5)
Greatest common factor (6.5)
Long division (6.4)
Parity (6.4)
Quotient (6.3)
Reciprocal (6.3)
Remainder (6.3)
Subtraction (6.1)

Self-Quiz for Chapter 6

1. Subtract:

   a. 15– (–42) b. –38– 85 c. –80– (–30)

2. Complete this addition box puzzle.

	5	-8
-1		13

3. Describe three different views of subtraction.

4. Subtract -47-(-51) and explain your result:

   a. With the missing addend view of subtraction.

   b. By adding the opposite.

   c. Using the take away view of subtraction.

5. Complete this box puzzle.

			7
	33		
-5		20	

6. Describe three different views of division.

7. Divide $38 \div 9$ and illustrate the work:
   a. As repeated subtraction
   b. Using algebra tiles and area

8. a. Divide $427 \div 35$ by long division.
   b. Picture the work with area.

9. Find the greatest common factor of 2,384 and 798.

Chapter Seven

Fractions

Now we widen the focus to include rational numbers, which can be written as fractions. After our review of integers, you may be pleased to find that fractions are not as difficult as you expected.

7.1 The Concept of Fraction

Counting questions can be answered with whole numbers, but measurement quickly leads to fractions. To introduce fractions, take a few minutes to make a list of where fractions are encountered in daily life. After you make your list, compare it with this one.

Area of life	**Example**
time	half an hour
money	quarter of a dollar
distance	three tenths of a mile: half an inch
cooking	one third of a cup: half a pint
music	half notes, eighth notes, etc.
area	a quarter of an acre
weight	quarter pound: half ton

You can easily add to the list. The point is that fractions turn up in all sorts of situations, so it is important to learn to work well with them and understand them. A secondary point is that fractions, like all mathematical ideas, are abstractions, which grow out of concrete cases, so the concept of fraction must be built through examples. In selecting examples, pay attention to variation in learning style. To use an extreme case, a blind student can learn from touching and hearing but not from seeing, so musical examples would be helpful, as would physical models that can be held, weighed, etc., but pictures would be of no use. More generally, each person has unique strengths and weaknesses in perception and learning. Therefore, the more varied the examples used to build an idea, the more readily students can learn it. Also, the more senses you can involve, the better. (Most adults rely mainly on sight and somewhat less on hearing. Children, however, use all their senses, often spending time touching, feeling, smelling, and even tasting when they meet new objects.) Now reconsider the list above (or your own list) and think of ways to use it to make fractions as concrete, natural, and appealing as possible.

There are unlimited opportunities to demonstrate fractions. Helping to cook by measuring a half cup of milk or a cup and a half of flour teaches about fractions. So does cutting a pizza or cake, or breaking a chocolate bar along the lines. Learning about money by making change, for example, teaches more than counting, adding, and subtracting. It also teaches the values of the coins, so that a nickel will eventually be seen as worth half as much as a dime – and there again fractions come in. A project, such as a bake sale, can make mathematical concepts concrete at many stages, from buying the ingredients, measuring and mixing them, right through to cleaning up, selling the final products, and figuring the profit. (Selling prepackaged goods, as is done today with Girl Scout Cookies, has far less educational value.)

As with any abstract concept, students will learn about fractions at different rates. Therefore, introduce fractions as an idea well in advance of when you need to teach arithmetic with fractions. That strategy puts time on your side and allows you to approach fractions in a low pressure way. When it comes time to teach adding or multiplying fractions, the students will be ready.

Models for Fractions

The word "fraction" is related to "fracture" and to the idea of breaking a whole number into equal pieces. When the number 1 (the **unit** or **whole**) is broken into equal parts, the number of equal parts is the kind of fraction. Thus, if the unit is broken into eight equal parts, the resulting fractions are eighths, and if the unit is broken into eleven equal parts, the resulting fractions are elevenths.

Fractions are written $\frac{a}{b}$, where b, the **denominator**, is the number of parts into which the unit is broken, and a, the **numerator**, is the number of those parts. Fractions may be modeled pictorially and physically with areas, sets, or lengths, as well as with time, money, weight, and volume. Whenever you model a fraction, be sure to specify the unit clearly.

Area model Circles, squares, rectangles, and pattern blocks work well for picturing fractions as areas. Pattern blocks come in different shapes – hexagons, trapezoids, parallelograms, and triangles.

If the hexagon 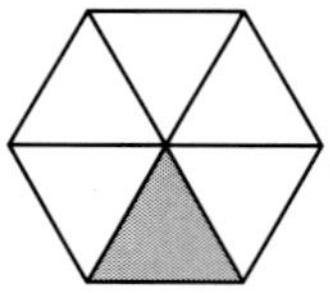 is the unit, then the triangle  is $\frac{1}{6}$,

because 6 of the triangles make up one whole. If two hexagons

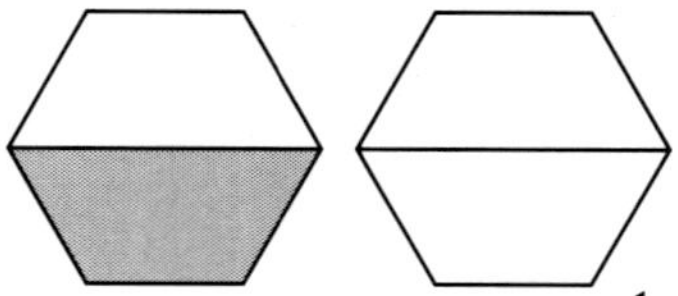 are taken as the unit, then the

trapezoid 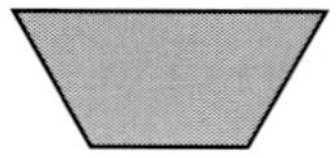 is $\frac{1}{4}$, because 4 of those trapezoids make up one unit.

If the triangle  is $\frac{1}{3}$, then the trapezoid 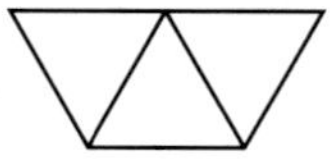 is the unit

since three of those triangles make the trapezoid.

Length Model A number line, a ruler, strips of paper, and Cuisenaire rods can all be used to model fractions as lengths. We use Cuisenaire rods here.

If the 5-rod  is the whole, then the 1-rod  is $\frac{1}{5}$,

since five 1-rods have the same length as the 5-rod. 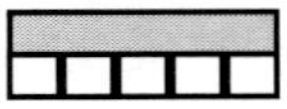

Again, if the 8-rod is the whole, then the 6-rod is $\frac{3}{4}$, as the following picture shows.

whole
whole cut into fourths
three fourths

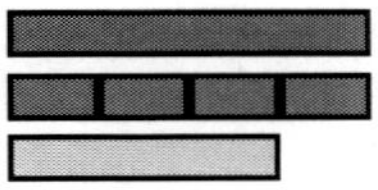

Set Model In set models, the whole is represented by a set of objects such as chips, beans, dots, etc., and the fractional parts are represented by subsets of the given set.

If a set of 12 eggs (a dozen) is the unit, then eight eggs is $\frac{8}{12}$.

Often we are given the whole and asked to find a fractional part, but sometimes we are asked to reverse the process and determine the whole from a given fractional part..

Example 1. (area model) If this rectangle 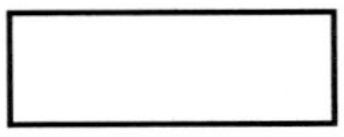 is $\frac{3}{4}$, find the whole.

Solution: Cut the given rectangle into three equal parts, each of which is $\frac{1}{4}$ (one fourth) of the whole:

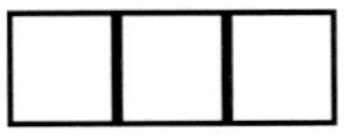

The whole is four of these fourths:

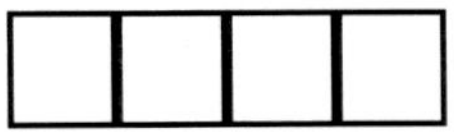 or 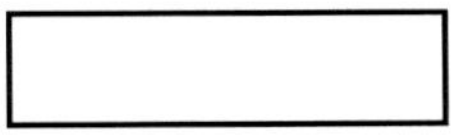

Example 2. (set model) If a set of 6 dots ○ ○ ○ ○ ○ ○ is $\frac{2}{5}$, find the whole.

Solution: Since the set of 6 dots is two parts (fifths) out of five, break the set

into two equal parts so that 3 dots is one part or $\frac{1}{5}$: ○ ○ ○|○ ○ ○

Therefore, the whole (five-fifths) is 15 dots:

○○○ ○○○ ○○○ ○○○ ○○○

In general, if you are given a fractional part $\frac{p}{q}$, first divide it by p to get $\frac{1}{q}$, then multiply the result by q to get the unit.

The conceptual understanding needed for the part-to-whole comparisons shown here is crucial for operations on fractions, decimals, and percents.

Once you have some feeling for fractions, you are ready to look more theoretically at them. Mathematically, a fraction is a number represented as a division.

Definition: The fraction $\frac{p}{q}$ means $p \div q$.[1]

The division sign even looks like a picture of a fraction. The bottom number, the **denominator**, is the divisor, while the top number, the **numerator**, is the dividend. If you find these names too long and clumsy, just call them the top number and the bottom number. The denominator shows how many parts are in the whole unit. The numerator shows how many of these parts are used.

Be careful with shilling fractions. Sometimes fractions are written with a slash instead of a horizontal bar, so that, for example, $\frac{2}{3}$ is written 2/3. This way of writing fractions is called **shilling fractions**, because English money was traditionally written that way. Shilling fractions are easy to type, but they can be confusing. For example, does 5/3+9 mean $\frac{5}{3}+9$ or $\frac{5}{3+9}$? By the convention on order of operations, the first interpretation is correct, since multiplication and division (in this case dividing 5 by 3) take precedence over addition and subtraction. To write $\frac{5}{3+9}$ with shilling fractions requires parentheses, like this: 5/(3+9). Confusion about order of operations never arises if the fractions are written with horizontal bars, whose lengths act like parentheses in determining order of operations. Therefore, **unless no confusion can possibly arise, avoid shilling fractions** and ask your students to do likewise.

No fraction can have a 0 denominator. Since you can't divide by 0, you can't have fractions with denominator 0. Of course, there is nothing to stop you from writing, for example, $\frac{2}{0}$, but, as explained in section 6.3, it has no meaning as a division. On the other hand, fractions with 0 in the numerator, such as $\frac{0}{2}$, do have meaning. All are equal to 0. In general, if $n \neq 0$, $\frac{0}{n} = 0$, since $0 = 0 \times n$.

Fractions which equal 1. Fractions which equal 1 are especially useful. You can easily see that $\frac{2}{2}=1$, since $2=1\times2$. Similarly, $\frac{3}{3}$, $\frac{4}{4}$, $\frac{5}{5}$, and, generally, any fraction of the form $\frac{y}{y}$, is equal to 1, since $y=1\times y$. Of course, y cannot be 0, since we can not divide by 0, but y can be any other number at all, positive or negative. An example of this, which we will use often, is $\frac{-1}{-1}=1$. The number y in a fraction $\frac{y}{y}$ can itself be a fraction so that, for example, $\frac{\frac{2}{3}}{\frac{2}{3}}=1$.

[1] Use of a line to separate numerator and denominator goes back at least to the eleventh century Arab writer al-Hassar and to Leonardo of Pisa (Fibonacci-see note, section 5.4), but it was first generally used in sixteenth century Europe.

Mixed numbers A number such as $1\frac{3}{4}$, which consists of an integer followed by a fraction, is called a **mixed number**. This one is read "one and three fourths," and it means $1+\frac{3}{4}$. A minus sign before a mixed number applies to the entire number. Thus, for example, $-1\frac{3}{4}$, shown in figure 7.1.1, means $-(1+\frac{3}{4})$ and not $-1+\frac{3}{4}$.

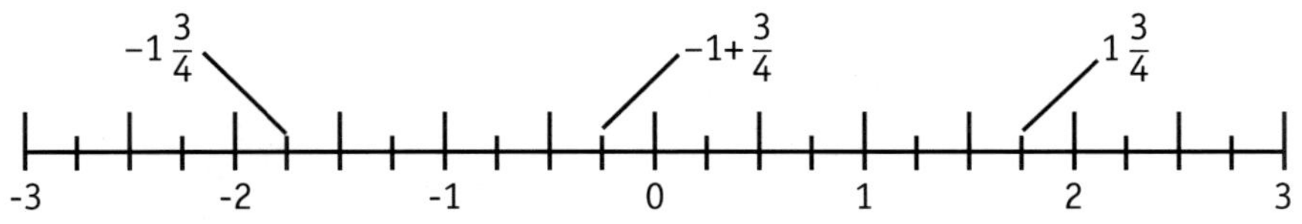

Figure 7.1.1

Measuring lengths with rulers and tape measures provides good practice with the use of mixed numbers in real world contexts.

Proper and improper fractions. A fraction such as $\frac{6}{3}$ or $\frac{9}{2}$, with numerator greater than denominator, is sometimes called an **improper fraction**, in contrast to fractions with numerator smaller than denominator, which are called **proper fractions**. For many years, elementary school students were told to shun improper fractions, except when they were needed for computation. In higher mathematics, the distinction between proper and improper fractions is less important, and the appearance of $\frac{22}{7}$ as an approximation for π has always been considered reasonable. The word "improper" sounds pejorative, but improper fractions are sometimes appropriate. Express numbers in whatever way is most useful and reasonable in each context.

Exercises 7.1

1. Test your feel for fractions. With straight lines from the center to the edge, divide the circles into the number of pieces asked, making the pieces equal in size, as if you were dividing a pie fairly.

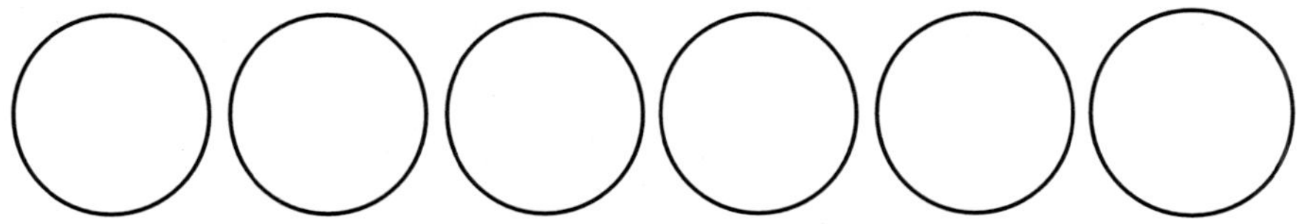

2 pieces 4 pieces 3 pieces 6 pieces 8 pieces 5 pieces

Most people find 2 and 4 relatively easy. They usually begin with a diameter (line through the center of the circle), thus cutting it in half, and then they draw a second diameter perpendicular to the first to get four equal parts. If asked to divide a pie for three people, some children cut it into four parts, but give out only three of them. Other children cut the pie into four parts, then make one piece twice as big as each of the others, like this:

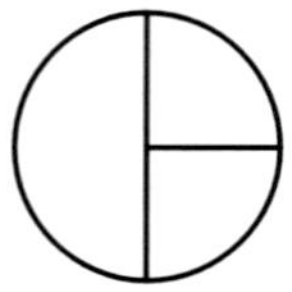

Children do this even when their family car is a Mercedes, whose logo is a circle divided neatly in thirds. If you can draw thirds, it is not hard to draw sixths. Can you do that? A child who can draw fifths accurately is unusual, and very few people, children or adults, can draw sevenths well.

2. **A kinesthetic approach to fractions.** Musicians think of fractions in terms of time. For example, they might think of $^1/_2$ by tapping one foot in a steady beat and tapping the other foot one every other beat. A more interesting challenge, which often arises, is to beat three beats against two. Explain how this picture can help visual learners with this.

3. Name at least one common example that could be used to illustrate each of these fractions.

   a. Thirds b. Sevenths c. Tenths
   d. Twelfths e. Sixteenths f. Sixtieths

4. There are four quarts to a gallon and two pints to a quart.
   a. How many pints to a gallon?
   b. What fraction of a gallon is a pint?
   c. What fraction of a gallon is three pints?
   d. What fraction of a gallon is a half pint?

5. There are twelve inches in a foot and three feet in a yard.
   a. What fraction of a yard is an inch?
   b. What fraction of a yard is 5 inches?
   c. What fraction of a yard is a quarter of an inch?

6. On the number line shown, estimate where each of these fractions should go, then mark the place.

   a. $\frac{1}{2}$ b. $\frac{1}{3}$ c. $\frac{1}{4}$ d. $\frac{3}{4}$ e. $\frac{2}{3}$

7. Children are sometimes surprised to learn that a fourth is smaller than a third, since four is larger than three. How would you help them understand this?

8. What fraction of each of these boxes is shaded?

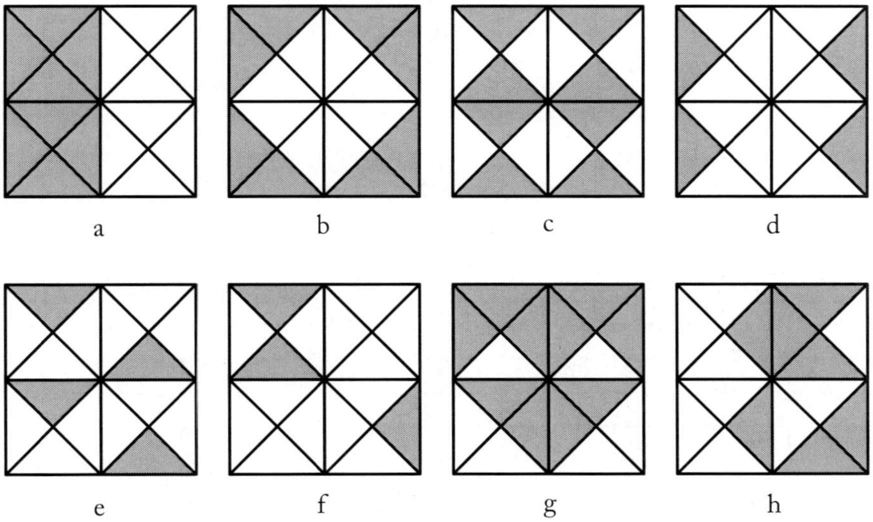

9. If each of these squares is one square inch, shade in 1/4 square inch in a dozen different ways as imaginatively as you can. (Make sure that in each case the area you shade really is 1/4 square inch.)

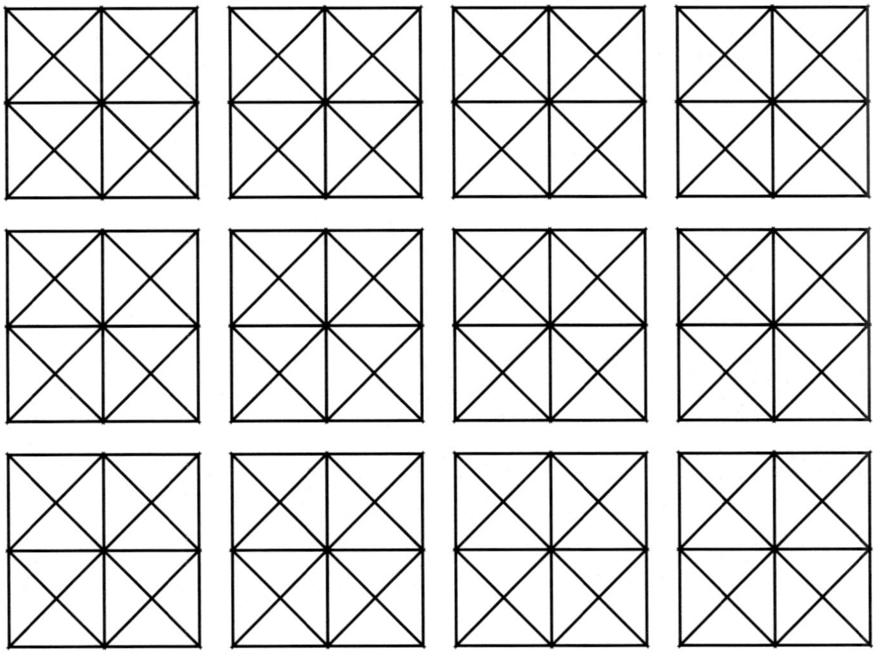

10. We often think of $\frac{3}{4}$ as what you get if you start with one fourth, then use three of them. However, by the definition, $\frac{3}{4}$ is the result of dividing 3 into four equal pieces. Describe each of these interpretations in terms of dollars.

11. What numbers on this ruler correspond to the points A, B, and C?

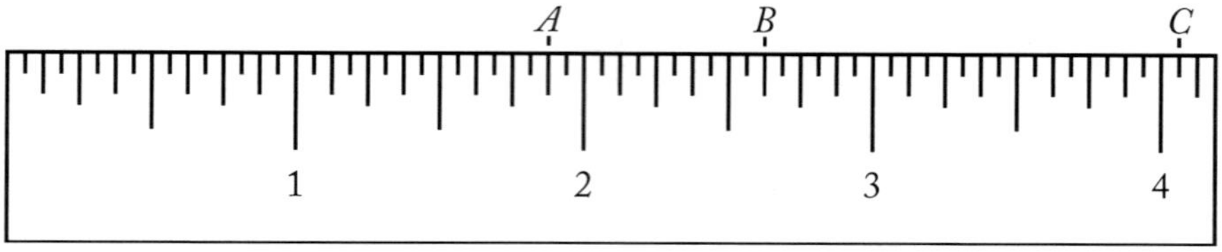

12. This picture can help you and your students to visualize fractions.

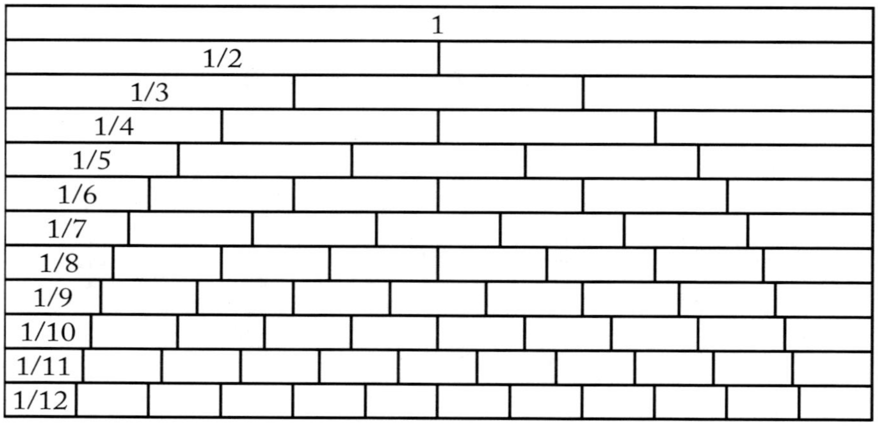

a. Coloring copies of this picture helps bring out the patterns. Try it.

b. You can see from the picture that $\frac{1}{2}$ is larger than $\frac{1}{3}$, which in turn is larger than $\frac{1}{4}$, etc. Make similar comparisons among $\frac{2}{3}$, $\frac{2}{4}$, $\frac{2}{5}$, etc.

c. Based on your observations above, generalize. If two fractions have the same numerator but different denominators, how does the fraction with the larger denominator compare to the fraction with the smaller denominator?

d. As in part b, compare $\frac{3}{4}$, $\frac{4}{5}$, $\frac{5}{6}$, $\frac{6}{7}$, etc.

e. If you compare $\frac{1}{2}$ with $\frac{2}{4}$, you will see they are the same size. What other fractions can you find in the picture that are the same size as $\frac{1}{2}$?

f. If you compare $\frac{1}{3}$ with $\frac{2}{6}$, you will see they are the same size. What other fractions can you find in the picture that are the same size as $\frac{1}{3}$?

g. Find two fractions in the picture, which are the same size as $\frac{2}{3}$.

7.2 Multiplying Fractions

We introduced multiplication of whole numbers as repeated addition, and therefore, for whole numbers at least, multiplication is more complicated than addition. Indeed, addition is used to do long multiplication, so it makes sense to teach addition before multiplication. For fractions the situation is reversed. Multiplication is simpler than addition, and multiplication (by fractions equal to 1) is often needed to add fractions. Therefore, for fractions it makes sense to teach multiplication before addition.

Using the area model of multiplication, students can discover for themselves how to multiply fractions. Begin with a square one unit long and one unit wide (figure 7.2.1a). Its area is, of course, one square unit. Now mark off as length and width of a new rectangle the fractions to be multiplied, and find the area of the resulting rectangle. Figure 7.2.illustrates the idea.

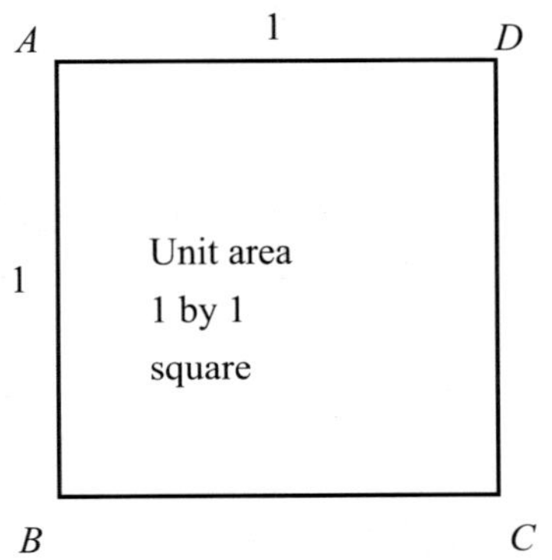

Figure 7.2.1a

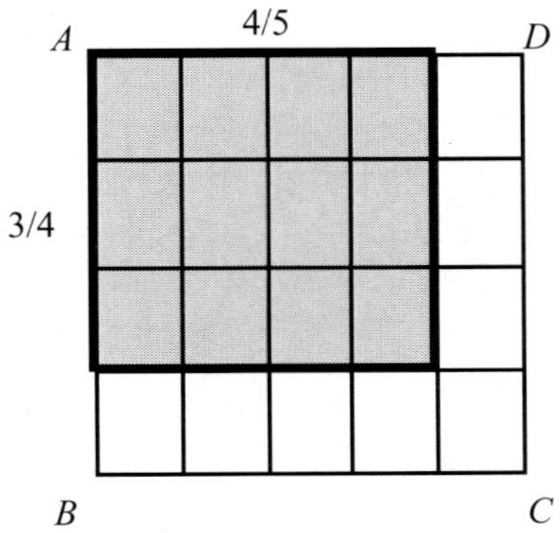

Figure 7.2.1b

Here, square *ABCD* has length and width 1, so its area is 1×1=1. The length and width of the shaded rectangle are $\frac{4}{5}$ and $\frac{3}{4}$ respectively , so the area of the shaded square is $\frac{4}{5} \times \frac{3}{4}$. But square *ABCD* is divided into 20 equal little rectangles, each of which must have area $\frac{1}{20}$, since the area of *ABCD* is 1. The shaded rectangle contains 12 of the little rectangles, so its area is $\frac{12}{20}$. Since area is length times width, $\frac{4}{5} \times \frac{3}{4} = \frac{12}{20}$.

With the above example in mind, try the next three as a discovery exercise, assuming in each case that the large square is 1 unit long, 1 unit wide, and so has area 1.

	Area of each small rectangle	Area of shaded rectangle	Length and width of shaded rectangle	Multiplication illustrated
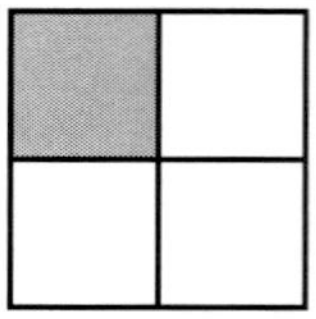				
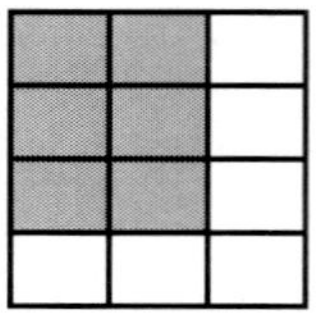				
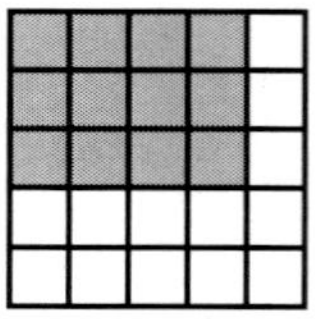				

From the four multiplications of fractions calculated so far, you can probably guess how to multiply fractions without pictures. However, if you want a few more examples of the same type, do the first problems in exercises 7.2 below.

How to multiply fractions

> 1. **Multiply numerators to get the numerator of the product.**
> 2. **Multiply denominators to get the denominator of the product.**
>
> **In shorthand,** $\frac{a}{b} \times \frac{c}{d} = \frac{ac}{bd}$

You can write any whole number as a fraction by putting it over a denominator of 1, so that, for example, $8 = \frac{8}{1}$. This turns out to be the key to seeing how to multiply a fraction by a whole number. For example, to multiply $\frac{3}{5}$ by 8, use the rule for multiplying fractions to multiply $\frac{3}{5} \times \frac{8}{1}$. Then $\frac{3}{5} \times 8 = \frac{3}{5} \times \frac{8}{1} = \frac{24}{5}$.

The rule in the box above applies to negative fractions as well as positive.

Example 1. Multiply $\frac{-5}{7} \times \frac{-4}{9}$.

Solution: $\frac{-5}{7} \times \frac{-4}{9} = \frac{(-5)(-4)}{7 \times 9} = \frac{20}{63}$. In practice, it may help to observe at the start that since both factors are negative, the product is positive. Then you can ignore the minus signs in carrying out the rest of the work.

Example 2. Multiply -7 by $\frac{2}{3}$.

Solution: $-7 \times \frac{2}{3} = \frac{-7}{1} \times \frac{2}{3} = \frac{-14}{3}$.

Don't mix units! Look again at figure 7.2.1, and also in the diagrams in exercise 7.2. The same unit of length is used for both height and width. This is no accident. **To express rectangular areas as square units, use the same unit for the length of a rectangle as for the width.**

Example 3. Find the area of a rectangle 2 yards long and 2 feet wide.
Solution: You cannot simply multiply 2 by 2 to get the area, since that would be neither in square yards nor square feet. There are 3 feet in a yard, so you could express the length as 6 feet, then multiply length and width to get the area as 6×2=12 **square feet.**

Second solution: Express both the length and the width in yards. The length is 2, the width $\frac{2}{3}$, and the area is $2 \times \frac{2}{3} = \frac{4}{3}$ **square yards.** (In practice, this would usually be expressed as a mixed number, $1\frac{1}{3}$ square yards.)

Example 4. Multiply $1\frac{1}{2} \times 2\frac{3}{4}$.

Solution: Write $1\frac{1}{2}$ as $\frac{3}{2}$ and $2\frac{3}{4}$ as $\frac{11}{4}$. Then $1\frac{1}{2} \times 2\frac{3}{4} = \frac{3}{2} \times \frac{11}{4} = \frac{33}{8}$.

In many contexts it would be appropriate to express this answer as $4\frac{1}{8}$.

Algebraic fractions are treated just like those in arithmetic.

Example 5. Multiply $\frac{2a^2b}{3c} \times \frac{4bd}{5c}$.

Answer: $\frac{8a^2b^2d}{15c^2}$

Exercises 7.2

1. Here are extra examples for the discovery exercise on multiplying fractions. Each outer square is 1×1.

Area of each small rectangle	Area of shaded rectangle	Length and width of shaded rectangle	Multiplication illustrated
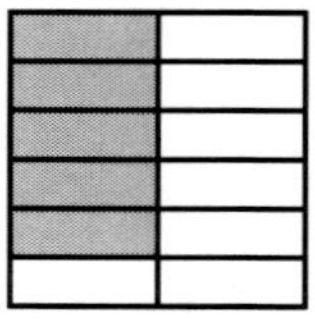			
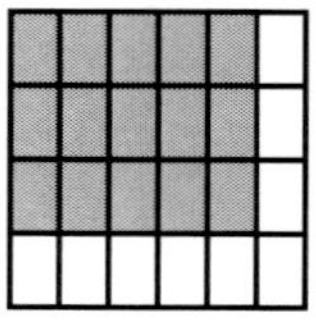			
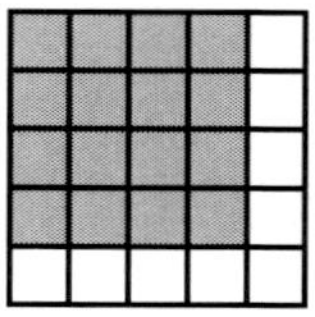			
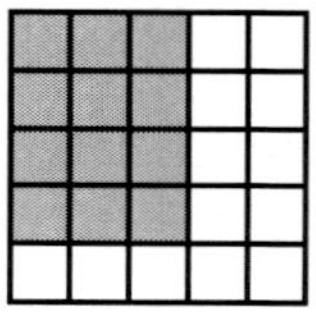			
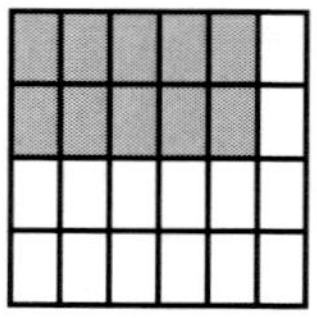			
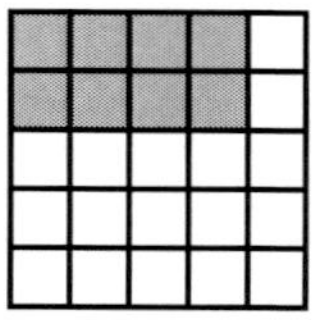			

2. Make drawings like those in question 1 to illustrate these multiplications.

a. $\frac{1}{6} \times \frac{1}{3}$ b. $\frac{1}{3} \times \frac{1}{4}$ c. $\frac{3}{5} \times \frac{3}{4}$

3. This question explores products with factors greater than 1. In each case, the rectangle whose area is to be found has at least one side longer than 1. Find the lengths of the sides and the areas just as you did in question 1.

	Area of each small rectangle	Area of shaded rectangle	Length and width of shaded rectangle	Multiplication illustrated
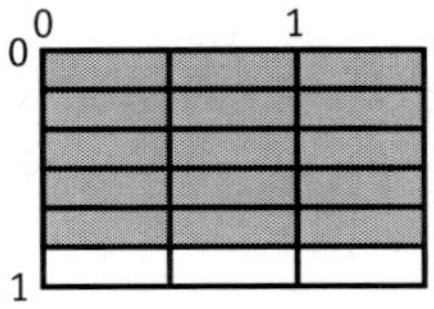				
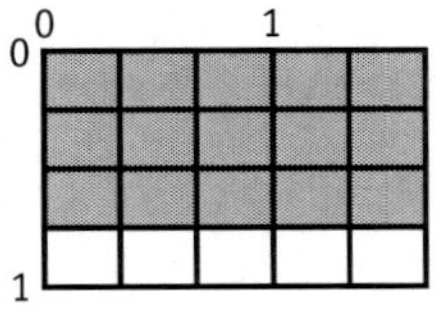				
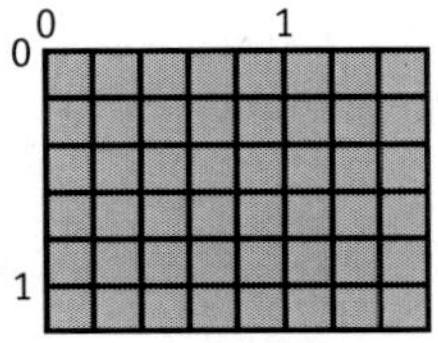				

4. Illustrate and compute as in question 3:

a. $\frac{1}{2} \times \frac{3}{2}$ b. $\frac{4}{3} \times \frac{1}{2}$ c. $\frac{3}{2} \times \frac{3}{4}$ d. $\frac{6}{5} \times \frac{4}{3}$

5. A highway proposal requires a rectangular strip of land 17 km long and 100 m wide. Calculate the area of this strip.

6. Multiply: a. $\frac{3rst}{5v} \times \frac{ps}{4u}$ b. $\frac{2xy}{3} \times \frac{x^2}{7}$

7. In English, the word "of" sometimes indicates multiplication (e.g. "Three groups of five").

a. Draw a picture that shows one half of one third. Is one half of one third the same as $\frac{1}{2} \times \frac{1}{3}$?

b. Draw a picture that shows two thirds of three fourths. Is two thirds of three fourths the same as $\frac{2}{3} \times \frac{3}{4}$?

7.3 Equal Fractions

Figure 7.3.1a shows three fractions which are equal. Figure 7.3.1b shows two more equal fractions. Equal fractions are simply different names for the same number.

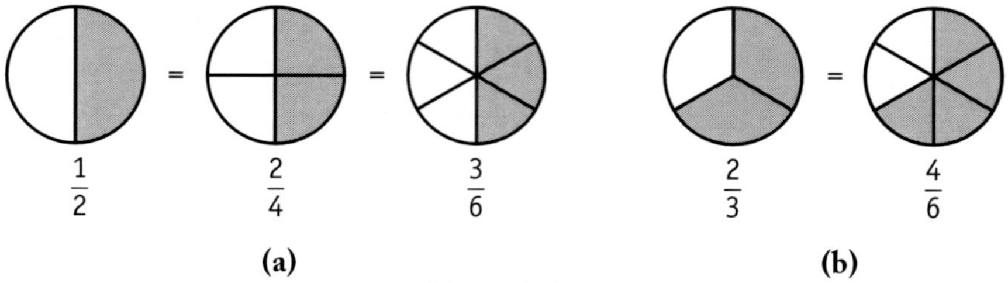

Figure 7.3.1

Just as 2×6 and 3×4 are different ways to express 12, equal fractions are different ways to express the same number as a quotient (result of division). As written symbols for numbers, fractions are numerals. Numbers that can be expressed as fractions are called **rational numbers**.[2] Understanding equal fractions is crucial to understanding how to add, subtract, and divide fractions. Unfortunately, many students learn rote methods for these processes without the necessary conceptual understanding. These students may perform adequately on simple mechanical computation, but when they encounter fractions in algebra, those trying to get by on rote alone will find tough going. With that in mind, let's take a careful look at equal fractions.

In section 7.1 you met fractions which equal 1. They are the fractions with numerator and denominator equal, since the division statement $x \div x = 1$ is equivalent to $\frac{x}{x} = 1$. Now, put together two observations:

1. Multiplying by 1 does not change any number.

2. Fractions equal to 1 have the form $\frac{x}{x}$.

Evidently, if you multiply any fraction by a fraction which equals 1, you will get a fraction equal to the one you began with. For example, multiply $\frac{1}{2}$ by $\frac{2}{2}$ and you get $\frac{2}{4}$, so $\frac{2}{4} = \frac{1}{2}$, since $\frac{2}{2}$ =1. Multiplying $\frac{1}{2}$ by 1 in the form $\frac{3}{3}$ yields $\frac{3}{6}$, which is also equal to $\frac{1}{2}$. Figure 7.3.1 shows that these results make sense.

In working with fractions, you often need to find a fraction equal to a given fraction, but with a specified numerator or denominator. That is not hard to find, once you understand equal fractions.

Example 1. Find a fraction equal to $\frac{3}{4}$ with a denominator of 12.

Solution: The plan is to multiply $\frac{3}{4}$ by a form of 1 to get a new fraction with a denominator of 12. Which form of 1 should we use? To decide, write $1 = \frac{y}{y}$, where $\frac{3}{4} \times \frac{y}{y} = \frac{?}{12}$. Multiplying out the left shows $\frac{3y}{4y} = \frac{?}{12}$, so $4y=12$, and $y=3$. Then the numerator is $3y$ or 9, and the desired fraction is $\frac{9}{12}$. Figure 7.3.2 pictures this equality in terms of a clock.

[2] Technically, we add (multiply, divide, subtract) rational numbers, not fractions. But it is pedantic to labor the distinction between number and numeral here.

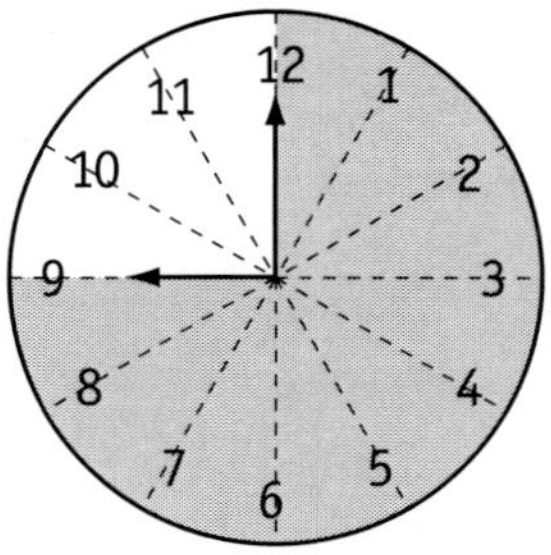

Figure 7.3.2

Example 2. Find a fraction equal to $\frac{3}{8}$ with a numerator of 36.

Solution: Since 3×12 = 36, multiply $\frac{3}{8}$ by 1 in the form $\frac{12}{12}$.

Then $\frac{3}{8} = \frac{3}{8} \times \frac{12}{12} = \frac{36}{96}$.

A spinoff from the above examples allows you to replace a fraction with a negative denominator by an equal fraction with a positive denominator whenever you find it convenient; just multiply by 1 in the form $\frac{-1}{-1}$.

The effect is to change the signs of both numerator and denominator. For example, $\frac{13}{-7} = \frac{13}{-7} \times \frac{-1}{-1} = \frac{-13}{7}$. Working mathematicians use this property without restating this justification, but we teachers owe our students a clear explanation of anything we do, and certainly we need to understand it ourselves.

Reducing and unreducing

A fraction is said to be in **lowest terms** if both numerator and denominator are integers and their greatest common factor is 1. For example, $\frac{3}{8}$ and $\frac{-16}{9}$ are in lowest terms but $\frac{3}{6}$ is not, since the greatest common factor of 3 and 6 is 3, not 1. For now we deal only with fractions in which both numerator and denominator are integers. Examples such as $\frac{\frac{3}{4}}{\frac{1}{2}}$ or $\frac{5}{\frac{1}{2}}$ are important for understanding division of fractions, but they will be considered later. If a fraction is not in lowest terms, then both numerator and denominator are multiples of some whole number greater than 1. To reduce the fraction, divide both numerator and denominator by a common factor. **To reduce the fraction to lowest terms, divide both numerator and denominator by their greatest common factor.**

Example 3. Reduce $\frac{6}{12}$ to lowest terms.

Solution: Dividing both numerator and denominator by the common factor 2 reduces $\frac{6}{12}$ to $\frac{3}{6}$, but this is still not in lowest terms, since the greatest common factor of numerator and denominator is 3. Divide both numerator and denominator by 3 to get $\frac{1}{2}$, which is in lowest terms. Of course, this could have been done in a single step, dividing numerator and denominator of $\frac{6}{12}$ by their greatest common factor, 6. It is a matter of convenience whether you reduce a fraction in several steps or all at once.

Reducing a fraction is particularly easy if you can see the factors common to the numerator and denominator, such as when they are both factored, as in example 4.

Example 4. Reduce $\frac{2\times 3\times 4\times 5}{2\times 3\times 3\times 10}$.

Solution: Removing any factor common to the numerator and denominator reduces the fraction. For example, reducing by 2 yields $\frac{3\times 4\times 5}{3\times 3\times 10}$. We can reduce further by a factor of 3, to get $\frac{4\times 5}{3\times 10}$. We eliminated only one of the two factors of 3 in the denominator, since there was only one factor of 3 in the numerator. Even further reducing is possible, as can be seen by factoring both numerator and denominator into primes: $\frac{4\times 5}{3\times 10}=\frac{2\times 2\times 5}{3\times 2\times 5}=\frac{2}{3}$

This is in lowest terms; no more reducing is possible.

Example 4 shows that factoring is a key step in reducing any fraction. Sometimes the factoring is so easy that you can do it in your head without even realizing what you are doing. In more complicated cases like example 5 you need to make the factoring explicit.

Example 5. Reduce $\frac{187}{121}$ to lowest terms.

Solution: Factor: 121=11×11 and 187=17×11. Then $\frac{187}{121}=\frac{11\times 17}{11\times 11}=\frac{17}{11}$.

As in example 4, you can reduce any fraction which is not in lowest terms by getting rid of factors common to numerator and denominator. When you do that, it may help to replace the factors you have lined out with 1s, to remind you that what remains is not "nothing," which might be incorrectly interpreted as 0. Lining out common factors and then dropping them is known as **canceling**. This is, of course, not a fundamental operation like adding or multiplying, but simply mathematical slang for a way of reducing fractions. Unfortunately, the idea of lining out the same number in numerator and denominator has such appeal that students sometimes incorrectly line out numbers which are not common factors. For example, in $\frac{3+5}{4+5}$ it would be wrong to line out the fives and eliminate them, because 5 is not a factor of the numerator and denominator. In fact, $\frac{3+5}{4+5}=\frac{8}{9}$, which is in lowest terms. Perhaps because of abuses like this, some teachers avoid the word "canceling." However, the problem is lack of understanding rather than vocabulary. Avoiding the word does not guarantee that people will avoid the mistakes.

Reducing fractions in algebra is much the same as in arithmetic, as example 6 shows.

Example 6. Reduce $\frac{8s^2t}{12st^3}$ to lowest terms.

Solution: Factor the 8 and the 12 to rewrite the fraction as

$$\frac{2\times 2\times 2\times s\times s\times t}{2\times 2\times 3\times s\times t\times t\times t}.$$

Now reduce by two factors of 2, one of s and one of t to get

$$\frac{8s^2t}{12st^3}=\frac{\not{2}\times\not{2}\times 2\times\not{s}\times s\times\not{t}}{\not{2}\times\not{2}\times 3\times\not{s}\times t\times t\times\not{t}}=\frac{2s}{3t^2}.$$

Sometimes we deliberately introduce common factors into the numerator and denominator of a fraction. This process has no simple, accepted name; we will call it **unreducing.** This is usually done in order to have a given denominator, as in example 1 above.

Reciprocals

The **reciprocal** of a fraction is the fraction obtained by turning it upside down.

The reciprocal of $\frac{3}{5}$, for example, is $\frac{5}{3}$. (Of course, since fractions with 0 denominators are not defined, fractions with 0 numerators, which is to say, fractions which equal 0, have no reciprocals.) If a fraction is multiplied by its reciprocal, the product is always 1, since $\frac{m}{n}\times\frac{n}{m}=\frac{mn}{nm}=1$.

Comparing fractions

Example 7. In one group, three out of seven students are boys. In a second group, four our of eleven students are boys. Which group has a larger proportion of boys?

Solution: Express both fractions with the same denominator. For this purpose, the least common multiple of 7 and 11, which is 77, is convenient.

$\frac{3}{7}=\frac{3\times 11}{7\times 11}=\frac{33}{77}$ and $\frac{4}{11}=\frac{4\times 7}{11\times 7}=\frac{28}{77}$. Since 33 is greater than 28, $\frac{3}{7}$ is greater than $\frac{4}{11}$, and the group with 3 boys out of seven has the greater proportion of boys.

You can see from example 6 that comparing fractions can be difficult, especially if there are many fractions to compare, and they are not very different in size. Ease of comparison is one of the main advantages of decimal fractions, which we will discuss in chapter 8. Question 10 in the exercises below is an excellent source of practice in comparing fractions.

Exercises 7.3

1. In parts a-d, express each fraction in lowest terms. (Try to avoid work.)

   a. $\frac{1\times 2}{2\times 3}$ b. $\frac{1\times 2\times 3}{2\times 3\times 4}$ c. $\frac{1\times 2\times 3\times 4}{2\times 3\times 4\times 5}$

   d. $\frac{1\times 2\times 3\times 4\times 5}{2\times 3\times 4\times 5\times 6}$ e. Generalize: $\frac{1\times 2\times 3\ldots(n-1)}{2\times 3\times 4\ldots n} =$

2. a. If $\frac{3}{5}\times\frac{n}{n}=\frac{6}{10}$, then n=   b. If $\frac{3}{5}\times\frac{m}{m}=\frac{-9}{-15}$, then m=

   c. If $\frac{3}{5}\times\frac{k}{k}=\frac{150}{250}$, then k=

3. Which, if any, of the following fractions is/are equal to $\frac{4}{7}$?

   a. $\frac{12}{21}$ b. $\frac{13}{22}$ c. $\frac{-40}{70}$

   d. $\frac{-76}{-133}$ e. $\frac{1548}{2709}$

4. Express each fraction in lowest terms. (It may help to think of money.)

   a. $\frac{25}{100}$ b. $\frac{10}{100}$ c. $\frac{75}{100}$ d. $\frac{90}{100}$ e. $\frac{50}{100}$

5. Express each fraction in lowest terms. (It may help to think of time.)

   a. $\frac{30}{60}$ b. $\frac{20}{60}$ c. $\frac{45}{60}$ d. $\frac{50}{60}$ e. $\frac{90}{60}$

6. Explain what is wrong with "canceling" 6 to reduce $\frac{16}{64}$.

Pattern hunts

7. In parts a-d, express each in lowest terms.

   a. $\frac{1+3}{5+7}$ b. $\frac{1+3+5}{7+9+11}$

   c. $\frac{1+3+5+7}{9+11+13+15}$ d. $\frac{1+3+5+7+9}{11+13+15+17+19}$

   e. Find a pattern in parts a-d, and use it to write and predict the next case. Then do the arithmetic and check your prediction.

f. (This elegant idea was suggested by Don Chakerian of U.C. Davis.) Use this drawing from figure 3.1.4 to explain case a.

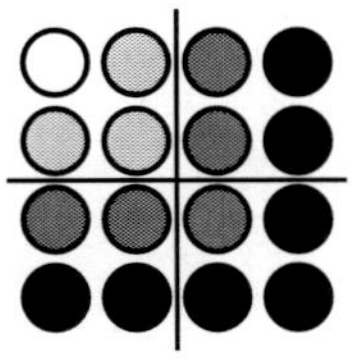

g. Make drawings like the one in part f to explain cases b, c, and d.

8. In parts a-c, fill in the blanks.

a. $\frac{1+2+3}{4+5+6} = \frac{}{5}$ b. $\frac{2+3+4}{5+6+7} = \frac{}{6}$ c. $\frac{3+4+5}{6+7+8} = \frac{}{7}$

d. Find the pattern, use it to predict the next case, then do the arithmetic to check your prediction.

9. In parts a-c, fill in the blanks.

a. $\frac{1-2+3}{4-5+6} = \frac{}{5}$ b. $\frac{2-3+4}{5-6+7} = \frac{}{6}$ c. $\frac{3-4+5}{6-7+8} = \frac{}{7}$

d. Find the pattern, use it to predict the next case, then do the arithmetic to check your prediction.

10. **A Game: Fraction War (suggested by Laura Pickett)**

This is like the card game of war, but here the cards have fractions on them instead of the usual numbers. The cards are dealt face down to each player. The players simultaneously turn up their top cards, and the one with the highest number wins all the cards turned up and places them face down at the bottom of his/her stack. A tie for the highest card creates a "war" between the tied players. Each tied player turns three cards face up, one at a time. The player with the highest third card wins all the turned up cards of the other players. (If the war produces a tie, the tied players have another war with three more cards each.) Any player who is out of cards drops out. The game continues until one player, the winner, has all the cards or until a set time limit, at which time the player with the most cards wins.

As teacher, you can vary the difficulty of the game by your choice of fractions written on the cards. Fractions which are nearly equal and have large denominators require the most skill. If you include many cards with equal fractions, the players will have lots of wars. Usually, the game is played with two players and an even number of fraction cards, but you can find ways to vary this.

11. Picturing equal fractions

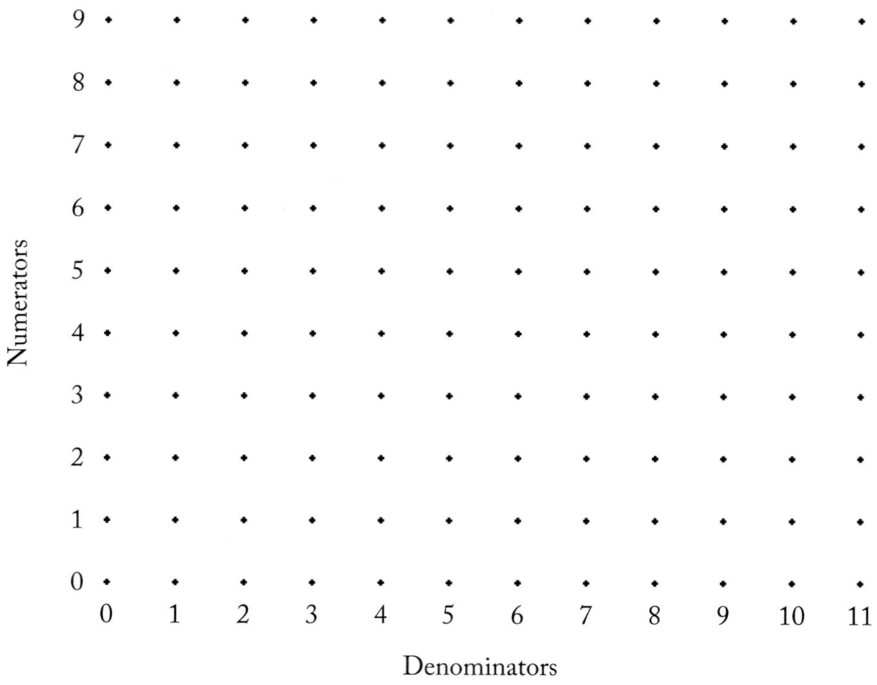

a. In the chart above, circle the dots corresponding to the fractions $\frac{2}{2}$, $\frac{3}{3}$, $\frac{4}{4}$, $\frac{5}{5}$, $\frac{6}{6}$, $\frac{7}{7}$, $\frac{8}{8}$, and $\frac{9}{9}$.

   What do you notice about these fractions?

b. On the same chart, but with a different color, circle the dots corresponding to $\frac{1}{2}$, $\frac{2}{4}$, $\frac{3}{6}$, $\frac{4}{8}$, and $\frac{5}{10}$.

   What do you notice about them?

c. In a third color, circle all dots for fractions which equal $\frac{1}{3}$.

d. With a fourth color, circle all dots for fractions which equal $\frac{2}{3}$.

e. With a fifth color, circle all dots for fractions which equal $\frac{3}{4}$.

f. With a fifth color, circle all dots for fractions which equal $\frac{3}{2}$.

g. On the basis of parts a-f, make a general statement about where the dots corresponding to equal fractions can be found.

h. Extend the diagram to include negative numerators and denominators. Where are all the dots corresponding to –1?

12. a. Rewrite this table with each fraction reduced to lowest terms.

Row number	**0**	**1**	**2**	**3**	**4**
			Column number		
0	1	$\frac{1}{1}$	$\frac{1\cdot 2}{1\cdot 2}$	$\frac{1\cdot 2\cdot 3}{1\cdot 2\cdot 3}$	$\frac{1\cdot 2\cdot 3\cdot 4}{1\cdot 2\cdot 3\cdot 4}$
1	1	$\frac{2}{1}$	$\frac{2\cdot 3}{1\cdot 2}$	$\frac{2\cdot 3\cdot 4}{1\cdot 2\cdot 3}$	$\frac{2\cdot 3\cdot 4\cdot 5}{1\cdot 2\cdot 3\cdot 4}$
2	1	$\frac{3}{1}$	$\frac{3\cdot 4}{1\cdot 2}$	$\frac{3\cdot 4\cdot 5}{1\cdot 2\cdot 3}$	$\frac{3\cdot 4\cdot 5\cdot 6}{1\cdot 2\cdot 3\cdot 4}$
3	1	$\frac{4}{1}$	$\frac{4\cdot 5}{1\cdot 2}$	$\frac{4\cdot 5\cdot 6}{1\cdot 2\cdot 3}$	$\frac{4\cdot 5\cdot 6\cdot 7}{1\cdot 2\cdot 3\cdot 4}$
4	1	$\frac{5}{1}$	$\frac{5\cdot 6}{1\cdot 2}$	$\frac{5\cdot 6\cdot 7}{1\cdot 2\cdot 3}$	$\frac{5\cdot 6\cdot 7\cdot 8}{1\cdot 2\cdot 3\cdot 4}$

b. Use the patterns to extend the table one more row down and one more column to the right. This pattern was discovered by Thomas Harriot.[3] Where have you seen it before?

c. Write an algebraic formula for the entry in the nth row down and the kth column across. (Count the row of 1's as the 0th row and the column of 1s at left as the 0th column.)

[3] Thomas Harriot (1560-1621) of Oxfordshire, graduated from Oxford in 1577 and then moved to London. Sir Walter Raleigh sent him to Virginia in 1585 as navigator. While in Virginia (today North Carolina) Harriot studied the local flora and fauna and invented a remarkable written alphabet for a local Native American language. On returning to England, he published his *True Report of the Newfound Land of Virginia* (1588) and continued to work on science and mathematics for his patrons, Raleigh and Henry Percy. Queen Elizabeth's successor James I imprisoned Raleigh in the Tower of London, and Harriot, too, was imprisoned there briefly in 1605. Harriot, who had learned to smoke tobacco in Virginia, died of cancer of the nose, having made extensive but largely unpublished contributions in mathematics and science. He introduced the convention that if two numbers are written next to each other and no other operation is indicated, they are to be multiplied. He also introduced the inequality signs < and >.

7.4 Adding Fractions

Fractions are easy to add, if they happen to have a common denominator. You can get the idea from the slide rule adder used in section 2.1, this time with fractions shown on it. Figure 7.4.1 shows a slide rule for adding sixths.

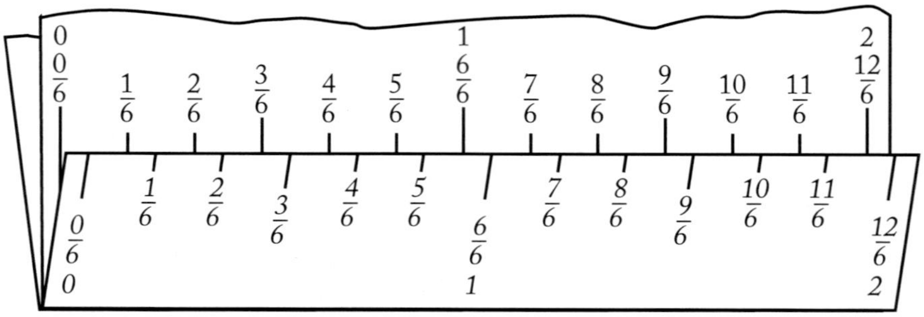

Figure 7.4.1

For example, slide the slider so that its 0 mark is by $\frac{2}{6}$ on the carrier. Then opposite the $\frac{3}{6}$ mark on the slider, you will find $\frac{5}{6}$ on the carrier. This shows that $\frac{2}{6}+\frac{3}{6}=\frac{5}{6}$, a result which may be checked with the distributive property:

$\frac{2}{6}+\frac{3}{6}=2\times\frac{1}{6}+3\times\frac{1}{6}=(2+3)\times\frac{1}{6}=5\times\frac{1}{6}=\frac{5}{6}$. This reasoning is entirely general.

How to add fractions

To add fractions with a common denominator, add the numerators, and write their sum over the common denominator.

In symbols, $\frac{a}{c}+\frac{b}{c}=\frac{a+b}{c}$ **(Of course, we assume c is not 0.)**

If the fractions don't have a common denominator, replace the given fractions with equal fractions that do have a common denominator. Then proceed as above.

Example 1. Add: a. $\frac{3}{5}+\frac{1}{5}$ and b. $\frac{1}{3}+\frac{1}{2}$.

Solution: a. $\frac{3}{5}+\frac{1}{5}=\frac{3+1}{5}=\frac{4}{5}$

b. First write $\frac{1}{3}$ as $\frac{2}{6}$ and $\frac{1}{2}$ as $\frac{3}{6}$. Then $\frac{1}{3}+\frac{1}{2}=\frac{2}{6}+\frac{3}{6}=\frac{5}{6}$.

You can see this is reasonable, if you use the slide rule in Figure 7.4.1, with $\frac{2}{6}$ relabeled as $\frac{1}{3}$ and $\frac{3}{6}$ as $\frac{1}{2}$. Figure 7.4.2 shows another way to picture this addition.

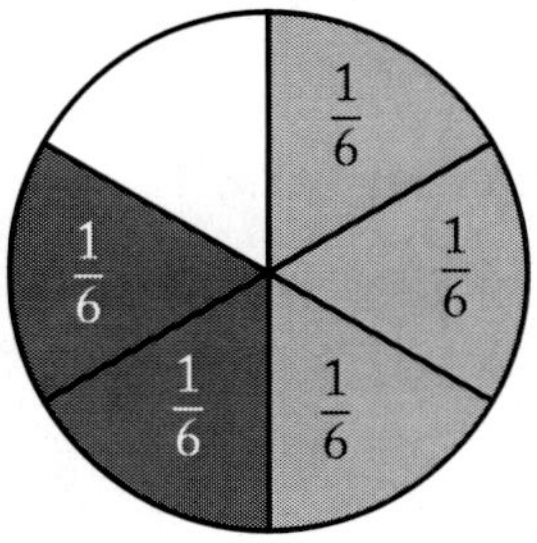

lightly shaded part: $\frac{1}{2}$ or $\frac{3}{6}$

heavily shaded part: $\frac{1}{3}$ or $\frac{2}{6}$

total shaded part: $\frac{1}{3} + \frac{1}{2} = \frac{2}{6} + \frac{3}{6} = \frac{5}{6}$

Figure 7.4.2

How was 6 chosen as common denominator? It is the least common multiple of the denominators 2 and 3. Any common multiple whatever would do, but the **least** common multiple keeps the numbers as small as possible. With 90 as the common denominator (90 is, after all, a common multiple of 2 and 3), we would have $\frac{1}{3}+\frac{1}{2}=\frac{30}{90}+\frac{45}{90}=\frac{75}{90}$. This reduces to $\frac{5}{6}$, so you get the same answer either way in the end.

Example 1 shows the steps of any addition of fractions, but the arithmetic can get complicated when the least common denominator is large. Another slight complication arises when any of the fractions to be added have negative denominators as in Example 2.

Example 2. Add $\frac{3}{4}+\frac{4}{-5}+\frac{1}{6}$.

Solution: Our discussion of the LCM in section 5.2 dealt only with positive numbers, so first rewrite the problem so that all the denominators are positive. The procedure for that, from section 7.3, is to multiply by 1 in the form $\frac{-1}{-1}$ whenever the given fraction has a negative denominator. In this case, $\frac{4}{-5}\times\frac{-1}{-1}=\frac{-4}{5}$, so rewrite the original question as $\frac{3}{4}+\frac{-4}{5}+\frac{1}{6}$. Now that all denominators are positive, proceed as before. The LCM of 4, 5, and 6 is 60, so unreduce each addend to an equal fraction with denominator 60.

$\frac{3}{4}=\frac{45}{60}$, $\frac{-4}{5}=\frac{-48}{60}$, and $\frac{1}{6}=\frac{10}{60}$.

$$\text{Then } \frac{3}{4}+\frac{4}{-5}+\frac{1}{6}=\frac{45}{60}+\frac{-48}{60}+\frac{10}{60}=\frac{45+(-48)+10}{60}=\frac{7}{60}.$$

Addition and multiplication of mixed numbers

If you know how to add fractions and how to add integers, then you already know how to add mixed numbers. For example, to add $2\frac{3}{4}+3\frac{5}{8}$, first change them to improper fractions, then add:

$$2\frac{3}{4}+3\frac{5}{8}=\frac{11}{4}+\frac{29}{8}=\frac{22}{8}+\frac{29}{8}=\frac{51}{8}.$$

You can change $\frac{51}{8}$ to a mixed number, $6\frac{3}{8}$, if you like.

Another way: Add the integers and the fractions separately.

$$2\frac{3}{4}+3\frac{5}{8}=\left(2+\frac{3}{4}\right)+\left(3+\frac{5}{8}\right)=\left(2+3\right)+\left(\frac{3}{4}+\frac{5}{8}\right)=5+\left(\frac{6}{8}+\frac{5}{8}\right)=5\frac{11}{8}.$$

Now change $\frac{11}{8}$ to a mixed number and add it to 5, getting $6\frac{3}{8}$ as before.

Example 3. Multiply $1\frac{1}{2}\times 2\frac{3}{4}$.

Solution: In section 7.2 we did this by rewriting the factors as improper fractions. Here is a second solution.

Write $1\frac{1}{2}$ as $1+\frac{1}{2}$, and write $2\frac{3}{4}$ as $2+\frac{3}{4}$. Then $1\frac{1}{2}\times 2\frac{3}{4}=\left(1+\frac{1}{2}\right)\times\left(2+\frac{3}{4}\right)$.

By the distributive property, this is $1\times 2+\frac{1}{2}\times 2+1\times\frac{3}{4}+\frac{1}{2}\times\frac{3}{4}$. This is

$2+1+\frac{3}{4}+\frac{3}{8}=3+\frac{9}{8}=3+1\frac{1}{8}=4\frac{1}{8}$. Question 10 below pictures this.

Comments about fractions in the curriculum

Example 2 shows why adding fractions is so complicated. To understand addition of fractions students need to know:

1. How to find the LCM of several numbers (this in turn requires skill in factoring and understanding of primes).
2. The meaning of fractions in general and especially of equal fractions (this in turn requires an understanding of multiplication by fractions equal to 1).
3. How to add fractions with a common denominator.

This has implications for the arithmetic curriculum sequence, because students who try to learn how to add fractions without the necessary base of skills and concepts are candidates for frustration and disappointment. They may blame themselves or think that mathematics is just too hard for them, when they might have caught on easily if the topics had been presented in a reasonable order.

The Curriculum Sequence for Fractions

1. If you plan to teach fractions, make sure the students understand prime numbers, factoring, and LCM. Address those areas before you need those skills. (You can discuss fractions any time; just avoid complicated computations with them until your students are up to speed on primes, factors, and LCM.)
2. Teach multiplication of fractions before addition. Why? Multiplication is easy and boosts students' confidence. Knowing how to multiply fractions is needed to understand equal fractions, which in turn is needed to teach addition of fractions with unlike denominators.
3. From multiplication of fractions, go on to fractions which equal 1, then to equal fractions, reducing and unreducing.
4. When your students can multiply, reduce and unreduce fractions, they are ready to add fractions. Start with fractions with common denominators, then go on to unlike denominators.
5. Finally, teach division and subtraction of fractions.

Treating fractions in the order suggested might help students to find fractions less confusing. Their weakness in this area may emerge only when they meet sums like $\frac{1}{x^2}+\frac{1}{xy}$ in algebra. The first step is to choose x^2y as common denominator. How does this relate to arithmetic? To see, let $x=2$ and $y=3$, so $x^2=4$ and $xy=6$. Then $\frac{1}{x^2}+\frac{1}{xy}$ becomes $\frac{1}{4}+\frac{1}{6}$, and the lowest common denominator, 12, is x^2y with $x=2$ and $y=3$. Students who understand arithmetic fractions find algebraic fractions no harder, but those who do not really understand fractions in arithmetic usually find fractions in algebra incomprehensible.

Exercises 7.4

1. a. Make a slide rule like the one shown in figure 7.4.1, but extending from –2 to 2 in eighths.
   b. Experiment with the slide rule you made in part a.

2. Add, and then use your slide rule from question 1 to check your answers.

   a. $\frac{3}{8}+\frac{5}{8}$ b. $\frac{-11}{8}+\frac{3}{8}$ c. $\frac{-7}{8}+\frac{13}{8}$ d. $\frac{-5}{8}+\frac{-9}{8}$

3. Add, and then use your slide rule from question 1 to check your answers.

   a. $\frac{2}{4}+\frac{5}{8}$ b. $\frac{1}{2}+\frac{-7}{8}$ c. $\frac{-7}{4}+\frac{13}{8}$ d. $\frac{3}{4}+\frac{-9}{8}$

4. a. How many eggs in a third of a dozen?
   b. How many eggs in a fourth of a dozen?
   c. A family ate a third of a dozen eggs on Saturday, and a fourth of a dozen eggs on Sunday. What fraction of a dozen eggs did they eat in all?

5. a. How many cents in two tenths of a dollar?
   b. How many cents in three quarters of a dollar?
   c. Amanda spent two tenths of a dollar for a pencil and three quarters of a dollar for a snack. What fraction of a dollar did she spend in all? Express your answer both in hundredths of a dollar (cents) and in lowest terms.

6. a. Draw a clock face like the one in figure 7.3.2.
   b. How many minutes are in two fifths of an hour? Shade $\frac{2}{5}$ of an hour on your picture from part a.
   c. How many minutes are in one sixth of an hour? Shade one sixth of an hour on your picture from part a, being careful not to re-shade any part of the picture already shaded in part b.
   d. What fraction of an hour is two-fifths plus one-sixth of an hour? Express your answer both in sixtieths of an hour (minutes) and in lowest terms.

7. Make up a problem like question 4, 5, or 6, which uses addition of fractions with unlike denominators in a familiar setting.

8. Do box puzzles work with fraction entries? Try these to see. The first is straightforward, and they grow successively more sophisticated.

a.

$\frac{1}{8}$	$\frac{3}{8}$	
$\frac{5}{8}$	$\frac{7}{8}$	

b.

$\frac{1}{3}$	$\frac{1}{4}$	
$\frac{5}{6}$	$\frac{7}{12}$	

c.

$\frac{3}{10}$	$\frac{1}{4}$	
$\frac{5}{8}$	$\frac{1}{5}$	

d.

$\frac{7}{15}$	$\frac{-3}{5}$	
$\frac{-5}{6}$	$\frac{1}{2}$	

9. a. On one job a painter used $5\frac{3}{4}$ gallons of paint, and on a second job she used $3\frac{1}{8}$ gallons. How many gallons did she use in all?

   b. Bob walked three and seven tenths miles one morning and another five and three fourths miles that afternoon. How far did he walk in all?

10. Illustrate the solution to example 3 of section 7.4 by drawing a rectangle $1\frac{1}{2}$ inches wide and $2\frac{3}{4}$ inches long on graph paper and finding its area.

11. If $\frac{1}{3}=\frac{1}{A}+\frac{1}{B}$, where A and B are different whole numbers, find the values of A and B. (Used, with permission, from Lenchner, *Math Olympiad Contest Problems for Elementary and Middle Schools,* MOEMS, 1997).

Pattern hunts with built-in practice in adding fractions

12. In parts a-d, add as indicated.

a. $\frac{1}{2}+\frac{1}{4}$ b. $\frac{1}{2}+\frac{1}{4}+\frac{1}{8}$ c. $\frac{1}{2}+\frac{1}{4}+\frac{1}{8}+\frac{1}{16}$

d. $\frac{1}{2}+\frac{1}{4}+\frac{1}{8}+\frac{1}{16}+\frac{1}{32}$

e. Use the pattern of answers to parts a-d to predict the sum

$$\frac{1}{2}+\frac{1}{4}+\frac{1}{8}+\frac{1}{16}+\frac{1}{32}+\frac{1}{64}$$

f. Carry out the addition from part e to check your answer.

g. In the above additions, how does the numerator of the answer relate to the denominator in each case? Use your observation to fill in this algebraic generalization:

$$\frac{1}{2}+\frac{1}{4}+\frac{1}{8}+\ldots+\frac{1}{2^n}=\frac{\quad}{2^n}$$

13. In parts a-d, add as indicated.

a. $\frac{1}{3}+\frac{1}{9}$ b. $\frac{1}{3}+\frac{1}{9}+\frac{1}{27}$ c. $\frac{1}{3}+\frac{1}{9}+\frac{1}{27}+\frac{1}{81}$

d. $\frac{1}{3}+\frac{1}{9}+\frac{1}{27}+\frac{1}{81}+\frac{1}{243}$

e. Use the pattern of answers to parts a-d to predict the sum

$$\frac{1}{3}+\frac{1}{9}+\frac{1}{27}+\frac{1}{81}+\frac{1}{243}+\frac{1}{729}$$

f. Carry out the addition from part e to check your answer.

g. In the above additions, how does the numerator of the answer relate to the denominator in each case? Use your observation to fill in this algebraic generalization:

$$\frac{1}{3}+\frac{1}{9}+\ldots+\frac{1}{3^n}=\frac{\quad}{3^n}$$

14. Use the patterns you found in questions 12 and 13 to predict the following sums without actually carrying out the additions:

a. $\frac{1}{4}+\frac{1}{16}$ b. $\frac{1}{4}+\frac{1}{16}+\frac{1}{64}$ c. $\frac{1}{4}+\frac{1}{16}+\frac{1}{64}+\frac{1}{256}$

d. $\frac{1}{4}+\frac{1}{16}+\frac{1}{64}+\frac{1}{256}+\frac{1}{1024}+\frac{1}{4096}$

15. In parts a-d, add as indicated.

a. $\frac{1}{2}+\frac{1}{3}$ b. $\frac{1}{3}+\frac{1}{4}$ c. $\frac{1}{4}+\frac{1}{5}$ d. $\frac{1}{5}+\frac{1}{6}$

e. Look at the pattern of answers to parts a-d, and use it to predict the sum $\frac{1}{8}+\frac{1}{9}$. Then do the arithmetic to check your answer.

f. Express the pattern in the general language of algebra:

$$\frac{1}{n}+\frac{1}{n+1}=$$

16. In parts a-e, add, and express each sum in lowest terms:

a. $\frac{1}{3}+\frac{1}{3\cdot 2}$ b. $\frac{1}{4}+\frac{1}{4\cdot 3}$ c. $\frac{1}{5}+\frac{1}{5\cdot 4}$

d. $\frac{1}{6}+\frac{1}{6\cdot 5}$ e. $\frac{1}{7}+\frac{1}{7\cdot 6}$

f. From the pattern of answers to parts a-f, predict the sums $\frac{1}{20}+\frac{1}{20\cdot 19}$ and $\frac{1}{50}+\frac{1}{50\cdot 49}$. Then check your predictions by carrying out the additions.

17. Add, expressing each answer with a numerator of 2 to bring out the pattern.

a. $\frac{1}{2\cdot 3}+\frac{1}{3\cdot 4}$ b. $\frac{1}{3\cdot 4}+\frac{1}{4\cdot 5}$ c. $\frac{1}{4\cdot 5}+\frac{1}{5\cdot 6}$

d. $\frac{1}{5\cdot 6}+\frac{1}{6\cdot 7}$ e. $\frac{1}{6\cdot 7}+\frac{1}{7\cdot 8}$

f. In general, what do you predict for $\frac{1}{(n-1)n}+\frac{1}{n(n+1)}$?

18. (From the *Mathematical Olympiad Contest Problems for Elementary and Middle Schools*, p11) Express the following sum as a simple fraction in lowest terms:

$$\frac{1}{1\times 2}+\frac{1}{2\times 3}+\frac{1}{3\times 4}+\frac{1}{4\times 5}+\frac{1}{5\times 6}+\frac{1}{6\times 7}$$

(Hint: What if you added just the first, the first two, or the first three fractions? Try it and find a pattern.)

19. The two smallest "perfect numbers" (defined in section 5.3) are 6 and 28. Their factors are, respectively, 1, 2, 3, 6, and 1, 2, 4, 7, 14, 28. With that in mind, compute these rather curious sums of fractions.

a. $\frac{1}{1}+\frac{1}{2}+\frac{1}{3}+\frac{1}{6}=$ b. $\frac{1}{1}+\frac{1}{2}+\frac{1}{4}+\frac{1}{7}+\frac{1}{14}+\frac{1}{28}=$

c. The next smallest perfect number, 496, has factors 1, 2, 4, 8, 16, 31, 62, 124, 248, and 496. On the basis of parts a and b, what do you predict is the sum of the reciprocals of these factors?

d. To check your prediction from part c, add

$$\frac{1}{1}+\frac{1}{2}+\frac{1}{4}+\frac{1}{8}+\frac{1}{16}+\frac{1}{31}+\frac{1}{62}+\frac{1}{124}+\frac{1}{248}+\frac{1}{496}=$$

Hint: The sum of the last five fractions may be written $\frac{1}{31}\left(\frac{1}{1}+\frac{1}{2}+\frac{1}{4}+\frac{1}{8}+\frac{1}{16}\right)$.

7.5 Subtracting Fractions

As with addition, subtraction with fractions is easy, if the fractions involved have the same denominator. For example, to subtract $\frac{3}{8}$ from $\frac{7}{8}$, view this as taking $\frac{3}{8}$ away from $\frac{7}{8}$, as in figure 7.5.1.

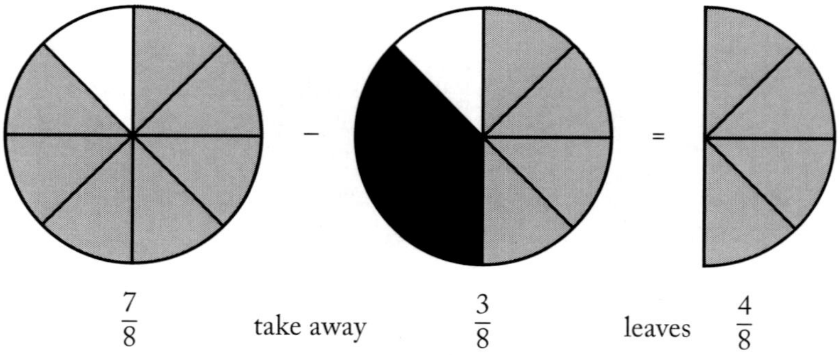

Figure 7.5.1

To view this subtraction as adding the opposite, think of it as $\frac{7}{8} + \frac{-3}{8}$ and apply the rules for adding fractions: $\frac{7}{8} + \frac{-3}{8} = \frac{7+(-3)}{8} = \frac{4}{8}$. Finally, to see this in terms of what must be added, note that if you start at $\frac{3}{8}$, you need to add $\frac{4}{8}$ to get $\frac{7}{8}$.

As with integers, the "take away" view of subtraction is not natural when negative fractions are involved. In those cases it is usually easier to model subtraction in another way, such as adding the opposite.

Example 1. Subtract $\frac{-5}{11}$ from $\frac{-2}{11}$.

Solution: Subtracting $\frac{-5}{11}$ is the same as adding the opposite, $\frac{5}{11}$. In this case, $\frac{-2}{11} - \frac{-5}{11} = \frac{-2}{11} + \frac{+5}{11} = \frac{-2+5}{11} = \frac{3}{11}$. To check, note that $\frac{-5}{11} + \frac{3}{11} = \frac{-2}{11}$, so $\frac{3}{11}$ is indeed the number that must be added to $\frac{-5}{11}$ to get $\frac{-2}{11}$.

What if the fractions in a subtraction problem have different denominators? Rewrite the problem using fractions that have the same denominator, just as you do in adding fractions.

Example 2. Subtract: $\frac{-3}{8} - \frac{5}{6}$.

Solution: Use the LCM of 8 and 6, which is 24, to rewrite $\frac{-3}{8}$ as $\frac{-9}{24}$ and $\frac{5}{6}$ as $\frac{20}{24}$. Then $\frac{-3}{8} - \frac{5}{6} = \frac{-9}{24} - \frac{20}{24}$.

Subtracting $\frac{20}{24}$ is the same as adding $\frac{-20}{24}$, so this is $\frac{-9}{24}+\frac{-20}{24}$ or $\frac{-29}{24}$.

To check this, add $\frac{5}{6}+\frac{-29}{24}$ as $\frac{20}{24}+\frac{-29}{24}$. The sum is $\frac{-9}{24}$ or $\frac{-3}{8}$.

When mixed numbers or integers are part of a subtraction problem, it may be convenient to modify the subtraction procedure as shown in example 3.

Example 3. Subtract $\frac{5}{12}$ from 15.

Solution: Think of 15 as $\frac{15}{1}$, and then replace $\frac{15}{1}$ with an equal fraction that has a denominator of 12.

$$15-\frac{5}{12}=\frac{15}{1}-\frac{5}{12}=\frac{180}{12}-\frac{5}{12}=\frac{175}{12}$$

We usually write this as a mixed number, $14\frac{7}{12}$.

A second solution: Think of 15 as 14 +1, and rewrite only the 1 as a fraction.

$$15-\frac{5}{12}=14+1-\frac{5}{12}=14+\left(\frac{12}{12}-\frac{5}{12}\right)=14+\frac{7}{12}=14\frac{7}{12}$$

You may find the second solution easier to picture or to do in your head. It is like borrowing. Imagine a bakery that sells cake by the piece and makes 12 slices per cake. If this bakery has 15 unsliced cakes and a customer wants 5 pieces, how much will the bakery have left after the sale? Solution 1 amounts to cutting all 15 cakes to make 180 pieces, then selling 5 of them. That leaves 175 pieces or $\frac{175}{12}$ cakes. The second solution is what you would probably do in that situation; "borrow" one cake, cut it into twelfths, and sell five of the pieces. That leaves 14 whole cakes and 7 pieces, or $14\frac{7}{12}$ cakes.

Exercises 7.5

1. Explain and complete the subtraction $\frac{11}{5}-\frac{3}{5}$ in terms of:

   a. Take away
   b. Adding the opposite
   c. What must be added

2. Explain and complete the subtraction $\frac{3}{8}-\frac{-7}{8}$ in terms of:

   a. Adding the opposite
   b. What must be added
   c. Why were you not asked to explain this subtraction in terms of take away?

3. Explain and complete the subtraction $2\frac{7}{10} - 1\frac{2}{3}$ in terms of:

   a. Take away
   b. Adding the opposite
   c. What must be added

4. Addition box puzzles with missing entries are a good source of subtraction practice, because, in effect, they check themselves. Try these.

a.

$\frac{3}{8}$		$\frac{9}{8}$
	$\frac{1}{8}$	
		$\frac{15}{8}$

b.

$\frac{2}{3}$		$\frac{3}{4}$
	$\frac{1}{12}$	
		$\frac{17}{12}$

c.

$\frac{-1}{6}$		$\frac{2}{3}$
	$\frac{-3}{4}$	
$\frac{-5}{12}$		

5. From a bolt of cloth 35 yards long Lotte used pieces that were $2\frac{2}{3}$ yards, $3\frac{1}{3}$ yards, and $5\frac{2}{3}$ yards long respectively. How many yards remained?

6. The fuel tank of a car holds $13\frac{5}{8}$ gallons. When the low fuel warning light came on, it was immediately refilled, taking $12\frac{4}{5}$ gallons to fill. How many gallons were in the tank when the warning light came on?

7. Ron and Chad set out to hike the $12\frac{3}{4}$ miles by trail from Hook Creek to Coxey Road. How far had they hiked along this trail when they came to a sign saying $3\frac{9}{10}$ miles to Coxey Road?

8. Make up additional questions using subtraction of fractions, if possible, based on your own experiences.

7.6 Dividing Fractions

"Yours is not to reason why, just invert and multiply."

A true story A teacher tried to explain division of fractions to a student who had been absent. At last the student exclaimed, "I get it. You turn over the one by the window. In my old school it was the one by the door!" Students who see math that way, as a set of arbitrary rules, eventually give it up; for them, it has no basis in reason or common sense.

No doubt you, too, were taught to invert and multiply when dividing by a fraction, and indeed it is a legitimate procedure. Our object here is to understand what is going on. The key is in the very definition of division. The three ways to think of division, discussed in section 6.3, are analogous to the three ways to think of subtraction.

3 views of subtraction	**3 views of division**
Take away	Repeated subtraction
Missing addend	Missing factor
Add the opposite	Multiply by the reciprocal

To divide by a fraction, multiply by its reciprocal; that is the third view of division and the only one we have not yet examined carefully. We will do so now.

We begin with division questions which we can do by other methods as well. For example, suppose we want to divide by a whole number, such as 3. To divide 12 by 3, we can subtract 3 repeatedly (4 times), or we can note that 4 is the factor by which 3 must be multiplied to get 12. What of inverting and multiplying in this case? The reciprocal of 3 is $\frac{1}{3}$, so multiplying 12 by $\frac{1}{3}$ should be the same as dividing 12 by 3. But $12 = \frac{12}{1}$, and therefore $12 \times \frac{1}{3} = \frac{12}{1} \times \frac{1}{3} = \frac{4 \times 3}{1 \times 3} = \frac{4}{1} = 4$. In this case, at least, multiplying by $\frac{1}{3}$ is the same as dividing by 3. Does this work the same way when we must divide one fraction by another? To find out, consider an example.

Example 1. Divide $\frac{3}{5} \div \frac{1}{5}$.

Solution: You can do this by repeated subtraction of $\frac{1}{5}$ from $\frac{3}{5}$ as shown in figure 7.4.1.

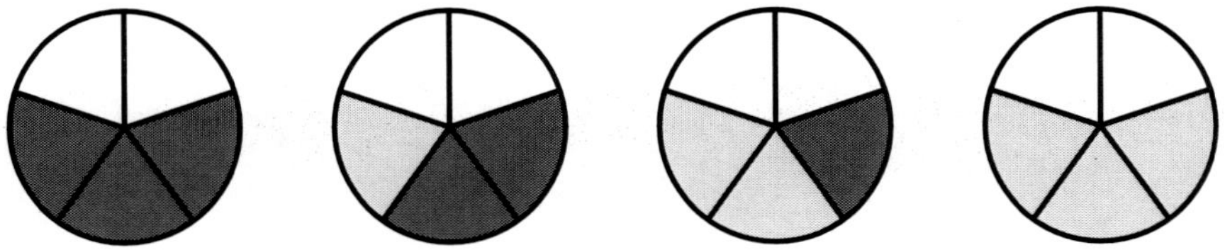

Figure 7.4.1

You can subtract $\frac{1}{5}$ three times from $\frac{3}{5}$ to get 0, so $\frac{3}{5} \div \frac{1}{5} = 3$.

Second solution: Multiply $\frac{3}{5}$ by the reciprocal of $\frac{1}{5}$. The reciprocal of $\frac{1}{5}$ is 5, so this amounts to calculating $\frac{3}{5} \times \frac{5}{1}$, which, when 5 is canceled, is 3. Let's reexamine this multiplication by the reciprocal, first writing $\frac{3}{5} \div \frac{1}{5}$ as a fraction of fractions, $\frac{\frac{3}{5}}{\frac{1}{5}}$. Multiply this by $\frac{\frac{5}{1}}{\frac{5}{1}}$, which is a cleverly chosen form of 1, to get $\frac{\frac{3}{5}}{\frac{1}{5}} \times \frac{\frac{5}{1}}{\frac{5}{1}}$. Now apply the rule for multiplying fractions to get $\frac{\frac{3}{5} \times \frac{5}{1}}{\frac{1}{5} \times \frac{5}{1}}$. The denominator is just $\frac{5}{5}$, which is a form of 1, so the thing boils down to $\frac{3}{5} \times \frac{5}{1}$, where the $\frac{5}{1}$ is the reciprocal of the original divisor.

The above steps can be used to justify inverting the divisor and multiplying whenever we divide by a fraction. We usually skip the justification to save time, but we and our students should be able to explain our reasoning in detail, even if we do so only rarely. We have found the number by which $\frac{1}{5}$ must be multiplied to get $\frac{3}{5}$, namely 3. Like any division, you can check it by multiplying: $\frac{1}{5} \times 3 = \frac{1}{5} \times \frac{3}{1} = \frac{3}{5}$.

A second example sheds further light on inverting and multiplying.

Example 2. Compute $1\frac{1}{2} \div \frac{1}{4}$.

Solution: Write this as $\frac{1\frac{1}{2}}{\frac{1}{4}}$ or $\frac{\frac{3}{2}}{\frac{1}{4}}$, and picture it this way, where a large square is the unit.

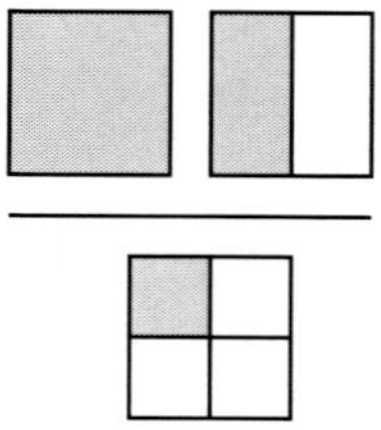

Now you can see that there are six fourths in the numerator, so $1\frac{1}{2} \div \frac{1}{4} = 6$.

Solution 2: To have a whole number in the denominator, multiply numerator and denominator by 4, as you would when you invert the denominator and multiply. The result may be pictured this way, showing again that the quotient is 6.

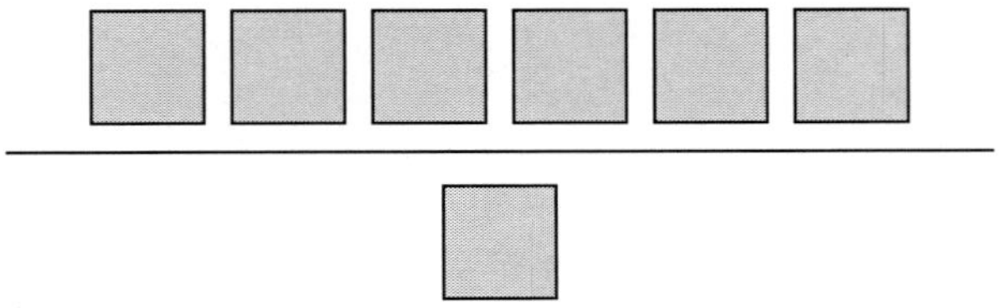

Example 3. A 2-acre piece of land is to be divided into $\frac{1}{2}$-acre lots to build homes. How many $\frac{1}{2}$-acre lots will there be, and what division does this question illustrate?

Solution: This problem illustrates $2 \div \frac{1}{2} = 4$. There are 4 lots, as shown.

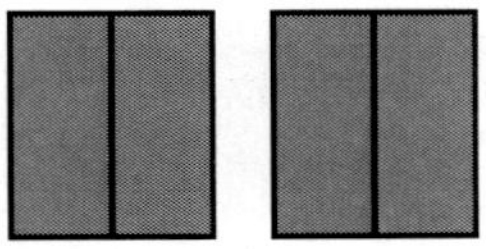

Another question that leads to the same division problem: How many half-dollars are in $2? Each dollar is 2 half-dollars, so there are 4 half-dollars in $2.

Example 4. Make up and solve a problem to illustrate $\frac{2}{3} \div 2$.

Solution: If the remaining two-thirds of a candy bar is split equally between two children, what portion of the candy bar will each child get? Divide $\frac{2}{3}$ into two equal parts by putting $\frac{1}{3}$ into each part.

$$\frac{2}{3} \div 2 = \frac{1}{3}$$

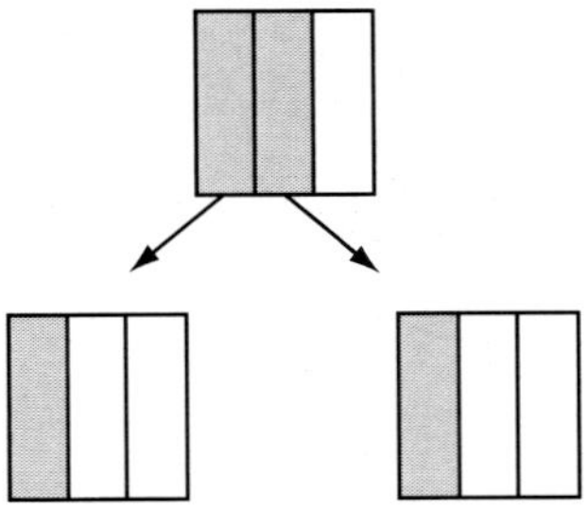

Example 5. Make up and solve a problem to illustrate $\frac{1}{2} \div \frac{2}{3}$.

Solution: (An example) A box of oatmeal contains $\frac{1}{2}$ cup of oatmeal. How many $\frac{2}{3}$-cup servings are in the box? There will be less than one whole $\frac{2}{3}$-cup serving in the box, since $\frac{2}{3}$ is more than $\frac{1}{2}$. What portion of the whole serving is there? Picture this with a measuring cup:

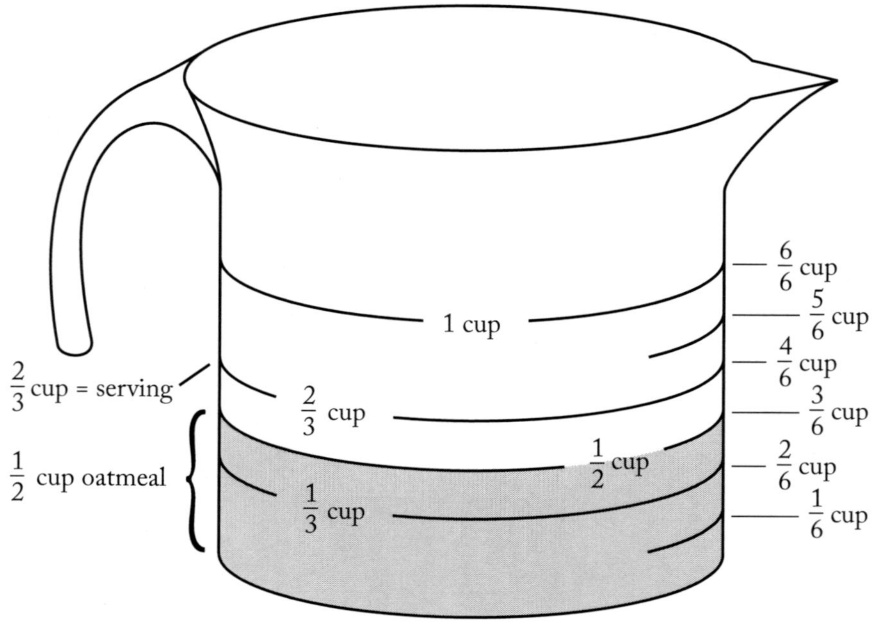

One way to picture this is to mark the cup in sixths of a cup. There are four of those sixths in $\frac{2}{3}$ cup and three of those sixths in $\frac{1}{2}$ cup. There is only $\frac{3}{4}$ of a $\frac{2}{3}$-cup serving in the box of oatmeal, showing that $\frac{1}{2} \div \frac{2}{3} = \frac{3}{4}$.

Exercises 7.6

1. Explain in terms of repeated subtraction how to find $6 \div \frac{2}{3}$.

2. Fill in the blanks.

   a. $\frac{3}{7} \div \frac{4}{5} = \dfrac{\frac{3}{7}}{\frac{\_}{5}} = \dfrac{\frac{3}{7} \times \frac{\_}{\_}}{\frac{4}{5} \times \frac{\_}{\_}} = \dfrac{\frac{3}{7} \times \frac{5}{4}}{\frac{\_}{20}} = \frac{\_}{1} = \frac{\_}{\_}$

   b. $\frac{5}{7} \div \frac{4}{9} = \dfrac{\frac{5}{\_}}{\frac{4}{9}} = \dfrac{\frac{5}{7} \times \frac{9}{\_}}{\frac{4}{9} \times \frac{9}{4}} = \dfrac{\frac{\_}{7} \times \frac{9}{\_}}{1} = \frac{\_}{\_}$

   c. $\frac{-3}{11} \div \frac{7}{15} = \dfrac{\frac{\_}{11}}{\frac{\_}{15}} = \dfrac{\frac{-3}{11} \times \frac{15}{7}}{\frac{7}{15} \times \frac{15}{\_}} = \dfrac{\frac{-3}{11} \times \frac{\_}{7}}{1} = \frac{\_}{\_}$

   d. $\frac{-17}{-4} \div \frac{-6}{5} = \dfrac{\frac{-17}{-4}}{\frac{-6}{5}} = \dfrac{\frac{-17}{\_} \times \frac{5}{\_}}{\frac{\_}{5} \times \frac{5}{-6}} = \dfrac{\frac{-17}{-4} \times \frac{\_}{-6}}{1} = \frac{\_}{\_}$

3. In parts a-f, divide and check by multiplying.

   a. $\frac{5}{8} \div \frac{1}{4}$ b. $\frac{1}{4} \div \frac{5}{8}$ c. $\frac{1}{2} \div \frac{1}{3}$

   d. $\frac{1}{3} \div \frac{1}{2}$ e. $\frac{-3}{4} \div \frac{4}{5}$ f. $\frac{4}{5} \div \frac{-3}{4}$

   g. Compare your answers for parts a and b, for parts c and d, and for parts e and f. Then make a general statement relating the quotients

   $\frac{p}{q} \div \frac{s}{t}$ and $\frac{s}{t} \div \frac{p}{q}$.

4. A certain recipe calls for $2\frac{1}{4}$ cups of flour.

   a. How many batches of that recipe can be made with 11 cups of flour?
   b. What division is needed to answer this question?

5. A rectangular field is $\frac{3}{4}$ mile long and has an area of $\frac{3}{10}$ square mile. How wide is the field?

6. At a steady $5\frac{1}{4}$ km per hour, how long will it take to walk 8 km? (Distance = rate × time.)

7. Zoning restrictions in a certain town require every new house to be built on a lot of at least 2/3 acre. Under this restriction, what is the maximum number of houses a builder could put on $7\frac{1}{2}$ acres?

7.7 Miscellaneous Exercises on Fractions

1. Add and express each sum as an improper fraction with a numerator of 5 (whether or not it is in lowest terms).

   a. $\frac{1}{1}+\frac{1}{2}+\frac{1}{6}$ b. $\frac{1}{1}+\frac{1}{5}+\frac{1}{20}$ c. $\frac{1}{2}+\frac{1}{3}+\frac{1}{6}$ d. $\frac{1}{2}+\frac{1}{4}+\frac{1}{12}$

   The Polish mathematician Sierpinski has conjectured that every fraction of the form $\frac{5}{n}$ with denominator n of at least 3 can be expressed as a sum of just three fractions, no two of which are the same, and all three of whose numerators are 1. Parts a-d illustrate the first four cases. This conjecture has not been proved, but it is known to be true for all values of n from 3 to 1,057,438,801.

2. In parts a-d, express each of these as a single fraction in lowest terms.

   a. $1-\frac{1}{2}$ b. $\left(1-\frac{1}{2}\right)\left(1-\frac{1}{3}\right)$ c. $\left(1-\frac{1}{2}\right)\left(1-\frac{1}{3}\right)\left(1-\frac{1}{4}\right)$

   d. $\left(1-\frac{1}{2}\right)\left(1-\frac{1}{3}\right)\left(1-\frac{1}{4}\right)\left(1-\frac{1}{5}\right)$

   e. Use your work from parts a-d to predict

   $\left(1-\frac{1}{2}\right)\left(1-\frac{1}{3}\right)\left(1-\frac{1}{4}\right)\left(1-\frac{1}{5}\right)\left(1-\frac{1}{6}\right)$ and

   $\left(1-\frac{1}{2}\right)\left(1-\frac{1}{3}\right)\left(1-\frac{1}{4}\right)\ldots\left(1-\frac{1}{20}\right)$.

3. In parts a-d, express each of these as a single fraction.

   a. $\left(1-\frac{1}{2^2}\right)$ b. $\left(1-\frac{1}{2^2}\right)\left(1-\frac{1}{3^2}\right)$ c. $\left(1-\frac{1}{2^2}\right)\left(1-\frac{1}{3^2}\right)\left(1-\frac{1}{4^2}\right)$

   d. $\left(1-\frac{1}{2^2}\right)\left(1-\frac{1}{3^2}\right)\left(1-\frac{1}{4^2}\right)\left(1-\frac{1}{5^2}\right)$

   e. Find a pattern in the answers to parts a-d and use it to predict the value of

   $\left(1-\frac{1}{2^2}\right)\left(1-\frac{1}{3^2}\right)\left(1-\frac{1}{4^2}\right)\left(1-\frac{1}{5^2}\right)\left(1-\frac{1}{6^2}\right)$ and

   $\left(1-\frac{1}{2^2}\right)\left(1-\frac{1}{3^2}\right)\left(1-\frac{1}{4^2}\right)\left(1-\frac{1}{5^2}\right)\left(1-\frac{1}{6^2}\right)\left(1-\frac{1}{7^2}\right)$.

Note: To see the pattern, first put each answer in lowest terms, then multiply those with odd denominators by 2/2.

f. Check your predictions for part e by doing the arithmetic.

4. **The four fours problem**[4] Make positive integers by these rules: You must use exactly four 4s, and you may not use any other digits. You may add, subtract, multiply, or divide, using parentheses as desired.

For starters, you can make 1 as $\frac{4+4}{4+4}$ or as $\frac{4+4-4}{4}$ and 2 as $\frac{4\times 4}{4+4}$ or as $\frac{4}{4}+\frac{4}{4}$ (Can you think of other ways?)

a. Make 3 with four 4s.
b. Make 4 with four 4s.
c. Make 5 with four 4s.
d. Make 6 with four 4s.
e. Make 7 with four 4s.
f. Make 8 with four 4s.
g. Make 9 with four 4s.
h. Make 10 with four 4s
i. Make the numbers from 11 to 18, each with four 4s. (Warning: This puzzle is addictive)

5. In parts a-e, express each of these as a single improper fraction.

a. $1+\cfrac{1}{1+1}$

b. $1+\cfrac{1}{1+\cfrac{1}{1+1}}$

c. $1+\cfrac{1}{1+\cfrac{1}{1+\cfrac{1}{1+1}}}$

d. $1+\cfrac{1}{1+\cfrac{1}{1+\cfrac{1}{1+\cfrac{1}{1+1}}}}$

e. $1+\cfrac{1}{1+\cfrac{1}{1+\cfrac{1}{1+\cfrac{1}{1+\cfrac{1}{1+1}}}}}$

f. Identify by name the set of integers which appear in these fractions as numerators and denominators. (Hint: They are mentioned in Exercises 5.1.)

4 This problem goes back at least to a letter to the editors which appeared in 1881 in the London publication, *Knowledge: An Illustrated Magazine of Science, Plainly Worded—Exactly Described.* The writer stated that any integer from 1 to 20 could be made this way except for 19. In 1964 Martin Gardner, in his *Mathematical Games* column of the *Scientific American*, stated that if additional notation, such as roots, exponents, decimals, etc. is allowed, all positive integers up through 112 can be represented by four 4s.

Vocabulary for Chapter 7

Canceling (7.3)
Common denominator (7.4)
Denominator (7.1)
Fraction (7.1)
Lowest terms (7.3)
Mixed number (7.1)
Numerator (7.1)
Proper and improper fraction (7.1)
Rational number (7.1)
Reciprocal (7.3)
Reducing and unreducing fractions (7.3)

Self-Quiz for Chapter 7

1. a. Make a rectangle to illustrate $\frac{2}{3} \times \frac{4}{5}$.

   (Use the 1×1 square at right.)

   b. $\frac{2}{3} \times \frac{4}{5} =$

2. a. Write a fraction equal to $\frac{3}{7}$ with a denominator of 35.

   b. Reduce $\frac{196}{238}$ to lowest terms.

3. Compute $2\frac{3}{4} + 5\frac{7}{10}$

4. a. Add: $\frac{-3}{4} + \frac{1}{-12} + \frac{5}{6}$ b. $1\frac{5}{8} + 3\frac{1}{12}$

5. Explain and complete the subtraction $\frac{7}{5} - \frac{3}{5}$ in terms of:

   a. Take away b. Adding the opposite

   c. What must be added

6. $\frac{5}{6} \div \frac{4}{7} =$

7. a. Explain the difference between dividing by 2 and dividing by $\frac{1}{2}$.

   b. Write a problem that can be understood by sixth graders and requires division by $\frac{1}{2}$.

8. Subtract $\frac{-1}{4}$ from $\frac{5}{6}$.

Chapter Eight

Decimals and Percents

This chapter is about the extension of decimal numeration from integers to fractions. How fractions are written may not initially seem very important, but it is. Writing fractions as decimals extends to fractions the algorithms for addition, subtraction, multiplication, and division that are used with whole numbers and also leads to new questions and a new view of the number line.

8.1 Decimal Fractions

The system of place value used for whole numbers is based on successive powers of ten. These underlying place values are usually not written explicitly, but they can be made explicit by writing the number in extended form. For example, 3,077 can be written as $3\times10^3+0\times10^2+7\times10^1+4\times10^0$. The place value of each digit is ten times the place value of the digit to its immediate right. As Aryabhata put it so eloquently, "From place to place, each is ten times the preceding." Then moving to the right, the place value of each digit is one-tenth the place value of the digit to its immediate left.

For whole numbers, the right most place value is $10^0=1$. Now, instead of stopping at 1, we extend the system of place values to the right by continuing to divide by ten.

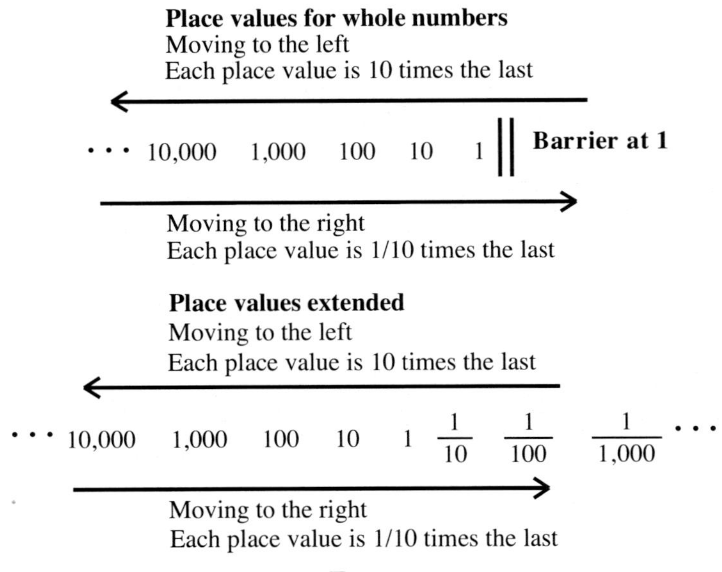

Figure 8.1.1

With whole numbers, the right most place value is understood to be 1, but in the extended decimal system there is no right most place value. Therefore, a period, called a **decimal point**, is placed between the ones place and the one-tenths place to separate the **integer part** of a number from the **fractional part** and thus signal the place values of all the digits. (The use of the decimal point is widely accepted, but occasionally you will see printed material from Europe that uses a comma as we use a decimal point.)[1]

Example 1. Write 56.34 in extended notation and express it as a mixed number.
Solution: The decimal point indicates that the digit to its immediate left has place value 1. Therefore, $56.34 = 5 \times 10 + 6 \times 1 + 3 \times \frac{1}{10} + 4 \times \frac{1}{100}$. To write this as a mixed number, note that $3 \times \frac{1}{10} + 4 \times \frac{1}{100} = \frac{3}{10} + \frac{4}{100} = \frac{30}{100} + \frac{4}{100} = \frac{34}{100}$. Then as a mixed number, $56.34 = 56\frac{34}{100}$.

The above example may be less abstract if modeled with money as \$56.34. The fractional part, 34 cents, can be thought of as three dimes and four pennies $\left(\frac{3}{10} + \frac{4}{100}\right)$ or as 34 pennies $\left(\frac{34}{100}\right)$.

Of course, decimal fractions may be modeled just like any other fractions. Among standard manipulatives, base-ten blocks and graph paper with one-inch squares subdivided into hundredths are especially well suited to modeling decimal fractions.

Locating decimal fractions on the number line can be instructive.

Example 2. Locate 24.079 on the number line.
Solution: In extended form, 24.079 is $2 \times 10 + 4 \times 1 + 0 \times \frac{1}{10} + 7 \times \frac{1}{100} + 9 \times \frac{1}{1000}$.

The decimal point separates the integer part, 24, from the fractional part, 079. To locate 24.079 on the number line, extend the process used for whole numbers to fractions, as follows:

1. The first digit, 2, is in the tens column, so 24.079 lies in the portion of the number line shown.

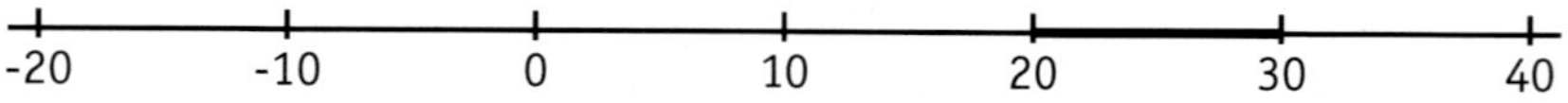

2. Magnify the interval from 20 to 30 and subdivide it as shown. The units digit, 4, puts 24.079 in the heavy interval.

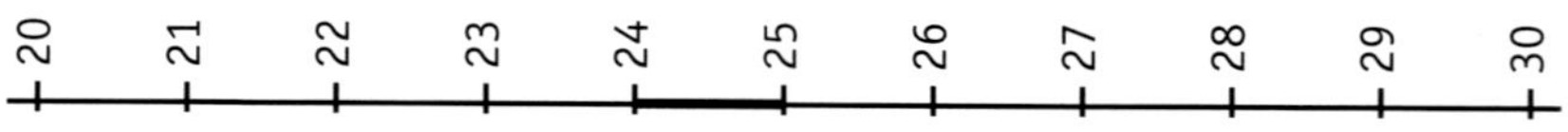

---

[1] Decimal fractions were used in ancient China and sporadically in the medieval Islamic world and Renaissance Europe. Simon Stevin (1548-1620), an engineer in Holland, wrote *De Thiende* (*On Tenths*), explaining decimal fractions and advocating their use in 1584. The decimal point was first used not by Stevin but either by Clavius (see note, section 10.4) in 1593 or in 1592 by G.A. Magini (1555-1617). Stevin's book was translated into French and later into English, where John Napier (see section 17.5) introduced decimal fractions and the decimal point in work published in 1614. The success of Napier's work led to widespread adoption of its decimal notation.

3. Magnify the interval from 24 to 25 and subdivide it as shown. The tenths digit is 0, so 24.079 lies in the heavy subinterval from 24.0 to 24.1.

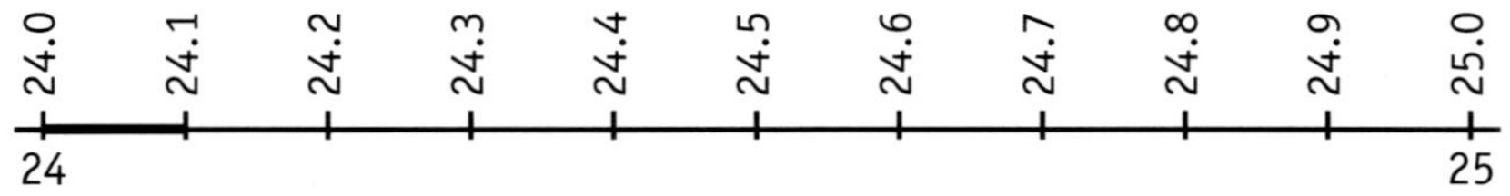

4. Magnify the shaded subinterval again and subdivide as shown. The hundredths digit places 24.079 still more precisely in the heavy subinterval from 24.07 to 24.08.

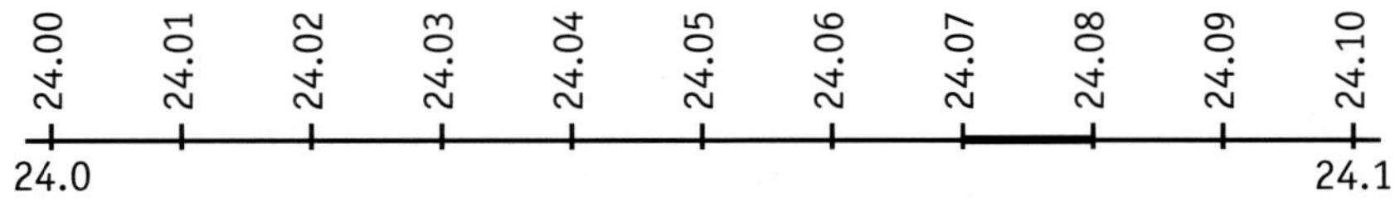

5. Repeat the process once more to reach the goal.

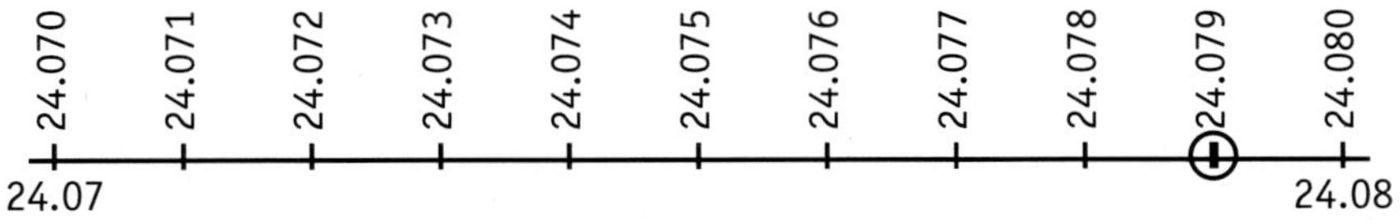

Locating a decimal on a number line is, as you can see, a process of repeatedly zooming in closer, much like locating a word in a dictionary.

Sometimes we do not zoom in all the way. We might say, for example, that to the nearest whole number, 24.079 is 24. This extends rounding from integers to decimal fractions.

Example 3. Round 3.14159 to: a) the nearest integer, b) the nearest tenth, and c) the nearest thousandth.
Solution: For part a), view number line through a relatively coarse lens:

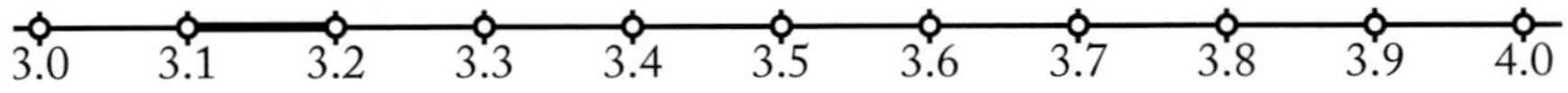

Since 3.14159 is in the heavy interval, which is closer to 3 than to any other integer, 3.14159 rounded to the nearest integer, is 3.

For part b), zoom in on the sub-interval between 3.1 and 3.2:

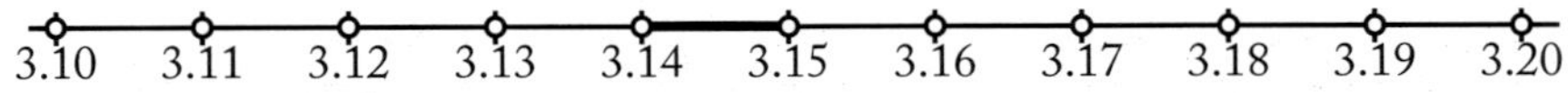

The heavy interval containing 3.14159 is closer to 3.1 than to 3.2, so, rounded to the nearest tenth, 3.14159 is 3.1.

For part c), zoom in even closer on the part of the number line containing 3.14159.

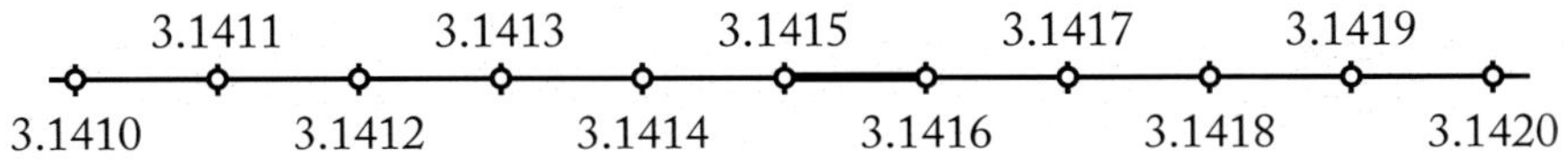

The heavy interval, which contains 3.14159, is closer to 3.142 than to 3.141, so, to the nearest thousandth, 3.14159 is rounded to 3.142.

Computing with Decimals

Extending to fractions the place value relationship of each column to its neighbors makes it possible to extend to decimal fractions the methods of computation developed for whole numbers.

Addition of decimals The vector addition model used for integers in chapter 2 also models addition of decimals, as shown in example 4.

Example 4. Add: a) 1.7 +1.2 and b) 3.2+(-2.6).
Solution: a. Make an arrow pointing to the right from 0 to 1.7. Then make a second arrow with length 1.2, also pointing to the right, beginning at 1.7. The tip of this second arrow is at 2.9, showing that 1.7 +1.2=2.9.

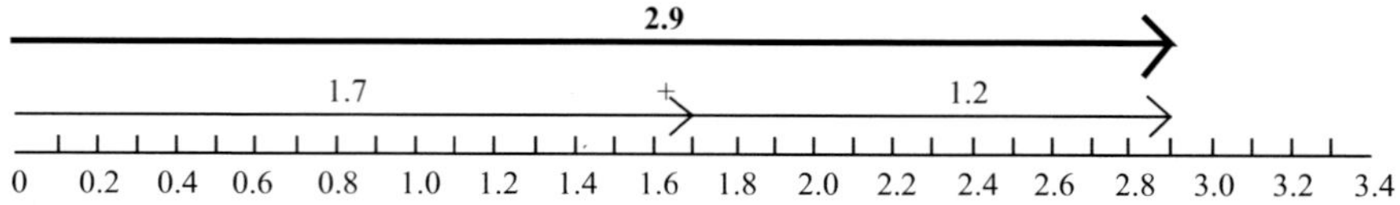

b. Picture a vector starting at 0 and ending at 3.2. Then, starting at 3.2, adjoin a vector 2.6 units long, pointing in the negative direction. The tip of this vector is at 0.6, showing that 3.2+(-2.6)=0.6.

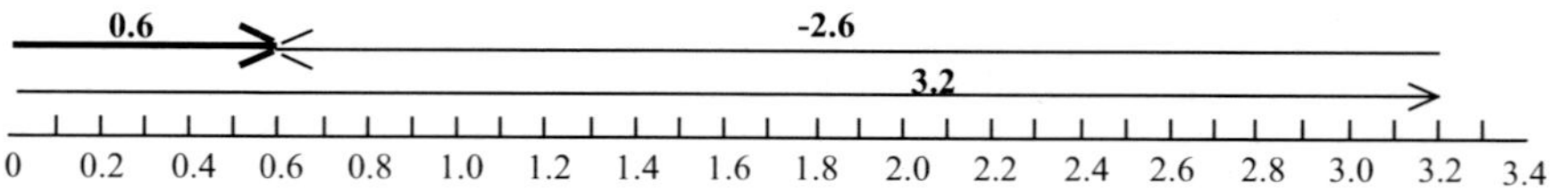

In question 1 below, you can check these solutions.

Diagrams of vector addition are very useful when one or more of the addends is negative, since the vectors, so to speak, point you in the right direction. But in practice the main issue in adding decimals is to organize the work efficiently.

Write the addends, much as you would write integer addends, with digits of like place value directly above each other. That happens automatically, if you write the numbers neatly, with the decimal points in a vertical column. (Students who have trouble writing numbers neatly enough may find this easier to do if they write on squared paper.) Then add as you would add whole numbers, putting the decimal point directly below the decimal points of the addends.

Example 5. Add 803.27 and 14.59.
Solution:

$$\begin{array}{r} 803.27 \\ +\underline{\ \ 14.59} \\ 817.86 \end{array}$$

Carrying, as must be done from the hundredths column to the tenths column, is the same as for whole numbers. To check this addition, write it as

$$803\frac{27}{100}+14\frac{59}{100}, \text{ which is } 803+14+\frac{27+59}{100}=817+\frac{86}{100}=817.86$$

To model this addition concretely, use money, adding $803.27 to $14.59.

	bills			coins	
	hundreds	tens	ones	dimes	pennies
	8	0	3	2	7
		1	4	5	9
	8	1	7	7	16
=	8	1	7	8	6

Example 6. Subtract 3.87 from 8.50.
Solution 1: **Missing addend** What number must be added to 3.87 to get 8.50? Model this with money. You leave home with $8.50 but spend $3.87. How much do you have left? Count up from $3.87. Add 13 cents to get to $4.00, and then add four dollars to get up to $8.00. Another 50 cents gets you to $8.50. In all, you have added $0.13+$4.00+$0.50=$4.63, Then 8.50-3.87=4.63.
Solution 2: **Take away** Write the number to be subtracted beneath the number being subtracted from, putting one decimal point directly below the other and thus vertically aligning digits with like place values:

$$\begin{array}{r} 8.50 \\ -\ \underline{3.87} \end{array}$$

Now subtract as you would with whole numbers. To make the place values of the digits of the difference the same as the place values of the digits directly above them, put the decimal point of the difference directly below those of the other numerals:

$$\begin{array}{r} 8.50 \\ -\ \underline{3.87} \\ 4.63 \end{array}$$

Solution 3: **Add the opposite**. 8.50-3.87=8.50+(-3.87). Carry out this addition as in example 4 above.
Solution 4: Express the numbers involved as mixed numbers and carry out the subtraction as in chapter 7.

$$8.50-3.87=8\frac{50}{100}-3\frac{87}{100}=8-3+\frac{50-87}{100}=5-\frac{37}{100}=4\frac{63}{100}=4.63$$

Multiplication with decimals: multiplying and dividing by 10

With whole numbers, students learn that multiplication by ten is easy: just add a 0 to the end of the number: e.g. 10×43=430. The reason for this is revealed by writing 43 in expanded notation.

$$10\times43=10\times(4\times10+3\times1)=10\times(4\times10)+10\times3=4\times100+3\times10+0\times1$$

Adjoining a 0 as a new right-hand digit has the effect of moving the other digits to the left, thus multiplying all their place values by ten.

The same idea works for the extended decimal system. To multiply 43.72 by ten, multiply all the place values by ten.

hundreds	tens	ones	tenths	hundredths
	4	3	7	2
4	3	7	2	

Then 10×43.72=437.2, and the effect of multiplying by ten is to move the decimal point one place to the right. The same reasoning shows why multiplying by 100 moves the decimal point two places to the right, dividing by 10 moves the decimal point one place to the left, etc.

You can visualize this with money. If you have $43.72 in the form of four ten-dollar bills, three ones, seven dimes, and two pennies, then a quick way to multiply this amount by ten is to trade each bill or coin for one worth ten times as much: trade four ten-dollar bills for four hundreds, trade three ones for three tens, trade seven dimes for seven dollars, and trade two pennies for two dimes. Multiplying the value of each coin and bill by ten multiplies the total value by ten.

hundreds	tens	ones	dimes	pennies
	4	3	7	2
4	3	7	2	

Moving the decimal point one place to the right has the same effect as moving each digit one place to the left.

Now that we can multiply by ten, we can multiply any decimals.

Example 7. Multiply 1.6×2.1.
Solution 1: Use the area model of multiplication as in figure 8.1.2:

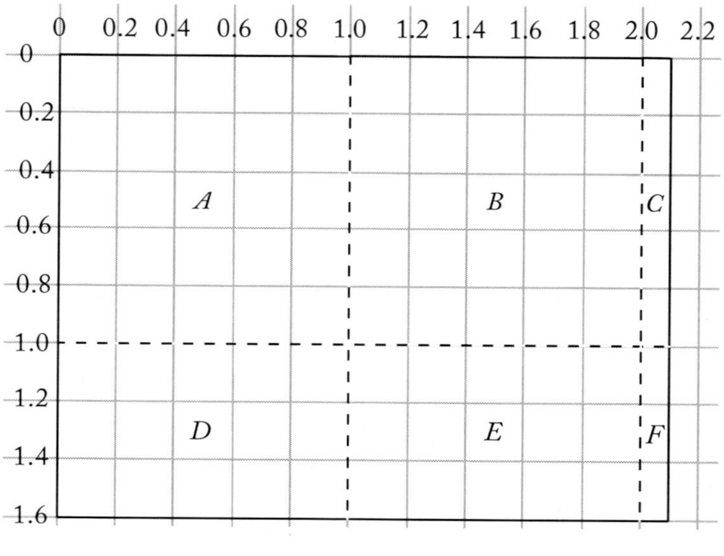

Figure 8.1.2

Now add up the areas of the pieces: A and B are each 1×1 units of area, while D and E are each 1×(0.6)=0.6. The area of C is 1×(0.1)=0.1, and the area of F is 0.6×0.1. The last of these products may be found, as described above, by moving the decimal point of 0.6 one place to the left, to get 0.06.
Then 1.6×2.1=1+1+0.6+0.6+0.1+0.06=3.36.
Solution 2: By the distributive property, 1.6×2.1=1.6×(2+0.1)= 1.6×2+1.6×0.1. Find 1.6×2 by repeated addition (1.6+1.6=3.2), and find 1.6×0.1 as described above, by moving the decimal point in 1.6 one place to the left to get 0.16, Then 1.6×2.1=1.6×(2+0.1)= 1.6×2+1.6×0.1=3.2+0.16=3.36.

Solution 3: Rewrite the problem without decimals and apply the methods of chapter 7.

$$1.6 \times 2.1 = \frac{16}{10} \times \frac{21}{10} = \frac{16 \times 21}{10 \times 10} = \frac{336}{100} = 3.36$$

Example 8. Multiply 3.21 by 5.04.
Solution: Think of this as the area, in square meters, of a rectangular room 3.21 meters (m) wide and 5.04 m long. Change the units to centimeters (cm). Since there are 100 centimeters in a meter, the area in square centimeters is 321×504, which can be found by multiplication of whole numbers to be 161,784 cm^2. Now there are 100×100=10,000 square centimeters in each square meter, so 161,784 cm^2=16.178,4 m^2.
As a check, note that, since the width and length of the room are slightly greater than 3 and 5 m, respectively, the area must be somewhat greater than 3x5=15.

Examples 7 and 8 suggest an easy way to multiply decimals.

1. Ignore the decimal points and multiply as you would integers.

   In example 8, begin by multiplying the whole numbers 321×504 to get 161,784.

2. Place the decimal point in the product. To do this correctly, students are often told a rule:

   Place as many digits to the right of the decimal point in the product as there are digits to the right of the decimal points in all the factors.

   Since there are two digits to the right of the decimal point in 3.21 and two in 5.04, place the decimal point in the product so it has 2+2=4 digits to its right. This is the case in the answer to example 8.

The rule for step 2 is correct (the rationale for it is explored in questions 8 and 9 below), but it seems arbitrary. More important, a simple counting error can easily lead to an absurdly large or small answer. Therefore, minimize reliance on the rule; help students develop the habit of estimating the approximate size of the answer. In example 8, since the width and length of the room are slightly greater than 3 and 5 m, respectively, the area must be somewhat greater than 3x5=15m^2. An answer a bit over 16 is reasonable. Miscounting digits to the right of the decimal point and thus misplacing the decimal point leads to numbers such as 161.784 or 1.6784, which are not even close to the correct product. Such errors, which can be important in real life situations, can be prevented by estimating the size of the product.

Division with decimals is discussed in section 8.2.

Exercises 8.1

1. Label the marked points with decimal fractions.

a. 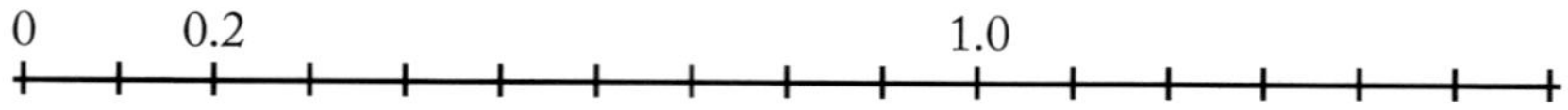

b. 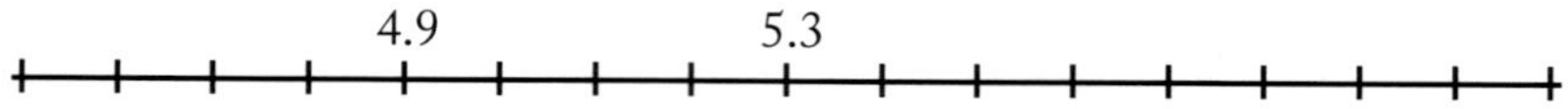

c. 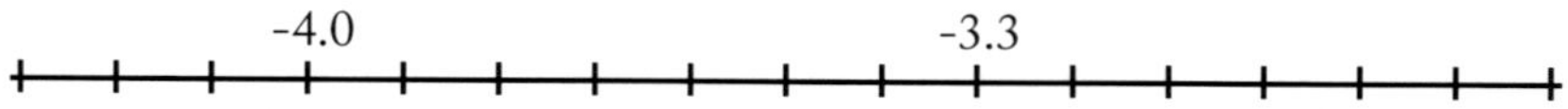

d. 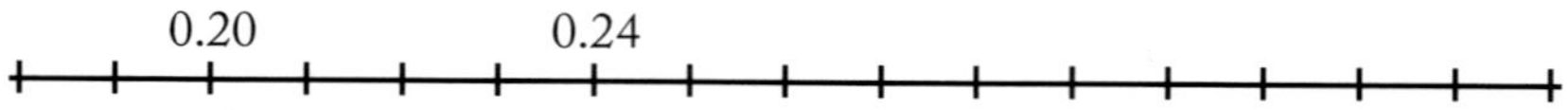

e. 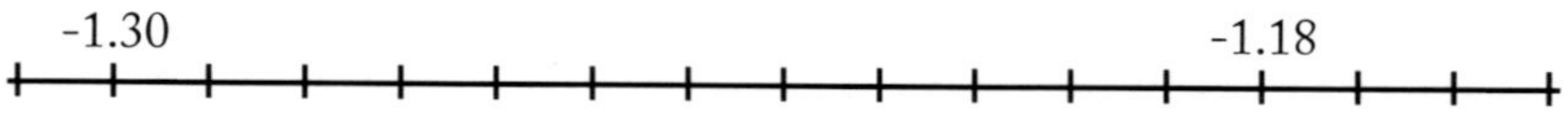

f. 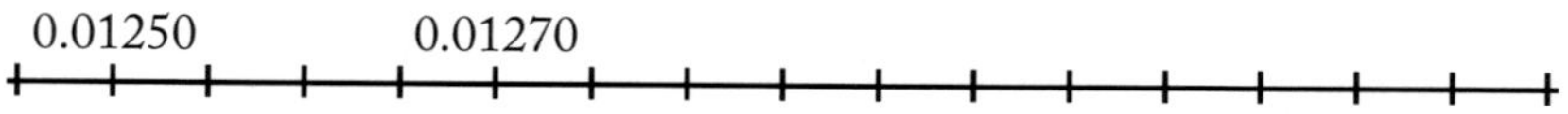

2. Which is greater, 21.2 or 21.05? Explain your answer.

3. Use successive enlargements of subintervals, as shown in the text for 24.079, to locate the following on a number line:
   a. 1.732 b. 2.718 c. 1.618 d. 2.059

4. Box puzzles are a good way to practice arithmetic with decimals, since they are self-checking. Here is a sampler; you can, of course, make up more for yourself, if you need the practice.

a.

3.1	2.8	
4.7	9.3	

b.

-0.4	8.01	
12.5	-7.6	

c.

5.01		
		3.7
2.78		6.0

d.

	1.1	0.4	
0.5			
3.1			

e.

	2.01	-0.21	
0.6			
-1.1			

f.

	0.7		
0.1			3.5
	-2.1		

5. As of April 1, 2009, according to the US Treasury web site (www.publicdebt.treas.gov/opd/opd.htm), the United States public debt stood at $11,110,654,357,209.33. Round this:
   a. To the nearest $ trillion
   b. To the nearest $ million.

6. Column-wise addition of decimal fractions, as described in the text, probably seems reasonable to you. This problem shows that the result of column-wise addition agrees with the result of addition of fractions as explained in chapter 7.
   a. Add 3.54+7.8 as described in the text.
   b. Write 3.54 and 7.8 as mixed numbers and add them that way.
   c. Model this addition using base ten blocks. (For this purpose, you will need to take a flat as your unit. In that case, how much is a rod? How much is a small cube?)

7. Picture these additions with vectors as in example 4.
   a. 7.2+5.3 b. −6.1+2.8 c. −4.8+(-3.1)

8. In parts a-e, write as ordinary fractions or mixed numbers, showing numerators and denominators:
   a. 6.02 b. 11.47 c. 20.1234
   d. 0.53426 e. 123.456789
   f. Based on examples a-e, if you write a decimal in fraction form, how will the number of 0s in the denominator be related to the number of digits to the right of the decimal point?

9. To check the rule for placing the decimal point in a product, consider an example, 4.032×31.87.
   a. How many digits are to the right of the decimal point in 4.032?
   b. How many digits are to the right of the decimal point in 31.87?
   c. According to the rule, how many digits are to the right of the decimal point in 4.032×31.87?
   d. Multiply 4,032×3,187, and then put the decimal point where it belongs for 4.032×31.87.

   Parts e-g check the placement of the decimal point by working the problem another way

   e. Write 4.032 and 31.87 as fractions. How many 0s are in each denominator?
   f. Multiply the fractions from part e. How many 0s are in the denominator of the product?
   g. Use the results of part f, together with problem 8f (above) to determine how many digits belong to the right of the decimal point in this product.
   h. Explain how to check the result of this multiplication by estimating.

10. If you buy 13.2 gallons of fuel at \$3.24 per gallon, it will cost you \$3.24×13.2.
    a. Estimate about how much money this is.
    b. Calculate 3.24×13.2 to determine your total cost, placing the decimal point by the rule given in the text. Do your answers from parts a and b agree?

11. If you are paid \$8.62 per hour and you work 42.3 hours, how much pay have you earned?

12. Make and complete a box puzzle to model the multiplication in example 7.

13. Evaluate $\frac{1}{2}+\frac{1}{0.2}$

8.2 Division Beyond the Decimal Point

Writing numbers in decimal form does not change their meaning or the basic meaning of the way we operate with them. However, in the case of long division, that requires some fine tuning.

Example 1. Divide 53 by 2.
Solution: Before we extended decimal numeration to fractions, the work for this question was written like this:

$$\begin{array}{r} 26 \\ 2\overline{)53} \\ \underline{4} \\ 13 \\ \underline{12} \\ 1 \end{array}$$

The 1 at the bottom was left as a remainder, because we could not divide further. With decimal numeration extended to fractions, we can continue to divide. Write 53 as 53.0, and put a decimal point in the numerator to indicate that the 6 has place value 1 and anything which follows it will have fractional place value. Then extend the line at the top over the new 0, and think of the remainder 1 as

$$\begin{array}{r} 26. \\ 2\overline{)53.0} \\ \underline{4} \\ 13 \\ \underline{12} \\ 1 \end{array}$$

1.0, which is the effect of copying the 0 next to the remainder ("bringing it down"). To continue the division, we must divide 1.0 by 2. Before we used decimals fractions, we would have written this as 1/2. Now, with decimal fractions available, we note that since 2×5=10, 2×(0.5)=1.0. This is reasonable, since $0.5=\frac{5}{10}=\frac{1}{2}$. The division is completed as shown below, ending, because the final subtraction yields 0.

```
   26.5
 2)53.0
   4
   13
   12
    1 0
    1 0
```

If you think of division as repeated subtraction, this result means that you can subtract 2 from 53 just 26.5 times. To check this answer, note that $2 \times 26.5 = 53$.

Any division that does not have a whole number quotient may be extended beyond the decimal point just as in example 1. Example 2 concerns the case where the dividend is not a whole number.

Example 2. Divide 20.724 by 6.
Solution: Write the division as usual, putting a decimal point in the quotient directly above the decimal point of the dividend.

```
     .
 6)20.724
```

Now divide as you would with whole numbers. Put a 3 in the quotient, then subtract $3 \times 6 = 18$.

```
    3.
 6)20.724
   18
```

Now we must divide 2.724 by 6, but in practice we can just bring down one digit, in effect to divide 2.7 by 6, and still get the next digit of the quotient correct.

```
    3.
 6)20.724
   18
    2 7
```

Continue as you would do long division with whole numbers.

```
    3.4               3.454
 6)20.724          6)20.724
   18                18
    2 7               2 7
    2 4               2 4
      3                 32
                        30
                         24
                         24
```

If the divisor is not a whole number, change the division question to an equivalent one that has a whole number divisor. For example, to divide $0.0075 \div 0.015$, think of this division as a fraction and multiply numerator and denominator by 1,000 to get an equivalent fraction with a whole number denominator.

$$\frac{0.0075}{0.015} = \frac{0.0075 \times 1{,}000}{0.015 \times 1{,}000} = \frac{7.5}{15}$$

Now divide as above.

Dividing beyond the decimal point is straightforward in principle, but beginners are sometimes surprised that the process may never end.

Example 3. Divide 40 by 11.
Solution:

$$\begin{array}{r@{}l}
 & \underline{\;3.636...} \\
11\big) & 40.000 \\
 & \underline{33} \\
 & \;\;7\,0 \\
 & \;\;\underline{6\,6} \\
 & \;\;\;\;40 \\
 & \;\;\;\;\underline{33} \\
 & \;\;\;\;\;\;70 \\
 & \;\;\;\;\;\;\underline{66}
\end{array}$$

Figure 8.2.1

Here successive subtractions alternately leave remainders of 7 and 4, which results in a quotient with endless alternation of the digits 3 and 6. The three dots in the quotient indicate, as usual, that the pattern continues. For infinite decimals, however, a more precise notation is needed to describe the unending patterns. To show unending repetitions in decimals, put a bar over the repeating digits. In the above example, for instance, write $3.\overline{63}$. To indicate that only the digit 3 repeats endlessly, write $3.6\overline{3}$.

Although practical applications rarely demand accuracy beyond a few decimal places, there are many interesting questions about infinite decimals. Some of those questions are considered in the exercises below.

The decimal equivalent of a fraction
To express a fraction as a decimal, carry out the division indicated by the fraction bar.

Example 4. Find the decimal equivalent for $\frac{3}{8}$.
Solution: Divide 3 by 8 as follows:

$$\begin{array}{r@{}l}
 & \underline{\;0.375\;} \\
8\big) & 3.000 \\
 & \underline{2\,4} \\
 & \;\;60 \\
 & \;\;\underline{56} \\
 & \;\;\;\;40 \\
 & \;\;\;\;\underline{40}
\end{array}$$

The result is checked in two different ways in question 4 below.

In example 4, the decimal equivalent for $\frac{3}{8}$ was found to be a finite decimal, 0.375. In contrast, some fractions have infinitely long decimal equivalents. A familiar example is $\frac{1}{3} = 0.\overline{3}$. Multiplying this equation by 3 yields $1 = 0.\overline{9}$, which they find puzzling. The exercises below explore additional puzzles and patterns involving infinite decimals.

The examples given above are entirely typical, both in how long division is done with decimal fractions and in the results. Specifically, when you divide one whole number by another, there are just three possibilities:

1. The quotient is an integer

2. The quotient is a decimal with a finite number of non-0 digits (a so-called "terminating" decimal).
3. The quotient is a decimal which, instead of ending, has a pattern of digits which repeats endlessly.

You might wonder about decimals that neither end nor have an endlessly repeating pattern of digits. One decimal in that category is 0.212 212 221 222 212 222 21. . ., where the strings of 2s grow steadily longer, always separated by a single 1. Any decimal in that fourth category does not represent a rational number (a number which can be expressed as a quotient of integers). Any such decimal is therefore an example of what is called an **irrational** (not rational) number. Irrational numbers were confusing and philosophically troubling to the ancient Greeks who first pondered them. Some of the numbers we meet in later chapters, notably $\sqrt{2}$ and π, are irrational.

Exercises 8.2

1. a. Divide 1.036 by 25.
   b. Check the result of part a by multiplication.
   c. Note that dividing by 25 is the same as multiplying by its reciprocal, $\frac{1}{25}$, which is, in turn, the same as multiplying by $\frac{4}{100}$ (since the two fractions are equal.) Multiplying by 4 is not hard to do in your head (double, then double again), and dividing by 100 is a snap (just move the decimal point two places to the left). This approach helps you divide by 25 in your head.

2. Use the idea from question 1 to divide in your head:
   a. $120 \div 25$ b. $230 \div 25$ c. $408 \div 25$

3. How can you use the idea from questions 1 and 2 to divide by 50 in your head? Try it on:
   a. $420 \div 50$ b. $720 \div 50$

4. a. Divide 3 by 8.
   b. Check your answer from part a by multiplication.
   c. As another check, write 0.375 as a fraction, and reduce it to lowest terms.

5. a. Find the decimal equivalent for $\frac{1}{7}$ by carrying out the division to at least 12 places.
   b. Look for patterns in your answer to part a before trying part c.
   c. Fill in the blanks appropriately, one digit for each dashed space.

$7 \times 2 =$ __ __
$7 \times 4 =$ __ __
$7 \times 8 =$ __ __
$7 \times 16 =$ __ __ __
$7 \times 32 =$ __ __ __
$7 \times 64 =$ __ __ __
$7 \times 128 =$ __ __ __
$7 \times 256 =$ __ __ __ __

---

d. Add the digits you entered in part c vertically in columns. (The first five columns have only one digit in each.)

e. Name the family of numbers which multiply 7 at the left side of the diagram in part c.

f. How are parts a and b of this question related to parts c and d? (Answer in a clearly written sentence or two.)

g. Extend the pattern from part c at least two more lines and check to see that the pattern observed in part d continues to hold.

6. In parts a-e, find decimal equivalents for the given fractions to at least six decimal places. Use a calculator if you wish.

a. $\frac{2}{7}$ b. $\frac{3}{7}$ c. $\frac{4}{7}$ d. $\frac{5}{7}$ e. $\frac{6}{7}$

f. Describe in a sentence or two how the decimal equivalents you found in parts a-e are related.

7. a. Add the infinite decimals for $\frac{1}{7}$ and $\frac{2}{7}$. Do you get the infinite decimal for $\frac{3}{7}$?

b. Add the infinite decimals for $\frac{1}{7}$ and $\frac{6}{7}$. Do you get the infinite decimal for $\frac{7}{7}=1$? (Note: Adding the decimals for just a few places allows you to see what would happen if the infinitely long decimals were added. However, if you add only a few digits, you get a sum only a few digits long. A **finite** string of 9s after the decimal point is **not** equal to 1, but an **infinite** string is.)

c. Add the decimals for $\frac{2}{7}$ and $\frac{5}{7}$. Do you get a decimal for 1?

d. Add the decimals for $\frac{3}{7}$ and $\frac{4}{7}$. Do you get a decimal for 1?

8. a. Find the decimal for $\frac{1}{14}$ by dividing.

b. Check your work in part a using the fact that $\frac{1}{14}+\frac{1}{14}=\frac{1}{7}$.

c. Find the decimal for $\frac{10}{14}$ by dividing.

d. Explain how you can use your work from question 6 to check your answer to part c.

e. As another check on your work, compare your answers to parts a and c. Are they related as they should be? Explain in a sentence or two.

9. A rectangular plot of land is surveyed and found to be 0.35 square kilometers in area and 0.67 km. long. How wide is it?

10. A town plans a tax to earmark 0.002 cents of every sales dollar to finance a $370,000.00 special project. How many dollars of sales revenue will be needed to raise $370,000.00 this way?

8.3 Exploring Decimal Equivalents of Fractions

You may have noticed some curious things about decimal equivalents for fractions. Some, like 0.5 for $\frac{1}{2}$, end, but others go on forever. Those which do not end have interesting patterns. For example, the decimal for $\frac{1}{7}$ endlessly repeats 142857. Which fractions have finite decimals? If a fraction has a finite decimal, how many digits long is it? If a fraction has an infinite decimal, must that decimal consist of endless repetitions of the same digits? If so, how long a string of digits repeats? This set of exercises explores these questions. For simplicity, we focus on fractions whose numerators are 1, however, fractions whose denominators are not 1, behave in the same way, provided they are in lowest terms.

1. To get some data, find decimal equivalents for each of these fractions. Carry each division until it ends or you can clearly see the digit(s) which repeat indefinitely.

$\frac{1}{2} =$

$\frac{1}{3} =$

$\frac{1}{4} =$

$\frac{1}{5} =$

$\frac{1}{6} =$

$\frac{1}{7} =$

$\frac{1}{8} =$

$\frac{1}{9} =$

$\frac{1}{10} =$

$\frac{1}{11} =$

$\frac{1}{12} =$

$\frac{1}{13} =$

$\frac{1}{14} =$

$\frac{1}{15} =$

$\frac{1}{16} =$

$\frac{1}{17} =$

$\frac{1}{18} =$

$\frac{1}{19} =$

$\frac{1}{20} =$

$\frac{1}{21} =$

$\frac{1}{22} =$

$\frac{1}{23} =$

$\frac{1}{24} =$

$\frac{1}{25} =$

$\frac{1}{26} =$

$\frac{1}{27} =$

$\frac{1}{28} =$

2. a. List all of the eight fractions from question 1 which have finite decimal expansions.
   b. In each fraction from part a, factor the denominator into primes.
   c. What do you notice about the prime factorizations in your answer to part b?
   d. Try to explain why your observation must fit any fraction that has a finite decimal equivalent.
   Does this fit your observations in part a?

3. Use your observation from question 2 to predict which of these fractions have finite decimal equivalents and which do not.

$$\frac{1}{31} \quad \frac{1}{32} \quad \frac{1}{33} \quad \frac{1}{35} \quad \frac{1}{36} \quad \frac{1}{40} \quad \frac{1}{48} \quad \frac{1}{55} \quad \frac{1}{60} \quad \frac{1}{75} \quad \frac{1}{80} \quad \frac{1}{120} \quad \frac{1}{125}$$

4. If a fraction has a finite decimal equivalent, can you predict how many digits long it will be? You may be able to reason directly from question 2. In the table below, some of the exponents in the denominators are omitted, as are the decimal equivalents.

	$\frac{1}{2} = \frac{1}{2^1 \times 5^0} =$	$\frac{1}{4} = \frac{1}{2^2 \times 5^0} =$	$\frac{1}{8} = \frac{1}{2^{\square} \times 5^0} =$
$\frac{1}{5} = \frac{1}{2^0 \times 5^1} =$	$\frac{1}{10} = \frac{1}{2^1 \times 5^1} =$	$\frac{1}{20} = \frac{1}{2^2 \times 5^1} =$	$\frac{1}{40} = \frac{1}{2 \times 5^{\square}} =$
$\frac{1}{25} = \frac{1}{2^0 \times 5^{\square}} =$	$\frac{1}{50} = \frac{1}{2^1 \times 5^{\square}} =$	$\frac{1}{100} = \frac{1}{2^2 \times 5^2} =$	$\frac{1}{200} = \frac{1}{2^3 \times 5^{\square}} =$
$\frac{1}{125} = \frac{1}{2^0 \times 5^3} =$	$\frac{1}{250} = \frac{1}{2^{\square} \times 5^3} =$	$\frac{1}{500} = \frac{1}{2^{\square} \times 5^3} =$	$\frac{1}{1000} = \frac{1}{2^{\square} \times 5^3} =$

   a. Write in the decimal equivalent for each fraction in the space provided, and fill in the missing exponents. Then look for patterns in the table.
   b. Find a pattern relating the number of decimal places in each decimal equivalent to the exponents on the factors of 2 and 5 in the denominator. Explain this pattern in a sentence or two.

5. If a fraction has an infinite decimal equivalent, must that decimal repeat the same pattern of digits forever? If so, how long is the repeating string of digits? To investigate this, consider in detail the process of expressing $\frac{1}{7}$ as a decimal.

```
                                          .1428571
subproblem 1:  10÷7  ——————————————   7 ) 1.0000000
                                            7
subproblem 2:  30÷7  ————————————————       30
                                            28
subproblem 3:  20÷7  —————————————————       20
                                             14
subproblem 4:  60÷7  ——————————————————       60
                                              56
subproblem 5:  40÷7  ———————————————————       40
                                               35
subproblem 6:  50÷7  ————————————————————       50
                                                49
subproblem 7:  10÷7  —————————————————————       10
                                                  7
```

Each subproblem comes from dividing 7 into a number whose last digit is 0 and whose first digit is the remainder from the previous step. When dividing by 7, the only possible remainders are 0, 1, 2, 3, 4, 5, and 6. Of course, a remainder of 0 would end the process and make the decimal finite. However, we know from question 2 above that this decimal does not end, so there are only 6 possible remainders, and they lead to the 6 subproblems above. Once a subproblem is encountered for a second time, the process recycles, as does the sequence of digits in the decimal.

This reasoning guarantees that the repeating sequence of digits in the decimal for (fill in the blanks)

a. $\frac{1}{3}$ is at most ___ digits long

b. $\frac{1}{11}$ is at most ___ digits long

c. $\frac{1}{29}$ is at most ___ digits long

d. $\frac{1}{107}$ is at most ___ digits long

e. $\frac{1}{n}$ is at most ___ digits long

6. According to question 5, the decimal for $\frac{1}{3}$ has a repeating pattern no longer than 2 places, but, in fact, the period of $0.\overline{3}$ is just one place long. Again, the result from question 5 guarantees that the period for $\frac{1}{9}$ is no more than 8 places long, but $\frac{1}{9} = 0.\overline{1}$ has a period of just one place.

   Evidently, infinite decimals may have periods shorter than the maximum guaranteed by question 5.

   a. Using your results from questions 1 and 5, fill in the table below to compare the actual length of the repeating cycle of digits with the maximum guaranteed in question 5.

n	3	6	7	9	11	12	13	14	15	17	18	19	21	22
Maximum length of period of $1/n$ by problen 5														
Actual length of period of $1/n$														

b. Based on the results in your table, how does the actual length of the period compare to the maximum length guaranteed by problem 5?

7. In cases where the denominator n is the product of two primes (other than 2 or 5), so that $n=pq$, where p and q are primes, Wallis observed that the period of $\frac{1}{n}$ is the least common multiple of the periods of $\frac{1}{p}$ and $\frac{1}{q}$.
   a. Check this observation when $n=21$, using the periods for $\frac{1}{3}$, $\frac{1}{7}$, and $\frac{1}{21}$ found in problem 1.
   b. Check Wallis' observation when $n=77$. Use the periods for $\frac{1}{7}$ and $\frac{1}{11}$ found from problem 1.

8. Problem 5 probably convinced you that **if** a fraction $\frac{1}{n}$ has an infinite decimal form, it involves cyclic repetition of at most n-1 digits. However, you may have noticed that in certain cases this repeating pattern does not begin right away. For example, the decimal for $\frac{1}{12}$ begins its endless repetition of 3s only after two preliminary digits (0 and 8), and the decimal for $\frac{1}{15}$ has an initial 0 before settling into its endless sequence of 6s.
   a. List at least four other examples like this (consult your answers from previous questions).
   b. Find and describe a rule relating the prime factorization of n to the number of digits in the decimal for n **before** the infinite repetitions begin. (Hint: This pattern is closely related to the one in question 4 above.)

9. Let us summarize what we know so far about the decimal for $\frac{1}{n}$.
   Case 1. If n has no prime factors other than 2 or 5, so we can write $n=2^r5^s$, then $\frac{1}{n}$ has a decimal form which ends after a number of digits equal to the larger of r and s. (Note: Either r or s may be 0.)

   Case 2. If n has no factors of either 2 or 5, n has a decimal form which is infinite and consists of endless repetitions of a sequence of, at most, n-1 digits, beginning immediately after the decimal point. The actual number of digits may be less than n-1.

   Case 3. If n has factors of 2 or 5 as well as other primes, we can write $n=2^r5^sm$, where m is not a multiple of either 2 or 5.

In this case, the decimal for n consists of a number of non-repeating digits (equal to the larger of r and s) followed by endless repetitions of a pattern of, at most, m-1 digits.

Is there a more precise rule for the number of repeating digits in cases 2 and 3? Yes, but the rule is not easy to guess.[2] If you write $n=2^r5^sm$, where m is not a multiple of either 2 or 5, with the understanding that in case 2, $r=s=0$, then the length of the periodic cycle of digits in the decimal for $\frac{1}{n}$ depends on m. The rule itself is found in problem 10 below.

10. Here is a rule for the length of the period of a repeating decimal. The period is the smallest exponent t for which 10^t is one more than a multiple of m. For example, for $\frac{1}{56}$, we have $56= 2^3 5^0 7$. The larger of the exponents (3) on the 2 and 5 is the number of places before the periodic behavior begins, and the fact that 10^6 is the smallest power of 10 which is one more than a multiple of 7 indicates that 6 is the length of the repeating cycle of digits.
    a. Check this by computing the first 10 digits of the decimal for $\frac{1}{56}$.
    b. Divide 10^1, 10^2, 10^3, 10^4, 10^5, and 10^6 by 7 to show that, indeed, 10^6 is the smallest power of 10 which is 1 more than a multiple of 7.

11. The repeating decimal for $\frac{1}{7}$ is six digits long. If you treat the first three digits as the number 142 and add it to the three-digit number formed from the last three digits, what do you get? Check for similar results with regard to:
    a. $\frac{1}{13}$ b. $\frac{1}{17}$ c. $\frac{1}{19}$ d. $\frac{1}{23}$

12. You know how to express a fraction as a decimal, but can you go the other way? If you start with a decimal, can you find an equivalent fraction? That is not hard to do in the finite case; here is how to find a fraction equal to a given infinite repeating decimal. To illustrate, look at 0.373737. . ..
    1. Assign a letter to stand for the number. We use N.
    2. Observe the length of the cycle of repeating digits in the decimal. In this case it is 2 digits long.
    3. Multiply the equation N=0.373737. . . by 10^2 (the exponent is the length of the cycle) to get

[2] Albert Beiler, in *Recreations in the Theory of Numbers: the Queen of Mathematics Entertains* (Dover, NY, 1964) gives some history of the exploration of patterns of decimal fractions. In 1823 Henry Goodwin published a table 107 pages long giving the decimal equivalent of every fraction with a denominator that has no factors of 2 or 5 and is less than 1,024. Later in the 19th century, William Shanks calculated the length of the period of every repeating decimal $1/p$, with p a prime number less than 120,000. Shanks' tables for p up to 3,000 are in the *Proceedings of the Royal Society of London* for 1873. The rest of his tables, handwritten and never published, remain in the archives of the Royal Society today.

$$\begin{array}{ll} & 100N = 37.373737 \\ \text{Subtract} & \underline{\quad N = .3737373737} \\ & 99N = 37 \\ \text{Solve for } N & N = \dfrac{37}{99} \end{array}$$

As you can check, the decimal for $\frac{37}{99}$ is indeed 0.37373737. . . .
Apply this process to find fractions whose decimals are:

a. 0.22222. . .

b. 0.451451451. . .

c. 0.999999. . .(Does this answer surprise you? Compare with section 8.2.)

d. 0.78343434 . . . (repeats 34 forever).

8.4 Percent

The word **percent**, denoted by the sign %, brings rates (e.g., interest rates) to mind, and you can indeed think of percents as rates. We will explore rates in detail in chapter 10. In percent, **per** means "by," as in "divided by," and **cent** is simply Latin for 100. Percent involves dividing by 100.

Which is greater, $\frac{5}{16}$ or $\frac{16}{50}$? If the fractions had the same denominator, we could compare them easily. The idea behind percent is to use 100 as the denominator for all fractions. For $\frac{16}{50}$ that is easy to do $\left(\frac{16}{50} = \frac{32}{100}\right)$, but it is a bit more awkward with $\frac{5}{16}$, which is $\frac{31\frac{1}{4}}{100}$. In any case, the two are now easy to compare, since, in the language of percents, $\frac{16}{50} = 32\%$, and $\frac{5}{16} = 31\frac{1}{4}\%$.

A number before a percent sign is a number of hundredths.

Thus, 5% means $\frac{5}{100}$; likewise, 117%= $\frac{117}{100}$, and so on.

To express a number as a percent, express it as a fraction with 100 as denominator. The numerator is the percent.

Some numbers, like $\frac{1}{2}$, are easy to express as a percent. Mentally, unreduce $\frac{1}{2}$ to $\frac{50}{100}$, showing right away that $\frac{1}{2}$=50%. This is pictured by shading half of one hundred squares, or 50 of them.

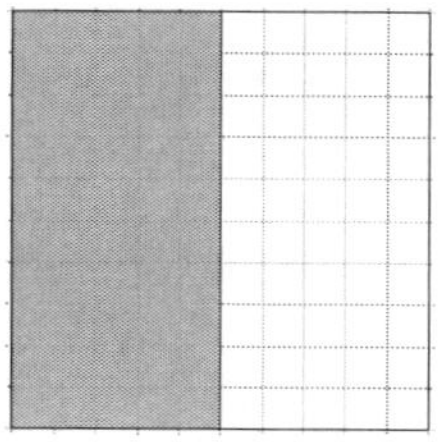

Other numbers, like $\frac{1}{3}$, are more awkward to unreduce; 3 is not a factor of 100, but if you notice that $3 \times 33\frac{1}{3} = 100$, you can see that $\frac{1}{3} = 33\frac{1}{3}\%$. In effect, possibly without knowing it, you solved the equation $\frac{1}{3} = \frac{?}{100}$ mentally. (You multiplied both sides of the equation by 100 to get $100 \times \frac{1}{3} = \frac{?}{100} \times 100$, then canceled 100 on the right to get $\frac{100}{3} = ?$ or $? = 33\frac{1}{3}$.) Again, you can picture this by shading one-third of the 100 squares shown.

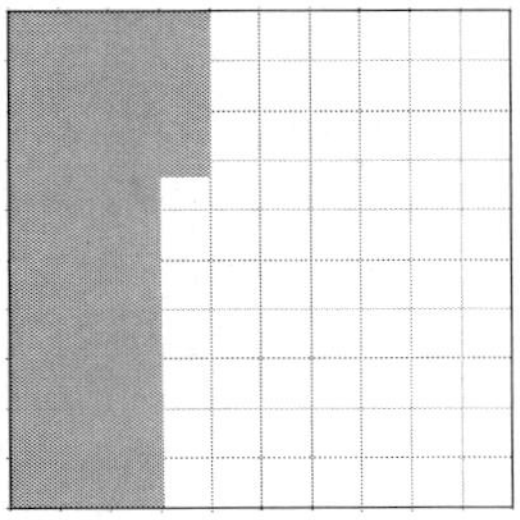

Another way to express a fraction as a percent: Express it as a decimal, then multiply the result by 100. As a decimal, $\frac{1}{3} = 0.3333...$, so $\frac{1}{3} = 33.33...\%$.

Example 1. There are 70 people in an orchestra, and 40% of them play wind instruments. How many people in the orchestra play wind instruments?

Solution: By the definition of percent, 40% is $\frac{40}{100}$ or $\frac{4}{10}$. Then 40% of 70 is $\frac{4}{10}$ of 70 or 28. We may picture this by shading 70 of the hundred squares, and then shading 40% of those squares darker. There are in all 28 darkly shaded squares.

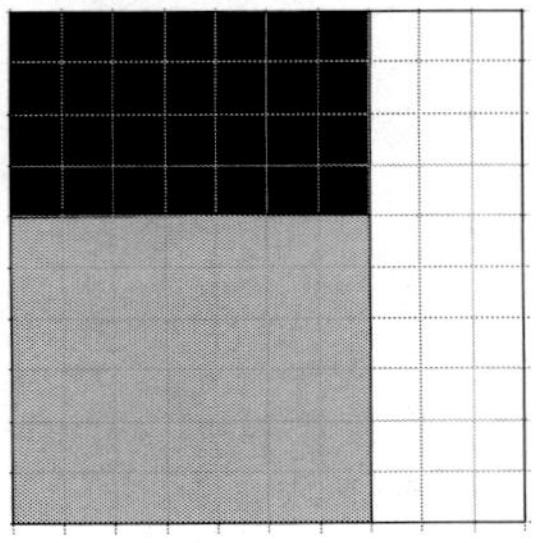

Example 2. Of the 420 Math 301A students last quarter, 40% had Professor Ahlgren. How many students had Professor Ahlgren?
Solution: Let the 420 students be represented by the whole strip below. When broken up into tenths (10%), each tenth represents 42 students. Then 40% is 4-tenths or 4×42=168 students.

42	42	42	42	42	42	42	42	42	42
10%	10%	10%	10%	10%	10%	10%	10%	10%	10%
40%									

Example 3. Janelle bought her desk at a 12 1/2 % discount. She paid \$140. How much money did she save? What was the original price of the desk?

Solution: Since $12\frac{1}{2}\% = \frac{1}{8}$, the sale price is $\frac{7}{8}$ of the original price.

Since 140÷7=20, each of the 7 eighths of the sale price is \$20, and 1/8 of the sale price is \$20. Therefore Janelle saved \$20, and the original price of the desk was \$160.

	\$20	\$20	\$20	\$20	\$20	\$20	\$20
Sale price (\$140 is $\frac{7}{8}$ of whole)							$\frac{1}{8}$
Original price							

Example 4. A box of crackers advertised that the reduced fat variety has 30% less fat per serving than the regular variety. If the reduced fat crackers have 3.5 grams of fat per serving, how many grams of fat are in a serving of regular crackers?
Solution: The reduced fat variety has 70% fat of the original variety. So each of the 7 equal parts (10% of the whole) for the reduced fat crackers has 3.5÷7=0.5 grams of fat. Therefore, the regular variety has 1.5 grams more fat or 5 grams of fat per serving.

0.5	0.5	0.5	0.5	0.5	0.5	0.5			
Reduced fat variety (3.5 grams) -70%							30%		
Regular variety (5 grams)									

Example 5. In an election, Bob won with 1,966 of the 5,314 votes. What percent of the vote did he get?

Solution: Express $\frac{1,966}{5,314}$ as a decimal fraction, 0.3699661272.... To express this as a percent, multiply by 100, getting $\frac{1,966}{5,314}$ =36.99661272...%. In this context, it makes sense to round to the nearest integer and say that Bob got 37% of the votes.

Example 6. A certain cheese contains 63% fat, according to the label. Teresa is concerned about eating too much fat but likes this cheese. How much of it can she eat without exceeding 50 grams of fat?
Solution: Let y be the number of grams of cheese that contains 50 grams of fat. Then 63% of y is 50 grams, or $0.63y=50$. This is the missing factor definition of division; $y=50\div0.63$, which is 79.36. Teresa should limit herself to 79 grams of this cheese.

Example 7. Jack makes \$7.85 per hour. What will he make after a 6.2% pay increase?
Solution: The increase will be 6.2% of \$7.85, or \$0.062×7.85. This is \$0.4867, or 48.67 cents an hour. Round this to the nearest cent and add the increase to what Jack already makes. He will get \$8.34 per hour.
Second solution: Jack will make 106.2% of his current pay, or \$1.062×7.85, which is, again, \$8.34 per hour.

Estimating percent visually
What percent of the square in figure 8.4.1 is shaded?

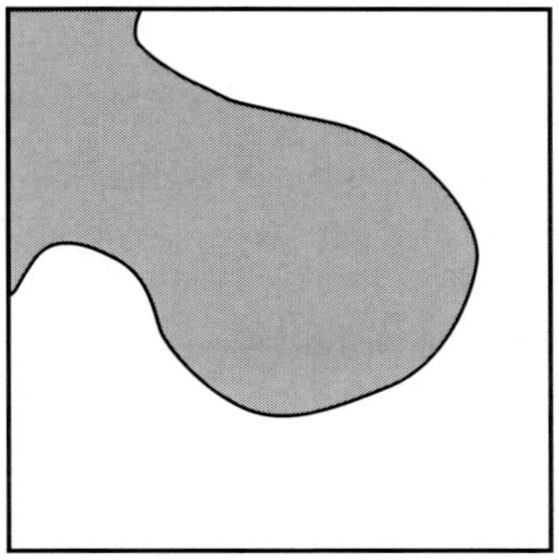

Figure 8.4.1

This is not easy to answer accurately, but trying to estimate it and then checking the answer builds intuition and confidence in using percents. After you have made a careful estimate, check it by superimposing a grid of 100 equal pieces, each of which is, of course, 1% of the whole. The result, in figure 8.4.2, allows a more accurate estimate.

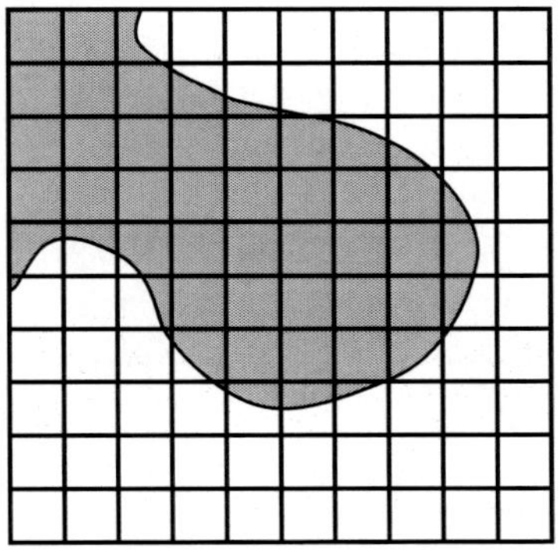

Figure 8.4.2

Counting shows 32 little squares are completely shaded, and 47 are not shaded at all, so between 32% and 53% of the area is shaded. An estimate of 42.5% would be reasonable, since it is at most 10.5% off. Estimating how much of each partially shaded square is shaded might sharpen this estimate.

Percents and mental arithmetic
It is often useful to be able to calculate percentages mentally, at least approximately. For example, Marge wants to leave a 15% tip on a restaurant bill of $32.45. To estimate this quickly, she takes 10% of $32.45, which is about $3.25, and adds half of that or about $1.63, which is another 5%. That would make a $4.88 tip, which she might round to a convenient number, like $5.00.

Mental calculations like the above are based largely on combining some easily calculated percentages like these:

50%=half
25%=one fourth (often found as half of half)
10%=one tenth, found by moving the decimal point one place to the left
5%=half of 10%
1%=one hundredth, found by moving the decimal point two places to the left, or by taking a tenth of 10%

When you need only an approximate answer, make life easy for yourself by using numbers you find easy to work with. To estimate 15% of $69.23, for example, just find 15% of $70.

Example 8. Chris bought a gift for $47.00 and was told there would be an additional charge of 7% for tax. How much tax would he pay?
Solution: First take 10% of $47.00, which is $4.70. Half of that, or $2.35, is 5% of $47. Now add 1%, or $0.47, twice to find the total tax of $2.35+$0.94=$3.29.

The methods used for mental calculation are often different from those used with pencil and paper. Still, if you try it, after a while you will be pleased to see your skill and confidence at mental arithmetic grow.

Exercises 8.4

1. Shade 75% of the circle shown.

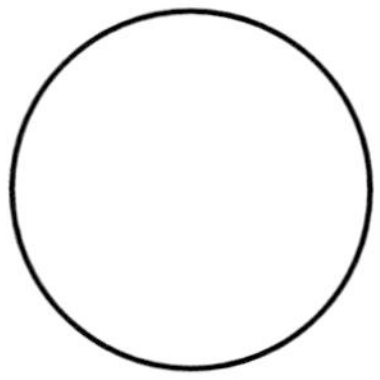

2. If this rectangle is 60% of something, sketch the whole thing.

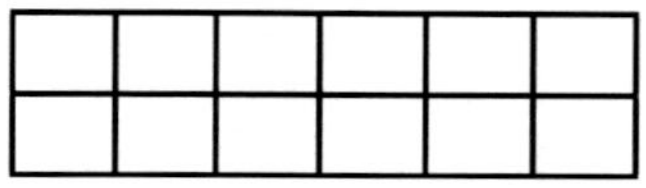

3. Of the 30 students in class, 60% wore tennis shoes. How many of the students wore tennis shoes?

4. A bank pays 4.53% interest for a year on a deposit of $8,000. How many dollars interest does the bank pay for a year on that deposit?

5. Three students did not bring their books to class. They represented 10% of the class. How many students were in the class?

6. What is a realtor's 6% commission on a $350,000 sale?

7. The school board election finished as follows:

candidate	**votes**
Jane St. John	1957
Molly Gallacher	1751
George Stehmeier	1166
Paul Muscorella	862

What percent of the vote went to Jane St. John?

8. Estimate, mentally, a 15% tip on a restaurant tab of:
   a. $16 b. $29.50 c. $63.47

9. A customer paid $3.28 per gallon for gasoline, and the sign at the pump indicated that the price includes 72 cents tax.
   a. What percent of the total cost is tax?
   b. What percent of the pretax price is the tax?

10. a. A store announced a 10% price reduction on some merchandise. When this didn't stimulate sales enough, prices were cut an additional 15%. By what percentage were prices reduced overall? Hint: The answer is not 25%; the 15% discount was applied to an already lowered price.
    b. Another store made the same price reductions in the opposite order, first cutting prices by 15% and then taking off 10% of the already reduced prices. By what percent did this store reduce prices overall?
    c. Compare your answers from parts a and b. What mathematical principle does this illustrate?

11. Every year, a mutual fund reports the increase or decrease in value of each share in the fund. The year after Martha bought some shares, they rose 50% in value. The next year, they fell 30%. What was the percent change in value for the two-year period?

12. A tire sale offers, "Buy three at the regular price, get the fourth tire for 50% off." What percent discount does this represent to a customer who buys a set of four tires?

13. Gold jewelry made in the United States is required to have the purity of the gold stamped on it as a number of carats: 24 carat gold is 100% pure gold, so that 50% gold is 12 carat. What percent gold is in jewelry that is:
    a. 18 carat gold b. 14 carat gold

14. According to the U. S. Census Report, based on data from the 2000 census, the US population in the year 2000 was 281.1 million, making the US the third largest national population, behind China and India, with populations estimated at 1.3 billion and 1 billion respectively. The total world population was estimated to be 6.1 billion.
    a. What percent of the world's population lived in China in the year 2000?
    b. What percent of the world's population lived in India in the year 2000?
    c. What percent of the world's population lived in the USA in the year 2000?

15. What percentage of the electric energy used by a 60 watt incandescent light bulb will be saved by switching to a 13 watt compact fluorescent bulb that puts out equivalent light?

16. Nils and Lin want to buy a house for $182,000. They want to make a down payment of 21% of the price. How much is their down payment?

17. A gold mine estimates its unmined ore to contain 4.0 grams of gold per ton of ore. By that estimate, what percent of the ore is gold?

18. In a ton of "super pure" aluminum (guaranteed to be at least 99.999% pure), what is the maximum weight of impurities?

19. The recommended daily dosage of vitamin D for adults was recently raised to 1000 international units, but doctors sometimes prescribe higher dosages on a temporary basis. What percent of the recommended daily dose is 50,000 international units?

20. How much water must be added to one liter containing 30% alcohol to get a mixture containing 17% alcohol?

21. The cooling system of a car holds 11 liters of a coolant that is 25% antifreeze. How much should be drained off and replaced by pure antifreeze to make the coolant 40% antifreeze?

22. A team's "winning percentage" is defined as the fraction of all the games it has played which it won, expressed as a decimal carried to three places. A team which has won half its games, or 50% of them, thus has a winning percentage of .500. (In colloquial usage, the decimal point is often ignored, and the team is said to be playing at the five hundred level. Also, in printing winning percentages, it is customary to drop the initial 0, writing .500, not 0.500.)
    a. What is the winning percentage of a team that wins 98 games and loses 70?
    b. The 1927 New York Yankees, playing the 154 game schedule that was standard for many years, had a .714 winning percentage. How many games did they win?

23. **A paradox**. A baseball player's **batting average** is defined as the percentage of times at bat in which the player makes a hit. Batting averages are typically expressed to three places, so that, for example, a player with 150 hits in 500 times at bat is said to have a batting average of .300 (an enviable figure in most circumstances). Because of the way curve balls curve, most batters fare best against pitchers of opposite handedness, and statistics on this are kept.

a. Larry has 11 hits in 26 times at bat against right-handed pitchers. What is his batting average against right-handers?

b. Larry has 2 hits in 10 times at bat against left-handers. What is his batting average against them?

c. Mo has 4 hits in 9 times at bat against right-handers. What is his batting average against them?

d. Mo has 5 hits in 22 times at bat against left-handers. What is his batting average against them?

e. Who has the better batting average against left-handers?

f. Who has the better batting average against right-handers?

g. On the basis of this evidence, whom do you think of as the stronger hitter?

h. Compute the overall (against both kinds of pitchers) batting average of each player.

i. Do the answers from parts g and h surprise you? They show you need to be a bit careful with percentages.

Vocabulary for Chapter 8

Decimal point (8.1)
Fractional part (8.1)
Integer part (8.1)
Irrational numbers (8.2)
Percent (8.4)

Self-Quiz for Chapter 8

1. Mark, as accurately as possible, the point 1.8816 on the number line shown.

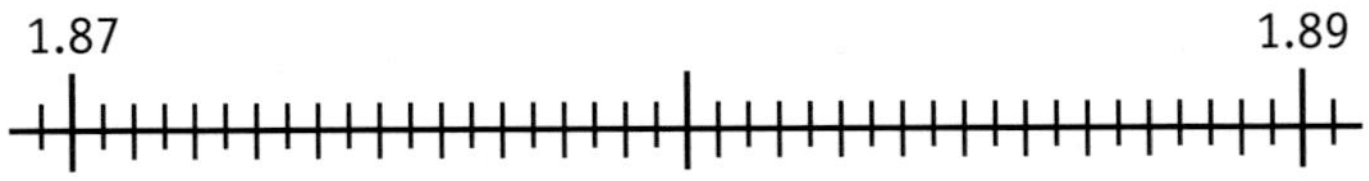

2. Express the decimal fraction 0.0723 as an ordinary fraction (with integer numerator and denominator). Do not reduce the answer.

3. Express $\frac{11}{21}$ as a decimal, carrying the work far enough to find and clearly identify the repeating pattern of digits.

4. Complete this box puzzle.

	-0.2		
0.03		0.9	
			59.6

5. An item that lists for $28.35 is on sale at a 20% discount. How much must you pay in all, if you must also pay a tax of 7.75% of the sale price?

6. A copier reduced a document to 78% of its original size. By what percent must the reduced copy be enlarged to be the same size as the original document?

Chapter Nine
Rules and Graphs

Relationships between numerical quantities are often more important and interesting than numbers themselves. This chapter introduces the study of such relationships, expressed as rules or, more formally, as functions. We will picture these relationships by means of graphs in the coordinate plane.

9.1 The Coordinate Plane

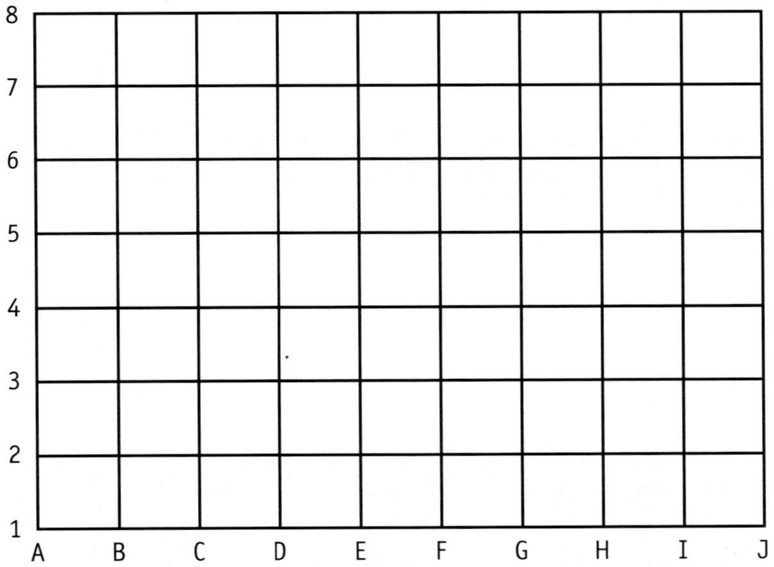

Figure 9.1.1

If you can find the point B5 in figure 9.1.1, you can play "battleship," a game for children as young as first grade. Each of the two sides writes down the coordinates of the points where it places its ships. The teams take turns guessing where each other's ships are. After each guess, the guessing team is told whether or not it has "hit" (correctly guessed a point on) an opponent's ship. The game introduces coordinates and also helps children learn letters and numerals.

For more sophisticated applications, we specify points with respect to a pair of perpendicular number lines called **axes** (singular: **axis**) which cross at their 0 points (figure 9.1.2). This crossing point, which has coordinates (0,0), is called the **origin**. The horizontal and vertical axes, sometimes called the ***x*-axis** and ***y*-axis**, respectively, divide the plane into four **quadrants** (figure 9.1.2). With reference to given axes, each point is specified by a pair of numbers called **coordinates**, which are written in parentheses, separated by a comma.[1]

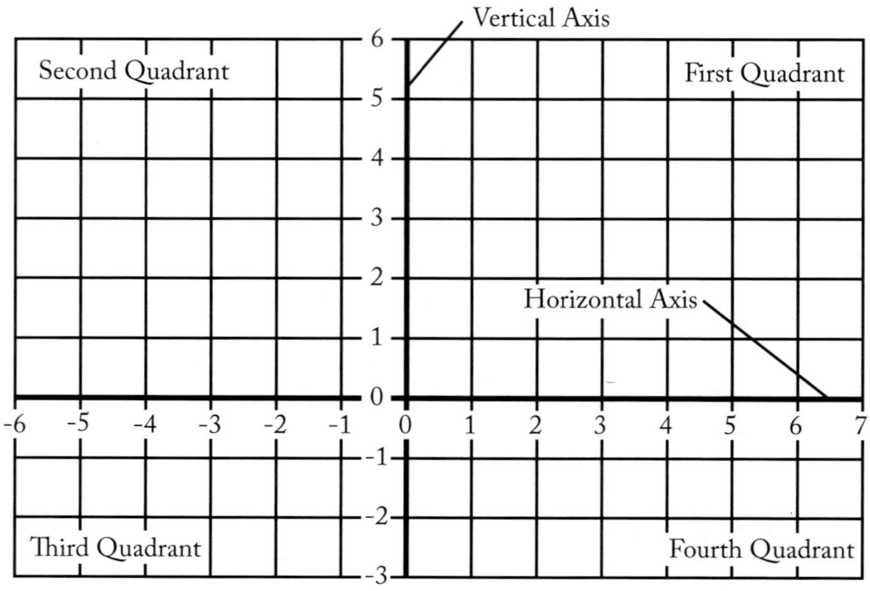

Figure 9.1.2

To locate or, as we say, **plot** a point, start where the axes cross (which is why that point is called the origin), and move horizontally to the first coordinate. Then, move vertically to the height of the second coordinate. For example, to plot (-2,3), start at the origin, move horizontally to -2, and then go up 3 units, as in figure 9.1.3.

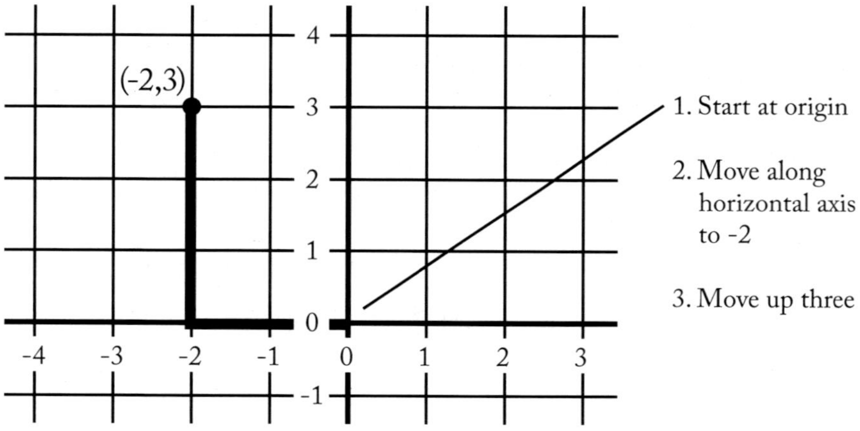

Figure 9.1.3

The fact that the first coordinate determines how to travel horizontally and the second vertically is mere convention. It could as well be the other way around. Like any convention, such as driving on the right side of the road, it is arbitrary but much better than having no convention at all.

[1] Coordinates were used implicitly by some ancient Greek mathematicians, but Descartes' *La Geometric* (1637) gave them a central role in mathematics. Referring to that pioneering work, the coordinate plane is sometimes called the Cartesian plane.

When students first draw and label coordinate axes, be sure that they number the vertical axis so that the lines split the numbers. Otherwise, they may want to make the numbers sit on the lines, as they are taught when they learn handwriting. The result, which looks something like the left hand part of figure 9.1.4, may mislead students to think they are numbering the spaces between the lines instead of the lines themselves.

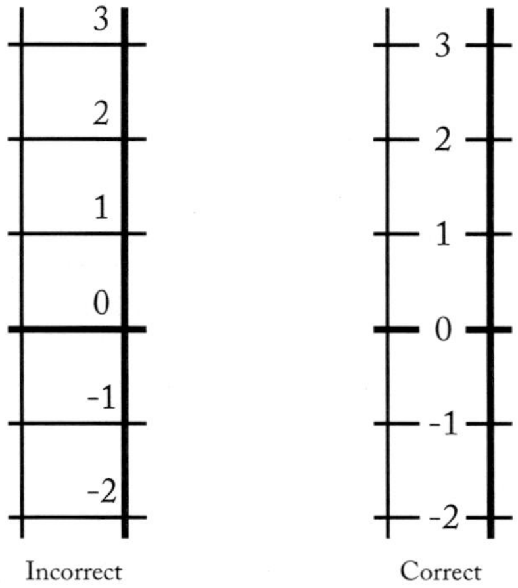

Figure 9.1.4 How to label the vertical axis.

Likewise, vertical grid lines should slice through the numbers on the horizontal axis as they do in figure 9.1.3.

Coordinates in higher dimensions

We have identified a single number with a point on a number line (one dimension), and we have identified ordered pairs of numbers with points in a coordinate plane (two dimensions). Similarly, ordered triples of numbers like (3,-2,7) can be regarded as points in three-dimensional space. In more advanced work, this is extended to even higher dimensions, where, for example, an ordered 6-tuple of numbers like (6,0,9,-4,-7,12) is thought of as a point in six-dimensional space. Even though nobody can visualize more than three dimensions physically, higher dimensional spaces are of immense value in solving very practical problems. Once again, elementary mathematics is the foundation that supports later work.

Exercises 9.1

1. a. Draw and number coordinate axes on squared paper.

   b. On your paper from part a, put dots at each of these points:

   $(-2,0), (-2,1), (-2,-2), (-2,2), (-2,-1), (-1,0), (0,0), (1,0),$

   $(1,2), (1,1), (1,-1), (1,-2), (3,2), (3,1), (3,0), (3,-1), (3,-2).$

   c. Find a coded message in the result of part b.

   The exercise in parts a-c is a good introduction to plotting points. It is particularly convenient for the teacher, because it is easy to see at a glance if all the points are correctly plotted. Students who consistently interchange the horizontal and vertical coordinates (a common error, no matter what the teacher does) will have the message reading backwards, from right to left, along the vertical axis. If only two or three points are plotted incorrectly, that, too, is easy to see.

2. Make up a short message and encode it as follows:
   a. Write the message, forming the letters from dots entered on a coordinate plane.
   b. Write down the coordinates of the dots from part a.

3. Plot these points in the given order, connecting each to the next by a straight line.

$$\left(\frac{1}{2},-3\right),\ \left(\frac{1}{2},-\frac{1}{2}\right),\ \left(5\frac{1}{2},-2\right),\ \left(1,1\frac{1}{2}\right),\ (4,0),\ \left(1,\ 2\frac{1}{2}\right),\ (3,2),$$

$$(1,4),\ \left(2,3\frac{1}{2}\right),\ (0,6),\ \left(-2,3\frac{1}{2}\right),\ (-1,4),\ (-3,2),\ \left(-1,2\frac{1}{2}\right),\ (-4,0),$$

$$\left(-1,1\frac{1}{2}\right),\ \left(-5\frac{1}{2},-2\right),\ \left(-\frac{1}{2},-\frac{1}{2}\right),\ \left(-\frac{1}{2},-3\right),\ \left(\frac{1}{2},-3\right).$$

4. In parts a-f, plot each pair of points and draw the line segment between them. Then, visually, determine the coordinates of the midpoint of that line segment and write them next to the coordinates of the endpoints.
   a. (2,3) and (4,6)
   b. (4,1) and (0,3)
   c. (5,8) and (11,4)
   d. (-4,-6) and (2,8)
   e. (-10,-1) and (-4,3)
   f. (3,1) and (4,3)
   g. How are the coordinates of the midpoints related to the coordinates of the endpoints? State this relationship in a carefully written sentence or two.

9.2 Functions or Rules

One of the most interesting and important uses of mathematics is to describe how quantities depend on each other. Here, informally, we introduce the mathematical relationships called **functions**. We begin with a game.

Guess the Rule Someone thinks of a rule, and the others try to guess it, something like this.

Jean: I'm thinking of a rule. Ask me a number.

Mike: 5

Jean: 11. Ask me another number.

Carlos: 9

Jean: 19

Lucy: I think I know your rule. Ask me a number.

Jean: OK, try 10.

Lucy: 21

Jean: You've got my rule.

(Other hands go up as more kids think they know the rule.)

Ben: I think I know.

Jean: Try 15.

Ben: 30

Jean: Not quite. Does anyone know?

Monique: 31?

Jean: You've got it.

(Now the teacher might invite Lucy or Monique to replace Jean as leader.)

Those who think they know the rule may not explain it; that would spoil everyone else's chance to discover the rule for themselves. This way their answers give everyone else a little more information to work with and may help them figure out the rule. The game continues until everyone guesses the rule.

Well, did **you** guess Jean's rule? Whatever number Jean was given, she multiplied by 2, then added 1. In the shorthand of algebra, Jean's rule is written $\mathbf{2}\boldsymbol{n}\mathbf{+1}$, meaning "twice the number, plus 1." This expression is not itself a number, but if n is given a particular value, such as 10, then $2n+1$ has a corresponding value 21, which the rule associates with 10. Here the letter n has no particular significance; any other letter, even from a different alphabet, or, for that matter, any symbol at all, would do. Thus $2n+1$, $2N+1$, $2a+1$, $2\theta+1$, and even 2☺ +1 all name the same rule.

Guessing the rule is an open-ended activity that can be played by an entire class, by a small group, or by a family riding in a car or bus. The rules themselves can be very simple or very complicated, depending on the sophistication of the participants. For first graders, just adding two might be a complicated enough rule, whereas for sixth graders, the rule might be something like "Square whatever number you are given, then add 7."

Rules are an informal introduction to what mathematicians call **functions**, which express relationships between quantities.

Example 1. This table relates children's sock sizes to shoe sizes.

Shoe Size	0 to $\frac{1}{2}$	1 to $2\frac{1}{2}$	3 to $3\frac{1}{2}$	4 to $5\frac{1}{2}$	6 to 7	$7\frac{1}{2}$ to 9
Sock Size	4	$4\frac{1}{2}$	5	$5\frac{1}{2}$	6	$6\frac{1}{2}$

For any shoe size from 0 to 9, this table names a corresponding sock size. We say sock size is a **function** of shoe size.

Example 2. The cost of developing a new car is estimated at $4.06 billion. (This includes market research, engineering, construction and testing of prototypes, design of machines to produce body parts, factory retooling, etc.). The actual costs of building the car (labor, factory overhead, materials and finished parts bought from other firms, etc.) are estimated to average about $6,700 per car. The total cost per car depends on how many are made. We say production cost is a **function** of how many cars are built.

Example 3. A dense ball, such as a marble, rolling down an inclined plane picks up speed, as Galileo showed by a simple experiment.[2] He put small metal bars across a board, somewhat like frets on a guitar fingerboard, so that he could hear a click each time the ball hit a bar. He observed that if the bars were placed at regular intervals, the clicks came faster and faster, showing that the ball sped up. He next varied the location of the bars, trying to make the clicks come at regular time intervals. He found this happens when the bars are spaced as in figure 9.2.1.

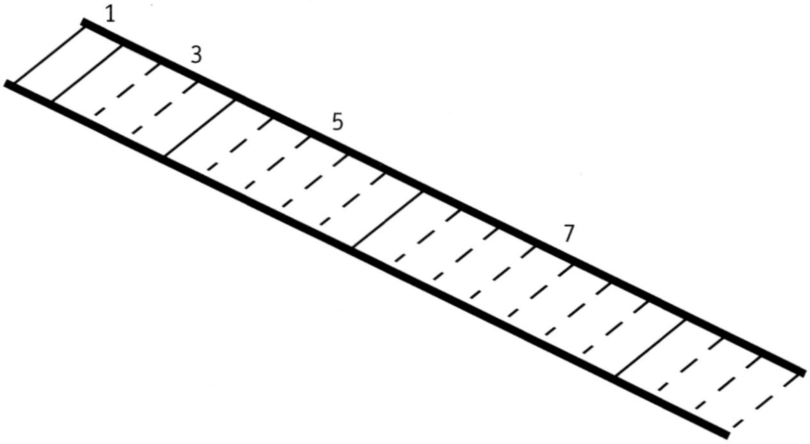

Figure 9.2.1

Since

$$1 = 1$$
$$1 + 3 = 4$$
$$1 + 3 + 5 = 9$$
$$1 + 3 + 5 + 7 = 16$$
$$1 + 3 + 5 + 7 + 9 = 25$$

and the sums on the right are perfect squares, Galileo concluded that the distance rolled is proportional to the square of the time, or, writing s for the distance and t for the time, $s=kt^2$. This expresses s as a **function** of t. The number k depends on how steeply the board is inclined and on the units of time and distance. Galileo's simple, but ingenious experiments required no accurate timepieces (they did not exist in his day), yet led to the discovery of a basic rule of nature. They also contributed to the feeling, rising at that time, that nature is not capricious but behaves according to laws. The search for such laws has been the focus of science ever since.

Each of these examples involves two sets of numbers; with each member of the first set the function associates a specific member of the second. If P and Q are sets, a **function** from P to Q associates a specific member of Q with each member of P. The set P is called the **domain** of the function. Figure 9.2.2 pictures the function in example 1, where P is the set of shoe sizes and Q is the set of sock sizes. Lines link each member of P with the corresponding member of Q.

[2] Galileo Galilei (1564-1642) gave up the study of medicine to go into science and mathematics. He was a popular and brilliant teacher, and he made many discoveries in physics. With a homemade telescope he discovered four of the moons of Jupiter, and watching them revolve led him to the idea that the earth moves around the sun. The Inquisition branded that idea heresy and imprisoned Galileo after he refused to say that it is false. He died blind, broken, and discredited by the church.

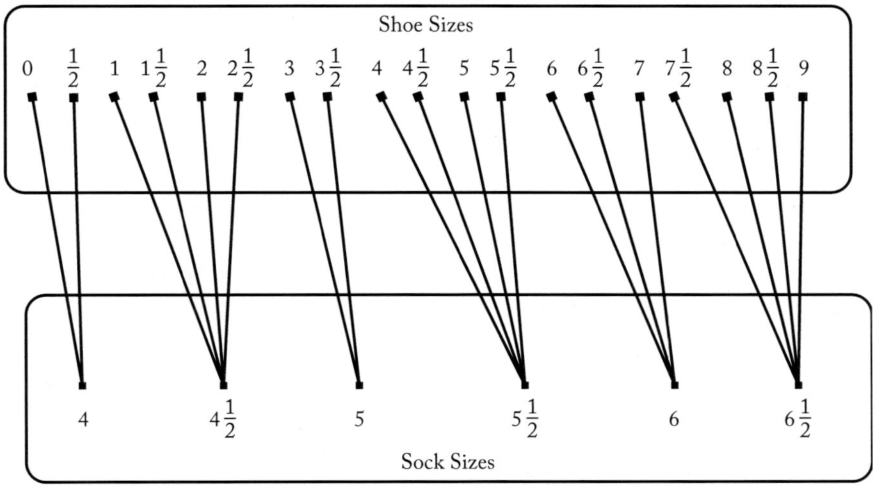

Figure 9.2.2

> **Definition of a Function**[2]
> If P and Q are sets, a **function** from P to Q is a rule which associates a member of Q with each member of P.

It common to use a letter, such as f or g to name a function. Then if x is a member of the domain P, $f(x)$ denotes the corresponding member of Q.

It sometimes helps to think of a function as a process symbolized by the machine in figure 9.2.3. If a member of P is put in at the top, the associated member of Q comes out at the bottom. The domain is the set of possible inputs.

Caution: The notation $f(x)$ can be confusing. It looks as if multiplication is intended, but that is not the case. $f(x)$ simply indicates the result of applying the rule f to x, a member of the domain.

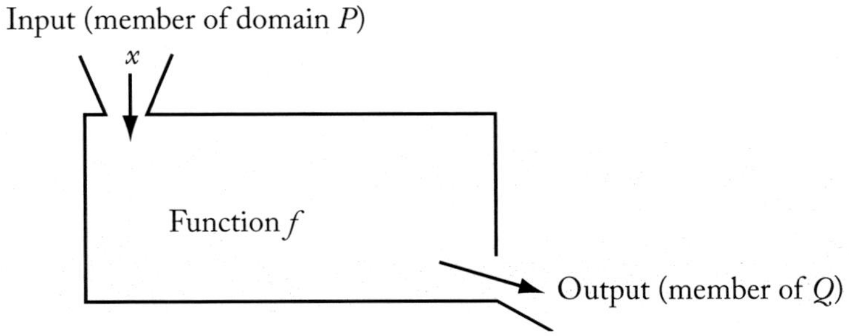

Figure 9.2.3

In our work we will not bother to identify the sets P and Q, which will usually be sets of numbers. Instead we will concentrate on the rule relating the input to the output.

[2] The word "function" was first used in mathematics by Leibniz (see note, section 1.3) but it was Euler (see note, section 6.4) who, in his *Intrdductio in Analysin Infinitorum (Introduction to the Analysis of the Infinite)* of 1748 made functions a central focus of study.

Exercises 9.2

1. Here are four rules for determining the drug dosage to be given to a child on the basis of A, the recommended adult dosage, and the child's age or size.[3]

   Clark's rule: Multiply A by the child's weight in pounds, and divide the product by 150.

   Cowling's rule: Multiply A by the child's age in years on the next birthday, and divide the product by 24.

   Dilling's rule: Multiply A by the child's age in years and divide the product by 20.

   Young's rule: Multiply A by the child's age in years and divide the product by 12 more than the child's age.

   a. For a six year-old weighing 45 pounds, what do each of these rules give for a medication with adult dosage of 100 mg?

   b. For a 10 year-old weighing 72 pounds, what do each of these rules give for a medication with adult dosage of 100 mg?

2. The computer at the science fair showed the following table of inputs and outputs.[4]

Inputs	3	7	12	10	-1
Outputs	6	14	24	20	-2

   a. Jenny said, "I know what it does. It multiplies your input by four, then divides the result by two." Does Jenny's idea work out?

Inputs	3	7	12	10	-1
Inputs × 4					
Half of line above					

   b. Kenny said, "I think it adds one to the input, doubles the result, and then subtracts two." Does Kenny's idea work out?

[3] Source: *The Arithmetic of Dosages and Solutions* by Laura K. Hard, 7th edition, 1987. C. V. Mosby & Co. St. Louis, MO.

[4] The idea for this problem comes from the *Math Workshop for Children.*

Inputs	3	7	12	10	-1
Inputs + 1					
Double line above					
Previous line minus 2					

c. Lenny said, "What it really does is add 5 to the input, multiply that by 6, subtract 9, divide the result by 3, and then subtract 7." Does Lenny's idea work out?

Inputs	3	7	12	10	-1
Inputs + 5					
Previous line times 6					
Previous line minus 9					
Previous line ÷ 3					
Previous line minus 7					

d. How would you describe the rule?

Exploration: Combining rules (questions 3-5)

3. Rule A doubles whatever number it is applied to. Rule B adds 9 to whatever number it is applied to. What if the outputs of rule A are used as inputs for rule B? Picture the rules hooked up like this.

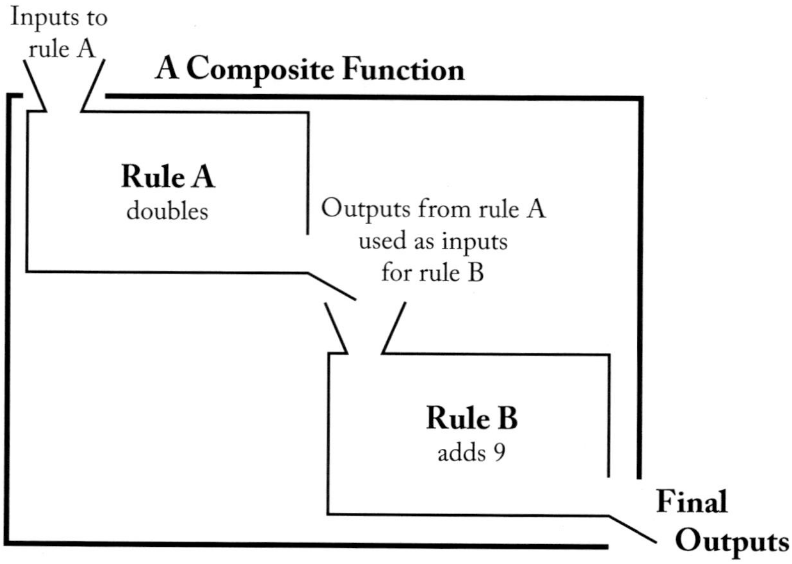

A cutaway view of a composite function

a. Trace the effect of the function pictured above on some inputs by filling in the following table. In a few cases the "table is turned," and you are given either the intermediate or final output and asked to find the other entries.

Inputs for composite rule = Inputs for Rule A	3	8	24	0		-6	
Outputs from Rule A = Inputs for Rule B	6	16			-6		
Outputs for composite rule = Outputs from Rule B	15	25					-30

b. Write an algebraic expression to describe the composite rule from part a.

4. This question is like question 3, but with rules A and B reversed.

a. Fill in the following table:

Inputs for composite rule = Inputs for Rule B	3	8	24	0		-6	
Outputs from Rule B = Inputs for Rule A	12	17			7		
Outputs for composite rule = Outputs from Rule A	24	34					-21

b. Write an algebraic expression for the composite rule in the table above.

5. a. Compare the composite rules from questions 3 and 4. Do these rules produce the same outputs for each input?

b. If a composite rule is made up by following one rule with another, does it matter which rule is applied first? Explain, using the information found above to support your answer.

6. Have you ever seen a flash of lightning and then waited to hear the thunder? Light travels about $\frac{1}{5}$ mile per second faster than sound, so you can estimate your distance in miles from the lightning by counting the number of seconds between when you see the lightning and you hear the thunder. State a rule relating the number of miles you are from the lightning to the number of seconds counted.

7. **Exploration: (Due to John Marks) Patterns formed by cutting a triangle.**

a. Lines from one corner of a triangle to the opposite side:

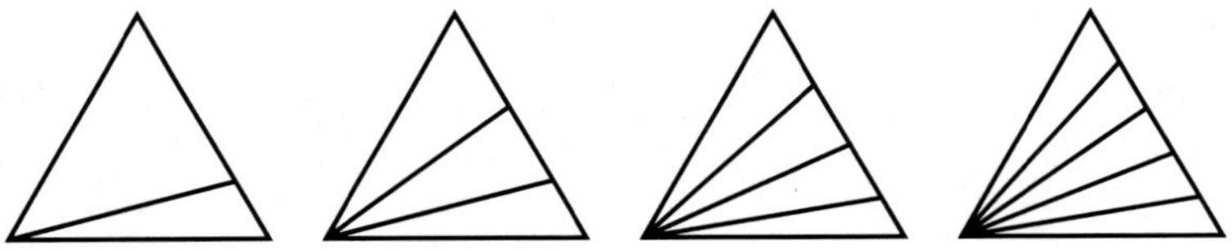

1 line	2 lines	3 lines	4 lines
__ pieces	__ pieces	__pieces	__ pieces

Guess nature's rule: n lines, all from the same corner, cut a triangle into ______ pieces.

b. What if lines are drawn from two corners?

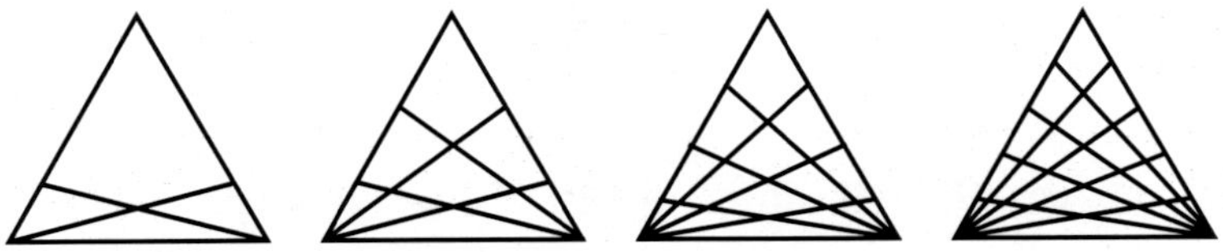

1 line	2 lines	3 lines	4 lines
__ pieces	__ pieces	__pieces	__ pieces

Guess nature's rule: n lines from each of two corners cut the triangle into ______ pieces.

c. With n lines from **1** corner (part a) you got $(n+1)^1$ pieces.

With n lines from **2** corners (part b) you got $(n+1)^2$ pieces.

Predict: With n lines from **3** corners you will get ______ pieces.

d. Count the pieces in these triangles to check your prediction from part c. Is your prediction correct?

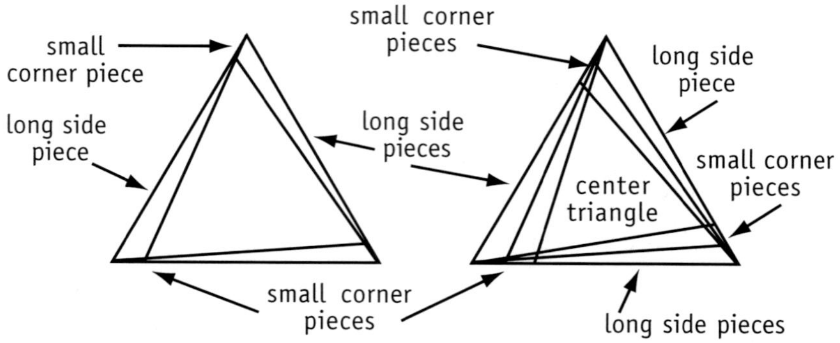

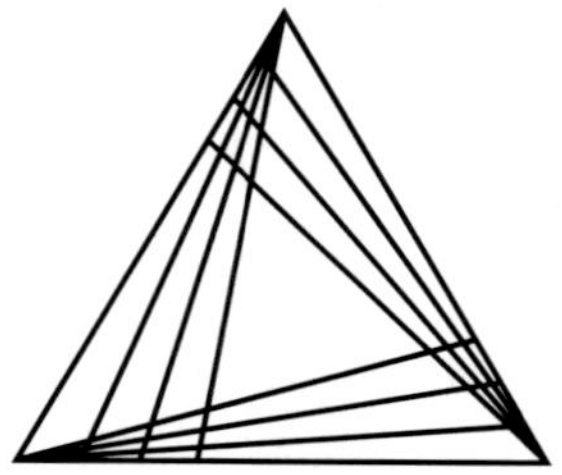

e. Referring to the pictures in part d, fill in the first four columns (headed by 0, 1, 2, and 3) of this table.

Number of lines from each corner	0	1	2	3	4
Number of small corner pieces					
Number of long thin pieces					
Number of center triangles					
Total number of pieces					

f. Guess the rules from the table. If n lines are drawn from each corner:

______ center triangles will be formed

______ long, thin pieces will be formed

______ small corner pieces will be formed

Total ____________ pieces will be formed in all

g. To test your work from part f, use it to predict the last column of the table in part e. Then count pieces in this triangle to check your predictions.

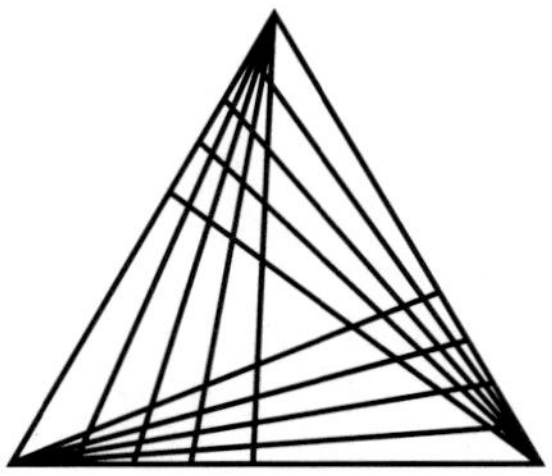

h. In part c you probably expected the number of pieces to be $(n+1)^3$, but in part d you saw that rule does not fit the data. However, the number of pieces is related to cubes after all. To see this, calculate:

$$1^3 - 0^3 =$$

$$2^3 - 1^3 =$$

$$3^3 - 2^3 =$$

$$4^3 - 3^3 =$$

$$5^3 - 4^3 =$$

i. You now have two ways to express the rule for the total number of pieces in the case of n lines from each corner. Show these two ways always produce the same number. To do this, use the fact, which can be shown with algebra, that $(n+1)^3=n^3+3n^2+3n+1$. (See question 30 of Exercises 13.3)

8. **Exploration: Diagonals**

A diagonal of a regular polygon is a line segment connecting vertices which are not adjacent. (A **regular** polygon is one with all sides equal and all angles equal. Polygons, and regular polygons in particular, are discussed in chapter 19.) The picture shows diagonals of a square and pentagon.

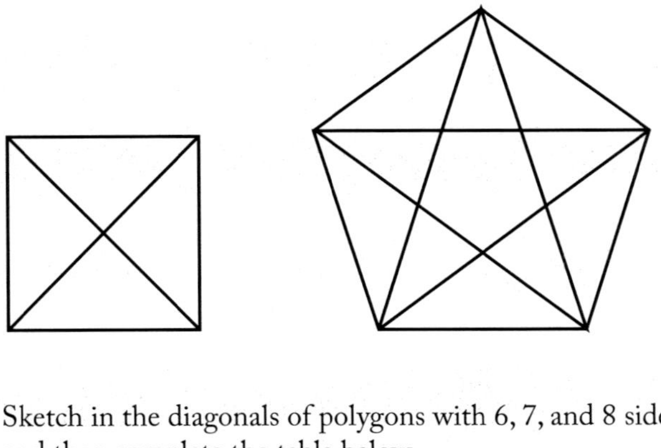

a. Sketch in the diagonals of polygons with 6, 7, and 8 sides, and then complete the table below.

Number of sides	3	4	5	6	7	8
Number of diagonals						

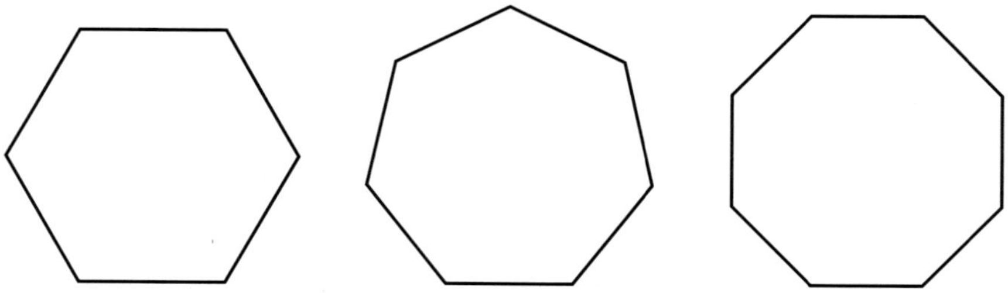

b. Find a rule for how many diagonals a regular polygon has if you know how many sides it has. To get started, calculate:

$0 \times 3 =$

$1 \times 4 =$

$2 \times 5 =$

$3 \times 6 =$

$4 \times 7 =$

Compare these products to the numbers of diagonals you observed in part a, and you can guess nature's rule for the number of diagonals.

c. To test your guess from part b, predict how many diagonals there are in a regular polygon with 10 sides. Then sketch a regular polygon with 10 sides and draw in all the diagonals, counting them as you go, to see if your prediction was correct.

d. Try to explain your pattern. Hints: If there are 5 sides, how many diagonals meet at each corner? What if there are 6 sides? In general, how does the number of diagonals meeting at each corner relate to the number of sides of the polygon? Finally, use the fact that each diagonal has 2 ends.

9. How many intersections can be made with n lines in a plane?

a. Use the sketch to make a table which shows the number of intersections you can have with 0, 1, 2, 3, 4, and 5 lines,

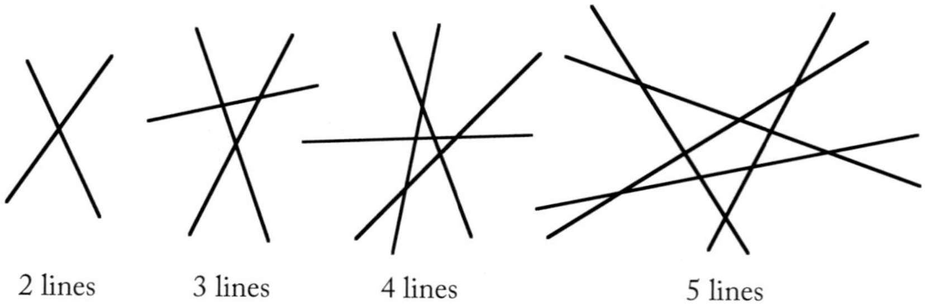

b. Find and state a rule expressing the number of intersections in terms of the number of lines.

10. Find and state the relation between the rules in problems 8 and 9.

11. If points on a circle are all connected by lines, the inside of the circle will be cut into regions. The number of regions that can be formed is a function of the number of points, as this picture shows.

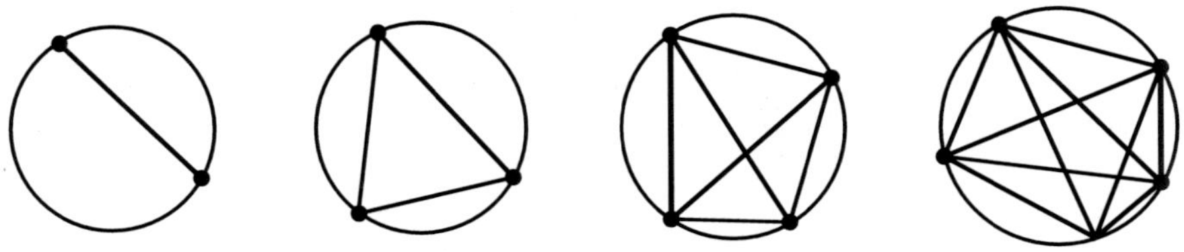

a. Make a table showing how many regions are possible when a circle is cut by lines between 2, 3, 4, and 5 points on it.

b. Guess a rule for the number of regions as a function of the number of points used.

c. Use your answer from part b to predict how many regions could be formed by connecting 6 points on the edge of a circle.

d. Check your prediction from part c with the sketch below. (This shows that it is important to test guesses and, where possible, prove them, before you can feel confident about them.)

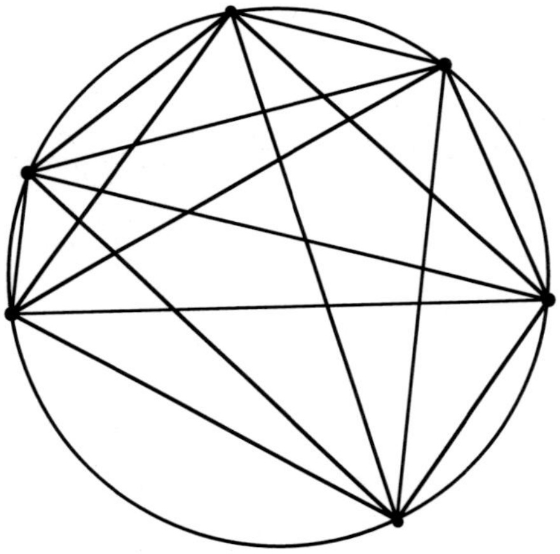

e. The correct function can be calculated by adding across rows in Pascal's triangle up to the line. Check that this correctly predicts the number of regions in all the cases examined so far.

1

1 1

1 2 1

1 3 3 1

1 4 6 4 1 /

1 5 10 10 5 / 1

1 6 15 20 15 / 6 1

1 7 21 35 35 / 21 7 1

f. The maximum number of regions that may be obtained with n points on the circumference is

$$\frac{n^4 - 6n^3 + 23n^2 - 18n + 24}{24}.$$ Check this formula for

n=1, 2, 3, 4, 5, 6, and 7. Use a calculator if you like. This formula is derived in *"Partitioning the Interior of a Circle with Chords"* by Dennis Parker. *The Mathematics Teacher,* Vol 99, No 2, September 2005.

12. **Kepler's Third Law**[5] This table shows each planet's mean distance from the sun and the length of its year. Mean distance from the sun is measured in **astronomical units**; one astronomical unit is defined as Earth's mean distance from the sun. Lengths of years are measured in Earth years.

Planet	Mean Distance from Sun	Length of Year
Mercury	0.387	0.2408
Venus	0.723	0.6152
Earth	1.000	1.0000
Mars	1.524	1.8808
Jupiter	5.203	11.862
Saturn	9.539	29.457
Uranus	19.191	84.013
Neptune	30.071	164.783
Pluto	39.517	248.420

a. For each planet, divide the square of the length of the year by the cube of the mean distance from the sun. (Use a calculator, a tool Kepler would have deeply appreciated. If your calculator has spreadsheet capabilities, enter the mean distances from the Sun in one column, and the length of the year in a second column. Then define the third column as the quotient of the first two.)

b. Express a relation, either in words or in algebra, between the length of the year of a planet and its mean distance from the sun.

[5] Source: H. N. Russell, R. S. Dugan, and J. Q. Stewart, *Astronomy, Revised Edition* Boston. Ginn. 1945. Johannes Kepler (1571-1630) was born in Weil der Stadt, in Southwestern Germany, an area torn in his day by strife between Catholics and Protestants. Kepler had an unhappy childhood; his father was often away as a military adventurer, and his mother, a mystical woman given to preparing herbal salves and believing in magic, did not always care for her children well. (She had been raised by a woman who was later burned at the stake for sorcery, and she herself was tried for witchcraft in her old age, but acquitted with Kepler's help.) Johannes' keen mind and religious interest won him a university scholarship, but once there his interest shifted to astronomy. After graduation, he found work as a teacher in a Protestant school in Graz. Always short of money, his circumstances grew dire when he and his wife, like all Protestants, had to flee Graz. He believed Copernicus' idea, that the planets orbit the sun in circles, but he worried about the reaction that heresy could arouse. Eventually, Kepler found work in Prague at the court of Emperor Rudolph II, where he got access to the meticulous observations of the planet's positions compiled by the Danish astronomer Tycho Brahe. After ten years trying to reconcile Copernicus' ideas with Brahe's data, Kepler gave up on circles and, in 1609, considered elliptical orbits. He then found the three laws for which he is still known. Newton later proved mathematically that Kepler's laws are the result of a force which he called gravity.

9.3 Graphs of Rules and Equations

Plotting points can help you visualize rules. To see how it works, let's try it with the rule $2n+7$. First, make up a table of inputs and outputs, much as we did before:

Inputs n	0	1	2	3	4	5	6	7
Outputs $2n+7$	7	9	11	13	15	17	19	21

Now use each input as the x-coordinate of a point whose y-coordinate is the associated output. The table above gives us

$$(0, 7), (1, 9), (2, 11), (3, 13), (4, 15), (5, 17), (6, 19), \text{ and } (7, 21).$$

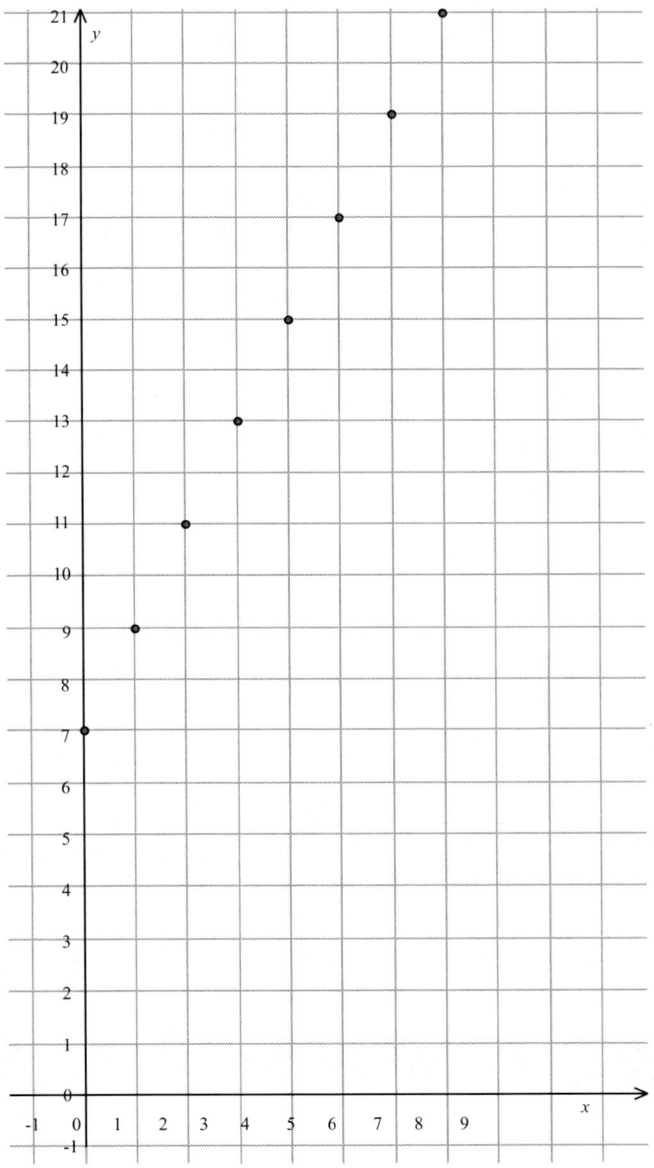

Figure 9.3.1

These points are not scattered at random. On the contrary, they line up perfectly. Try some different inputs and outputs for this same rule. Do they always give you points on this line? What if you use a negative input, such as –3? What if you use a fraction input, like 2/3? Trying this for many different inputs is more than just practice in plotting points. It shows that for this rule, **no matter what number you use as input, you get a point on the line.** You can also reverse the process. Every point on the line corresponds to an input (the x-coordinate) and its output (the y-coordinate) for this rule. The line is, therefore, a picture, or, as we say, a **graph**, of the rule. Therefore, picture not only isolated points, as in figure 9.3.1, but the solid line, as in figure 9.3.2. More formally, the **graph** of a function f is the set of points (x,y) for which $y=f(x)$. In this case we say that y depends on x, so we call x the **independent variable** and y the **dependent variable**.

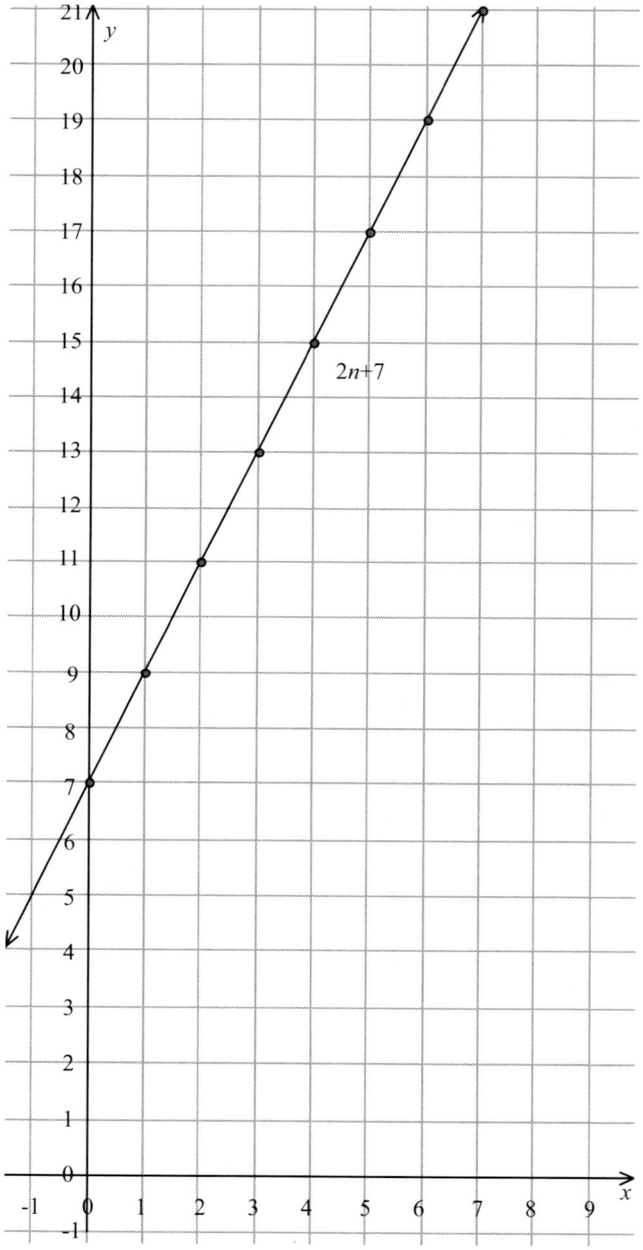

Figure 9.3.2

Graphing rules can be taken on many levels. At its most basic, this activity provides practice in arithmetic (plugging inputs into a rule and computing the outputs) and plotting points. The work is easy to check, since you can see at a glance if the points don't line up. Soon students will use graphs to check their arithmetic. If they notice that all but one of their points lie on a line, for example, they will check the work that led to that point. Learning to assess one's own work is a major step in becoming an independent learner.

Graphs of Equations

So far we have generated graphs by starting with a function and considering all the points (x,y), where $y=f(x)$. If we begin not with a function but with an equation in two unknowns, x and y, such as $y=2x+7$, then we regard the graph as coming not from a function but from an equation. The graph of an equation in two unknowns x and y is the set of points (x, y) in the plane whose coordinates satisfy the equation. For example, the point (1, 2) is not on the graph of $y=2x+7$, because 2 is not 2(1)+7, but the point (1,9) is on the graph, since 9=2(1)+7. Indeed, the points on the graph of $y=2x+7$ are the points on the line plotted in figure 9.3.2.

Graphing functions and graphing equations are so closely related that the difference between those two ideas is often just a slight change in point of view. However, graphing equations turns out to be a slightly broader idea. Every function has a graph, but you will see that there are graphs of equations that are not graphs of functions. An example of such an equation is $x=1$. This equation is satisfied by every point (x, y) with $x=1$, regardless of what number y is. The points that have 1 as their x-coordinate lie on a vertical line, plotted in figure 9.3.3.

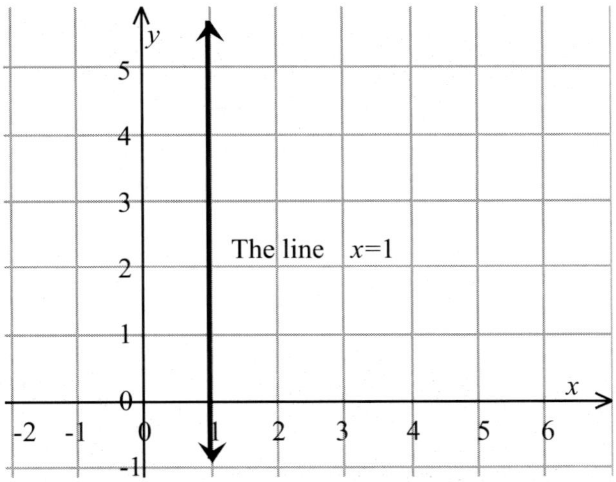

Exercises 9.3

1. a. Fill in this table for the rule $3n-7$.

Inputs n	6	5	3	1	0	-1	-2
Outputs $3n-7$							

b. Plot the points resulting from part a.

c. Examine your results from part b. The points should all lie on a line. If they do not, check your work. When you have fixed it up so that all your points do indeed lie on a line, draw the line and go on to part d.

d. Use your graph from parts b and c to predict the output of the rule for n=4. Then check your prediction by substituting n=4 in the rule $3n$-7.

e. Use your graph from parts b and c to predict the output of the rule for $n=\frac{1}{2}$. Then check your prediction by substituting $n=\frac{1}{2}$ in the rule $3n$-7.

f. Use your graph from parts b and c to predict the output of the rule for $n=-\frac{1}{3}$. Then check your prediction by substituting $n=-\frac{1}{3}$ in the rule $3n$-7.

2. a. Fill in this table for the rule $\frac{2}{3}n-1$.

Inputs n	-2	$\frac{-3}{2}$	-1	$\frac{-1}{2}$	0	1	$\frac{3}{2}$
Outputs $\frac{2}{3}n-1$							

b. Plot the points resulting from part a. To estimate thirds of squares is difficult. Using three squares as the unit avoids the difficulty.

c. Examine your results from part b. The points should all lie on a line. If they do not, check your work. When you have fixed it up so that all your points do indeed lie on a line, draw the line and go on to part d.

d. Use your graph from parts b and c to predict the output of the rule for n=2. Then check your prediction by substituting n=2 in the rule.

e. Use your graph from parts b and c to predict the output of the rule for $n=\frac{1}{2}$. Then check your prediction by substituting $n=\frac{1}{2}$ in the rule.

f. Use your graph from parts b and c to predict the output of the rule for $n=-\frac{5}{2}$. Then check your prediction by substituting $n=-\frac{5}{2}$ in the rule.

3. a. Fill in this table for the rule n^2.

Inputs n	-3	-2	-1	0	1	2	3
Outputs n^2							

b. Plot the points determined from part a. Are they on a line?

c. To learn more about the graph of n^2, complete this extra table and plot the new points along with those already plotted.

Inputs n	$\frac{-5}{2}$	$\frac{-3}{2}$	$\frac{-1}{2}$	$\frac{1}{2}$	$\frac{3}{2}$	$\frac{5}{2}$
Outputs n^2						

d. Sketch a smooth curve through the points plotted in parts a – c.

The points you have plotted lie along a graceful curve called a **parabola**. Ancient Greek mathematicians studied parabolas geometrically, but these curves took on added importance some 1900 years later with the discovery that, neglecting air resistance, the path through the air of an object such as a thrown stone or cannonball is a parabola.

Exploration: Linear rules (questions 4-7)

Any rule, such as $2n+7$, which can be expressed in the form $An+B$, is said to be **linear**, since its graph is a line. To graph a linear rule, just use two convenient inputs (such as 0 and 1), plot the two resulting points, and draw the line through those points. As a check against errors, either in arithmetic or in plotting the points, try a third value of n. If all three points do not lie on a line, that is a signal to recheck your work. The goal of this exploration is to find out how the numbers A and B in the rule $An+B$ determine just which line is the graph of a given linear rule.

4. Graph each of these rules using a full-sized piece of graph paper and the same set of coordinates for all eight lines.

a. $n+3$ b. $2n+3$ c. $3n+3$ d. $\frac{1}{2}n+3$

e. $-n+3$ f. $-2n+3$ g. $-3n+3$ h. $-\frac{1}{2}n+3$

5. The rules graphed in problem 4 all fit the general format $An+3$.

a. What common feature do all these lines have that reflects the fact that these rules all end in 3?

b. The rule $2n+7$, graphed in the text above, has $A=2$, as does $2n+3$, graphed in question 4b above. How is that common value of A reflected in their graphs?

c. How is the value $B=7$ reflected in the graph of $2N+7$? Explain.

d. Predict where the graphs of $-2n+7$ and $-n+7$ are in the plane.

e. To check your predictions from part d, graph the rules $-2n+7$ and $-n+7$.

6. A point and direction determine a line. For the graph of $An+B$, the number B determines a point on the line, (seen in question 5) and the number A determines the direction. Here is how A determines the direction:

Each time you increase n by 1, you move up or down by A units. If A is positive, you move up; if A is negative, you move down. For example, when $A=2$, as in question 4b, each time you move 1 unit to the right, you move up 2 units to stay on the line, like this:

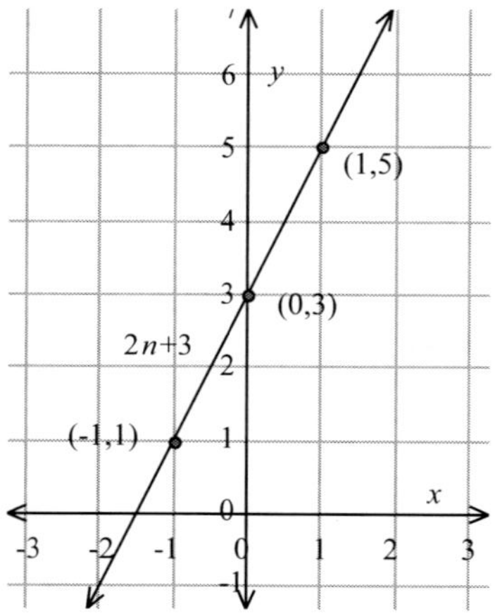

For example, from the point (0,3), moving 1 to the right and 2 up gets you to (1,5), which is also on the graph. You can also reverse the process, moving one step to the left and down 2, so that if you begin at (0,3), you get to (–1,1).

Similarly, for the rule $-\frac{1}{2}n+3$, every time you move 1 unit to the right you move **down** (because of the minus sign) a half unit, like this:

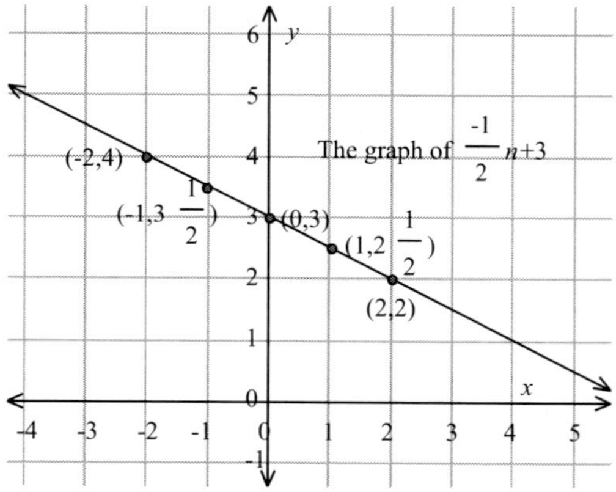

a. Check that this interpretation of the effect of A fits with the other lines graphed in question 4.

b. The number A is called the **slope** of the line $An+B$. Explain why that is a reasonable name for A.

7. Which of these lines could be the graph of $3n+\frac{1}{2}$? Explain what features of the line are determined by the numbers 3 and $\frac{1}{2}$.

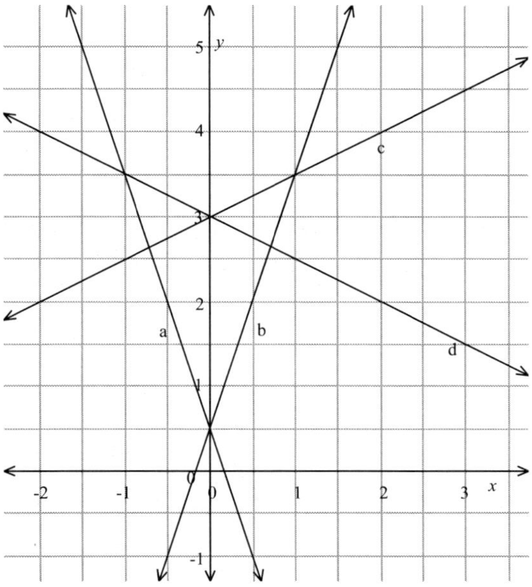

9.4 Graphs of Linear Functions

Functions or rules that can be expressed in the form $mx+b$ are called **linear functions**, because their graphs $y=mx+b$ are straight lines. Just which line is the graph of a particular linear function depends on the numbers m and b as you can discover for yourself by examining figures 9.4.1 and 9.4.2.

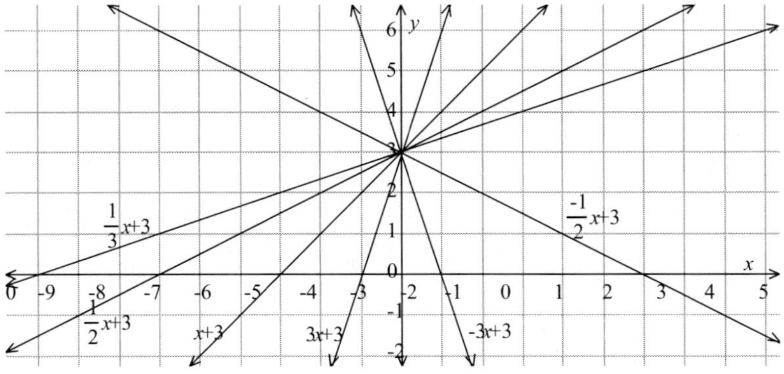

Figure 9.4.1

Figure 9.4.1 shows the graphs of several functions of the form $mx+3$. All the lines cross the y-axis at $y=3$; the different values of m are reflected in the different directions of the lines. Figure 9.4.2 shows the graphs of several functions of the form $2x+b$. They all have a common value of 2 for m, but they all cross the y-axis at different points, since they have different values of b.

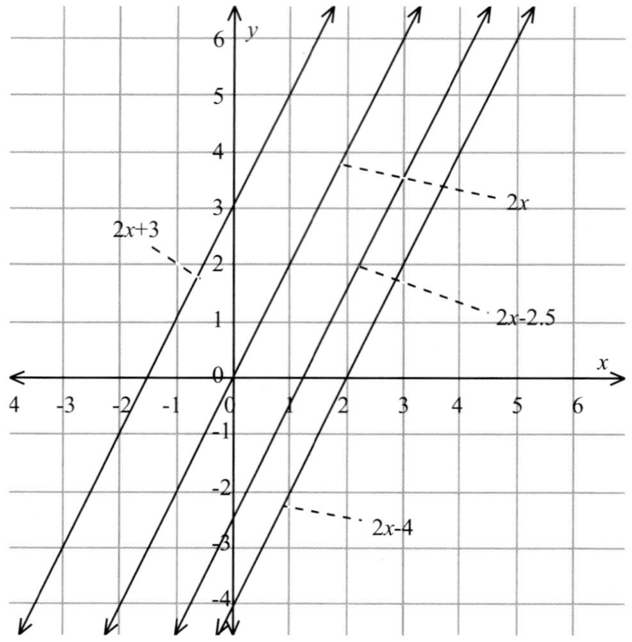

Figure 9.4.2

Evidently, b determines a point on the line (the point where the line crosses the y-axis), and m determines the direction of the line. Together, these two pieces of information determine which line is the graph of the function $mx+b$. The number b is the ***y*-intercept**, the coordinate at which the line crosses the y-axis. The number m, called the **slope** of the line, determines its direction, not as an angle but as a ratio, much as a helicopter pilot above a city might report flying two blocks north for each block east. To see how this works on graphs, look at figure 9.4.3, which shows the graph of $2x-1$ passing through the points (-1, -3), (1,1), and (3,5).

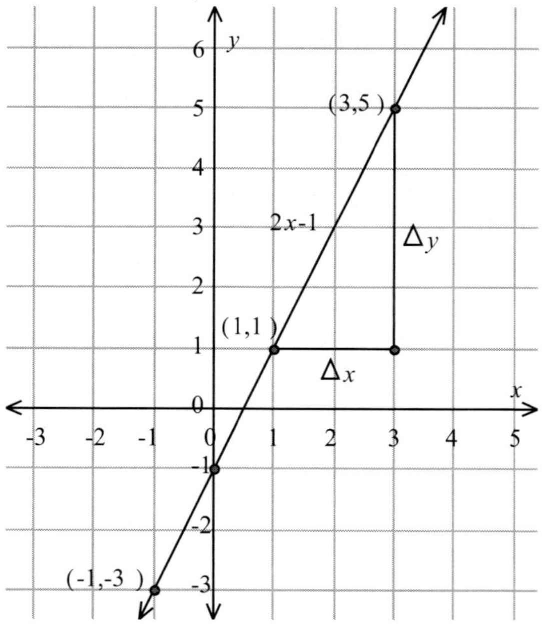

Figure 9.4.3

To move from (1,1) to (3,5) you have to move two units to the right and four units up, so the x coordinate increases by 2 (from 1 to 3) and the y coordinate increases by 4 (from 1 to 5). In this case the ratio $\frac{y\text{-change}}{x\text{-change}} = \frac{4}{2} = 2$.

Similarly, in moving from (3,5) to (-1,-3) the x-coordinate changes by –4 (from 3 to –1), and the y-coordinate changes by –8 (from 5 to –3). In this case the ratio $\frac{y\text{-change}}{x\text{-change}} = \frac{-8}{-4} = 2$. It is not a coincidence that the ratio $\frac{y\text{-change}}{x\text{-change}}$ is the same in both cases. Indeed, this ratio is the same for any pair of points on the line $2x$-1. For a different line, the ratio $\frac{y\text{-change}}{x\text{-change}}$ might have a different value, but again it would be the same for any two points on that line. It is customary to denote the y-change as Δy (read "delta y") and the change in x as Δx (read "delta x"). With this notation, the ratio $\frac{y\text{-change}}{x\text{-change}}$ is written $\frac{\Delta y}{\Delta x}$.

The ratio $\frac{\Delta y}{\Delta x}$ is the same for all pairs of points on the graph of a linear function $mx+b$. This ratio is called the **slope** of the line, and it is the same number as m. If you do not have a formula for the linear function, you can find the slope of its graph by computing $\frac{\Delta y}{\Delta x}$, which is equal to $\frac{y_2 - y_1}{x_2 - x_1}$ for any two points (x_1, y_1) and (x_2, y_2) on the line.

Example 1. Find the slope of the line through (-1,2) and (4,9).
Solution:

$$m = \frac{\Delta y}{\Delta x} = \frac{9-2}{4-(-1)} = \frac{7}{5}.$$

Example 2. Find the slope of the line through (-5,5) and (-3, 3).
Solution:

$$m = \frac{\Delta y}{\Delta x} = \frac{3-5}{-3-(-5)} = \frac{-2}{2} = -1.$$

Here the minus sign indicates that an increase in x goes with a decrease in y, so that as you move to the right along the line, you descend (see figure 9.4.4).

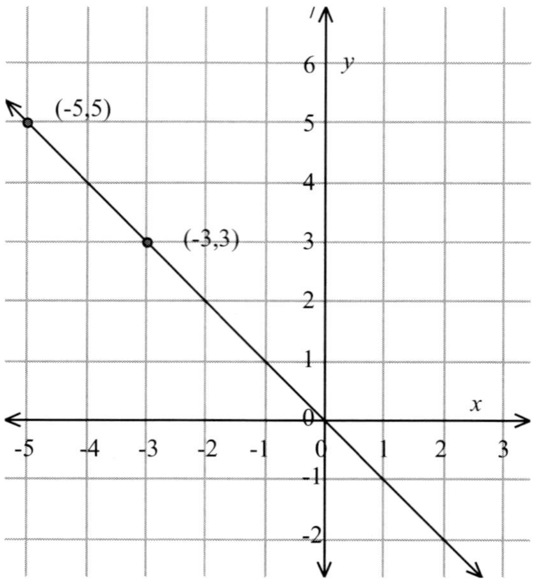

Figure 9.4.4

Example 3. Graph $\frac{1}{2}x+4$.

Solution: The y-intercept is 4, so the point (0,4) is on the graph. The slope, $\frac{1}{2}$, is the ratio $\frac{\Delta y}{\Delta x}$ for any two points on the line, so select any convenient size change as the increase in y, and make the increase in x twice as large. You might chose Δy=1 and Δx=2, for example. From (0,4) an increase of 1 in the y-coordinate and 2 in the x-coordinate gets you to (2,5). Further increases of 1 in the y-coordinate and 2 in the x-coordinate take you to (4,6), (6,7), and (8,8). Plot these points and draw the line through them (Figure 9.4.5).

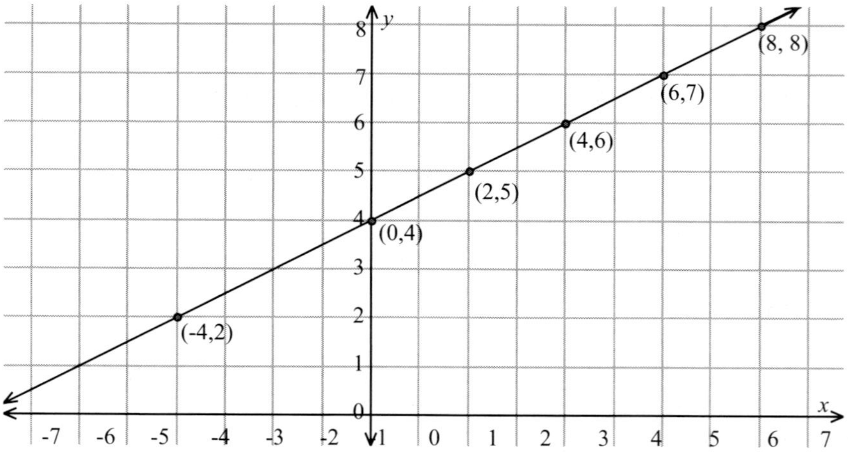

Figure 9.4.5

Among all linear functions $mx+b$, those with $m=0$ deserve special mention. If $m=0$, the function is $0x+b$, or just b. It is called a **constant function**, since no matter what value of x is used as input, the output is always b. The graph of this function is a horizontal line with y-intercept b.

You may wonder if vertical lines are also special, and indeed they are. A vertical line cannot have a slope at all, since for points directly above or below each other, the x-coordinates are the same, and $\Delta x=0$, and then $\frac{\Delta y}{\Delta x}$ is not defined because of the 0 denominator.

The slope of a line

A line with positive slope rises toward the right.
A line with negative slope falls toward the right.
A line with slope 0 is horizontal.
The closer to 0 the slope, the less steeply the line rises or falls.
A vertical line has no slope; its slope is undefined.

Figure 9.4.6 shows lines with various slopes.

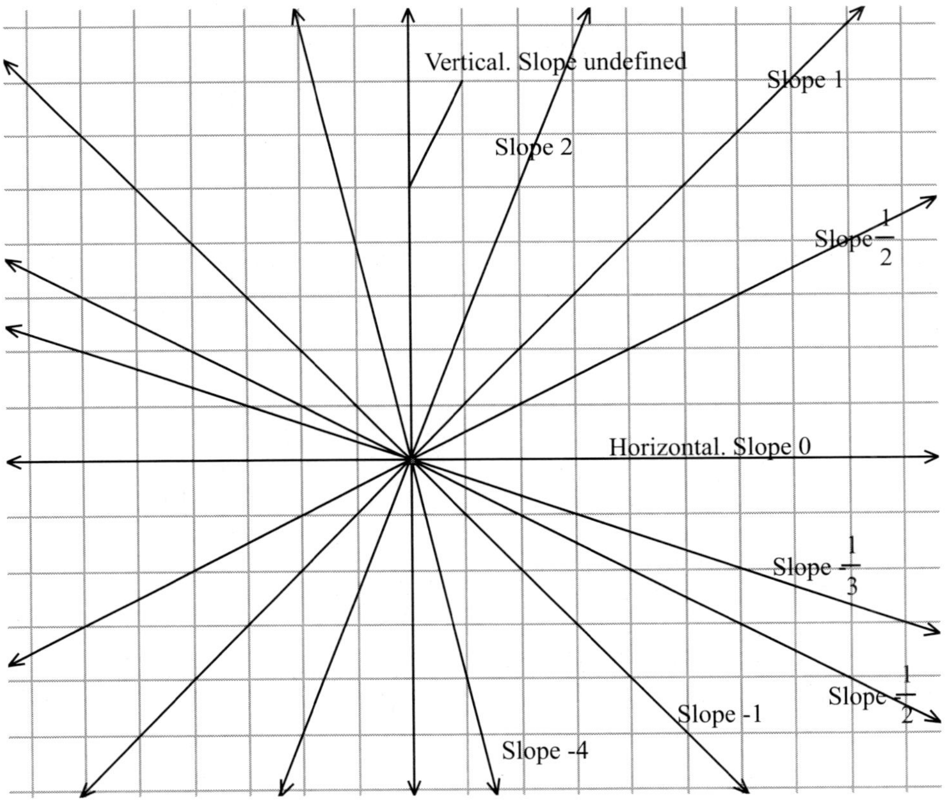

Figure 9.4.6. Slopes of lines

The above remarks apply to lines in general, whether or not they are obtained as graphs of linear functions. If a line is not vertical, it has a slope, which is the ratio $\frac{\Delta y}{\Delta x}=\frac{y_2-y_1}{x_2-x_1}$, and it has a y-intercept, which is the y-coordinate of the point where the line crosses the y-axis. The slope and y-intercept together determine the line.

Exercises 9.4

1. Find the slope of the line through the given points
   a. (2,4) and (1,6)
   b. (-1,5) and (0,4.5)
   c. (10,10) and (40,50)
   d. (-8,4) and (-2,0)
   e. (0, b) and (4,4)
   f. (a, 0) and (0, b)

2. Use the slope and intercept to graph each of these functions
   a. $3x-6$
   b. $-3x+4$
   c. $\frac{1}{2}x+1$
   d. $-\frac{1}{3}x-1$
   e. $\frac{5}{2}x-4$
   f. $-x+\frac{1}{3}$

3. The **glide ratio** of an airplane is the horizontal distance it can glide (without power) divided by the vertical distance it drops as it glides. If an airplane with a glide ratio of 15 loses power at a height of 8,000 feet, can it glide to an airport 16 miles away? Explain.

4. Scientific farming techniques in California produce unusually high yields of rice per acre. The techniques involve grading the land so that the maximum slope in any direction is 0.002. What is the maximum possible difference in elevation between two points on such a field, if they are 4,000 feet apart?

5. A line with slope 2 passes through the point (5,8). Find its y-intercept.

6. In parts a-d, graph the given pair of functions on the same set of coordinate axes.
   a. $\frac{1}{2}x+1$ and $\frac{1}{2}x-1$
   b. $4x-5$ and $4x-3$
   c. $-2x+1$ and $-2x+2$
   d. $\frac{2}{3}x+1$ and $\frac{2}{3}x-4$
   e. What is true of each pair of graphs that is related to their slopes?

7. In parts a-d, graph the given pair of functions on the same set of coordinate axes.
   a. $2x+1$ and $-\frac{1}{2}x+3$
   b. $3x+2$ and $-\frac{1}{3}x+1$
   c. $\frac{2}{3}x+4$ and $-\frac{3}{2}x$
   d. $-\frac{3}{5}x+1$ and $\frac{5}{3}x-2$
   e. What is true of each pair of graphs that is related to their slopes?

8. Determine the slope and the y-intercept of lines a-f graphed here.

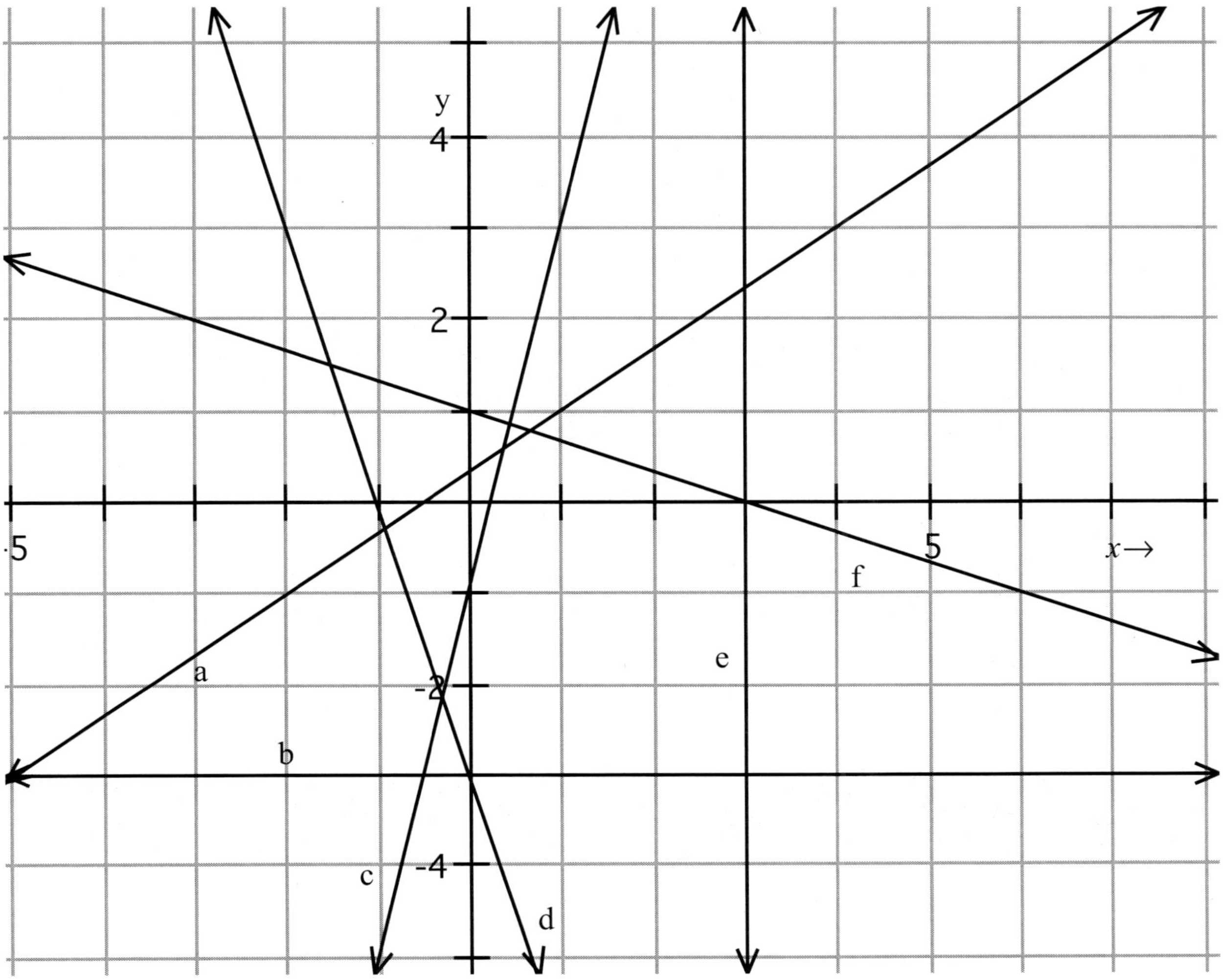

Vocabulary for Chapter 9

Axis (9.1)
Constant function (9.4)
Coordinates (9.1)
Data point (9.5)
Dependent variable (9.3)
Domain (9.2)
Function (9.2)
Graph (9.3)
Independent variable (9.3)
Linear function (9.4)
Parabola (9.3)
Plot (9.1)
Quadrant (9.1)
Slope (9.4)
y-intercept (9.4)

Self-Quiz for Chapter 9

1. Draw and number the coordinate axes on a piece of graph paper. (Save this for question 5.)

2. a. Describe in words what the rule $2n–4$ does to whatever number is given.
   b. Describe in words what the rule $2(n–4)$ does to whatever number is given.
   c. Explain why the rules in parts a and b are not the same.

3. A rule multiplies whatever number is given by 3, then adds 5 to the result. Write an algebraic expression for that rule.

4. Is the point (5,13) on the graph of the rule $3(n-\frac{2}{3})$? Explain.

5. Graph the rule $-2n+\frac{1}{3}$.

6. Explain in a sentence or two what is meant by saying that a rule is "linear."

7. Explain in a sentence or two what is meant by saying that a line has a slope of –3?

8. What is the slope of the line through the points (-3, 8) and (4, 7)?

Chapter Ten

Transition to Algebra

Chapters 1-9 built the arithmetic foundation for algebra. Now the spotlight turns to algebra itself. We will treat algebra as generalized arithmetic. This teaching approach eases the abrupt jump in abstractness that is difficult for beginning algebra students.

10.1 Take a Number

Here is a little trick for anyone who can add, subtract, multiply, and divide with simple numbers:

1. Take a number (remember it, but don't say what it is).
2. Add three.
3. Multiply by four.
4. Add eight.
5. Divide by four.
6. Subtract the number you began with.

At this point, remarkably, you should have five, regardless of what you began with. In practice, however, if you try this with a class, some will not have five, because they did not follow the directions carefully or because of faulty arithmetic. To make sure everyone understands the process, make a table like figure 10.1.1, starting with various numbers. Have students put their own numbers at the top and then complete the columns. Soon they will see that they always end up with five.

1. Take a number	3	10	15		
2. Add 3	6	13	18		
3. Multiply by 4	24	52	72		
4. Add 8	32	60	80		
5. Divide by 4	8	15	20		
6. Subtract your original number	5	5	5		

Figure 10.1.1

So far, this trick has involved only arithmetic and care in following directions, but you can do much more with it. Before you read further, experiment a bit with this trick. Then think of some interesting questions that students might ask about it and how you might use them in the classroom.

Some of these questions concern (and will be raised by a class) relate to what numbers can be used at the start. Does the trick work if you start with a very big number like a million? Does it work if you start with 0? What if you start with a fraction? How about a decimal? What about a negative number? Experiment with these questions, if you have not already done so.

As was the case with box puzzles, these questions are an excellent source of arithmetic practice, with each student making up his or her own questions and working at a level of challenge he or she selects. You, as teacher, need not correct or even read most of the work. Only when a student finds a number for which the trick appears not to work might you look over the work carefully, and even then a student might check it for you. Just circulate around the classroom, encouraging those who are doing interesting work and helping those who need it.

After a period of experimentation, it is time to move on to other questions, such as why the trick works and whether other tricks like it can be made up. These questions cannot be answered experimentally, but they do lead to new insights. A first step in answering them is an elegant pictorial explanation developed by the late Robert Wirtz, one of the authors of the *Math Workshop for Children*.

Imagine the original number as a number of stones in a bag. We cannot see inside the bag (maybe it is tied shut), so we just picture it like this . Now, if we add three more stones but cannot open the bag, we will picture the result like this . The next step in the trick is to multiply by four. How are we to picture that if we do not know how much is in the bag? Like this:

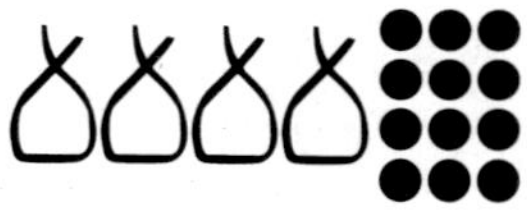

Now we must add eight, and we picture the sum like this

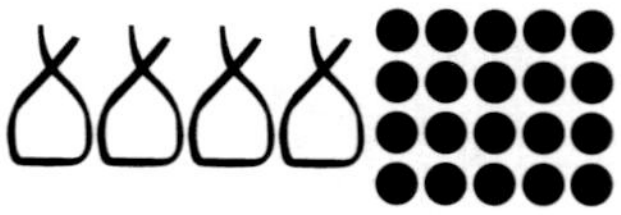

Next, divide by four, which we picture, even without knowing what number is in the bag, by dividing the bags by four and the "stones" by four to get 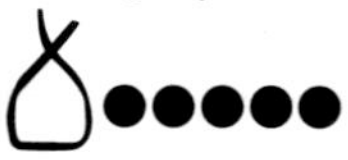. Finally, subtract the number we began with, which is the number of objects in a bag, leaving only the five stones . The number of objects in the bag turns out to have no influence on the final answer of five, and that is why the trick works regardless of what number you begin with. The entire explanation is summarized in figure 10.1.2.

The trick is like starting with a number, adding 2, and subtracting the number you began with, leaving 2 regardless of what you began with. The smokescreen of additions and multiplications hides what is going on.

1. Take a number.
2. Add three.
3. Multiply by four.
4. Add eight.
5. Divide by four.
6. Subtract the number you began with.

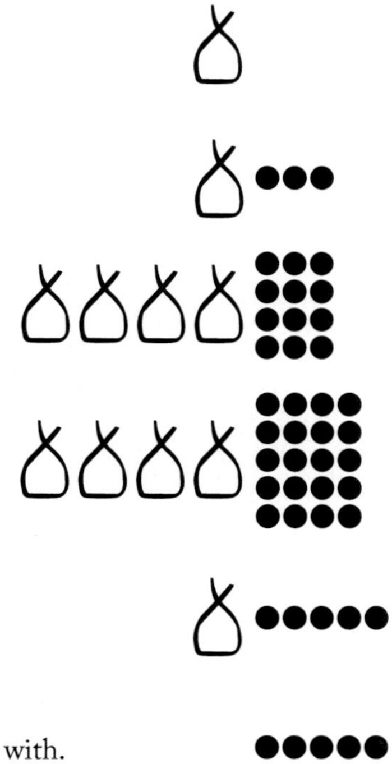

Figure 10.1.2

Initially, even if you can follow this explanation, you may feel you could not have made it up or explained it to others. However, if you go through this kind of explanation for similar tricks, perhaps over a period of several weeks, you will find it becomes easier and more natural for you, and more and more students will catch on, too. Understanding here comes in stages. Stage 1 is to be able to match the picture steps with the instructions. Exercises 1-5 below provide practice in that skill. When you are reasonably good at that, you are ready for stage 2, to begin making up similar tricks on your own. At first you may want to write them down, but with a little practice, you can move to stage 3, which is to make up tricks as you go along, without writing anything down. Your students will be impressed, and in time they, too, will reach that stage. The time needed for this transition varies widely, but if you return to this kind of trick regularly, many students will reach stage 3 by the end of grade seven.

To reach this goal, however, we need to consider a few fine points, which are best illustrated by actually making up a number trick together. First, ask people to take a number (which we think of as a bag), and then build up what they began with by a series of additions and multiplications. Let's add five to the original number (think: one bag and five stones), and then multiply by five (think: five bags and 25 stones). Now, we could simply reverse, dividing by five and then subtracting the original number, but that would be too transparent. Instead, let's add the original number (think: six bags and 25 stones). At this point, we have created enough of a smokescreen, and it is time to begin simplifying things and bringing everyone back to a common outcome. One way to simplify would be to divide by the current number of bags, six. However, dividing six bags and 25 stones by six will yield one bag and $4\frac{1}{6}$ stones, which is clumsy, because of the fraction. A smoother set of instructions, because it avoids fractions, would first be to subtract one (think: six bags and 24 stones) and then divide by six (think: one bag and four stones). Now, subtract the original number, and the result is 4, regardless of the choice of original number. This trick is summarized in figure 10.1.3.

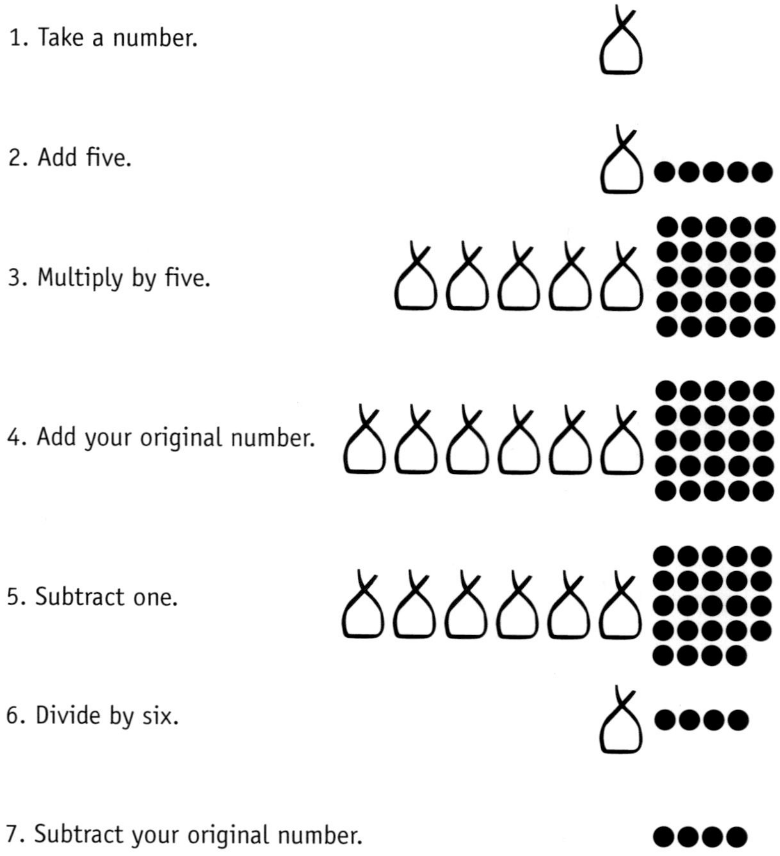

Figure 10.1.3

Our explanation in terms of bags and stones is relatively concrete, and it can be used not only to understand these tricks, but also to make them up. However, the picture notation, for all its intuitive appeal, gets clumsy when large numbers or fractions are involved. Therefore, we introduce a shorthand. For example, to represent the picture in step 4 of figure 10.1.3, write 6+25. With this shorthand, we no longer need to make elaborate pictures, and we no longer need count either bags or stones when reading the picture writing. With this notation, the trick in figure 10.1.3 is described at the right-hand side of figure 10.1.4.

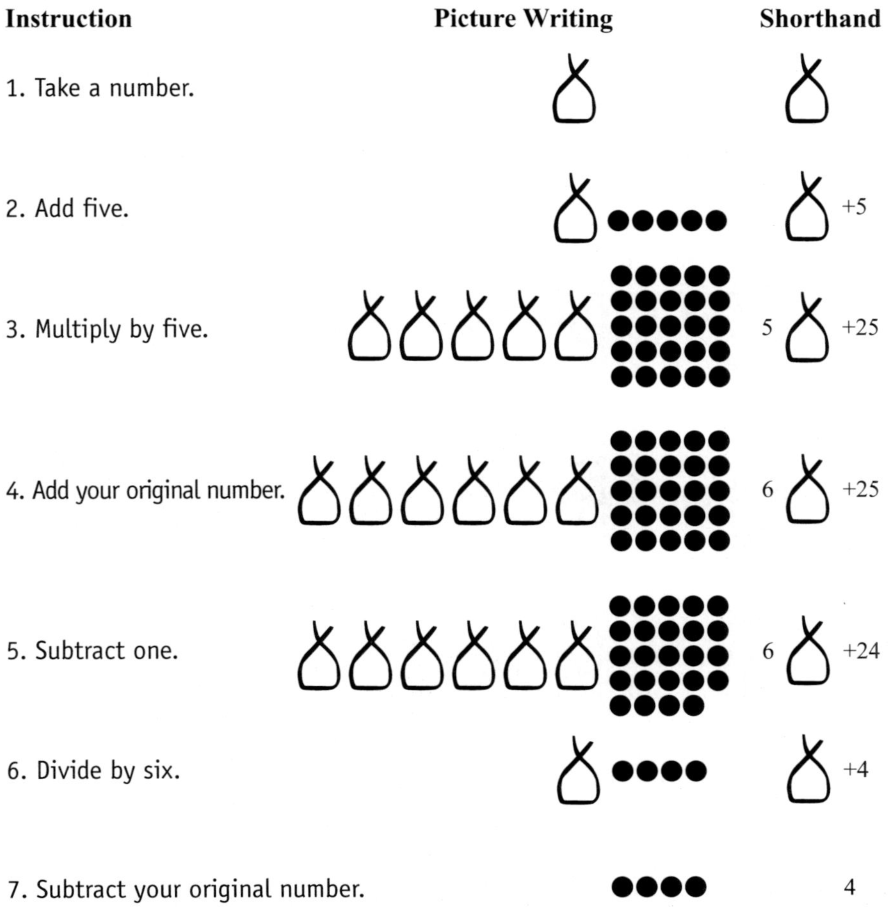

Figure 10.1.4

Our shorthand improves on the picture writing, but it is more abstract. Indeed, it is every bit as abstract as the notation of algebra, in which numbers are often represented with letters instead of pictures of bags. To see this, erase the bottom of the symbol for a bag. What remains looks much like the letter x.

Look where these number tricks have led us! They are an interesting source of arithmetic practice, but they are also a stepping stone to the abstract language of algebra. Many youngsters find the transition from arithmetic to algebra difficult and confusing, but these number tricks allow for a gradual transition using picture writing as a bridge to algebraic notation.

Of course, number tricks are not the only way to introduce algebraic notation, nor should they be. We have already used a little algebra in our shorthand for rules and given examples of algebra computations that parallel those in arithmetic. The Mason Mathematics Program, shown in the film *"Mathematics for Five-Year Olds"*, introduces algebraic notation to preschoolers with "Mrs. A" and "Mr. B" even as it helps teach the alphabet, asking such questions as, "If A+3=5, what number is Mrs. A hiding?" Again, children in Russia are regularly exposed to exercises like that in Figure 10.1.5, which is from a 1978 version of a third grade text. All of these introductions to algebra are simple and natural, and they allow students to get used to using letters for numbers long before this abstraction is used formally. If such activities are deliberately incorporated into the mathematics curriculum from the beginning, the transition to algebra will be easier than it usually is today.

b	40		18	54	12
c		24	1		
b ÷ c	8	3		9	1

Figure 10.1.5

Exercises 10.1

In questions 1–5, use bags and stones to explain the tricks.

1. Take a number
   Add 3
   Multiply by 2
   Add your original number
   Divide by 3
   Subtract your original number

2. Take a number
   Add 2
   Multiply by 4
   Add 1
   Subtract the number you began with
   Divide by 3
   Subtract the number you began with

3. Take a number
   Add 7
   Multiply by 3
   Subtract 1
   Subtract the number you started with
   Divide by 2
   Subtract the number you started with

4. Take a number
   Add 7
   Multiply by 4
   Subtract the number you began with
   Add 2
   Divide by 3
   Subtract the number you began with

5. Take a number
   Add 6
   Multiply by 5
   Add 2
   Subtract your original number
   Divide by 4
   Subtract your original number

6. Write instructions for the number trick pictured here.

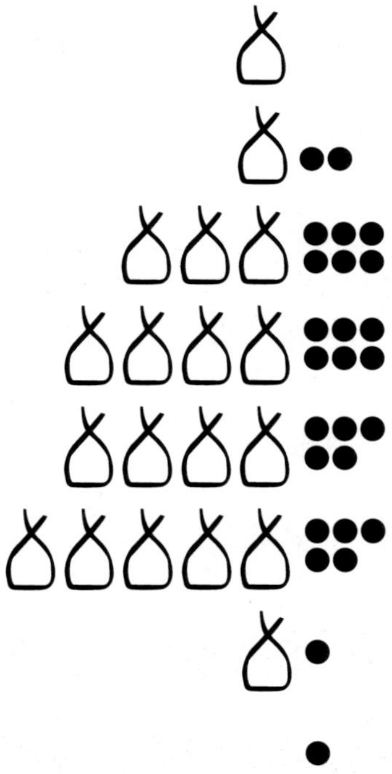

7. Write instructions for the number trick pictured here.

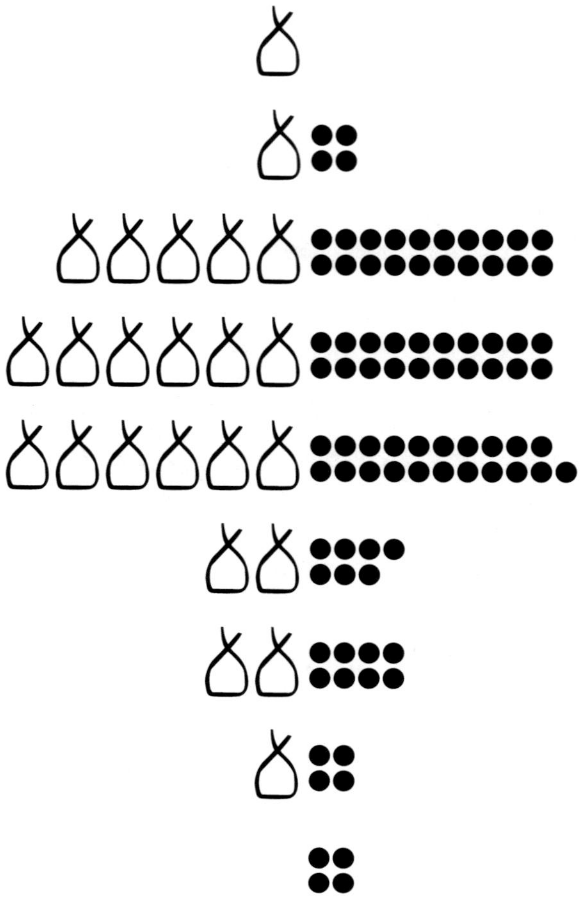

8. Write algebra shorthand equivalents for the pictures in question 6.

9. Write algebra shorthand equivalents for the pictures in question 7.

10. Write algebra shorthand equivalents for the pictures you made in question:

    a. 1 b. 2 c. 3 d. 4 e. 5

11. Make up a number trick and show it actually works by means of bags and stones or algebra.

12. View the movie (available also as a videotape) *Mathematics for Five Year Olds*. It may still be available from The Newgrange School, 52 Lafayette Avenue, Trenton, NJ 08610. Price unknown.

10.2 Equations

Numbers and algebraic expressions are important in their own right, but often we are interested in relationships among them. Some of the most important of these relationships are expressed as equations.

An **equation** is a statement that two objects are the same, or, as we say, equal. The verb of an equation is written with the sign =, which is read "equals" or "is equal to."[1] Its meaning is even clearer in German, where it is simply read "is." (In English, we read 3+5=8 as "Three plus five **equals** eight," whereas in German it is "Three and five **is** eight.") Equations may be used to state equality of many kinds of mathematical objects, such as numbers, sets, or angles, but here we shall confine our attention to equations about numbers.

Like any statement, an equation may be true or false. For example, 3+4=7 is true, since both sides do indeed represent the same number, but 3-5=7 is false. An equation like $x+4=9$ cannot be classified as true or false, since that depends on what number x represents. This is called a **conditional** equation, since it represents a condition that may or may not be true for a particular number x. Conditional equations are among the most interesting of all, since each one implicitly poses a problem of finding which numbers, if any, make it true. A number which satisfies (makes true) a conditional equation is called a **root** of that equation; to **solve** a conditional equation is to find all of its roots. Some conditional equations have no roots, and others have more than one, but most of the equations we will work with have just one root. Depending on the particular equation, finding solutions can be very difficult. On the other hand, it is relatively straightforward to check whether or not a number is a root of an equation. Simply substitute the number in question for the letter in the equation, then do the arithmetic to see if both sides are the same, as in example 1 below.

When a conditional equation has not been solved, we do not know yet what its roots are, so we call the letter an **unknown**. This letter, incidentally, is not used as part of an alphabet but merely as a symbol for a number. You can use any letters from any alphabet, or even symbols like a bag or a happy face as unknowns in equations.

Example 1. Which of the numbers, 2, 3, and –5 is (are) a root of $x + 3 = \dfrac{10}{x}$?

Solution: Substitute 2 for x. $\quad 2 + 3 = \dfrac{10}{2}$

Both sides are 5, so 2 is a root.

Substitute 3 for x. $\quad 3 + 3 = \dfrac{10}{3}$

Here the left side is 6, but the right side is $\dfrac{10}{3}$, so 3 is not a root.

Substitute –5 for x. $\quad -5 + 3 = \dfrac{10}{-5}$

Both sides are –2, so –5 is a root.

[1] This use of two parallel marks goes back to Robert Recorde (1510-1558), who wrote in his *Whetstone of Witte* (1557) that "noe 2 thynges, can be moare equalle." Recorde's sign for equality was adopted by several prominent English mathematicians in the early 17th century and was introduced to continental Europe by Leibniz in the second half of that century.

Example 1 shows the procedure for checking whether or not a number satisfies an equation, and it shows that an equation can have more than one root. Finding roots of equations may be difficult, and the search for ways to solve difficult equations has long been a major quest in mathematics. In contrast, checking to see whether a number is, in fact, a root is, at least in theory, a purely mechanical question, but the arithmetic involved can be sophisticated, as in example 2.

Example 2. Is $\frac{-1}{3}$ a root of $3(x\text{-}2) = \frac{2}{x}\text{-}1$? Show why or why not.

Solution: Substitute $\frac{-1}{3}$ for x, and evaluate both sides. If they are the same, then $\frac{-1}{3}$ is a root of the equation, otherwise not.

The left side is $3(\frac{-1}{3}\text{-}2)$, which can be evaluated either by means of the distributive property, $3(\frac{-1}{3}) + 3(-2) = -1 + (-6) = -7$, or by first evaluating the factor in parentheses and then multiplying that by 3:

$$3(\frac{-1}{3}\text{-}2) = 3(-\frac{7}{3}) = -7.$$

On the right hand side, $\frac{2}{\left(-\frac{1}{3}\right)} - 1 = 2\left(-\frac{3}{1}\right) - 1 = -6 - 1 = -7$. Since both sides are -7, $\frac{-1}{3}$ is indeed a root of the given equation.

Example 2 shows eloquently that skill in arithmetic is fundamental for algebra. Students need to be able to recognize the correct order of operations, and they need to be reasonably reliable at carrying out those operations with whatever numbers are involved.

Here are some examples of how to solve simple equations.

Example 3. Solve $3x+11=5x+3$.
Solution: Think of x as representing the number of stones in a bag, to picture

$3x+11$ as 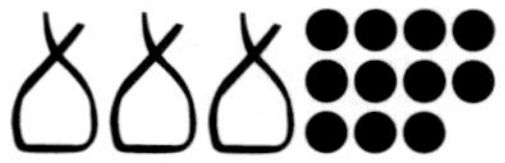

and $5x+3$ as 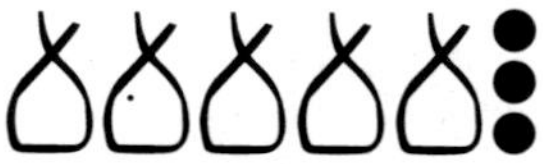

The equation states that both of these pictures have the same number of stones, although some of the stones are hidden inside the bags. Pictorially,

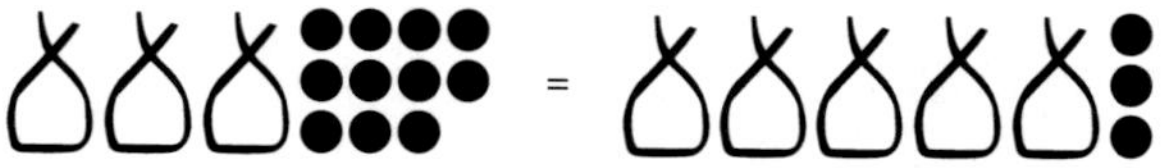

Solving this equation amounts to figuring out how many stones must be in each bag. If you remove three stones from each side, each side has fewer stones than before, but still there are as many stones on the left side as on the right. Pictorially,

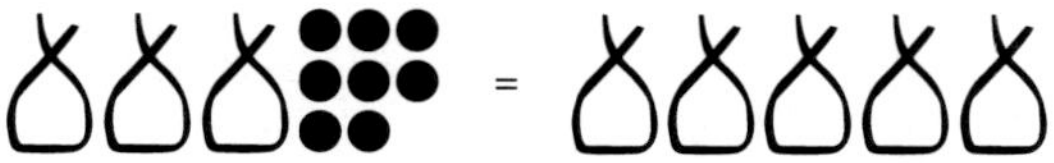

Now, the idea of removing an equal number of stones from each side can also be applied to stones in bags, even though we don't know how many stones there are in each bag. If you remove three bags worth of stones from both sides, then

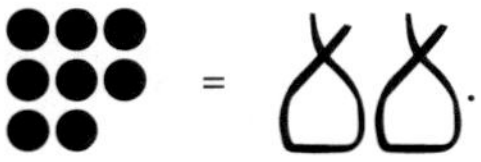. Now, each of the bags has the same number of stones, so it is easy to see that this last picture equation will be true provided that each bag has four stones. To show that the original equation has the root x=4, substitute 4 for x, or, equivalently, imagine four stones in each bag. Then, if we look into the bags, the original picture equation looks like this

Here it is easy to count up the stones to see that there are 23 on each side of the picture, so that 4 does indeed satisfy the original equation. The entire process is summed up in figure 10.2.1.

Instruction	**Picture Writing**	**Algebra**
Solve.	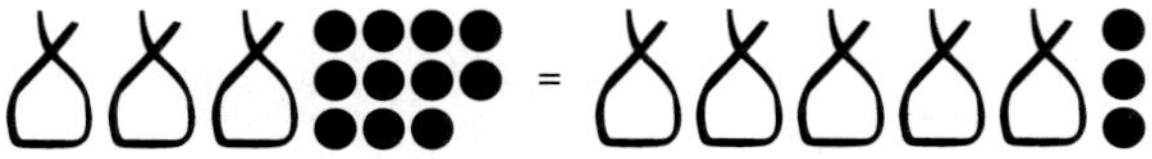	$3x+11=5x+3$
Subtract 3.	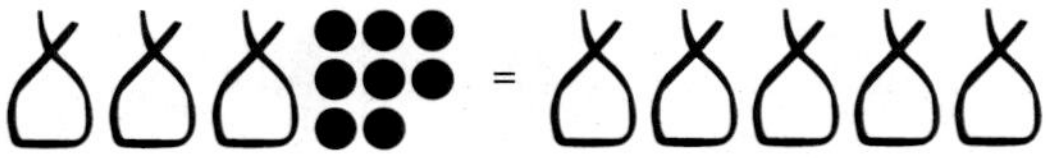	$3x+8=5x$
Subtract 3x.	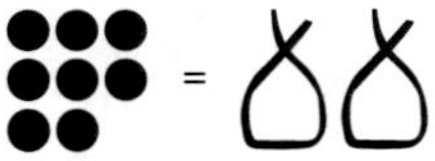	$8=2x$
Divide by 2.	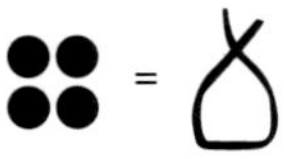	$4=x$
Check.		$3(4)+11=5(4)+3$

Figure 10.2.1

Pictures of bags and stones are helpful as an introduction to algebra, but students cannot rely on the bags and stones forever, because they offer no easy way to represent fractions or negative numbers. If we think of an equation as balancing weights, we can think of negative numbers of bags and stones as having negative weight, pulling upward like helium balloons. We could, for example, picture a negative stone as hollow circle, like this: O. This vision would make physical sense of a zero pair, which would consist of a stone whose weight pulls downward and a hollow stone with negative weight, upward pull equal but opposite to the downward pull of a positive stone. The weight of the pair would thus be zero.

Algebra tiles can help us picture equations with negative numbers of bags and stones. Consider two kinds of tiles, rods, which represent x (or any other unknown letter we please) and unit squares, which represent 1. When shaded, we will view them as positive, and when turned over, which we will picture as unshaded, we will consider them negative.

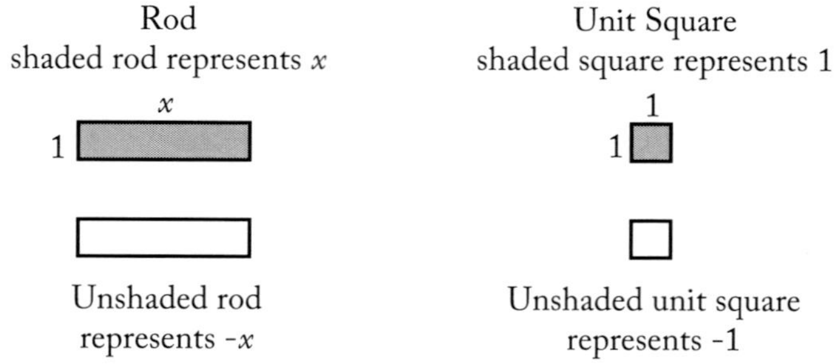

Example 4. Picture the equation x-4=3 and solve it with algebra tiles.
Solution:

Add 4 to both sides

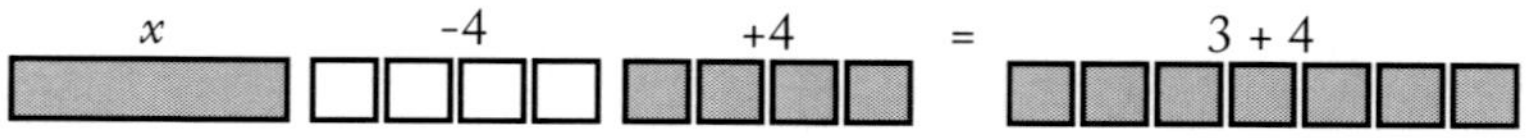

Simplify, using the zero pairs on the left.

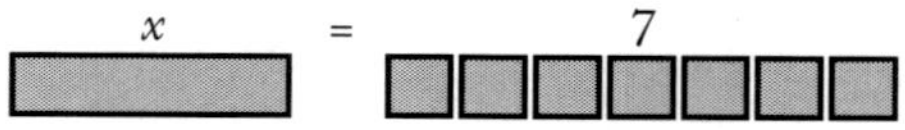

Example 5. Picture the equation $2x$-5=7-4x with algebra tiles and solve it.
Solution:

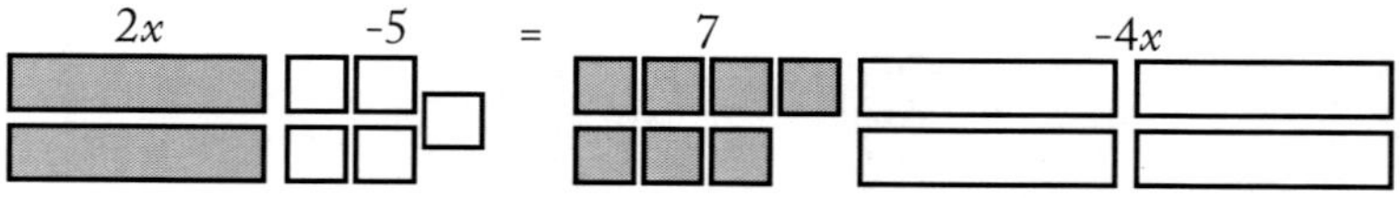

Operate on this equation, doing the same thing to both sides, to get all the rods (positive and negative) on one side of the equation and all the unit squares (again, positive and negative) on the other side. As a first step, add 5.

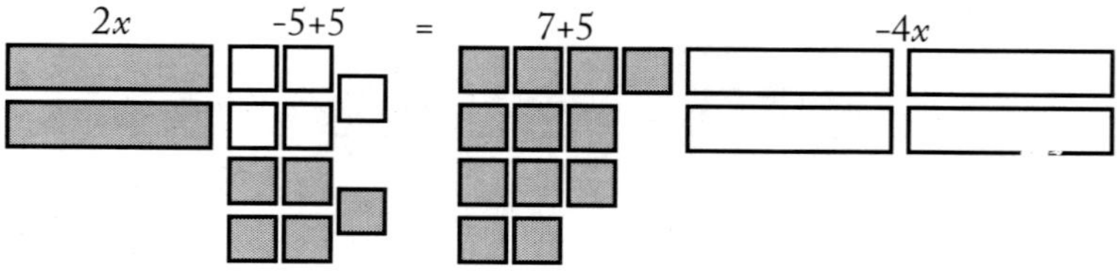

On the left-hand side there are five zero pairs, corresponding to the fact that the equation $2x$-5+5=7+5-4x can be simplified, using the associative property of addition, as $2x$+(-5+5)=(7+5)-4x, or $2x$=12-4x.

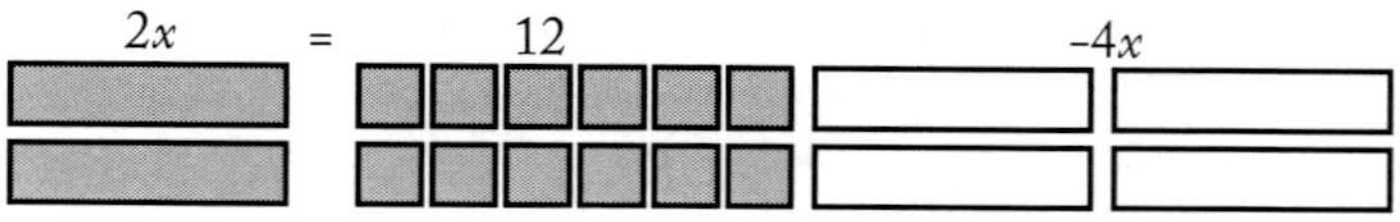

Now, to get all the rods on one side, add $4x$ to both sides.

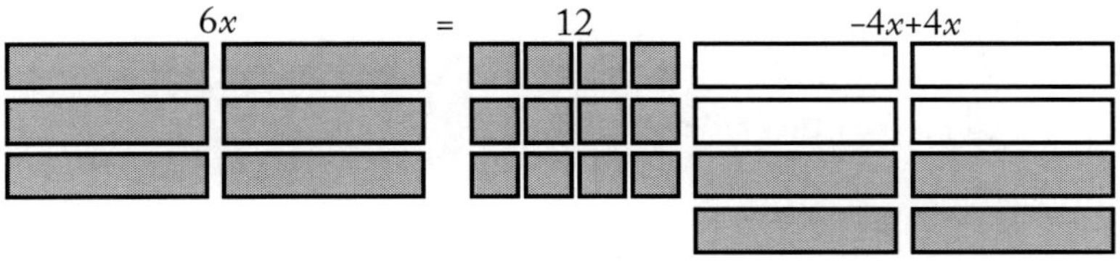

On the right-hand side, $-4x+4x=0$. You can see this from the distributive property, which applies, even though we do not know what number x stands for. Thus $-4x+4x=(-4+4)x=0x$, which you can envision as the result of four zero-pairs for rods. Then the equation becomes

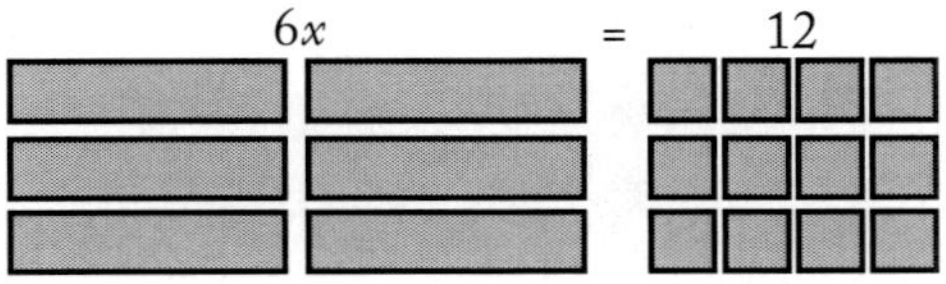

Now that all the rods are on one side of the equation and all the unit squares are on the other side, divide both sides by the number of rods, 6, to get

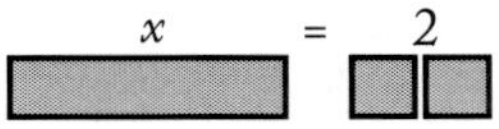

You can check for yourself that 2 is indeed a root of the original equation. There are other sequences of operations that work as well in solving this equation. For example, you could first add $4x$ and then add 5, reversing the order of our first two steps, or you could combine them by adding $4x+5$ in a single step. Alternatively, you could subtract $2x$ from both sides to get $5=7-6x$, then subtract 7 to get $-12=-6x$. Now divide by the number of rods, -6, to get $2=x$.

The general idea for solving simple equations is to perform a succession of simplifications until we reach an equation like $2=x$, which explicitly names a root. In making each equation from the one before, we are bound by one simple but very important rule.

Basic rule for equations:

Anything done to change the value of one side of an equation must also be done to the other side.

This rule is inherent in the logic of solving equations. If, for some number x, the equation is true, then for that value of x, the two sides of the equation represent the same number. Consequently, if you alter each side in the same way, the two sides, though different from before, will still be equal to each other. To make this idea graphic, imagine yourself as the blindfolded justice pictured in traditional paintings, holding a balanced scale. (The blindfold signifies that you are not biased and so will weigh the two sides fairly.) In this case the pans hold stones of uniform weight, but some of the stones are hidden inside bags. The bags themselves have negligible weight, and all bags hold the same number of stones. Both sides have the same number of stones, so you do not feel one side as heavier than the other. If someone alters both sides in exactly the same way, you will feel a change, but you will not feel one pan as heavier than the other.

The basic rule for equations does not help you decide what operations will actually help to simplify a given equation. For this there is no simple rule, but there is a general strategy that works in many cases.

Strategy For Solving Simple Equations

> **1. Add or subtract to get an equation which on one side has only multiples of the unknown, such as $5x$ or $-3y$, and on the other side no unknowns.**
> **2. Combine like terms.**
> **3. Divide by the number which multiplies the unknown.**

The details of applying this strategy can vary considerably, as illustrated in problems 1 and 2 below.

Exercises 10.2

1. In example 3 we first subtracted 3, then subtracted $3x$ to get $8=2x$. Then, following the basic strategy, we divided by 2, the number which multiplies the unknown. In parts a-f, write out the sequences of equations, that come from applying the following sequences of operations to the same equation, $3x+11=5x+3$. Illustrate, where possible, with bags and stones or algebra tiles.

   a. 1. Subtract $3x$
      2. Subtract 3
      3. Divide by 2

   b. 1. Subtract 1
      2. Subtract x
      3. Divide by 2
      4. Subtract 1
      5. Subtract x

   c. 1. Subtract $3x$
      2. Subtract 1
      3. Divide by 2
      4. Subtract 1

   d. (Use algebra, not bags and stones.)
      1. Subtract $5x$
      2. Subtract 11
      3. Multiply by -1
      4. Divide by 2

   e. (Use algebra, not bags and stones.)
      1. Subtract $5x$
      2. Subtract 11
      3. Divide by -2

   f. (Use algebra, not bags and stones.)
      1. Divide by 2
      2. Subtract 1
      3. Subtract x
      4. Subtract 1/2
      5. Subtract $x/2$

   g. Which of the ways in parts a-f seem the simplest and most direct to you?

2. Use algebra or algebra tiles to solve:

   a. $5t+1=2t+10$

   b. Check your solution to part a.

3. a. Use algebra or algebra tiles to solve $10y+3=9y+4$.
   b. Check your solution and picture the check as in the last line of figure 10.2.1, with bags and stones.

4. Some equations look more complicated than those we have solved so far, only because they are not written in simplest form. For example, the equation

$$x+5+x+4+2x+2=2+3x+1+4x$$

may look intimidating at first, but if you figure out how many xs and other numbers are on each side, you will find the equation is easy to solve. Rewrite the equation as simply as possible and solve it.

5. Rewrite the equation $t+5+4t-4=2+t+3+t+4$ as simply as possible and solve it.

6. Solve these equations, picturing the work with bags and stones if you like, but in any case writing out each step. Check each solution.

   a. $3a+8=4a+7$ b. $2b+6=4b+10$ c. $4c+4=2c+10$
   d. $3d+9=d+13$ e. $5e+3=2e+12$ f. $6f+3=2f+15$
   g. $g+9=5g+6$ h. $2h+11=5h+6$

7. Refer to question 2 in Exercises 9.2. Use algebra to show that Kenny's and Lenny's rules always give the same results as Jenny's rule.

8. **A teaching question**

   Some algebra texts teach four rules for solving equations.

   Rule 1. You can add the same number to both sides of an equation.

   Rule 2. You can subtract the same number from both sides of an equation.

   Rule 3. You can multiply both sides of an equation by the same number.

   Rule 4. You can divide both sides of an equation by the same number, provided that the number you divide by is not 0.

   a. Why is rule 2 the same rule as rule 1?
   b. Why is rule 4 simply a case of rule 3?
   c. We did not present these rules, because they are all examples of the fundamental idea that doing the same thing to both sides of a true equation yields a true equation. What are the pros and cons of teaching the four rules?

10.3 Language and Mathematics

Two of the traditional three Rs are language skills. The complexity of modern civilization is reflected in its language, with many new words and specialized jargons. We create specialized languages, such as codes and computer languages, for particular purposes, and we often think of music and other arts as languages of a different kind.

Mathematics, too, has a linguistic component, and using mathematics to work on any problem, no matter how simple, involves translating the problem into mathematical language. Algebra, one of the earliest and most important languages created for a special purpose, is a powerful tool for problem solving. The general approach to using mathematics, outlined in figure 10.3.1, is to translate a problem into mathematical language, solve the translated problem, and then translate the solution back to answer the original problem.

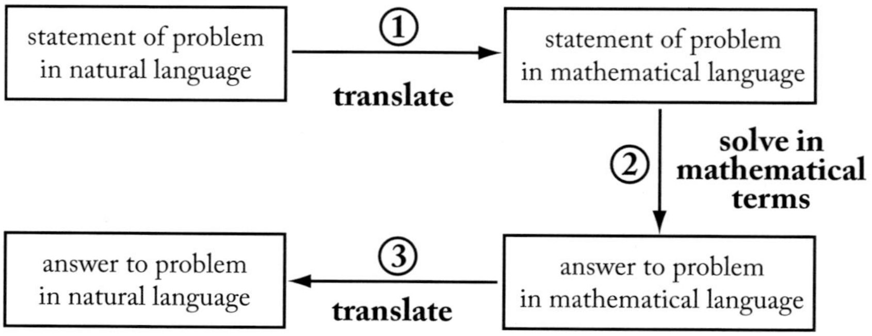

Figure 10.3.1

In practice the process outlined in figure 10.3.1 is not at all cut and dried. Steps 1 and 3, the translation steps, can be especially tricky, since they require knowledge of two languages, as well as the ability to relate them to each other. As an example, consider the question of translating the English phrase "ten less than twice a number" into algebra. Letting x represent the (unspecified) number, students often come up with three possibilities:

a. $10<2x$ b. $10-2x$ c. $2x-10$.

The first proposed translation looks reasonable enough, if you read the sign $<$ as "less than." However, $10<2x$ is a complete sentence, corresponding to the English statement, "Ten is less than twice a number," whereas, grammatically, "ten less than twice a number" is not a sentence at all, since it lacks a verb. Evidently, the sign $<$ corresponds to the English verb phrase "is less than" and not to "less than."

The absence of the small word "is" in the phrase to be translated forces us to consider other uses of the word "less." This word is sometimes used to indicate subtraction, as in "forty less three" for 40–3. This leads to the second option, $10–2x$. However, we have ignored the word "than" in this translation.

We use the phrase "less than" in subtraction examples like "seven less than twelve," which translates as 12–7. Here the order in which the numbers 12 and 7 appear is reversed from the English. Thus the third option, $2x–10$, is the correct translation of "ten less than twice a number."

Translation is an art, not a science, and it requires precise understanding of two languages. In this case, one of the languages, English, is notorious for its quirks and inconsistencies. Each of the three steps in figure 10.3.1 requires higher order thinking and may be difficult. Therefore students often find verbal problems (sometimes called "word problems" or "story problems") time consuming and difficult. The resulting frustration discourages many students, and it is especially acute for those whose English is not standard. No wonder students often dislike word problems.

One way to avoid this frustration is to avoid word problems, but in the long run that only guarantees failure in this area. Instead we must confront the issues and develop strategies for dealing with them. With this in mind, we now examine the three step process from figure 10.3.1 more closely.

Step 2 is in many ways the easiest to deal with, because it uses technical skills from within mathematics. These skills depend on the nature of the problem. They may require nothing more than simple arithmetic, but they could involve algebra, calculus, or other mathematical techniques. This step will not be the main difficulty with word problems, if you select problems for which the mathematical techniques needed are not difficult for your students. In this section, for example, the word problems are selected to use only the mathematical skills which we have covered so far. This allows us to focus on steps 1 and 3, which usually involve little formal mathematics but may require considerable linguistic skill. The exercises below illustrate this point. They include several problems which you will find challenging, even though the technical level of mathematics involved is relatively low.

Since translation requires knowledge of two languages, it is good to break the problem into smaller parts by means of exercises which involve only one of the languages. That is the purpose of the following exercise, taken from the *Math Workshop for Children*, in which the only responses required are to identify statements as true or false.

Based on the given information, decide whether or not each statement must be true. Circle **T** if it must be true, circle **F** otherwise.

I have 3 pockets. In them I have a total of 4 keys. No two pockets have the same number of keys.

a.	All the keys might be in one pocket.	T	F
b.	One pocket might have twice as many keys as another.	T	F
c.	One pocket has to be empty.	T	F
d.	There must be 3 keys in one of the pockets.	T	F
e.	There must be a single key in one of the pockets.	T	F

This exercise is not meant to be difficult, but it is by no means obvious. (In question a, for example, people may overlook the fact that zero is a number, so all the keys cannot be in the same pocket.) What mathematical skills are used in the exercise? None! All you need is to picture the situation, which amounts to nothing more than understanding it. This is an exercise in reading comprehension, not in mathematics. The sentences are short, simple and use only basic words; the reading level should not tax a fourth grader, yet the exercise is not easy for literate adults. Why? You can understand each word and each sentence at some level but still not "get the picture." You need to go beyond the words to make your own mental picture. Surprisingly, this key aspect of reading comprehension is neglected by many tests of reading comprehension. How can we help people learn to visualize?

1. Practice reading comprehension.

Exercises like the one above can both test reading comprehension and improve it. For those who do not "get the picture," you might make a concrete representation of the situation. For the example above, get a garment with three pockets, get four keys, and ask students to put keys in the pockets in various ways, using their observations to answer the true-false questions. The next step would be to represent the situation with other objects, such as envelopes for pockets and paper clips for keys. The question of keys in pockets is **generalized** to a question of objects in containers. **Generalization**, the process of fitting particular examples into a wider context, underlies all abstraction. From an abstract point of view, the question above is not about keys or pockets. It is about representing four as a sum of three different numbers. Seeing the pattern as independent of the context allows us to ignore the context and keep only the pattern itself. The mathematician Jean Dieudonné has called abstraction "the art of selective ignorance."

Students who can generalize from pockets and keys to objects in containers are ready for the next step, which is to use pictures instead of actual objects. Pictorial representations, as illustrated by hieroglyphics, are the original forms of most alphabets and Chinese characters. They represent a big step in linguistic development. Algebra can be thought of as growing out of mathematical picture writing.

To help students along this path of linguistic development, first have them make concrete models of situations they read about, then encourage them to make models which use different objects to represent a given abstract situation, and, later, have them make more and more abstract pictures to represent situations.

Another way to practice reading comprehension is to have students scrutinize advertisements for what they really say. See exercises 8-10 below.

2. Write problems to fit equations.

Build the bridge between English and mathematical language from both ends. Start with some mathematical statements, and have students write English statements to fit them. This should begin at a very elementary level (e.g. Write a brief story to fit 2+3=5.) and continue as needed through the mathematics curriculum. This works on step 3 of figure 10.3.1.

3. Arrange situations which help students to link mathematics to reading.

In recent years "writing across the curriculum" has incorporated the teaching and practicing of language skills into other areas, such as social studies, science, and mathematics. This should, naturally, include "reading across the curriculum." It is not easy to find material which fits the students' level in both reading and mathematics and engages them. Games involving both mathematics and reading are ideal for this purpose, because students get deeply involved in both subjects and check each other zealously. Mathematical problems that merit study as literature are woefully rare, but they do exist. Books of puzzles by Martin Gardner are excellent. Those by Sam Lloyd, Hubert Phillips ("Caliban") and Lewis Carroll are interesting both as mathematics and as literature. However, for all their wit, some old classics contain material that would be inappropriate or even offensive in today's classrooms, so be sure to check it yourself before using it in a classroom.

Realism in problems. Many of the problems found in books are not realistic to the students who are asked to solve them. How often, outside the classroom, are you asked to find someone's age, given that seven years ago she was one fourth as old as she will be fourteen years from now? Ultimately, no problem is realistic unless it is your problem. Time, rate, and distance problems were never realistic to me until I took flying lessons. Suddenly they were very realistic. One way to make word problems more realistic is to involve the class in activities and projects which naturally lead to word problems. This is an excellent approach and should be used as much as possible. However, it may be difficult or impossible to arrange for class activities that generate the full range of problems you would like the class to consider, so you may have to use problems which are clearly not realistic. This is not as bad as you might at first think.

Realism in problems, while sometimes desirable, is not necessary. A problem is a puzzle, and solving puzzles is an activity which appeals to large numbers of people the world over, as is shown by the popularity of jigsaw puzzles, crossword puzzles, and many other kinds of puzzles which have little or no practical value. A good problem, like any good puzzle, "hooks" people. Usually the question is easy to understand but seems challenging. There is an art to posing such problems in appealing ways.

Exercises 10.3

1. (From the *Math Workshop for Children*) There are 12 goats on a hill. Fifty percent of them are black. Each of the others is either white or brown. A third of the goats are eating daisies. Two thirds of the goats have bells tied around their necks.

   What can you say for sure? Mark each statement T (true) or F (false), using the T only if you can be sure the statement is true.

   a. There are six black goats. T F
   b. Four goats are eating daisies. T F
   c. Those not eating daisies have bells. T F
   d. Some black goats have bells. T F
   e. Some white goats have bells. T F
   f. Two thirds of the black goats are eating daisies. T F
   g. There are half as many white goats as black goats. T F
   h. If all the goats eating daisies are brown, then there are fewer white goats than brown goats. T F

2. (From the *Math Workshop for Children*)

   Artist's Supplies for Sale
   Pencils 7¢ or 3 for 20¢
   Boards 9¢ or 3 for 25¢

   (Note: No person bought more than 9 of any one item.)

What Can You Say?

a. Five pencils cost 35¢.* T F

b. Five pencils and five boards cost a total of 77¢. T F

c. No one can spend exactly 26¢ for pencils and boards. T F

d. There are at least two ways to spend exactly 43¢ at this sale — 5 pencils and 1 board or 5 boards and 0 pencils. T F

e. There are at least two ways to spend exactly 88¢ at this sale. T F

f. There are at least three ways to spend exactly 54¢ at at this sale. T F

* *Assume all pencils are bought by one person.*

3. (By Diane Melvin) Andrea left her math homework on her desk yesterday, and this morning it is gone. Andrea was the last one out of the room yesterday. This morning three students had already shown up for class before Andrea. Todd came in just before Andrea. Muggsie came in before Andrea. Cindy came in right after Muggsie.

   a. Cindy came in before Todd. T F

   b. Andrea was the last student to enter the classroom today. T F

   c. Todd came in just after Muggsie. T F

   d. If one of the three students took Andrea's homework, then Muggsie took it. T F

4. (By Teresa Feger) Jenny has gone out with Kevin for six days. Samantha has gone out with Matt for five times as many days as Loren has gone out with Kim. Loren and Kim just started going out with each other yesterday.

   a. Jenny and Kevin have gone out together longer than Loren and Kim. T F

   b. Matt and Samantha have gone out together longer than Jenny and Kevin. T F

   c. Loren and Kevin were flirting behind Jenny's and Kim's backs. T F

   d. Samantha and Matt have gone out together five days. T F

   e. Loren and Kim are going to break up, because Kim saw Loren and Kevin flirting. T F

   f. Jenny and Kevin have been together for less time than Samantha and Matt but more than Loren and Kim. T F

5. Make up a problem like questions 1-4 above. Keep the vocabulary and sentence structure simple. Make the questions focus on understanding the situation rather than on using clues to solve a mathematical problem.

6. Write a brief story (at most two sentences) corresponding to each of the following mathematical sentences:

   a. $3+5=8$ b. $4<11$ c. $500-238=262$

   d. $13\times11=143$ e. $91=10\times9+1$ f. $3+n=7$

7. Examine some reading comprehension tests used in schools today. To what extent do they test visualization? To what extent do they test precise understanding of every word and phrase (as opposed to the less specific understanding needed to deal with items like, "In the paragraph above, how did Amanda feel as she left town?" or, "Write an ending to the story that fits the part you are given above.")

8. What is the catch in this offer?

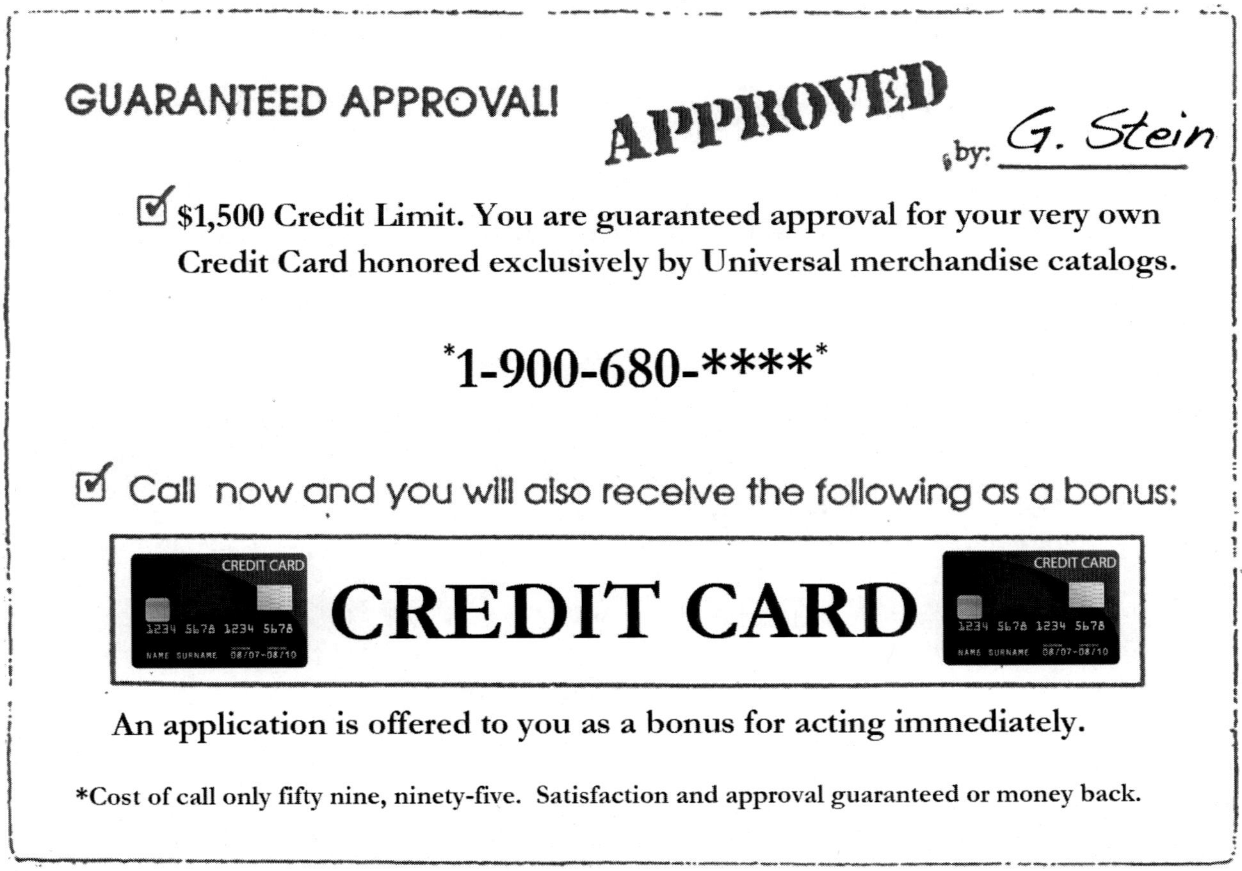

9. What is the catch in this offer?

**CR CR60
MRS ROBERT G STEIN
5500 UNIVERSITY PKWY
SN BERNARDINO CA 92407

ID: SLR10W0919D

DETERMINATION:
Mrs. STEIN did
not claim gift.

EXPIRES: OCTOBER 6, 1989

NUMBER OF CLAIMS: TWO (2)
MAXIMUM REDEMPTION VALUE: $21,500 in gifts to Mrs. STEIN
GIFT CLAIM CONTINGENCY: Mrs. STEIN to receive a 1989 Ford Taurus Sedan, and $5,000 Cash.

INTENTION TO CLAIM GIFT: Notify within 72 hours by calling toll-free: 1-800-456-7868, EXT. SLR.

Mrs. STEIN, our records confirm that you have definitely been chosen, & are going to receive, upon qualifying, two of the four gifts. These gifts include: 1)1989 Ford Taurus; 2)$5,000 Cash; 3)Zenith VCR/TV Remote complete with speed search, channel scanner and AB selection features; 4)AM/FM Multiplex Stereo. We have determined that you have not claimed your TWO stated gifts. Find attached a Certificate of Disbursement which entitles you to your TWO (2) of the four stated gifts. This is your final opportunity to claim these gifts.

To make an appointment, CALL TOLL FREE: 1-800-456-7868 and ask for EXT. SLR. Operators are on duty from 9AM to 7PM Mon. thru Fri. and 9AM to 1PM on Saturday.

IMPORTANT: This offer expires on the date shown above. The actual gifts that you will receive are determined by matching the below stated claim numbers with the official Master List at The Inn at Silver Lakes.
SPONSORED IN CA BY: THE INN AT SILVER LAKES, 14818 CLUB HOUSE DR., HELENDALE, CA 92342.

DETACH HERE — DETACH HERE

CERTIFICATE OF DISBURSEMENT

THIS DISBURSEMENT HAS BEEN ORDERED AND SHOULD BE EXECUTED IN ACCORDANCE WITH THE RULES STATED IN THIS OFFER AS A QUALIFIED RESPONDENT. WHEN YOU MATCH THE CLAIM NUMBERS ENTITLING YOU TO ITEMS 1 & 2. THEN

REGISTERED CERTIFICATE OF DISBURSEMENT NUMBER
RV00960-04574606
VALUE OF ABOVE STATED MERCHANDISE
$21,500.00
MERCHANDISE CLAIM NUMBERS
1467327 AND 1985358

AS A DULY APPOINTED REPRESENTATIVE OF THE SWEEPSTAKES DIVISION, I HEREBY ENDORSE THIS ORDER IN FAVOR OF THE ABOVE NAMED RECIPIENT

1-31-89
DATE

DIRECTOR

THIS HEREBY ENTITLES
MRS. STEIN OF SN BERNARDINO, CA
TO RECEIVE A NEW FORD
TAURUS AND $5,000 CASH.

SCHEDULE YOUR APPOINTMENT
TO VISIT US ON MONDAY
& YOU WILL ALSO RECEIVE IN
ADDITION TO YOUR GIFTS LISTED
ABOVE, A CATALINA CRUISE
FOR 2. RETAIL VALUE $49.95.

10. What is the catch in this offer?

Fisherman's Retreat

32300 SAN TIMOTEO CANYON RD., REDLANDS, CA 92373

CAR-RT SORT ** CR60

MRS. ROBERT G STEIN
5500 UNIVERSITY PKWY
SN BERNARDINO, CA 92407 (EXTENSION 514) AWARD NUMBER **45228**

EXP. DATE MAY 21, 1989
FILE NO: 15485-03-45228
ATTENTION: MRS. STEIN

MRS. STEIN AND FAMILY CHOSEN AS SOUTHWESTERN RECIPIENTS IN NATIONAL GIVEAWAY/STOP/YOU ARE SELECTED TO RECEIVE ATARI CC 1900 WITH DISPLAY, MEMORY AND KEYBOARD OR $1,995.00 CASH/STOP/THIS IS DEFINITELY AN URGENT NOTICE/STOP/ PLEASE CALL TOLL FREE 1-800-234-5632 IMMEDIATELY TO CLAIM YOUR AWARD/STOP/ALL MERCHANDISE MUST BE GIVEN AWAY TO COMPLY WITH REGULATIONS/STOP/YOU WILL RECEIVE THE CASH OR ACTUAL ATARI DESKTOP UNIT ON THE SPOT/STOP/

PROGRAM WINNERS INCLUDE:

MCGIBBON	WISC.	FORD THUNDERBIRD OR $10,000 CASH-CLAIMED
MCCALL	CALIF.	HONDA A.T.V. OR $1,499 CASH-CLAIMED
BIRCHEM	TENN.	RCA 25" COLOR TV OR $598 CASH-CLAIMED
MRS. STEIN	CALIF.	DESKTOP ATARI OR $1,995 CASH ***UNCLAIMED***
HOPPIE	UTAH	$500.00 U.S. SAVINGS BOND-CLAIMED

MERCHANDISE (OR CASH) MUST BE AWARDED/STOP/FAILURE TO RESPOND IMMEDIATELY WILL RELEASE YOUR MERCHANDISE (OR CASH) TO OTHER PERSONS IN YOUR REGION/STOP/ YOU MAY BE ELIGIBLE FOR OTHER BONUS MERCHANDISE/STOP/REMEMBER THIS IS A SWEEPSTAKES IN WHICH YOU DEFINITELY ARE A SIGNIFICANT WINNER AS HAS BEEN DETERMINED BY YOUR AWARD NUMBER/STOP/SO PLEASE CALL IMMEDIATELY/STOP/PLEASE SEE BELOW/STOP/

TO CLAIM YOUR AWARD SIMPLY MEET THE ELIGIBILITY AND REDEMPTION RULES LISTED ON THE REVERSE SIDE AND CALL THE AWARD CLAIM HOTLINE TOLL FREE AT 1-800-234-5632 BETWEEN 9:00 AM AND 7:00 PM MONDAY THROUGH FRIDAY FRIDAY AND 9:00 AM TO 5:00 PM SATURDAY (EASTERN TIME)/STOP/

YOUR BONUS: CALL WITHIN 72 HOURS AND RECEIVE A $100.00 GIFT CERTIFICATE.

11. A fifth grader once showed me her homework, which she was proud to have done in just a few minutes. The first dozen or so questions were simply multiplication drill. Then, at the end, there were several word problems. Surprised at her speed, I asked if she had actually read the problems. She said she had not, and added that to do so would be a waste of time. Knowing that the set of problems was about multiplication, she would just find the numbers in the short paragraph and multiply them. To my dismay, that worked in all cases. She had learned not to read or think. She did not understand why I was upset. When I looked further, I found that although her math book had many word problems, none required reading! It seemed that those who produced the textbook had deliberately arranged this so that the teachers and students who used the book would feel successful with word problems without really learning to connect language and mathematics.

a. How could you, as a classroom teacher, deal with teaching word problems if the textbook your class was issued was of the type described above?

b. If you were on a committee involved with selection of textbooks in mathematics, what are some of the things you would look for in the area of word problems?

12. The game Chutes and Ladders originated in India centuries ago as Snakes and Ladders.
    a. What mathematical skills are required?
    b. What reading skills are required?
    c. What skills in mathematics and reading are practiced in playing Chutes and Ladders?

13. For the game of Monopoly:
    a. What mathematical skills are required?
    b. What reading level is required?
    c. What skills in mathematics and reading are practiced in playing?

14. In an old riddle, a man speaks about a second man, saying,

    Brothers and sisters I have none,
    But that man's father is my father's son.

    a. How is the speaker related to the person spoken about?
    b. The difficulty of the riddle lies entirely in the language. Can you find a way to picture the situation?

10.4 Solving Problems with and without Algebra

Section 10.3 emphasized the role of language in understanding and solving problems. Now we will use language skills and some special techniques to solve problems. Consider, for example, this question.

Example 1. A man took some money to a fair but spent a third of it to get in, and then lost a third of the rest. At that point he had \$24 left. How much money did he take to the fair?
Solution: Experiment, starting with various numbers. To keep the numbers simple, start with multiples of 3. Suppose you start out with \$300. You spend a third of it, or \$100, on admission, leaving \$200. Next you lose a third of this, or

$\$66\frac{2}{3}$, leaving you with $\$133\frac{1}{3}$.

What if you start with less? Let's try \$90. Spend a third (that's \$30) and you have \$60 left. Now lose a third of that (\$20) and \$40 is left.

This is much closer, so start with even less. Try \$27. Spend a third of it (\$9) to get in, leaving \$18, then lose a third of that (\$6), leaving only \$12 at the end. Since \$27 seems to be too low, \$90 too big, try something in between, such as \$60. We can sum up the work in a table much like figure 9.1.1, where we put various numbers through a process of subtracting a third, then subtracting a third of what is left, hoping to come out with a "target" value of \$24.

Number of dollars at start (guess)	300	90	27	60
Number of dollars left after admission	$\frac{2}{3}(300) = 200$	$\frac{2}{3}(90) = 60$	$\frac{2}{3}(27) = 18$	$\frac{2}{3}(60) = 40$
Number of dollars left after losing a third	$\frac{2}{3}(200) = 133\frac{1}{3}$	$\frac{2}{3}(60) = 40$	$\frac{2}{3}(18) = 12$	$\frac{2}{3}(40) = 26\frac{2}{3}$

The last result, $26\frac{2}{3}$ is close to \$24, so \$60 is not all that far off. Try \$54.

Number of dollars at start (guess)	300	90	27	60	54
Number of dollars left after admission	200	60	18	40	$\frac{2}{3}(54) = 36$
Number of dollars left after losing a third	$133\frac{1}{3}$	40	12	$26\frac{2}{3}$	$\frac{2}{3}(36) = 24$

You hit the target. The initial number of dollars was \$54.

Algebraic solution: Let x be the starting number of dollars (right-hand column).

Number of dollars at start (guess)	300	90	27	60	54	x
Number of dollars left after admission	200	60	18	40	36	$\frac{2}{3}x$
Number of dollars left after losing a third	$133\frac{1}{3}$	40	12	$26\frac{2}{3}$	24	$\frac{2}{3}\left(\frac{2}{3}x\right)$

Now this last is supposed to be \$24, so $\frac{2}{3}\left(\frac{2}{3}x\right) = 24$

Multiply out the left side $\quad \frac{4}{9}x = 24$

Multiply by $\frac{9}{4}$ $\quad x = \frac{9}{4}(24) = 54$

The algebra flows naturally from trial and error, and we get the same result as we did before.

A third solution: Work backwards. If he had \$24 after losing a third of his money, then \$24 was $\frac{2}{3}$ of what he had just before, so he must have had \$36 before he lost any money. That in turn is $\frac{2}{3}$ of the money he had when he first got to the fair, so multiply again by $\frac{3}{2}$.

Then he must have had \$54 before paying admission.

A fourth solution: Picture the initial sum of money like this:

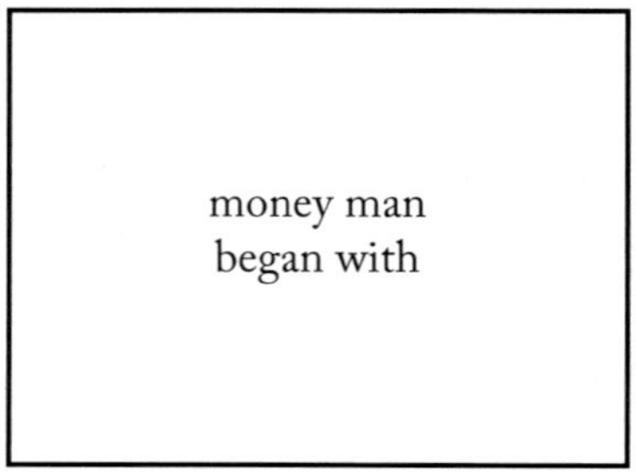

Now picture what happened to the money.

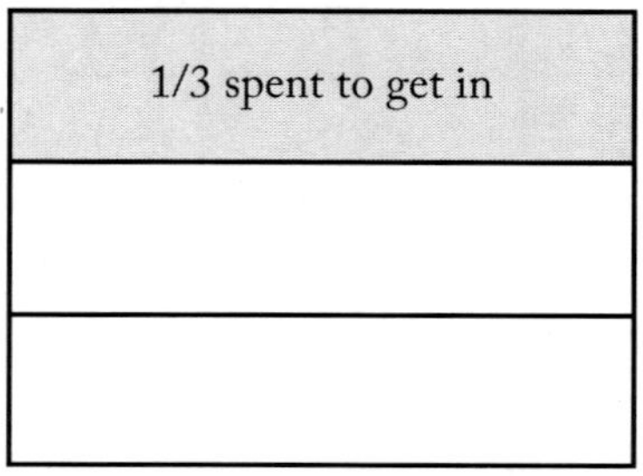

The money remaining after paying admission to the fair is represented above by the unshaded portion of the big rectangle. The man now loses a third of that, (darker shading).

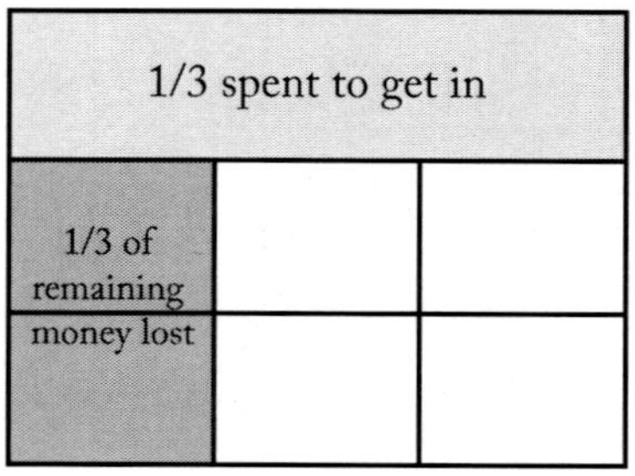

Completing the vertical lines shows that the man has $\frac{4}{9}$ of the money he started out with.

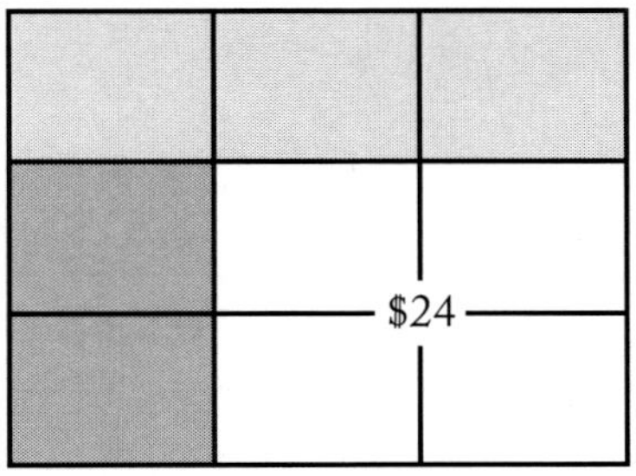

Evidently, $\frac{1}{9}$ of the original money was a fourth of \$24, or \$6, so the man set out with \$9×6 or \$54.

Example 2. I have three times as many quarters as nickels, and my quarters and nickels together are worth \$8.80. How many of each have I?
Solution: Give a name to one of the numbers sought, say, the number of nickels. If n is the number of nickels, then $3n$ is the number of quarters, and their total value in dollars is $0.05n+0.25(3n)$, which we are told is \$8.80.

Then	$0.05n+0.25(3n)=8.80$
Multiply by 100	$5n+25(3n)=880$
Simplify	$80n=880$
Divide by 80	$n=11$

There were 11 nickels and 33 quarters.

Another solution: Experiment. You could make \$8.80 with \$7 in quarters and \$1.80 in nickels, but that uses 28 quarters and 36 nickels. You need more quarters and fewer nickels. With \$8 in quarters and \$.80 in nickels, you use 32 quarters and 16 nickels. Try still more quarters and fewer nickels. Trading in 5 nickels for a quarter, you could make \$8.80 with 33 quarters and 11 nickels. This uses three times as many quarters as nickels, as required.

As you can see, experimenting with a problem can give you a feel for what is needed, and might even lead you to a solution. Algebra enables you to handle problems for which trial and error would be impractical, but do not dismiss non-algebraic methods out of hand.

Example 3. A woman is three times as old as her son. Five years ago her age was four times his age. How old are they?
Solution: Begin by experimenting. If the son is ten, then the woman must be 30 (three times as old), but then five years ago the child was five, and the woman was 25, which was five times his age.
Try fifteen for the son's age. Then the mother is 45, and five years ago they were ten and 40, which fits the conditions of the problem.

Another solution:
Organize the information in a table:

	Now	5 Years Ago
Woman's Age		
Son's Age		

The first sentence, "A woman is three times as old as her son" concerns their ages now. If s is the son's present age, then $3s$ is the woman's present age.

	Now	5 Years Ago
Woman's Age	$3s$	
Son's Age	s	

Five years ago the son's age was s-5 and the woman's was $3s$-5.

	Now	5 Years Ago
Woman's Age	$3s$	$3s-5$
Son's Age	s	$s-5$

Now the second sentence of the problem says that $3s-5=4(s-5)$

Solve: $3s-5=4s-20$

$15=s$

Then the son is 15, and the woman is 45. As a check, 5 years ago the woman, at 40, was indeed four times her son's age, which was 10. Had we chosen the woman's age as the unknown, the solution would have involved fractions, but it would not have been fundamentally different (see question 10 below.)

Here trial and error was easy. Algebra is often best when the numbers involved are hard to guess, as might happen if the answer involved clumsy fractions. In such cases setting up equations can be greatly facilitated by organizing the information. Charts, tables, and sketches are all useful for this purpose.

Example 4. A box contained apples. Alpha took half the apples, then took one more. Next, Bravo took half of what was left. Then there were four apples in the box. How many were there at first?
Solution: Make a chart and try a few guesses, as in the first three columns of the chart below. These guesses did not lead to the target value of 4 in the bottom row, but additional guesses would probably solve the problem. Instead, however, consider an algebraic approach. Since you are looking for the original number of apples, use that as your unknown; call it s (for starting number), and follow down the table as in the right-hand column:

Number of apples at start	50	42	30	s
After Alpha takes half	$\frac{50}{2} = 25$	$\frac{42}{2} = 21$	$\frac{30}{2} = 15$	$\frac{s}{2}$
After Alpha takes one more	25–1 = 24	21–1 = 20	15–1 = 14	$\frac{s}{2}-1$
After Bravo takes half	$\frac{24}{2} = 12$	$\frac{20}{2} = 10$	$\frac{14}{2} = 7$	$\frac{1}{2}\left(\frac{s}{2}-1\right)$

At the end there were four apples left, so $\frac{1}{2}\left(\frac{s}{2}-1\right)=4.$

To solve, you could first multiply out the left side, but here you don't need to do that.

Multiply by 2 $\qquad \frac{s}{2}-1=8$

Add one $\qquad \frac{s}{2}=9$

Multiply by 2 $\qquad s=18$

Starting with 18 apples does indeed satisfy the problem. Once again the steps used to solve the equation suggest a way to solve the problem without algebra by working backwards. If there were four apples left at the end, there must have been 8 before Bravo took half. Then there must have been 9 before Alpha took an extra apple, and 18 before Alpha took half, which is to say, at the start.

Helpful though these examples might be, they cannot substitute for tackling problems on your own. Problem solving is an art, not a science; to learn it takes practice. A certain amount of frustration is inherent in problem solving. Anything you can solve in a few minutes hardly counts as a problem, but perseverance in problem solving must be built up carefully. One tool for doing that is group problem solving.

Group problem solving

In group problem solving, no single individual carries the entire burden for coming up with the solution, and members of the group can build on each other's ideas and correct each other's misperceptions. That way, working with others can ease the frustration inherent in problem solving and encourage perseverance. Group problem solving also encourages participation by students who would be afraid to take on difficult-looking problems by themselves. Although some problems yield to a flash of insight, it may help to think of problem solving as more like putting together a picture puzzle. Work at it bit by bit, often with others, until the answer emerges. For fun and practice, a collection of word problems follows. You may find some of these straightforward, but others may stump you, at least for a while. When that happens, the following suggestions may help.

Problem-Solving Checklist
1. Reread the problem until you understand it clearly and can restate it in your own words.
2. Take an inventory of the clues and information given.
3. Identify what is wanted and give it a name. (In algebra what is wanted is usually a number, which can be named by a letter.)
4. Organize the information (a table or sketch often helps). That may lead you to break the problem into smaller problems and try to solve them separately.
5. Ask what additional information would be needed to solve the problem.
6. Look for similar problems. Have you solved one in an earlier exercise or seen one worked in a book?
7. If you have worked a long time on a problem and feel stuck, let it rest; it may seem easier when you return to it another time.

Presenting your solution

A solution must be clear and convincing, both inside the classroom and out. If, for example, you bid on a job, say to build or repair something, your bid must include more than simply the total cost. You should explain, in a way the customer can understand, what materials you will use, when you expect to complete the job, etc. It is your job to convince the customer, not the customer's job to understand you. Therefore, do not overwhelm the customer with technical jargon but present your work in a way he/she can understand.

That is the attitude to model, when you ask students for clear and complete solutions. Avoid assigning large numbers of routine exercises and having the students list their answers at the margin of the page (so that you can read the answers without looking at the supporting work). This practice sends exactly the wrong message-that the final, numerical answer is all that matters. With that in mind, here are some tips for presenting solutions.

Presenting solutions to word problems

- **Organize the paper neatly and write legibly. Crowding the work saves paper but wastes time.**
- **Make sure the reader knows what question is being addressed. Either restate it, or, if it comes from a book, identify it clearly.**
- **If you use a letter for a number, introduce it to your reader by defining it, with units where appropriate. Right: "Let *x* be the number of dollars. . . ." (An equation that would be correct if *x* is a number of dollars might well be incorrect if *x* is a number of pennies, for example.) Wrong: "Let *x* be the amount of money. . ."**
- **Never use the same letter for more than one number in a given problem.**
- **In a sequence of equations, line up the equals signs vertically to make your work easy to read.**
- **If erasures are making the work look messy, just line out the work you don't want and rewrite it correctly in another place.**
- **State your conclusions in clear English. If *w*, for example, is the width of a rectangle, in feet, that you were supposed to find, do not end the work with "*w*=17" but write a clear sentence: "The rectangle is 17 feet wide."**

Exercises 10.4

In all questions, if you use a new letter, introduce it with a sentence of the form "Let *x* be the number of. . . ." Do not hurry into writing equations!

1. Suppose pencils cost c cents each.
   a. How much would 9 pencils cost?
   b. If I have 3 cents more than I need to buy 9 pencils, express in terms of c the amount of money I have.
   c. I need 15 more cents to buy 12 pencils. Express in terms of c, the amount of money I have.
   d. Suppose both parts b and c are true. Find the price of each pencil and how much money I have.

2. The sum of two thirds of a number and one fourth of that number is 33. What is the number?

3. The sum of the weights of two students is 138 pounds, and one is 34 pounds heavier than the other. How much does the heavier student weigh? (From George Lenchner, *Creative Problem Solving in School Mathematics.*)

4. Two years ago I was twice as old as Maria will be ten years from now. Now I am 44. How old is Maria?

5. My wife is six times as old as my son. In four years his age will be one fourth of hers. How old are they?

6. I'm thinking of a number. If I add 5 to this number, then triple that, the result is nine less than five times the original number. What is the number? State this problem as an equation and solve it.

7. Terry made some money baby sitting. She spent half of it on a new pair of shoes, and then she spent one third of the rest on a gift for a friend. That left her with $12 less than she earned. How much did she earn?

8. I want to give each child here nine cookies, but I lack eleven cookies for that. If I gave four cookies to each child, I would have nine left over.
   a. How many children are here?
   b. How many cookies have I?

9. If you multiply a certain number by 5, then subtract 3, you get 11 more than the original number. What is the original number?

10. Rework example 3, using the woman's present age as the unknown.

11. In fourteen years Tarzan's parrot will be two years older than three times its present age. How old is it now?

12. Dear Action Reporter:

    If I give each of my children 11 cents, I'll have 9 cents left. If I had 47 cents more, I could give 15 cents to each. I can't count my money or my kids, and my pocket calculator is on the fritz. I don't know what do to.

    Old Woman in Shoe

    How much money and how many children has the Old Woman?

13. I had some money, and I earned $17 more. Then I spent one third of my money. At that point I had $42 left. How much did I begin with?

14. One day at the races I lost $10 on the first race. I bet all my remaining money on the second race and doubled my money. That left me with $50 more than I had brought to the races. How much did I begin with?

15. A boy is half as old as his sister and two years younger than his brother. The sum of their ages is 38. How old is the boy?

16. In one game team A doubled team B's score. In a return match team B won by as many points as it had lost by in the first game, even though team A got as many points in the second game as both teams together scored in the first game. In the second game a total of 42 points were scored. What were the scores of each game?

17. One day I went to the races. On the first race I won \$20. Then I spent half my money buying a ticket on a horse that, if he won, would pay off \$4 for each dollar bet. Luckily, I won. After spending \$20 to celebrate, I went home with four times as much money as I had when I first got to the races. How much did I start with?

18. Three merchants at sea suddenly experience a terrible storm. To stay afloat, they have to throw 100 guilders worth of their cargo overboard. After they reach port, they have to divide the loss in proportion to the values of their cargos. Originally, the first merchant had 300 guilders worth of cargo on the ship, the second 400, and the third 500. What should be each merchant's share of the remaining cargo? (Adapted from Peter van Halle's arithmetic text of 1568[2])

19. On the second week in November, a family spent \$2 more for groceries than they had spent in the first week that month. Their bills for the third week (which included a holiday party) totaled just \$5 less than the sum of their grocery bills for the first two weeks. For the last week in November, they spent \$1 less on groceries than they did in the first week. How much did they spend in each of the four weeks if the four bills totaled \$197?

20. Rollins Stables bought two horses for \$360 at an auction. One cost \$80 more than the other. How much did each horse cost?

21. Two numbers total 999. One-ninth of one of them added to one-sixth of the other yields 138. What are the numbers?

22. "A man met several beggars and wished to give 25 pence to each; but, on counting his money, he found that he had 10 pence too little for that; and then he made up his mind to give each 20 pence. After doing this he had 25 pence over. What had he at first, and how many beggars were there?" (From Chrystal's *Textbook of Algebra*)

2 Problem 18 is from Marjolein Kool, "*Two Pots with One Lid: The first arithmetic textbooks in the Netherlands before 1600*," Proceedings of the HPM Conference on History in Mathematics Education, Department of Mathematics, National Taiwan Normal University, 2000.

23. "To encourage his son to learn arithmetic, a father agrees to pay his boy 8 cents for every problem correctly solved and to fine him 5 cents for each incorrect solution. At the end of 26 problems neither owes anything to the other. How many problems did the boy solve correctly?" (From an algebra text of 1608 by Christoph Clavius[3])

24. If I were to give 7 cents to each of the beggars at my door I would have 24 cents left. I lack 32 cents of being able to give them 9 cents apiece. How many beggars are there, and how much money have I? (Adapted from Clavius)

25. A servant is promised $100 and a cloak as his wages for a year. After 7 months he leaves this service and receives the cloak and $20 as his due. How much is the cloak worth? (Adapted from Clavius)

26. A merchant visited three fairs. At the first he doubled his money and spent $30, at the second he tripled his money and spent $54, at the third he quadrupled his money and spent $72, and then he had $48 left. How much money had he at the start? (Adapted from *Triparty en la science des nombres*, a text by Nicolas Chuquet, 1484)

27. A carpenter agrees to work under the conditions that he is to be paid $5.50 every day he works, while he must pay $6.60 every day he does not work. At the end of 30 days he finds he has paid out as much as he received. How many days did he work? (Note. The carpenter is paid for partial work days and is docked for partial days he does not work.) (Adapted from Chuquet)

28. A woman at Las Vegas wins $50 at roulette. Then she plays half of her money (including the roulette winnings) and loses it. Then she wins $10, after which she has the same amount she began with. How much was that?

29. A man spends one third of his money and loses two thirds of the remainder. He then has $12. How much money had he at first?

30. If 6 oranges cost as much less than 80¢ as 14 cost more than 80¢, how much do they cost each?

---

[3] Christoph Clavius (1537-1612) was a Jesuit from Germany. His textbook was widely used, including at La Flèche, where Descartes studied. At the request of Pope Gregory XIII he proposed a revision of the Julian calendar (used since the time of Julius Caesar). This proposal was eventually adopted, (though some Protestant regions delayed for many years) and we now use the "Gregorian" calendar. The change to the new calendar required the day after 4 October 1582 to be numbered 15 October, causing riots in Frankfort against this robbery of ten days.

31. At the market I bought plums, peaches, and pears. I bought twice as many peaches as pears and $4\frac{1}{2}$ times as many plums as pears. In all, I bought 45 pieces of fruit. How many of each did I buy?

32. I have 56 coins, some half dollars, the rest nickels. Their total value is \$10. How many of each have I?

33. Little is known of the ancient mathematician Diophantus except for this puzzle,[4] which is said to come from his epitaph.
"This tomb holds Diophantus. Ah, how great a marvel! The tomb tells scientifically the measure of his life. God granted him to be a boy for the sixth part of his life, and adding a twelfth part to this, He clothed his cheeks with down; He lit him the light of wedlock after a seventh part, and five years after his marriage He granted him a son. Alas! Lateborn wretched child; after attaining the measure of half his father's life, chill Fate took him. After consoling his grief by this science of numbers for four years, Diophantus ended his life."
How long did Diophantus live?

34. (An old favorite) Three travelers share equally the cost of a \$30 hotel room, and each pays \$10 at the desk before going to the room. The desk manager notices that the room really costs only \$25, so he sends the bell man to refund the room with \$5. But the bellman returns just \$1 to each traveler and pockets the other \$2. Now the travelers have paid \$27 in all (\$10 each, less \$1 refund), and the bellboy has \$2. But 27+2=29. Where's the other dollar?

35. If to my Age there added be
One Half, one Third, and three times Three
Six score and ten the Sum you'll see
Pray find out what my Age may be?
(From the *Ladies Diary,* in 1707, founded in 1704 by Charles Tipper.)

10.5 Linear Equations in Two Unknowns

Sometimes we use linear equations in more than one unknown. For example, to say that Xavier has three times as many marbles as Yasmin, we might write $x=3y$. More generally, a **linear equation** in two unknowns, x and y, is one which may be written

[4] This problem is from the *Greek Anthology*. For many years Heidelberg's Palatine library had the only copy of a large collection of short epigrams known as the *Greek Anthology* or the *Palatine Anthology*. It is a revision from the twelfth or thirteenth century of a ninth century work which drew on many ancient Greek and Roman sources. It includes 46 numerical puzzles, thought to be mainly of ancient Greek origin, gathered by Metrodorus, a grammarian, probably around 500 C.E. A five volume set containing both the Greek text and an English translation by W. R. Paton was published in 1916-18 by William Heinemann, Ltd., London and reprinted in 1979 by the Harvard University Press. Here, as in many other modern books, Paton's translation is used.

in the form $ax+by+c=0$. (We assume, of course, that a and b are not both 0, since, if they are, the equation boils down to $c=0$, which is either true or false, depending on the value of c, but there is not much to say about it either way. If just one of the numbers a and b is is 0, we have an equation in one unknown, which we have already discussed.) A linear equation of this type is satisfied by infinitely many pairs of numbers. If neither a nor b is 0, you can find a solution by selecting a value for one unknown, say x, substituting that value into the equation, and then solving for the second unknown.

Example 1. Find solutions of $3x + 2y = 7$
a. If $x=1$ and b. If $y=5$.

Solution: a. Substitute 1 for x in the equation. $3\cdot 1 + 2y = 7$

Now solve this equation for y

$$2y = 4$$
$$y = 2$$

Check: $3 \cdot 1 + 2 \cdot 2 = 7$

b. Substitute 5 for y in the equation. $3x + 2\cdot 5 = 7$

Now solve this equation for x

$$3x = 7 - 2\cdot 5$$
$$3x = -3$$
$$x = -1$$

Check: $3(-1) + 2 \cdot 5 = 7$

The desired solutions are $x=1, y=2$ for part a and $x=-1, y=5$ for part b.

Many problems involve finding a single pair of numbers which satisfy each of two equations in two unknowns. Such equations are called **simultaneous equations**. Example 2 shows how to solve a system of two of simultaneous equations.

Example 2. Solve the simultaneous equations $\begin{cases} 2x - y + 7 = 0 \\ 3x - y + 5 = 0 \end{cases}$

Solution: First solve each equation for y in terms of x

$$\begin{cases} y = 2x + 7 \\ y = 3x + 5 \end{cases}$$

Now if the same x and y satisfy both equations, since both $2x+7$ and $3x+5$ equal y,

$$3x + 5 = 2x + 7$$

Solve for x

$$x = 2$$

Substitute this value of x into either of the given equations

$$2\cdot 2 - y + 7 = 0$$

Solve for y

$$11 = y$$

Check the solution $x=2, y=11$ in **both** original equations.

$$\begin{cases} 2\cdot 2 - 11 + 7 = 0 \\ 3\cdot 2 - 11 + 5 = 0 \end{cases}$$

Example 3. Some thieves, who had stolen silk cloth, gathered together under a bridge to decide how to distribute their ill-gotten gains among them. Some pedestrians who happened to pass the bridge overheard their loud arguments. They were saying that if each one took 12 *tan*, then 12 *tan* would be left, and if each one took 14 *tan*, then 6 *tan* would be short. How many thieves were there and how much cloth was there?[5]

Solution: Evidently the *tan* is a unit of measure for cloth. Let c be the amount of cloth, measured in *tan*, and let t be the number of thieves. Then

$$\begin{cases} 12t = c - 12 \\ 14t = c + 6 \end{cases}$$

Solve each equation for c in terms of t

$$\begin{cases} 12t + 12 = c \\ 14t - 6 = c \end{cases}$$

Equate the expressions for c

$$12t + 12 = 14t - 6$$

Solve for t

$$18 = 2t$$

$$9 = t$$

Substitute t=9 into either original equation and solve for c.

$$12 \cdot 9 = c - 12$$

$$108 + 12 = c$$

$$120 = c$$

Check the solution t=9, c=120 in both original equations.

$$\begin{cases} 12 \cdot 9 = 120 - 12 \\ 14 \cdot 9 = 120 + 6 \end{cases}$$

The arithmetic is left to you.

Solving each of the two equations for the same unknown and equating the results is a way to extract a single equation in one unknown from the system of two equations in two unknowns. This is just a special case of solving one equation for one unknown in terms of the other, and substituting that into the second equation, as will be shown in example 5. It is also possible to eliminate one of the unknowns by adding a multiple of one equation to a multiple of the other, as in example 4.

5 (Adapted with permission from *Jinkoki,* November, 1641 edition, translated into English by Osamu Takenouchi and published by Wasan Institute, 2000. *Jinkoki* is representative of the "mathematics for ordinary people" developed in Japan in the period 1603-1867.)

Example 4. Solve the system of equations
$$\begin{cases} 4p + 3m = 11 \\ 2p - 9m = 37 \end{cases}$$

Solution: Multiply both sides of the first equation by 3
$$\begin{cases} 12p + 9m = 33 \\ 2p - 9m = 37 \end{cases}$$

Add the two equations to eliminate m $$14p = 70$$
Solve for p $$p = 5$$

Substitute this value of p in either original equation
$$2 \cdot 5 - 9\text{m} = 37$$
$$10 - 9m = 37$$
$$10 = 37 + 9m$$

Solve for m.
$$10 - 37 = 9m$$
$$-27 = 9m$$
$$-3 = m$$

Check the solution $m = -3, p = 5$ in both original equations

$$\begin{cases} 4(5) + 3(-3) = 11 \\ 2(5) - 9(-3) = 37 \end{cases}$$

Again, the arithmetic is left to you to complete.

In example 4 we eliminated m by arranging multiples of the two equations that contained $9m$ and $-9m$, respectivey, so that when the two equations were added, there were no m-terms left. Often, when using this method, it helps to multiply both of the two equations by carefully chosen numbers.

For example, to eliminate r in the system
$$\begin{cases} 3r + 2s = 7 \\ 4r + 3s = 1 \end{cases}$$
you could multiply the top equation by 4 and the bottom one by 3 to get
$$\begin{cases} 12r + 8s = 28 \\ 12r + 9s = 3 \end{cases}$$

Then subtract one equation from the other to eliminate r and solve for s.
If you prefer, you could multiply the bottom equation of the original system by -3 and then add the two equations.

To Solve a System of Two Linear Equations in Two Unknowns

1. **Use any of the methods demonstrated above to get an equation in one unknown.**
2. **Solve for the unknown that was not eliminated.**
3. **Substitute the value of the unknown found in step 2 into one of the original equations and solve for the other unknown.**

The next examples show that the above plan can yield surprises.

Example 5. Solve the system $\begin{cases} y = 7x + 3 \\ x = \dfrac{y}{7} + 1 \end{cases}$

Solution: Since the second equation is already solved for x, substitute that expression for x to get an equation involving y alone. $y = 7\left(\dfrac{y}{7} + 1\right) + 3$

On the right-hand side, use the distributive property. $y = y + 7 + 3$
Subtract y $0 = 10$, which is false!

How can our method lead to a false equation? Our method was based on the assumption that there is a pair of numbers that satisfy both equations. But in this case there are no such numbers. Simultaneous equations such as these are said to be **inconsistent**, because any pair of numbers x and y which satisfies one of them does not satisfy the other.

A related difficulty occurs when the two equations are **equivalent**, or have exactly the same solutions, as are, for example,

$$2x + 3y = 7$$
$$\frac{-2x}{3} + \frac{7}{3} = y$$

If you solve the first equation for y in terms of x, you get the second equation. Every pair of numbers x and y that satisfy one of these equations also satisfy the other. This system of equations is said to be **indeterminate**, because they do not determine a value of x or a value of y.

Possibilities for a System of Two Linear Equations in Two Unknowns

1. **Exactly one pair of values for x and y satisfies both equations. This pair of numbers is the unique solution.**
2. **Trying to solve leads to a false equation. The system of equations is inconsistent, and no pair of numbers satisfies it.**
3. **Trying to solve leads to a true equation. The system of equations is indeterminate. Every pair of numbers satisfying one equation also satisfies the other.**

Graphical interpretation

If we associate a linear equation with its graph, which is a line, then two linear equations in two unknowns specify two lines, and a pair of numbers x and y which satisfies both equations corresponds to a point (x,y) which is on both lines. There are three possible cases:

1. The graphs of the equations are two lines that intersect. Then the coordinates of the point of intersection are the unique x and y values that satisfy the system of linear equations.
2. The graphs of the equations are two lines that are parallel and so do not intersect. Then no pair of numbers x and y satisfies the system of linear equations. The system of equations is inconsistent.
3. The graphs of the equations are the same line, so that any pair of numbers (x,y) that satisfies one equation also satisfies the other equation. In this case the system of equations is indeterminate.

Example 6. The figure below shows the graphs of three equations.

a. Based on the graphs, which of these systems of linear equations is inconsistent?

b. Find the solutions of the other systems.

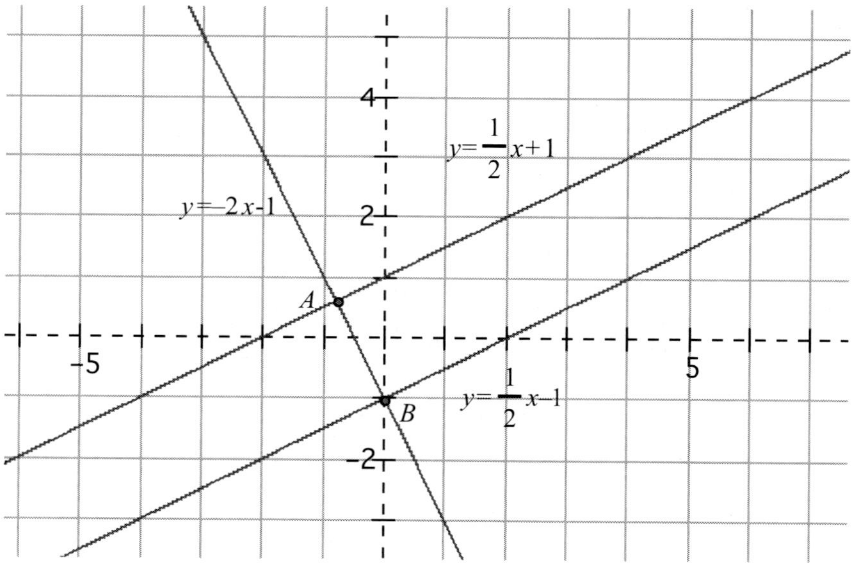

Solution: a. The first system of equations corresponds to the parallel lines in the graph and is therefore inconsistent.
b. The second system corresponds to the lines whose graphs meet at point A in the graph above. This system has a unique solution, which may be found by

equating the values of y to get $-2x-1=\frac{1}{2}x+1$ and solving for x to get $x=-\frac{4}{5}$.
Substitute this in either equation to get $y=\frac{3}{5}$. Thus the point A is $\left(-\frac{4}{5},\frac{3}{5}\right)$.

c. The third pair of lines meets close to (0,-1). In this case, the graphs of both

$y=-2x-1$ and $y=\frac{1}{2}x-1$ have -1 as y-intercept, so (0.-1) is the exact location of point B.

Reading a solution to a system of linear equations from their graphs is usually only approximate. As in the point A above, it is hard to read the coordinates exactly. Even in the case of point B, it was necessary to check.

Exercises 10.5

In questions 1-12 solve the pairs of simultaneous equations. Check your solutions both algebraically and by sketching the graphs of the equations. Identify any inconsistent or indeterminate systems.

1. $\begin{cases} x+y=11 \\ x=3 \end{cases}$

2. $\begin{cases} r=1.8s-2.1 \\ r=s \end{cases}$

3. $\begin{cases} 8m+5p=7 \\ p+1=-\frac{8}{5}m+\frac{15}{8} \end{cases}$

4. $\begin{cases} t+2v=-11 \\ 2t+v=5 \end{cases}$

5. $\begin{cases} 3w+7z=48 \\ w+15z=16 \end{cases}$

6. $\begin{cases} s+t=19 \\ s-t=-3 \end{cases}$

7. $\begin{cases} 4v+3z=17 \\ -v+\frac{5}{2}z=12 \end{cases}$

8. $\begin{cases} 5x+6y=-1 \\ 3x+2y=-4 \end{cases}$

9. $\begin{cases} 2u+5v=3 \\ -u+10v=-2 \end{cases}$

10. $\begin{cases} 9s+4t=11 \\ t+2\frac{1}{4}s=3 \end{cases}$

11. $\begin{cases} 2q+3w=2 \\ 6q-12w=-1 \end{cases}$

12. $\begin{cases} 4z+5y=2 \\ -12z+20y=9 \end{cases}$

13. Show whether or not $x=-\frac{11}{8}$, $y=\frac{67}{22}$ is a solution of $\begin{cases} 3x+4y=10 \\ -4x+2y=9 \end{cases}$

14. Show whether or not $x=\frac{32}{17}$, $y=\frac{7}{17}$ is a solution of $\begin{cases} 2x+3y=5 \\ 3x-4y=4 \end{cases}$

15. Stefan is one year older than Phyllis. Seven years ago Stefan's age was one third of Phyllis' present age. How old are they now?

16. Abe had nine acres of land more than Bill. But after Abe sold Bill 23 acres, he had only half as much land as Bill. How many acres did each have before the sale?

17. The sum of two numbers is 73, and their difference is 15. Find them.

18. My son's age is one seventh my wife's age. Next year he will be one sixth of her age. How old are they now?

19. One number is 16 more than another. Three times the larger number equals seven times the smaller number. What are the numbers?

20. Sally is half as old as Tammy. In four years Tammy will be three times as old as Sally was six years ago. How old are they?

21. A mule and an ass were carrying heavy burdens. The ass complained of his load, and said to the mule: "I need only one hundred pounds of your load, to make mine twice as heavy as yours." The mule answered, "Yes, but if you gave me a hundred pounds of your load, I would carry three times as much as you would." How much did each carry?[6]

22. Someone has two pots and one lid, which fits either pot. The price of the lid is 16 guilders. The lid and first pot together cost four times as much as the second pot. The lid and second pot together cost three times much as the first pot. What is the price of each pot? (This problem appears in several Northern European texts of the 16th century.)[7]

23. "The mixed price of 9 citrons and 7 fragrant wood-apples is 107; again the mixed price of 7 citrons and 9 fragrant wood-apples is 101. O you arithmetician, tell me quickly the price of a citron and of a wood-apple here, having distinctly separated those prices well." (From Mahavira, about 850 C.E., in India. Quoted by D.E. Smith in *History of Mathematics*.)

---

[6] Adapted from G. Polya, *Mathematical Discovery*, vol 1, who gives this as a problem from Euler's *Elements of Algebra*.

[7] Marjolein Kool, op.cit.

10.6 Rates and Units

Rates are important in practical applications but can be a source of confusion.

Speed is the most familiar example of a rate. Speed can be expressed in miles per hour, meters per minute, or furlongs per fortnight, but all the units of speed fit the format

units of speed = units of distance per unit of time

Of course, speed is not the only rate. Medics think of heart rates, economists think of interest rates and inflation rates, demographers think of birth and death rates, and nearly everyone thinks of pay rates and tax rates. What, then, is a rate? One dictionary defines a rate as "a quantity or degree of one thing measured per unit of something else." The following table illustrates the idea.

Type of rate	Typical units of measure
Speed	miles per hour, feet per second
Exchange rate	dollars per Euro, pesos per yen
Rate of pay	dollars per hour, dollars per month
Water flow rate	gallons per minute, acre-feet per minute
Heart rate	beats per minute
Price of vegetables	dollars per pound
Fuel economy	miles per gallon

The word **per** indicates a rate. Even concepts that you might not think of as rates, such as pressure (typical unit: pounds **per** square inch) or density (typical unit: grams **per** cubic centimeter) are best understood and treated as rates. From this point of view, percent is also a rate.

Here, as with percent, the Latin word **per** means **by** and indicates division as in "divided by." For example, if you go 100 miles in 4 hours, your average speed, by definition, is found by dividing distance (miles) by (**per**) time (hours). In this case $\frac{100 \text{ miles}}{4 \text{ hours}}$=25 miles per hour. Knowing the units of a rate and the role of the word **per** helps you work with rates.

Example 1. If you earn \$420 in 50 hours, what is your average rate of pay?
Solution: The pay rate will be expressed in dollars per hour, so divide the number of dollars by the number of hours: $\frac{\$420}{50 \text{ hours}}$=\$8.40 per hour.

Example 2. How much money will you earn, if you work 38 hours at \$7.00 per hour?
Solution: The pay rate is expressed in dollars per hour and found by dividing $\frac{\text{dollars}}{\text{hours}}$. Think of this as an ordinary fraction, and multiply it by hours to get

$$\frac{\text{dollars}}{\text{hours}} \times \text{hours} = \text{dollars}$$

Now put in the numbers.

$$\frac{\$7}{\text{hour}} \times 38 \text{ hours} = \$7 \times 38 = \$266$$

You can think of the hours as cancelling. Multiplying and dividing units this way may look strange to you at first. However, it is a legitimate technique, which is sometimes given the pretentious name of **dimensional analysis**.

Example 3. How many inches are there in a mile?
Solution: There are 12 inches per foot and 5,280 feet per mile. Treat these as rates and multiply, $\frac{\text{inches}}{\text{foot}} \times \frac{\text{feet}}{\text{mile}}$. Simplify by cancelling feet (disregard singular or plural) to get inches/mile. Finally, put in the numbers:

$$\frac{12 \text{ inches}}{\text{foot}} \times \frac{5280 \text{ feet}}{\text{mile}} = 63{,}360 \text{ inches per mile.}$$

Example 4. If you row 6 miles upstream at 2 miles per hour and 6 miles back downstream at 4 miles per hour, what is your average speed for the trip?
Solution: Careful! It is tempting but incorrect to average the speeds, 2 and 4. The total distance rowed is 12 miles (6 each way). Divide that by the total time, since average rate for the round trip is, by definition, total distance divided by total time. Going 6 miles upstream at 2 miles per hour takes 3 hours, and 6 miles downstream at 4 miles per hour takes $1\frac{1}{2}$ hours, so the total time for the round trip is $3 + 1\frac{1}{2} = 4\frac{1}{2}$ hours.

$$\text{average speed} = \frac{\text{total number of miles}}{\text{total number of hours}} = \frac{12 \text{ miles}}{4\frac{1}{2} \text{ hours}} = 2\frac{2}{3} \text{ miles per hour.}$$

In this example the upstream and downstream times were easy to find, but in more complicated problems, the two may be found from the units. To find the upstream times, start with the basic relationship:

$$\frac{\text{distance}}{\text{time}} = \text{rate}$$

Multiply by time to get $\quad \text{distance} = \text{rate} \times \text{time}$

Divide by rate to get $\quad \frac{\text{distance}}{\text{rate}} = \text{time}$

Put in the upstream numbers: $\quad \frac{6 \text{ miles}}{4 \text{ miles per hour}} = \frac{6 \text{ miles}}{\frac{4 \text{ miles}}{\text{hour}}}$

Now carry out the division by inverting and multiplying. The miles cancel, leaving $\frac{6}{4}$ hours.

Here the numerical part, $\frac{6}{4}$, is clear. Simplify the units as if they were numbers; i.e., divide by the fraction $\frac{\text{miles}}{\text{hour}}$ by multiplying by its reciprocal, $\frac{\text{hour}}{\text{miles}}$ and then cancel miles to get hours.

Example 5. Ken bought gasoline in Sweden for 9.0 Swedish kroners per liter, and he wondered what that would be in dollars per gallon. He knew there are 0.95 liters per quart and 4 quarts per gallon, and at the time the exchange rate was 7.45 kroners per dollar. Help him out.
Solution: As before, the units help us convert both kroners to dollars and liters to gallons.

For the money,
$$\text{dollars} = \frac{\text{dollars}}{\text{kroner}} \times \text{kroners}$$

For the volume,
$$\begin{aligned}\text{gallons} &= \text{liters} \times \frac{\text{gallons}}{\text{liter}} \\ &= \text{liters} \times \frac{\text{quarts}}{\text{liter}} \times \frac{\text{gallons}}{\text{quart}}\end{aligned}$$

Manipulate units as you would numbers, dividing the first equation by the second to get

$$\frac{\text{dollars}}{\text{gallon}} = \frac{\frac{\text{dollars}}{\text{kroner}} \times \text{kroners}}{\text{liters} \times \frac{\text{quarts}}{\text{liter}} \times \frac{\text{gallons}}{\text{quart}}}$$

Invert the denominator and multiply
$$\frac{\text{dollars}}{\text{gallon}} = \frac{\text{dollars}}{\text{kroner}} \times \frac{\text{kroners}}{\text{liter}} \times \frac{\text{liters}}{\text{quart}} \times \frac{\text{quarts}}{\text{gallon}}$$

Put in the numbers
$$\frac{\text{dollars}}{\text{gallon}} = \frac{1}{7.45} \times 9.0 \times 0.95 \times 4 = \$4.59 \text{ per gallon}.$$

Then 9.0 kroners per liter is $4.59 per gallon.

Exercises 10.6

1. Name typical units for each of these rates
   a. Population birth rate
   b. Rate of climb in an airplane
   c. Concentration of impurities in water
   d. Retail price of potatoes
   e. Sales tax
   f. Property tax

2. Occasionally you may encounter a rate that is not expressed with the word per. One example, often used for the speed of a boat or aircraft, is knots. Look up the definition of a knot, and use it to express speed in knots in terms of distance and time.

3. The slope of a line was encountered in chapter 9. Explain why this may be viewed as a rate.

4. The density of a material is defined as its mass per unit volume. Find the mass of 12 cubic centimeters of a material with a density of 8.1 grams per cc. Show that the units in your calculation yield an answer which is a number of grams.

5. A certain man drinks a barrel of beer in 20 days. If his wife drinks with him, they will empty the barrel in 14 days. How long would the woman take to drink the barrel alone? (from a 1558 text by Peter van Halle).[8]

6. A German brewery sells beer for 0.85 euros per liter. At 1.32 euros to the dollar and 0.95 liters per quart, find this price in dollars per quart.

7. A pound of meat sells for 5.80 British pounds per kilogram, What is the price in dollars per pound, if there are 2.2 pounds per kilograms and 1.81 US dollars to a British pound?

8. A motorist thought she had just enough cash to buy 12 gallons of gas. But the price was 30 cents per gallon higher than she had expected, so she could afford only 10 gallons. How much did gas cost per gallon?

9. (A challenge) At what time between 7:00 and 8:00 do the hands of a clock coincide? (It is easy to find an approximate answer by looking at a watch as you wind it. What is wanted here, however, is an answer that is accurate to the precise fraction of a second.)

10. An airplane on a search has fuel for a safe flight of 4 hours and 24 minutes. Because of wind, its ground speed will be 250 km/hr outbound and 300 km/hr on the return trip. How far out can the plane fly safely?

11. A swimming pool has a drain system that can drain it in 6 hours and a filling system that can fill it in 4 hours. After draining the pool, an attendant turns on the filling system but neglects to close the drain. If the flow in both systems is steady, how long will it take to fill the pool?

12. At 37 cents each, Maya could sell her entire supply of mangos today at the fruit stand. At 45 cents each she would sell 3 fewer mangos, but she would take in $1.05 more. How many mangos has Maya?

13. Americans who are used to the Fahrenheit scale of temperatures sometimes convert local weather forecasts from Celsius to Fahrenheit by he "tourist's approximation:" "Double the Celsius temperature and add 30." This is easier to perform mentally than

[8] Marjolein Kool, op. cit.

multiplying the Celsius temperature by nine fifths and adding 32, which is the accurate rule.

a. The world record high temperature recorded in shade is 58° C, at Al Azziyah, Libya. Find the Fahrenheit equivalent and the tourist's approximation for this. (The North American record is 57.6° C, at Death Valley, CA.)

b. The world record low temperature is -83.3° C, at Vostok Station in Antarctica. Find the Fahrenheit equivalent and the tourist's approximation for this temperature.

c. At what Fahrenheit temperature is the tourist's approximation correct?

14. (A challenge) A man was $\frac{2}{5}$ of the way across a narrow railroad bridge when a train appeared, headed for the bridge. The man ran as fast as he could (13 km/hr) and got to the end of the bridge at the same instant as the train, avoiding being hit by a hairsbreadth. Interestingly, in view of his speed and that of the train, which neither sped up nor slowed down, the same thing would have happened at the other end of the bridge–man and train would have reached an end of the bridge simultaneously, regardless of which way he ran. The bridge was 30 meters long.

a. How fast was the train going?

b. How far from the bridge was the train when the man first began to run?

15. (A challenge) **The milk and coffee problem**

You have two large (8 ounce) glass tumblers. One holds six ounces of milk, and the other holds six ounces of coffee. First an ounce is poured from the coffee glass into the milk glass, where it mixes thoroughly. Then an ounce of the milk-coffee mixture is poured from the milk glass into the coffee glass. The situation is pictured this way:

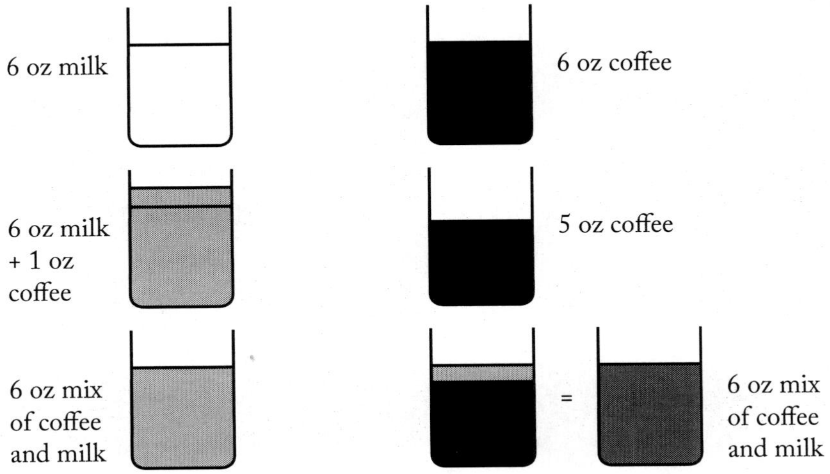

At this point there is some milk "contaminating" the coffee, and there is coffee "contaminating" the milk. How do the amounts of contamination compare at this stage? Specifically, is there more milk contaminating the coffee than coffee contaminating the milk? -or is it the other way around? -or are the amounts equal? Explain your reasoning in a convincing way.

16. The Holy Father sent a courier from Rome to Venice, commanding him to reach Venice in 7 days. And the most illustrious Signoria of Venice also sent another courier to Rome, who should reach Rome in 9 days. And from Rome to Venice is 250 miles. In how many days will the couriers meet, and how many miles will each have traveled? (Assume the travelers leave at the same time.)[9]

17. A family owns a car, which gets an average of 24 miles per gallon of gas, and a small truck, which gets an average of 12 miles per gallon of gas. They drive each vehicle 10,000 miles per year.
    a. How many gallons of fuel does the truck use each year?
    b. How many gallons of fuel does the car use each year?
    c. For all 20,000 miles that the family drives their vehicles each year, what is their overall fuel economy in miles per gallon?
    d. If the family keeps the car they have but replaces their truck with one that gets 24 miles per gallon, what will be the average fuel economy for the two vehicles?
    e. If the family replaces their car with a new car that gets 48 miles per gallon and then drives their car and truck each 10,000 miles, how many gallons of fuel will they use over the 20,000 miles?
    f. Under the assumptions of part f, what will be the combined fuel economy of the new truck and old car?

18. Sam and Janet Evening each drive 10,000 miles per year. Sam traded in his car, which got 20 miles per gallon, for a new car that gets 30 miles per gallon. At the same time, Janet traded her car, which got 30 miles per gallon, for a new car that gets 50 miles per gallon.
    a. By what percent did Sam's gas mileage increase?
    b. By what percent did Janet's gas mileage increase?
    c. Over 10,000 miles, how much fuel did Sam save by switching to his new car?
    d. Over 10,000 miles, how much fuel did Janet save by switching to her new car?

19. A man canoeing on a quiet lake in the northland saw a fish jump directly ahead of him. As he paddled toward the place

[9] From the *Treviso Arithmetic* (1478). The *Treviso* is the first printed mathematics book. Its author is unknown.

where the fish had jumped, he met the expanding circle of ripples after just 12 strokes of his paddle. In another 12 strokes he had completely crossed the expanding ring of ripples. When the fish jumped, how many strokes was it from the man? (Disregard the length of the canoe and that of the fish.) This problem was told to me by Gordon "Andy" Anderson, who liked fishing and puzzles.

10.7 Inequalities

Sometimes we care more about the relative sizes of numbers than whether or not they are equal. You might care, for example, about whether the amount in your checking account is sufficient to cover a certain check without wanting to know exactly how much is in the account. For such comparisons we use **inequalities**. These come in four varieties, all closely related.

1. The inequality $x>y$, read "x **is greater than** y," means that $x-y$ is positive. Geometrically, this means x is to the right of y on the number line in figure 10.7.1.
2. The inequality $x<y$, read "x **is less than** y," means that $x-y$ is negative. Geometrically, this means x is to the left of y on the number line in figure 10.7.1.
3. The inequality $x\geq y$, read "x **is greater than or equal to** y," means that $x-y$ is either positive or 0. Geometrically, this means that x is either to the right of y or at the same point as y on the number line in figure 10.7.1.
4. The inequality $x\leq y$, read "x **is less than or equal to** y," means that $x-y$ is either negative or 0. Geometrically, this means that x is either to the left of y or at the same point as y on the number line in figure 10.7.1.

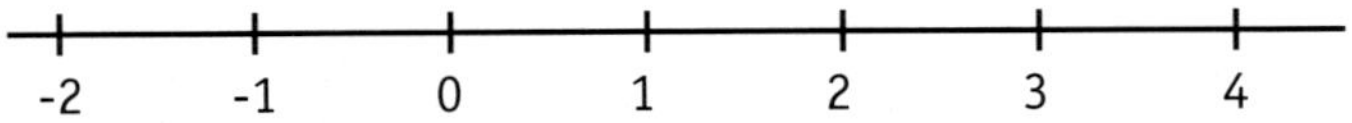

Figure 10.7.1

Inequality signs are much like the crescendo and decrescendo signs in music; the small, pointed end goes at the lesser number (softer sound), and the large, open end goes at the greater number (louder sound).

Like equations, inequalities are statements. Some, like $3<5$, are true, whereas others, like $-4>0$ and $-1\leq-2$, are false. Still other inequalities, like $x+3>1$, may be true or false, depending on the number x. These are **conditional inequalities**. To solve a conditional inequality is to find any and all values of the unknown which make it true.

To solve a conditional inequality, you can add, subtract, multiply or divide both sides by the same number much as you did with linear equations. However, if you multiply or divide both sides by a negative number, you must reverse the direction of the inequality sign. This reversal comes from the fact that multiplying by a negative number reflects about 0. For example, 3 is to the left of

5, but if both these numbers are multiplied by –1 and the results are compared, -5 is to the left of –3.

Example 1. Solve $3x+4\leq 0$, and picture the solution on a number line.
Solution: Subtract 4 from each side $3x\leq -4$

Divide both sides by 3 $x\leq -\frac{4}{3}$

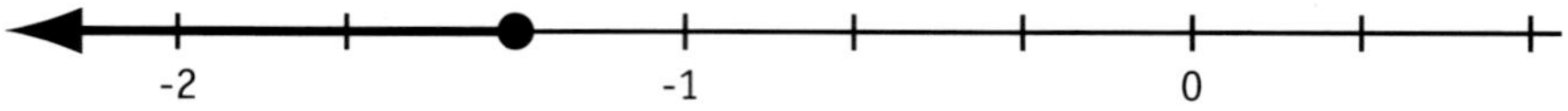

To indicate $\frac{-4}{3}$ on a number line is itself a solution, the dot at $\frac{-4}{3}$ is filled in.

Otherwise a hollow dot would be used.

Example 2. Solve $2(t+3)>5t-4$, and picture the solution on a number line.
Solution: Multiply out the left-hand side $2t+6>5t-4$
Subtract $5t+6$ (from both sides) $-3t>-10$

Divide by –3 (and reverse the inequality) $t<\frac{-10}{-3}$

Simplify $t<\frac{10}{3}$

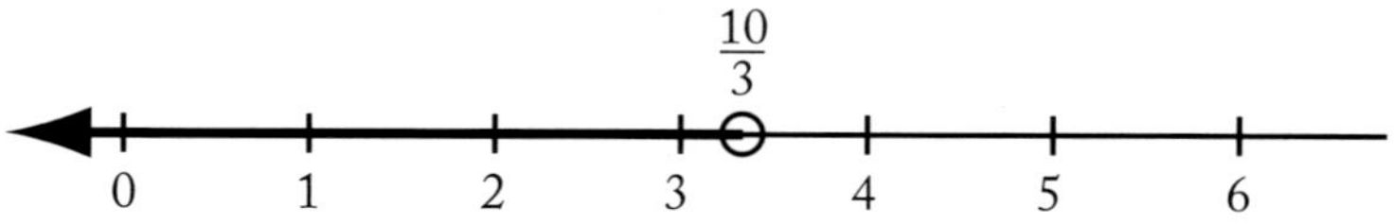

To indicate that $\frac{10}{3}$ is not itself a solution, the dot at $\frac{10}{3}$ is hollow.

Inequalities typically have infinitely many solutions, so there is no way to check them all. Still, substituting one or more strategically chosen values can help detect gross errors, such as forgetting to reverse the inequality sign after multiplying or dividing by a negative number.

We sometimes combine two equalities $a<b$ and $b<c$ and write the double inequality as $a<b<c$. Similarly, we might write $a>b>c$, and we can also use ≤ or ≥ in double inequalities. For example, figure 10.7.2 shows $2<x\leq 3$. Double inequalities are never used unless all the inequality signs have the same direction.

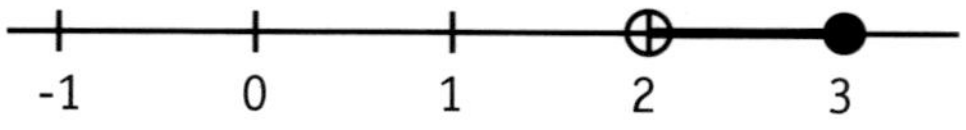

Figure 10.7.2

Example 3. A meteorologist forecasts temperatures in the 80's in Sacramento but only in the 60's in San Francisco. Write a double inequality expressing the difference in predicted temperatures in the two cities.
Solution: If it is 60° in San Francisco and almost 90° (it could be 89.9° and still be "in the 80's") in Sacramento, the difference can be up to 30°. The minimum would occur if it is nearly 70° in San Francisco (for example 69.9°) and 80 in Sacramento. Then 10°<temperature difference<30°.

Exercises 10.7

In questions 1–6, write inequalities to express the given statements.

1. The number r is greater than a number s.
2. The number w is less than half of the number z.
3. The number x is at least 3 greater than the number y.
4. If 2 is subtracted from y, the result is no greater than 6 times x.
5. The sum of c and d is less than one third the square of c.
6. Twice the product of s and t is at least seven times the sum of their squares.

In questions 7–17, solve for the letter and picture the solution on a number line.

7. $5a>20$
8. $4b<-12$
9. $7c\geq 84$
10. $3.1d\leq -15.5$
11. $-6\frac{1}{2}e>65$
12. $-11f<-143$
13. $5g-4<g+3$
14. $10h+12\leq -4h-2$
15. $\frac{1}{5}+7k<2k+\frac{1}{3}$
16. $73t-18\leq 89t+144$
17. $6(8-2y)>5y-4$
18. I bought five items at the store, of which the least expensive cost \$2.25. Write an inequality about my total expenses.
19. I drove for 2 hours at speeds which varied between 45 and 65 mph. Write a double inequality for the number of miles m traveled in the two hours.
20. Every week for seven weeks in a row Mary saved 10 cents of her allowance in a piggy bank, while Juan saved 15 cents. Every week in that time Pierre saved more than Mary, but he never saved as much as Juan. Write a double inequality for how much Pierre saved.

Vocabulary for Chapter 10

Conditional equation (10.2)
Dimensional analysis (10.6)
Equals (10.2)
Equation (10.2)
Equivalent equations (10.5)
Generalization (10.3)
Inconsistent equations (10.5)
Indeterminate system (10.5)
Inequality (10.7)
Less than, greater than (10.7)
Rate (10.6)
Root of equation (10.2)
Simultaneous equations (10.5)
Solution (10.2)
Unit (10.6)
Unknown (10.2)

Self-Quiz for Chapter 10

1. Use pictures with bags and stones to show why this trick always yields the same number in the end, regardless of what number you begin with.
   a. Take a number
   b. Add three
   c. Multiply by 5
   d. Add three
   e. Add the number you began with
   f. Divide by six
   g. Subtract your original number

2. Is $\frac{-2}{3}$ a root of $\frac{3}{4}(x-1)=\frac{1}{2x}+\frac{1}{4}$? Justify your answer by evaluating each side when $x=\frac{-2}{3}$.

3. Solve $3(t-2)=5t+4$.

4. I bought 9 *varas* of velvet for as much more than 40 pesos as 12 *varas* at the same time is less than 70 pesos. How much did a *vara* cost? (From Juan Diez, *Sumario Compendioso de las Cuentas*[10]).

5. Alexa walked four kilometers up a hill at a steady 3 kilometers per hour, and then she immediately jogged down the same hill at a steady 6 kilometers per hour. What was her average walking speed for the whole trip? Careful. The answer is **not** $4\frac{1}{2}$ kilometers per hour.

6. Solve $t+1<3t-5$, and picture the solution on a number line.

7. Solve the system of equations $\begin{cases} 4x-3y=18 \\ 2x-7y=20 \end{cases}$.

[10] The *Sumario* (*Brief Summary of Reckoning*), published in Mexico City in 1560, translated by Ed Sandifer and Shirley Gray, is the earliest mathematics book printed in the New World.

Chapter Eleven

Area, Perimeter, and Volume

In chapter 1 we defined the area of a rectangle as the product of its length and width, and we later we used that definition to figure out how to multiply fractions. In this chapter we will consider areas of other shapes, as well as the related concepts of perimeter and volume.

11.1 Exploring Area

Polygons, a few of which are shown in Figure 11.1.1, are figures made of line segments joined end to end.

Figure 11.1.1

Many students think of areas in terms of formulas, but we begin with an exploration which focuses on area as the size of a region and brings out basic properties of area.

Taking the square at the upper left of figure 11.1.2 as the unit of area, find the areas of the other polygons. Work through them in alphabetical order, using earlier answers to help with later ones.

[1] For this work it is sometimes handy to use rubber bands and geoboards. Geoboards were used as early as the first half of the 18th century (the blind mathematician Nicholas Saunderson made one) but became popular in schools only when inexpensive, light weight ones made of plastic became available.

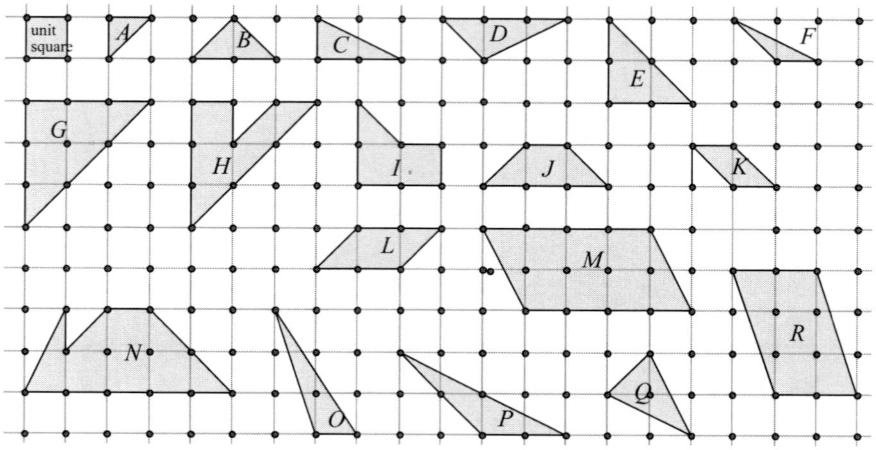

Figure 11.1.2

Finding the areas in figure 11.1.2 develops intuition about area, and it brings out basic ideas which are often used but not stated explicitly in elementary work:

> 1. **If a figure is made up of pieces which do not overlap, its area is the sum of the areas of the parts.**
> 2. **The area of a figure does not change if the figure is moved.**

It is interesting to see the varied ways that students can see the same thing. Here are two ways fifth graders handled triangle *F* (figure 11.1.1). One student first found area *A* as $\frac{1}{2}$ and then found area *C* this way:

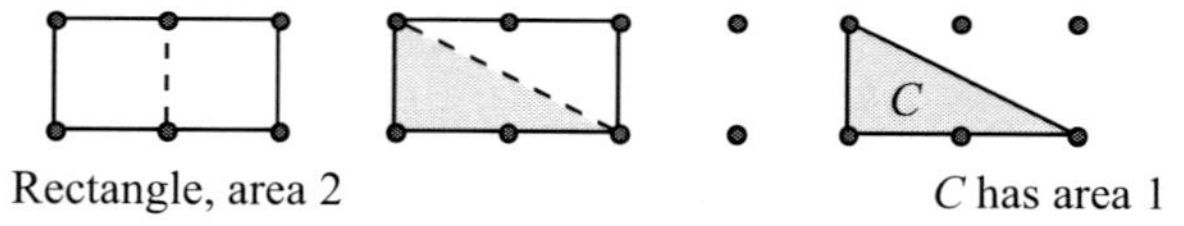

Next, she thought of *C* as made up of two shapes, one like *F* and the other like *A*, like this:

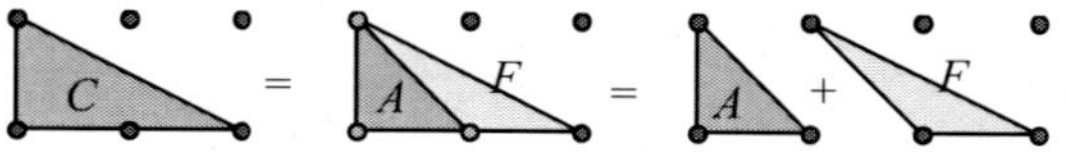

Area of C=1=area of A+ area of F=$\frac{1}{2}$ + area of F, so area of F= $\frac{1}{2}$.

A second student found *F* hard, so she skipped it and came back to it later. She found a clue in *K*, which she saw as a rectangle with area 2 minus two triangles like *A*, one of which is upside down, like this:

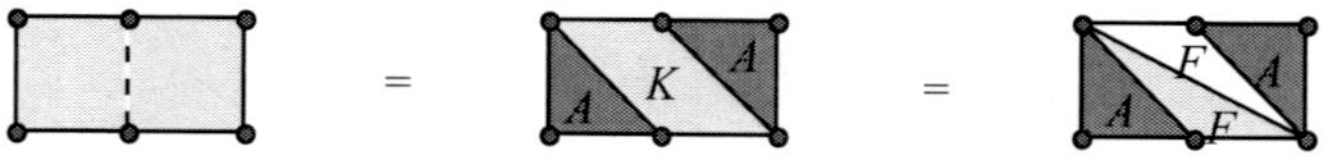

Area of rectangle = 2 = area A + area K + area A = $\frac{1}{2}$ + area K + $\frac{1}{2}$ = 1+2 area F

She then saw K as made of two copies of F, one of which is upside down. She reasoned that the area of K must be 1, so that the area of F must be $\frac{1}{2}$. Other students found still more ways to conclude that the area of F is $\frac{1}{2}$.

Exercises 11.1

1. **The area of a parallelogram.** In figure 11.1.2, shapes K, L, M, and R are parallelograms, i.e., they are four-sided figures, each with two pairs of parallel sides. This exercise helps you discover a rule for the area of a parallelogram in terms of the length of the **base** and the **height** of the parallelogram. The **base** of a parallelogram is a side on which it is seen as resting. The **height** of a parallelogram is the length of an **altitude,** a line segment perpendicular to the base (possibly extended) from a point on the opposite side of the parallelogram. Which side you take as base is up to you, but you must measure the height perpendicular to your chosen base.

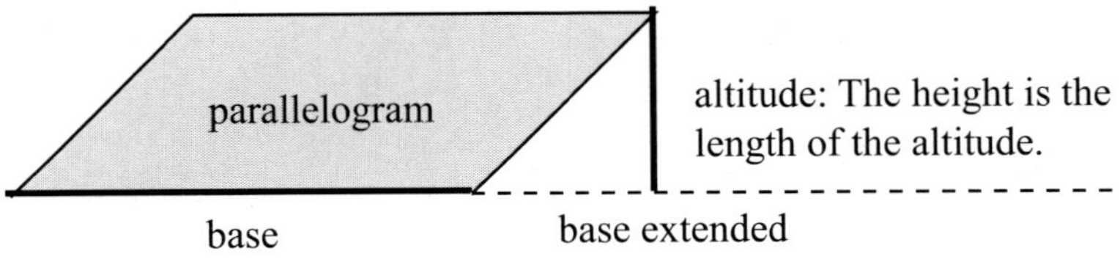

a. Fill in this table with information from parallelograms K, L, M, and R.

parallelogram	K	L	M	R
base				
height				
area				

b. From the table in part a, guess the rule expressing the area of a parallelogram in terms of its base and height. Hint: In these pictures the parallelogram at left has the same area as the rectangle at right:

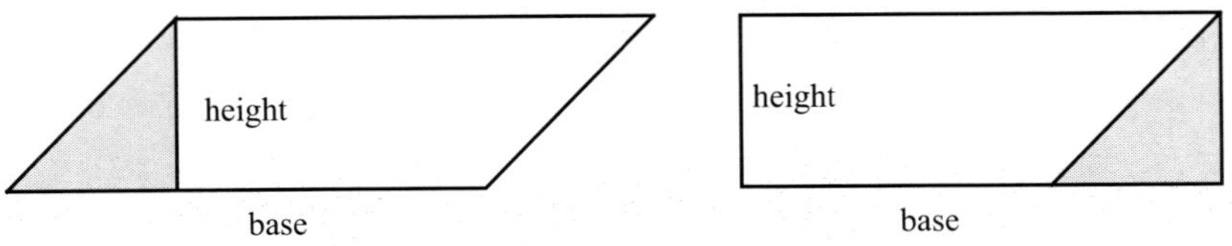

2. **The area of a triangle** The area of a triangle, like that of a parallelogram, depends on the base and the height. For a triangle, as for a parallelogram, the **base** is a side, on which you can think of the triangle as resting, and the **height** is the length of a line from the third vertex (the one not on the base) perpendicular to the base. (Sometimes, as in two of the three pictures below, the base must be extended beyond the triangle in order to draw the height.)

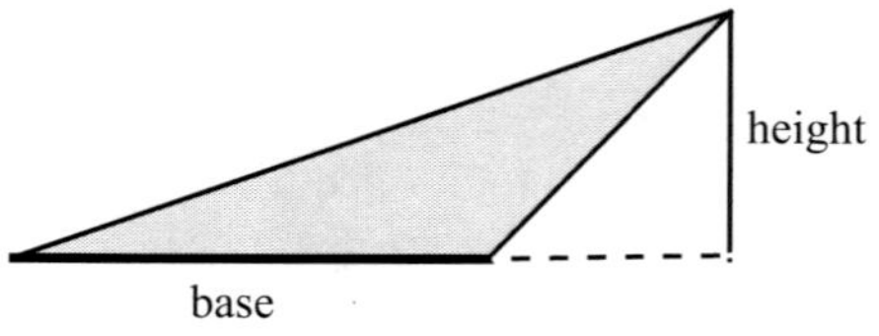

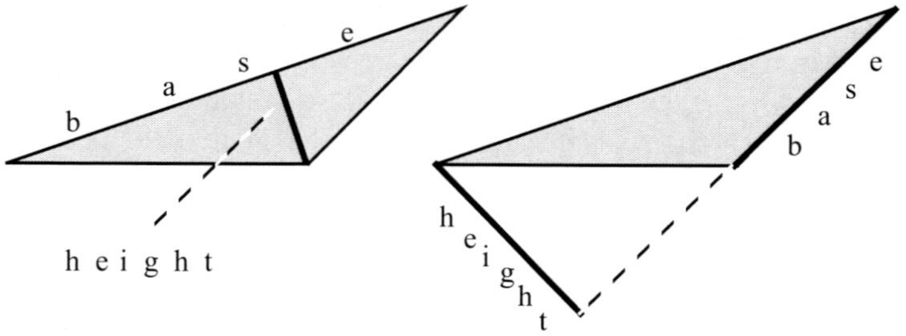

a. For each triangle in figure 11.1.2, except Q, fill in this table. Choose as base whatever side is convenient.

triangle	*A*	*B*	*C*	*D*	*E*	*F*	*G*	*O*	*P*
base									
height									
area									

b. From the table in part a, guess the rule expressing the area of a triangle in terms of its base and height.

c. Explain why the rule from part b is correct. Hint: If you put a copy of any triangle upside down next to itself, as shown here, what kind of shape do you get? Now, refer to question 1 above.

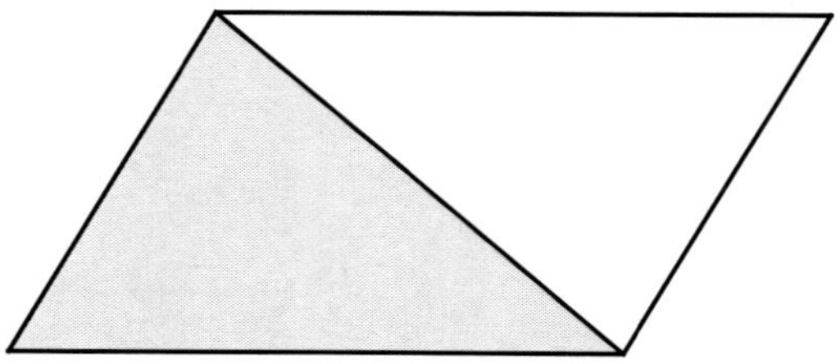

3. **Exploration: Pick's rule** In 1899, G. Pick[2] found a curious relation about areas of figures bounded by straight edges connecting dots at lattice points. Pick's rule links the area of a figure to the number of lattice points on its boundary and the number of lattice points in its interior. In figure 11.1.2, for example, triangles *A* and *F* each have three lattice points on their boundaries (at their vertices) and none in their interiors, while triangle *G* has nine lattice points on its boundary and one in its interior, and figure *N* has 11 lattice points on its boundary and two in its interior.

   a. Draw at least four figures which have three lattice points on their boundaries and no lattice points in their interiors, and find the area of each of them. Be sure the figures you draw are truly different from each other and are not just the same ones moved to new positions. A grid of lattice points is supplied for your convenience, but you can do this exercise with graph paper just as well. You should find that the four figures you made all have the same area.

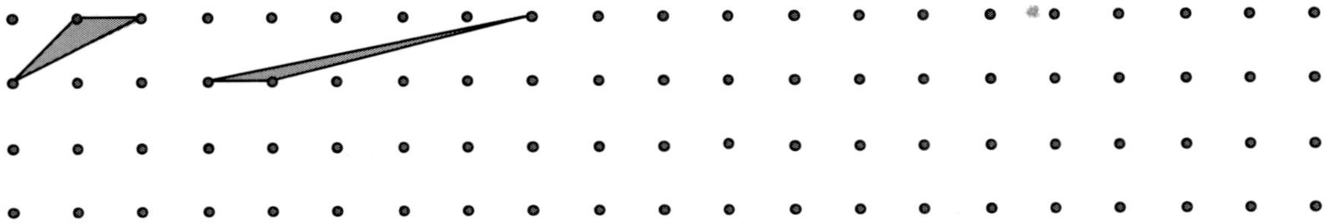

   b. Draw at least four figures, all truly different, which have four lattice points on their boundaries and none in their interiors, and find the area of each of figure you draw. All four of your figures should have the same area.

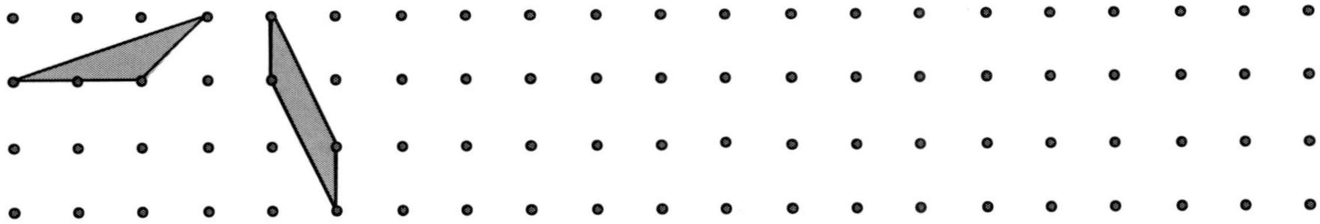

   c. Draw at least four figures, all truly different, which have three lattice points on their boundaries and one lattice point in their interiors, and find the area of each figure. All four of your figures should have the same area.

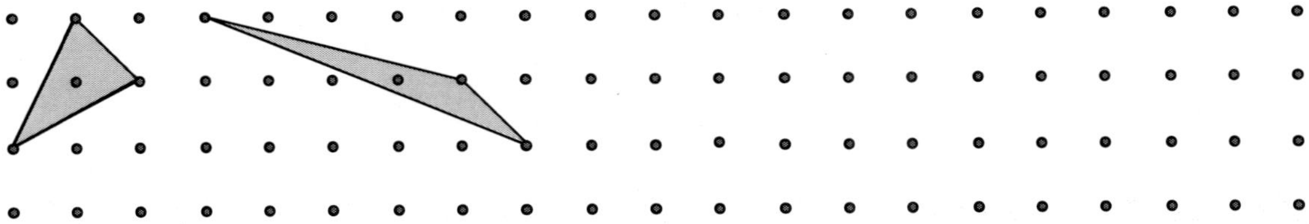

[2] Georg A. Pick (1859-1942) was born in Vienna, where he earned his doctorate in 1880. He taught at the Prague German University from 1880 to retirement in 1929, published 67 papers and served briefly as a dean. He died in the Theresienstadt concentration camp.

d. Draw at least four figures, all truly different, which have four lattice points on their boundaries and one lattice point in their interiors, and find the area of each figure. All three of your figures should have the same area.

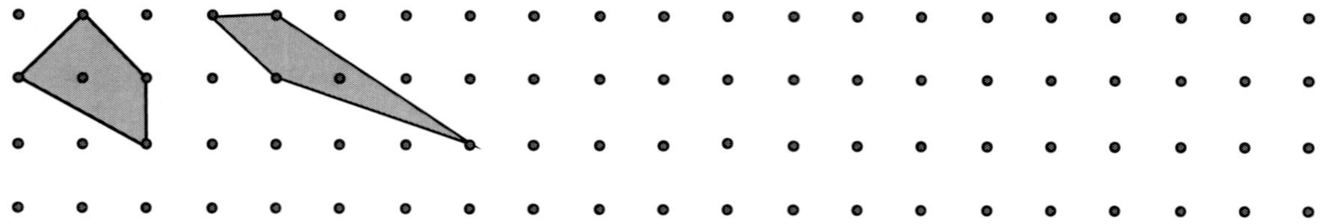

e. Draw at least four figures, all truly different, which have five lattice points on their boundaries and no lattice point in their interiors, and find the area of each figure. All four of your figures should have the same area.

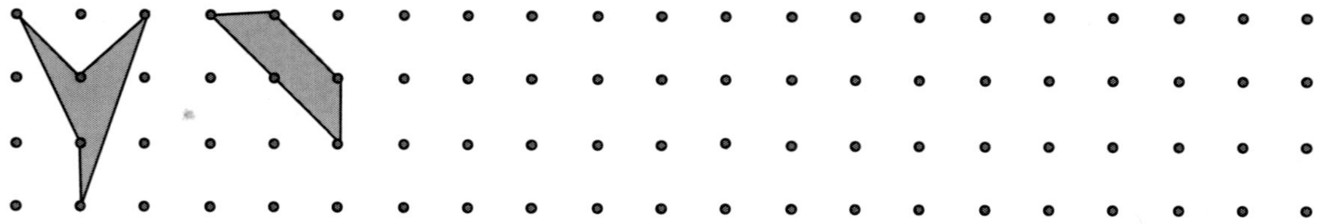

f. Draw at least four figures, all truly different, which have four lattice points on their boundaries and two lattice point in their interiors, and find the area of each figure. All four of your figures should have the same area.

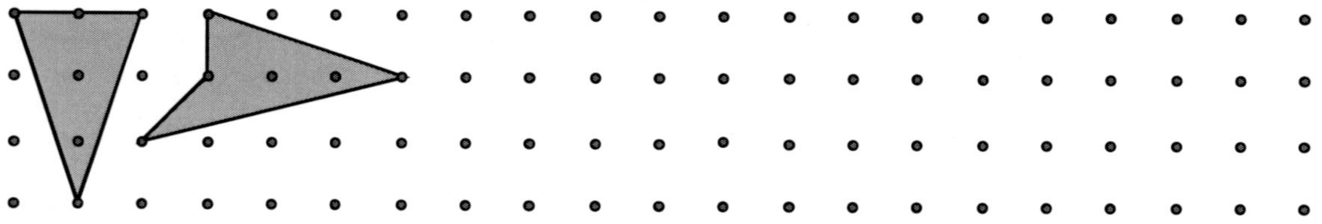

g. Enter the results from parts a-f in the table below, then make more sketches to fill in the table with the areas you have found for various figures.

		Number of dots on boundary				
		3	4	5	6	7
Number of dots in interior	0					
	1					
	2					
	3					
	4					

h. From the entries in the table, how does the area seem to change if the number of dots in the interior stays the same, while the number of dots on the boundary increases by 1?

i. From the entries in the table, how does the area seem to change if the number of dots on the boundary stays the same, while the number of dots in the interior increases by 1?

j. Put your answers from parts g, h, and i together to state Pick's rule.

4. In this picture, line *AB* is parallel to line *CDE*.

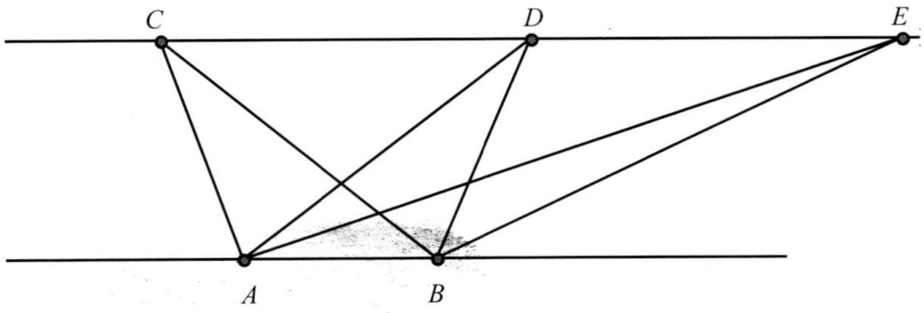

a. How do the areas of triangles *ABC*, *ABD*, and *ABE* compare?

b. In a sentence or two, explain the reasoning behind your answer to part a.

c. How do the areas of triangles *ACE* and *BCE* compare?

d. In a sentence or two, explain the reasoning behind your answer to part c.

e. If distance *CE* is twice distance *DE*, how do the areas of triangles *ACD* and *ACE* compare?

f. In a sentence or two, briefly justify your answer to part e.

5. **Area in the coordinate plane** Find the areas of the polygons (shapes with line segments as sides) with vertices at the points given, in order.

a. (4,2), (6,2), (4,8) b. (1,1), (6,1), (9,3), (4,3)

d. (-1,1), (-1, 3), (3, 6) d. (1,2), (6,1), (6,5), (1,5), (2,3), (4,3)

Exploration: Areas of Lattice Squares (problems 6-10)

Figure 11.1.3 is based on an array of dots, or **lattice points**, which we assume extends without end over the entire plane. Square *A* is assumed to be one unit long and one unit wide, so its area is 1. Question: What areas can squares have, if all four of their corners are lattice points?

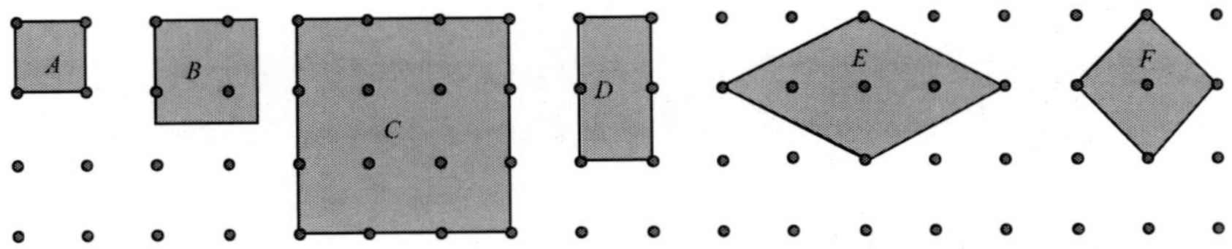

Figure 11.1.3

6.a. Begin by experimenting, sketching lattice squares for yourself and finding their areas. Such experimenting often defines a problem more clearly, and it can help you or your students get into it.

b. What exactly are we looking for? Is *B* the kind of square asked for? No, because some of its vertices are not lattice points. *C* is the kind of square required, and so is *F*. But neither of *D* and *E* are squares. Students are usually slower to come up with examples like *F* than with examples like *C*, because they assume, incorrectly, that the sides of a square must be horizontal and vertical on the page.)

c. Now that you understand what is wanted, try to find lattice squares with various areas up to 25. Whether you do this with others or just by yourself, allow plenty of time for this. Graph paper works as well as dot paper.

7. In each case where a lattice square is claimed to have a particular area, give a persuasive argument to support that claim, along the lines of the following example, which shows that the square in figure 11.1.4 has area 5.

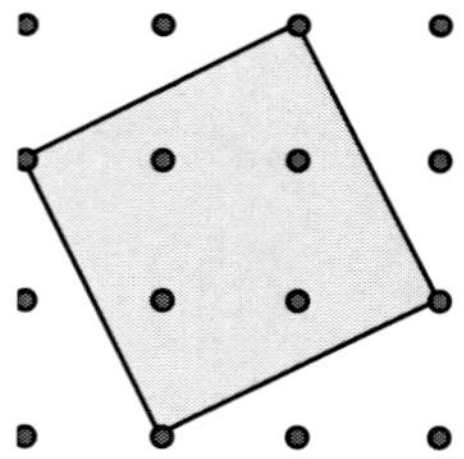

Figure 11.1.4

a. First method: What is the area of each shaded triangle in figure 11.1.5? What is the area of the four shaded triangles plus the center square?

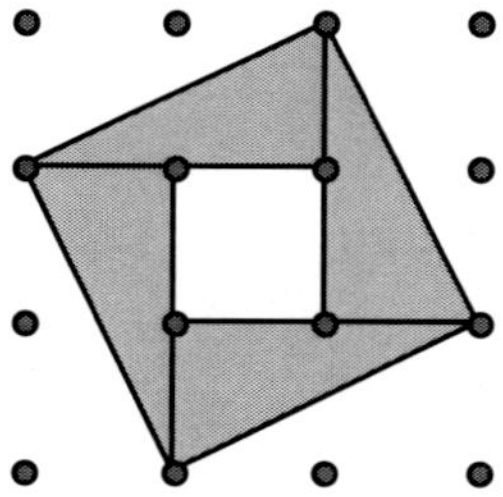

Figure 11.1.5

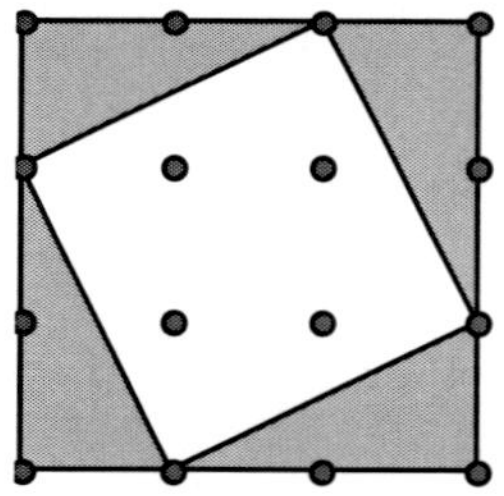

Figure 11.1.6

b. Second method: What is the area of the large outer square in figure 11.1.6? Subtract the area of the four shaded triangles from the area of the large square to get the area of the tilted square.

c. If you did problem 3 above, check that Pick's rule agrees with your calculations from parts a and b in this case.

8. All the areas you made so far are whole numbers. Now organize them; put the numbers from 1 to 25, in order, into a table like this. (For example, list 5 as an area you can make, in view of problem 7.)

Can Make																									
Can't Make																									

9. To find the pattern in the table from question 3, go back to the geometry. In question 1, you probably found that one way to make a lattice square is to put a corner at any dot, then choose another dot to be a second corner of the square. Then, it is not hard to complete the square, if you keep track of how far over and up you move from one dot to the next. For example, look again at the square in question 2, now in figure 11.1.4. If you start at *P*, you get to *Q* by going 2 units to the right and 1 unit up like knight's move in chess. From *Q*, go 1 to the right and 2 down to vertex *R*, then 1 down and 2 to the left to vertex *S*. Evidently, the square is determined by the "key move" from *P* to *Q*, which we write (2,1), meaning 2 over, 1 up. A move, such as this, in a plane, is called a **two-dimensional vector**.

   a. Complete this table. If a square cannot be made, leave the key move blank. Look for a rule relating the area of the square to its key move.

Area of lattice square	1	2	3	4	5	6	7	8	9	10	11	12	13
Key move (vector)	(1,0)	(1,1)		(2,0)	(2,1)								

Area of lattice square	14	15	16	17	18	19	20	21	22	23	24	25	26
Key move (vector)													

   b. State the rule in words or algebra: If the key move for a lattice square is (s,t), then the area of the square is ______ ______ .

10. What distinguishes those **prime** numbers which are areas of lattice squares from those prime numbers which are not areas of lattice squares? Hint: See problem 9 of exercises 6.3.

11.2 Perimeter

The **perimeter** of a figure such as a polygon is the distance you would travel if you went once around its boundary. For a figure with line segments as sides, the perimeter is, of course, the sum of the lengths of the edges. Although perimeter and area are very different concepts, many students confuse them with each other. The two concepts should be introduced separately, as they have been here, giving you plenty of opportunity to get acquainted with area before meeting perimeter. Perimeter is measured in units of length, whereas area is measured in square units.

People sometimes assume that shapes with larger areas automatically have larger perimeters, but, in fact, area and perimeter are independent of each other. Indeed, of the polygons in figure 11.2.1, the square at the left has the most area and the least perimeter, and the polygons to its right have successively less area and more perimeter. How much fence is used does not measure how much area is enclosed.

Figure 11.2.1

A practical question, for example, by someone building a garden with a limited amount of fencing material, is what shape encloses the most area for a given perimeter. This question is explored in the exercises below for rectangles. If there is no restriction on the type of shape used, the answer turns out to be a circle, as you might guess. We will consider circles in sections 11.3 and 11.5 and focus this section on polygons.

Example 1. Forty square tables, each two feet on a side, are arranged as in figure 11.2.2a. What is the perimeter of that shape?
Solution: Count the number of table edges along one fourth of the figure, shown in figure 11.2.2b. Do not count the two edges in parentheses, because they are not part of the edges of figure 11.2.2a. The result is 20. Multiply that by four to get the number of table edges in the perimeter of figure 11.2.2a, 80. Since each table edge is two feet long, the figure has a perimeter of 160 feet. You can no doubt think up alternative ways to find this perimeter, which is essentially a counting problem.

Figure 11.2.2a

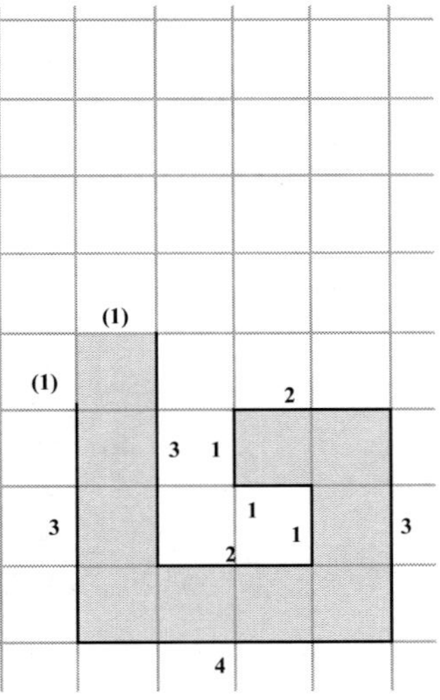

Figure 11.2.2b

Example 2. A rectangle with area $\frac{3}{4}$ square inches is three times as long as it is wide. What is the perimeter of the rectangle?

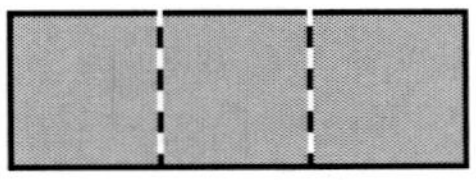

Solution: Picture the rectangle, as at right, partitioned into three squares. Each square has area $\frac{1}{4}$ square inch, so each square must be $\frac{1}{2}$ inch long and wide. Then the rectangle is $\frac{1}{2}$ inch wide, $\frac{3}{4}$ inch long, and has perimeter $\frac{1}{2}+\frac{3}{4}+\frac{1}{2}+\frac{3}{4}=2\frac{1}{2}$ inches.

Example 3. A certain field has an area of 40 acres, which is $\frac{1}{16}$ square mile. The field is $\frac{3}{8}$ mile long. How long would it take to walk once around this field at 3 miles per hour?

Solution: Recall fromSection 10.6: average rate $=\dfrac{\text{distance}}{\text{time}}$. Then

time $=\dfrac{\text{distance}}{\text{average rate}}$. We know the rate, 3 miles per hour (m/h). To find the time needed to walk around the field, first find the distance around the field, i.e., its perimeter. The field is a rectangle. Its perimeter is twice the length, $\frac{3}{8}$ mile (m), plus twice the width, which is not given. We are, however, given the area, $\frac{1}{16}$ square mile (m²), and we know that length × width = area. Then

width $=\dfrac{\text{area}}{\text{length}}=\dfrac{\frac{1}{16}\text{ m}^2}{\frac{3}{8}\text{ m}}=\dfrac{1}{16}\times\dfrac{8}{3}\text{ m}=\dfrac{1}{6}$ m. Perimeter is then $\frac{1}{6}+\frac{3}{8}+\frac{1}{6}+\frac{3}{8}$ m or $\frac{13}{12}$ m, and the time needed to walk that far at 3 miles per hour is

$\dfrac{\frac{13}{12}\text{ m}}{3\text{ m/h}}=\dfrac{13}{36}$ hours, or about 22 minutes.

Exercise 11.2

1. The U-shaped figure below contains 11 squares of the same size. The area of the U-shaped figure is 176 square inches. How many inches are in the perimeter of the U-shaped figure?[3]

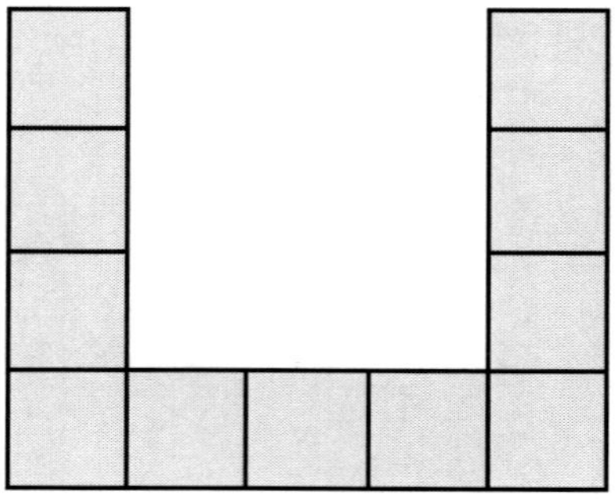

2. Rectangles are to be enclosed with 16 inches of string.
   a. If the length and the width of the rectangles are to be whole numbers of inches, what are the possible lengths and widths of the rectangles?
   b. For each rectangle found in part a, fill in the following table:

Length	Width	Perimeter	Area

3. Repeat question 2 with 24 inches of string, filling in this table.

Length	Width	Perimeter	Area

[3] From George Lenchner: *Math Olympiad Contest Problems for Elementary and Middle Schools.* MOEMS, 1997.

4. Repeat question 2 with 32 inches of string, filling in this table.

Length	Width	Perimeter	Area

5. a. Based on questions 2-4, how many different rectangles do you think there are with integer length sides and a perimeter of 40?
   b. Of all rectangles with a perimeter of 40 inches, which do you think has the largest area?
   c. Make a table like those in questions 2-4 and use it to check your answers from parts a and b.

6. Ancient Greeks proved that of all the rectangles with a given perimeter, the square has the largest area. The proof compares a square *ABCD* with a rectangle *AHGE*, made by shortening height *AD* by *DE* and lengthening width *AB* by the same amount, so that *DE*=*CF*=*FG*=*BH*.

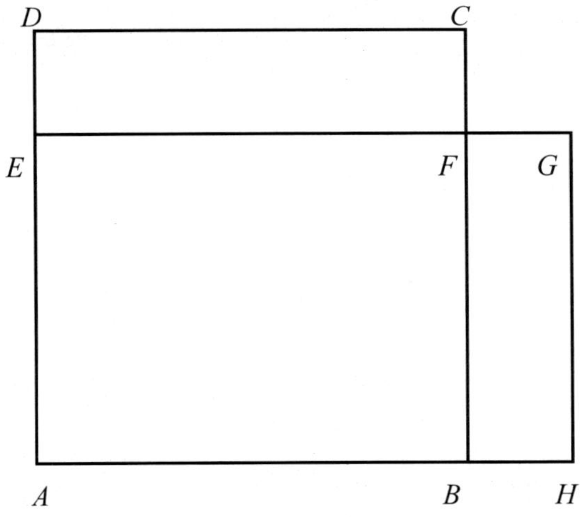

   a. Explain why rectangle *AHGE* has the same perimeter as square *ABCD*.
   b. Explain why rectangle *CDEF* has a larger area than rectangle *BHGF*.
   c. Show that rectangle *AHGE* has a smaller area than does square *ABCD*.

Exploration: Toothpick triangles (Problems 7 and 8)
Make triangles, using toothpicks of a standard length to make the sides (Make triangles with integer length sides, where the unit of length is the toothpick.). You can make one triangle from three toothpicks, one on each side, but there is no way to make a triangle from four toothpicks.

7. How many triangles can you make with longest side of a given length? Discovering nature's rule for this is fun and not too hard.

8. How many different triangles can be made with n toothpicks? (For a given n, how many triangles with integer sides have perimeter n?)
   a. Experiment, using actual toothpicks (paper clips of a uniform length work just as well), and enter your results in a table like this:

number of toothpicks	1	2	3	4	5	6	7	8	9	10	11	12
number of triangles	0	0	1	0								

   b. (Hard) Extend the table up to 21 toothpicks, and try to find patterns.

11.3 Similarity, Perimeter, and Area

. . . no building can be said to be well designed which wants symmetry and proportion. In truth they are as necessary to the beauty of a building as to that of a well formed human figure. . .

Vitruvius, *On Architecture*[4]

Which of the objects in figure 11.3.1 are alike?

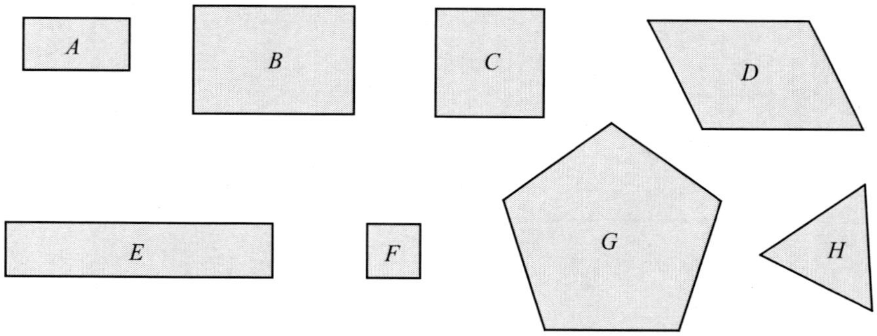

Figure 11.3.1

It all depends, you might say, on what you mean by "alike." All the objects are alike in that they are closed polygons. *C, F, G,* and *H* are alike in being regular

[4] Vitruvius, *On Architecture*, quoted from http://penelope.uchicago.edu/Thayer/E/Roman/Texts/Vitruvious/3*.html as of May 5, 2009.

(all sides and all angles equal in each case), and *A, B, C, D, E,* and *F* are alike in having four sides. Here we are concerned with two particular ways of being alike, known as congruence and similarity.

Congruence and Similarity

Objects are said to be **congruent** if they have the same shape and size. Objects mass produced in the same production run, such as bolts, books, pencils, chairs, or coffee cups, are congruent for all practical purposes. If a pair of 2-dimensional geometric objects, such as polygons, are congruent, one of them may be placed over the other so that the two objects match exactly.

Geometric objects are said to be **similar** if they have the same shape, regardless of whether or not they have the same size. Of course, congruent objects are also similar. All circles are similar to each other, but only those of any one size are also congruent to one another. The same can be said for all squares, for all cubes, and for all spheres. A scale model of a boat, house, or airplane is similar to the real one. An enlarged picture is similar to the original.

If a scale model car is one-tenth as long as the original, then it must also be one-tenth as wide and one-tenth as tall. In fact, every distance between points on the scale model is then one-tenth the distance between the corresponding points of the real car. We say the model is built on a scale of one-tenth. This is sometimes written as a ratio, 1:10. Curiously, there is no standard name for this ratio. You can find it in various books as the **ratio of similarity**, the **ratio of similitude**, and **stretching factor**. Taking Mark Twain's advice to avoid a $64 word when a two-bit word will do, we simply call this ratio the **scale**.

If M_1 and M_2 are similar objects and s is the scale, then the distance between two points A_2 and B_2 on M_2 is s times the distance between the corresponding points A_1 and B_1 on M_1 (see figure 11.3.2). Now, the distance from B_2 to C_2 is also s times the corresponding distance from B_1 to C_1, so the ratio of distances A_2B_2 to B_2C_2 on M_2 is the same as the ratio of distances between corresponding points on M_1.

$$\frac{\text{Distance } A_2B_2}{\text{Distance } B_2C_2} = \frac{s\text{Distance } A_1B_1}{s\text{Distance } B_1C_1} = \frac{\text{Distance } A_1B_1}{\text{Distance } B_1C_1}$$

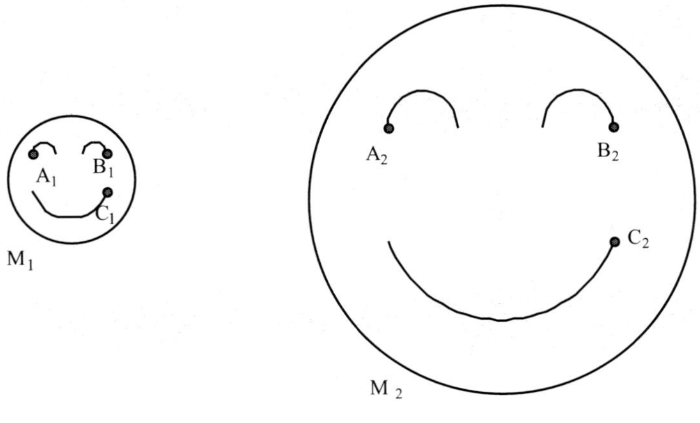

Figure 11.3.2

On similar objects, the ratios of distances between corresponding points are equal. We say that similar objects are **proportional** or that they "have the same proportions."

Example 1. Vitruvius, the Roman architect quoted above, stated various human proportions, such as:

The length of the foot is one sixth the height of the body

The length of the forearm is one fourth the height of the body.

He asserted that by using these rules, "the famous painters and sculptors of antiquity attained to great and endless renown." If these rules are used on a ten-foot tall statue of a person, how long would each foot be?

Solution: The length of each foot would be $\frac{10}{6}$ of a foot, or 1 foot 8 inches.

Exploration 1: Areas of similar objects

a. Triangles

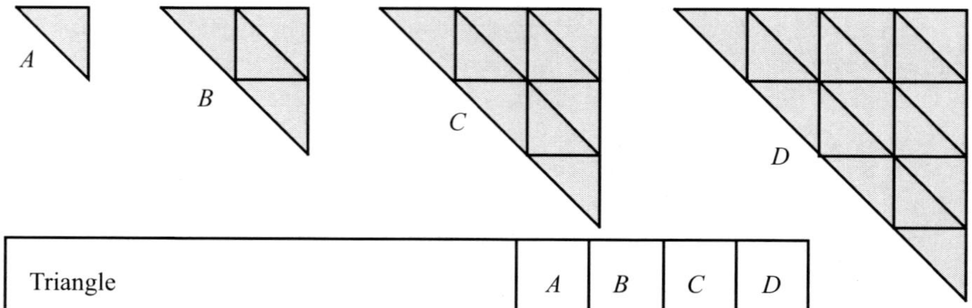

Triangle	*A*	*B*	*C*	*D*
Factor by which A is scaled up	1	2	3	4
Factor by which area is multiplied	1	4	9	16

b. Another polygon

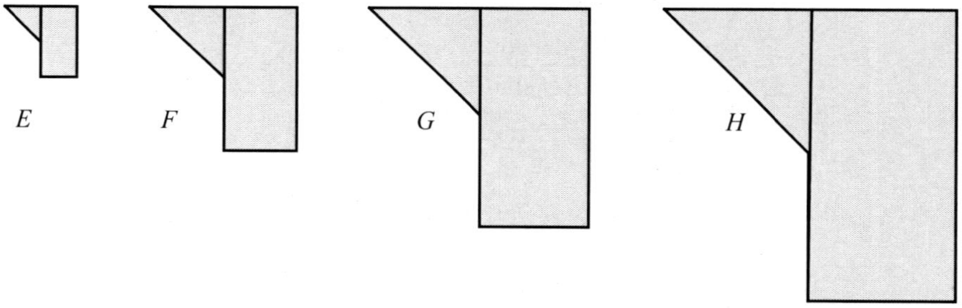

Shape	*E*	*F*	*G*	*H*
Factor by which E is scaled up	1	2	3	4
Factor by which area is multiplied	1	4	9	16

From the pictures above, you can see the general rule:

The ratio of areas of similar figures is the square of the scale.

In other words, if you begin with a figure F that has area A, and you scale the figure by a factor of n, the result will be a figure with area n^2A (figure 11.3.3).

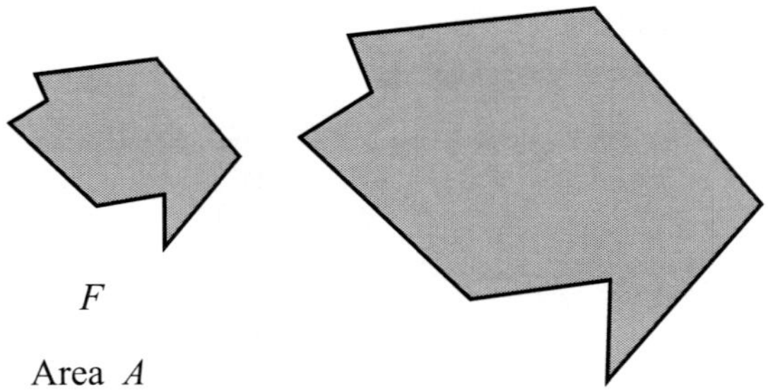

F

Area A

F scaled up by 2

Area $4A$

Figure 11.3.3

For simple shapes, whose areas are described by formulas, it is easy to see why areas behave this way. Consider, for example, a parallelogram, whose area is given by $A=bh$, where b is the length of the base and h is the height. Scaling the triangle by a factor k produces a new parallelogram with base and height each k times as long as those of the original triangle. Therefore the new triangle has area $(kb)(kh)$, which, by the associative and commutative properties of multiplication, is just k^2bh, or k^2 times the original area. The fact that the scale k enters the formula two times reflects the fact that the region whose area is involved here is two-dimensional. In much the same way, volumes of three-dimensional figures depend on the cube of the scale factor. We will explore this section 11.4.

Example 2. The area of a circle
All circles have the same shape and so are similar. Scaling a circle by a factor n multiplies the length of the radius, the diameter, and the circumference all by n, and it multiplies the area by n^2. Therefore, if C is the area of a circle with radius 1, then a circle with radius 3, for example, has area $9C$, and the area of a circle with radius 7 is $49C$. In general, if C is the area of a circle with radius 1, then the area of a circle with radius r is Cr^2. We will see in section 11.5 that C is the famous number π.

Exercises 11.3

1. Before constructing a new kind of cargo ship, which is to be 320 meters long, a scale model is built and tested in a wave tank. The model is to be built on a scale of 1/80.
    a. How long will the model be?
    b. If the model is 0.29 meters wide, how wide will the real ship be?

2. A chart is printed on a scale of 1:500,000.
    a. What does this mean? Explain in a clear and careful sentence or two.
    b. A 1:500,000 chart shows two towns are shown 23 cm apart. How many kilometers are the actual towns from each other?

c. A 1:500,000 chart of Washington State shows Anacortes and Everett $5\frac{5}{8}$ inches apart. How many miles is it from Everett to Anacortes? (There are 12 inches to a foot, 5,280 feet to a mile.)

3. A map of West Africa (scale, 1:4,000,000) shows Abidjan and Accra $4\frac{3}{4}$ inches apart. How many miles apart are Abidjan and Accra? (There are 12 inches to a foot, 5,280 feet to a mile.)

4. A photo shop sells 5×7 enlargements for $1.75 each. If the price of an enlargement is determined by the cost of the materials alone, what would expect to be the cost of a 10×14 enlargement?

5. Telescopes and binoculars are classified by magnification and light gathering power. A popular binocular size is 7×50, meaning that it magnifies by a factor of 7 and has objective lenses 50 mm in diameter. (The area of the objective lens determines the light gathering power of the binoculars.)
   a. Compared to a pair of 7×20 binoculars, how many times as much light is gathered by a pair of 7×50s?
   b. Does a scene viewed through 20×100 binoculars appear brighter or dimmer than the same scene viewed through 7×50s? Explain your answer.

11.4 Volume

Just as length and area measure size in one and two dimensions, respectively, **volume** measures size in three dimensions. The usual units of volume are cubes, whose length, width, and height are each one unit of length (figure 11.4.1). For example, the cubic centimeter, cubic inch, cubic foot, and cubic meter are common units of volume.

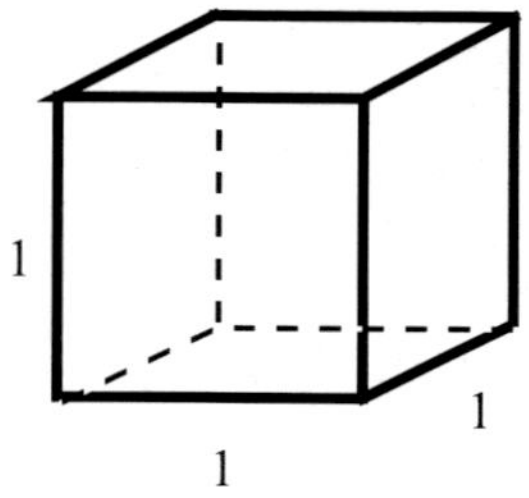

Figure 11.4.1 A unit cube

The volumes of certain other shapes are easy to measure in terms of the unit cube. In particular, if unit cubes are arrayed in a rectangle, and then one or more of these rectangles are piled on top of each other, the result is a **rectangular solid**, two of which are shown in figure 11.4.2.

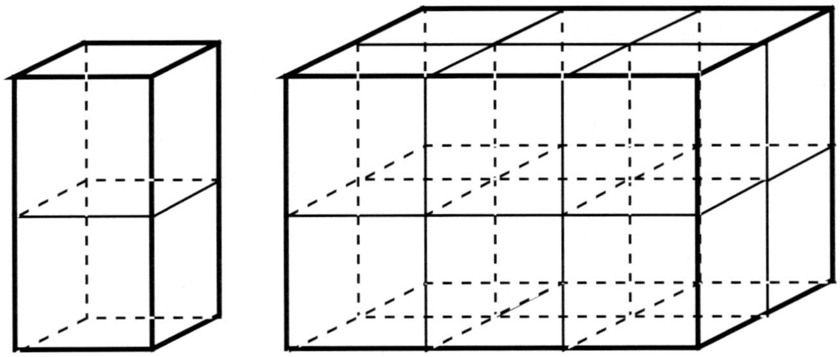

Figure 11.4.2 Rectangular solids

Pictures like these are helpful, but to really get a feeling for volumes of rectangular solids it is best to actually make some from unit cubes. Such cubes are widely available, usually made of plastic or wood. For whole number length, width, and height, the volume of a rectangular solid is found by multiplying length, width, and height. With that in mind, we define the volume as that product, whether or not the dimensions are whole numbers.

> **Definition: The volume of a rectangular solid is the product of its length, width, and height. If *l*, *w*, and *h* are the length, width, and height, the volume *V* is *lwh*.**

If one side of a rectangular solid is chosen as its base, then the area of that base is the product of its length and width. In that case, the height of the rectangular solid is the third dimension, and the volume may be expressed as Bh, where B is the area of the base, and h is the height.

Two key properties of volume are analogous to those for area.

Two key properties of volume

1. **If a three-dimensional figure is made up of pieces that do not overlap, its volume is the sum of the volumes of the pieces.**

2. **The volume of a three-dimensional figure does not change if the figure is moved.**

These properties, together with formula for the volume of a rectangular solid, help to find volumes of other three-dimensional figures. For example, if you cut a rectangular solid along a plane through opposite vertical edges, you can rearrange the pieces as in figure 11.4.3 to form a four-sided **prism**. (Rectangular solids are four-sided prisms with rectangular base and top.) This prism was made by rearranging the pieces of a rectangular solid, so its volume is the same as the volume of the rectangular solid.

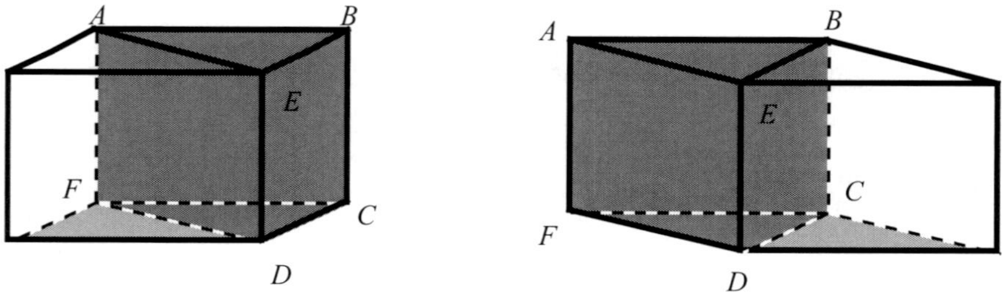

Figure 11.4.3

The result pictured in figure 11.4.3 holds even for prisms like the one shown in figure 11.4.4, whose base is not perpendicular to the adjoining edges.

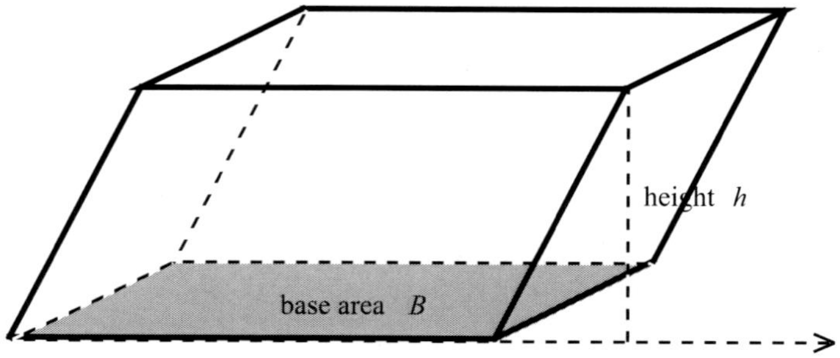

Figure 11.4.4

For a prism with a parallelogram base, the volume *V* is the base area *B* times the height *h*. In algebra *V*=*Bh*.

Because two copies of any triangle can be put together to make a parallelogram, Any prism with a parallelogram base may be cut into two triangular prisms with equal volumes, as in figure 11.4.5.

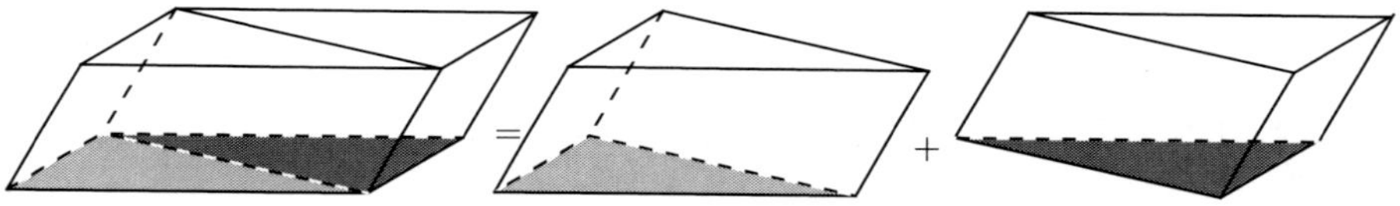

Figure 11.4.5

Now the area of the base of the triangular prism is half that of the base of the prism with a parallelogram base, and the volume of the triangular prism is half the volume of the parallelogram prism. Therefore,

The volume of a triangular prism is the area of its base times its height. *V*=*Bh*

The same reasoning extends to prisms with more complicated polygons as bases, since these can be cut up into triangular prisms, as in figure 11.4.6.

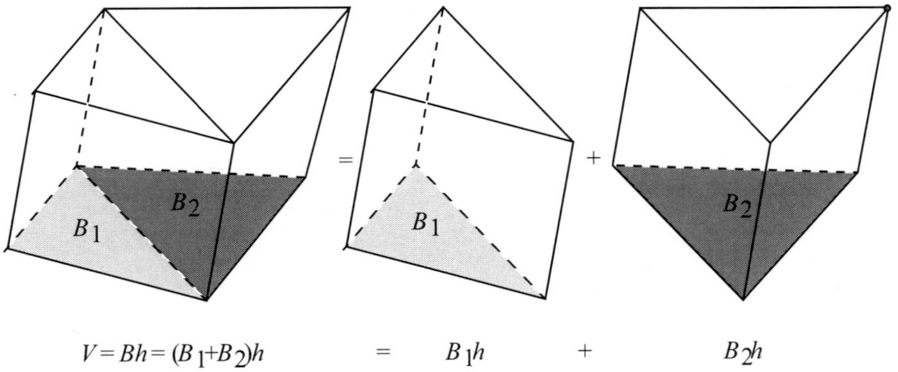

Figure 11.4.6

The volume of any prism is the area of its base times its height. *V=Bh*

Knowing the volume of a triangular prism is the key to finding the volume of a triangular pyramid. To see how this works, take a careful look at figure 11.4.7.

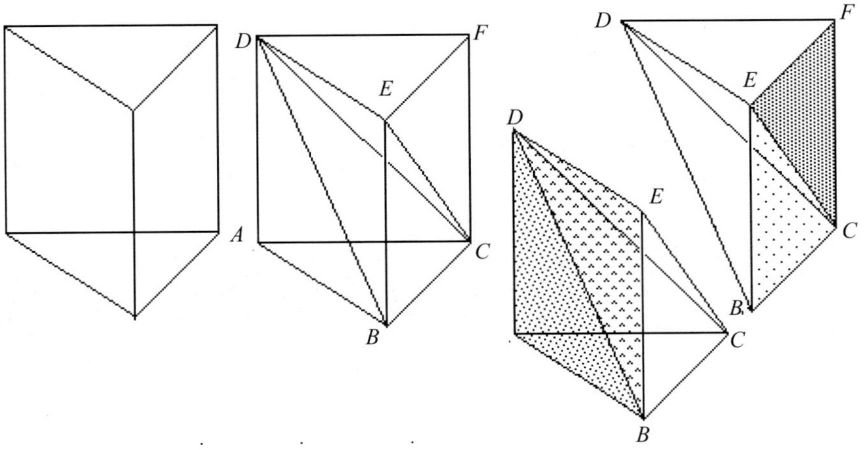

Figure 11.4.7

The idea is to cut the prism into three triangular pyramids and then show that the three triangular pyramids all have equal volumes. Since *DB* is the diagonal of rectangle *ABED*, triangles *ABD* and *BED* have equal areas. Then, pyramids *ABDC* and *BCDE* have equal volumes, since the height from *C* "above" plane *ABCD* is the height of both those pyramids. Similar reasoning shows that pyramids *BCED* and *CEFD* have equal volumes. Then the three triangular pyramids have equal volumes and together make up the triangular prism with volume *Bh*. Therefore,

The volume of a triangular pyramid is $\frac{1}{3}Bh$, where *B* is the base area and *h* is the height.

Figure 11.4.8 shows how this reasoning extends to pyramids with more complicated bases.

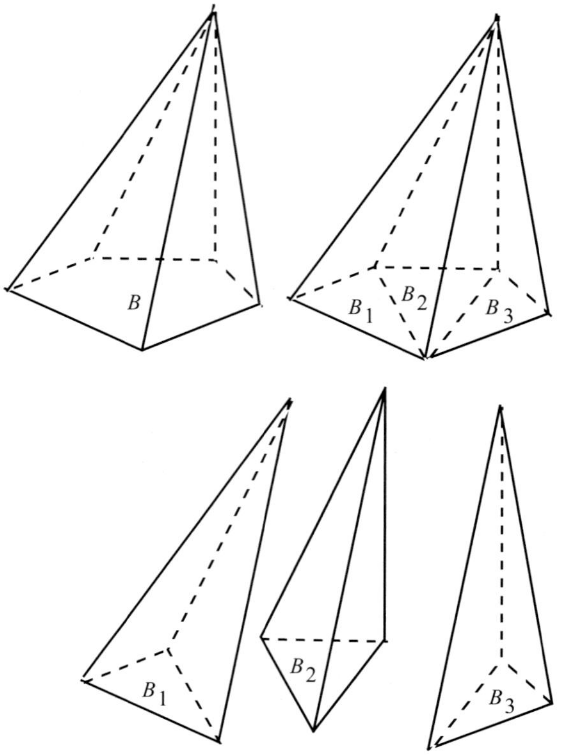

Figure 11.4.8

Here, the triangular pyramids all have the same height, h, as the pentagonal pyramid. If the base areas of the triangular pyramids are B_1, B_2, and B_3, then the volumes of the triangular pyramids are $\frac{1}{3}B_1h$, $\frac{1}{3}B_2h$, and $\frac{1}{3}B_3$h. Since the volumes of the triangular pyramids add up to that of the pentagonal pyramid $\frac{1}{3}B_1h + \frac{1}{3}B_2h + \frac{1}{3}B_3h = \frac{1}{3}(B_1 + B_2 + B_3)h = \frac{1}{3}Bh$.

The volume of a pyramid with base area B and height h is $\frac{1}{3}Bh$.

Volumes of similar figures

To explore volumes of similar objects, you need a collection of congruent three-dimensional objects that can be assembled into larger objects. Identical cubes suit this purpose well. (Such cubes, sold under various commercial names, are in many school classrooms). Congruent lego pieces, Cuisenaire rods, dominoes, or sugar cubes can also serve for this exploration.

Begin with a simple three-dimensional shape, such as, for example, the L-shape in figure 11.4.9. If each little cube has unit volume, then this shape has volume 3. Now make a similar shape twice as long, twice as high, and twice as thick. How many unit cubes does the scaled up shape use? Enter the volume of the scaled up shape in the table of volumes below

Now scale up the original shape by a factor of 3, and, again, count the cubes and enter the volume in the table. Repeat this for the original shape scaled up by a factor 4.

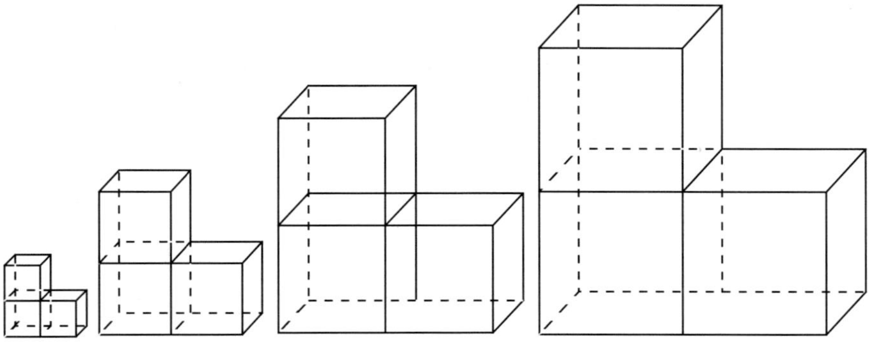

Figure 11.4.9

In a classroom, have students do this work in pairs, so they can help each other. This cuts down the number of occasions where a student does not scale up in all three dimensions (a common source of error), and it also reduces the number of sets of manipulatives needed.

Now complete a second row of the table of volumes, based on a different starting object, which you make up. Again, enter your results in the table below.

Table of volumes

	Original	scale x 2	scale x 3	scale x 4
L-shape	3			
Your shape				

Based on the evidence in the table, you can see that

The ratio of the volumes of similar objects is the cube of the scale.

This conclusion is built into volume formulas. For example, the volume of a rectangular solid with length l, width w, and height h, is lwh. If all three dimensions are multiplied by k, the new volume will be $(kl)(kw)(kh)$, which, by means of the associative and commutative properties of multiplication, may be written k^3lwh, which is the original volume multiplied by k^3. The exponent 3 here corresponds to the fact volume is a measure of size in three dimensions.

Our results on the area and volume of similar shapes have important practical implications. For example, you cannot design a supersonic airliner by scaling up a supersonic fighter plane. If you build a plane, say, ten times as long, wide, and high as the fighter, the wing area (which determines how much lifting force is available at a given airspeed) is multiplied by 10^2=100, but the volume, and hence the weight, of the plane, is multiplied by 10^3=1,000. Each square foot of wing will have to lift 10 times the weight it lifts on the fighter plane, making it hard to get the scaled up airplane to fly.

These considerations apply to biological design as well. Spiders and insects have exoskeletons, which give them structural strength and protect them from drying out. However, exoskeletons have significant drawbacks for larger animals. The larger the creature, the stronger (and therefore thicker and heavier) an exoskeleton must be, until the exoskeleton is more of a disadvantage than an advantage. In water, exoskeletons can support animals like crabs that are much larger than any land animals with exoskeletons. Huge insects make scary science fiction, but in reality they would be quite helpless. A fascinating article on this is "*On Being the Right Size*" by J. B. S. Haldane, reproduced in Nagel and Newman, *World of Mathematics*, New York, Simon and Schuster, 1956.

Exercises 11.4

1. If there are 1.1 yards to a meter, how many cubic yards are in a cubic meter?

2. The sum of the perimeters of two squares is 800 inches, and the area of the larger square is 16 times the area of the smaller square. Find the length of the side of the smaller square.

3. The area of one square face of a cube is 36/49 the area of the square face of a second cube. What is the ratio of the volume of the first cube to the volume of the second cube?

4. If the edge of a cube is increased by 10%, by what percent is the volume increased?

5. In the figure shown below, tetrahedron $ABCD$ is part of a unit cube. What is the volume of tetrahedron $ABCD$?

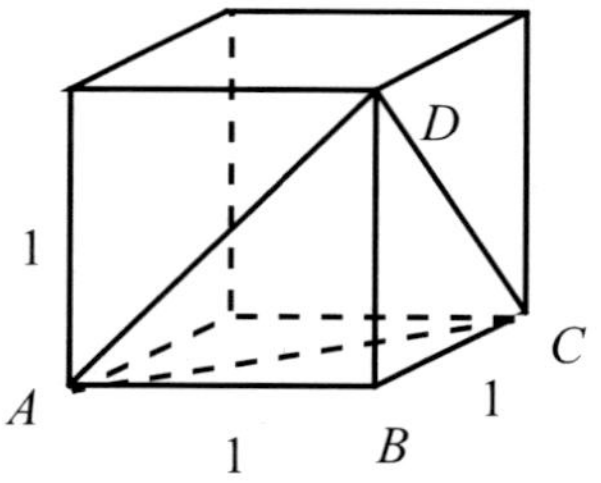

Exploration: Surface to Volume Ratio (problems 6-8)

6. If you make a rectangular solid or box from six unit cubes, the dimensions of the box are either $1 \times 1 \times 6$ or $1 \times 2 \times 3$.
   a. Make your own models of these with plastic cubes.
   b. Find the surface areas (the total number of unit squares showing on the top, bottom, and four sides) of these two boxes.

7. There are four ways to make a rectangular box from 12 unit cubes, and they are $1 \times 1 \times 12$, $1 \times 2 \times 6$, $1 \times 3 \times 4$, and $2 \times 2 \times 3$.
   a. Make your own models of these boxes.
   b. Find the surface areas (the total number of unit squares showing on the top, bottom, and four sides) of these four boxes.

8. (This question builds on questions 6 and 7)
   a. List all eight possible ways to make a rectangular box from 36 identical cubes.
   b. Find the surface areas of each of the boxes from part a.
   c. Which of the boxes from part a has the smallest surface area?

9. a. Cut out the following models, then fold and tape them each into a triangular pyramid.
   b. Assemble the three triangular pyramids from part a into a triangular prism.
   c. Explain why all three of the triangular pyramids have equal volumes. This shows why the volume of a triangular pyramid is one third that of a prism with the same base area and height.

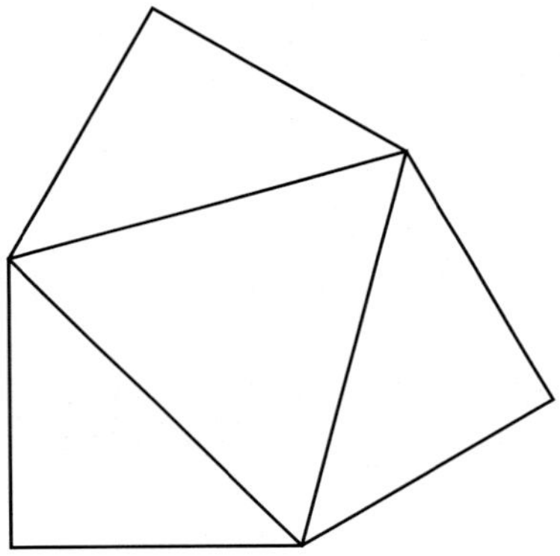

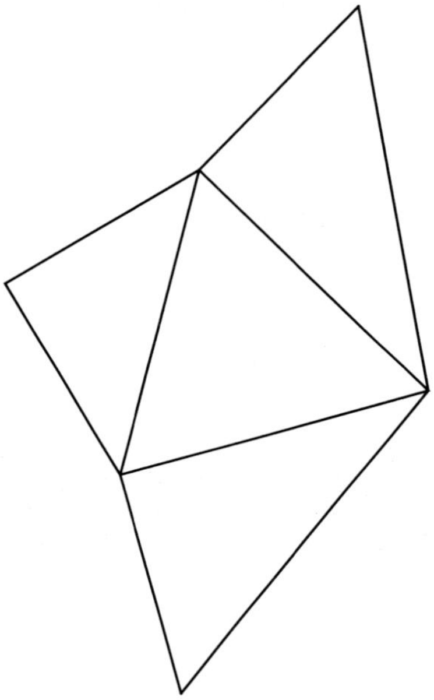

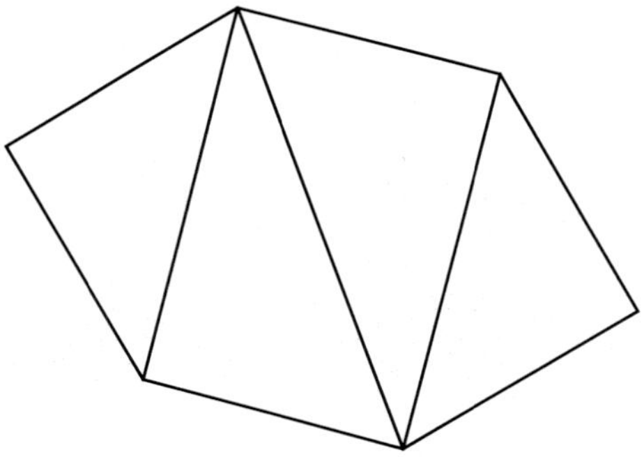

11.5 Circles

Shapes bounded not by line segments but by other curves are more difficult to work with than polygons. We will confine our attention to circles, exploring their perimeter and area through hands on explorations.

First some basic definitions: A **circle** consists of all the points in a plane which are at a fixed distance from a specified point, called the **center.** A line segment from one point on a circle to another is called a **chord** of the circle. A chord through the center is called a **diameter.** The distance around a circle is called its **circumference**. Figure 11.3.1 pictures these terms on a circle with center *O.*

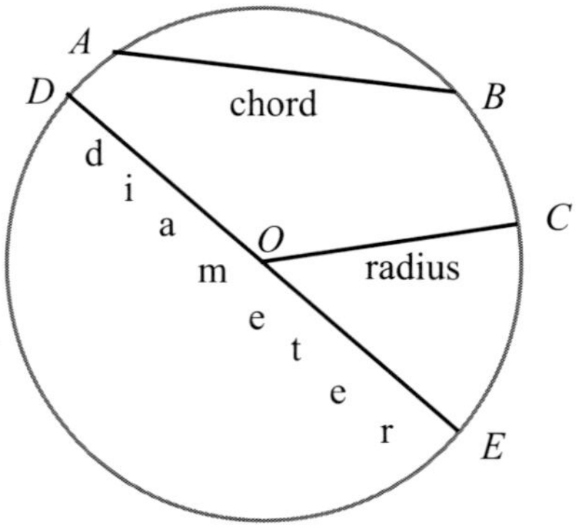

Figure 11.5.1

Relating circumference and diameter: a measurement approach

Get a tape measure and circular objects, such as dishes, pans, flower pots, lampshades, tires (bicycle wheels are excellent), cans, and hoops. Do not use small objects, such as coins, because they are hard to measure precisely enough. Measure each circumference and diameter as accurately as possible, and then enter the measurements in a table like the one below. This is a good classroom activity for students with some skill in using and reading a tape measure, but even then it takes supervision. Some students are happy to sacrifice precision in order to avoid having to work with awkward fractions. Others may not be used to reading a tape measure to the degree of accuracy needed. When all the measurements have been made and recorded, lead the students in a comparison of their tables. A discussion of the differences will show the difficulties in making the measurements and bring out the need for care and precision. Among the difficulties mentioned may be that it is hard to locate the center of some circular objects and therefore to be sure you are measuring the diameter. Also, no tape measure is perfectly accurate, and on some objects it is hard to hold it in just the right position for the needed measurements. If students record markedly different numbers for the same object, have them measure again, checking and helping each other. Ideally, dozen or more circular objects will be measured, and the class will agree on their circumferences and diameters, at least approximately. Now have the students calculate the ratio of circumference to diameter for each object, to two or three decimal places, and ask them what they observe about the various ratios $\frac{c}{d}$ calculated in the right-hand column.

	circumference c	diameter d	ratio c/d
lampshade			
pie dish			
bycicle tire			
wastebasket			
garbage can			
hoop			

The class will observe that in each case the circumference is a bit more than three times the diameter. In fact, the ratio, denoted by the Greek letter π (pronounced like "pie"), is independent of the size of the circle.[5] This number is approximately 3.14, but you may find considerable variation in the ratios calculated in the table. Take this as an opportunity to discuss measurement error and how to minimize it.

Accurate determination of π by measuring is not easy, and over the centuries it was abandoned in favor of theoretical methods, some of which are mentioned in the exercises below. A noteworthy early pioneer of such methods was Archimedes, who estimated the circumference of a circle by the following ingenious process:

Relating circumference and diameter: a theoretical approach

All circles are similar, so the ratio of two one-dimensional features, such as circumference to diameter, is the same for all circles. If we call that ratio π, then for all circles $c=\pi d$. The question, then, is to accurately determine the number called π, which we know from our measurements above is a little more than 3.

Archimedes first considered a hexagon inscribed in a circle (figure 11.5.2). If the diameter of the circle is 1, then the inscribed hexagon has perimeter 3, and the circumference of the circle must be greater than 3. On the other hand, the distance around the outer hexagon was, he found, about 3.464, so he concluded that π, the circumference of a circle with diameter 1, is between 3 and 3.464. Archimedes next trapped the same circle between an inner and an outer dodecagon (12-sided polygon) as in figure 11.5.3, and repeated the calculations.

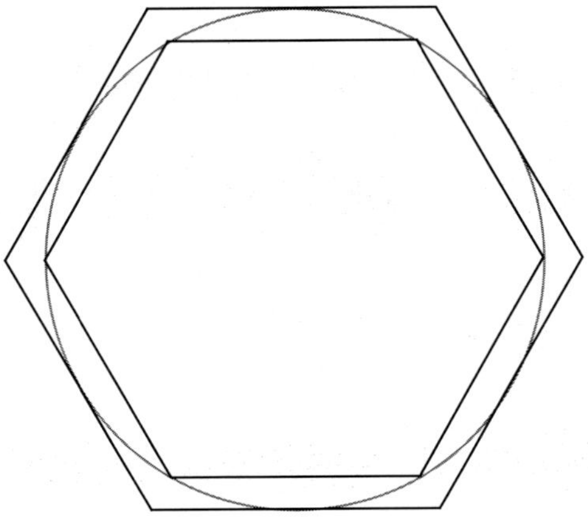

Figure 11.5.2

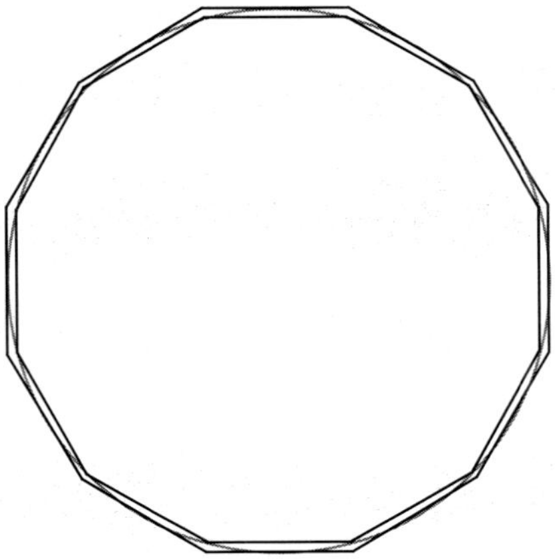

Figure 11.5.3

[5] William Jones (1675-1749) introduced this use of the symbol π in his *Synopsis palmariorum matheseos: or, a New Introduction to the Mathematics* of 1706. However, that notation was widely used only after Euler (see note, section 6.4) reintroduced it in 1737.

From this he pinned down the value of π more accurately than was possible with hexagons. He then repeated the work, successively, with polygons of 24 sides, 48 sides, and 96 sides, each time determining the value of π with greater accuracy than before. With 96-gons, he found that $3\frac{10}{71} < \pi < 3\frac{10}{70}$. Several centuries later, in China, Liu Hui applied the same method to polygons with even more sides to estimate π more accurately as $\frac{355}{113}$, a fraction which is easy to remember (think 1, 1, 3, 3, 5, 5). Beginning in the 15th century, new methods were developed, which allowed the mysterious number π to be calculated more and more accurately. By the early 18th century, the first 100 decimal places of π had been calculated, and by the end of World War II π was known to more than 600 decimal places. Since then, as more and more powerful techniques and computers appeared, π has been calculated to ever more astonishing numbers of decimal places. At this writing, in 2009, the record is held by two computer scientists in Japan, who in 2005 calculated more than 1.2 trillion digits of π.

The public has sometimes been confused about π. In 1897, the Indiana state legislature passed a bill, enthusiastically backed by the state superintendent of public instruction, to legislate the value of π as 3, revealing a complete misunderstanding of the meaning of π. Today π is better understood, and schools even celebrate π day on March 14 (3/14). A web search for "digits of pi" yields thousands of replies, showing worldwide fascination with this number. The first 30 digits of π are, incidentally, 3.14159265357979323856264338269.

To sum up:

If *d* is the diameter of a circle, then its circumference *c* is given by πd.

The area of a circle: a measurement approach

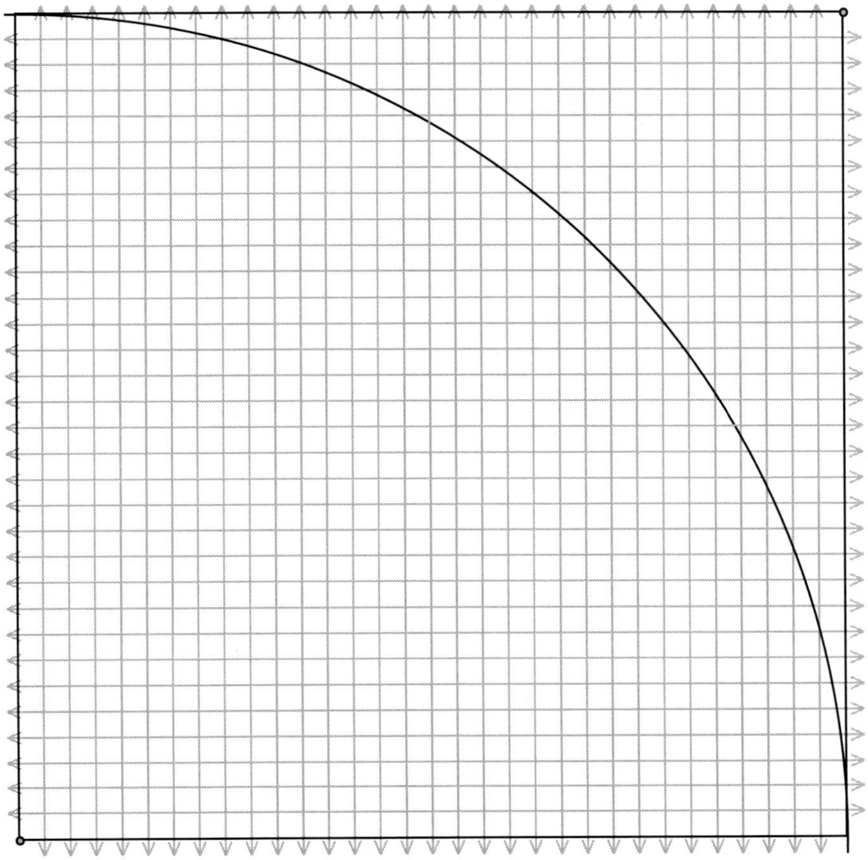

Figure 11.5.4

Figure 11.5.4 shows a quarter circle. If the radius of this circle is taken as 1, then the parallel lines in this grid are $\frac{1}{32}$ apart, and each little square has area $\frac{1}{32^2} = \frac{1}{1024}$, while the big outer square has area 1. Counting the number of squares inside the quarter circle and estimating the partial squares in it as well, gives an estimate for the area of a quarter of the circle. Multiplying this estimate by 4 gives an estimate of the area of the whole circle. For example, if you estimate $\frac{3}{4}$ of the area of the square lies inside the circle, then, multiplying that by 4, you would estimate the area of the entire circle as 3.

Unfortunately, to make this measurement yield much more accurate results take increasingly fine graph paper and skillful estimation. Again, a theoretical approach is more powerful, and again Archimedes played a leading role.

Area of a circle: a theoretical approach
Since all circles are similar, scaling a given circle up or down produces any other circle. Scaling a circle by a factor n multiplies the length of the radius, the diameter, and the circumference all by n, and it multiplies the area by n^2. Therefore, the area of a circle with radius r can be expressed as kr^2, for some

constant k. This area can also be expressed as hd^2 for some constant h, where d is the diameter, or as jc^2, for some constant j, where c is the circumference.

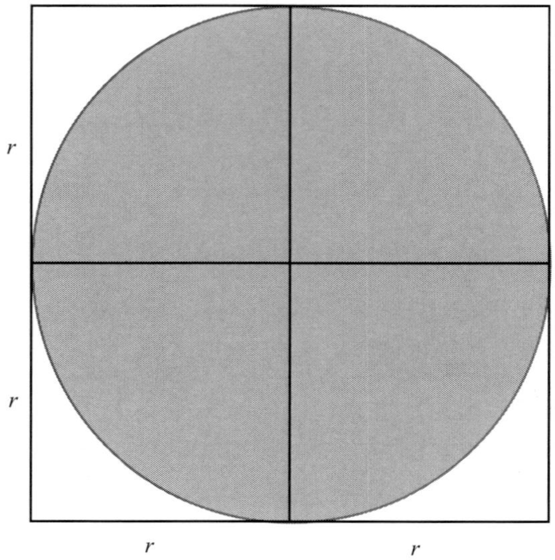

Figure 11.5.5

Figure 11.5.5 shows that the area of a circle is less than $4r^2$, or, equivalently, that the area of the circle is less than the square of the diameter, since $2r=d$. To sum up, we know at this point that the area of a circle is kr^2, for some constant k, and figure 11.5.5 shows that the constant k is less than 4. Can we determine k more accurately? Yes. More than 2,000 years ago Archimedes settled the question. He proved that the area of a circle is the same as the area of a right triangle whose height is the radius of the circle and whose base is the circumference of the circle, as in figure 11.5.6.

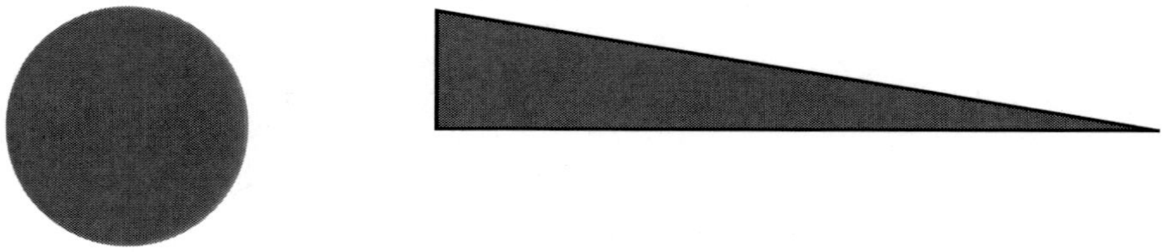

Figure 11.5.6
The base of the right triangle is the circumference of the circle
The height of the triangle is the radius of the circle.
The triangle and the circle have equal areas.–*Archimedes*

Since the circumference c of a circle is $c=\pi d$, where the diameter d is twice the radius r, the area may be expressed as $A=\frac{1}{2}cr=\frac{1}{2}(\pi d)r=\frac{(\pi 2r)r}{2}=\pi r^2$.

Archimedes' theorem

> **A circle of radius r has area πr^2, where π is the circumference of a circle with diameter 1.**

Then in figure 11.5.5, the area of the circle is πr^2, while the area of the outer square is $4r^2$, so the fraction of the square that lies inside the circle is $\pi/4$.

Exercises 11.5

1. a. If the earth is a sphere whose diameter is 8,000 miles, how long is the equator? (For this problem, ignore mountains on the equator.)

   b. If a string 100 feet longer than the equator were stretched around the earth at a constant height above the equator, how much clearance would there be between the earth's surface and the string?

2. If a bicycle tire is a perfect circle two feet in diameter, how many turns will it make in a hundred-mile trip? (Note: One mile is 5,280 feet. Each time a tire makes one complete turn, if it does not slip or skid, it rolls a distance equal to its circumference.)

3. The largest hailstone ever measured, according to aviation writer Barry Schiff, fell on June 22, 2003 in Aurora, Nebraska.[6] Its circumference was measured at 18.75 inches, even after losing an estimated 40% of its size when it hit a gutter. What would be the diameter of a spherical hailstone with a circumference of 18.75 inches?

4. a. Express $\frac{355}{113}$ as a decimal, carrying the result to 8 decimal places.

   b. To how many places does this agree with the true value of π?

5. a. Martin Gardner, in *Hexaflexagons and Other Mathematical Diversions* (University of Chicago Press, 1988), gives the following quotations.

   What is the common purpose of these quotations

   i. May I have a large container of coffee?

      - Anonymous

   ii. How I want a drink, alcoholic of course, after the heavy chapters involving quantum mechanics.

      -Sir James Jeans, physicist

   iii. Now I – even I – would celebrate
      In rhymes unapt the great
      Immortal Syracusan rivaled nevermore
      Who in his wondrous lore,
      Passed on before,
      Left men his guidance
      How to circles mensurate.

      -Adam C. Orr, in *The Literary Digest*, January 20, 1906

   b. Who was the "immortal Syracusan?"

---

[6] "*Test Pilot*," by Barry Schiff, *AOPA Pilot*, July, 2004, p 130.

6. This exercise shows a different approach to finding the area of a circle.

   a. Cut the circle into 8 pieces and reassemble them as shown. The area of the shape at right is equal to the area of the circle.

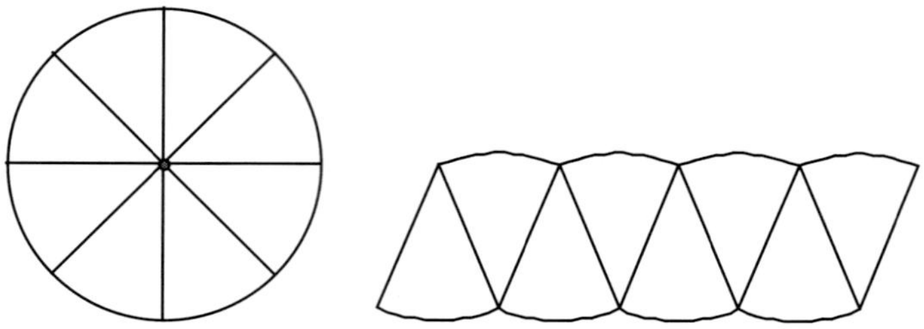

   b. Cut the circle into 16 pieces and re-assemble them as shown. The area of the shape at right is equal to the area of the circle.

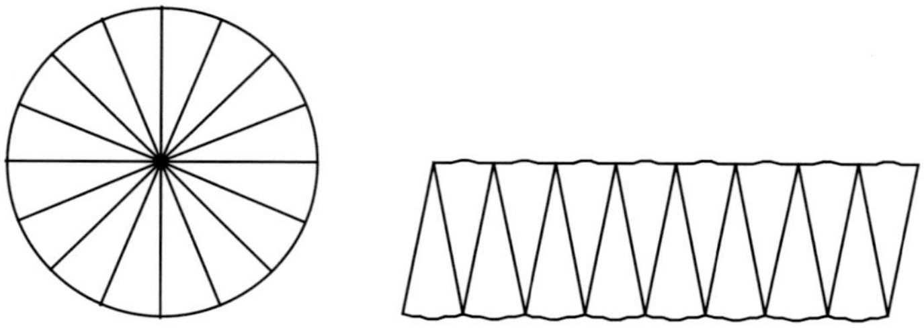

   c. Cut the circle into 32 pieces and re-assemble them as shown. The area of the shape at right is equal to the area of the circle.

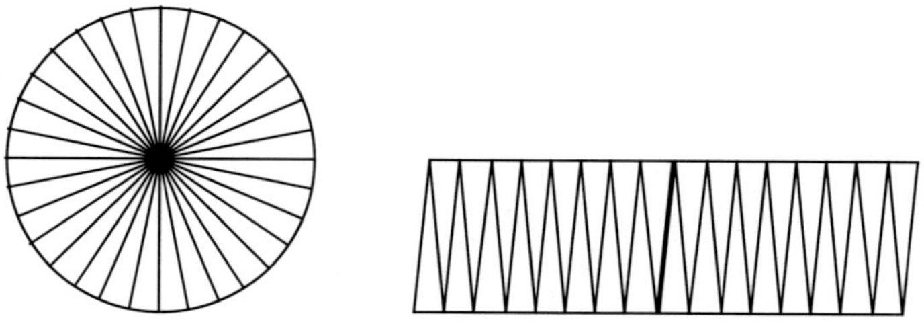

   d. Continuing this process re-assembles the area of the circle into shapes that look more and more rectangular. Find the length, height, and area of the limiting rectangle.

11.6 Cylinders, Cones, and Spheres

When Archimedes approximated the circumference of a circle with the perimeters of polygons, he used polygons with many small sides to approximate a circle. The larger the number of sides, the better the approximation. The volume of a prism is given by $V=Bh$, where B is the area of the base, and h is the height. If B is nearly the area of a circle, then V is approximately $V=\pi r^2h$. When B really is a circle, we have a cylinder instead of a prism, and the volume of the cylinder is given by $V=\pi r^2h$.

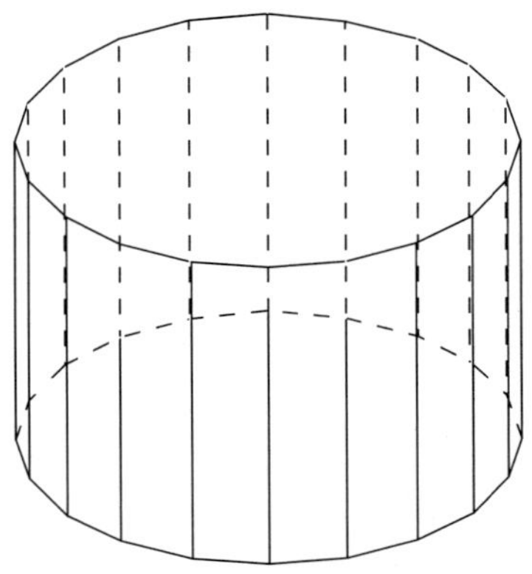

Figure 11.6.1

In direct analogy with approximating cylinders by prisms, cones can be approximated by pyramids. Figure 11.6.2 shows a 20-sided pyramid, which has a volume of $\frac{1}{3}Bh$, where, again, B is the area of the base, and h is the height. If B is nearly the area of a circle, then V is approximately $V=\frac{1}{3}\pi r^2h$. When B really is a circle, we have a cone instead of a pyramid, and the volume of the cone is given by $V=\frac{1}{3}\pi r^2h$.

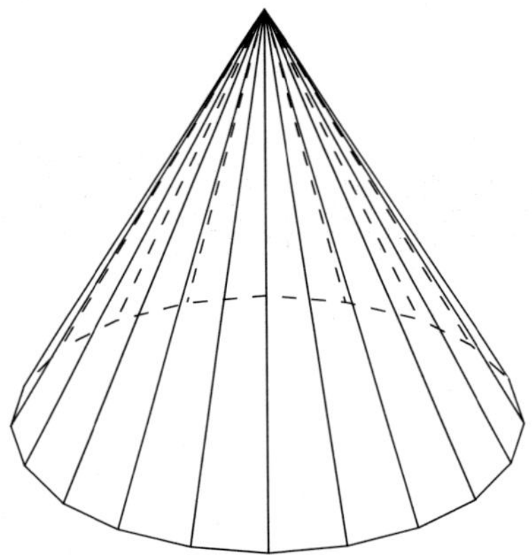

Figure 11.6.2

Spheres

All spheres are similar to each other, and since volumes of similar objects are in proportion to the cube of the scale, it is not surprising that for a sphere, $V=Kr^3$. The value of K was found by Archimedes, who proved that if you have a right circular cylinder with radius r and height $2r$, a right circular cone with the same base and height as the cylinder, and a sphere of radius r, then

Volume of the sphere + Volume of the cone = Volume of the cylinder.

Archimedes gave an elegant proof of this beautiful relationship, which we will not present. However, we will show how this relationship determines the formula for the volume of a sphere.

Archimedes knew the volume of a circular cone with radius r and height h is $\frac{1}{3}\pi r^2 h$, and he knew the volume of a circular cylinder with radius r and height h is $\pi r^2 h$. If both the cone and the cylinder have height $2r$, then the relationship found above becomes: Volume of the sphere $+\frac{1}{3}\pi r^2(2r) = \pi r^2(2r)$. Subtract $\frac{1}{3}\pi r^2(2r)$ from both sides of this equation to get the volume of the sphere as $\pi r^2(2r) - \frac{1}{3}\pi r^2(2r)$. This is $2\pi r^3 - \frac{1}{3}2\pi r^3$. Subtracting $\frac{1}{3}$ of $2\pi r^3$ from all of it leaves $\frac{2}{3}2\pi r^3$, which is $\frac{4}{3}\pi r^3$.

Formulas for the volume of a cylinder, cone, and sphere.

> 1. **The volume of a cylinder is Bh, where B is the area of the base and h is the height. If the base is a circle of radius r, the volume is $\pi r^2 h$.**
> 2. **The volume of a cone is $\frac{1}{3}Bh$, where B is the area of the base and h is the height. If the base is a circle of radius r, the volume is $\frac{1}{3}\pi r^2 h$.**
> 3. **The volume of a sphere of radius r is $\frac{4}{3}\pi r^3$.**

Example 1. Jakob, my old Volvo, has a four cylinder engine. Each piston moves up and down 80 millimeters (mm) inside a right circular cylinder with a diameter of 84 millimeters (mm). The volume of air/fuel taken in when the piston goes down and expelled as exhaust when the piston comes up is called the **displacement** of that cylinder. What is the total displacement of this four cylinder engine?

Solution: The displacement of each cylinder is the volume of a circular cylinder with radius 42 mm (half of the diameter, 84) or 4.2 cm, and height 80 mm or 8 cm. This volume is $\pi(4.2)^2 \times 8 = 443.34$ cubic centimeters (cc). Since there are four cylinders, the displacement of the engine is $443.34 \times 4 = 1773.36$ cc. (The service manual approximates this as 1.75 liters.)

Example 2. An ice cream cone is sold frozen and pre-packed, filled solid with ice cream and topped with a hemisphere (half-sphere) of ice cream. If the cone has a height of four inches and its top is a circle with a diameter of 2 inches, what is the volume of ice cream? (Disregard the thickness of the cone for this approximate calculation.)

Solution: The volume of the cone is $\frac{1}{3}\pi r^2 h$, where r=1 and h=4. This is $\frac{4}{3}\pi$ or about 4.2 cubic inches. A sphere of radius 1 inch has volume $\frac{4}{3}\pi \times 1^3 = \frac{4}{3}\pi = 4.2$ cubic inches as well, so the hemisphere has volume 2.1 cubic inches, and the total volume of ice cream is $4.2 + 2.1 = 6.3$ cubic inches.

Example 3. A farmer wants to replace a cylindrical silo with one that is scaled up by 50%. How will the volume of the new silo compare with that of the old one?

Solution: As with any three-dimensional objects, the the ratio of the volumes of similar objects is the cube of the scale. In this case the scale is 1.5, and new silo has a volume 1.5^3 times the volume of the old one. The new silo will hold 3.375 times as much as the current one.

Exercises 11.6

1. A cylindrical tank with a circular base stands on level ground. The circumference of the base is 25 feet, and the height is 8 feet.
   a. Find the volume of the tank.
   b. If the tank is initially empty and is filled at a rate of 10 cubic feet per minute, how long will it take to fill completely?

2. A circular cylindrical can has an inner diameter of 69 mm, which is just large enough to hold a tennis ball. The height of the can is just enough to hold three tennis balls.
   a. What is the volume of the can?
   b. Compare the circumference of the can with its height.
   c. When there are three tennis balls in the can, what fraction of the volume of the cylinder do the three balls occupy?

3. a. How much beer fits in a conical glass 15 inches tall and 4.5 inches in diameter at the top?
   b. The above beer glass was full, when a ball with a 2 inch radius was put in and floated just half submerged, having caused some of the beer to overflow. How much beer remained in the glass?
   c. How much beer is in the glass if it is filled to a depth of just five inches? Hint: Use similarity.

4. A dozen tennis balls are sold as a four pack of 3-ball cans, wrapped in plastic in a package that looks from above like this:

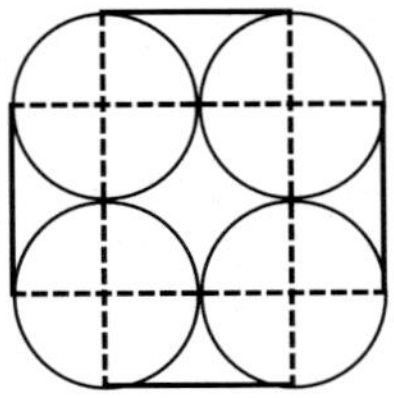

   If the cylinders are 210 cm tall and 70 cm in diameter, what is the volume of the package?

5. The earth is not perfectly spherical, largely because it is flattened at its poles. The diameter between the poles is about 7,901 miles, and through the equator it is closer to 7,926 miles. On a pool ball 2.25 inches in diameter, what would be a proportional difference in diameters?

Vocabulary for Chapter 11

Altitude (of a parallelogram 11.1)
Base (11.1)
Chord (11.5)
Circumference (11.5)
Cone (11.6)
Congruent (11.3)
Cylinder (11.6)
Diameter (11.4)
Height (11.1)
Lattice point (11.1)
Parallelogram (11.1)
Perimeter (11.2)
Prism (11.4)
Proportion (11.3)
Pyramid (11.4)
Rectangular solid (11.4)
Scale (11.3)
Similar (11.3)
Two-dimensional vector (11.1)
Volume (11.4)

Self-Quiz for Chapter 11

1. What is the area of triangle *A*? Explain your answer,

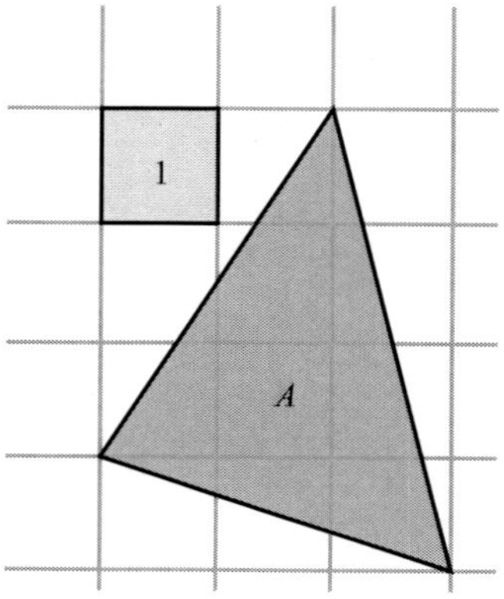

2. Explain why the larger square shown has area 13.

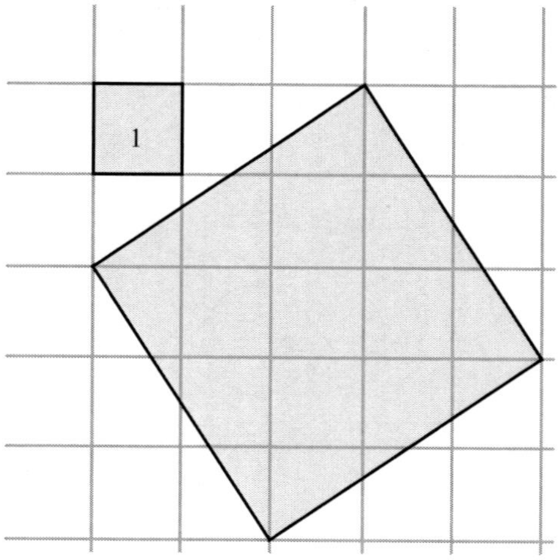

3. Which of these numbers can be areas of lattice point squares?

   a. 5 b. 7 c. 10 d. 37 e. 63 f. 65

   Explain your answer.

4. Explain why the area of a parallelogram is equal to its base times its height.

5. Explain why the area of a triangle is half its base times its height.

6. A room has a rectangular floor, 18 feet long and 12 feet wide. If the ceiling is 10 feet above the floor, what is the volume of the room in cubic feet?

7. A shop sells a souvenir model of a famous statue, scaled to 1/20 original size but made of identical stone. If the original weighs 800 pounds, what does the souvenir weight? Hint: Weight is proportional to volume.

8. An old pine by my house had a straight trunk that was very nearly circular in cross section. At a height of about 5 feet above the ground, its circumference was 17 feet. How thick was the tree trunk at that height?

9. The Great Pyramid of Khufu, in Egypt, has a square base. Each side of the square is about 756 feet long. The original height of the pyramid (it has worn down some 25 feet today) was about 481 feet. Find the original volume of the Great Pyramid in cubic feet.

10. Three tennis balls are packed snug in a circular cylinder can. What fraction of the volume of the can is occupied by the tennis balls?

11. A scale model of a sculpture weighs 43 pounds. If the scale model and the original are made of the same material, and the scale model is one sixth the height of the original, how much does the statue weigh?

12. Three circular holes, each 2 inches in diameter, ar drilled out of a circular steel plate 8 inches in diameter. If the original plate is of uniform thickness and weighs 30 pounds, how much does the plate with three holes in it weigh?

Chapter Twelve
The Pythagorean Theorem

This chapter is about one of the most important relationships in all of mathematics. You probably have at least a passing acquaintance with it, but here we study it in some detail and consider some of the many interesting questions related to it.

12.1 Exploration

Part 1. Gathering evidence

On squared paper with lines $\frac{1}{4}$ or $\frac{1}{5}$ inch apart, place coordinate axes along a pair of lines, as in figure 12.1.1, dividing the page into four parts which are approximately equal.

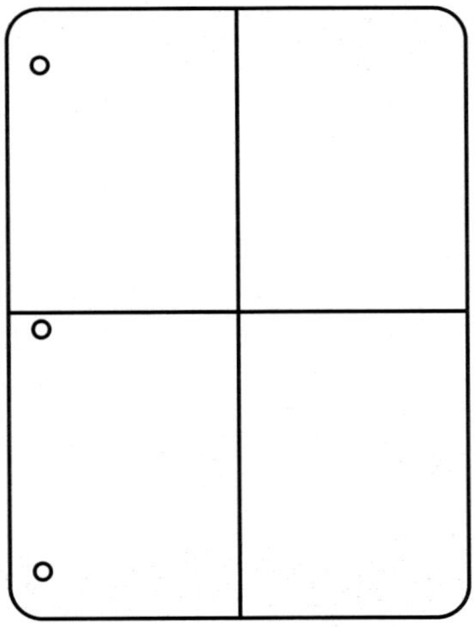

Figure 12.1.1

Next, plot the points given below, and connect them to form triangles.

Triangle A: (-13, 14), (-13, 19), (-1, 19)
Triangle B: (-6, 3), (0, 3), (0, 11)
Triangle C: (3, 1), (3, 16), (11, 1)
Triangle D: (7, 16), (11, 16), (11, 13)
Triangle E: (-12, 0), (12, 0), (-12, -7)
Triangle F: (12, -2), (-12, -12), (12, -12)

The triangles must be plotted accurately to bring out the pattern to be discovered. Using graph paper and coordinates ensures that all students fit the desired triangles onto their pages, while practicing plotting points. The paper should now look something like figure 12.1.2.

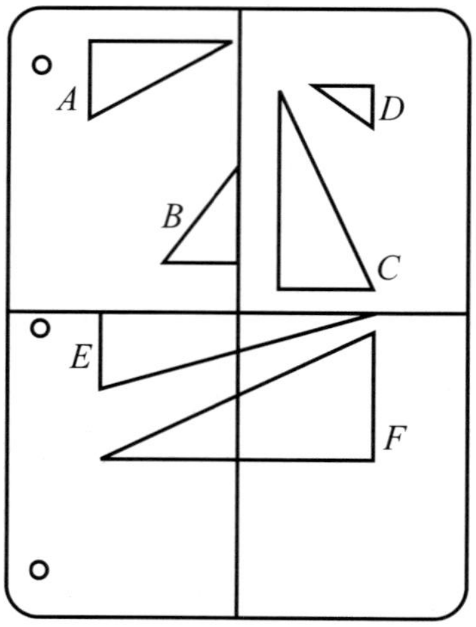

Figure 12.1.2

Now, measure the lengths of all three sides of each triangle, using one square on your graph paper as one unit. The horizontal and vertical sides are easy to measure by just counting squares, but the sloping side is more difficult. To measure it, tear off an edge of your graph paper and use it as a ruler (figure 12.1.3).

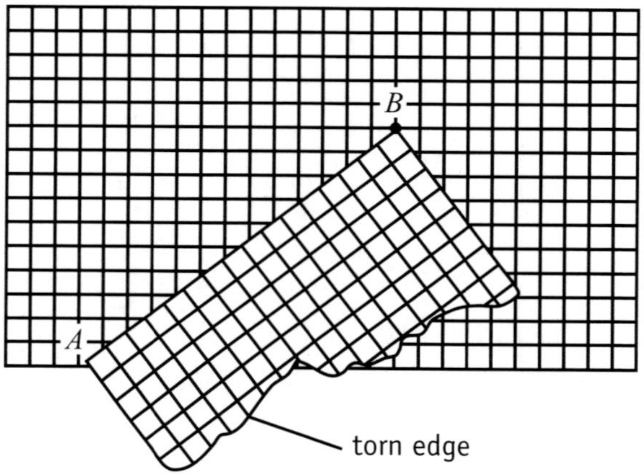

Figure 12.1.3 A home-made ruler

Summarize your findings in the table on the next page.

Triangle	Length of shortest side	Length of next-to-shortest side	Length of longest side
A			
B			
C			
D			
E			

To measure the longest side of each triangle, your home-made graph paper ruler is better than a commercial ruler, because its unit is precisely the same as used on the horizontal and vertical sides. From the classroom point of view this has an additional advantage: one less piece of equipment needed.

Part 2. Finding the rule

Can you find nature's rule for the length of the longest side? This rule is subtle, and if you were playing guess the rule you probably would not use it, because it would be too difficult to guess. Still, you may enjoy pondering it a while. Hint: Make a table much like the one above, but replace all lengths by their squares. When you are ready, read on.

In this case the key to finding nature's rule is an indirect approach that grows out of the work on area in chapter 11.

Figure 12.1.4 shows four copies of triangle *A* arranged to make a large square and a tilted inner square. Finding the areas of these squares will make it possible to find the length of the sloping side of triangle *A*. Try to figure out the answers to the three questions below the figure on your own before reading on.

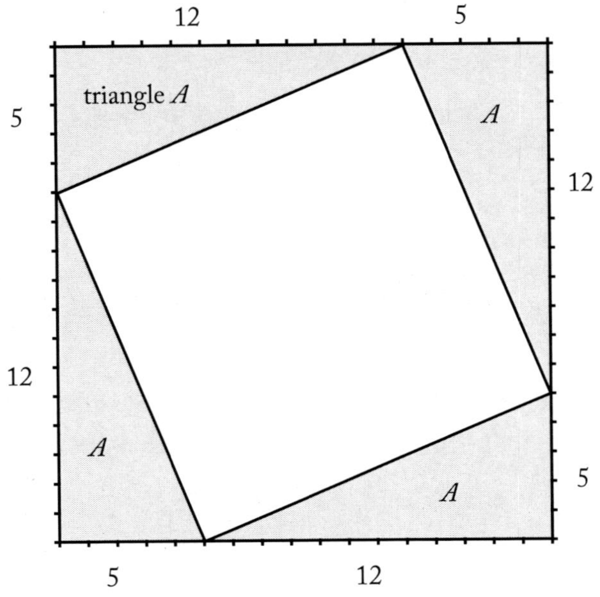

Figure 12.1.4

The area of the large outer square is ______
The total area of the four copies of triangle *A* is ______
Subtract to find the area of the tilted inner square. ______

Now check your work:

1. The outer square is 5+12=17 units on edge, so its area is $17^2=289$.
2. Two copies of triangle *A* fit together to make a rectangle 5×12 with an area of 60, so the four copies of triangle *A* in figure 12.1.4 have a total area of 120.
3. Then the area of the tilted inner square is 289-120=169.

Another way to find the area of the tilted white square in figure 12.1.4 is to slide the copies of triangle *A* around to get figure 12.1.5. The white areas in figure 12.1.5 equal the area of the tilted white square in figure 12.1.4, which must be 5^2+12^2 or 169.

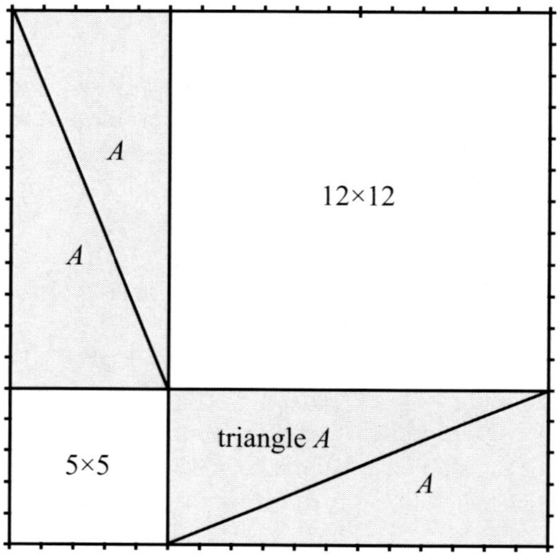

Figure 12.1.5

We have seen, by two different lines of reasoning, that the area of the tilted inner square in figure 12.1.4 is 169.

Given the area of the tilted inner square, can you find the length of its sides? This number is called the **square root** of the area. Since 13×13=169, the square root of 169 (written $\sqrt{169}$) is 13. Then, the tilted inner square must have sides 13 units long, which is the length of the sloping side of triangle *A*. Does that agree with your measurement of the longest side of triangle *A*?

The reasoning used on triangle *A* applies to find the length of the longest side of triangle *B*. Figure 12.1.6 shows four copies of triangle *B* arranged to make a larger square and a tilted inner square.

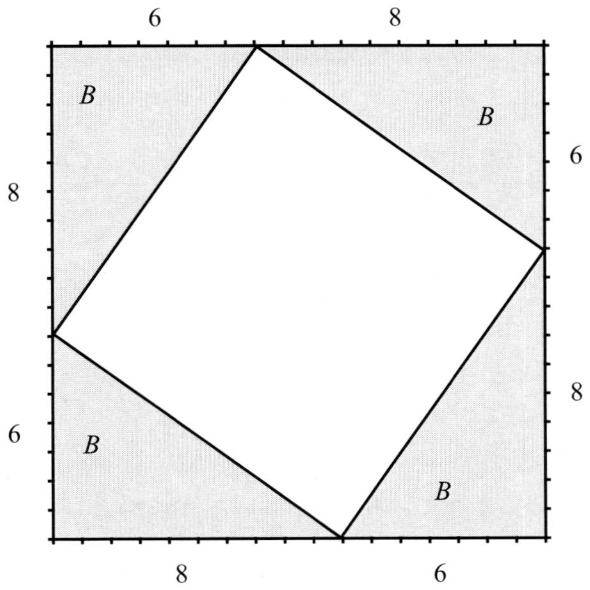

Figure 12.1.6

The area of the large outer square is	______
The total area of four copies of triangle *B* is	______
Subtract to find the area of the tilted inner square.	______

Another way to find the area of the tilted white square in figure 12.1.6 is to slide the copies of triangle *B* around to get figure 12.1.7. The white areas in figure 12.1.7 equal the area of the tilted white square in figure 12.1.6, which therefore must be 6^2+8^2. Then, the edge of the tilted square has length $\sqrt{6^2+8^2}$, which is $\sqrt{36+64} = \sqrt{100} = 10$.

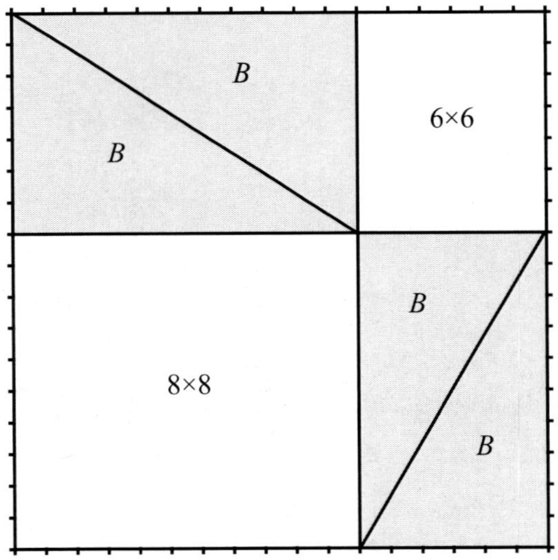

Figure 12.1.7

The reasoning presented above for triangles *A* and *B* holds for any right triangle whatever. The general statement of the resulting relationship between the short sides (called the **legs**) and the long side of any right triangle (called the **hypotenuse**) is known as the **Pythagorean theorem**.

Pythagorean Theorem

> **In a right triangle, if a and b are the lengths of the legs and c is the length of the hypotenuse, then $a^2+b^2=c^2$ or $c=\sqrt{a^2+b^2}$.**

Figure 12.1.8, based on the examples above, is one way to see why the Pythagorean theorem must be true.

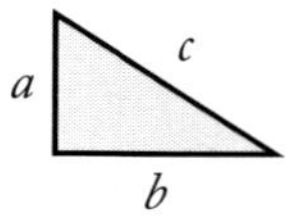

Figure 12.1.8a

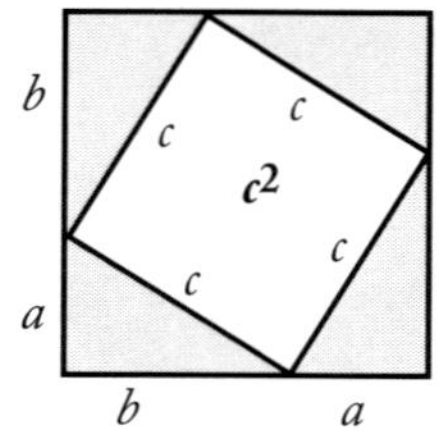

Figure 12.1.8b

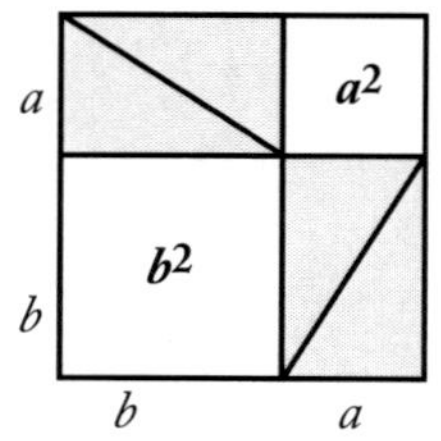

Figure 12.1.8c

Put four copies of the right triangle together to make a square with a tilted inner square (figure 12.1.8b). Then slide the copies of the triangle to make figure 12.1.8c. The area of the tilted inner square, c^2, must be the sum of the areas of the two smaller squares, $a^2 + b^2$. Then $a^2 + b^2 = c^2$. Take square roots: $c = \sqrt{a^2 + b^2}$.

The reasoning motivated by figure 12.1.8 can be restated algebraically, based on the two ways of viewing the area of a square with sides $a+b$. If we view the large square as formed by four triangles in the corners plus a tilted inner square of side c, its area is c^2+2ab. If we view the same area as shown in the figure 12.1.8c, it is a^2+b^2+2ab. Since both expressions represent the same area,

$$a^2+b^2+2ab = c^2+2ab$$

Subtract $2ab$ from each side to get $\quad a^2+b^2 = c^2$

Therefore $\quad \sqrt{a^2 + b^2} = \sqrt{c^2}$

With the Pythagorean theorem, finding the hypotenuse of a right triangle is routine if you know the lengths of the sides, as in example 1.

Example 1. A rectangular table top is 21 inches long and 20 inches wide. How far apart are diagonally opposite corners?

Solution: $\sqrt{20^2 + 21^2} = \sqrt{400 + 441} = \sqrt{841} = 29.$

How to find square roots is discussed in section 12.2, but in the exercises below, you can find all needed square roots by guessing. For example, to find $\sqrt{841}$ just experiment; $30^2=900$, which is too big. Trying smaller numbers, you will find $29^2=841$, so $\sqrt{841}=29$.

You can also use the Pythagorean theorem to find the length of one leg of a right triangle, if you know the length of the other leg and the hypotenuse. Example 2 illustrates the idea.

Example 2. One leg of a right triangle is 16 feet long, and the hypotenuse is 34 feet long. How long is the second leg of the triangle?
Solution: If the length of the second leg of the triangle is b, then, by the Pythagorean theorem, $16^2+b^2 = 34^2$
Subtract 16^2 from both sides $b^2 = 34^2-16^2$
$= 1156-256$
$= 900$
Take the square root of both sides $b = 30$
Then the second leg of the right triangle is 30 feet long.

Exercises 12.1

1. Here are sketches of triangle D from the exploration above and four copies of D arranged in a square.

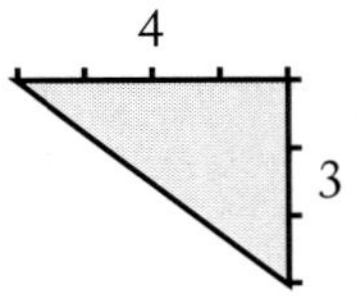

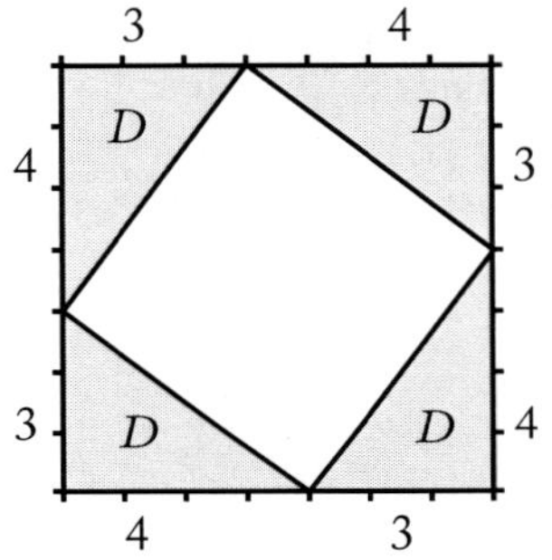

a. Find the area of the large square.
b. Find the area of the four copies of triangle D.
c. Find the area of the tilted inner square.
d. Find the length of the hypotenuse of triangle D.
e. Here is the same large square as sketched above, without the copies of triangle D.

Sketch how to put in four copies of triangle D in a way that leaves two squares, and show that those squares have areas 9 and 16.

f. Here is a diagram based on the ancient Chinese *Zhoubi suanjing*, from perhaps 200 B.C.

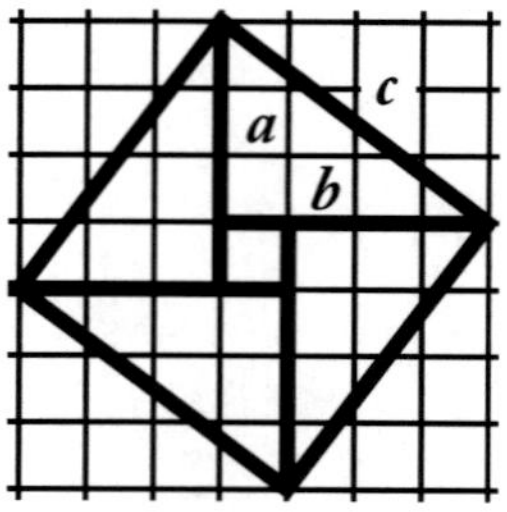

Based on this diagram, do you think that the writers knew the Pythagorean theorem? Explain the reasons for your answer.

2. a. Put four copies of triangle C from figure 12.1.2 into this picture in a way that shows a tilted inner square on the hypotenuse, as in figure 12.1.8b.

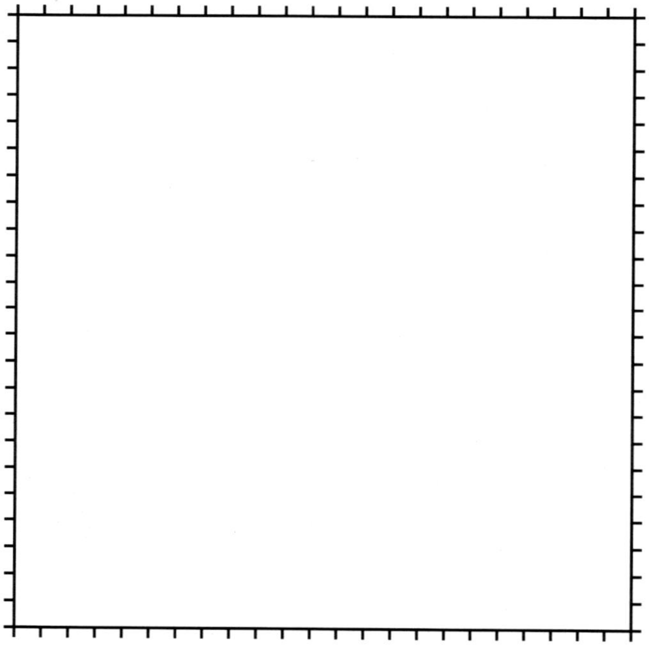

b. In a second copy of the same square, put in copies of triangle C in a way that shows the big square as made up of two smaller squares plus four copies of triangle C, as in figure 12.1.8c.

c. Use the sum of the areas of the two squares in part b to find the area of the tilted square in part a.

d. Use a square root to find the length of the hypotenuse of triangle C.

e. Check your result against the one you obtained by measuring in the discovery exercise.

3. Repeat the instructions from question 2 on triangle E from the text, this time making your own sketches on a piece of graph paper.

4. Find the length of the hypotenuse of a right triangle with legs of these lengths. (In all cases you can find the necessary square roots by guessing.)

   a. 16 and 30   b. 12 and 35   c. 24 and 45

5. A rectangular lot is 55 yards long and 48 yards wide. How far apart are diagonally opposite corners of the lot?

6. A cable secures a vertical pole which stands on flat land. The cable is attached 96 feet up the pole and is anchored in the ground 40 feet from the base. How long is the cable?

7. A ship is 27 kilometers (km) due east of a lighthouse in the Caribbean, while a second is 120 km due north of the same light. How far apart are the ships?

8. View *The Theorem of Pythagoras*, a video about 20 minutes long from *Project MATHEMATICS!* This video has been widely distributed to schools and school districts. It may be reproduced at no copyright cost for non-commercial, educational use. You can also get it from the NCTM, 1906 Association Drive, Reston, VA 22091.

9. If you know the lengths of the hypotenuse and one leg of a right triangle, the length on the other leg may be found as in example 2. Describe this procedure in general.

10. The length of the hypotenuse and one leg of a right triangle are given. Use the Pythagorean theorem to find the length of the other leg.

    a. Hypotenuse 169, leg 119 b. Hypotenuse 195, leg 189

11. A 130 foot cable is attached to a building 120 feet above flat ground. If the cable is taut, how far from the building will it touch the ground?

12. A 25-foot ladder leans against a vertical wall with its foot on level ground 7 feet from the wall. If the foot is moved 8 feet further from the wall, how many feet will the top of the ladder slide down the wall?

13. (The Pythagorean theorem in three dimensions.) Find the distance between diagonally opposite corners of the rectangular solid (brick shape) shown here. Note that both triangles ABD and DBF are right triangles. Thus, for example, if $AB=4$, $AD=3$, and $BF=12$, how long is DF? One way to do this is in stages:

    a. Use the Pythagorean theorem on triangle ABD to find length BD.

    b. Use the Pythagorean theorem on triangle DBF to find length DF.

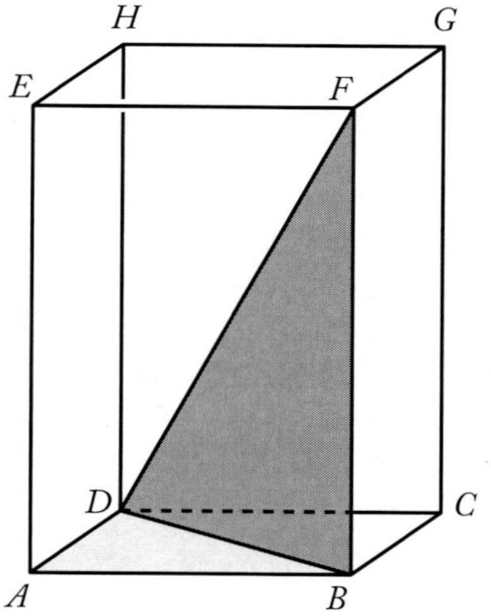

14. A shipping company refuses to take any package that is more than 1 meter in length, width, or height. What is the longest straight object that can be shipped under these rules?

12.2 Square Roots

It is time for a careful look at square roots, a topic treated rather casually so far.

A square root of a number *n* is, by definition, a number whose square is *n*.

For example, 3 is a square root of 9. We say **a** square root, because -3 is also a square root of 9, as you can check by squaring -3. This is typical. Both 2 and -2 are square roots of 4, and both 7 and -7 are square roots of 49. In general, positive numbers have two square roots, one positive and the other negative. Negative numbers, on the other hand, have no square roots at all, and 0, which is neither positive nor negative, has only one square root , 0. [1]

The square root sign $\sqrt{\ }$ denotes the positive square root or 0.

How can you actually calculate a square root? The naïve approach is trial and error. Often, it will do the job.

Example 1. Find $\sqrt{56{,}169}$.
Solution: Experiment, trying to hit the target 56,169 by squaring numbers. Start with numbers that are easy to square, to save work.

Try 100.	100^2=10,000.	That's less than 56,169, so 100 is too small.
Try 200.	200^2=40,000.	That's less than 56,169, so 200 is too small.
Try 300.	300^2=90,000.	That's more than 56,169, so 300 is too big.

Evidently $\sqrt{56{,}169}$ is somewhere between 200 and 300. The next step is to search numbers in this range.

Try 250.	250^2=62,500.	That's more than 56,169, so 250 is too big.
Try 240.	240^2=57,600.	That's more than 56,169, so 240 is too big.
Try 230.	230^2=52,900.	That's less than 56,169, so 230 is too small.

The search is narrowed to the range from 230 to 240.

Try 235.	235^2=55,225.	That's less than 56,169, so 235 is too small.
Try 237.	237^2=56,169.	Success! $\sqrt{56{,}169}$ =237.

Example 1 worked out nicely, because 56,169 happens to be a square number. Example 2 shows what happens when trial and error is used to find the square root of a number which is not a perfect square.

[1] Square roots of negative numbers were rejected for many years as preposterous, but today they are accepted as part of an expanded system of numbers called complex numbers. See section 14.2.

Example 2. Find $\sqrt{2}$.
Solution: Once again, use trial and error. Since $1^2=1$ and $2^2=4$, the search is narrowed to the interval from 1 to 2, shown in figure 12.2.1.

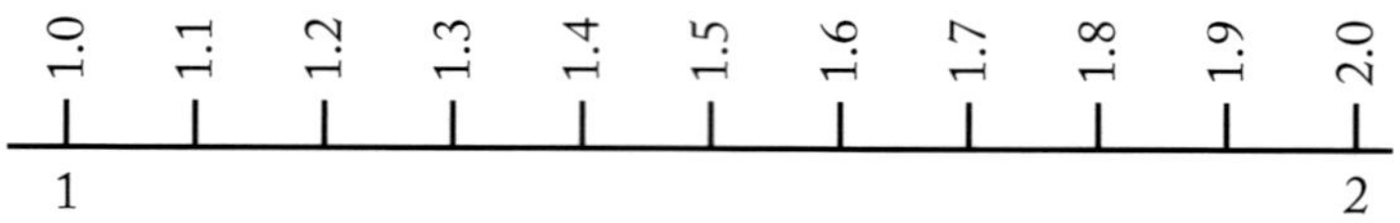

Figure 12.2.1

Continue the trial and error procedure. $1.5^2=2.25$, so 1.5 is too big, but $1.4^2=1.96$, so 1.4 is a bit too small. Next, search the interval from 1.4 to 1.5, shown in figure 12.2.2.

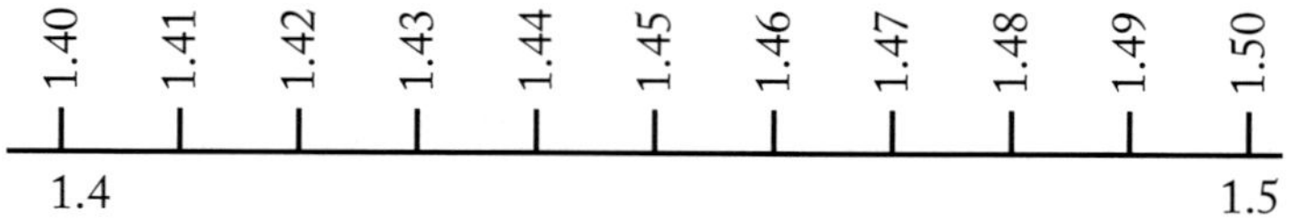

Figure 12.2.2

Try 1.45.	$1.45^2=2.1025$.	That is more than 2, so 1.45 is too large.
Try 1.42.	$1.42^2=2.0164$.	That is more than 2, so 1.42 is too large.
Try 1.41.	$1.41^2=1.9881$.	That is less than 2, so 1.41 is too small.

The next step is to search the interval from 1.41 to 1.42, shown in figure 12.2.3.

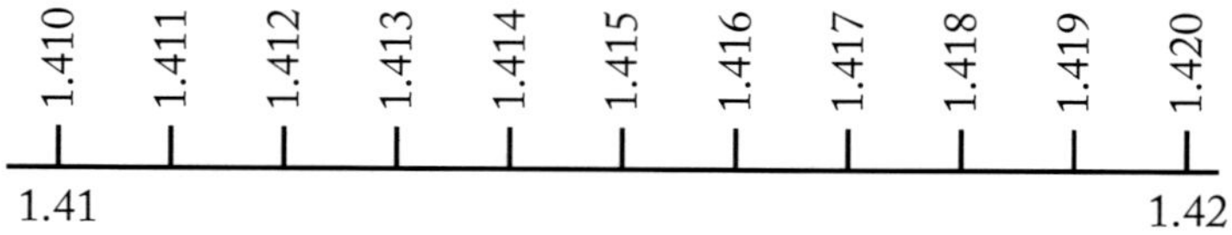

Figure 12.2.3

At this stage Vicki, a sixth grader, said, "We'll never get to $\sqrt{2}$." When asked to explain, she pointed out a problem with the last digit. For example, squaring 1.415 amounts to multiplying it by itself, as in figure 12.2.4.

```
      1.415
    × 1.415
   --------
      7075
     1415|
    5660 |
   1415  |
  ---------
         5
```

Figure 12.2.4

The last digit in this case is a 5, but if we had an exact decimal for $\sqrt{2}$, then after squaring, the last digit should be 0. Vicki observed that the only last digit that would produce a 0 would itself be 0, but then the digit just before would cause the same trouble. Since $\sqrt{2}$ is not an integer, you will never get it exactly by calculating more digits.

Vicki's observation and reasoning are remarkable for a 12-year old. You can calculate more and more decimal places by trial and error, and for many students, this turns into a valuable lesson about the number line. Indeed, they often get very excited about getting decimals whose squares are closer and closer to 2, but no matter how hard they try, they will never get exactly 2. Not only is there no finite decimal which is exactly equal to $\sqrt{2}$, there is also no fraction, even one with an infinite decimal, which is exactly equal to $\sqrt{2}$. This discovery, which is attributed to the ancient Pythagoreans, may seem strange to you. Because $\sqrt{2}$ cannot be represented as a fraction or ratio of integers, it is an example of an irrational number. In contrast, numbers that can be represented as ratios of integers are called rational numbers. The discovery of irrational numbers was philosophically troubling to Pythagoreans, who had developed an elaborate theory of proportions based on ratios of whole numbers.

Calculation of square roots by trial and error is a good place to use calculators or computers to save work and improve accuracy. Do not, however, use a calculator with a square root key until the students have explored extensively, computing several square roots by trial and error. Then they will appreciate the square root key. More important, they will know what a square root is in a way they could never have learned without actually calculating a few by trial and error.

Isaac Newton[2] improved on the above method of calculating square roots. To find $\sqrt{n}$, begin with a guess, and divide n by that guess. If you are lucky enough to guess $\sqrt{n}$ perfectly, the quotient will be the same as the guess. Otherwise, $\sqrt{n}$ will lie between the guess and the quotient. For example, to find $\sqrt{90}$, you might start with a guess of 9. Divide 90 by 9, and the quotient is 10. Then, $\sqrt{90}$ lies between 9 and 10. The idea to Newton's method is to average your guess and the quotient to get a better guess. For $\sqrt{90}$, the next guess would be the average of 9 and 10, which is 9.5. Now, divide and average again, repeating the process until the square root is approximated as accurately as needed.

```
        9.4736
9.5 ) 90.0000
      85 5
       4 50
       3 80
         700
         665
          350
          285
           650
           570
```

[2] Isaac Newton (1642-1727) was born shortly after the death of his father. When Isaac was three, his mother remarried and left him in the care of his grandmother. Very much a solitary child, he soon developed a remarkable ability for sustained, concentrated thought. His great scientific talents led him to Cambridge, where he soon became interested in mathematics and read the works of most of the leading mathematicians up to that time. When the university was closed in 1665-1666 because of the plague, Newton returned to his home. There he made discoveries in physics and mathematics, which place him among the greatest thinkers of all time. Among his great achievements, he proved that a force which he called gravity would cause the planets to travel in elliptical orbits about the sun and to obey Kepler's laws. In later life, he did little mathematics, as he was preoccupied with his duties as director of the royal mint. The method for finding square roots discussed in the text is known as Newton's method, but its beginnings can be traced all the way back to ancient Mesopotamia.

Average once again to get a new guess: $\frac{9.4736 + 9.5}{2} = 9.4868$. You can check with a calculator that this is remarkably close to $\sqrt{90}$, but if you need still more accuracy, use 9.4868 as your next guess and carry the process one more round. Newton's method uses division, which is more tedious than multiplication, but it homes in on the actual square root faster than simple trial and error.

Simplifying square roots

> **For the remainder of this section, we will assume that numbers appearing inside square root signs are all positive. For positive numbers x and y, $\sqrt{xy} = \sqrt{x}\sqrt{y}$.**

To see why this is so, square both sides. On the left, $\left(\sqrt{xy}\right)^2$ is, by definition, xy, and the right-hand side is $\left(\sqrt{x}\sqrt{y}\right)\left(\sqrt{x}\sqrt{y}\right)$, which, by the associative and commutative properties of multiplication, is $\left(\sqrt{x}\sqrt{x}\right)\left(\sqrt{y}\sqrt{y}\right) = xy$.

The above rule is useful in simplifying square roots. For example, $\sqrt{20} = \sqrt{4}\sqrt{5} = 2\sqrt{5}$. Again, $\sqrt{\frac{3}{4}} = \sqrt{3}\sqrt{\frac{1}{4}} = \sqrt{3}\left(\frac{1}{2}\right) = \frac{\sqrt{3}}{2}$.

Generally, if $a \geq 0$, $\sqrt{a^2 b} = a\sqrt{b}$, and a square root is considered simplified if this rule is applied to the point where b has no square factors. For example, you could simplify $\sqrt{80}$ part way as $\sqrt{80} = \sqrt{4 \times 20} = \sqrt{4}\sqrt{20} = 2\sqrt{20}$, but $\sqrt{80}$ is not entirely simplified yet, because $\sqrt{20}$ can be simplified further to $2\sqrt{5}$. Completely simplified, $\sqrt{80} = 2\sqrt{20} = 2 \times 2\sqrt{5} = 4\sqrt{5}$.

Example 3. Simplify $\sqrt{\frac{98ab}{9c^4}}$ completely.

Solution: $\frac{7\sqrt{2ab}}{3c^2}$.

Example 4. Simplify $\sqrt{2 \cdot 5^2 \cdot 9^2 \cdot 29^2 \cdot 79^2}$

Solution: In this case, if you tried to compute the square root with a small calculator by first multiplying out the number under the square root sign and then pushing the square root button, you might get a poor approximation, because the number under the square root sign might have more digits than the small calculator can handle. Instead, since $\sqrt{5^2} = 5$, $\sqrt{9^2} = 9$, etc.,

$$\sqrt{2 \cdot 5^2 \cdot 9^2 \cdot 29^2 \cdot 79^2} = 5 \cdot 9 \cdot 29 \cdot 79\sqrt{2}$$

Square root signs (and also signs for cube roots and higher roots) are sometimes called **radical** signs. The word is related to "radish" and comes from an old word for root. Its use in politics refers to policies, which, in contrast to superficial measures, aim at the root of the situation.

Exercises 12.2

1. Compute the positive square roots of these numbers by trial and error. (All answers are whole numbers.)

   a. 8,649 b. 5,776 c. 91,204
   d. 23,409 e. 81,796

2. Find $\sqrt{2}$ to three decimal places:

   a. By continuing the trial and error process begun in the text.
   b. By Newton's method.

3. Find $\sqrt{3}$ to the nearest hundredth:

   a. By trial and error. b. By Newton's method.

4. Find $\sqrt{5}$ to the nearest hundredth:

   a. By trial and error. b. By Newton's method.

5. a. Find $\sqrt{8}$ to the nearest hundredth:
   b. Compare your answer from part a with the value of $\sqrt{2}$ found in question 2 above. What do you notice?

6. A baseball diamond is a square, 90 feet on an edge. To the nearest tenth of a foot, how long is the diagonal of that square? (This is approximately the distance the catcher must throw when a runner tries to steal second base.)

7. A rectangular display screen measures 19 inches between diagonally opposite corners. If the display is 15 inches wide, how high is it?

8. To compute a square root by trial and error, it helps to make your first guess a good one. One way to do that is based on a picture. For example, to find $\sqrt{149}$, make a square array of 144 dots, and put the remaining five dots along the border, beginning to extend the array to a 13× 13 square. Since 149 is more than a 12×12 square and less than a 13×13, $\sqrt{149}$ is between 12 and 13. More precisely, since 5 of the 25 dots in

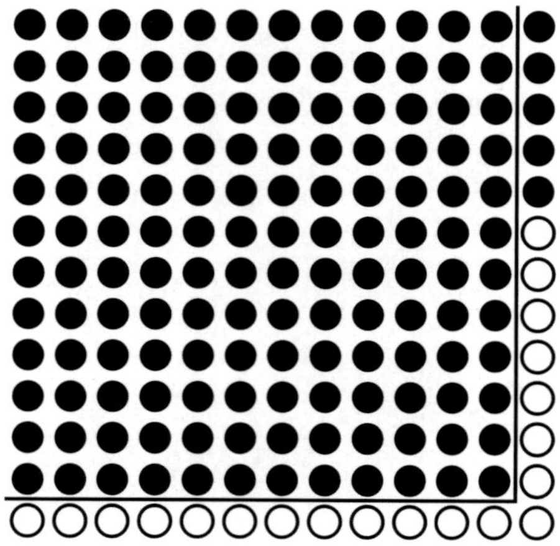

the outer row are shaded, $\sqrt{149}$ is about $12\frac{5}{25}$ or 12.2, as you can easily check.

Make similar picture estimates and checks for the square roots of:

a. 27 b. 85 c. 128 d. 69

9. Simplify as much as possible:

a. $\sqrt{12}$ b. $\sqrt{72}$ c. $\sqrt{99}$ d. $\sqrt{288}$

10. Imagine a two-mile piece of straight railroad track, rigid except for a hinge in the middle allowing it to flex due to expansion in heat. If *AC* and *BC* remain straight, and the two-mile section of track expands a total of two feet, how far will point *C* rise above the center of line segment *AB*? (Use a calculator for the actual computations, with 5,280 feet to a mile.) The surprisingly large answer shows why builders of railroads, highways, and bridges must plan ahead for expansion, which is typically much more than an inch per mile.

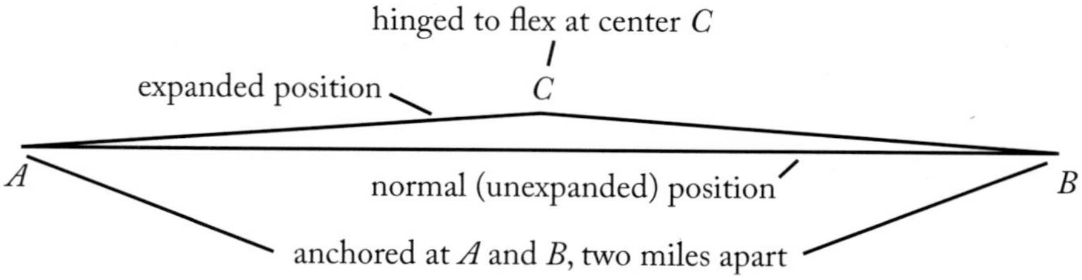

11. Calculate the perimeter of triangle *Q* in figure 11.1.2 to the nearest hundredth.

12. If you are one mile above sea level, and the weather is perfectly clear, how far away is an object, such as a boat, at sea level on your horizon? (For this problem, assume the Earth is a sphere with a radius of 4,000 miles.)

13. The Korean mathematician Hong Kil-ju (1786-1841) invented the following procedure for finding the square root of a square number. It is remarkable in that, unlike other methods, it is based on subtraction rather than multiplication or division. We illustrate with 49.

Step 1. Take half of the given number.	Half of 49 is 24.5	
Step 2. Subtract 1	24.5-1=	23.5
Step 3. Subtract 2	23.5-2=	21.5
Step 4. Subtract 3	21.5-3=	18.5
Step 5. Subtract 4	18.5-4=	14.5
Step 6. Subtract 5	14.5-5=	9.5
Step 7. Subtract 6	9.5-6=	3.5

The next step would be to subtract 7, but it is to be done only if doing so leaves a positive number. In this case it will not, so the procedure moves to its final step, doubling the last difference, 3.5. The result, 7, is $\sqrt{49}$.

In general, subtract one higher number at each step as long as you can do so without getting a negative answer. Twice the result of your final subtraction is the square root desired.

a. Try this procedure on a few numbers of your choice. (Suggestions: 169 and 441).

b. Try to explain why this procedure will always work. (Consult Exercises 3.1, question 4.)

12.3 Distance, Circles, and Absolute Value

The Pythagorean theorem allows us to find the distance between two points in the coordinate plane, as the following example shows.

Example 1. Find the distance from (2,5) to (7,1).
Solution: The distance is the length of line segment *AB* in figure 12.3.1. First form a right triangle by means of a point *C*, whose coordinates are (2,1). Point *C* takes its *x*-coordinate from point *A* and its *y*-coordinate from point *B*. (The point (7,5), which takes its *x*-coordinate from *B* and its *y*-coordinate from *A*, could also be used). The points *A, B*, and *C* form a right triangle with *AB* as hypotenuse. By the Pythagorean theorem, the length of *AB* is $\sqrt{(7-2)^2+(5-1)^2}$. This is $\sqrt{5^2+4^2}=\sqrt{41}$.

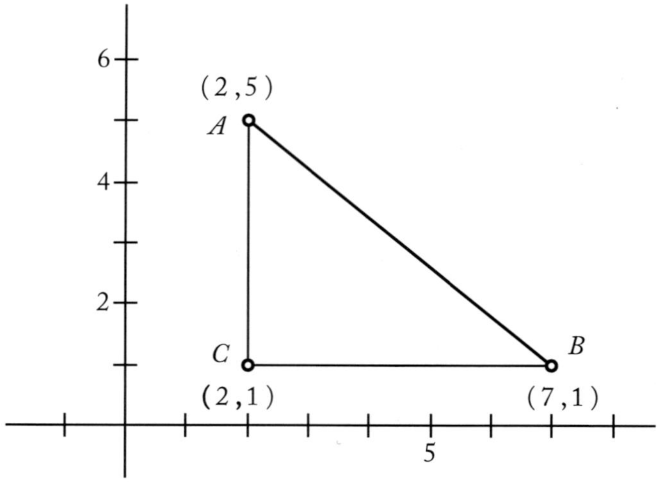

Figure 12.3.1

The method of Example 1 may be used to find the distance between two arbitrary points $P_1(x_1,y_1)$ and $P_2(x_2,y_2)$. First form a right triangle (figure 12.3.2) by means of a third point $P_3(x_1,y_2)$. The distance between P_3 and P_2 is the same as the distance between x_1 and x_2 on the x-axis, and, similarly, the distance between P_1 and P_3 is the same as the distance between y_1 and y_2 on the y-axis. Then by the Pythagorean theorem the distance from P_1 to P_2 is $\sqrt{(x_2 - x_1)^2 + (y_2 - y_1)^2}$. In this context, the Pythagorean theorem is called the **distance formula.**

Distance Formula

The distance between points $P_1(x_1,y_1)$ and $P_2(x_2,y_2)$ is $\sqrt{(x_2 - x_1)^2 + (y_2 - y_1)^2}$.

Remember, the square root sign indicates the non-negative square root. **Distance is never negative.**

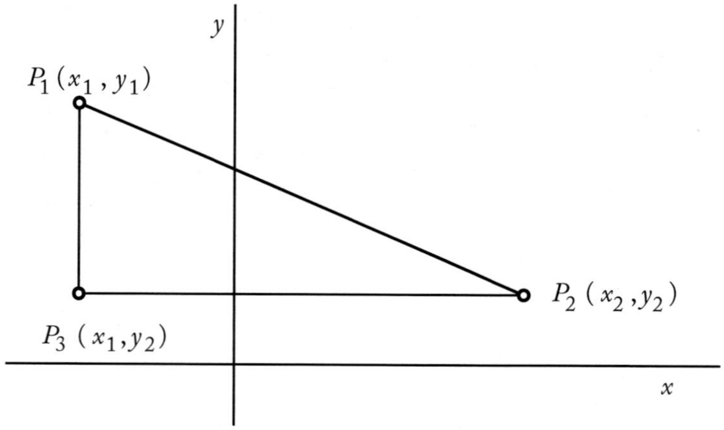

Figure 12.3.2

Example 2. Find the distance between (-1,5) and (3,2).

Solution: The distance formula gives $\sqrt{(3-(-1))^2+(2-5)^2} = \sqrt{25} = 5$. From the sketch (figure 12.3.3), this answer looks reasonable. The line segment appears to be about 5 units long.

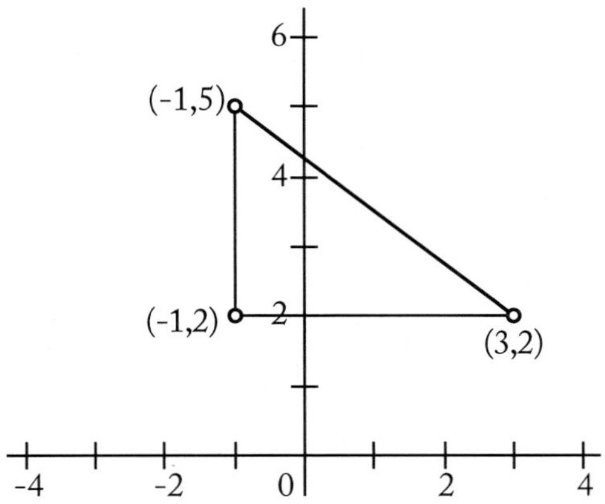

Figure 12.3.3

Example 3. Write an equation which states that the distance from the point (x,y) to the point (3,4) is 6 units.
Solution: The distance from (x,y) to (3,4) is $\sqrt{(3-x)^2+(4-y)^2}$, so the desired equation is $\sqrt{(3-x)^2+(4-y)^2} = 6$. This equation expresses the algebraic condition that (x,y) must satisfy if it is 6 units from (3,4). The set of all such points is a circle with center (3,4) and radius 6, so $\sqrt{(3-x)^2+(4-y)^2} = 6$ is said to be an equation for that circle. It is often convenient to write an equation that does not have a square root sign by squaring the last equation to get

$$(3-x)^2+(4-y)^2 = 6^2$$

More generally, to find an equation for a circle with center (p,q) and radius r, note that a point (x,y) is on the circle if and only if its distance from (p,q) is r. In terms of the distance formula, this condition is $\sqrt{(p-x)^2+(q-y)^2} = r$, or, squaring both sides, $(p-x)^2+(q-y)^2 = r^2$. Conversely, every equation which can be put in the form $(p-x)^2+(q-y)^2 = r^2$ with $r > 0$ represents a circle with center (p,q) and radius r.

Equation of a Circle

The equation $(p-x)^2+(q-y)^2 = r^2$ with $r > 0$ represents a circle with center (p,q) and radius r.

Note: The equation $(p-x)^2 + (q+y)^2 = 0$ is satisfied by a single point, (p,q). You may think of this as a circle with radius *0*.

Example 4. Describe the set of points (x,y) for which $x^2+(y-3)^2=49$.
Solution: Since $49=7^2$ and $x^2=(x-0)^2$, we can rewrite the given equation as $(x-0)^2+(y-3)^2=7^2$. This is a circle with center (0,3) and radius 7.

Example 5. Describe the set of points for which $\sqrt{(x-2)^2+(y-3)^2} < 2$.
Solution: This is much like example 4, but in this case the distance from (x,y) to (2,3) is **less than** 2, which is the radius of the circle. This inequality is satisfied not by points on a circle itself but by every point **inside** the circle (figure 12.3.4).

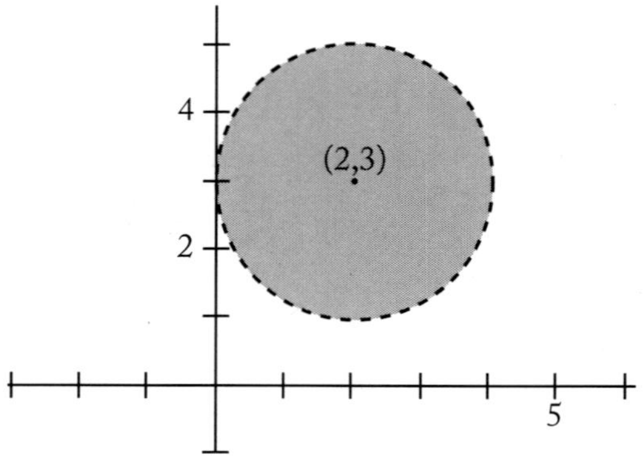

Figure 12.3.4

Example 6. Describe the set of points which satisfy the equation $(x-2)^2+(y-3)^2=-2$.
Solution: No pair of real numbers (x,y) satisfies this equation, since the left side, a sum of squares of real numbers, is at least zero, but the right side is negative.

Distance along a number line: Absolute value
The distance between two numbers x_1 and x_2 along the number line is written $|x_2-x_1|$. Since one way to find this distance is to subtract the smaller number from the larger one, $|x_2-x_1|$ is either x_2-x_1 or its opposite, x_1-x_2, whichever is positive. In either case, the square of the distance is $(x_2-x_1)^2$, and the distance itself is the non-negative (positive or zero) square root of $(x_2-x_1)^2$, or

$\sqrt{(x_2-x_1)^2}$. This is just what you would expect if you think of the number line as the x-axis, on which all points have y-coordinate 0, and apply the distance

formula $\sqrt{(x_2-x_1)^2+(y_2-y_1)^2}$. In that case the formula becomes

$\sqrt{(x_2-x_1)^2+(0-0)^2}$, which simplifies to $\sqrt{(x_2-x_1)^2}$.

The distance between x and 0, written $|x-0|$ or simply as $|x|$, is called the **absolute value** of x. From computational point of view you can think of $|x|$ as

$\sqrt{x^2}$. For example, $\sqrt{(-3)^2}=\sqrt{9}=3$.

Definition of absolute value

The absolute value of a number x is written $|x|$. If $x \geq 0$, $|x| = x$, but if $x < 0$, then $|x| = -x$.

Since $-x$ and x are the same distance from 0, $|x| = |-x|$. Therefore, the equation $|x| = k$ has two solutions, $x = k$ and $x = -k$, when $k > 0$ and only the single solution $x = 0$ when $k = 0$. When $k<0$, the equation has no solution.

Example 7. Graph the inequality $|x| \le 3$.
Solution: Interpret $|x|$ as $|x-0|$, the distance from x to 0. Then the given inequality is satisfied by all points which are three units or less from 0. If you think of a goat tethered to 0 by a rope three units long, these are the points where the goat could graze. They are graphed in figure 12.3.5.

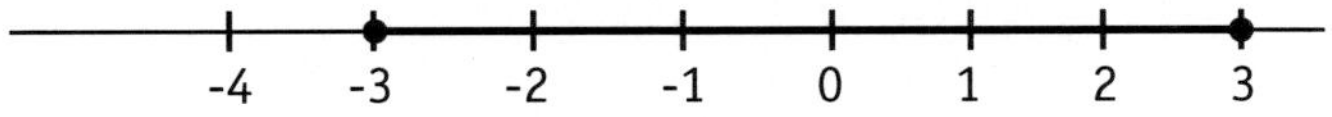

Figure 12.3.5

Example 8. Solve and graph $|x-3|>5$.
Solution: Interpret $|x-3|$ as the distance between x and 3. The inequality is satisfied by any number x which is more than five units from 3. These are pictured in figure 12.3.6.

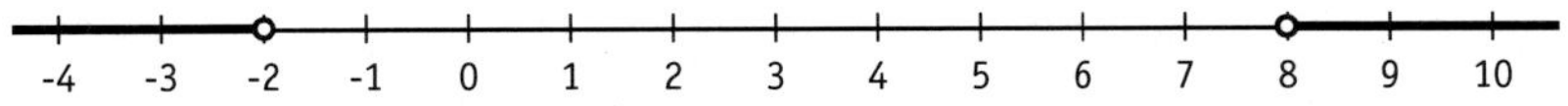

Figure 12.3.6

Another use of absolute value is in simplifying radical expressions.

Example 9. Simplify $\sqrt{9a^4b^2}$.

Solution: $\sqrt{9a^4b^2} = \sqrt{9}\sqrt{a^4}\sqrt{b^2} = 3a^2|b|$. The absolute value sign is used, since b may be negative. This is not the case for a^2, which is not negative.

Exercises 12.3

In questions 1-3, find the distances between the given pairs of points.

1. a. (3,0) and (0,-4) b. (1,4) and (6,16) c. (-6,3) and (0,11)

2. a. (3,16) and (11,1) b. (7, 16) and (11, 13) c. (12, 0) and (-12,7)

3. a. (3,1) and (1,3) b. (-4,2) and (3, -6) c. $\left(\frac{-1}{2}, \frac{\sqrt{3}}{2}\right)$ and (0,0)

In questions 4-10 write an equation for each circle.

4. Center (0,0), radius 9
5. Center (0,0), radius 25
6. Center (4,0), radius 4
7. Center (0,7), radius $\sqrt{11}$
8. Center (-5,8), radius 13
9. Center $\left(\frac{1}{2}, \frac{1}{3}\right)$, radius $\frac{1}{5}$
10. Center $(3\sqrt{2}, 1.3)$, radius 3.7

In questions 11-15, write an equation or an inequality to express the given conditions.

11. The point (x,y) is 5 units from (4, 7).

12. The point (r,s) is less than 4 units from (-1, 8).

13. The point (t,u) is more than 70 units from (25, 23).

14. The distance from (x,y) to (0,1) equals the distance from (x,y) to (3,10).

15. The point (w,z) is further from (-2,10) than from (4,14).

In questions 16-19, write an English sentence about the distance from point (x,y) to one or more other points that says the same thing as the given equation or inequality.

16. $\sqrt{(x-2)^2+(y-3)^2}=8$

17. $\sqrt{(x-3)^2+(y-4)^2} > \sqrt{(x-1)^2+(y-1)^2}$

18. $\sqrt{(x-7)^2+(y-4)^2} > 5$

19. $\sqrt{(x+3)^2+(y-9)^2} < 2$

In questions 20-21, write an inequality to describe algebraically the points which satisfy the given conditions.

20. The interior of a circle with center (0,0) and radius 1

21. The exterior of a circle with center (2,5) and radius 12

23. Show that the points $\left(\frac{-1}{2},\frac{\sqrt{3}}{2}\right),\left(\frac{-1}{2},\frac{-\sqrt{3}}{2}\right),(1,0)$ are the vertices of an equilateral triangle — i.e., show that the distances between all three pairs of these points are the same.

24. a. Write an inequality stating that the number x is closer to 11 than to 0.

    b. Solve your inequality from part a, and graph the solution on a number line.

In questions 25-28, write equations in terms of absolute value which express the given conditions.

25. The distance from a number y to 4 is twice the distance from y to 1.

26. The number z is three times as far from 7 as it is from -2.

27. The distance from a number t to 7 is half that from t to 13.

28. Match each inequality in the left-hand column with its equivalent in the right-hand column.

a. $\lvert x+3\rvert<11$	(1) $-5<x<5$
b. $\lvert x+5\rvert\geq\lvert x-2\rvert$	(2) $x<-5$ or $x>9$
c. $\lvert x\rvert<5$	(3) $1\leq x<9$
d. $\lvert x-2\rvert>7$	(4) $-14<x<8$
e. $\lvert 5-x\rvert\leq 4$	(5) $x\geq\frac{-3}{2}$

29. a. Evaluate $\frac{p+q}{2}+\frac{\lvert p-q\rvert}{2}$ for at least five pairs of numbers, including some pairs where p and q have the same sign and others where p and q have opposite signs.

b. Show that $\frac{p+q}{2}+\frac{\lvert p-q\rvert}{2}$ is the greater of the two numbers p and q.

c. Write a similar expression that always yields the lesser of the two numbers p and q.

30. Simplify these radical expressions.

a. $\sqrt{x^6}$ b. $\sqrt{t^4y^8}$ c. $\sqrt{27w^2t^5}$

d. $\sqrt{32r^3s^5}$ e. $\sqrt{\frac{49p^2}{z^3}}$ f. $\sqrt{\frac{0.01n^2a^6}{b^{12}}}$

12.4 Pythagorean Triples and Fermat's Last Theorem — from Mesopotamia to Andrew Wiles

Three positive integers are called a **Pythagorean triple**, if the squares of two of them add up to the square of the third. Interpreted as lengths, they are whole-numbered sides of a right triangle. Pythagorean triples you met in section 12.1 are: $(3, 4, 5), (5, 12, 13), (7, 24, 25), (6, 8, 10), (8, 15, 17)$, and $(10, 24, 26)$.

The study of Pythagorean triples is very old. Euclid wrote about them in his *Elements*, as did many much earlier writers. A remarkable example is a Mesopotamian clay tablet dating from 1800-1750 B.C.E. called Plimpton 322—it is in the George A. Plimpton Collection at Columbia University's rare book library.

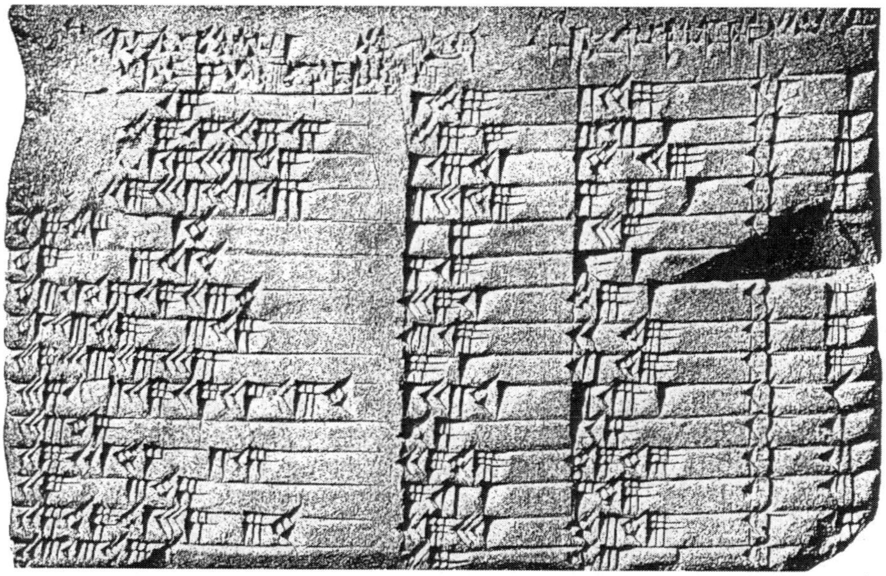

Figure 12.4.1 Plimpton 322. From Neugebauer and Sachs, *Mathematical Cuneiform Texts*, with permission of the American Oriental Society.

In 1945, Otto Neugebauer and Abraham Sachs scrutinized this tablet, whose lists of numbers had previously been assumed to be commercial records, like many other Mesopotamian tablets.[3] They translated the tablet as follows:

[3] Otto Neugebauer, a promising young German mathematician in the 1930's, was noted for his linguistic skills and his interest in bridging the gulf between mathematics and more humanistic studies. He found history of mathematics an excellent vehicle for this. He left Germany rather than submit to Nazi rulings about who could serve on the editorial board or write articles for a journal he edited. Eventually he came to the United States, where he developed an Assyrian dictionary and translated many ancient tablets. His book, *The Exact Sciences in Antiquity*, New York, Harper, 1957, remains a classic.

solving number of the width	**solving number of the diagonal**	**its name**
119	169	1
3,367	11,521*	2
4,601	6,649	3
12,709	18,541	4
65	97	5
319	481	6
2,291	3,541	7
799	1,249	8
541*	769	9
4,961	8,161	10
45	75	11
1,679	2,929	12
25,921*	289	13
1,771	3,229	14
56	53*	15

Neugebauer and Sachs recognized the entries in the first two columns as parts of Pythagorean triples. For example, 119 and 169 in the top line is part of the Pythagorean triple (119, 120, 169). This observation fits all the entries except those with stars. Using the other numbers in those four cases, it was possible to correct the four starred numbers, which are clearly errors (for example, there is no way a rectangle with width 25,921 could have a diagonal of 289, but if 161 is put in place of 25,921 the result is indeed part of Pythagorean triple, and $\sqrt{25,921} = 161$). Evidently, ancient Mesopotamians knew a lot of Pythagorean triples. Did they know the Pythagorean theorem itself? The best evidence to date is surprisingly ambiguous. In 1999, Eleanor Robson, a specialist in Mesopotamian mathematics, reexamined Plimpton 322. She was able to read it more accurately than had been done before, and she traced the origin of the tablet, both in place and time, far more accurately than ever before, showing it was used not in connection with right triangles but rather to solve problems about numbers and their reciprocals, which were studied at that time and place and were interpreted geometrically.[4] It seems that Mesopotamians were aware of the relationship now known as the Pythagorean theorem, but they may not have viewed it as we do today.

Ancient knowledge of Pythagorean triples set the stage for an intriguing saga, ignited by Pierre de Fermat (1601-1665). Fermat was a lawyer from Toulouse, France, and member of parliament, who paid little attention to mathematics until he was in his thirties. From then on, he made huge contributions to mathematics, though always as an amateur, who usually communicated his findings in letters. Fermat's contributions to the development of analytic geometry and calculus were immense, but perhaps his most remarkable achievements were in the study of numbers, which is where this tale begins. Fermat had a copy of *Arithmetica*, part of an ancient classic by Diophantus of Alexandria.[5] In the section on Pythagorean triples, Fermat wrote a note in the margin, in Latin:

4 *"Neither Sherlock Holmes nor Babylon: a reassessment of Plimpton 322,"* *Historia Mathematica*, volume **28**, (2001) pages 167-206

5 Little is known of the life of Diophantus, save what appears in an epigram in the *Greek Anthology* (see question 33 of Exercises 10.4). He lived in the third century and wrote in Greek, but most of his writings, including seven of the 13 books of his classic *Arithmetica*, have not survived. His surviving work shows a mastery of algebraic techniques more advanced than any other work known from ancient Greece.

> *On the other hand, it is impossible to separate a cube into two cubes or a fourth power into two fourth powers or generally any power but a square into powers with the same exponent. I have discovered a truly remarkable proof of this, which, however, the margin is too narrow to contain.*

Did Fermat really have a proof? Had he made a mistake? Nobody knows, but his assertion became a challenge. It was proved that there were no positive integers for which $a^4+b^4=c^4$, and $a^3+b^3=c^3$ fell soon after that, as did the result for some other exponents. In 1823 and again in 1850, the Academy of Science in Paris offered prizes for a general proof that $a^n+b^n=c^n$ has no solutions with a, b, and c positive integers if n is greater than 2, as did the Academy of Brussels in 1883. Later, F. F. Wolfskehl of Darmstadt, left 100,000 German marks for the same purpose, but its value was drastically reduced in Germany's post-World War I inflation. The prizes were not awarded, but not for lack of effort. Over 2,000 incorrect "proofs" of this result were published between 1908 and 1918 alone, part of an avalanche of effort which included brilliant work as well as a lot of trash. By 1992 it was known that if there are any exponents $n > 2$ for which $a^n+b^n=c^n$ is satisfied by positive integers a, b, and c, then $n > 4{,}000{,}000$.

In the meantime, theoretical advances were made, which did not at first seem related to this question. In the summer of 1993, Andrew Wiles found a way to use these new results to outline a proof of "Fermat's last theorem," as it had long been known. Initial reaction was enthusiastic but cautious, but by 1999 the issue at last appeared settled, and Wiles was awarded the Wolfskehl prize, its value having recovered considerably.

Exercises 12.4

1. Although there are no positive numbers a, b, and c for which $a^3+b^3=c^3$, there are positive integers a, b, c, d, for which $a^3+b^3+c^3=d^3$. In particular, show that $3^3+4^3+5^3=6^3$.

2. a. The Pythagorean triples (3, 4, 5) and (6, 8, 10) are related. Show that if k is any positive integer $(3k, 4k, 5k)$ is a Pythagorean triple.
   b. If (a, b, c) is a Pythagorean triple, show that $(2a, 2b, 2c)$ is a Pythagorean triple.
   c. If (a, b, c) is a Pythagorean triple and k is a positive integer, show that (ka, kb, kc) is a Pythagorean triple.

12.5 Exploration: Continued Fractions and Square Roots

Continued fractions[6] are a bit off the beaten track, but they are a fine subject for classroom use. This investigation relates fractions and decimals to square roots, to reveal some remarkable patterns.

[6] Continued fractions were first studied by John Wallis and Lord Brounkner in the mid 17th century. For more about continued fractions, see C. D. Olds, *Continued Fractions.* Mathematical Association of America, 1963.

1. Simplify in parts a-d, showing the intermediate steps in each case. Leave each answer as an improper fraction.

a. $1+\frac{1}{2}$ b. $1+\cfrac{1}{2+\cfrac{1}{2}}$ c. $1+\cfrac{1}{2+\cfrac{1}{2+\cfrac{1}{2}}}$ d. $1+\cfrac{1}{2+\cfrac{1}{2+\cfrac{1}{2+\cfrac{1}{2}}}}$

Part a is just $\frac{3}{2}$, but later parts look like the work of a fevered brain. The key to dealing with them is that the fraction bar, much like parentheses, is a grouping symbol. In part b, for example, it would be no problem to add 1, but to what?

To find out, first evaluate $\cfrac{1}{2+\cfrac{1}{2}}$.

That, in turn, depends on dividing 1 by something, which can be found by adding 2 and $\frac{1}{2}$ to get $\frac{5}{2}$. Evidently, continued fractions must be evaluated from the bottom up. In this case, $1+\cfrac{1}{2+\cfrac{1}{2}}=1+\cfrac{1}{\frac{5}{2}}$. Now, dividing by $\frac{5}{2}$ is the same as multiplying by the reciprocal, $\frac{2}{5}$, so $1+\cfrac{1}{2+\cfrac{1}{2}}=1+\cfrac{1}{\frac{5}{2}}=1+1\times\frac{2}{5}=1+\frac{2}{5}=\frac{7}{5}$.

Do parts c, d, and e the same way, from the bottom up. Here is a skeleton for part c; you can just fill in the blanks.

$$1+\cfrac{1}{2+\cfrac{1}{2+\cfrac{1}{2}}}=1+\cfrac{1}{2+\cfrac{1}{\frac{5}{2}}}=1+\cfrac{1}{2+\frac{\underline{\quad}}{5}}=1+\cfrac{1}{\frac{\underline{\quad}}{5}+\frac{\underline{\quad}}{5}}=1+\cfrac{1}{\frac{12}{\underline{\quad}}}=1+\frac{\underline{\quad}}{12}=\frac{\underline{\quad}}{\underline{\quad}}$$

Try part d on your own. You should come out with $\frac{41}{29}$.

2. a. Write your answers from parts a-d of question 1 next to each other, and look for a pattern: $\frac{3}{2}, \frac{7}{5}, \frac{\underline{\quad}}{12}, \frac{41}{29}, \ldots$

(From each fraction, it is easy to find the denominator of the next. That denominator can then help you find the numerator.) Describe the pattern with words or algebra, then use it to predict the value of

$$1+\cfrac{1}{2+\cfrac{1}{2+\cfrac{1}{2+\cfrac{1}{2+\cfrac{1}{2}}}}}$$

b. Check your prediction from part a by actually working this out.

3. a. A remarkable property of these continued fractions emerges if you square each answer from questions 1 and 2. Fill in the blanks as fractions (not as decimals).

$$\left(\frac{3}{2}\right)^2 = \quad , \left(\frac{7}{5}\right)^2 = \quad , \left(\frac{\quad}{12}\right)^2 = \quad , \left(\frac{41}{29}\right)^2 = \quad , \left(\frac{\quad}{\quad}\right)^2 =$$

b. The squares in part a are alternately greater than and less than a certain number, and they are getting closer and closer to it. What number do the squares from part f seem to be approaching?

c. Part b suggests a relatively easy way to get good approximations to a number we have found elusive. After all, if you know what the **squares** are approaching, you can figure out what number the fractions themselves are approaching. What number is that?

4. Do the results of questions 1-3 give us a way to calculate other square roots? As a first step, express each of the following as an improper fraction, much as in question 1. Show all intermediate steps.

a. $1+\frac{1}{3}$ b. $1+\cfrac{1}{3+\cfrac{1}{3}}$ c. $1+\cfrac{1}{3+\cfrac{1}{3+\cfrac{1}{3}}}$ d. $1+\cfrac{1}{3+\cfrac{1}{3+\cfrac{1}{3+\cfrac{1}{3}}}}$

e. Do the answers to a-d above approach $\sqrt{3}$ as the answers in question 1 approach $\sqrt{2}$? Justify your answer with a clear sentence or two.

5. Evaluate, again showing all intermediate steps.

a. $2+\frac{1}{4}$ b. $2+\cfrac{1}{4+\cfrac{1}{4}}$ c. $2+\cfrac{1}{4+\cfrac{1}{4+\cfrac{1}{4}}}$ d. $2+\cfrac{1}{4+\cfrac{1}{4+\cfrac{1}{4+\cfrac{1}{4}}}}$

e. Find a pattern in your answers to parts a-d, and describe it. (What is twice the denominator plus the numerator? What is twice that number plus the original denominator?)

f. Use your answers from parts a-e to predict the value of

$$2+\cfrac{1}{4+\cfrac{1}{4+\cfrac{1}{4+\cfrac{1}{4+\cfrac{1}{4}}}}}.$$

g. Carry out the arithmetic to check your prediction from part f.

h. Square your answers to parts a-f.

i. What number do the squares seem to approach?

j. What square root do the answers to parts a-f seem to approach?

6. Repeat **all parts** of questions 1 and 2, first for part a, then for part b.

a. $3+\frac{1}{6}$, $3+\cfrac{1}{6+\cfrac{1}{6}}$, $3+\cfrac{1}{6+\cfrac{1}{6+\cfrac{1}{6}}}$, $3+\cfrac{1}{6+\cfrac{1}{6+\cfrac{1}{6+\cfrac{1}{6}}}}$, ...

b. $4+\frac{1}{8}$, $4+\cfrac{1}{8+\cfrac{1}{8}}$, $4+\cfrac{1}{8+\cfrac{1}{8+\cfrac{1}{8}}}$, $4+\cfrac{1}{8+\cfrac{1}{8+\cfrac{1}{8+\cfrac{1}{8}}}}$, ...

7. a. Using questions 1-4 as a guide, what square root do you expect the following expressions to approximate?

$$10+\frac{1}{20},\quad 10+\cfrac{1}{20+\cfrac{1}{20}},\quad 10+\cfrac{1}{20+\cfrac{1}{20+\cfrac{1}{20}}},\quad 10+\cfrac{1}{20+\cfrac{1}{20+\cfrac{1}{20+\cfrac{1}{20}}}}$$

b. By arithmetic or algebra, test your conclusions from part a.

8. Generalize the pattern from questions 1-7.

Vocabulary for Chapter 12

Absolute value (12.3)
Continued fraction (12.5)
Distance formula (12.3)
Hypotenuse (12.1)
Leg (12.1)
Pythagorean theorem (12.1)
Pythagorean triple (12.4)
Radical (12.2)
Square root (12.2)

Self-Quiz for Chapter 12

1. A pole 84 feet tall is to stand vertically on level ground, held in place by taut wires from the top of the pole to anchors on the ground. If the wires are 85 feet long, how far will the anchors be from the foot of the pole?

2. Find $\sqrt{7.5}$ to the nearest hundredth by trial and error or by Newton's method (your choice). Show all your work.

3. Explain the difference between the square of a number and the square root of a number.

4. In the coordinate plane, what is the distance between (5, -4) and (10,8)?

5. Write an equation for a circle with radius 7 and center (-5,2).

6. Simplify completely: a. $\sqrt{\dfrac{96}{125}}$ b. $\sqrt{192x^3y}$

7. Evaluate: $1+\cfrac{1}{3+\cfrac{1}{5+\cfrac{1}{7}}}$

8. What is the maximum length of any object that can be packed without bending or breaking in a rectangular box whose interior is 2 feet long, 1 foot high, and 1 foot wide?

9. Write an inequality stating that the distance from a number x to 7 is greater than 5. Then solve your inequality and picture it on the number line.

Chapter Thirteen
Polynomials and Rational Expressions

Q: How much is 5q and 5q?
A: 10q.
Questioner: You're welcome.

–old algebra joke

Algebraic notation arose as a shorthand for expressing relationships among numerical quantities. In time, however, algebraic expressions came to be objects of study in their own right. Here we will study polynomials and rational expressions, which play roles in algebra analogous to the roles of the integers and fractions in arithmetic.

13.1 Polynomials: Addition and Subtraction

The introduction of letters as symbols for unspecified numbers led naturally to algebraic expressions such as $3x^3 - 5x + 4$, $u^7 - 27u^6 - 11$, $\sqrt{x+7}$, 2^x, and $t^2 + \frac{2}{t}$. The first two of these are examples of polynomials. More generally, if x is an unspecified number and n is a whole number, a **polynomial in x** is a sum of the form

$$a_0 + a_1x + a_2x^2 + \dots + a_nx^n$$

(Note: The subscript numbers are used only for identification purposes. Thus, a_2 means the second a. Unlike superscript numbers, which are used as exponents, subscripts do not indicate a mathematical operation.) The numbers, $a_0, a_1, a_2, \dots$ are called **coefficients**. Most of our work will focus on polynomials with integer coefficients, but sometimes we allow other numbers as coefficients. The letter x itself is referred to as a **variable**, an **unknown**, or an **indeterminate**. Here x^n denotes the product of n factors of x. Each item added, such as a_0, a_1x, or a_2x^2, is called a **term** of the polynomial. Polynomials with only one term, such as $3x^4$ or $-2x$, are called **monomials**. Similarly, those with two terms, such as $6x - 3$ or $3x^7 + 2x^5$, are called **binomials**, and those with three terms are called **trinomials**. A number such as 3 may be regarded as a polynomial whose only nonzero coefficient is $a_0 = 3$. Such a polynomial is called a **constant polynomial.**

The **degree** of a polynomial in one variable is the largest exponent of the variable on a term whose coefficient is not zero. For example, the degree of

$3x^7 + 2x^5$ is 7. The degree of a constant polynomial is zero.[1]

You can use any letters, from any alphabet, even symbols you invent, as the variables of polynomials. For example, $2y^2 - 3y + 4$ is a polynomial in y of degree 2, and $\frac{1}{2}t^4 - 3$ is a polynomial in t of degree four.

You can have polynomials in more than one variable.

Definition: The degree of a polynomial in more than one variable is the greatest of the sums of exponents of the variables of terms with nonzero coefficients.

Example 1. What is the degree of the polynomial $5s^4 + 2^7s^2t^3 - s^3t^3$?
Solution: For the term $5s^4$ the sum of the exponents of the variables is 4. For the term $2^7s^2t^3$ the sum of the exponents of the variables is 5 (the exponent 7 belongs to the coefficient, not a variable). For the term $-s^3t^3$ the sum of the exponents of the variables is 6. Then the degree of this polynomial is 6.

Polynomials are algebraic expressions, but if you **substitute** numbers for the letters in a polynomial, you get a number. The process of substituting numbers for the letters in a polynomial and calculating the resulting value is called **evaluating** the polynomial at that number. For example, to evaluate $x^2 - 1$ at $\frac{2}{3}$, substitute $\frac{2}{3}$ to get $\left(\frac{2}{3}\right)^2 - 1 = \frac{4}{9} - 1 = \frac{-5}{9}$.

Any algebraic operation with polynomials becomes an arithmetic operation if numbers are substituted for the letters in the polynomial. For example, if you substitute 10 for x, the polynomial multiplication $(x - 1)(x + 1)$ becomes the arithmetic multiplication 9×11.

You can model polynomials of degree two or less with **algebra tiles**, a versatile, low-tech manipulative consisting of the following kinds of tiles:

1. **Unit squares** 1 unit on each side. Each unit square has area 1.
2. **Rods** 1 unit wide and x units long, where x is not specified. Each rod has area x.
3. **Flats** x units long by x units wide. Each flat has area x^2.

These basic tiles may be turned face down to represent their negatives. (See figure 13.1.1.)

[1] The constant polynomial 0 has no non-zero coefficients. We will not define its degree.

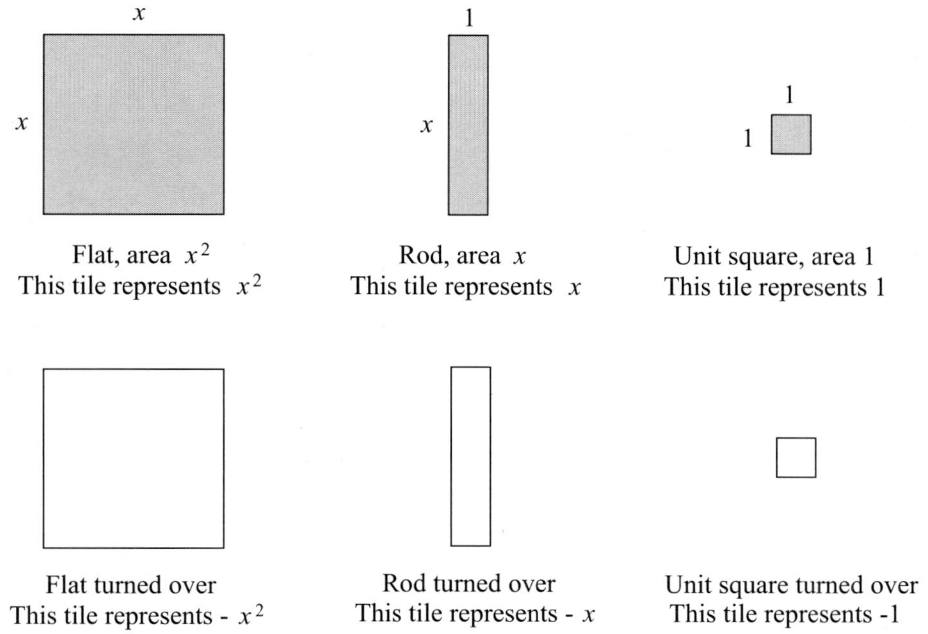

Figure 13.1.1 Algebra tiles

Addition of Polynomials

You have already added simple polynomials in the process of solving equations in chapter 10. Now we will extend the work to more general polynomials. To add polynomials, add each kind of term separately, just as you would add any lists of unlike objects. For example, if you had 3 apples and 2 bananas, and a friend gave you 5 apples and 10 bananas, you would have (3 + 5) apples and (2 + 10) bananas, or 8 apples and 12 bananas.

In algebraic shorthand,

$$\begin{array}{r} 3a + 2b \\ 5a + 10b \\ \hline 8a + 12b \end{array}$$

To add polynomials, add like terms, paying attention to signs.

To keep track of the work, it sometimes helps to write the polynomials so that like terms are under each other, especially when there are many terms.

Example 1. Add $3x^2 + 3x + 2$ to $2x^2 + 5$.
Solution: Write the polynomials with like terms directly beneath each other and add.

$$\begin{array}{rrr} 3x^2 + & 3x & + 2 \\ 2x^2 & & + 5 \\ \hline 5x^2 + & 3x & + 7 \end{array}$$

Here both polynomials have been written in descending powers of x. While not strictly necessary, a systematic arrangement clarifies the work. This addition may be pictured with algebra tiles like this:

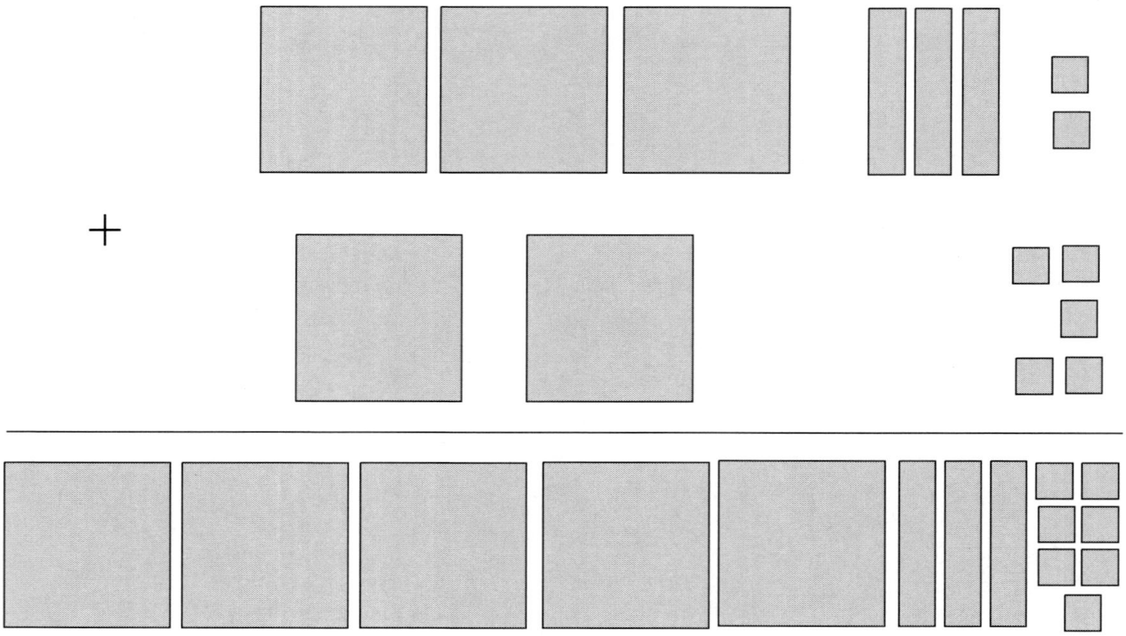

Example 2. Add $x^2 - 3x + 2$ to $x^2 - x - 3$.
Solution: Write the polynomials with like terms directly below each other and add column-wise.

$$\begin{array}{r} x^2 - 3x + 2 \\ x^2 - x - 3 \\ \hline 2x^2 - 4x - 1 \end{array}$$

The work is pictured with algebra tiles below.

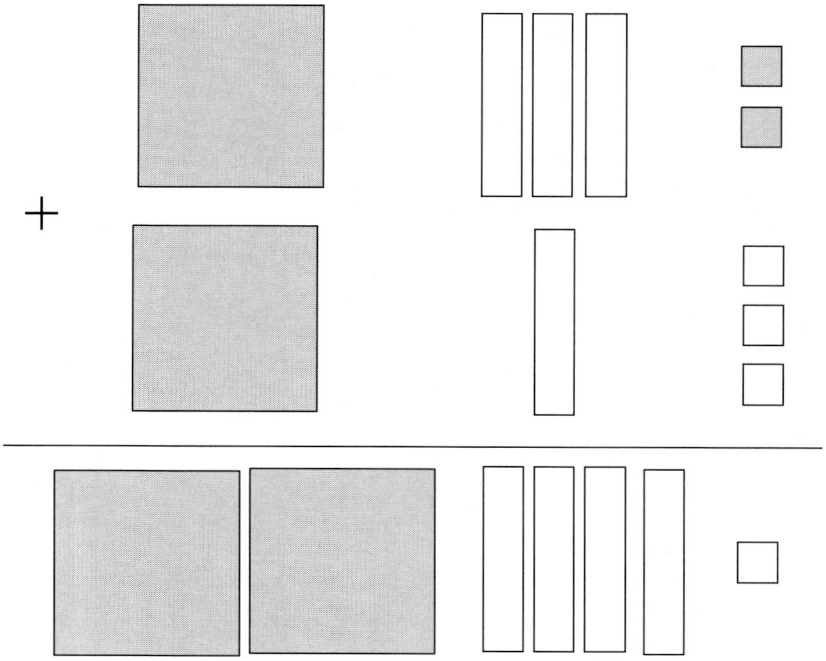

Box puzzles can help organize the work and at the same time show that this part of algebra is very much like arithmetic.

Example 3. Add $3t^2 + 5t + 7$ and $6t^2 + 9t + 1$.
Solution: Make a box puzzle and complete it as shown.

$3t^2$	$5t$	7	$3t^2 + 5t + 7$
$6t^2$	$9t$	1	$6t^2 + 9t + 1$
$9t^2$	$14t$	8	$9t^2 + 14t + 9$

Does this look like arithmetic? If $t = 10$, then this is addition of 357 and 691 in extended notation, except that with polynomials there is no carrying, because the coefficients need not be digits from 0 through 9. Therefore, leave $14t$ as a term in the sum, where with numbers you would have carried to the hundreds column.

Subtraction of Polynomials

The constant polynomial 0 plays a role in the set of polynomials just like that of the number 0 in the set of numbers. Specifically, 0 is the **additive identity** for polynomials, meaning that 0, and only 0, has the property that adding it to any polynomial leaves that polynomial unchanged. Extending the analogy with arithmetic, note that each polynomial has an opposite polynomial (also referred to as its **negative** or **additive inverse**), which, if added to that polynomial, yields 0. For example, the opposite of $3x - 7$ is $-3x + 7$, as you can easily check by adding. In general, to find the opposite of a polynomial, change the sign of each coefficient.

The **additive inverse** (or **opposite**, or **negative**) of the polynomial $a_0 + a_1x + a_2x^2 + \ldots + a_nx^n$ is $-(a_0 + a_1x + a_2x^2 + \ldots + a_nx^n)$ or, equivalently, $-a_0 - a_1x - a_2x^2 \ldots - a_nx^n$. As with numbers, the opposite of the opposite is the original polynomial.

With numbers, we used three views of subtraction: Take away, the missing addend, and adding the opposite. These three views also apply to polynomials. In very simple cases, such as $7y + 5 - (3y + 4)$, it is natural to use the take-away model. Start with 7 rods and 5 unit squares.

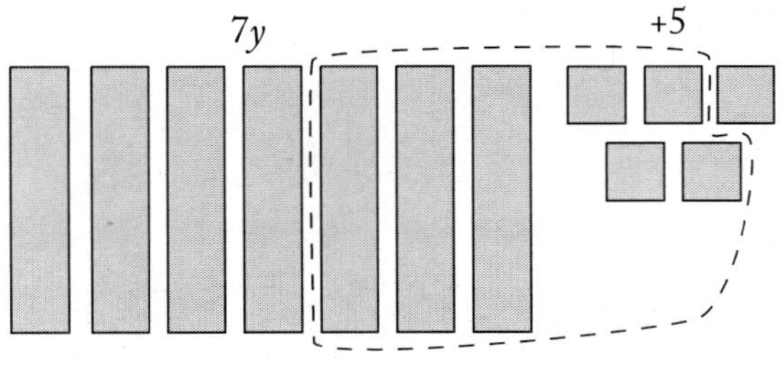

Now take away 3 rods and 4 unit squares, enclosed in the dashed region, leaving 4 rods and 1 unit square. Then $7y + 5 - (3y + 4) = 4y + 1$.

For more complicated examples students and practicing mathematicians alike usually find that adding the opposite is the simplest and most reliable way to subtract polynomials.

> **To subtract a polynomial, add its additive inverse. That is, change the sign of each coefficient of the polynomial to be subtracted and add the resulting polynomial.**

Example 4. Subtract $x^2 - 3x + 1$ from $x^2 - 2x + 4$.
Solution: First, write the polynomial to be subtracted below the one to be subtracted from, with like terms directly below each other.

$$\begin{array}{r} x^2 - 2x + 4 \\ - \quad x^2 - 3x + 1 \\ \hline \end{array}$$

Next, change the sign of each term of the polynomial to be subtracted, and add.

$$\begin{array}{r} x^2 - 2x + 4 \\ + \quad -x^2 + 3x - 1 \\ \hline x + 3 \end{array}$$

Pictorially

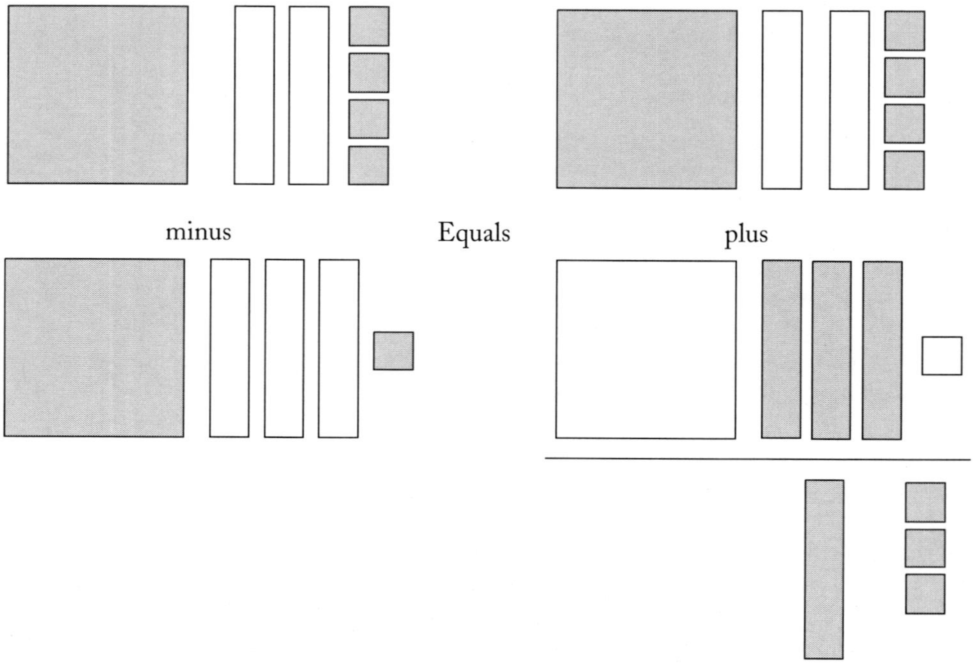

The examples above use polynomials with integer coefficients and degrees no more than 2, because they can be modeled with algebra tiles. You could augment algebra tiles with cubes to represent third degree polynomials, but physical representations are not possible for polynomials of degree higher than three. To work with higher degree polynomials, or those with non-integer coefficients, you need to rely on algebra, following the same procedures shown above.

Exercises 13.1

1. Use algebra tiles to model these polynomial additions, then compute the sums:
   a. $3x + 2$ plus $5x + 1$
   b. $2x - 5$ plus $4x - 2$
   c. $x^2 + 3x - 4$ plus $x^2 - 7x + 3$
   d. $5x^2 - 2x + 6$ plus $-2x^2 - x + 6$

2. Use algebra tiles to picture these polynomial subtractions, and then compute the differences:
   a. $5x - 12$ minus $7x + 4$
   b. $-3x - 6$ minus $5x - 1$
   c. $2x^2 + 4x + 1$ minus $x^2 + 5x + 3$
   d. $x^2 - 2x + 6$ minus $-2x^2 - x + 7$

In questions 3-9, add.

3. $$\begin{array}{r} 3a^2 + 4ab + b^2 \\ -6a^2 - 2ab - b^2 \\ 5a^2 + ab - b^2 \\ \hline \end{array}$$

4. $$\begin{array}{lrr} m & - 3n & + 5p \\ 6m & & - 2p \\ & 4n & - 8p \\ \hline \end{array}$$

5. $y^3 + 4y^2 - 8y + 7$, $4y - 2$, and $9y^3 + 3y + 4$

6. $t^3 - s^3$, $2ts^2 - 4s^3 + t^3$, and $4t^2s + 3ts^2$

7. $x^4 - 3x^3y$, $4xy^3 + 2x^3y + x^2y^2$, and $y^4 - x^4$

8. $\frac{1}{4}c^3 - \frac{3}{5}c^2 + 4$, $\frac{1}{2}c^3 - 4c + 1$ and $\frac{-1}{5}c^2 - \frac{1}{2}$

9. $0.8d^2 - 0.03d - 1.1$, $3.7d - 1$, and $0.4d - 3.2d^2 - 3.8$

In questions 10-12, subtract, and check your answers by addition.

10. $$\begin{array}{r} a - b + c + 4 \\ \underline{-a + b + c } \end{array}$$

11. $$\begin{array}{r} x + 2y - z \\ \underline{u + x + y - z} \end{array}$$

12. $$\begin{array}{r} x^3 + 3x^2 - 2x + 7 \\ \underline{4x^3 - 7x^2 - 9} \end{array}$$

13. $$\begin{array}{r} 5y^2 - 4y - 3 \\ \underline{y^3 - y^2 - y + 5} \end{array}$$

14. Subtract $x^2 - 8x + 4$ from $2x^2 - 5x + 2$.

15. What must be added to $3y^2 - 4y - 1$ to get $-5y^3 + y^2 - 6y + 5$?

13.2 Multiplication of Polynomials

Multiplication of polynomials is much like multiplication of numbers. In particular, the distributive property applies to polynomials, and we can often model multiplication of polynomials in terms of area. Also, the role of 1 as multiplicative identity for numbers ($1 \times n = n$ for any number n) extends to polynomials ($1 \times P = P$ for any polynomial P). We will first multiply monomials and then use the distributive property to multiply polynomials with more terms.

Multiplication of monomials

The polynomial $3x$ is the product of the polynomials 3 and x, so it may be regarded as the area of a rectangle with length 3 and width x. Here is just such a rectangle, pictured with algebra tiles.

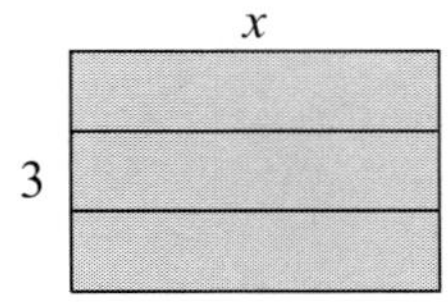

What if you multiply this by 4? You could interpret this as repeated addition of $3x$ to get

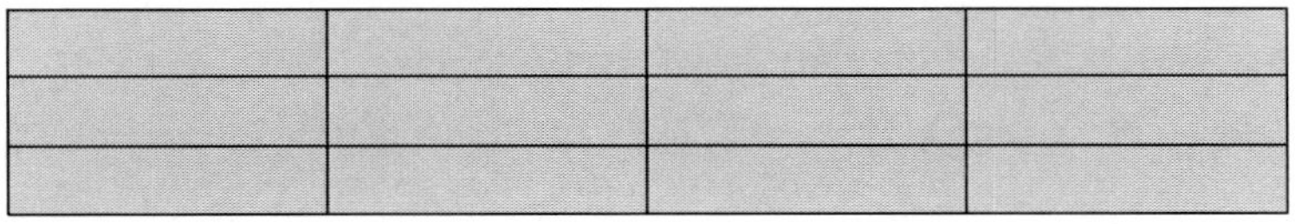

Again, you can think of $4(3x)$ as the area of a rectangle with dimensions 4 and $3x$, like this.

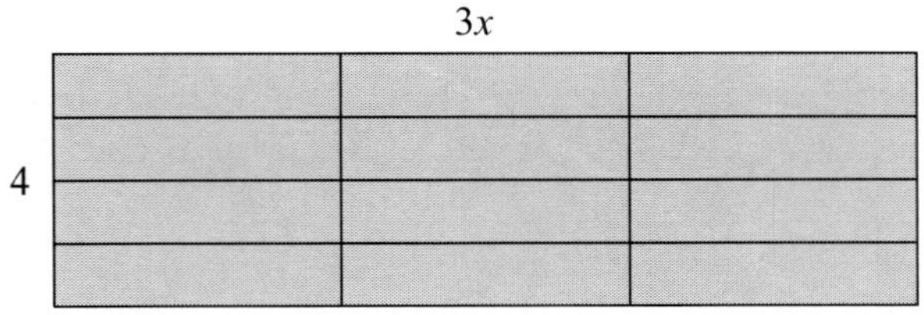

Either way, 4 times $3x$ is $12x$. This result makes sense. If you substitute a number for x, then $4(3x) = (4 \times 3)x$ is just a case of the associative property of multiplication for numbers. In fact, multiplication of polynomials is also associative.

Example 1. Multiply $2x$ by $5x$.
Solution: The product is the area of a rectangle with dimensions $2x$ by $5x$. That rectangle is shown in below. Evidently, $(2x)(5x) = 10x^2$, which is what you would expect, since, by the associative and commutative properties, $(2x)(5x) = (2)(5)(x)(x)$.

Even when we cannot picture the factors or the product with algebra tiles, the associative and commutative properties of multiplication are the basis for multiplying any monomials whatever.

Example 2. Multiply $-3x^2y^3$ by $10x^4yz^5$.
Solution: By the associative and commutative properties of multiplication,
$$(-3x^2y^3)(10x^4yz^5) = -3 \times 10x^2x^4y^3yz^5$$
$$= -30x^{(2+4)}y^{(3+1)}z^{(0+5)}$$
$$= -30x^6y^4z^5.$$

Multiplication of polynomials in general

Multiplication of polynomials with more than one term involves the distributive property. Our first examples are modeled with algebra tiles.

Example 3. Multiply $3x + 1$ by 4.
Solution: Consider a rectangle with dimensions $3x + 1$ and 4.

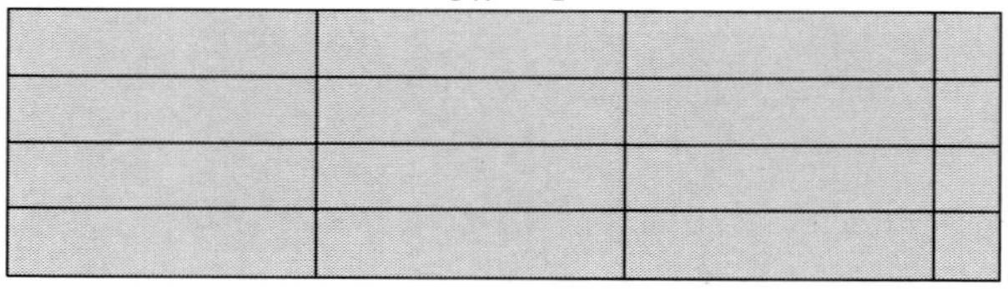

Its area is $4(3x + 1)$ or, as you can see from the picture, $12x + 4$. This is the same result you would get if you applied the distributive property: $4(3x + 1) = 4(3x) + 4(1) = 12x + 4$. The distributive property holds for polynomials as well as for numbers.

Example 4. Multiply $2x + 3$ by $3x + 1$.
Solution: Make a rectangle with dimensions $2x + 3$ by $3x + 1$ as shown.

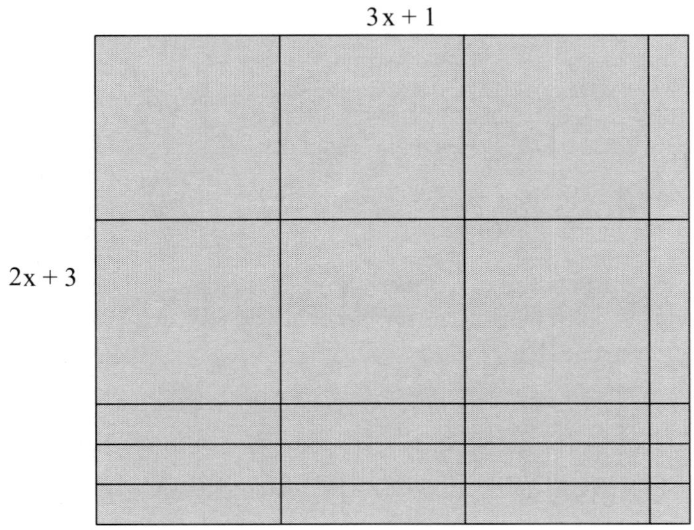

You can see from the picture that $(2x + 3)(3x + 1) = 6x^2 + 11x + 3$. Question 5 below shows how you could obtain the same result from a box puzzle based on the distributive property.

Any multiplication of polynomials is based, like the above examples, on the distributive property, together with the commutative and associative properties, even in cases too complicated to model easily with algebra tiles. It does not even matter if some or all of the coefficients are themselves letters, as in example 5.

Example 5. Multiply ax by $3x^3 + bx - 1$.
Solution: By the distributive property,
$ax(3x^3 + bx - 1) = ax(3x^3) + ax(bx) + ax(-1)$.
Now use the associative and commutative properties of multiplication to simplify each term, which yields $3ax^4 + abx^2 - ax$.

We have no algebra tiles that would allow us to think of $3x^3 + bx - 1$ as a length, but you can model the process with a box puzzle.

	$3x^3$	$+bx$	-1	$3x^3+bx-1$
ax	$3ax^4$	abx^2	$-ax$	$3ax^4+abx^2-ax$

To multiply polynomials with more terms, extend the work from examples 4 and 5. However, for complicated examples, even box puzzles can be cumbersome, and ultimately you would like your students to be able to multiply polynomials without them, as in example 6. Arrange the work systematically, with the exponents in both factors either increasing or decreasing, and proceed much as you would with long multiplication in arithmetic.

Example 6. Multiply $4y^3 + 2y + 1$ by $y^2 - 2y - 3$.

Solution:

$$
\begin{array}{rrrrrrl}
 & & 4y^3 & & +2y & +1 & \\
 & & & y^2 & -2y & -3 & \\
\hline
 & & -12y^3 & & -6y & -3 & \text{The partial product } (4y^3 + 2y + 1)(-3) \\
 & -8y^4 & & -4y^2 & -2y & & \text{The partial product } (4y^3 + 2y + 1)(-2y) \\
4y^5 & & +2y^3 & +y^2 & & & \text{The partial product } (4y^3 + 2y + 1)(y^2) \\
\hline
4y^5 & -8y^4 & -10y^3 & -3y^2 & -8y & -3 & \text{Product} = \text{sum of partial products}
\end{array}
$$

The analogy with arithmetic is even clearer if we write only the numerical coefficients and not the powers, which serve primarily as place holders. The zero in the top line indicates $0y^2$.

$$
\begin{array}{rrrrrr}
 & & 4 & 0 & 2 & 1 \\
 & & & 1 & -2 & -3 \\
\hline
 & & -12 & 0 & -6 & -3 \\
 & -8 & 0 & -4 & -2 & \\
4 & 0 & 2 & 1 & & \\
\hline
4 & -8 & -10 & -3 & -8 & -3
\end{array}
$$

This is like long multiplication, except that there is no carrying from one column to the next, and entries need not be integers from 0 to 9.

Common Patterns

In multiplying binomials, it helps to be familiar with certain common patterns. The table below summarizes them.

General Pattern	Specific Example
1. $(ax + b)(cx + d) = acx^2 + (ad + bc)x + bd$	$(2x + 5)(3x - 1) = 6x^2 + (-2 + 15)x - 5$ $= 6x^2 + 13x - 5$
2. $(a + b)(a + b) = a^2 + 2ab + b^2$	$(3y + 2)(3y + 2) = 9y^2 + 12y + 4$
3. $(a - b)(a - b) = a^2 - 2ab + b^2$	$(3z - 2)(3z - 2) = 9z^2 - 12z + 4$
4. $(a - b)(a + b) = a^2 - b^2$	$(2r - t)(2r + t) = 4r^2 - t^2$

These patterns may be easily checked by multiplying out the left-hand sides of the equations. Pattern #1 underlies the FOIL (First, Outer, Inner, Last) mnemonic that algebra students sometimes use. There is nothing wrong with FOIL, but if you understand how to multiply polynomials, you should not need it. You met the second pattern in section 12.1, and the third arises if you substitute $-b$ for b in the second. These are pictured below with algebra tiles.

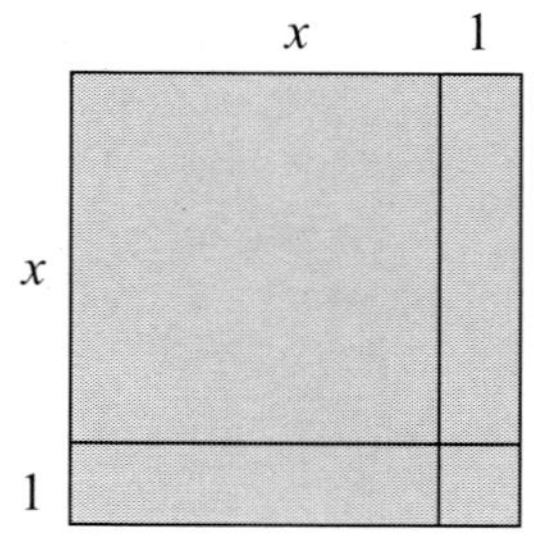

$(x + 1)^2 = x^2 + 2x + 1$

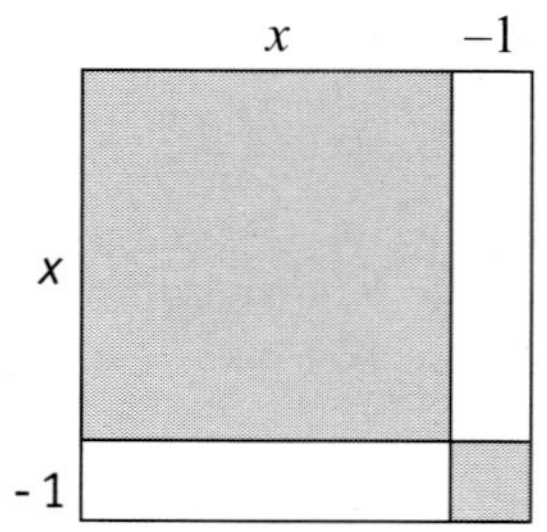

$(x - 1)^2 = x^2 - 2x + 1$

The last pattern is explored concretely in questions 9-13 of Exercises 4.1.

Exercises 13.2

In questions 1-4, use the length and width of the rectangles pictured to state what polynomials are being multiplied. Then read off the product as the area of the rectangle.

1. 

2. 

3. 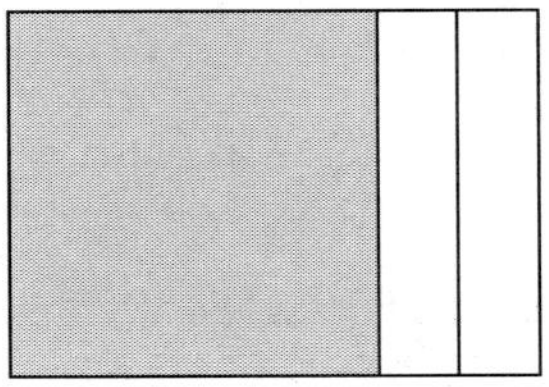

4. 

5. Redo example 4 of the text, first by completing this box puzzle and then by writing out the work with algebra.

	$3x$	1	$3x + 1$
$2x$			
3			
$2x + 3$			

In questions 6-9, multiply the given monomials.

6. $(-4x^3)(-3x^5)$
7. $(-7t^3)(-9t^5s)$
8. $(5xy^4z)(-11x^3z^5)$
9. $(-12p^3r^5q)(23p^2q^7)$

In questions 10-13, multiply the given polynomials, and illustrate your work with algebra tiles where you can.

10. $3x(5x + 2)$
11. $(3x + 2)(3x + 2)$
12. $(-2x + 1)(3x - 4)$
13. $(-x + 5)(-x - 1)$

In questions 14-17, multiply the given polynomials, and illustrate your work with box puzzles.

14. $(3t^2)(5t^2 - 4t + 7)$
15. $(4w^3 - 2)(5w^2 + w)$
16. $(4y^2 - 3)(4y^2 - 3)$
17. $(10z^2 + 1)(10z^2 - 1)$

In questions 18-29, multiply any way you please.

18. $(x^2 + x + 1)(x - 1)$
19. $(y^3 - 2)(y^3 - 2)$
20. $(de + 5)(de - 5)$
21. $(fg - 5)(fg + 8)$
22. $(h^2 - a)(h^2 + a)$
23. $(j^2 - 2j + 1)(j^2 + 3)$
24. $(k - g)(k^2 + kg + g^2)$
25. $(m^2 - m + 1)(m^2 + m + 1)$
26. $(n^2 + p^2 + np)(n^2 + p^2 - np)$
27. $(q - 5)^2$
28. $(2r + 1)^2$
29. $(s^2 - 1)^2$

Exploration: Binomial powers Questions 30-33 explore a pattern that occurs in many parts of mathematics. They are a good source of practice in multiplying polynomials, because the patterns guide the students.

30. Compute the following and arrange each answer in ascending powers of x. Note that multiplying each answer by $a + x$ gives the next. If that leads you to the last answer, which is given, the prior multiplications must be correct.
    a. $(a + x)^2$
    b. $(a + x)^3$
    c. $(a + x)^4$
    d. $(a + x)^5 = a^5 + 5a^4x + 10a^3x^2 + 10a^2x^3 + 5ax^4 + x^5$

31. The answers from parts a, b, c, and d of question 30 involve successive powers of x, always preceded by numerical coefficients.
    a. Identify this array of numbers by name. (See Exercises 5.1, problem 19)
    b. Use the pattern to write out $(a + x)^6$ without actually multiplying.

32. a. In problem 30, let $a = x = 1$. How does that help explain the pattern in part a of problem 19, Exercises 5.1, that the rows of the triangle add up to powers of 2?
    b. In problem 30, let $a = 1$ and $x = -1$. How does that help explain the pattern in part b of problem 19, Exercises 5.1, that with alternating signs, the rows of Pascal's triangle add up to 0?

33. Use the pattern from questions 30 and 31 to carry out these calculations (Do NOT use a calculator, except to check your final answers.)
    a. $\dfrac{\left(1+\sqrt{5}\right)-\left(1-\sqrt{5}\right)}{2\sqrt{5}}$
    b. $\dfrac{\left(1+\sqrt{5}\right)^2-\left(1-\sqrt{5}\right)^2}{2^2\sqrt{5}}$
    c. $\dfrac{\left(1+\sqrt{5}\right)^3-\left(1-\sqrt{5}\right)^3}{2^3\sqrt{5}}$
    d. $\dfrac{\left(1+\sqrt{5}\right)^4-\left(1-\sqrt{5}\right)^4}{2^4\sqrt{5}}$
    e. $\dfrac{\left(1+\sqrt{5}\right)^5-\left(1-\sqrt{5}\right)^5}{2^5\sqrt{5}}$

f. Remarkably, each answer from parts a-e is a whole number. Do you recognize the series of numbers being generated? Test your answer by evaluating

$$\frac{\left(1+\sqrt{5}\right)^6-\left(1-\sqrt{5}\right)^6}{2^6\sqrt{5}}.$$

34. Do these multiplications, and find the underlying pattern.

a. $(a - x)(a + x)$

b. $(a - x)(a^2 + ax + x^2)$

c. $(a - x)(a^3 + a^2x + ax^2 + x^3)$

d. $(a - x)(a^4 + a^3x + a^2x^2 + ax^3 + x^4)$

e. Predict the product $(a - x)(a^5 + a^4x + a^3x^2 + a^2x^3 + ax^4 + x^5)$, then check your prediction by multiplying.

Exploration: A polynomial pattern hunt: Bernoulli polynomials. (Problems 35-42)

35. In section 3.1 it was seen that the sum of the first n whole numbers is the triangular number $\frac{n(n+1)}{2}$, which may also be written as a polynomial, $\frac{1}{2}n^2+\frac{1}{2}n$. Verify this formula for the cases:

a. n=7 b. n=100

36. Problems 9 and 10 of Exercises 3.2 lead to a formula for the sum of the first n squares, which may be written as the polynomial $\frac{1}{3}n^3+\frac{1}{2}n^2+\frac{1}{6}n$. Verify this formula for the cases:

a. n=3 b. n=5 c. n=6

37. Problems 15-21 of Exercises 4.1 involve the pattern $1^3+2^3+3^3+\cdots+n^3=(1+2+3+\cdots+n)^2$. Using the result of question 1 above, this leads to the formula

$$1^3+2^3+3^3+\cdots+n^3=\left[\frac{n(n+1)}{2}\right]^2.$$

a. Show that the above expression may be written as the polynomial $\frac{1}{4}n^4+\frac{1}{2}n^3+\frac{1}{4}n^2$.

b. Add the first seven cubes and show that the sum is the same number as the value of the polynomial in part a when n=7.

38. The results of problems 35-37 were all known in ancient times. About a thousand years ago the Arab mathematician Alhazen found a formula for the sum of the first n fourth powers, which can be written as the polynomial $\frac{1}{5}n^5+\frac{1}{2}n^4+\frac{1}{3}n^3+0n^2-\frac{1}{30}n$. Check this formula for the cases n=:

a. 1 b. 2 c. 3 d. 4

17th century mathematicians tried to find formulas like those of questions 35-38 for higher powers. However, finding such formulas was a lot of work, and eventually one mathematician, Jakob Bernoulli[2], partly in hopes of easing this work, compiled a table of the formulas for the first kth powers, for k=1, 2, 3, . . , 10, and looked for patterns in the table. His table is reproduced below. Here the symbol $S_6(n)$, for example, is the formula for $1^6 + 2^6 + 3^6 + ... + n^6$. Explore this table on your own, looking for patterns.

$$S_1(n)=\frac{1}{2}n^2+\frac{1}{2}n$$

$$S_2(n)=\frac{1}{3}n^3+\frac{1}{2}n^2+\frac{1}{6}n$$

$$S_3(n)=\frac{1}{4}n^4+\frac{1}{2}n^3+\frac{1}{4}n^2+0n$$

$$S_4(n)=\frac{1}{5}n^5+\frac{1}{2}n^4+\frac{1}{3}n^3+0n^2-\frac{1}{30}n$$

$$S_5(n)=\frac{1}{6}n^6+\frac{1}{2}n^5+\frac{5}{12}n^4+0n^3-\frac{1}{12}n^2+0n$$

$$S_6(n)=\frac{1}{7}n^7+\frac{1}{2}n^6+\frac{1}{2}n^5+0n^4-\frac{1}{6}n^3+0n^2+\frac{1}{42}n$$

$$S_7(n)=\frac{1}{8}n^8+\frac{1}{2}n^7+\frac{7}{12}n^6+0n^5-\frac{7}{24}n^4+0n^3+\frac{1}{12}n^2+0n$$

$$S_8(n)=\frac{1}{9}n^9+\frac{1}{2}n^8+\frac{2}{3}n^7+0n^6-\frac{7}{15}n^5+0n^4+\frac{2}{9}n^3+0n^2-\frac{1}{30}n$$

$$S_9(n)=\frac{1}{10}n^{10}+\frac{1}{2}n^9+\frac{3}{4}n^8+0n^7-\frac{7}{10}n^6+0n^5+\frac{1}{2}n^4+0n^3-\frac{3}{20}n^2+0n$$

$$S_{10}(n)=\frac{1}{11}n^{11}+\frac{1}{2}n^{10}+\frac{5}{6}n^9+0n^8-1n^7+0n^6+1n^5+0n^4-\frac{1}{2}n^3+0n^2+\frac{5}{66}n$$

Then try the exercises below, which will guide you in further exploration.

2 Jacob Bernoulli (1654-1705), first of a family of outstanding mathematicians and scientists, wrote his *Ars Conjectandi (The Art of the Conjecture)*, which included the work above on sums of powers, between 1680 and his death. Its publication in 1713 effectively established the probability as a branch of mathematics.

39. Consider the numerical array of coefficients of the above polynomials:

$$
\begin{array}{ccccccccccc}
\frac{1}{2} & \frac{1}{2} & & & & & & & & & \\
\frac{1}{3} & \frac{1}{2} & \frac{1}{6} & & & & & & & & \\
\frac{1}{4} & \frac{1}{2} & \frac{1}{4} & 0 & & & & & & & \\
\frac{1}{5} & \frac{1}{2} & \frac{1}{3} & 0 & \frac{-1}{30} & & & & & & \\
\frac{1}{6} & \frac{1}{2} & \frac{5}{12} & 0 & \frac{-1}{12} & 0 & & & & & \\
\frac{1}{7} & \frac{1}{2} & \frac{1}{2} & 0 & \frac{-1}{6} & 0 & \frac{1}{42} & & & & \\
\frac{1}{8} & \frac{1}{2} & \frac{7}{12} & 0 & \frac{-7}{24} & 0 & \frac{1}{12} & 0 & & & \\
\frac{1}{9} & \frac{1}{2} & \frac{2}{3} & 0 & \frac{-7}{15} & 0 & \frac{2}{9} & 0 & \frac{-1}{30} & & \\
\frac{1}{10} & \frac{1}{2} & \frac{3}{4} & 0 & \frac{-7}{10} & 0 & \frac{1}{2} & 0 & \frac{-3}{20} & 0 & \\
\frac{1}{11} & \frac{1}{2} & \frac{5}{6} & 0 & -1 & 0 & 1 & 0 & -\frac{1}{2} & 0 & \frac{5}{66}
\end{array}
$$

a. Extend the leftmost column down three more entries

b. Extend the second column (next to the left) down three more entries.

c. Which other columns can you extend?

40. a. Express each entry in the third column as a number of twelfths.

b. Find a pattern in your answer to part a and use it to predict the next three entries of the third column.

41. a. Find the lowest common denominator of the fractions in the fifth column.

b. Express each number in the fifth column as a fraction with the denominator found in part a.

c. Find the pattern of the numerators of the fractions found in part b and use that pattern to predict the next three entries in the fifth column.

42. a. Add the fractions in each horizontal row. Show all the work.

b. Based on your results from question 3, make a general statement about the sum of the entries in each row of this array.

13.3 Division of Polynomials

Division in algebra is based on the same ideas as division in arithmetic. It may be viewed as repeated subtraction, as finding a missing factor, or as multiplication by the reciprocal. The last of these will be postponed until we discuss algebraic fractions. Here we focus on repeated subtraction and finding the missing factor. We begin by exploring simple examples.

Example 1. Divide $3x + 6$ by $x + 2$.
Solution: **Repeated subtraction** (using take away):

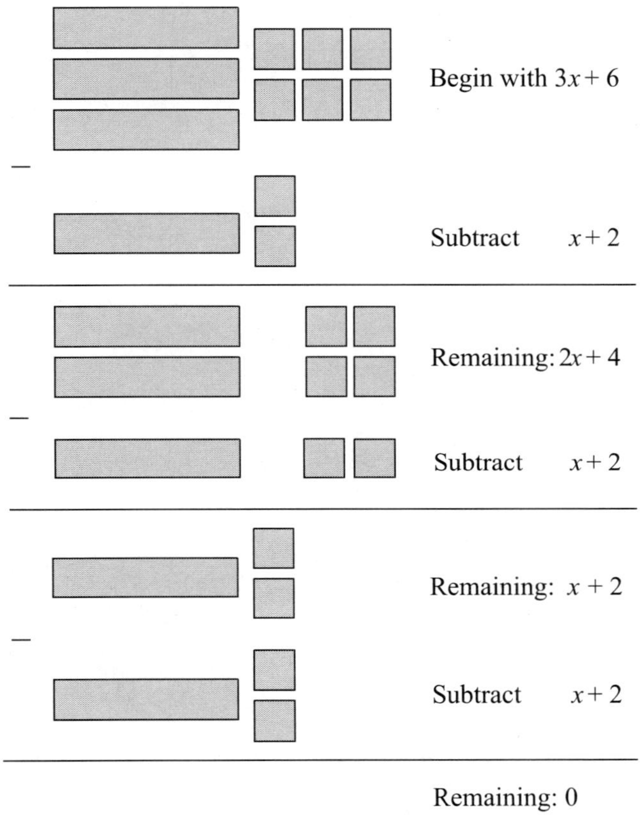

Figure 13.3.1

The process ends when there is nothing left to subtract.
We subtracted $x + 2$ three times, so $(3x + 6) \div (x + 2) = 3$.

A second solution: **Missing factor.** By what must we multiply $x + 2$ to get $3x + 6$? To see the answer, form a rectangle of area $3x + 6$ with one dimension $x + 2$ as shown.

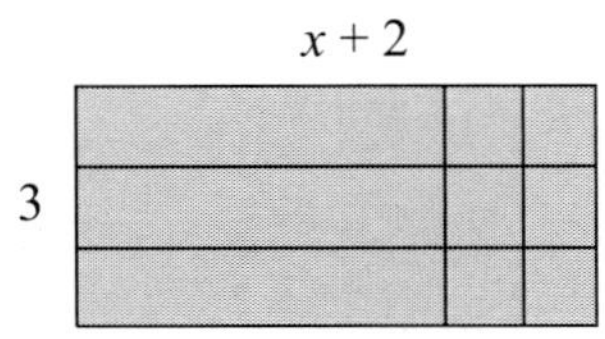

The width of this rectangle, 3, is the quotient $(3x + 6) \div (x + 2)$.

Example 2. Divide $x^2 + 3x + 2$ by $x + 1$.
Solution: In this case you can see that simply subtracting $x + 1$ does not affect the x^2 term of the dividend. The key new idea is to subtract $x + 1$ not a whole number of times but x times. That involves subtracting $x(x + 1)$ or $x^2 + x$.

Begin with	$x^2 + 3x + 2$
Subtract $x(x + 1)$ or $x^2 + x$	$\underline{x^2 + x}$
Remaining	$2x + 2$
Subtract $x + 1$	$\underline{x + 1}$
Remaining	$x + 1$
Subtract $x + 1$	$\underline{x + 1}$

Since $x + 1$ was subtracted first x times and then twice more, it was subtracted $x + 2$ times in all. Therefore, the quotient is $x + 2$. To check this, conclusion, note that $(x + 2)(x + 1) = x^2 + 3x + 2$. Figure 13.3.2 shows this solution with algebra tiles.

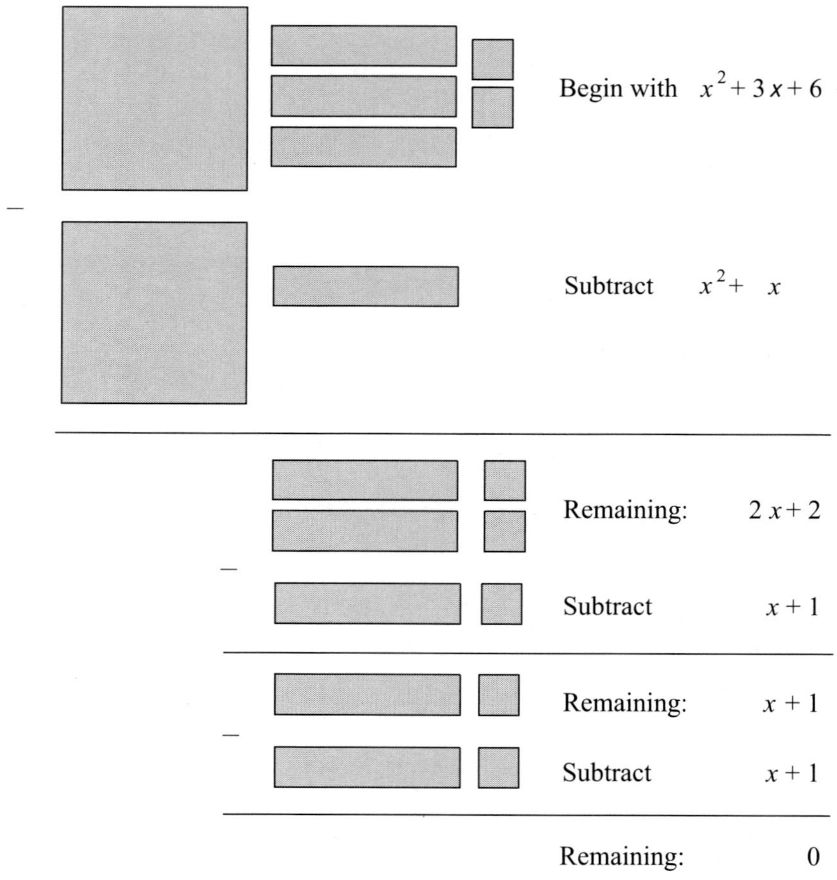

Figure 13.3.2

A second solution: **Missing factor**. Make a rectangle with width $x + 1$ and area $x^2 + 3x + 2$. You can see from the picture that its length is $x + 2$, so $(x^2 + 3x + 2) \div (x + 1) = x + 2$.

Long Division with Polynomials

Repeated subtraction of polynomials can be refined into polynomial long division.

1. Arrange both polynomials in descending powers of the variable.
2. Divide the leading term of the **dividend** by the leading term of the **divisor**. Write the result above as the first term of the quotient.
3. Multiply the entire divisor by this result and subtract the product from the dividend.
4. Using the remainder from step 3 as a new dividend, repeat steps 2 and 3 until the remainder has lower degree than the divisor.

Example 3. Divide $4x^2 + 6x - 5$ by $2x + 1$.
Solution:

1. The divisor, $2x + 1$, and the dividend, $4x^2 + 6x - 5$, are both arranged in descending powers of x.
2. Divide the leading term of the dividend, $4x^2$, by the leading term of the divisor, $2x$. The result, $2x$, is the first term of the quotient. Write it above.
3. Multiply the entire divisor by $2x$.
   $2x(2x + 1) = 4x^2 + 2x$
4. Subtract the product from the dividend and continue as shown.

$$\begin{array}{r|l} & 2x + 2 \\ \hline 2x + 1 & 4x^2 + 6x - 5 \\ & 4x^2 + 2x \\ \hline & 4x - 5 \\ & 4x + 2 \\ \hline & -7 \end{array}$$

The quotient is $2x + 2$, and -7 is the **remainder**.

Some students may find a negative remainder puzzling. Model the division with algebra tiles, as in figure 13.3.3, to explain the negative remainder.

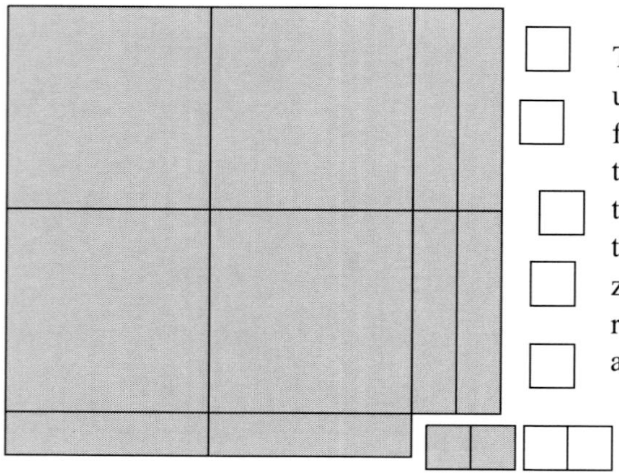

Try to form a rectangle $2x+1$ wide, using four x^2 flats, six x rods, and negative five unit squares. The flats and rods force the length to be $2x+2$, but then you need two positive unit squares to complete the rectangle. To get them, add two zero pairs. That completes the rectangle and leaves -7 unit squares as remainder.

Figure 13.3.3

To check a numerical division with remainder, multiply the divisor by the quotient, then add the remainder. If the work is correct, this yields the dividend. For example, dividing 13 by 5 yields quotient 2 and remainder 3. To check it, multiply 2 by 5 and add the remainder 3 and note that this yields the dividend 13. The procedure is just the same for polynomials. In the example above,

1. Multiply the quotient by the divisor.
   $(2x + 2)(2x + 1) = 4x^2 + 6x + 2$
2. Add the remainder, -7.
   $4x^2 + 6x + 2 + (-7) = 4x^2 + 6x - 5$

The result of step 2 is the dividend, so the division checks. When the remainder is 0, we say the divisor is a **factor** of the dividend.

Example 4 will show that polynomial long division allows you to divide polynomials of degree greater than two, even though you cannot use algebra tiles to picture them. That example also shows that when the dividend lacks one or more terms, you need to leave space when you set up the division.

Note: Incorrect subtraction is a major source of error in algebraic long division. A reliable way to carry out the subtractions is to add the opposite; change each sign of the polynomial you are subtracting, then add the result.

Example 4. Divide $y^4 - 16$ by $y - 2$.
Solution: Step 1. Write the division as shown, with space for missing y^3, y^2, and y terms.
Step 2. Since $y^4 \div y = y^3$, write y^3 as the first term of the quotient. Multiply $y - 2$ by y^3 and subtract as shown.
Step 3. Since $2y^3 \div y = 2y^2$, write $2y^2$ as the second term of the quotient. Multiply $y - 2$ by $2y^2$ and subtract as shown.
Step 4. Since $4y^2 \div y = 4y$, write $4y$ as the third term of the quotient. Multiply $y - 2$ by $4y$ and subtract as shown.
Step 5. Since $8y \div y = 8$, Write 8 as the last term in the quotient. Multiply $y - 2$ by 8 and subtract as shown.

$$
\begin{array}{r|l}
 & y^3 + 2y^2 + 4y + 8 \\
\hline
y - 2 & y^4 \qquad\qquad\qquad - 16 \\
 & \underline{y^4 - 2y^3} \\
 & 2y^3 \qquad\qquad - 16 \\
 & \underline{2y^3 - 4y^2} \\
 & 4y^2 \qquad - 16 \\
 & \underline{4y^2 - 8y} \\
 & 8y - 16 \\
 & \underline{8y - 16}
\end{array}
$$

Note: To make the repeated subtraction more transparent in the work above, -16 is shown ("brought down") after each subtraction. In practice, many people skip repeated copies such as those.
To check this division, verify that $(y - 2)(y^3 + 2y^2 + 4y + 8) = y^4 - 16$.

Exercises 13.3

In questions 1-4, divide and illustrate your solution with algebra tiles.

1. $x^2 + 5x$ by x
2. $x^2 + 6x + 8$ by $x + 4$
3. $2x^2 + 6x + 8$ by $x + 4$
4. $6x^2 + 13x + 6$ by $3x + 2$

In questions 5-18, divide, but do not necessarily illustrate with tiles.

5. $x^2 + x - 30$ by $x + 5$
6. $y^2 - y - 56$ by $y - 8$
7. $a^2 - a - 20$ by $a - 5$
8. $b^2 + b - 42$ by $7 + b$
9. $c^2 - c - 72$ by $c - 9$
10. $d^2 + d - 56$ by $d + 8$

11. $22 + e^2 - 9e$ by $e - 4$

12. $f^2 + 7f + 11$ by $2 + f$

13. $g^3 - 4g^2 - 23g$ by $3 + g$

14. $h^3 + 9h - 2$ by $h - 1$

15. $j^3 - 17j^2 - 18j + 3$ by $j + 1$

16. $k^4 - 32k^2 + 60$ by $k^2 - 2$

17. $8z^6 - 27$ by $z^2 - 3$

18. $m^4 + 15m^2 + 36$ by $m^2 + 3$

Exploration: The remainder theorem (questions 19-22)

19. Divide $x^2 - x - 20$ by each divisor, and note the remainder each time. Then evaluate the dividend at the given value by letting x equal that value. Compile the results in the following table.

Remainder when $x^2 - x - 20$ is divided by	Value of $x^2 - x - 20$ when
$x - 1$	$x = 1$
$x - 2$	$x = 2$
$x - 3$	$x = 3$
x − 4	$x = 4$
$x - 5$	$x = 5$
x	$x = 0$
$x + 1$	$x = -1$
$x + 2$	$x = -2$
$x + 3$	$x = -3$

The answers in the table should reveal a pattern. Evaluate $x^2 - x - 20$ at $x = -10$, and then use the pattern from the table to predict the remainder when $x^2 - x - 20$ is divided by $x + 10$.
Note $x + 10 = x - (-10)$. Check your prediction by dividing.

20. What do you predict will be the remainder if $x^2 - x - 20$ is divided by $x - a$? Carry out the division to check your prediction from question 19.

21. a. Complete this table by doing the divisions and substitutions.

Remainder when $x^3 - 4$ is divided by	Value of $x^3 - 4$ when
$x - 1$	$x = 1$
$x - 2$	$x = 2$
$x - 3$	$x = 3$
$x - 4$	$x = 4$
$x - 5$	$x = 5$
x	$x = 0$
$x + 1$	$x = -1$
$x + 2$	$x = -2$
$x + 3$	$x = -3$

b. Based on the results of part a, what do you predict for the remainder if $x^3 - 4$ is divided by $x - b$?

c. Divide $x^3 - 4$ by $x - b$ to check your prediction from part a.

22. Use problems 20 and 21 to complete this statement of the general rule known as the **remainder theorem**. If a polynomial in x is divided by $(x - a)$, the remainder is ____________________.

13.4 Factoring Polynomials

Here we extend the analogy between polynomials with integer coefficients and the integers. Polynomials, like integers, may be factored, which is useful in finding least common multiples. In chapter 14 you will see how factoring polynomials can help in solving equations.

A polynomial is **factored**, if it is expressed as a product of polynomials, which are called **factors** of the given polynomial. For example, the polynomial $x^2 + x$ may be factored as $x(x + 1)$.

For polynomials, as for integers, factoring is usually more difficult than multiplying. Any polynomial, like any integer, can be factored as 1 times itself, but that "trivial" factorization is of little practical value, as is multiplying a polynomial by (-1)(-1).

Factoring a number as a product of two factors was modeled by making rectangles with given areas and whole number sides. In factoring 12, we could make rectangles that are 3×4, 2×6, and, trivially, 1×12. In factoring 7 we could only make a 1×7 rectangle, because 7 is a prime number. As a concrete introduction to factoring polynomials, try arranging various selections of the algebra tiles from figure 13.4.1 to make rectangles. For example, can you arrange one flat and two rods into a single rectangle? How about one flat, two rods and one unit square? How about one flat, five rods, and six unit squares? Students can work in groups at this, recording their results in a table like the one below, and occasionally sharing results between groups.

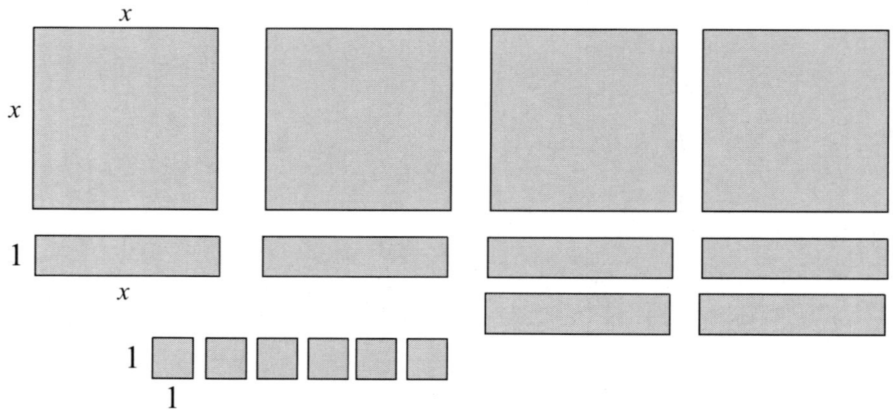

Figure 13.4.1

Make this an open-ended activity. Students will soon find that some sets of tiles, such as one flat and one unit square, simply cannot be formed into a single rectangle. That is because $x^2 + 1$ can only be factored trivially. A polynomial which can only be factored trivially is called a **prime polynomial**. On the other hand, one flat, five rods, and six unit squares can be arranged to make the rectangle shown in figure 13.4.2, which shows that $x^2 + 5x + 6$ can be factored as $(x + 3)(x + 2)$. This is recorded in the top line of the table on the next page.

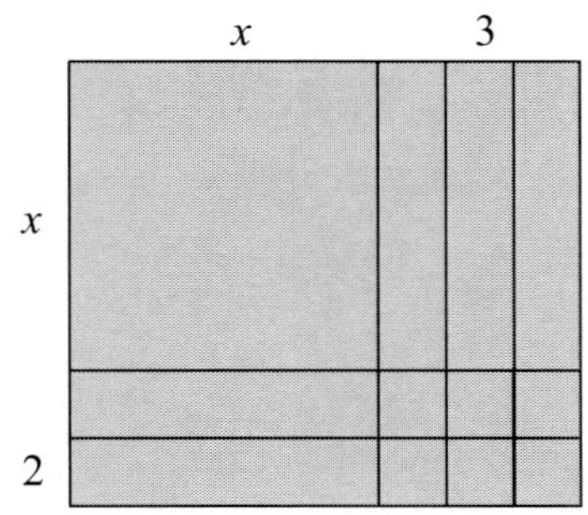

Figure 13.4.2

Number of x^2	Number of x	Number of 1s	Length	Width	Area
1	5	6	$x+3$	$x+2$	x^2+5x+6

Although factoring polynomials can be difficult, the basic factoring techniques shown below will give you a good start.

First technique: Common factor
If all terms of a polynomial have a common factor, you can factor by using the distributive property in reverse.

Example 1. Factor $6x^2 + 12x$.
Solution: Both terms are multiples of $6x$. Factor out the common factor to get $6x(x + 2)$.

Example 2. Factor $3(z + d) + 5(z + d)$.
Solution: The factor $(z + d)$ is common to both terms. Therefore, $3(z + d) + 5(z + d) = (3 + 5)(z + d) = 8(z + d)$.

Second technique: Difference of squares
A polynomial of the form $a^2 - b^2$ is called, naturally, a **difference of squares.**
You can use $a^2 - b^2 = (a + b)(a - b)$ as a template for factoring any difference of squares.
Note: There is no corresponding way to factor a sum of squares.

Example 3. Factor $49x^2 - 16$.
Solution: Write this as a difference of squares, $(7x)^2 - 4^2$.
By the pattern above, this factors as $(7x - 4)(7x + 4)$, which you can check by multiplying.

Third technique: Square of a binomial
A polynomial of the form $a^2 + 2ab + b^2$ is called a **perfect binomial square,** since $a^2 + 2ab + b^2 = (a + b)^2$. You can use this as a template to factor any polynomial of this form.

Example 4. Factor $y^2 - 6y + 9$.
Solution: This fits the perfect square template with $a = y$ and $b = -3$, so $y^2 - 6y + 9 = (y - 3)^2$. You can check this by multiplying out the right-hand side of the equation.

Example 5. Factor $25x^4 + 10x^2 + 1$.
Solution: This fits the binomial square template with $a = 5x^2$ and $b = 1$. Therefore, $25x^4 + 10x^2 + 1 = (5x^2 + 1)^2$.

Fourth technique: Sum or difference of cubes
You can check by multiplying that: $a^3 + b^3 = (a + b)(a^2 - ab + b^2)$ and $a^3 - b^3 = (a - b)(a^2 + ab + b^2)$.
These templates allow you to factor any sum or difference of cubes.

Example 6. Factor $t^3 - 27w^3$.
Solution: Since $27w^3 = (3w)^3$, the given polynomial is a difference of cubes and $t^3 - 27w^3 = (t - 3w)(t^2 + 3tw + (3w)^2) = (t - 3w)(t^2 + 3tw + 9w^2)$.
Although our focus is on polynomials in one variable, this example shows that much of the thinking that we use in the one-variable case applies as well to polynomials in two variables.

Factoring general trinomials

Recall that $(ax + b)(cx + d) = acx^2 + (ad + bc)x + bd$. Some trinomials, such as $x^2 + 5x + 4$ may be factored using this template, but others, such as $x^2 + x + 1$, are prime. The most practical approach to factoring in such cases is trial and error, guided by observations about the coefficients of the trinomial.

Example 7. Factor $y^2 - 3y - 10$.
Solution: The terms have no common factor, so if $y^2 - 3y - 10$ is not prime, it must be factored as a product of binomials. Therefore, begin by writing $(y \_\_\_ p)(y \_\_\_ q)$, where p and q are factors of 10 and the blank spaces are for plus or minus signs. Since -10 is negative, one sign is positive and the other negative, so $y^2 - 3y - 10 = (y + p)(y - q)$, where the product of p and q is 10. If you try 5 in the first factor and 2 in the second and multiply out $(y + 5)(y - 2)$ you get $y^2 + 3y - 10$.
Then try 2 in the first factor and 5 in the second.
At last, $(y + 2)(y - 5) = y^2 - 3y - 10$.

Example 8. Factor $6y^2 - 5y + 1$.
Solution: There is no common factor, and this does not fit one of our special forms, so try factoring as a product of binomials. You could factor $6y^2$ either as $y(6y)$ or as $(3y)(2y)$ so the factorization will be either $(y \_\_\_)(6y \_\_\_)$ or $(3y \_\_\_)(2y \_\_\_)$.

The constant term $+1$ in $6y^2 - 5y + 1$ indicates that the constant terms of the factors are either both 1 or both -1. Then there are only four possibilities:

a. $(y - 1)(6y - 1)$
b. $(y + 1)(6y + 1)$
c. $(3y - 1)(2y - 1)$
d. $(3y + 1)(2y + 1)$

Each of these cases will have the right first and last terms, so multiply mentally to check the coefficient of y in the middle term. Case a gives -7, and cases *b* and *d* yield positive middle terms, so the only hope is case c. Checking, you do indeed find $(3y - 1)(2y - 1) = 6y^2 - 5y + 1$.

Example 9. Factor $2x^2 + xy + 3y^2$.
Solution: There is no common factor, and the special forms do not fit, so try factoring as a product of binomials. Since all coefficients are positive, any factorization has the form $(2x + \_\_\_ y)(x + \_\_\_ y)$, where the blanks must be filled with positive integers whose product is 3. The only possibilities are $(2x + 3y)(x + 1y)$ and $(2x + 1y)(x + 3y)$, but neither product is $2x^2 + xy + 3y^2$. Therefore, $2x^2 + xy + 3y^2$ is a prime polynomial.

Some technical comments

In chapter 4 we classified whole numbers as primes, composite numbers, and, in a class by itself, the unit 1. We saw that each composite number can be factored completely in essentially one way. For this purpose, we ignored the order of the factors, and we ruled out the use of the unit 1 as a factor, since any number n may be factored "trivially" by expressing it as $1 \times n$.

In the larger arena of non-zero integers, we would encounter a more complex situation. For factoring purposes, we divide the non-zero integers into three categories:

Prime numbers –e.g. $2, -2, 3, -3, 5, -5, 7, -7, 11, -11$, etc.

Composite numbers –e.g. $4, -4, 6, -6, 9, -9$, etc.

Units (integers with multiplicative inverses) 1 and –1.

We count 5 and –5 as primes, because they cannot be factored in the system of integers without using units. In the non-zero integers, we count two factorizations of the same number as essentially the same, if the only difference between those factorizations is factors of –1. With that qualification, factorization into primes in the non-zero integers is unique. This means that any two factorizations of the same prime are the same, with the possible exceptions of order and factors of 1 or –1. For example, we count $(-3)(-5)$ and $(5)(3)$ as basically the same factorization of 15, since $(-3)(-5) = (-1)(3)(-1)(5) = (-1)^2(3)5 = 1(5)(3) = (5)(3)$.

The same considerations apply to the system of non-zero polynomials with integer coefficients. Each of these polynomials belongs to one of three categories:

Prime polynomials –e.g. $x + 4$, 7, $x^2 + 3$

Composite polynomials -e.g. $2x$, -6, $4x + 10$, $3x^2 - 27$

Units (polynomials with multiplicative inverses) 1 and -1.

We count –6 as composite, since it may be factored as $(2)(-3)$. Among one-variable polynomials with integer coefficients, factorization is unique up to order and factors of 1 and –1. For $3x^2 - 27$, the prime factorizations $3(x - 3)(x + 3)$ and $(-x + 3)(-3)(x + 3)$ are essentially the same, since $(-x + 3) = (-1)(x - 3)$ and $-3 = (-1)3$, so $(-x + 3)(-3)(x + 3) = (-1)(x - 3)(-1)(3)(x + 3) = 3(x - 3)(x + 3)$.

When we factor polynomials, we will generally do so in the context of polynomials in one variable with integer coefficients. However, when we solve quadratic equations, we will expand the arena to polynomials with coefficients which are not integers. For example, $x^2 - 3$ is prime in the set of polynomials with integer coefficients, but in the larger arena of polynomials with real coefficients it may be factored as $(x - \sqrt{3})(x + \sqrt{3})$.

Skill in factoring is an art, not a science, and it requires practice. The exercises below begin with straightforward examples, but among the higher numbered questions you will find some challenges.

Exercises 13.4

1. a. Students found two ways to show the factorization in example 1 above. Are they both correct? Explain.

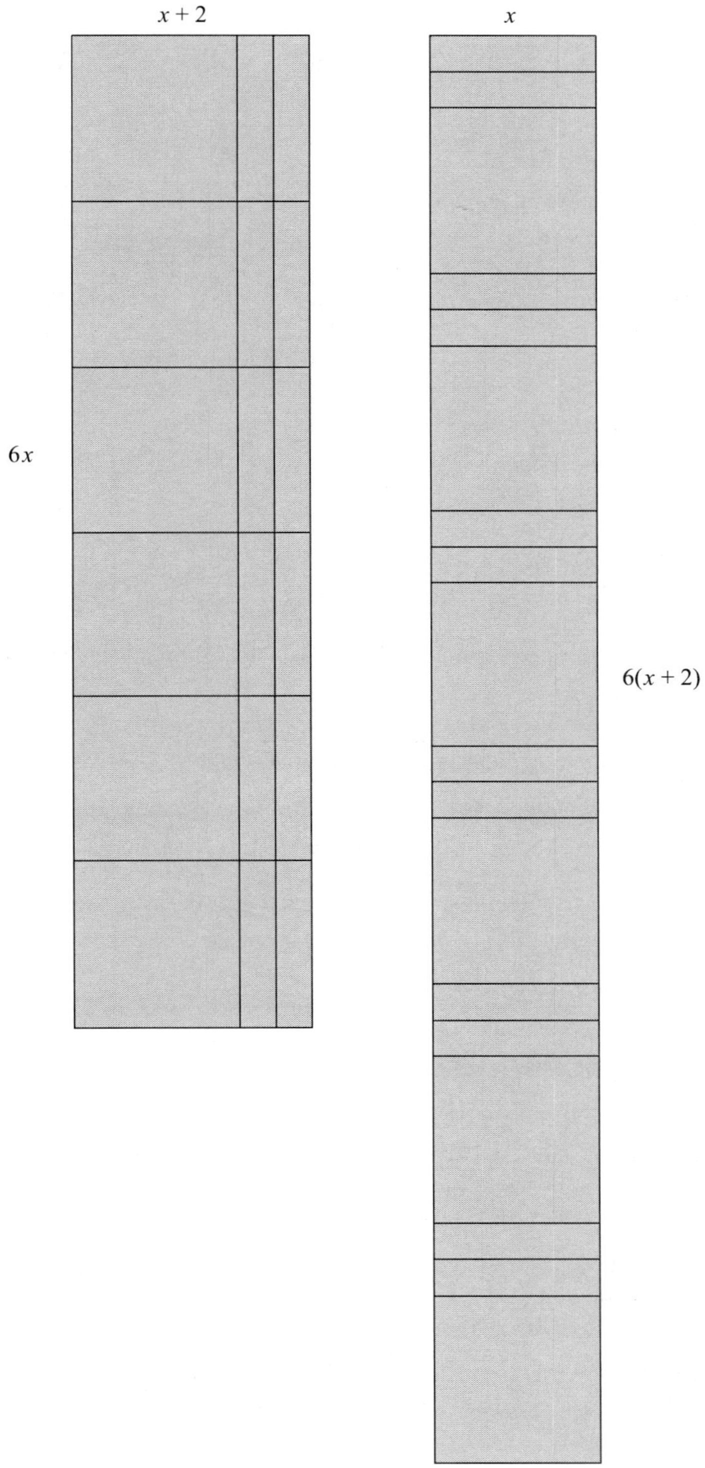

b. Other students tried to picture the same polynomial as the area of a rectangle with one dimension 6. What would the other dimension of this rectangle be? Why did they have trouble picturing this rectangle?

The polynomials in problems 2-5 all fit the common factor pattern $ab + ac = a(b + c)$. For each polynomial:

a. State what plays the role of a.
b. State what plays the role of b.
c. State what plays the role of c.
d. Factor the polynomial.

2. $7x-21xy$
3. $10sp-25p^2$
4. $-9y+12yz$
5. $30k+50kt$

The polynomials in problems 6-11 are differences of squares a^2-b^2. In each case:

a. State what plays the role of a.
b. State what plays the role of b.
c. Factor the polynomial.

6. $4x^2 - 81y^2$
7. $9r^2 - 49s^2$
8. $-25z^4 + 49$
9. $144x^2y^6 - 1$
10. $36c^2d^6 - 121e^4$
11. $81 - 256s^4$

The polynomials in problems 12-17 are squares of binomials $a^2+2ab+b^2$. In each case:

a. State what plays the role of a.
b. State what plays the role of b.
c. Factor the polynomial.

12. $x^2 - 8x + 16$
13. $t^2 + 10t + 25$
14. $z^2 - 12z + 36$
15. $4y^2 - 4y + 1$
16. $w^4 + 20w^2 + 100$
17. $9s^6r^2 + 6s^3r + 1$

The polynomials in problems 18-24 are sums or differences of cubes a^3+b^3 or a^3-b^3. In each case:

a. State what plays the role of a.
b. State what plays the role of b.
c. Factor the polynomial.

18. $27s^3 - 8$
19. $64x^3 + 125y^3$
20. $z^6 - 64$
21. $t^3s^3 - 8k^3$
22. $125x^3y^6 + z^9$
23. $-8 + 216s^9$

For each of the following polynomials, identify which, if any, of the patterns above applies and how. For example, if you are given $9x^2 - 16y^2$, identify it as a difference of squares, $(3x)^2 - (4y)^2$ and then factor it as $(3x + 4y)(3x - 4y)$.

24. $8s^3 - 27$
25. $4t^2 + 12t + 9$
26. $86z^3 + 3z$
27. $x^2 - 12x + 36$
28. $27t^6 + 1$
29. $144 - 25z^4$

In questions 30-39, factor by finding a factor common to all terms.

30. x^2+6x
31. y^2+3y
32. $2z^2+z$
33. $5w^2-w$
34. $8t^2+4t$
35. $10s^2-40s$
36. r^3-r
37. x^3+3x^2+x
38. $8q^4-4q^3-q$
39. x^5+x^6

In questions 40-43, factor the differences of squares.

40. t^2-81
41. $4a^2-25b^2$
42. $16c^2-d^2$
43. $144r^6-81$

In questions 44-47, factor the perfect binomial squares

44. $q^2+10q+25$
45. $36-12v+v^2$
46. $64-16r+r^2$
47. $16s^2+56st+49t^2$

In questions 48-51, factor the sums and differences of cubes.

48. $125q^3+27$
49. $1+m^3$
50. $8-27x^3$
51. $64y^3-1,000$

In questions 52-55, factor the polynomial if possible as a product of polynomials with integer coefficients. If the polynomial can only be factored trivially, state that it is prime.

52. $3a^2-3a-1$
53. $3j^2+8j+4$
54. b^2-b-6
55. $10w^3-12w^2+4w$

13.5 Prime Factorization and Least Common Multiples

Factoring into Primes

A polynomial can be factored as a product of prime polynomials, much as a positive integer can be factored as a product of prime numbers. As with integers, you factor a polynomial into primes by factoring, then factoring the factors, and continuing this way until all factors are primes. The polynomial is then said to be **completely factored** or **factored into primes**. In this section we will restrict attention to polynomials with integers and/or letters as coefficients.

Example 1. Factor $2ax^4 - 2a^5$ completely.
Solution: First factor out the common factor $2a$ to get $2a(x^4 - a^4)$

Now factor $x^4 - a^4$ as a difference of squares $2a(x^2 - a^2)(x^2 + a^2)$

Factor $x^2 - a^2$, too, as a difference of squares $2a(x - a)(x + a)(x^2 + a^2)$

Now $x^2 + a^2$ is prime, so the complete factorization of $2ax^4 - 2a^5$ is $2a(x - a)(x + a)(x^2 + a^2)$. Figure 13.5.1 shows the process as a factor tree, with the prime factors in bold type at the ends of the branches.

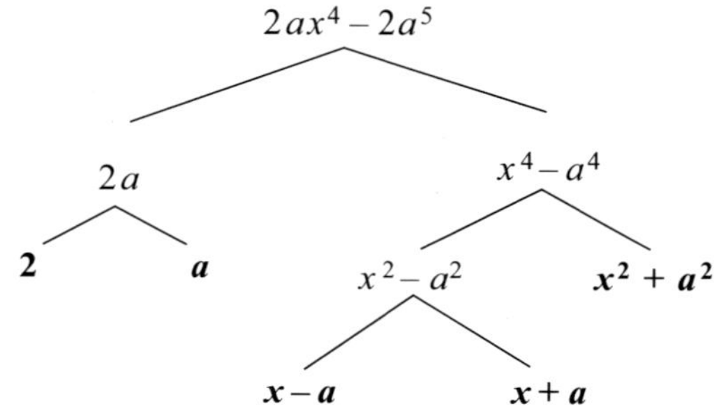

Figure 13.5.1

Example 2. Factor $x^6 - 1$ completely.
Solution: Here each term may be regarded either as a square or as a cube, since $x^6 = (x^2)^3 = (x^3)^2$, and $1 = 1^3 = 1^2$.

If we factor first as a difference of cubes,
$x^6 - 1 = (x^2)^3 - 1^3 = (x^2 - 1)(x^4 + x^2 + 1)$

Now factor $x^2 - 1$ as a difference of squares,
$x^6 - 1 = (x - 1)(x + 1)(x^4 + x^2 + 1)$

Further factorization depends on factoring $x^4 + x^2 + 1$, which is not obvious.

Start over, factoring first as a difference of squares.

$x^6 - 1 = (x^3)^2 - 1^2 = (x^3 - 1)(x^3 + 1).$

Now think of 1 as 1^3 and use the formulas for the difference and sum of cubes.
$x^6 - 1 = (x - 1)(x^2 + x + 1)(x + 1)(x^2 - x + 1).$

Here all factors are prime, and, as you can see by comparing the two factorizations, $x^4 + x^2 + 1 = (x^2 + x + 1)(x^2 - x + 1)$.

For polynomials, as for integers, factorization into primes is unique only up to order and factors which differ by units. This means, for example, that $(x - 1)(2x + 3)$ and $(1 - x)(-2x - 3)$ are not viewed as different factorizations, just as, in the integers, (3)(5) is viewed as fundamentally the same as $(-3)(-5)$.

Least Common Multiple

Often when we work with algebraic fractions, we need a common denominator. This denominator must be a multiple of each of the denominators involved, so it is said to be a **common multiple** of them. The common multiple that has the fewest factors is called the **least common multiple (LCM)**. The following parallel examples show that the procedure for finding the LCM is essentially the same for polynomials with integer coefficients as for positive integers (section 5.2).

Example 3. Find the LCM of:

a. 8, 9, and 30.

b. x^3, $x^2 + 2x + 1$, and $5x^2 + 5x$.

Solutions (note the similarities between the two):

Step 1:	**Step 1:**
Factor each number into primes. $8 = 2^3$ $9 = 3^2$ $30 = 2(3)(5)$	Factor each polynomial into primes. x^3 $x^2 + 2x + 1 = (x + 1)^2$ $5x^2 + 5x = 5x(x + 1)$
Step 2:	**Step 2:**
A number is a multiple of 8 if its prime factorization includes three factors of 2; it is a multiple of 9 if its prime factorization includes two factors of 3, and it is a multiple of 30 if its prime factorization has factors of 2, 3, and 5. Then the LCM is $2^3(3^2)(5)$.	A polynomial is a multiple of x^3 if its prime factorization includes three factors of x; it is a multiple of $(x + 1)^2$ if its prime factorization includes two factors of $x + 1$, and it is a multiple of $5x^2 + 5x$ if its prime factorization has factors of x, $x + 1$,and 5. Then the LCM is $5x^3(x + 1)^2$.

Note: If $x = 2$, the algebraic example above becomes the arithmetic example.

The above method may be summarized as follows:

To find the least common multiple (LCM) of several polynomials

> 1. **Factor each polynomial into primes.**
> 2. **Build a new polynomial (the LCM) from the prime factors found in step 1, using each factor the greatest number of times that it occurs in any of the given polynomials.**

Example 4. Find the LCM of $4p^2$, $6pq$, and $10q^3$.
Solution: Factor: 2^2p^2, $2(3)pq$, and $2(5)q^3$ respectively. The LCM must have two factors of 2, two factors of p, one factor of 3, one factor of 5, and three factors of q. Therefore, the LCM is $2^2(3)(5)p^2q^3 = 60p^2q^3$. It is worth taking a moment to make sure that this is indeed a multiple of each of the three polynomials given. Figure 13.5.2 summarizes the process. The prime factors of each polynomial are shown at right, and for each prime, the case with the most factors is circled. The circled entries are listed in the bottom line, and their product is the LCM.

polynomials	prime factors				
$4p^2$	2^2			p^2	
$6pq$	2	3		p	q
$10q^3$	2		5		q^3
LCM: $60p^2q^3$	2^2	3	5	p^2	q^3

Figure 13.5.2

Example 5. Find the LCM of $3y^3 - 27y$, $6y^2 + 9y - 27$, and $y^2 - 6y + 9$.
Solution: Factor each of these polynomials into primes.

$$3y^3 - 27y = 3y(y^2 - 9) = 3y(y - 3)(y + 3)$$
$$6y^2 + 9y - 27 = 3(2y^2 + 3y - 9) = 3(2y - 3)(y + 3)$$
$$y^2 - 6y + 9 = (y - 3)^2$$

Then the LCM has one factor of 3, one factor of y, one factor of $2y - 3$, one factor of $y + 3$, and two factors of $y - 3$. Therefore, the LCM is $3y(2y - 3)(y + 3)(y - 3)^2$. Figure 13.5.3 summarizes the work.

polynomials	prime factors				
$3y^2$-27y	3	y	y-3	y+3	
$6y^2$+9y-27	3			y+3	2y-3
y^2-6y+9			$(y\text{-}3)^2$		
LCM: $3y(y\text{-}3)^2(y\text{+}3)(2y\text{-}3)$	3	y	$(y\text{-}3)^2$	y+3	2y-3

Figure 13.5.3

Exercises 13.5

In questions 1-28, factor completely.

1. $3a^2 - 12$
2. $b^3x - x^3b$
3. $72 - 2c^2$
4. $5d^4 - 20$
5. $6e^3 + 2e^2$
6. $10f^2 - 90f^4$
7. $7g^2 + 42 - 35g$
8. $4h^2 + 24h + 27$
9. $y^2k - 121k$
10. $4j^3 - 4j^2 + 4j$
11. $3m^2 - 24mn + 48n^2$
12. $10p^2 - 15pq - 25q^2$
13. $3r^2 - 42s^2 - 15rs$
14. $3tu^2 + 5ut + 2t$
15. $5y^5 + 20y^4 + 25y^3$
16. $81r^9 - 3$
17. $16 - x^4$
18. $7a^2(b-1) - 28b^2(b-1)$
19. $c^2(x+3) + d^2(x+3)$
20. $h^2 - 16 - y^2(h^2 - 16)$
21. $4j^3 + 108$
22. $(p-q)^2 - 4r^2$
23. $-u^2 + (s+t)^2$
24. $4(v+w)^2 - (x-3)^2$
25. $49 - y^2 - 2Ay - A^2$
26. $z^2 + 2zb + b^2 - M^2$
27. $1 - c^2 - 25d^2 + 10cd$
28. $x^6 - 64$

In questions 29-42 find the LCM.

29. $2abc$, $6ab^2c$, and $4ab^3$

30. $18g^2$, $12h^2$, and $8gh$

31. $k^2 - 1$ and $k^2 - 3k - 4$

32. $4 - h^2$ and $4 + 2h$

33. $10m^2 - 10$ and $15m - 15$

34. $p^2 - 10p + 16$ and $p^2 - 5p + 6$

35. $5(r^4 - z^4)$ and $2(r^2 + z^2)$

36. $s^3 - t^3$ and $s^2 - 2st + t^2$

37. $u^2 + 1$, $u^3 - 1$, and $u^4 - 1$

38. $6 + 2x$, $x^2 - 9$, and $15x^2 + 20x + 2$

39. $z^2 - z - 20$ and $z^2 + z - 12$

40. $w^2 - 2w + 1$ and $w^2 - 5w + 4$

41. $X^2 + 5X + 6$ and $X^2 + 6X + 8$

42. $4y^2 - 16$, $6y - 12$, and $32 - 8y^2$

13.6 Algebraic Fractions

Just as dividing integers leads to numerical fractions, dividing polynomials leads to algebraic fractions and extends the analogy between arithmetic and algebra. Study of algebraic fractions doubles as review of arithmetic fractions.

Definition of algebraic fractions

Just as with numbers, the bar in an algebraic fraction indicates division.

Definition: If a and b are algebraic expressions and $b \neq 0$, then $\frac{a}{b}$ means $a \div b$.

For example, $\frac{x^2-9}{x+3}$ means $(x^2 - 9) \div (x + 3)$. Since division by 0 is not defined, an algebraic fraction is regarded as meaningless for any values of letters which make the denominator 0, as would $x = -3$ in the fraction $\frac{x^2-9}{x+3}$. We will assume (often without stating it explicitly) that the denominators of fractions are not 0, and we will avoid substituting values that lead to 0 denominators. We will focus on fractions in which the numerator and denominator are polynomials, but much of the discussion pertains as well to other algebraic fractions such as $\frac{\sqrt{a^2+b^2}}{2}$ or $\frac{\frac{3-x}{2x}}{\frac{2}{x+7}}$.

Fractions equal to 1

Any fraction with the same numerator and denominator is equal to 1, as you can check by multiplying. For example, $\frac{3y^3}{3y^3} = 1$, of course for $y \neq 0$, and $\frac{x-4}{x-4} = 1$, again, provided $x \neq 4$.

How to multiply algebraic fractions

1. **Multiply numerators to get the numerator of the product.**
2. **Multiply denominators to get the denominator of the product.**

In shorthand, if a, b, c, and d are polynomials and neither b nor d is 0,

$$\frac{a}{b} \times \frac{c}{d} = \frac{ac}{bd}$$

You can see this makes sense, because if numerical values are substituted into the polynomials, the result must be a correct multiplication of numerical fractions.

Example 1. Multiply $\frac{6a}{49}$ by $\frac{5b^2}{8y}$.

Solution: $\frac{6a}{49}\times\frac{5b^2}{8y}=\frac{30ab^2}{392y}$

As in arithmetic, you can often cancel to simplify a product of fractions in advance. That is discussed below in connection with equal fractions.

Just as with numbers, you can express any polynomial as a fraction by writing it over 1. This is useful for purposes of multiplication and division.

Example 2. Multiply $3x + 4$ by $\frac{2x+1}{x-2}$.

Solution: $(3x+4)\times\frac{2x+1}{x-2}=\frac{3x+4}{1}\times\frac{2x+1}{x-2}=\frac{(3x+4)(2x+1)}{1(x-2)}=\frac{6x^2+11x+4}{x-2}$

The algebraic equivalent of a mixed number is the sum of a polynomial and a fraction, such as $2y^2+7+\frac{3y+2}{y+1}$. To multiply an expression of this type by a fraction or by another polynomial, you can use the distributive property directly, or you can first combine the mixed expression into a single fraction.

Example 3. Multiply $x+1+\frac{1}{x+2}$ by $x + 2$.

Solution: By the distributive property

$$\begin{aligned}\left(x+1+\frac{1}{x+2}\right)(x+2)&=(x+1)(x+2)+\left(\frac{1}{x+2}\right)(x+2)\\&=x^2+3x+2+1\\&=x^2+3x+3\end{aligned}$$

Second solution: First combine as a single fraction

$$\begin{aligned}\left(x+1+\frac{1}{x+2}\right)(x+2)&=\left(\frac{(x+1)(x+2)}{x+2}+\frac{1}{x+2}\right)(x+2)\\&=\left(\frac{(x+1)(x+2)+1}{x+2}\right)(x+2)\\&=x^2+3x+3\end{aligned}$$

Equal fractions

The polynomial 1 is the multiplicative identity. ($1x = x$ for any number x, and $1P = P$ for any polynomial P, and $1Q = Q$ for any algebraic fraction Q). Two fractions are equal, if one can be multiplied by a form of 1 to get the other.

For example, $\frac{8x}{10}=\frac{4x}{5}$, because $\frac{8x}{10}=\frac{2}{2}\times\frac{4x}{5}$.

Again, $\frac{3a-4b}{3a+4b}=\frac{9a^2-16b^2}{(3a+4b)^2}$, because, provided the denominator is not 0, $\frac{3a-4b}{3a+4b}\left(\frac{3a+4b}{3a+4b}\right)=\frac{9a^2-16b^2}{(3a+4b)^2}$, and the fraction in parentheses is a form of 1.

As in arithmetic, two fractions are equal if you can multiply numerator and denominator of one fraction by the same thing to get the other fraction.

Reducing and unreducing fractions

If you multiply a fraction by a form of 1 and get fewer factors in the numerator and denominator, you have **reduced** the fraction. If, on the other hand, you multiply a fraction by a form of 1 that increases the number of factors in the numerator and denominator, you have **unreduced** the fraction.

For example, multiplying $\frac{4x}{5}$ by 1 in the form $\frac{2}{2}$ unreduces it to $\frac{8x}{10}$, and multiplying $\frac{8x}{10}$ by 1 in the form $\frac{\frac{1}{2}}{\frac{1}{2}}$ reduces it to $\frac{4x}{5}$. Often we reduce and unreduce fractions without explicitly noting the multiplication by 1, simply multiplying or dividing numerator and denominator by the same thing. For example, you can reduce $\frac{\pi r^2}{2\pi r}$ by dividing numerator and denominator r to get $\frac{\pi r}{2\pi}$. In turn, you can reduce this fraction by dividing numerator and denominator by π to get $\frac{r}{2}$. At this point the fraction cannot be reduced further, because the numerator and denominator have no common non-trivial factors. Such a fraction is said to be **lowest terms**.

For example $\frac{5}{7}$, $\frac{-3x}{4}$, and $\frac{x+2}{(x+1)^2}$ are in lowest terms, but $\frac{2a}{4a(x+1)}$, $\frac{5b^2}{4by}$, and $\frac{x^2-1}{(x+1)^2}$ are not.

The last fraction is not in lowest terms, because the numerator factors as $(x - 1)(x + 1)$, so that

$$\frac{x^2-1}{(x+1)^2}=\frac{(x-1)(x+1)}{(x+1)(x+1)}=\frac{(x-1)}{(x+1)}$$

Any fraction that is not already in lowest terms is equal to some fraction that is in lowest terms; to reduce a fraction to lowest terms, divide numerator and denominator by their common factors, until they have no non-trivial common factors left.

Example 4. Reduce $\frac{2a}{4a(x+1)}$ to lowest terms.

Solution: The factors common to numerator and denominator are 2 and a.

Thus $\frac{2a}{4a(x+1)}=\frac{2a(1)}{2a(2)(x+1)}=\frac{1}{2(x+1)}$.

Example 5. Reduce $\frac{c^2+cd-20d^2}{2c-8d}$ to lowest terms.

Solution: Factor numerator and denominator: $\frac{(c+5d)(c-4d)}{2(c-4d)}$

Then $\frac{c^2+cd-20d^2}{2c-8d}=\frac{c+5d}{2}$.

Example 6. Reduce $\frac{x^2+5x+6}{(x+1)(x+4)}$ to lowest terms.

Solution: Factor the numerator as $(x + 2)(x + 3)$, and you see that numerator and denominator have no common factors. This fraction is already in lowest terms.

Just as with numerical fractions, sometimes you can see before actually multiplying that numerator and denominator will have common factors. Then you can eliminate them ahead of time, in effect avoiding needless multiplication and division. As with numerical fractions, this is known in mathematical slang as **canceling**. Many educators have wished to replace this word with a more accurate description of this process, such as "reducing in advance," but the term "canceling" survives. The problem is not the word so much as the temptation it seems to bring students to line out parts of complicated fractions without understanding what they are doing.

Example 7. Multiply $\frac{5b^2}{x^2-y^2}$ by $\frac{x^2+2xy+y^2}{20ab}$.

Solution: Before multiplying numerators and multiplying denominators, notice the 5 in the first fraction will be offset by a 5 in the denominator of the second, since $20 = 5 \times 4$. Again, the denominator of the first fraction can be factored as $(x - y)(x + y)$, and the numerator of the second fraction factors as $(x + y)(x + y)$. Cancel one factor of $x + y$ in each. Then, thinking of $5b^2$ as $5bb$, you can cancel one factor of b in the numerator and one in the denominator.

$$\frac{5b^2}{x^2-y^2} \times \frac{x^2+2xy^2+y}{20ab} = \frac{\cancel{5}b\cancel{b}}{(x-y)\cancel{(x+y)}} \times \frac{(x+y)\cancel{(x+y)}}{4\cdot\cancel{5}\,a\cancel{b}} = \frac{b(x+y)}{(x-y)4a} = \frac{bx+by}{4ax-4ay}$$

Dividing algebraic fractions is the same as dividing fractions in arithmetic, but, like most division in algebra, it is not generally easy to visualize or explain in terms of a model based on sharing. In this context, it is simplest easiest to think of division by a fraction as multiplication by its reciprocal. Plenty of practice appears at the end of this section.

Example 8. Divide $\frac{q^2+tq}{4t^2}$ by $\frac{1-tq}{3tq}$.

Solution: Invert the divisor, $\frac{1-tq}{3tq}$, and multiply $\frac{q^2+tq}{4t^2} \times \frac{3tq}{1-tq}$.

One factor of t cancels, and we are left with $\frac{3q^3+3tq^2}{4t-4t^2q}$.

Adding or subtracting fractions is also the same for algebraic fractions as for arithmetic fractions.

$$\frac{-3}{b(x-t)} + \frac{c}{4(x-t)}$$

Example 9. Add .

Solution: Use $4b(x - t)$ as common denominator, and unreduce each addend to give it that denominator. That entails multiplying the first fraction by $\frac{4}{4}$ and the second by $\frac{b}{b}$.

Then $\frac{-3}{b(x-t)} + \frac{c}{4(x-t)} = \frac{-3}{b(x-t)} \times \frac{4}{4} + \frac{c}{4(x-t)} \times \frac{b}{b}$

$$= \frac{-3 \times 4}{4b(x-t)} + \frac{bc}{4b(x-t)}$$

$$= \frac{bc - 12}{4b(x-t)}$$

Example 10. Subtract: $\frac{7}{2x-y} - \frac{5}{xy}$.

Solution: Change the sign of $\frac{5}{xy}$ to $\frac{-5}{xy}$ and add, using $xy(2x - y)$ as common denominator.

$$\frac{7}{2x-y} - \frac{5}{xy} = \frac{7}{2x-y} + \frac{-5}{xy} = \frac{7}{(2x-y)} \times \frac{xy}{xy} + \frac{-5}{xy} \times \frac{2x-y}{2x-y}$$

$$= \frac{7xy}{(2x-y)\,xy} + \frac{-5(2x-y)}{(2x-y)\,xy}$$

$$= \frac{7xy + 5y - 10x}{2x^2y - xy^2}$$

Simplifying rational expressions

Sometimes you may find yourself with an expression, built up from polynomials and algebraic fractions, but more complicated looking. You can apply the basic rules of algebra to simplify such **rational expressions**.

Example 11. Simplify $\frac{1-\frac{2}{y}}{1+\frac{2}{y}}$.

Solution: Write the numerator as a single fraction $1 - \frac{2}{y} = \frac{y}{y} - \frac{2}{y} = \frac{y-2}{y}$.

Write the denominator similarly as a single fraction $1 + \frac{2}{y} = \frac{y}{y} + \frac{2}{y} = \frac{y+2}{y}$.

Then $\frac{1-\frac{2}{y}}{1+\frac{2}{y}} = \frac{\frac{y-2}{y}}{\frac{y+2}{y}}$. Carry out this division by inverting the denominator and multiplying: $\frac{y-2}{y} \times \frac{y}{y+2}$. Cancel factors of y to get $\frac{y-2}{y+2}$.

Second solution: Multiply by 1 in the form $\frac{y}{y}$.

$$\frac{1-\frac{2}{y}}{1+\frac{2}{y}} = \frac{1-\frac{2}{y}}{1+\frac{2}{y}} \times \frac{y}{y} = \frac{\left(1-\frac{2}{y}\right) \times y}{\left(1+\frac{2}{y}\right) \times y} = \frac{1 \cdot y - \frac{2}{y} \cdot y}{1 \cdot y + \frac{2}{y} \cdot y} = \frac{y-2}{y+2}$$

Exercises 13.6

In questions 1-10, multiply. Express your answers in lowest terms.

1. $\frac{4ab}{3cd} \times \frac{2ac}{5bd}$
2. $\frac{x-1}{a} \times \frac{x+1}{b}$
3. $\frac{6ax}{5by} \times \frac{4cy}{3ad} \times \frac{5bd}{2cx}$
4. $\frac{y+1}{4} \times \frac{8m}{y^2-1}$
5. $\frac{2m^2}{5pq} \times \frac{3pr}{4am} \times \frac{5p^2}{6mr}$
6. $\frac{b+2}{b-1} \times \frac{b^2-1}{b^2-4}$
7. $\frac{c+d}{c-d} \times \frac{c^2-d^2}{(c+d)^2}$
8. $\frac{e^2-9}{e^2+4e} \times \frac{e^2-16}{e^2-3e}$
9. $\frac{3f^2}{f-2} \times \frac{f-1}{2f+6f^2}$
10. $\frac{g-x}{g^2+2gx} \times \frac{g^2-4x^2}{g^2-gx}$

In questions 11-14, supply the missing numerator or denominator.

11. $\frac{p-q}{q-r} = \frac{\qquad}{r-q}$
12. $\frac{m+n}{m-n-1} = \frac{-(m+n)}{\qquad}$
13. $\frac{(a-b)^2}{3a^3b} = \frac{\qquad}{6a^4b(X-B)}$
14. $\frac{a^3y(x+2)}{3(x-5)^2y} = \frac{\qquad}{18(x-5)^3y^2}$

In questions 15-20, divide. Express your answers in lowest terms.

15. $\frac{8a^2x}{5b^2y} \div \frac{4ax}{5by}$
16. $\frac{a-1}{b} \div \frac{1-a^2}{ac}$
17. $\frac{6x^2}{7by^2} \div \frac{3ax}{4by}$
18. $\frac{1-c}{c} \div \frac{b}{c+1}$
19. $\frac{x+2}{x+1} \div \frac{x^2-4}{x^2-1}$
20. $\frac{a+Y}{a-c} \div \frac{aY+a^2}{a^2-c^2}$

In questions, 21-28, add or subtract as indicated.

21. $\frac{a+x}{4} + \frac{a-x}{5}$
22. $\frac{e+f}{e} + \frac{e-x}{f}$
23. $\frac{j-y}{3} - \frac{j+y}{5}$
24. $\frac{k+x}{3} - \frac{k-x}{6} + \frac{x-k}{2}$
25. $\frac{1}{1+v} + \frac{1}{1-v}$
26. $\frac{5}{n+x} + \frac{3}{n-x} + \frac{4}{n+x}$
27. $\frac{t+8}{2} - \frac{t+9}{3} + \frac{t+4}{4}$
28. $\frac{y+z}{y} + \frac{y-z}{z} + \frac{y^2-z^2}{yz}$

Exploration: Rules with a curious property (questions 29-35)

29. a. Fill in the following table for the rule $\frac{3n+6}{2n-3}$.

Inputs n	0	1	2	3	4	5	-2	12	-9
Outputs $\frac{3n+6}{2n-3}$									

b. On the basis of the table, what do you find peculiar about the rule?

c. Predict the rule's output if the input is $\frac{18}{5}$.

d. Check your prediction by evaluating $\frac{3n+6}{2n-3}$ when $n = \frac{18}{5}$.

e. To see if the rule behaves this way for all numbers or only for certain ones, first apply it to n to get $\frac{3n+6}{2n-3}$ then use that output as the new input.

The result is $\dfrac{3\left(\frac{3n+6}{2n-3}\right)+6}{2\left(\frac{3n+6}{2n-3}\right)-3}$.

To simplify this, first work separately with the top and the bottom. The top is (fill in the blanks)

a form of 1

$$3\left(\frac{3n+6}{2n-3}\right)+6 = \frac{3}{1}\left(\frac{3n+6}{2n-3}\right)+\frac{6}{1}\left(\frac{\quad}{\quad}\right) = \frac{\quad}{2n-3}+\frac{\quad}{2n-3} = \frac{\quad}{2n-3}$$

The bottom is (fill in the blanks)

a form of 1

$$2\left(\frac{3n+6}{2n-3}\right)-3 = \frac{2}{1}\left(\frac{3n+6}{2n-3}\right)-\frac{3}{1}\left(\frac{\quad}{\quad}\right) = \frac{\quad}{2n-3}-\frac{\quad}{2n-3} = \frac{\quad}{2n-3}$$

Divide the top fraction by the bottom fraction by inverting the bottom fraction and multiplying. You should be able to cancel $2n - 3$, and what is left should simplify to n.

30. a. The rule $\dfrac{2n+6}{n-2}$ is similar to the first. Does it have a property like the first rule? To get some idea, complete this table:

Inputs n	-8	-3	0	1	3	4	7	12
Outputs $\dfrac{2n+6}{n-2}$								

b. Try the method from question 29 on this new rule, evaluating and simplifying $\dfrac{2\left(\dfrac{2n+6}{n-2}\right)+6}{\dfrac{2n+6}{n-2}-2}$.

To express the top as a single fraction, fill in these blanks:

$$2\left(\frac{2n+6}{n-2}\right)+6=\frac{2}{1}\left(\frac{2n+6}{n-2}\right)+\frac{6}{1}\overset{\text{a form of 1}}{\left(\frac{\quad}{\quad}\right)}=\frac{\quad}{n-2}+\frac{\quad}{n-2}=\frac{\quad}{n-2}$$

To express the bottom as a single fraction fill in these blanks:

$$\frac{2n+6}{n-2}-2=\frac{2n+6}{n-2}-\frac{2}{1}\overset{\text{a form of 1}}{\left(\frac{\quad}{\quad}\right)}=\frac{\quad}{n-2}-\frac{\quad}{n-2}=\frac{\quad}{n-2}$$

Now divide the top by the bottom, inverting and simplifying as in the first example, and see what you get.

31. Does the rule $\dfrac{2n+6}{5n-2}$ have the same curious property as the first rule? To find out, follow steps similar to those on earlier rules to simplify

$$\frac{2\left(\dfrac{2n+6}{5n-2}\right)+6}{5\left(\dfrac{2n+6}{5n-2}\right)-2}$$

32. Does $\frac{3n+4}{4n-2}$ have the same curious property as the first rule? To find out, simplify

$$\frac{3\left(\frac{3n+4}{4n-2}\right)+4}{4\left(\frac{3n+4}{4n-2}\right)-2}$$

33. By the same method as used above, find out which of these rules have the curious property of rule 1 and which do not. SHOW ALL WORK.

a. $\frac{n+3}{5n-1}$

b. $\frac{5n+1}{3n-5}$

c. $\frac{4n+2}{n-5}$

d. $\frac{6n-1}{n+6}$

e. $\frac{2n-1}{3n-1}$

f. $\frac{3n+5}{n-2}$

g. $\frac{2n+7}{7n-2}$

h. $\frac{3n-4}{6n-3}$

i. $\frac{n-1}{n+1}$

j. $\frac{2n+1}{n-2}$

34. Organize the results of your work so far, to see if you can find the rule for determining which rules do have the property and which do not.

a. List all the rules from questions 29-33 that **do** have the property.

b. List all the rules from questions 29-33 that **do not** have the property.

35. a. On the basis of your work above, state a criterion involving the numbers A, B, C, and D which is satisfied by those rules (and only those rules) of the form $\frac{Ax+B}{Cx+D}$ which have the curious property.

b. How could you test the criterion you stated in part a?

Vocabulary for Chapter 13

Binomial (13.1)
Coefficient (13.1)
Constant polynomial (13.1)
Degree of a polynomial (13.1)
Evaluate (13.1)
Factor (13.3)
Lowest terms (13.6)
Monomial (13.1)
Polynomial (13.1)
Rational expression (13.6)
Reducing (13.6)
Remainder theorem (13.3)
Substitute (13.1)
Term (13.1)
Trinomial (13.1)
Variable (13.1)

Self-Quiz for Chapter 13

1. Evaluate $t^3 - 2t^2 + 1$ at $t = -\frac{2}{3}$.
2. Add $3x^2 - 7x^5 + 24x - 7$ to $4x^3 + 9x^2 - 45x - 11$.
3. Subtract $5y^3 - 3y + 8$ from $10y^4 + 6y^3 - 3y^2 - 4y + 1$.
4. Multiply $5z^2 - 4z + 2$ by $z^4 - 5z + 1$.
5. Divide $8w^3 + 2$ by $2w - 1$.
6. Factor completely: a. $s^4 - 16t^4$ b. $6x^2 + 27x + 21$.
7. Find the LCM of $6x^2 + 12x + 6$ and $8x^2 + 12x + 4$.
8. Multiply $\dfrac{3w+1}{2w^2+4} \times \dfrac{w^4-4}{1+6w+9w^2}$.
9. Divide $\dfrac{4x^2+7x}{4x-3}$ by $\dfrac{x}{16x^2-24x+9}$.
10. Add $\dfrac{1}{yz} + \dfrac{2y}{z^2} + \dfrac{3}{y^3}$.
11. Subtract $\dfrac{1}{a^2+2ab+b^2} - \dfrac{1}{a^2-b^2}$.
12. Simplify $1 - \dfrac{1}{1-\dfrac{1}{x}}$.

Chapter Fourteen
Quadratic Equations

Here we will consider quadratic equations in one unknown and how to solve them. Quadratic equations are a step up in complexity from the linear equations we have studied so far. We will begin with a simple example and gradually extend the methods to where you can solve any quadratic equation in one unknown.

14.1 Basic Methods: an Exploration

A quadratic equation is a polynomial equation of degree two, such as $3x - 7 = 5x^2 + x + 1$. You can operate on quadratic equations, as on any equations, by adding, subtracting, multiplying, or dividing both sides of the equation equally, but doing so will not, by itself, solve a quadratic equation as example 1 shows.

Example 1. Solve $3x^2 - 5 = x^2 - 4x + 8$.
Solution (attempted):

Subtract x^2 from both sides	$2x^2 - 5 = -4x + 8$
Add $4x$ to both sides	$2x^2 - 5 + 4x = 8$
Add 5 to both sides	$2x^2 + 4x = 13$

Now all terms with x are on the left-hand side and the constant term is on the right-hand side. If this were a linear equation, that would nearly solve it, but this equation is evidently more difficult. Examples 2 and 3 introduce the new ideas needed to solve quadratic equations.

Example 2. Solve $x^2 = 16$.
Solution 1: **Square roots**

Start with	$x^2 = 16$
Take the square roots of both sides	$x = 4$ or -4

This is usually written $x = \pm 4$, which is read, "plus or minus 4."

You might wonder why you need to consider both square roots of 16, while not doing the same for x^2 to get x and $-x$. Must we not treat both sides of

an equation alike? If you consider both square roots on both sides, you get $\pm\, x = \pm\, 4$, which is really four equations:
$x = 4$, $x = -4$, $-x = 4$, and $-x = -4$. However, if you multiply the first of these equations by -1 on both sides, you get the fourth, and if you multiply the second by -1 on both sides, you get the third. Therefore, all roots of the original equation may be found by taking both square roots of one side of $x^2 = 16$ and equating them to just one square root of the other side.

Solution 2: Factoring

Subtract 16 from both sides of the equation $\qquad x^2 - 16 = 0$

Any difference of squares can be factored, as figure 14.1.1 shows.

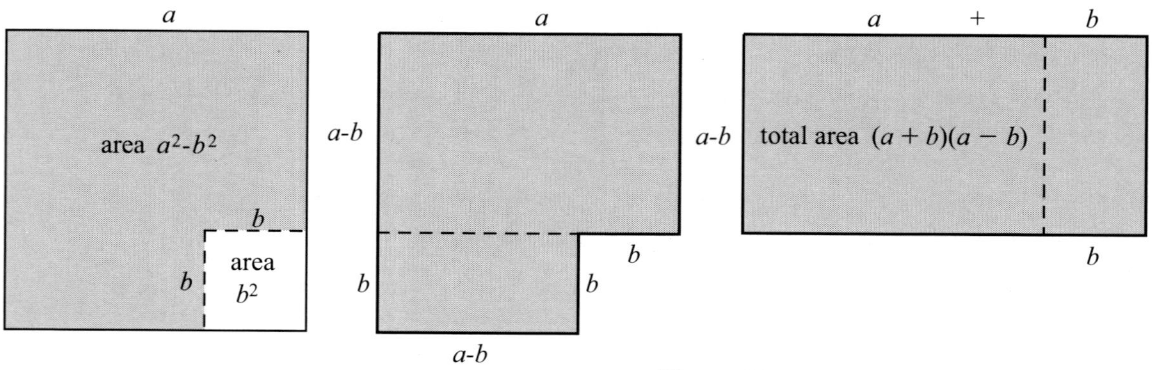

Figure 14.1.1

Then factor the left-hand side of the equation $\qquad (x - 4)(x + 4) = 0$

For a product to equal 0, at least one factor must be 0.

If $x + 4 = 0$, then $x = -4$.

If $x - 4 = 0$, then $x = 4$.

You can easily check that 4 and -4 both satisfy the original equation.

The basic approaches of factoring and taking square roots can be extended to solve any quadratic equation whatever in one unknown. The remainder of this section explores those extensions.

Example 2. Solve $y^2 + 2y + 1 = 25$.

Solution 1: Square roots

Rewrite the equation $\qquad (y + 1)^2 = 25$

Take square roots $\qquad y + 1 = \pm\, 5$

This last equation is two equations written in one.

With the plus sign on the right hand side $\qquad y + 1 = +\, 5$

With the minus sign on the right hand side $\qquad y + 1 = -\, 5$

Each of these may be solved by subtracting 1 to get, respectively, $y = 4$ and $y = -6$.

Solution 2: Factoring

From the original equation, subtract 25 $\qquad y^2 + 2y - 24 = 0$

Factor the left-hand side $\qquad (y - 4)(y + 6) = 0$

For a product to equal 0, at least one factor must be 0.

If $y - 4 = 0$, then $y = 4$. If $y + 6 = 0$, then $y = -6$.

Check the solution: Substitute both 4 and -6 into the original equation to verify that they are indeed roots. For $y = 4$ substitution yields $4^2 + 2(4) + 1$ on the left-hand side, which is 25, as it should be. The check for -6 is left for you to do.

Examples 1 and 2 above are somewhat special, because the square roots were easy to find. What if the right-hand side is not a square number?

Example 3. Solve $y^2 + 4y + 4 = 13$.
Solution: Factoring (attempted)
Subtract 13 $\qquad y^2 + 4y - 9 = 0$
The polynomial $y^2 + 4y - 9$ cannot be factored using polynomials with integer coefficients, so the solution by factoring cannot be completed at this stage.
Square roots
Rewrite the equation as $\qquad (y + 2)^2 = 13$
Take square roots $\qquad y + 2 = \pm\sqrt{13}$
This last equation is really two equations in one.
With the plus sign on the right hand side, it is $\qquad y + 2 = +\sqrt{13}$
With the minus sign on the right hand side, it is $\qquad y + 2 = -\sqrt{13}$
Each of these may be solved by subtracting 2 to get, respectively, $y = -2 + \sqrt{13}$ and $y = -2 - \sqrt{13}$.
Check the solution: You may substitute either of these numbers into the original equation to verify that it is indeed a root. For $y = -2 + \sqrt{13}$, substitution yields $(-2 + \sqrt{13})^2 + 4(-2 + \sqrt{13}) + 4$. Is this really 13?
Here the arithmetic is more complicated than before.
As a first step, calculate $(-2 + \sqrt{13})^2$, using $(a + b)^2 = a^2 + 2ab + b^2$ with $a = -2$ and $b = \sqrt{13}$.
This is $(-2)^2 + 2(-2)\sqrt{13} + (\sqrt{13})^2 = 4 - 4\sqrt{13} + 13 = 17 - 4\sqrt{13}$
Add to this $4(-2 + \sqrt{13}) + 4 = -8 + 4\sqrt{13} + 4 \qquad = -4 + 4\sqrt{13}$

The sum is 13, which shows that $-2 + \sqrt{13}$ is in fact a root.
The check for $y = -2 - \sqrt{13}$ is similar and is left for you to do.

Evidently, checking a solution to a quadratic equation can be as demanding as finding the solution in the first place.

Example 4. Solve $y^2 + 4y - 5 = 10$.
Solution: As in example 3, if you subtract 10, you get a polynomial on the left-hand side, which is prime in the realm of polynomials with integer coefficients. Therefore, try a solution by square roots.
The fact that 10 is not a square number is not a major obstacle, as seen in example 3, but here the left-hand side is not the square of a binomial. However, comparing examples 3 and 4 can lead to a bright idea. If you add 9 to both sides of the equation you get $y^2 + 4y + 4 = 19$, which may be solved like example 3. Rewrite the equation as $\qquad (y + 2)^2 = 19$
Take square roots $\qquad y + 2 = \pm\sqrt{19}$
This is really two equations, $\quad y + 2 = \sqrt{19}$ and $y + 2 = -\sqrt{19}$.
Subtract 2 from each equation to get $y = -2 + \sqrt{19}$ and $y = -2 - \sqrt{19}$.

The checks of these roots are left to you.

As the above examples show, solution by factoring depends on your being able to factor the quadratic polynomial in question. If you cannot factor that polynomial, this method of solution does not work. The method of square roots can be adapted, as in example 4, by adding a constant to both sides of the equation so as to make the side with the unknown a perfect square. Example 5 shows how to do this. The method is called **completing the square**.

Example 5. What number must be added to $x^2 + 6x + 1$ to make a perfect square binomial?
Solution: The problem amounts to finding a way to make a square from the tiles pictured, plus as many unit squares as you need.

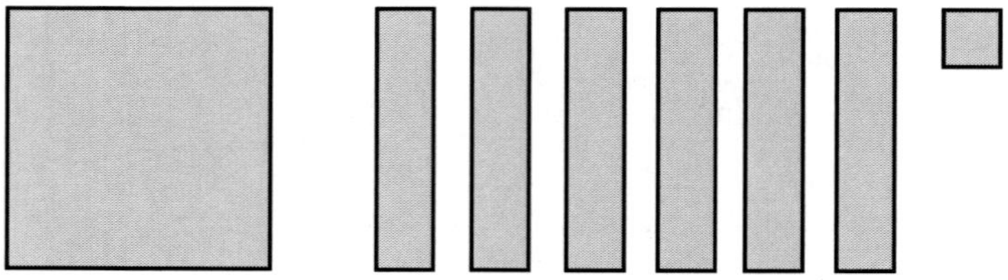

Put half the rods along one edge, and the other half along a perpendicular edge, like this:

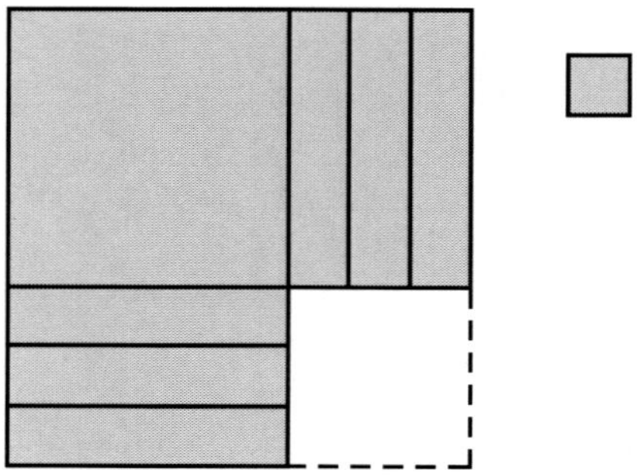

The result is an L-shape, which you can see as a large square, from which a smaller square is missing at the lower right-hand corner. This smaller square could be filled with nine unit squares, but we had only one to begin with. Therefore, add 8 to complete the square. Algebraically,

$$x^2 + 6x + 1 + 8 = x^2 + 6x + 9 = (x + 3)^2.$$

The thinking behind example 5 generalizes. If you have one flat and several rods, (figure 14.1.2) put half the rods on one side of the flat and half the rods on a perpendicular side of the flat. The result will be an incomplete square-one with a square corner missing. Add enough unit squares to fill in the missing corner and thus "complete the square." This is pictured in more detail below.

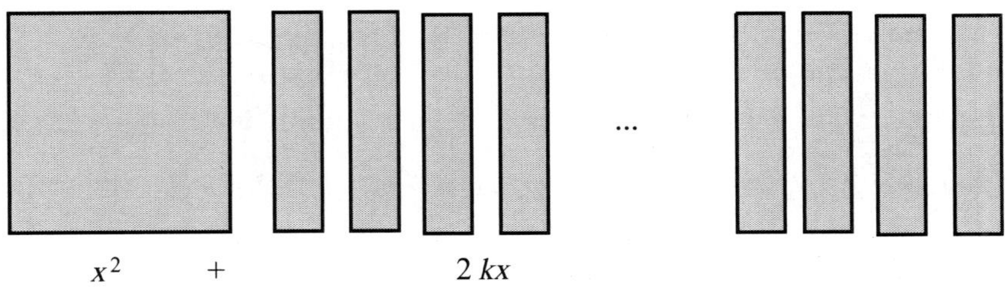

Figure 14.1.2

One flat and $2k$ rods may be rearranged to form parts of a square, which is $(x + k)$ on a side. To complete the square, add k^2 unit squares.

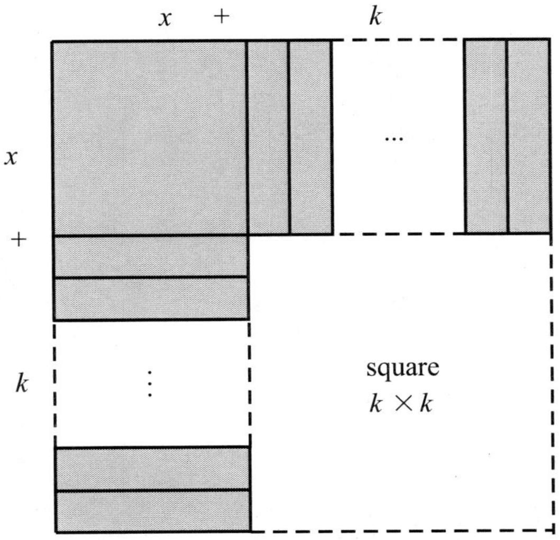

Figure 14.1.3

If the coefficient of x is odd, the same reasoning works, but it cannot be modeled with algebra tiles. For example, to complete the square with $x^2 + 3x + 4$, you must still use half the x-coefficient (3) on each side.

The square is then $\left(x + \frac{3}{2}\right)^2 = x^2 + 3x + \left(\frac{3}{2}\right)^2 = x^2 + 3x + \frac{9}{4}$,

so the number to add to complete the square is $\frac{9}{4} - 4 = \frac{-7}{4}$.

How to complete the square

> **A quadratic polynomial $x^2 + 2kx + h$ is a perfect square $(x + k)^2$ if and only if h is the square of half the coefficient of x — i.e., if and only if $h = k^2$. Therefore, to complete the square, add $k^2 - h$.**

Example 6 shows this reasoning works, even if the original number of rods is negative.

Example 6. Add a constant to $z^2 - 20z + 1$ to complete the square.
Solution: The coefficient of z is -20. Half of that, squared is $(-10)^2 = 100$, so $z^2 - 20z + 100$ is a perfect square, $(z - 10)^2$. Add 99 to $z^2 - 20z + 1$ to complete the square.
Solution 2: Think of figure 14.1.3 with $k = -20$. Put half of the rods (-10 of them) on each side of the flat to begin making a square $z - 10$ on a side. To complete this square will require 100 unit squares. Since $(z - 10)^2 = z^2 - 20z + 100$, add 99 to $z^2 - 20z + 1$ to complete the square.

Example 7. Solve $w^2 - 4w + 10 = 3$.
Solution: Complete the square.
Half of the w coefficient (-4) is -2,
so the square we seek is $(w - 2)^2 = w^2 - 4w + 4.$
Then, in the original equation,
subtract 6 to get $w^2 - 4w + 4 = -3$
or, in more convenient form, $(w - 2)^2 = -3$
Take square roots $w - 2 = \pm\sqrt{-3}$
But no real number is the square root of -3!
Therefore, no real number w satisfies the original equation.

Evidently, some quadratic equations have two real roots, but others have none, and we will see some quadratic equations that have just one real root.

Exercises 14.1

1. Solve each equation two ways, by factoring and by taking square roots:
   a. $t^2 = 9$
   b. $w^2 = 25$
   c. $y^2 = 49$
   d. $x^2 = a^2$

2. Solve by taking square roots. Do NOT multiply out the binomials.
   a. $(t - 1)^2 = 25$
   b. $(u + 2)^2 = 36$
   c. $(r + 5)^2 = 81$
   d. $\left(v + \frac{1}{2}\right)^2 = 121$
   e. $(x - 7)^2 = c^2$
   f. $(z - 6)^2 = 49$
   g. $\left(s - \frac{1}{9}\right)^2 = 25$

3. Use algebra tiles to picture completing the squares:
   a. $x^2 - 2x$
   b. $y^2 + 8y$
   c. $z^2 - 6z + 3$
   d. $t^2 + 10t + 11$
   e. $u^2 - 12u - 40$
   f. $v^2 + 6v + 1$

4. Solve by the method of square roots, completing the square where necessary to do so.
   a. $x^2 - 2x + 1 = 144$
   b. $y^2 - 2y - 3 = 140$
   c. $z^2 + 4z + 10 = 175$
   d. $t^2 - 6t + 11 = 83$
   e. $u^2 + 8u - 40 = -55$
   f. $v^2 + 20v + 1 = -50$

5. Solve by factoring:
   a. $x^2 - 8x - 153 = 0$
   b. $y^2 - 7y + 10 = 0$
   c. $z^2 + 8z + 7 = -5$
   d. $t^2 - 12t + 5 = -27$
   e. $u^2 + 9u - 1 = -15$
   f. $v^2 - 8v + 25 = 13$

6. A common error: To solve $x^2 - 6x + 5 = 1$, a student factors the left-hand side as $(x - 5)(x - 1)$ and then says: If $x - 5 = 1$, then $x = 6$. If $x - 1 = 1$, then $x = 2$.
   a. Show that 6 and 2 are not roots of $x^2 - 6x + 5 = 1$.
   b. Explain what the student has done wrong.
   c. Solve the equation correctly by completing the square.

7. In example 3 above, it is asserted that $y^2 + 4y - 9$ is a prime polynomial in the realm of polynomials with integer coefficients.
   a. Verify that this is the case.
   b. Multiply out $(y + 2 - \sqrt{13})(y + 2 + \sqrt{13})$. Comment on your result.

In part a of questions 8-11, square as indicated. Then use the results to solve the equation in part b.

8. a. $\left(x-\frac{1}{2}\right)^2$

b. $x^2 - x + 2 = 5$

9. a. $\left(x+\frac{3}{2}\right)^2$

b. $x^2 + 3x + 1 = 2$

10. a. $\left(x-\frac{5}{2}\right)^2$

b. $x^2 - 5x + 4 = -1$

11. a. $\left(x+\frac{9}{2}\right)^2$

b. $x^2 + 9x + 20 = -1$

14.2 More about Completing the Square

Now we will extend the method of completing the square to solve any quadratic equation whatever.

When a quadratic equation arises in a problem, it is not usually in the convenient form $x^2 + 2kx + h = 0$. To solve it, first move all terms to the left-hand side, then collect terms of like degree to get an equation of form $ax^2 + bx + c = 0$, where $a \neq 0$. (If $a = 0$, the equation has no second degree term and is not really a quadratic equation). Then divide by a to get an equation of the form $x^2 + 2kx + h = 0$, with x^2 coefficent 1. Here we write $2k$ for the coefficient of x, but we do not assume that k is an integer.

Thus, in the equation $x^2 + 7x + 12 = 0$, we let $7 = 2k$, so that $k = 3\frac{1}{2}$.

Example 1. Put the equation $5x + 2 - x(1 - 3x) = x^2 - 4(x - 7)$ into the form $x^2 + 2kx + h = 0$.

Solution: Use the distributive property to rewrite the equation as $\quad 5x + 2 - x + 3x^2 = x^2 - 4x + 28$

Now subtract $x^2 - 4x + 28$ to get

$$5x + 2 - x + 3x^2 - x^2 + 4x - 28 = 0$$

Use the commutative property of addition to rearrange the order of the terms, writing this as

$$3x^2 - x^2 + 5x - x + 4x + 2 - 28 = 0$$

Use the distributive property to write this

$$(3 - 1)x^2 + (5 - 1 + 4)x + 2 - 28 = 0$$

or

$$2x^2 + 8x - 26 = 0$$

This is the form $ax^2 + bx + c = 0$ with $a = 2$, $b = 8$, and $c = -26$.

Divide by 2, the coefficient of x^2, to put the equation in the desired form

$$x^2 + 4x - 13 = 0$$

From here on, we will assume that any quadratic equation can be rewritten in the manner of example 1 to put it in the form $ax^2 + bx + c = 0$, with $a \neq 0$. Equivalently, we assume that any quadratic equation can be rewritten in the form $x^2 + 2kx + h = 0$.

Now you can solve any quadratic equation whatever by the method of square roots.

Example 2. Solve $x^2 + 5x + 4 = 0$.
Solution: Compare the identity $x^2 + 2kx + k^2 = (x + k)^2$ with the expression on the left-hand side of the equation.

Evidently, $2k = 5$, and $k = \frac{5}{2}$. Then $k^2 = \left(\frac{5}{2}\right)^2 = \frac{25}{4} = 6\frac{1}{4}$.

The expression on the left would be a perfect square, if the constant term were $6\frac{1}{4}$. Therefore, add $2\frac{1}{4}$ to both sides of the equation, to get

$$x^2 + 5x + 6\frac{1}{4} = 2\frac{1}{4}$$

Rewrite this as

$$\left(x + \frac{5}{2}\right)^2 = \frac{9}{4}$$

Take square roots

$$x + \frac{5}{2} = \pm\frac{3}{2}$$

This is really two equations. Subtract $\frac{5}{2}$ to solve them.

With the positive root, this yields

$$x = -\frac{5}{2} + \frac{3}{2} = \frac{-2}{2} = -1$$

With the negative root, this yields

$$x = -\frac{5}{2} - \frac{3}{2} = \frac{-8}{2} = -4$$

The checks are left for you to do.

As you can see, the method of square roots, as outlined above, sometimes involves polynomials with coefficients that are not integers, even when the original equation has integer coefficients.

Example 3. Solve $3x^2 + 5x - 2 = 0$.
Solution: Divide by 3, to make the x^2 coefficient 1 $x^2 + \frac{5}{3}x - \frac{2}{3} = 0$

Add $\frac{2}{3}$ to both sides $x^2 + \frac{5}{3}x = \frac{2}{3}$

Complete the square on the left-hand side with $\frac{5}{3} = 2k$.

Add $\left(\frac{1}{2} \times \frac{5}{3}\right)^2 = \left(\frac{5}{6}\right)^2$. $x^2 + \frac{5}{3}x + \left(\frac{5}{6}\right)^2 = \frac{2}{3} + \left(\frac{5}{6}\right)^2$

Since $\frac{2}{3} + \frac{25}{36} = \frac{49}{36} = \left(\frac{7}{6}\right)^2$, we may rewrite the equation as $\left(x + \frac{5}{6}\right)^2 = \left(\frac{7}{6}\right)^2$

Take square roots $x + \frac{5}{6} = \pm\frac{7}{6}$

Subtract $\frac{5}{6}$ $x = -\frac{5}{6} \pm \frac{7}{6}$

Since $-\frac{5}{6} + \frac{7}{6} = \frac{2}{6} = \frac{1}{3}$ and $-\frac{5}{6} - \frac{7}{6} = \frac{-12}{6} = -2$,

the roots are $\frac{1}{3}$ and -2.

The checks are left for you to do.

Example 4. Solve $5x^2 - 2x + 1 = 0$.
Solution: Divide by 5 to make the leading coefficient 1 $x^2 - \frac{2}{5}x + \frac{1}{5} = 0$

Subtract $\frac{1}{5}$ $x^2 - \frac{2}{5}x = -\frac{1}{5}$

Complete the square on the left-hand side with $-\frac{2}{5} = 2k$

Add $k^2 = \frac{1}{25}$ to both sides $x^2 - \frac{2}{5}x + \frac{1}{25} = -\frac{1}{5} + \frac{1}{25}$

Rewrite this $\left(x - \frac{1}{5}\right)^2 = \frac{-4}{25}$

Next, we should take square roots, which yields $x - \frac{1}{5} = \pm\sqrt{\frac{-4}{25}}$

Solving, $x = \frac{1}{5} \pm \sqrt{\frac{-4}{25}}$.

These are not real numbers, since no real number has a negative square. Therefore, no real numbers satisfy $5x^2 - 2x + 1 = 0$.

Complex numbers Evidently, the system of real numbers is inadequate for the problem at hand. When the system of counting numbers could not handle certain subtraction problems such as $3 - 8\ (= -5)$, it was extended to include negative integers. The system of integers does not have numbers to express certain quotients, such as $3 \div 4\ (= 3/4)$, so the system of integers was further expanded to the system of rational numbers, which includes all numbers which can be expressed as fractions of integers, positive, negative, and 0. The system of rational numbers, adequate for the mathematics of daily life, was enlarged once again to include irrational numbers such as $\sqrt{2}$, resulting in the system of **real numbers**. Now the system of real numbers must be expanded to include square roots of negative numbers. The result of this expansion, known today as the system of **complex numbers,** consists of all expressions of the form $a + b\sqrt{-1}$, where a and b are real numbers. Complex numbers with $b = 0$ are real numbers, while those with $a = 0$ are called **imaginary numbers**. The adjectives "real" and "imaginary" are still used, but they are misleading. Numbers are abstract ideas, like truth and beauty. Whether or not abstract ideas are real is a philosophical question that goes beyond this book, but square roots of negative numbers are neither more real nor less real than other numbers. Square roots of negative numbers are among the imaginary numbers, since, for example, $\sqrt{-9} = 3\sqrt{-1}$

If you find square roots of negative numbers strange, you are not alone. Rafael Bombelli (1526–1572), a nobleman and civil engineer from Bologna, who pioneered the use of complex numbers in his *Algebra* of 1572, stated that the idea was a "wild thought" and did not pursue it very far. For nearly two centuries after Bombelli, the very idea of square roots of negative numbers was viewed as absurd and rejected by leading mathematicians. The enlarged number system of complex numbers gained grudging acceptance, legitimized especially by Leonhard Euler, who used them to obtain remarkable new discoveries, but even as late as the 19th century in England, many mathematicians rejected all negative numbers and complex numbers as well.

The next example shows that completing the square has uses other than solving quadratic equations.

Example 5. The equation $x^2 + 2x + y^2 - 4y = 44$ represents a circle in the coordinate plane. Find its center and radius.
Solution: Complete squares in x and y

$$x^2 + 2x + 1 + y^2 - 4y + 4 = 44 + 1 + 4.$$

Rewrite the left-hand side

$$(x + 1)^2 + (y - 2)^2 = 49$$

This circle has its center at the point (-1,2), and its radius is 7, as discussed in section 12.3.

The quadratic formula Rather than complete the square each time you solve a quadratic equation, you can do it once and for all on the general equation $ax^2 + bx + c = 0\ (a \neq 0)$ to derive a formula to solve all quadratic equations. This derivation is the subject of question 13 below.

The Quadratic Formula

If $ax^2 + bx + c = 0$ and $a \neq 0$, then $x = \dfrac{-b \pm \sqrt{b^2 - 4ac}}{2a}$.

The quadratic formula reduces the problem of solving a quadratic equation to numerical calculation, and it also brings out some relationships between the roots of quadratic equations and the factors of quadratic polynomials, which will be considered in the following section.

Example 6. Solve $x^2 + 11x + 28 = 0$.
Solution: Use the quadratic formula with $a = 1$, $b = 11$, and $c = 28$.

Then $x = \dfrac{-11 \pm \sqrt{11^2 - 4 \times 1 \times 28}}{2 \times 1} = \dfrac{-11 \pm \sqrt{121 - 112}}{2} = \dfrac{-11 \pm \sqrt{9}}{2} = \dfrac{-11 \pm 3}{2}$

This is really two equations, $x = \dfrac{-11-3}{2}$, which is -7 and $x = \dfrac{-11+3}{2}$, which is -4.

Checking the solutions is the same as before.

A note on teaching quadratic equations

Students and teachers sometimes wonder why they should study factoring and completing the square, when they could simply memorize the quadratic formula and use it to solve any quadratic equation. If the goal of instruction is limited to solving quadratic equations their point is well taken. However, factoring and completing the square are basic techniques, used in many parts of mathematics. The study of quadratic equations is an excellent opportunity to introduce these techniques.

Using the Quadratic Formula to Factor

In previous examples, we have factored quadratic polynomials as a way to solve quadratic equations. However, you can also reverse this process to factor $ax^2 + bx + c$ by first using the quadratic formula to find the roots of the equation $ax^2 + bx + c = 0$.

Roots of $ax^2 + bx + c = 0$ and factors of $ax^2 + bx + c$ ($a \neq 0$)

1. **If r_1 and r_2 are the roots of the equation $ax^2 + bx + c = 0$, then $x - r_1$ and $x - r_2$ are factors of $ax^2 + bx + c$.**
2. **If $x - r_1$ and $x - r_2$ are factors of $ax^2 + bx + c$, then r_1 and r_2 are the roots of the equation $ax^2 + bx + c = 0$.**

Example 7. Factor $x^2 - 17x + 11$.
Solution: This polynomial is not factorable by the methods of chapter 13, and indeed $x^2 - 17x + 11$ is a prime in the set of polynomials with integer coefficients. Instead, solve the equation $x^2 - 17x + 11 = 0$ by means of the quadratic formula. The roots turn out to be $\dfrac{17 \pm 7\sqrt{5}}{2}$.

Then $x - \frac{17+7\sqrt{5}}{2}$ and $x - \frac{17-7\sqrt{5}}{2}$ are factors of $x^2 - 17x + 11$.

That is, $x^2 - 17x + 11 = \left(x - \frac{17+7\sqrt{5}}{2}\right)\left(x - \frac{17-7\sqrt{5}}{2}\right)$

Without solving the equation these factors would be hard to discover.

Example 8. Factor $3x^2 + 2x - 5$.
Solution: By the quadratic formula, the equation $3x^2 + 2x - 5 = 0$ has roots

$$\frac{-2 \pm \sqrt{2^2 - 4(3)(-5)}}{2(3)} = \frac{-2 \pm \sqrt{4 + 4(3)(5)}}{2(3)} = \frac{-2 \pm \sqrt{64}}{6} = \frac{-2 \pm 8}{6} = \frac{-5}{3}$$

and 1. Then $x + \frac{5}{3}$ and $x - 1$ are both factors of $3x^2 + 2x - 5$.

Indeed, $3x^2 + 2x - 5 = 3\left(x + \frac{5}{3}\right)(x - 1) = (3x + 5)(x - 1)$.

Exercises 14.2

In problems 1-6, solve each equation and check each solution.

1. $z^2 + 10z = 39$

2. $2t^2 + 3t + 3 = 3t^2 + 4t + 1$

3. $(x - 1)^2 + 3(x - 1) = 0$

4. $(z - 2)(z - 7) = 6z$

5. $6x^2 - 7x - 7 = -4x^2 + 4x - 1$

6. $10y^2 - 7y - 12 = 0$

7. Bombelli's discovery of complex numbers came out of a question that led him to wonder whether or not
$$4 = \sqrt[3]{2 + \sqrt{-121}} + \sqrt[3]{2 - \sqrt{-121}}\ .$$
   a. Cube $2 + \sqrt{-1}$.
   b. Cube $2 - \sqrt{-1}$.
   c. Use parts *a* and *b* to calculate $\sqrt[3]{2 + \sqrt{-121}}$ and $\sqrt[3]{2 - \sqrt{-121}}$.
   d. Use parts *a*, *b*, and *c* to verify that $4 = \sqrt[3]{2 + \sqrt{-121}} + \sqrt[3]{2 - \sqrt{-121}}$.

In problems 8-11, find the center and the radius of the circles.

8. $x^2 + y^2 - 4y = 1$

9. $x^2 + 3x + y^2 - 7y = 0$

10. $y^2 - 8y + x^2 - 12x + 20 = 0$

11. $x^2 + y^2 + 10x = 41$

12. In Exercises 12.5 the sequence of continued fractions

$$1+\frac{1}{2} \quad 1+\cfrac{1}{2+\cfrac{1}{2}} \quad 1+\cfrac{1}{2+\cfrac{1}{2+\cfrac{1}{2}}} \quad 1,+\cfrac{1}{2+\cfrac{1}{2+\cfrac{1}{2+\cfrac{1}{2}}}} \quad ,\ldots$$

was seen as getting closer and closer to $\sqrt{2}$. Show that to be the case as follows. Suppose the infinite continued fraction is equal to a number x. Then by imagining the continued fraction as unending, we have

$$x = 1+\cfrac{1}{2+\cfrac{1}{2+\cfrac{1}{2+\cfrac{1}{2}\ddots}}} = 1+\frac{1}{1+x}$$

a. Equate the x on the left to the expression at the far right and solve.

b. Is the solution what you expected? What about the second root?

In problems 13-16, use the quadratic formula to factor the given polynomials.

13. $t^2 - 15t + 9$

14. $u^2 + 16u + 3$

15. $v^2 - 13v + 11$

16. $w^2 - 18w + 4$

17. Omar Khayyam,[1] around 1100 A.D. wrote (not, of course, in modern notation) that

$$x = \sqrt{\frac{s^2}{4} + t} - \frac{1}{2}s$$

is a root of $x^2 + sx = t$. Substitute Khayyam's alleged root in the equation to see if his statement is correct.

18. A rhyme by Lewis Carroll[2]

    Yet what are such gaieties to me,
    Whose thoughts are full of indices and surds?
    $x^2 + 7x + 53$
    $= -11/3$

    Solve Lewis Carroll's rhyme equation. A **surd** is a number expressed as a sum involving roots.

19. Use the quadratic formula to show that
    a. The sum of the roots of $ax^2 + bx + c = 0$ is $-b/a$.
    b. The product of the roots of $ax^2 + bx + c = 0$ is c/a.

20. **Exploration: Deriving the quadratic formula.**
    Begin with the general quadratic equation $ax^2 + bx + c = 0$ $(a \neq 0)$.
    a. Subtract c.
    b. Divide by a.
    c. Complete the square by adding $\left(\frac{b}{2a}\right)^2$.
    d. Rewrite the left-hand side of the equation as the square of a binomial. Carry out the addition on the right hand side, using the common denominator $4a^2$.
    e. Take square roots.
    f. Subtract $\frac{b}{2a}$ from both sides.
    g. Add the fractions on the right-hand side.

---

1 Umar Ibn Ibrahim al-Khayyami (1048-1131), known in the west as Omar Khayyam, spent many years working on calendar reform at the observatory in Isfahan, in his native Persia. Widely known today for *The Rubaiyat*, a collection of poems, he was also an outstanding astronomer and mathematician. Although Omar's thinking often differed from the orthodoxy of his day, he was usually able to maintain the support of powerful patrons.

2 Lewis Carroll is the pen name of Charles L. Dodgeson, writer of *Alice's Adventures in Wonderland* and *Through the Looking Glass*. Dodgeson was well known as a mathematician, with a particular interest in mathematical logic.

14.3 Problems Involving Quadratic Equations

Here is a sampler of word problems that lead to quadratic equations.

Example 1. A ship's captain tells a passenger to pay for fare the number of pesos which, when added to its square, yields 1,260. What is the fare? (This is from the first mathematics book published in the New World, the *Sumario Compendioso* by Juan Diez, Mexico City, 1556.)
Solution: Begin by guessing, filling your answers into a table like this:

Guess fare (in pesos)	50	40	30	p
Square of guess	$50^2=2{,}500$	$40^2=1{,}600$	$30^2=900$	p^2
Fare plus square	$50^2+50=2{,}550$	$40^2+40=1{,}640$	$30^2+30=930$	p^2+p
Check: Sum=1260?	$>1{,}260$	$>1{,}260$	$<1{,}260$	Equate: $p^2+p=1{,}260$

From the table $p^2 + p = 1{,}260$
Solve by the formula or by factoring $p^2 + p - 1{,}260$ as $(p - 35)(p + 36)$ to get the roots -36 and 35. For this problem, -36 is not a reasonable solution, for it would require the captain to pay the passenger 36 pesos! However, a fare of 35 pesos appears to be a reasonable answer.

In general, if a problem leads you to a quadratic equation, take extra care to interpret the roots. As in example 1, a number might satisfy the equation and still not fit the conditions of the problem that led to the equation.

Example 2. A furniture dealer paid $1,680 for a shipment of identical tables. She priced them each $30 more than she had paid for them, and after a week she had sold all but two of them for a total of $1,800. How many tables did she buy to begin with?
Solution: Let t be the number of tables in the shipment. Since the total bill was $1,680, the cost per table paid by the dealer was $\frac{1{,}680}{t}$ dollars.

The dealer sold the tables for $30 more per table than they had cost, or for $\frac{1{,}680}{t} + 30$ dollars each. The dealer's income, in dollars, from the sale of all but 2 of the tables was then $\left(\frac{1{,}680}{t} + 30\right)(t - 2)$.

We are told this income was $1,800.

Therefore,

$$\left(\frac{1{,}680}{t} + 30\right)(t - 2) = 1{,}800.$$

To solve this equation, divide by 10 and multiply by t to get

$$(168 + 3t)(t - 2) = 180t$$

Multiply out the left-hand side: $3t^2 + 162t - 336 = 180t$
Subtract $180t$ $3t^2 - 18t - 336 = 0$
Divide by 3 $t^2 - 6t - 112 = 0$
Factor $(t - 14)(t + 8) = 0$
Then $t = 14$ or $t = -8$
A negative root is meaningless in this problem.
The dealer bought 14 tables.

Example 3. Fourteen years ago Bob's age was the square of a certain number. The square of three more than that number was his father's age then. Bob is now half his father's age. How old are they each now?
Solution: As a start, organize the information in a table like this:

	Bob's Age	Father's Age
14 years ago		
Now		

How shall we fill in the table? Rereading the problem, we see that fourteen years ago "Bob's age was the square a certain number." If y is that number, then Bob's age 14 years ago was y^2. Then the age of Bob's father 14 years ago was $(y + 3)^2$. Adding 14 to these ages gives their ages now.

	Bob's Age	Father's Age
14 years ago	y^2	$(y + 3)^2$
Now	$y^2 + 14$	$(y + 3)^2 + 14$

We have not yet used the information in the third sentence of the problem, "Bob is now half his father's age."
Use this sentence to write $2(y^2 + 14) = (y + 3)^2 + 14$
Carry out the indicated multiplications and additions

$$2y^2 + 28 = y^2 + 6y + 23$$

Subtract everything on the right-hand side $y^2 - 6y + 5 = 0$
Factor $(y - 5)(y - 1) = 0$
If $y - 5 = 0$, then $y = 5$. If $y - 1 = 0$, then $y = 1$.

Interpretation: If $y = 5$, the ages of Bob and his father fourteen years ago were 25 and 64, and their current ages are 39 and 78. The root $y = 1$ yields ages of 1 and 16 fourteen years ago and present ages of 15 and 30.
Comment: The statement of the problem seems to imply that the ages are whole numbers of years. If so, you can solve the problem by guessing and checking. If Bob's age is a perfect square and within a believable human life span, it must be one of the numbers 1, 4, 9, 16, 25, 36, 49, 64, 81, and 100. Checking these numbers will reveal all the solutions.

Example 4. A rectangular sheet of metal is 20 cm longer than it is wide. A square 4 cm on edge is cut from each corner and the remainder folded to form a rectangular bin with no top. The volume of the bin is 3,444 cubic centimeters (cc). Find the original length and width of the piece.
Solution: We seek both the original length and width, but if we can find one of those numbers, we can get the other as well. To visualize the question, make a sketch like figure 14.3.1.

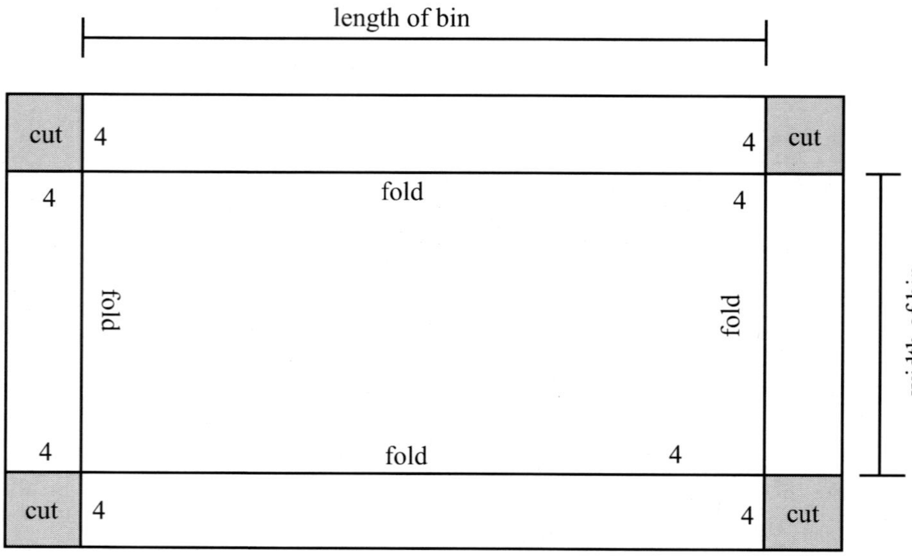

Figure 14.3.1

The volume of this rectangular solid, 3,444 cc, is the product of its length, width, and height. What are the dimensions of the bin? Clearly, the height will be 4 cm. The length and width of the bin are each 8 cm less than those of the original sheet of metal.

Then: (4)(original length − 8)(original width − 8) = 3,444.

Now if w = original width in centimeters,
then $w + 20$ is the original length, and $4(w + 20 - 8)(w - 8) = 3{,}444$
Divide by 4 $(w + 12)(w - 8) = 861$
Multiply out and simplify $w^2 + 4w - 957 = 0$
Complete the square $w^2 + 4w + 4 = 961$
Rewrite $(w + 2)^2 = 31^2$
Take square roots $w + 2 = \pm 31$
If $w + 2 = 31$, then $w = 29$. If $w + 2 = -31$, then $w = -33$.
Discard the negative root, because a width must be positive, but keep $w = 29$. If the sheet was 29 cm wide, it was 49 cm long. Then the volume of the bin is $21 \times 41 \times 4$ cubic centimeters, which is indeed 3,444 cm^3.

Example 5. On a round trip of 400 km, a pilot wants to determine the airspeed of a light plane and flies the entire distance at the same airspeed. On the outbound leg of 200 km a steady headwind slows the plane by 30 km/hr, but on the return trip the wind, now at his tail, speeds up the plane by 30 km/hr. The entire trip takes an hour and a half. Find the plane's airspeed.

Solution: If s is the plane's airspeed (the speed with respect to the moving air) in km/h, then on the outbound trip, its ground speed (the speed with respect to the ground) was $s - 30$, and on the return trip the ground speed was $s + 30$. Then the outbound travel time, in hours, was $\frac{200}{s-30}$, and the time needed for the inbound leg was $\frac{200}{s+30}$ hours.

Since the entire flight took an hour and a half,

$$\frac{200}{s-30} + \frac{200}{s+30} = \frac{3}{2}$$

Multiply by $2(s - 30)(s + 30)$ to clear out the fractions.

$$400(s + 30) + 400(s - 30) = 3(s + 30)(s - 30)$$

Simplify

$$3s^2 - 800s - 2{,}700 = 0$$

By the quadratic formula

$$s = \frac{800 \pm \sqrt{800^2 - 4 \times 3(-2{,}700)}}{6}$$

Do the arithmetic

$$s = 270 \text{ or } \frac{-10}{3}.$$

Check that 270 and $\frac{-10}{3}$ actually satisfy the original equation.

Finally, interpret the answer; 270 km/hr is a reasonable answer in this context, but $\frac{-10}{3}$ is not.

Example 6. Find the points, if any, where the line $-x + 2y = 5$ and the circle $x^2 + y^2 = 25$ intersect.

Solution: Write the equation of the line as $x = 2y - 5$. Then substitute this expression for x into the equation for the circle and solve the resulting quadratic equation.

$$x = 2y - 5$$
$$x^2 = (2y - 5)^2$$
$$x^2 = 4y^2 - 20y + 25$$

Substitute for x^2

$$x^2 + y^2 = 25$$
$$4y^2 - 20y + 25 + y^2 = 25$$

Simplify

$$5y^2 - 20y = 0$$
$$y^2 - 4y = 0$$
$$y(y - 4) = 0$$
$$y = 0 \text{ or } y = 4$$

If $y = 0$, then $x = -5$. If $y = 4$, then $x = 3$. Then the line and circle meet at (-5,0) and (3,4), as shown in figure 14.3.2.

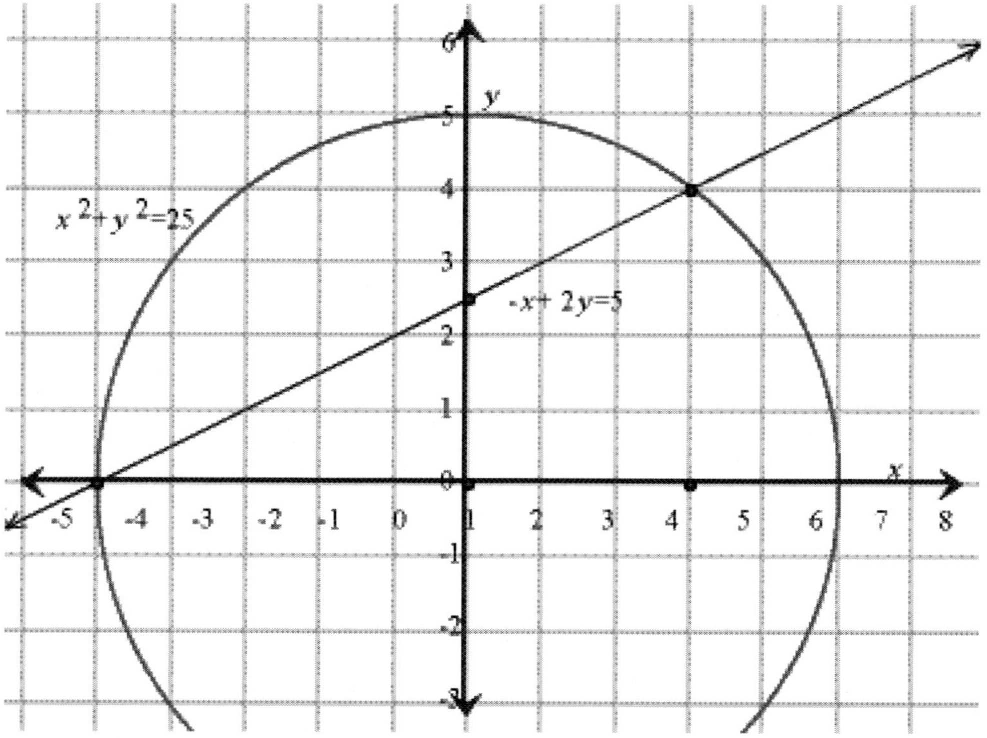

Figure 14.3.2

Exercises 14.3

1. The product of a number and six more than the number is 891. Find the number.

2. If a number is added to twice its square, the result is 595. What is the number?

3. The sum of two numbers is 53, and their product is 612. What are the numbers?

4. Divide 4.5 by a certain number, and the result is the same as subtracting that number from 4.5. What is the number?

5. A rectangle is 4/3 as long as it is wide. A second rectangle is 4 meters longer and 2 meters wider than the first, and its area is twice that of the first. Find the dimensions of the rectangles.

6. On all sides of a rectangular flower bed is a border of uniform width. The flower bed is 3 meters by 4 meters, and the area of the border is ten times the area of the bed. How wide is the border?

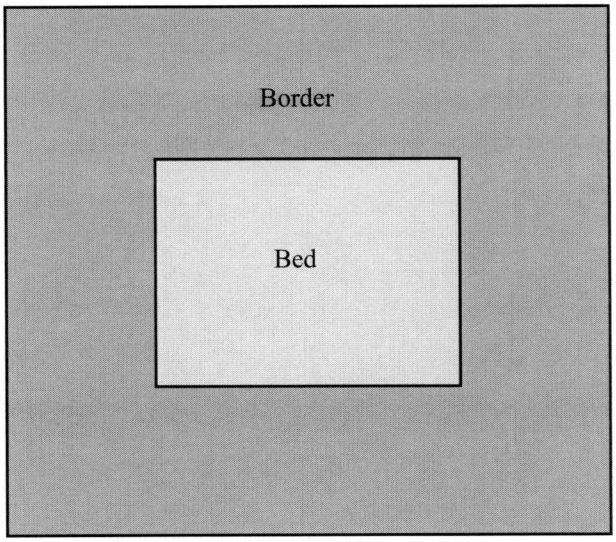

7. A man spends $4 to buy lemons. If the price of each lemon were 4 cents less, he could buy 50 more for the same amount of money. What was the price of each lemon?

8. From 100 kg of wheat, some was removed and an equal weight of rye was mixed in. Then the same amount of the mixture was removed, and it too was replaced by rye. At that point there were 64 kg wheat and 36 kg rye. How much rye was in the first mixture?

9. A rectangular sheet of steel is twice as long as it is wide. From each corner a 2 cm-square is cut. Then the sides are folded to make a box (without top) which holds 60 cubic centimeters. What were the original dimensions of the steel sheet?

10. To raise $1,000, the members of the Nice People Club pledged to chip in equal amounts. However, five of the Nice People did not contribute, so to make the $1,000 the rest of the members had to pay an extra $10 each. How much did each paying member contribute?

11. A rancher bought a certain number of sheep of equal value for $700. She lost five to disease but sold the others for $12 per head more than she paid for them. Her profit was $730. How many sheep did she buy?

12. In the figure below AB and AC are tangents to a circle, B and C are the points of contact, and AB and AC are perpendicular. Point E lies on segment AB, 9 units from A, and EP is perpendicular to AB and is 8 units long. Find the radius of the circle.

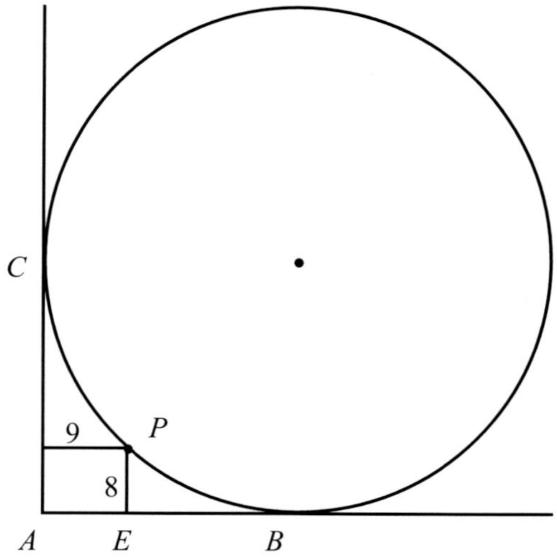

13. Juan and his assistant can paint a house in 36 hours. By himself, Juan could do the job in 30 hours less than his assistant would need to do the job alone. How long would Juan need?

14. In an electrical system, it is necessary to wire each of a set of points directly to every other point in the system. If 300 wires are used, how many points are in the system?

15. A plane flew 150 kilometers with a 30 km/hour tailwind, then turned into the wind and flew another 65 km. If the wind and the plane's airspeed were constant and the entire trip took 1/2 hour, what was the plane's airspeed? (Neglect the time needed to turn the plane around.)

16. An investment account grew by a substantial percentage in its first year. It grew by the same percentage the second year, and the original $2,000 investment was then worth $2,784.80. By what percent did the account grow each year?

17. An automobile cooling system was mistakenly filled to its 7.2-liter capacity with pure antifreeze. An amount was drained off and replaced with pure water, but the mixture still had too much antifreeze, so the same amount of the mixture was drained off and replaced with pure water. The mixture then contained 5 liters of antifreeze. How much liquid was drained each time?

18. Sinbad's boat accumulated a full load of bilge after every storm, and he used to dislike having to bail it out by hand. One day he bought a pump, which could do the job in 7 hours less than it took him to do it by hand. With the pump working and Sinbad bailing at the same time, the boat could be entirely pumped out in 2 hours and 56 minutes, if no more water came in. How long would the pump take to do the job alone?

19. A vat can be filled through two pipes. One pipe can fill the vat in 5 hours more than the other pipe can. Working together, they fill the vat in 6 hours. How long would it take with each pipe alone?

20. A store manager bought a shipment of DVD recorders for $13,500. Six of them were damaged and had to be scrapped, but the rest each sold with a $60 markup. The store profited $1,020 from these sales (excluding overhead and taxes). How many recorders were sold, and for what price?

21. "The length of a rectangle exceeds the width by 7, and the area is 60. What are the length and width?" (From a Babylonian clay tablet more than 3,500 years old. Quoted in "*Babylonian Quadratics,*" by Robert D. McMillian, *The Mathematics Teacher*, January, 1984.)

22. A rectangular garden plot is 2/3 as wide as it is long. If part of the plot is used for a grass border 1 meter wide on all sides, then the border will use 2/3 of the original rectangular area. Find the length and width of the plot.

23. Brahmagupta, an astronomer of central India, posed this problem around 630. A vertical bamboo stalk 18 cubits tall boke in the wind so that its top touchted the ground, which was level, 6 cubits from the base of the stalk. How high up did the bamboo break?

24. At what points, if any, do these lines meet the circle $x^2+y^2=25$?

a. $y=x+1$

b. $y=\frac{3}{4}x+6\frac{1}{4}$

Vocabulary for Chapter 14

Completing the square (14.2)
Complex numbers (14.2)
Quadratic equation (14.1)
Quadratic formula (14.2)

Self-Quiz for Chapter 14

1. Solve $x^2 + 3x = 40$

2. Solve by factoring: $2w^2 + 25w + 33 = 0$

3. What constant must be added to $y^2 + 9y$ to get the square of a binomial?

4. Use the quadratic formula to factor $3x^2 + 10x - 1$.

5. A rectangular sheet of copper is twice as long as it is wide. A four cm square is cut out of each corner, and the remaining copper is folded to form a rectangular box without a top, much as in figure 14.3.1. If the volume of the rectangular box is 360 cubic centimeters, what are the length and width of the original sheet of copper?

6. Solve and check: $(y - 1)^2 - 8(y - 1) + 12 = 0$

Chapter Fifteen
Data and Statistics

We live in a data-driven world. Airplanes carry flight data recorders, schools base decisions on enrollment and testing data, financial markets are influenced by data on prices, earnings, employment, etc, and everyone is concerned with the security of their personal data. Statistics is, broadly speaking, the study of data. The use of statistics typically begins with obtaining data and ends when that data has been studied and used as desired. Intermediate steps may include: coding and transmitting data, organizing it, analyzing, interpreting, and displaying it. Each step in the process can be expensive, complicated, and sophisticated, as it is for data gathered by space probes, but the steps themselves are the same ones as used in low tech examples like this:

Ian's Pocket Money[1]

Ian went home from school one day with an almost legible list of the pocket money each child in his class received each week.

Name	Amount in P	Name	Amount in P	Name	Amount in P
Tom	75	Sandra	100	Sara	100
Mark	100	Jacky	200	Martha	200
Philip	75	Jane	150	Ian	75
Julliet	125	Boose	150	Lillian	220
Dean	200	Lawrence	75	Amanda	150
Jan	125	Noel	200	Joan	200
Paul	225	Raul	100	Philipa	100
Richard	100	Luigi	100	Robert	125
Emma	100	Karen	125	Margaret	100
Kathy	250	Sharon	125	Antony	125

He had even added a histogram [the kind of graph shown] to reinforce the point. (Figure 15.0.1)

[1] From Hugh Burkhardt, *The Real World and Mathematics*, Blackie and Son, London, 1981.

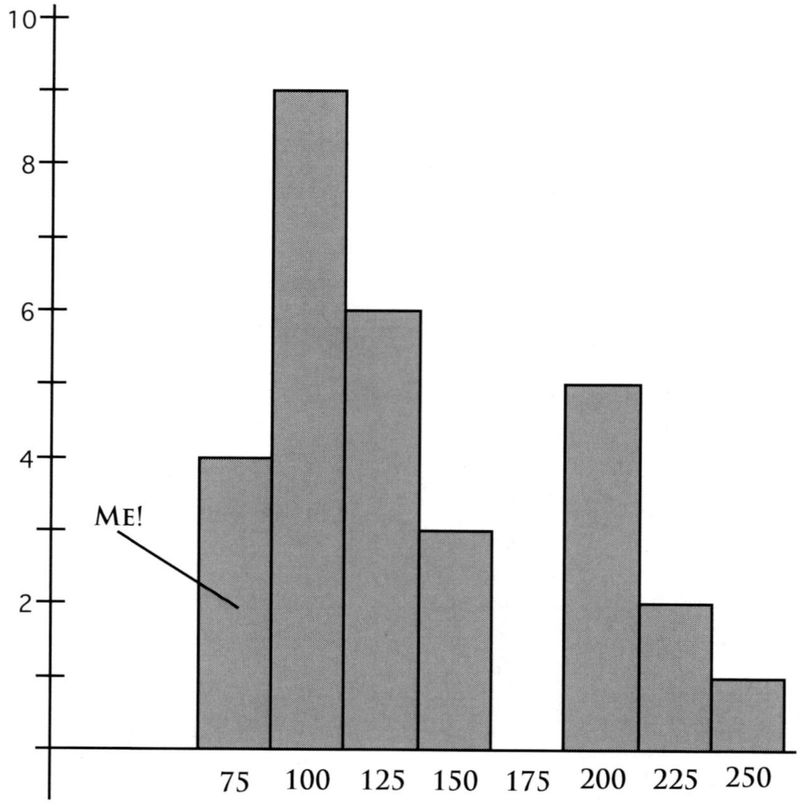

Figure 15.0.1

He resisted his parents' suggestion that the figures were not comparable—"Are you sure that the others don't have to buy dinners out of theirs?"—and persisted through some passive resistance. In the end the parents retreated in fairly good order and he carried his elder brother and sister along with him to victory, and, apparently, £ 1.00 per week each.

In this example Ian gathered data, organized it, displayed it, and used it to support his case. That is the essence of statistics.

15.1 About Data

Data originates as observations, possibly aided by instruments. Data about weather, for example, might consist of qualitative descriptions such as "clear," "cloudy," "foggy," "rainy," "hot," or "cold." It might also be quantitative measurements, such as temperatures, atmospheric pressures, wind velocities, and relative humidity readings. Because the type of data affects how it should be treated, statistical data is classified into different kinds. One common classification of data is as categorical, ordinal, or ratio data.

Categorical data typically involves categories for which there is no natural order, such as:

Country of birth
Religious affiliation
Make of car

Ordinal data can be meaningfully put into order, such as:

Freshman, sophomore, junior, senior
Cold, cool, tepid, warm, hot
Grade point average (GPA)
Private, Private First Class, Lance Corporal, Corporal, Sergeant

Data based on rank ordering is ordinal data. For example, a questionnaire presents a statement and asks respondents to choose which of these options they consider most appropriate[2]:

Strongly disagree	Mildly disagree	Undecided	Mildly agree	Strongly agree

These options are ordered, but there is no natural way to compare them for different people. One person might mark "strongly agree" where another, with essentially the same attitude but a tendency toward understatement, might mark "mildly agree." Again, two people two people who choose the same option may really feel quite differently about the statement.

Sometimes ordinal data is coded numerically. For example, one might code:

Strongly disagree 0
Mildly disagree 1
Undecided 2
Mildly agree 3
Strongly agree 4

Be careful with numbers used to code ordinal data! Their order corresponds to the order in the data, but the numbers themselves are arbitrary. The responses could as well be coded:

Strongly disagree -30
Mildly disagree -10
Undecided 0
Mildly agree 10
Strongly agree 50

Ratio data measures something that can be meaningfully compared by numerical ratios. Length, for example, is ratio data. It makes sense to speak of a ratio of lengths, as we do in geometry. Contrast that to ordinal data, where, for example, it makes no sense to speak of the ratio of "strongly agree" to "agree." Some examples of ratio data:

Mass
Area
Volume
Weight
Age
Velocity
Hourly pay

2 This is known as a Likert scale and is named for psychologist Rensis Likert (1903-1981), who invented it as a way to measure attitudes in his doctoral thesis of 1932.

Why classification of data matters
Different kinds of data allow different kinds of treatment. For example, an average of a ratio data quantity, such as the ages of people in a study, is meaningful, but it makes no sense to average ordinal data (e.g., what is the average of freshmen and juniors?), or to calculate its standard deviation (a measure of variability discussed in section 15.4), but it might make sense to calculate means and quartiles. None of these make sense for categorical data such as gender or political affiliation.

Confusion about the difference between ordinal and ratio data dates back at least to Plato, who, more than 2,000 years ago concluded that a truly good ruler is 729 times as happy as a despot. That kind of nonsense continues in the modern world, exemplified by an advertising jingle, "Double your pleasure, double your fun, with Wrigley's Doublemint gum." There is simply no way to tell whether you have doubled your pleasure or multiplied it by only 1.9. It is as meaningless to speak of doubling your pleasure as it is to imagine one man 729 times as happy as another.

Example 1. Classify the following data as categorical, ordinal, or ratio.

a. Crops grown commercially
b. Population (number of residents)
c. Telephone numbers
d. Number of questions answered correctly on a test
e. Student grades (A, B, C, D, F)

Answers, with explanations

a. Crops are categorical data. There is no reason to list tomatoes before (or after) corn or apples. Any ordering is either entirely arbitrary or is in accordance with an additional criterion, such as the alphabet, which is not inherent to the categories.
b. Population density is ratio data. It makes sense to compare populations numerically, to say, for example that one country has twice the population of another.
c. Telephone numbers are categorical data. The fact that they are numbers is irrelevant, since there is no interest or point in comparing phone numbers by size.
d. Number of questions answered correctly is ratio data. The numbers can be the basis for meaningful comparison, such as one person getting 20% more questions right than a second person.
e. Grades are ordinal data. An A is higher than a B, which in turn is higher than a C, but there is no basis for comparing how much higher an A is than a B or a C. Often grades are coded numerically and, and students are compared on the basis of grade point average (GPA). There is, of course, nothing to prevent such calculations, but it is a mistake to take the resulting numbers seriously, as is the case when GPA is used to determine scholarship awards. Grades and GPA reflect grading methods at least as

much as they reflect scholarly achievement. Systems for awarding grade points are really Borda systems for voting (see problems 3 and 4 below).

Exercises 15.1

1. Classify each of the following data as categorical, ordinal, or ratio.
   a. Atmospheric pressure
   b. School subjects
   c. Batting averages
   d. Calories per serving
   e. Favorite kind of music
   f. Birth weight
   g. Fuel economy (miles per gallon)
   h. Customer satisfaction rating (1 to 5, with 5 highest)
   i. Credit rating

Voting: An exploration of ordinal data (problems 2-4)
Voter preferences are ordinal data. The following exercises illustrate the problems that arise in manipulating ordinal data.

2. Suppose fifteen people have to elect one candidate, Al, Bo, or Cy, and it turns out that:
   Six people prefer Al, with Bo their second choice and Cy last.
   Five people prefer Cy, with Bo their second choice and Al last.
   Four people prefer Bo, with Cy their second choice and Al last.
   a. Who wins if only first place votes are counted?
   b. How many people would prefer Bo to Al?
   c. How many people would prefer Cy to Al?

   This example is not merely theoretical. Most elections in the US go to the highest vote getter, even if that candidate does not get the majority of the votes. Jesse Ventura won the 1998 governor's race in Minnesota, for example, when his 37% of the votes were the highest of the three candidates.

3. Instead of giving the election to the top vote getter, even without a majority, Borda[3], in 1770, proposed a scheme of weighted voting, where each first place vote gets a certain number of points, each second place vote gets a smaller number, and so on, and the candidate with the

[3] Jean-Charles de Borda (1733-1799) was a French naval officer as well as a military engineer and a prominent scientist. His numerous contributions included improved pumps, waterwheels, and chronometers, and other instruments of navigation. He made the standard meter. When the French backed the American Revolution against the English, Borda saw action but was captured and returned to France. He invented his voting system in 1770 for the French Academy of Sciences, of which he was a member.

most points wins. Determine the outcome in problem 2 if the points for the various votes are:

a. First place 2, second place 1, third place 0.

b. First place 20, second place 1, third place 1.

4. In a four-way election between A, B, C, and D, seven voters preferences are as follows:

Voter	First Choice	Second Choice	Third Choice
1	A	B	C
2	A	B	C
3	A	B	C
4	A	B	C
5	C	B	A
6	C	B	A
7	C	B	A

Suppose that a first place vote gets 3 points, a second place vote gets 2 points, and a third place vote gets no points at all. Show that candidate B wins the election, even though candidate A won a more than half of the first place votes.

5. A problem with rating teachers. On a standardized test, teacher A's class averaged 62 in the fall and 69 in spring. Teacher B's class averaged 73 in the fall and 79 in the spring. Explain why that does not necessarily mean that teacher A's class improved more than teacher B's class.

15.2 Obtaining Data: Sets and Samples

Once you have decided what data you need, you must get it. That may be as simple as making a few observations and tallying them or as complex as conducting a national census. It is often easier to use data someone else has gathered, but you need to be very cautious about doing so.

Ever since the origin of statistics as a subject, people have gathered data in order to make a point, and there are always those who will distort or even shred the truth to serve their own purposes. The best defense against that is to check up on the original source of the data, but that is not easy to do today, where, for example, on the internet, anyone can post information, true or not, and postings move easily from site to site and user to user. Regard with deep suspicion any data whose source is an authoritarian government, an extreme political or religious group, or anyone trying to sell something. Even if you are confident about the source of data, retain a healthy skepticism about the data itself.

Example 1. A pharmaceutical company, before being allowed to market a new drug, must demonstrate that it is safe and effective. To this end, imagine

a clinical trial in which the company's researchers give the drug to ten patients (the **treatment group**), and do not give the drug to ten other patients (the **control group**). At the end of the study, the health of patients in the two groups is compared. Are there grounds to question this procedure?
Solution: As described, this trial is deeply flawed. The fact that the company's researchers run the clinical trial is grounds for suspicion. Experience has shown that if the researchers or the patients know which patients are in the treatment group and which are in the control group, that alone can influence the outcome of the study. (The researchers and patients both may react differently in such circumstances.) This should be avoided by using a **double blind** design, in which the treatment group and the control group appear to be treated identically, but the treatment given to the control group is known to have no effect on the matter at hand. In a double blind study, neither the subjects nor those who come into contact with them know who is in the treatment or the control group. Data from a poorly designed study such as this may be biased.

Example 2. A commercial some years ago stated that 90% of the Volvos sold in America in the ten previous years were still in use.

a. Assuming the 90% figure is correct, what does it imply about durability of Volvos?
b. What additional information would you need to be able to assess the durability of Volvos sold in America in the last ten years?

Solution: a. The advertisement is designed to associate the numbers 90% and "ten years," with the Volvo name, but the commercial makes no claim about the durability of Volvos. (It is like saying that most of the hybrid powered cars built in the last 500 years are still on the road. True or not, it has nothing to do with the cars lasting 500 years.) This data cited, accurate or not, is useless. The advertisement implies absolutely nothing about the durability of Volvos.
b. To assess the durability of Volvos, you would need to know how long the cars lasted, but that information is not given in this commercial.

Example 3. In the 1950s, market researchers for Ford asked people what kind of car they wanted, while market researchers from General Motors asked people what kind of car their neighbors wanted. Ford respondents stressed safety, while GM respondents stressed style and power. Ford built the safe cars its respondents said they wanted, while GM built the powerful, stylish cars that people said their neighbors wanted. GM resoundingly outsold Ford. Conclusion: You cannot necessarily rely on people to be candid in replying to questions.

Example 4. Even the process of counting, which is completely straightforward in principle, is almost impossible to do accurately when the job is a large one. To count the students in a classroom is not difficult, but to count the people in a large country is very difficult, to say nothing of sorting them by citizenship status. The well-publicized problems in tallying close elections, like the 2000 Presidential election and the 2008 Minnesota senatorial race, have unique circumstances, but problems are virtually certain to arise in any really large counting process. (Just imagine trying to tally the items in the inventory of a chain of stores!)

Sampling and Sets A practical alternative to large data collection is to study a small but representative sample. The entire set you hope to learn about is called the **population**, a term used even for inanimate objects. A subset selected for study called a **sample**.

Sets

Before getting into sampling, we need a brief look at sets. Mathematically, a set is any collection of objects whatever[4]. A set might include a car, a baseball, a dollar bill, and a unicorn. Two sets are said to be equal if they have the same members. Usually we denote a set by a capital letter, and we describe it either by listing the objects in the set (called **members** or **elements** of the set) in curly braces or by giving a rule that distinguishes members of the set from non-members. For example, the set A=[2, 3, 5, 7} can also be described as the set of prime numbers less than ten, which is written A={x:x is a prime number less than ten}. Here the x before the colon stands for a possible member of the set A, and the expression after the colon states the defining condition that x must satisfy to be a member of the set. We can state this condition without knowing whether or not the condition of membership is actually satisfied by anything. Thus, we can write the set P={x:x is an airplane that seats 20,000 people} without knowing whether or not there are any such airplanes. Again, we can speak of the set of odd perfect numbers, though, as mentioned in chapter 5, nobody has any idea if there are any odd perfect numbers. If there are no members of a set, it is said to be **empty**. There is, in effect only one empty set (all empty sets are equal, having the same members, which is to say no members at all), and it is denoted by the symbol $\varnothing$, which is a bit like the number 0.

Sets are often pictured as circles or other shapes with the members inside and everything else outside. For example, M={x:x is a factor of 27} is pictured as in figure 15.2.1.

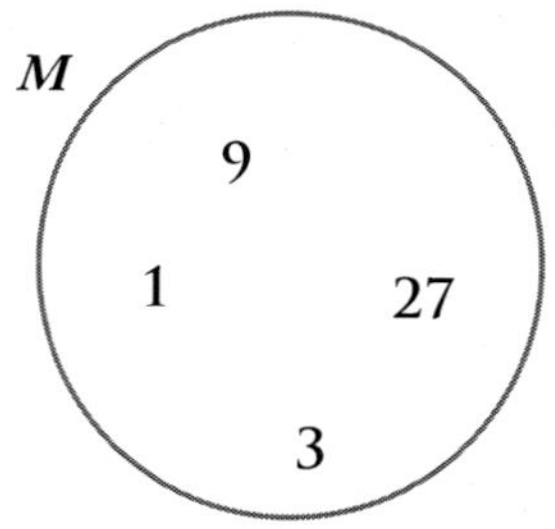

Figure 15.2.1

Anything not in M, if it is shown at all in this picture, would lie outside of the circle. Again, if N={x:x is a factor of 75}, then N={1,3,5,15,25,75}, which is pictured in Figure 15.2.2.

[4] This is a slight overstatement. Our discussion does not apply to sets that are members of themselves, such as the set of all sets.

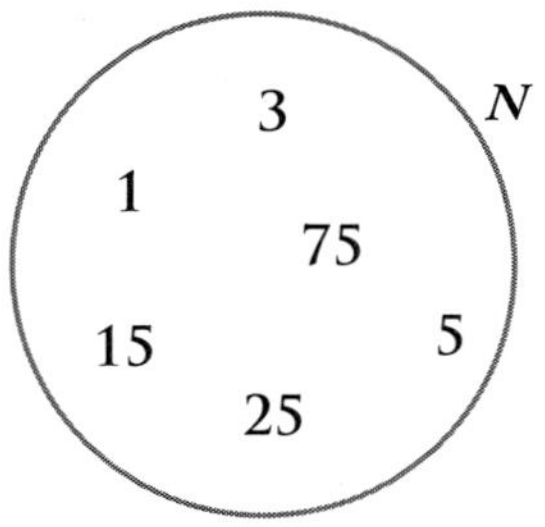

Figure 15.2.2

To show both sets in the same picture, the circles must overlap, because elements such as 1 and 3 are in both sets and therefore must lie in both circles at once. On the other hand, the circles should not overlap completely, because there are other elements that are in one set but not the other. Figure 15.2.3 shows the resulting picture.

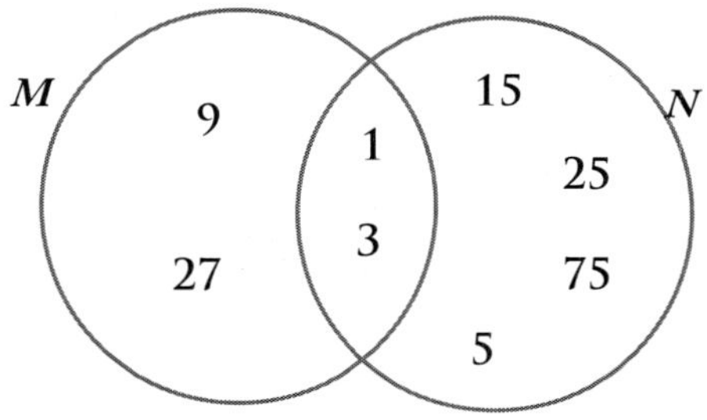

Figure 15.2.3

A **subset** of a set *A* is a set, all of whose members are also members of *A*. For example, in figure 15.2.3, the set $C=\{1,3\}$ is a subset of both *M* and *N* (it is the set of common factors of 27 and 75). The set $D=\{1,3,5,75\}$ is a subset of *N* but not of *M*. The members of a set that are not in a given subset comprise the **complement** of that subset. Thus the complement of *C* in *M* is {9, 27}, because 9 and 27 are all the members of *M* that are not members of *C*. Again, the complement of *C* in *N* is {5,15,25,75}.

We write $B \subset A$ to say *B* is a subset of *A*, and we write $C \not\subset A$ to say that *C* is not a subset of *A*. To indicate that *c* is a member of *A* we write $c \in A$, and to indicate that *c* is a not member of *A* we write $c \notin A$. Every set is a subset of itself, and the empty set $\varnothing$ is a subset of every set.

Set pictures like figures 15.2.1-15.2.3 are called **Venn diagrams**[5]. They are often extended to three sets, but for more than three sets they are clumsy and rarely used.

Sets and Samples

Every sample is a subset of the population. The goal in sampling is to use a sample that is representative of the population, yet small enough to be practical. In general, the larger the subset, the more likely it is to be repre-

[5] John Venn (1834-1923) was a British mathematician. Others, notably Euler, made similar diagrams before Venn did, but Venn gave them a prominence through his book *Symbolic Logic* (1881).

sentative of the population, but the more cumbersome it is to deal with, especially if the population is large. To know for sure whether a subset is representative of the population, one would need to know the same information that the sampling is supposed to help answer. For example, to know how well a sample of voters represents the population as a whole you need to know that information about the whole population, but the reason for using sampling in the first place was to avoid having to gather all that information! In practice, then, you never know for certain how representative your sample is, but you can avoid some of the errors that are most likely to produce non-representative samples. These errors involve **bias** in the sampling procedure itself.

An obvious but classic example of a biased sample occurs when, after difficulties finding a time when a sufficient number of committee members could meet, the chair of the meeting asks those assembled to vote on whether the same hour and day of the week would be convenient would be convenient for weekly meetings. Clearly, the people available to answer the question are the ones that had that time available now, and that fact alone makes them more likely to be able to meet at that time and day in future weeks than those who did not get to the meeting.

Many sources of sampling bias, such as "non-response bias," are discovered by hard experience. Non-response bias was discovered after a political polling company made embarrassingly inaccurate predictions about an election. They had based their predictions on the responses to written poll they had mailed to a sample of voters that had been carefully chosen as representative. They assumed that the attitudes of those who did not answer the poll were the same as those who did answer it, but that turned out not to be the case. It turned out that people in the lowest and highest strata of US society are less likely to return written questionnaires than are those in the middle class. The polling company had tried to sample the entire society, but the respondents were mostly middle class.

To avoid bias in sampling, first identify clearly the population of interest. Then be sure that the sampling procedure treats all population members the same way. For example, if a sample of size three is to be selected from the population $\{a, b, c, d, e\}$, then all of the samples: $\{a, b, c\}$, $\{a, b, d\}$, $\{a, b, e\}$, $\{a, c, d\}$, $\{a, c, e\}$, $\{a, d, e\}$, $\{b, c, d\}$, $\{b, c, e\}$, $\{b, d, e\}$, $\{c, d, e\}$ must have the same chance of being selected. Only then can the sampling procedure be considered unbiased.

Random sampling

Sampling is sometimes done by chance, using a device like tossing a coin or rolling a die in a way that gives equal chances to all population elements to be included in the sample. Coins and dice are examples of what are sometimes called **randomizing devices**. In classrooms a favorite randomizing device is a spinner (figure 15.2.1). To use it, put the spinner on a flat, level surface with a pencil point or similar pointed object at the center and a paper clip around it. Spin the paper clip with a flick of the finger and see which number it stops on. The chance the spinner will land on a given sector of the circle is proportional to the size of that sector. For each spinner in Figure 15.2.1 the sectors are all the same size, making all outcomes equally likely.

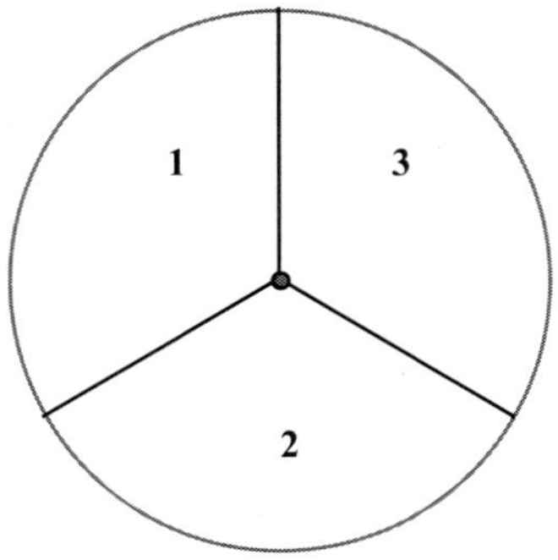

3-spinner

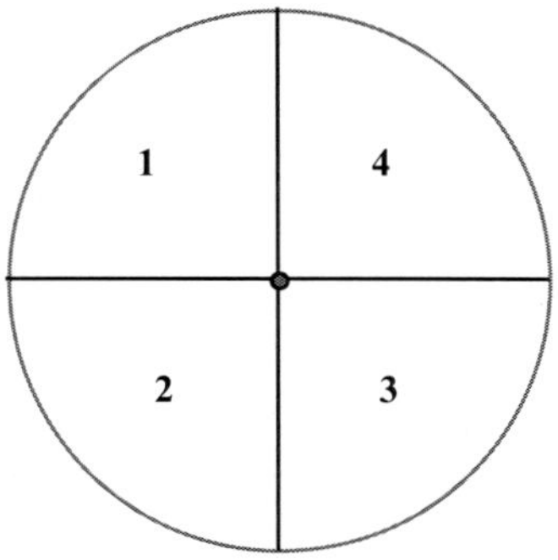

4-spinner

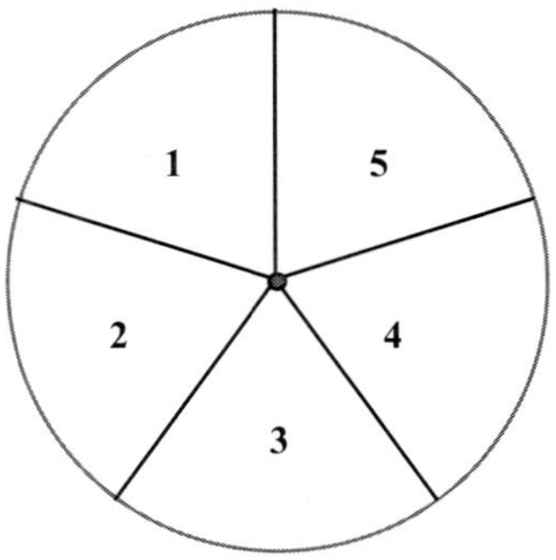

5-spinner

Figure 15.2.1

Another model for random selection is based on reaching into a pot and selecting from numbered but otherwise identical objects, much as the caller at a bingo game does.

Student projects with statistics

A good way to engage students in statistics is to have them follow Ian's example and investigate a question that involves gathering data. They may need help in selecting a topic that they can deal with in a fairly short time, without imposing unduly on others. Some possibilities:

1. Observe and record the temperature outside the school (or home) at a certain time of day for a convenient period, such as a month.
2. Record how many hours each day someone (most likely the student doing the study) engages in a designated activity, such as eating, traveling, doing homework, or watching TV.
3. Find out how old each US President was when he assumed office.
4. Check breakfast cereal labels to see how much carbohydrate each one has in a serving.
5. Have classmates shoot 10 free throws each and record how many of the free throws each person made.
6. Survey classmates as to what kind of music (if any) they listen to while studying.
7. Investigate how many games each World Series has lasted.
8. Investigate the batting averages of the Major League batting champions for the last 50 years.
9. Investigate people's birth dates to find whether they are related to the phases of the moon.

Exercises 15.2

1. The a set with two elements is $\{a, b\}$. It has the following subsets: $\varnothing$, $\{a\}$, $\{b\}$, $\{a, b\}$. Likewise, a set with just one element {a} has two subsets, $\varnothing$, and $\{a\}$.
   a. List the subsets of a set with three elements, $\{a, b, c\}$
   b. List the subsets of a set with four elements, $\{a, b, c, d\}$
   c. Based on your results from parts a and b, predict how many subsets a set with five elements has.
   d. List all the subsets of the five-element set $\{a, b, c, d, e\}$.

2. Why might each of the following polling methods be biased?
   a. Contact people using their telephone numbers from a phone book.
   b. Contact people at shopping malls on Saturday.
   c. Contact people by e-mail.
   d. Contact people by sending personal representatives door to door.

3. Bill Williams, in his wonderful book, *A Sampler on Sampling*, describes "the disappearing children" puzzle that arose in the course of conducting a poll. The polls took the form of interviews of adult residents in a certain neighborhood. To avoid missing anyone, the interviewers, who had initially visited the neighborhood during the day, returned later to those homes where nobody had responded the first time. On one of the questions in the poll, about the number of children in the household, the numbers of children per family reported in the follow-up interviews were substantially lower than those reported in the initial interviews. Can you suggest a plausible explanation?

4. At an elite private university in a small town, a few local high school students took calculus at the university, along with the freshmen. In those circumstances, the high school students usually were among the highest performers in the class. Why?

5. **The placebo effect**. In a study to determine the effectiveness of a drug, sometimes the treatment group of patients is given the drug in pill form, while the control group is given an inert pill, called a **placebo**, on a similar schedule. This avoids the patients or the experimenters knowing who is getting the drug and who is not in the study. Members of both groups are then assessed to see if the drug has the expected effect on patients. In some cases the patients on the placebo also show improvement! This is called the **placebo effect**. Try to explain what might cause the placebo effect.

15.3 Displaying Data

After data is gathered, it must be arranged and displayed in a way that brings out salient features. Here is a typical grade school activity, a class poll on favorite ice cream cone flavors. If the favorites are chocolate (9 students), pistachio (3 students), and strawberry (8 students), the data might be graphed as in figure 15.3.1. This kind of graph, like Ian's graph of spending money, is called a **bar graph** or **histogram**.

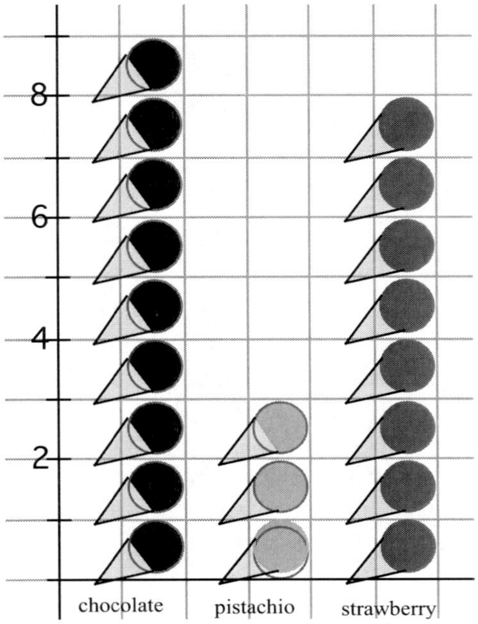

Figure 15.3.1

Because this is categorical data, the order in which the flavors appear on the horizontal axis has no significance. When ordinal data is presented in a bar graph, it is arranged along the horizontal axis so as to reflect the order of the data.

Categorical data is also often displayed in pie charts. A **pie chart** or circle plot is often used to picture the how something is divided into various categories. Pie charts are particularly appropriate for illustrating the relative size of various sources of income or expenditure, as in a budget. The ice cream data could be represented on the pie chart in figure 15.3.2.[6]

[6] Florence Nightingale (1820-1910) pioneered the field of nursing. In 1854 she led of a team of 38 nurses to treat soldiers injured in the Crimean War. Her compassion, energy, knowledge, and administrative skill quickly made her famous. Because of her insistence on sanitary measures, the mortality rate decreased sharply from the shocking 42.7% of cases that she found on arrival. To present her results to others, and build support at home for her work, Nightingale invented "polar-area charts," which, with minor modifications, became what we now call pie charts. This is an early example of the use of statistics to convince others. Nightingale was a lifelong leader in hospital design, public health, and sanitation.

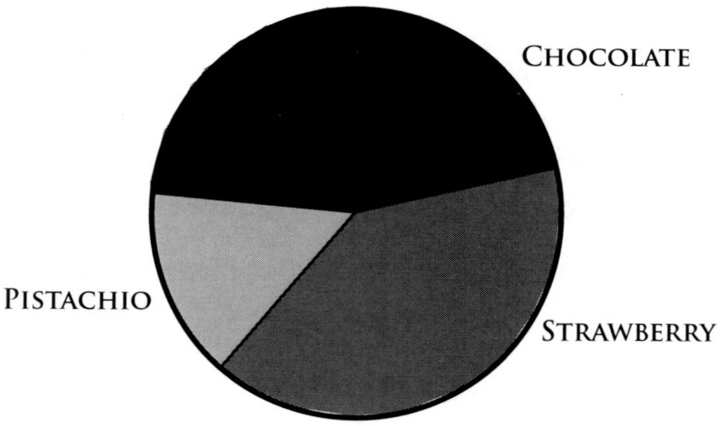

Figure 15.3.2

Example 2. Figure 15.3.3 is a bar graph showing the percentage of hockey players in three levels of Canadian youth hockey leagues that were born in each quarter of the year.[7] For the bottom (youngest) tier the percentages of players born in each quarter are about equal, but for the oldest, "top tier" leagues, participation drops as the birthdate advances. This trend, clear from the graph, might easily be missed in a table of data. The graph incorporates another feature that is worth pointing out. You might have expected birthdays to be plotted individually by day of the year. Instead, birthdays in figure 15.3.3 are grouped into three month periods. Such grouping of data can, as in this case, bring out patterns in a histogram that might otherwise be less clear.

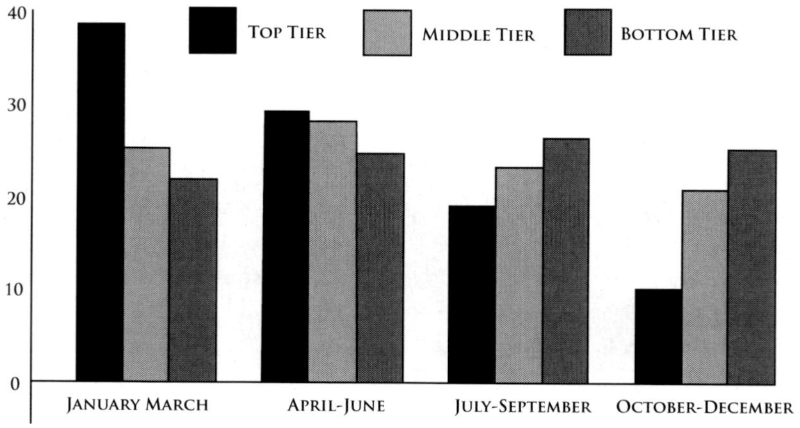

Figure 15.3.3

Line graphs

To display data that is obtained at intervals over a period of time, it is often helpful to plot the points represented by that data (these points are called

7 Barnsley, Roger H. and Thompson, A.H., *Birthdate and success in minor hockey: The key to the NHL* Canadian Journal of Behavioral Science, vol 20, #2 (1988), available at www.socialproblemindex.ualberta.ca/RelAgeMinorHockeyCJBS.pdf

data points) and then connect consecutive data points with line segments. The result is a **line graph**. An example is figure 15.3.4, which graphs US resident population, as tallied by the US Census every ten years for a span covering the twentieth century.

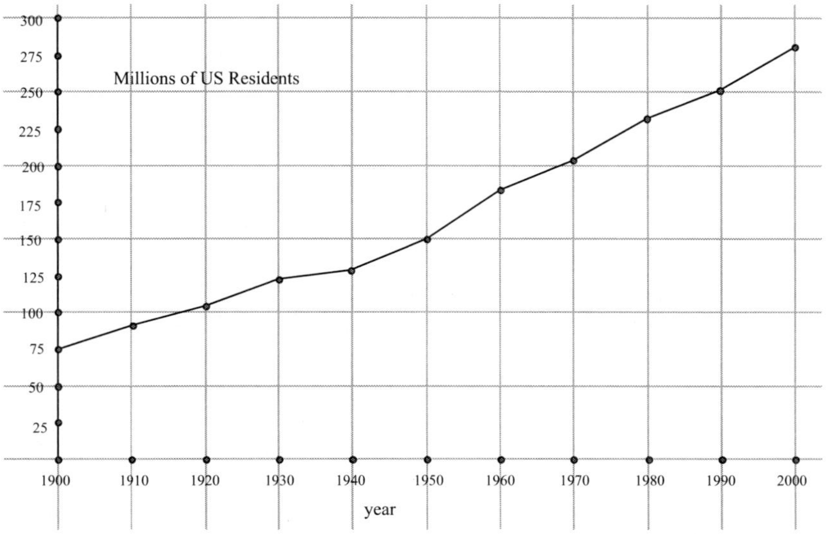

Figure 15.3.4

Example 3. A dramatic example of data display is the L Curve, which shows US income distribution. Some highlights:[8]

The US population is represented along the length of the football field, arranged in order of income.

Median US family income (the family at the 50 yard line, with half the families below it and half above it) is about $40,000 (a stack of $100 bills 1.6 inches high.)

—The family on the 95 yard line earns about $100,000 per year, a stack of $100 bills about 4 inches high.

—At the 99 yard line the income is about $300,000, a stack of $100 bills about a foot high.

—The curve reaches $1 million (a 40 inch high stack of $100 bills) one foot from the goal line.

—The graph continues to rise steeply. The yearly income of the very richest families is equivalent to a stack of $100 bills some 30 miles high!

Scatter Plots

Scatter plots are used to display data and may reveal relationships between two variables that are related, but perhaps not related as tightly as they would be if one were a function of the other.

8 Source: http://www.lcurve.org May 15, 2009.

Example 4. The Third International Mathematics and Science Study (TIMSS) revealed large differences in achievement levels of eighth grade mathematics students in different countries. Students from certain Asian countries performed particularly well on TIMSS tests. The cultures of these countries have traditions of hard work and thrift, related to their tradition of rice cultivation, which requires great diligence and effort. Prior to being assessed in mathematics, students in the TIMSS study were given a long questionnaire. Some students did not bother to finish the questionnaire, but others did so. Erling Boe and his team plotted the percentage of questions answered on this questionnaire against national scores on the TIMSS mathematics assessment to get the scatter plot below.

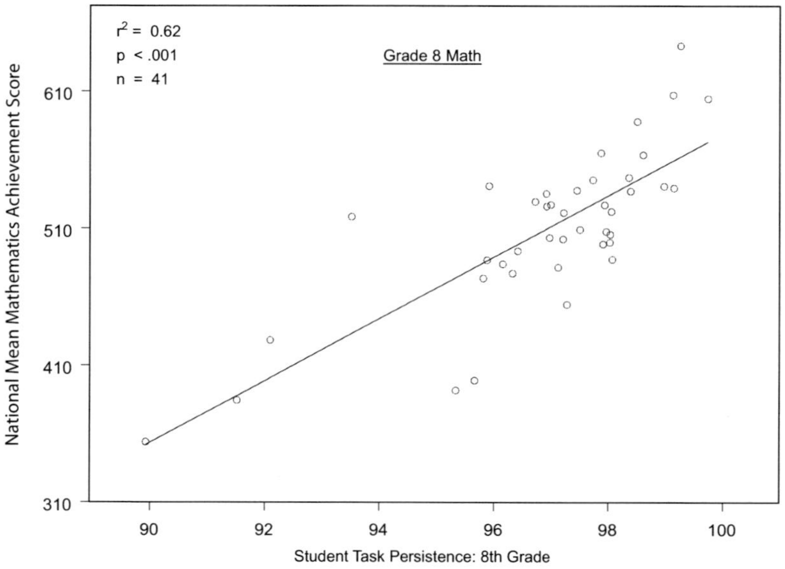

Source: Boe, Erling E.; May, Henry; Boruch, Roert F. *Student Task Persistence in the Third International Mathematics and Science Study: A Major Source of Achievement Differences at the National, Classroom, and Student Levels.* Center for Research and Evaluation in Social Policy, Graduate School of Education, University of Pennsylvania. Available on the Web at Education Resources Information Center (ERIC) as document ED478493.

The sloping line is not part of the scatter plot itself but is the line, which, statistically, best fits the data points plotted.

The types of data displays discussed here are just a few of those in common use today. Presentation of information has developed into a major field of expertise, of intense interest to designers of websites, cell phones, vehicle dashboards and cockpits, and equipment for hospitals and factories. A leader in the field is Edward Tufte, who, not surprisingly, has a beautiful website: http://www.edwardtufte.com/tufte/. In schools, we should encourage students to pay attention to presenting information well. This should apply to research projects, mathematics problems, and, indeed, coursework of all kinds. We should also help students develop judgment and taste about what makes a presentation of information, in a given context, either particularly effective, or, conversely, what makes it ineffective or clumsy.

Exercises 15.3

1. In a poll, prospective voters were asked to identify their most important concern from a list of four areas. The results of the poll were presented in the pie chart below.

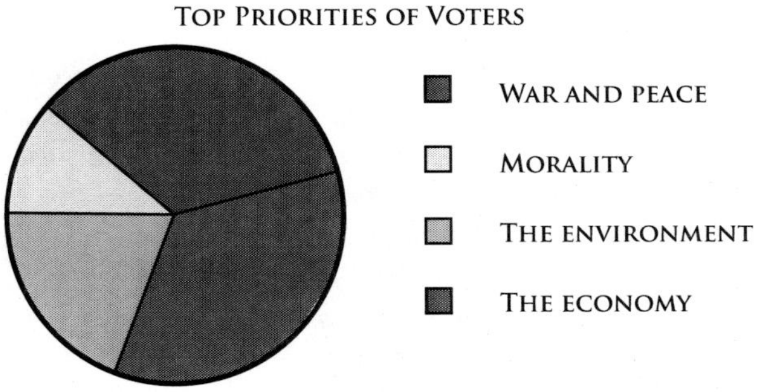

Explain why it would be inappropriate to present the results of this survey in a line graph.

2. When the differences between data points are small compared to their overall sizes, part of the vertical axis is sometimes omitted from a graph. For example, a management of a company whose annual earnings grew to \$373,000 from \$370,000 a year earlier might choose the upper graph over the lower (see below) in presenting their annual earnings. What might be some reasons for their choice?

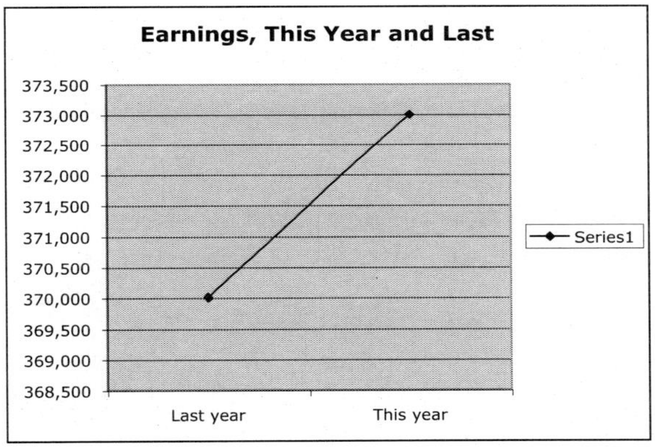

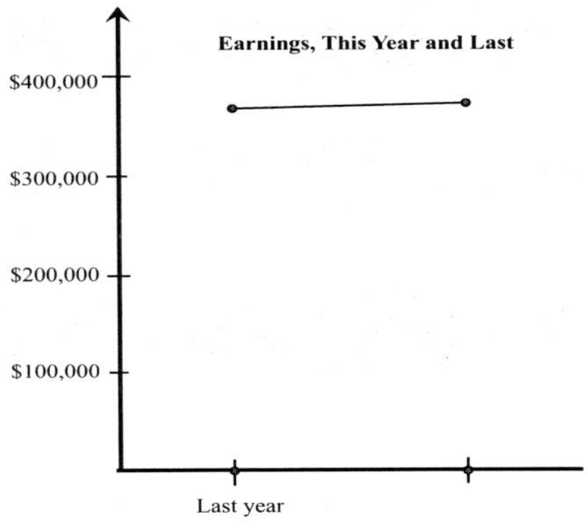

3. Suppose you had a line graph of temperatures taken daily at noon at a weather station. Would you expect the midpoints of the line segments between the data points to give useful estimates of midnight temperatures? Explain.

4. In "The Climate of Man–III" (*The New Yorker*, May 9, 2005) Elizabeth Kolbert wrote:

   In the seventeen eighties, ice-core records show, carbon-dioxide levels stood at about two hundred eighty parts per million. Give or take ten parts per million, this was the same level that they had been at two thousand years earlier, in the era of Julius Caesar, and two thousand years before that, at the time of Stonehenge, and two thousand years before that, at the founding of the first cities. When, subsequently, industrialization began to drive up CO_2 levels, they rose gradually at first–it took more than a hundred and fifty years to get to three hundred and fifteen parts per million–and then much more rapidly. By the mid-nineteen seventies, they had reached three hundred and thirty parts per million, and by the mid-nineteen nineties, three hundred and sixty parts per million.

   Graph these annual averages of atmospheric CO_2 measurements, made at the observatory on Mauna Loa, Hawaii. Use a line graph, and choose your y-axis scale with attention to the issues raised in problem 2.

Year	1960	1965	1970	1975	1980	1985	1990	1995	2000
CO_2	316.91	320.09	325.65	331.16	338.67	345.86	354.19	360.90	375.61

5. Below are the world records for the mile run (from Wikipedia, *World record progression for the mile run*, April 21, 2009).

Year	**Time**	**Runner**	**Location**
1865	4:36.5	Richard Webster	England
1868	4:29.0	William Chinnery	England
1868	4:28.8	Walter Gibbs	England
1874	4:26.0	Walter Slade	England
1875	4:24.5	Walter Slade	London, England
1880	4:23.2	Walter George	London, England
1882	4:21.4	Walter George	London, England
1884	4:18.4	Walter George	Birmingham, England
1894	4:18.2	Fred Bacon	Edinburgh, Scotland
1895	4:17.0	Fred Bacon	London, England
1895	4:15.6	Thomas Conneff	Travers Island, New York
1911	4:15.4	John Paul Jones	Cambridge, MA
1913	4:14.4	John Paul Jones	Cambridge, MA
1915	4:12.6	Norman Taber	Cambridge, MA
1923	4:10.4	Paavo Nurmi	Stockholm, Sweden
1931	4:09.2	Jules Ladoumegue	Paris, France
1933	4:07.6	Jack Lovelock	Princeton, NJ
1934	4:06.8	Glenn Cunningham	Princeton, NJ

1937	4:06.4	Sydney Wooderson	London, England
1942	4:06.2	Gundar Hägg	Goteborg, Sweden
1942	4:06.2	Arne Andersson	Stockholm, Sweden
1942	4:04.6	Gundar Hägg	Stockholm, Sweden
1943	4:02.6	Arne Andersson	Goteborg, Sweden
1944	4:01.6	Arne Andersson	Malmo, Sweden
1945	4:01.4	Gundar Hägg	Malmo, Sweden
1954	3:59.4	Roger Bannister	Oxford, England
1954	3:58.0	John Landy	Turku, Finland
1957	3:57.2	Derek Ibbotson	London, England
1958	3:54.5	Herb Elliott	Dublin, Ireland
1962	3:54.4	Peter Snell	Wanganui, New Zealand
1964	3:54.1	Peter Snell	Aukland, New Zealand
1965	3:53.6	Michel Jazy	Rennes, France
1966	3:51.3	Jim Ryun	Berkeley, California
1967	3:51.1	Jim Ryun	Bakersfield, California
1975	3:51.0	Filbert Bay	Kingston, Jamaica
1975	3:49.4	John Walker	Goteborg, Sweden
1979	3:49.0	Sebastian Coe	Oslo, Norway
1980	3:48.8	Steve Ovett	Oslo, Norway
1981	3:48;5	Sebastian Coe	Zurich, Switzerland
1981	3:48.4	Steve Ovett	Koblenz, West Germany
1981	3:47.3	Sebastian Coe	Brussels, Belgium
1985	3:46.3	Steve Cramm	Oslo, Norway
1993	3:44.4	Nouredine Morceli	Rieti, Italy
1999	3:43.1	Hicham El Guerrouj	Rome, Italy

a. Putting the years, from left to right, on the horizontal axis and the record mile times on the vertical axis, construct a line graph of the years and mile records given above, using the first record and every fourth one thereafter.

b. How might you use the graph from part a to predict when, if ever, someone will first run a mile in 3:30? Describe your method and use it to come up with a year for your estimate.

c. If you are asked to predict the year in which a mile will first be run in 3 minutes flat, would you have as much confidence in your prediction as in the prediction from part b? Explain the reasoning behind your conclusion.

6. This problem relates to the bar graph from Example 2 in the text. On January 9, 1998, *The Hockey News* published its list of the "Top 50 NHL Players of All-Time," which is reproduced below with the players' birth dates.

Syl Apps	Jan 18	Gilbert Perrault	Nov 13
Jean Beliveau	Aug 31	Red Kelly	Jul 9
Max Bentley	Mar 1	Guy Lafleur	Sep 20
Mke Bossy	Jan 22	Newsy Lalond	Oct 31*
Ray Bourque	Dec 28	Mario Lemieux	Aug 4
Johnie Bucyk	May 12	Ted Lindsay	Jul 29
Chrid Chelios	Jan 25	Frank Mahovlich	Jan 10
Dit Clapper	Feb 9	Joe Malone	Feb 28
Bobby Clarke	Aug 13	Mark Messier	Jan 18
Paul Coffey	Jun 1	Stan Mikita	May 20
Charlie Conacher	Dec 20	Dickie Moore	Jan 6
Bill Cook	Oct 9	Howie Morenz	Jun 21
Marel Dionne	Aug 3	Bobby Orr	Mar 20
Ken Dryden	Aug 8	Brad Park	Jul 6
Bill Durnan	Jan 22	Gilbert Perrault	Nov 13
Phil Esposito	Feb 20	Jaques Plante	Jan 17
Bernie Geoffrion	Feb 16	Denis Potvin	Oct 29
Wayne Gretzky	Jan 26	Henri Richard	Feb 29
George Hainsworth	Jun 26	Maurice Richard	Aug 4
Glenn Hall	Oct 3	Larry Robinson	June 2
Doug Harvey	Dec 19	Patrick Roy	Oct 5
Tim Horton	Jan 12	Terry Sawchuck	Dec 28
Gordie Howe	Mar 31	Milt Schmidt	Mar 5
Bobby Hull	Jan 3	Eddie Shore	Nov 25*
Jaromir Jagr	Feb 15*	Bryan Trottier	Jul 18

a. All of these players, as youngsters, played in the network of Canadian youth hockey teams, except for Jagr, who grew up in the Czech Republic, and Shore and Lalond, who grew up before Canadian youth hockey was organized. Omit those three players, and, for the other players, make a histogram of the number of players born in each quarter.

b. How does your histogram relate to Example 2 of the text?

7. Geert Hofstede, using data from an immense statistical study, identified five "dimensions of culture" that lie at the heart of cultural differences. Two of those dimensions are what he calls Power Distance and Individualism-Collectivism. He developed the Power Distance Index (PDI) as a measure of how uncomfortable members of a society feel in dealing with those in higher authority, and the Individualism Index (IDV) as a measure of how much people in a society regard themselves as individuals as opposed to members of a group, such as a family, clan, or tribe. Here is Hofstede's data for a few of the countries surveyed.[9]

[9] Source: Hofstede, Geert, and Hofstede, Gert Jan. *Cultures and Organizations. Software of the Mind.* 2nd edition. McGraw-Hill, New York, 2005

Country	PDI	IDV	Country	PDI	IDV
Arab Countries	80	38	Japan	54	46
Argentina	49	46	Malasia	104	26
Australia	36	90	Mexico	81	30
Austria	11	55	Morocco	70	46
Brazil	69	38	Norway	31	69
China	80	20	Pakistan	55	14
Costa Rica	35	15	Poland	68	60
Equador	78	8	Slovakia	104	52
France	68	71	South Korea	60	18
Guatemala	95	6	United States	40	91
Great Britain	35	89	Venezuela	81	12
India	77	48	West Africa	77	20
Israel	13	54			

a. Make a scatter plot of this data, putting the Power Distance Index along the horizontal axis and the Individualism Index along the vertical axis.

b. Hofstede got his data from questions on a questionnaire. He averaged the responses for each country, then scaled the averages to make the lowest near zero and the highest near 100 on each index. What kind of data are these numbers?

8. Look up the *L curve* on the internet and view the movie.

15.4 Location and Variability of Data

Because a large amount of data is usually complex, it is often desirable to summarize it concisely. One way to do this, at least for ratio data, is to state the **average**, or **mean**, which is found by adding up all the numbers and dividing by the number of data points.

Example 1. Find the mean of the weights, in pounds, of players listed as linemen (not including wide receivers) for the Green Bay Packer's roster, as of May, 2009.

Randolph	300	Soi	334	Dietrich-Smith	305
Wells	303	Hartline	297	Muhtadi	295
Tanker	316	McCaskill	315	Giacomini	311
Meredith	304	Lang	316	Spitz	302
Colledge	308	Moll	306	Clifton	320
Jenkins	305	Barbre	305	Pickett	330
Raji	337	Harrell	320	Toribio	304
Wynn	277	Tally	283	Montgomery	273
Jolly	320	Malone	312		

Solution: The sum of all 26 weights is 7,998 pounds. Divide that by the number of players, 26, to get a mean weight of about 307.6 pounds.

The mean has an interesting physical interpretation. If one unit of mass is placed along the horizontal axis at each data point, the mean is the point about which those masses balance. Figure 15.4.1 pictures this for the numbers 2, 4, 6, 12, whose mean is $\frac{2+4+6+12}{4} = \frac{24}{4} = 6$.

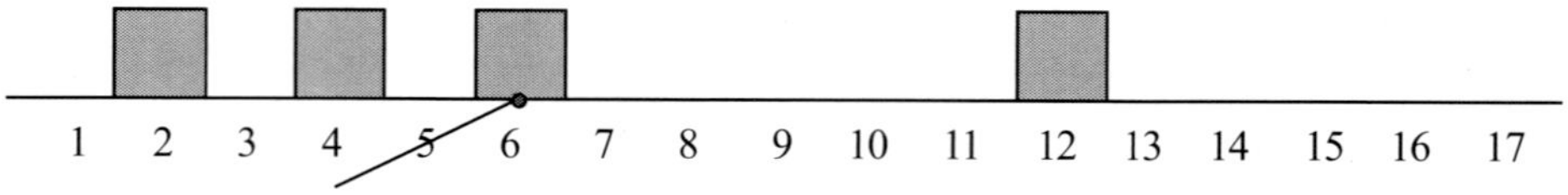

Mean of 2, 4, 6, and 12 is center of mass–the point about which unit weights at 2, 4, 6, 12 balance.

Figure 15.4.1

The mean is sometimes presented as a number that is somehow typical of the data set from which it is calculated. In example 1, that is true. The mean of 307.6 pounds is a typical weight for the Packers linemen listed. However, there are cases where the mean is not at all representative of the data. This happens, for example, when most of the data points are tightly bunched, but a few of them are far away from the rest. Those remote data points, called **outliers**, can disproportionately influence the mean. Example 2 shows that in those cases the mean may not be the best single-number to represent the data.

Example 2. The real estate sales at a certain agency last month were for properties with the following values, in thousands of dollars: 162, 180, 135, 190, 175, and 2,200. The mean sale price, $\frac{162+180+135+190+175+2{,}200}{6} = \frac{3{,}042}{6} = 507$, is not at all typical of the actual sales, because of the outlier sale for \$2,200,000. That is why real estate figures, for example, for the values of homes in a particular neighborhood, mention the median price more often than the mean.

Definition: The **median** of a set of numbers is the middle number, when the set of numbers is arranged in increasing (or decreasing) order, and including all repetitions. If there is an even number of numbers, the median is the average of the middle two numbers.

Half of the data set is below the median, and half above it.

In the set of numbers 3, 8, 10, 12, -5, the median is 8, because if the five numbers are arranged in increasing order, as -5, 3, 8, 10, 12, then the middle number is 8. For the numbers 3, 5, 3, 8 the median is found by arranging them in increasing order: 3, 3, 5, 8 and averaging the middle two, which are 3 and 5. Then the median is 4.

Sometimes it is useful to divide the data set into more than just two equal parts. If the data set is arranged in order and then divided into four equal parts, they are called **quartiles**.

Example 3. For the Green Bay lineman's weights in example 1, calculate the median and the quartiles.

Solution: Arranged in ascending order, the weights of the linemen are

273, 277, 283, 295, 297, 300, 302, 303, 304, 304, 305, 305, 305,
306, 308, 311, 312, 315, 316, 316, 320, 320, 320, 330, 334, 337.

Since there are an even number (26) weights, there is no middle number. Therefore, the median is the mean of the 13th and 14th numbers, or 305.5. To further divide the group into quartiles, find the median of each half. These are the 7th numbers in each list, which are 302 and 316. Then the quartiles are bounded by the numbers 273, 302, 305.5, 316, 337.

While users may find an entire data set a cumbersome information overload, they may find that a single number, such as the mean or the median, is an oversimplification. A middle ground that is easy to use is to provide, in increasing order, the smallest number in the data set, the first quartile, the median, the third quartile, and the largest number in the data set. For the linemen above this would be: 273, 302, 305.5, 316, 337. This is sometimes printed as a **box and whiskers** or **box plot**, as shown in figure 15.4.2.

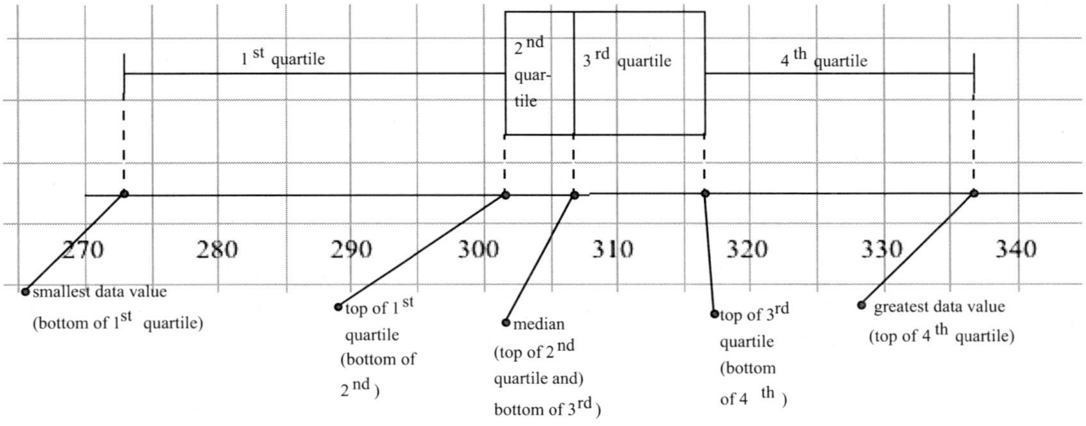

Figure 15.4.2 Box Plot

In this box plot the scale for the weights is horizontal, and the ends of the box and the whiskers, together with the location of the median, are shown on the horizontal scale. The vertical scale is irrelevant to this problem. Sometimes, for convenience or to make a nice-looking display, the horizontal and vertical axes are interchanged to show the box plot going up and down instead of sideways.

Where there are large numbers of data points, scores are sometimes stated in terms of **percentiles**. For percentiles, the ranked scores are divided into 100 equal groups (instead of into four equal groups, as is be the case with quartiles). The terminology in the use of percentiles differs from that of quartiles in a small but significant way: Percentiles name the percent of scores that are in **lower** hundredths. Thus, the top percentile is called the 99th percentile, and the lowest is the 0th. Contrast this with quartiles, where the lowest is called the first and the highest is called the 4th. Many health statistics are reported in terms of percentiles, as are some educational statistics.

The mean and the median are sometimes viewed as measures of location, because each in its own way is central to the data. If the data is categorical, the mean and median cannot be calculated, and the best that can be cited as typifying the data is the **mode**, which is the most frequent data value. For example, if five people prefer vanilla ice cream, three prefer chocolate, and two prefer strawberry, then the mode is vanilla (a fact unrelated to pie a la mode). If there is a tie for the most frequent data value, as there would be if five preferred vanilla, five chocolate, and two strawberry, then there is no unique mode. The greater the number of possibilities compared to the number of responses, the less likely the mode is to be of interest. The mode, cannot be considered to be an indicator of location for the data set as a whole when it is used with categorical data.

Measures of variability: Range, standard deviation and variance

In many practical situations, the most important feature of a data set is its variability. Indeed, if there is little or no variability, there is little or no need for statistics. If you survey dogs to find out how many legs they have, the answer, 4, occurs with so little variability that there is no need for statistical analysis. A naïve measure of variability is the difference between the largest and the smallest data points, which is called the **range**. For the Green Bay linemen in Example 3, the range is the largest number, 337, minus the smallest, 273, or 64. By itself, however, the range is a dubious measure of data variability, because if there are outliers, they alone determine the range, regardless of how much variability there is in all the other data points. A more balanced and commonly used measure of variability in data is the **standard deviation**.

To find the **standard deviation** of a set of numbers:

1. Find their mean
2. Subtract the mean from each number in the set and square each result
3. Find the mean of the squares found in step 2
4. Take the square root of the result in step 3

Example 4. Find the standard deviation of the data set -2, 0, 4, 5, 8.

Solution: 1. Find the mean of these numbers: $\frac{-2+0+4+5+8}{5} = 3$.

2. Subtract the mean from each data point and square: $(-2-3)^2 = 25$

$$(0-3)^2 = 9$$
$$(4-3)^2 = 1$$
$$(5-3)^2 = 4$$
$$(8-3)^2 = 25$$

3. Take the mean of these squares: $\frac{25+9+1+4+25}{5} = \frac{64}{5}$

4. Take the square root: $\sqrt{\frac{64}{5}} = \frac{8}{\sqrt{5}} = \frac{8\sqrt{5}}{5}$ or about 3.578.

In practice, standard deviations are calculated electronically. The only reason for calculating a standard deviation by hand is to be sure you understand the procedure.

If the standard deviation sounds a bit far fetched, here is a bit of the thinking behind it. In a data set with no variability, all the data points are the same, and they all equal the mean. It is reasonable to say that the variability of that data set is 0. The standard deviation measures the "distance" between the given data and data which has no variability at all. That distance is based on the Pythagorean theorem, extended to higher dimensions, averaged for the number of data points. Problems 9-11 below explore this in more detail.

Exercises 15.4

1. For each kind of data (categorical, ordinal, ratio) explain why it does or does not make sense to calculate:

   a. A mean b. A standard deviation.

2. A gambler at a roulette wheel bet $1 on red 38 times in a row. She won $36 dollars once but got no payoff at all on the other 37 plays. Considering that she paid $1 to play the game each time, what was her average payoff for each play?

3. a. An airline sold seats on a certain flight for three different prices. If the plane has 82 passenger seats, and the airline sells 31 seats at $179, 25 seats at $217, and 13 seats at $309, what is their mean revenue per seat on this flight? (Note: Not all seats were sold.)

   b. If the airline instituted a 20% price increase, but the number of each kind of tickets sold remained the same, what would be the mean revenue per coach seat on this flight?

4. Lorenzo's quiz scores to date are 80, 75, 72, 83, 68, 73, and 72. He has three more quizzes to take. What must be his average score on those three quizzes if he is to average of 80 on all ten quizzes?

5. How does the mean of 18, 27, and 30 compare with the mean of:

   a. 118, 127, and 130? b. 518, 527, and 530?

   c. If k is a number, what is the mean of k+18, k+27, and k+30?

   d. Complete the following generalization: If m is the mean of a set of numbers and k is added to each of the numbers, then ____________.

6. Compute the standard deviations for the data used in problem 5
   a. For 18, 27, and 30 b. For 118, 127, and 130
   c. For 518, 527, and 530
   d. Compare your answers from parts a, b, and c, and explain why what you have found is reasonable.

7. a. For which percentile is the median is the upper boundary point?
   b. Explain your answer to part a.

8. Here are five box and whiskers plots. Arrange them in ascending order of variability.

a.
b.
c.
d.
e.

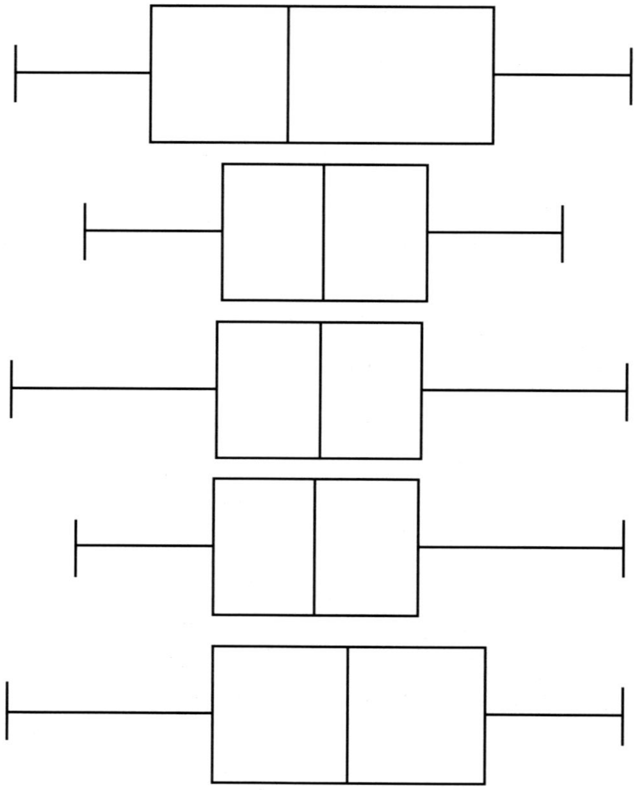

Exploration. The rationale for the standard deviation (problems 9-11)

9. a. On a number line plot the data sets $S=\{5, 8, 20\}$ and $T=\{5, 5, 7, 19\}$.
   b. Which of the two sets S and T seems to you to be the more spread out –i.e, the more variable? Explain your reasoning.
   c. Compute the means of sets S and T.
   d. Compute the standard deviations for the sets S and T.
   e. When S and T are plotted on the number line, does the set which the smaller standard deviation appear to be the less spread out set?

10. The Pythagorean theorem for the distance between two points (x_1,y_1) and (x_2,y_2) in the coordinate plane is $\sqrt{(x_2 - x_1)^2 + (y_2 - y_2)^2}$. In a similar way, the distance between two points (x_1,y_1,z_1) and (x_2,y_2,z_2) in three-dimensional space is $\sqrt{(x_2 - x_1)^2 + (y_2 - y_2)^2 + (z_2 - z_1)^2}$. Even though we cannot visualize higher dimensional spaces, four numbers (x, y, z, w) can be interpreted as a point in four-dimensional space, and the distance between two such points (x_1,y_1,z_1,w_1) and (x_2,y_2,z_2,w_2) is $\sqrt{(x_2 - x_1)^2 + (y_2 - y_2)^2 + (z_2 - z_1)^2 + (w_2 - w_1)^2}$.

    a. Calculate the distance between (5, 8, 20) and (11, 11, 11).

    b. Calculate the distance between (5, 5, 7, 19) and (9, 9, 9, 9).

11. Using distance to measure variability is a good start, but it is not adequate, because it assigns $\sqrt{138}$, to 5, 5, 7, 19, but only $\sqrt{134}$ to 5, 8, 20, even though 5, 5, 7, 19 is slightly less variable (more tightly bunched) than 5, 8, 20. Evidently, four data points put four numbers under the square root sign, but three data points puts only three numbers under the square root sign. That makes sets with more data points appear more variable than sets of fewer data points. To neutralize the effect of the number of data points, divide under the square root signs by the number of data points. For the three points 5, 8, 20, use $\sqrt{\frac{(5-11)^2 + (8-11)^2 + (20-11)^2}{3}}$, and for the four points, 5, 5, 7, 9 use $\sqrt{\frac{(5-9)^2 + (5-9)^2 + (7-9)^2 + (19-9)^2}{4}}$.

    These are the standard deviations for the three and four point data sets used here.

Vocabulary for Chapter 15

Bar graph (15.3)
Bias in sampling (15.2)
Box and whiskers (Box plot)(15.4)
Categorical data (15.1)
Control group (15.2)
Data point (15.4)
Double blind (15.2)
Element of a set (15.2)
Empty set (15.2)
Equal sets (15.2)
Histogram (15.3)
Line graph (15.3)
Mean (15.4)
Median (15.4)
Member of a set (15.2)
Mode (15.4)
Ordinal data (15.1)
Outlier (15.4)
Pie chart (15.3)
Percentile (15.4)
Placebo (15.2)
Placebo effect (15.2 exercises)
Population (15.2)
Quartile (15.4)
Range (of data) (15.4)
Ratio data (15.1)
Random sampling (15.2)
Randomizing device (15.2)
Sample (15.2)
Scatter plot (15.3)
Set (15.2)
Standard deviation (15.4)
Subset (15.2)
Treatment group (15.2)
Venn diagram (15.2)

Self-Quiz for Chapter 15

1. Explain the difference between categorical, ordinal, and ratio data.

2. For the data points plotted below, find:
   a. The range b. The mean
   c. The median d. The mode

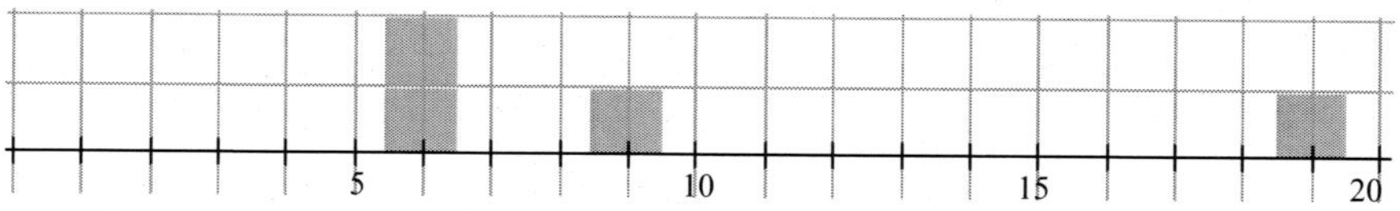

3. Write each element of sets *P*, *Q*, and *R* in the appropriate region of this Venn diagram.
   P={*a*, *b*, *e*, *g*} *Q*={*a*. *e*, *f*, *h*} *R*={*a*, *g*, *h*, *m*}

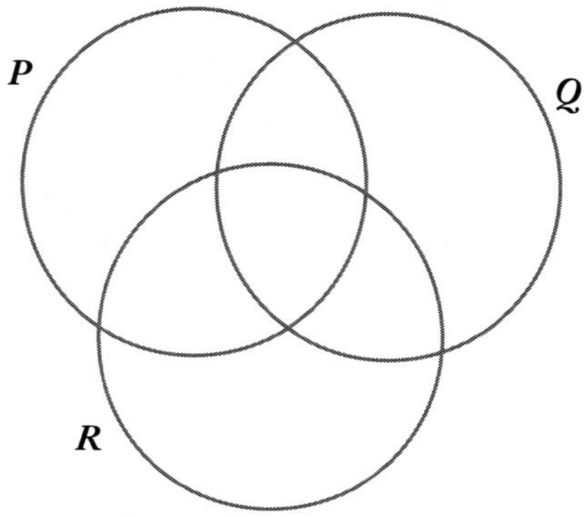

4. Here are the scores of a class on a test.

78	48	87	85
86	83	91	96
59	84	58	80
92	95	81	74
84	63	88	90
92	69	72	88
100	74	88	100

a. Calculate the range.

b. Calculate the median.

c. Construct a box plot for these scores, giving calculated values for the high and low scores, the median, and the quartile boundaries.

d. Construct a histogram of the grades from this test, if:

Scores 90 or higher are As

Scores from 78 to 89 are Bs

Scores from 65 to 77 are Cs

Scores from 50 to 65 are Ds

Scores less than 50 are Fs

5. Make a line graph of the annual net income of ExxonMobil Corp, using the figures below.

Year	2002	2003	2004	2005	2006	2007	2008
Net income in $billions	11.5	21.5	25.3	36.1	39.5	40.6	45.2

(Source Exxon web site, August, 2009.)

Chapter Sixteen
Counting and Probability

Here we will look at some basic counting techniques and then apply them to a new approach to the mathematics of chance, different in spirit but complementary to the observational point of view of chapter 15.

16.1 Basic Counting Tools

Example 1. Suppose someone has three shirts and two pairs of pants. In how many ways can that person choose a shirt and a pair of pants to wear?
Solution: The simplest idea is to list the possibilities. Suppose the shirts are red (R), green (G) and white (W), and the pants are black (B) and tan (T). Then the possibilities are RB, RT, GB, GT, WB, WT, and the person has six possible ways to choose shirt and pants. The **tree diagram** in figure 16.1.1 pictures the solution. The shirt can be selected in thee ways, and for each

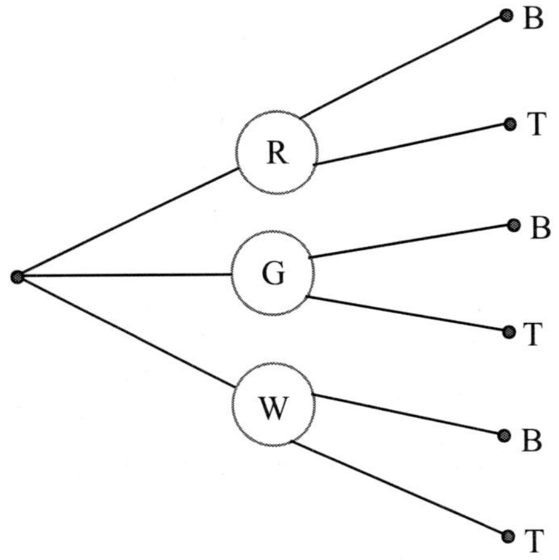

Figure 16.1.1

shirt there is a choice of two pairs of pants. Starting at the left in the diagram, each path through the tree corresponds to a choice of shirt and pants. Selecting the pants first and the shirt second would give us a different looking tree diagram, but the ends of the respective paths would match the same six options for selecting shirt and pants. Either way,

The number of choices = (number of shirts)$\times$(number of pants).

This solution suggests a more general fact. The number of ways to make successive choices is the product of the numbers of ways to make the individual choices. If two selections are to be made and the first can be made in n_1 ways and the second in n_2 ways, then the number of ways to make the two successive choices is $n_1 n_2$. This is called the **multiplication principle**.

> ***The Multiplication Principle:***
> If two successive choices are to be made, where the first choice can be made in n_1 ways and the second in n_2 ways, then the number of ways to make the two successive choices is $n_1 n_2$.

Tree diagrams and the multiplication principle apply to just as well when there are more than two choices to be made.

Example 2. A fast food chain offers 6 kinds of sandwiches and 4 kinds of drinks in 3 sizes: large, medium, and small. As a sales promotion, a wheel-of-fortune is set up showing all possible combinations of one sandwich and one drink of each size. A customer orders, spins the wheel, and gets the meal free if the wheel matches the order. How many outcomes are listed on the wheel?
Solution: There are 6 ways to choose a sandwich: $n_1 = 6$
There are 4 ways to choose a kind of drink: $n_2 = 4$
There are 3 ways to choose a drink size: $n_3 = 3$
Then $n_1 n_2 n_3 = 6 \cdot 4 \cdot 3 = 72$ is the number of outcomes shown on the wheel.

In general, the number of ways to make that sequence of choices is the product of the number of ways each of the choices can be made. If the first choice can be made in n_1 ways, the second in n_2 ways, and so on until the last can be made in n_t ways, then the total number of possible alternatives is $n_1 n_2 \cdots n_t$.

This property is called the **fundamental counting principle**.

> ***The Fundamental Counting Principle:***
> If a sequence of t choices is to be made and there are n_1 ways to make the first choice, n_2 ways to make the second choice, and so up to n_t ways for the last choice, then the number of ways to make the sequence of choices is $n_1 n_2 \cdots n_t$.

Example 3. A history quiz consists of 10 true-false questions. In how many different ways can the quiz be completed?
Solution: Think of the quiz as consisting of ten choices to be made successively, each of which can be made in two ways. Then there are $2 \times 2 \times 2 \cdots 2 = 2^{10} = 1{,}024$ ways to complete the quiz.

Exercises 16.1

1. Aline has 5 skirts, 8 blouses, and 4 scarves. In how many ways can she select a skirt, a blouse, and a scarf?

2. An automobile manufacturer makes three different styles of car, each of which comes in three trim levels and 12 different colors. How many different varieties of car does this company make?

3. In how many ways can first, second, and third place occur in a horse race with 10 entries?

4. There are no nonstop commercial flights between Portland, Oregon and Fairbanks, Alaska. I can fly from Portland to Seattle, then from Seattle to Fairbanks, or from Portland to Anchorage and then from Anchorage to Fairbanks. On the day I will travel there are three flights from Portland to Anchorage, and any of them could connect with any of 5 flights from Anchorage to Fairbanks. Alternatively, there are 8 flights from Portland to Seattle, each of which can connect with with four flights from Seattle to Fairbanks. Using the flights mentioned, in how many ways can I fly from Portland to Fairbanks?

5. Sharon she has won a dream trip. She can choose any one of 9 cities, any one of 6 dates, and any one of 3 hotels in each city. How many possibilities does Sharon have to choose from?

6. Mary wants to redecorate her kitchen. She wants wallpaper on one wall, paint on the other three walls (all three walls the same), and vinyl covering on the floor. She has a choice of 5 vinyl patterns, 4 paint colors, and 12 wallpaper patterns. Assuming any of the combinations would match, how many combinations does Mary have to choose from?

7. How many license plates are possible if each consists of 2 letters followed by 4 digits?

8. In some states, license plates use a county designation. Suppose a state uses the county designation and then 6 digits.
   a. How many plates are possible for each county?
   b. If there are 105 counties in the state, how many plates are possible for the state?

9. In a certain state, all license plates consist of 2 letters followed by 4 digits, 3 letters followed by 3 digits, or 5 digits. How many different license plates are possible?

10. How many different license plates can be made if the first 2 symbols are letters and the last 3 digits are digits, the first of which cannot be zero?

11. In how many ways can a true-false test consisting of 20 questions be completed?

12. How large a population is necessary to assure that at least 2 people have exactly the same 3 initials in the same order? Assume that all people have three initials.

13. Each day Roberto is allowed to watch cartoons for an hour. During the first day he has a choice of 3 cartoon shows, the second day only 2 cartoon shows, but the third day he has a choice of 4 cartoon shows. In how many different ways can he watch for the 3 days?

14. Pierre took a 10-question, multiple-choice quiz in which each question had 5 possible answers. In how many ways could he answer the quiz?

16.2 Permutations

Have you ever rearranged the letters of a word to make another word, for example, changing POTS to STOP or to OPTS? To put things in a different order is to **permute** them, and the different orderings are called **permutations.**

Example 1. In how many orders can the letters of the word MORN be arranged?
Solution: Each arrangement results from successive choices of the four letters. Any of the four letters can be first and any of the remaining three can be second, so by the fundamental counting principle there are 4×3 ways to choose the first two letters. Continuing this way, there are $4 \times 3 \times 2$ ways to choose the first three letters, and for each of these there is just one letter left to choose for the fourth position, making $4 \times 3 \times 2 \times 1 = 24$ possible orders or permutations of these four letters. Figure 16.2.1 shows these permutations.

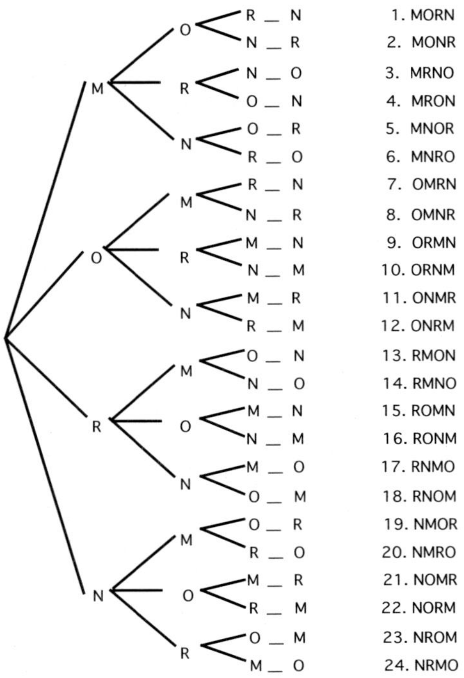

Figure 16.2.1

Example 1 generalizes to the arrangement of any number of n objects. If there are n objects, there are n choices for the first position, n-1 for the second, and so on, until thee is only 1 object left to place in the last position. Then, by the fundamental counting principle, there are $n(n-1)(n-2)\cdots 2\cdot 1$ permutations of n objects.

The fact that there are $n(n-1)(n-2)\cdots 2\cdot 1$ permutations of n objects makes long products of consecutive numbers turn up frequently in problems involving permutations and prompts special attention to them. The product of consecutive numbers from 1 to n is called $n!$ and is called ***n* factorial**. In that notation, the answer to example 1 would be written $4\times 3\times 2\times 1 = 24 = 4!$. Though the first few factorials are small (1!=1, 2!=2. 3!=6, 4!=24), the numbers $n!$ increase rapidly with n.

Example 2. How many different batting orders can be made up for a nine-person baseball team?
Solution: Any of the nine can bat first, any of the remaining 8 can bat second, and so on, yielding 9!=362,880 different batting orders.

Some problems about arrangements do not involve all the objects available. Example 3 is a case in point.

Example 3. A club with ten members will elect a slate of officers made up of a president, vice president, and secretary. If no member may hold more than one of the positions, how many different slates of officers can be nominated?

Solution: Again we may view this in terms of successive choices. There are ten individuals who may be elected president, and for each of these there are nine others who may be chosen as the vice president, making 10×9=90 combinations to fill the first two offices. For each of these 90 ways there are 8 possible secretaries, making 10×9×8=720 different slates.

If, instead of ten members, there were a different number, say n of them, the number of possible slates of three would be $n(n-1)(n-2)$.

Example 4. There will be four prizewinners in a contest. First prize is $100, second is $50, third is $20, fourth is $10. If 25 people enter the contest and nobody can win more than one prize, in how many ways can the prizes be awarded?
Solution: There are 25 possibilities for the first prize, 24 for the second, etc., or 25×24×23×22 ways in all, or 303,600 ways.

Example 5. How many different-looking orderings can be made with the letters of the word MOON?
Solution: In example 1 we found arrangements of the letters in the word MORN. Changing the R to a second O yields 24 arrangements of the letters of the word MOON, but since the two Os are identical, those arrangements do not all look different. For example, changing R to O changes both RMON and OMRN to OMON. In this way, two arrangements for MORN lead to each arrangement for MOON, so there are $\frac{24}{2} = 12$ different-looking arrangements of the letters in the word MOON.

To check this solution, use a tree diagram (figure 16.2.3).

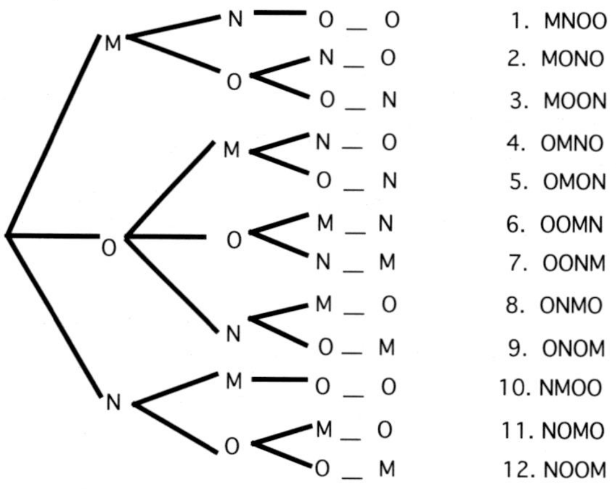

Figure 16.2.3

The basic idea in the solution to example 5 is: To count the number of different looking arrangements that can be made when objects are not all different, first count the number of different looking arrangements there would be if the objects were all different, then divide the result to account for the identical objects. In example 4 we divided by 2 because there are two ways to arrange the identical objects. If there were three identical objects, we would have to divide by 3! = 6, because that is how many ways you can arrange three identical objects.

Example 6. How many different looking arrangements can be made with the letters of the word SASS?
Solution 1: Count with a tree diagram, as in figure 16.2.4. This amounts to reasoning as follows: there are four places to put the A, and once the A is placed, put an S in each of the other three places. Hence there are four possible arrangements.

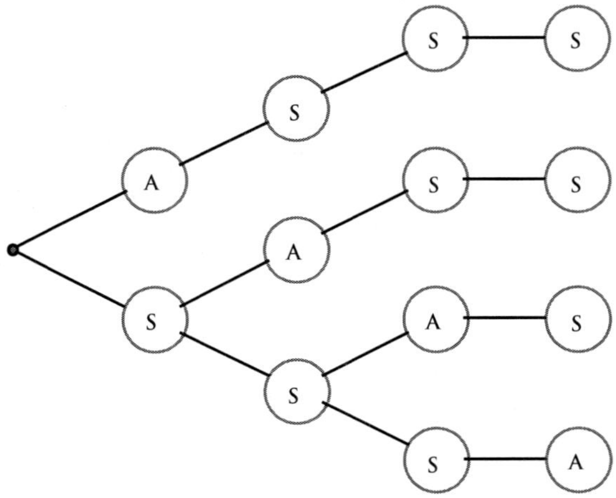

Figure 16.2.4

Solution 2: First label the Ss as S_1, S_2, S_3, so we can distinguish them from each other. With distinguishable letters, there are 24 different orderings of the letters in the word $S_1AS_2S_3$. Figure 16.2.5 shows all of these arrangements, as well as how they would look if the tag on every S were removed. Evidently, with 3 indistinguishable objects, we must divide by 3! = 6 to get $\frac{24}{6}$ = 4 distinguishable permutations of 4 objects.

The fraction encountered above as $\frac{4!}{3!}$ is relatively common in this part of mathematics. Such fractions can look intimidating, but they can be reduced a great deal, which saves a lot of work. For example, $\frac{10!}{8!}$ is really

$\frac{10!}{8!} = \frac{10 \cdot 9 \cdot 8 \cdot 7 \cdot 6 \cdot 5 \cdot 4 \cdot 3 \cdot 2 \cdot 1}{8 \cdot 7 \cdot 6 \cdot 5 \cdot 4 \cdot 3 \cdot 2 \cdot 1}$. To multiply out either the numerator or the denominator would be a lot of work, but it can be avoided by canceling all the factors from 1 up to 8, leaving only 10•9=90.

What if more than one kind of letter is repeated? As before, we first pretend all the letters are different, then divide to make up for the multiple counting caused by identical letters. Now, however, we divide by a product of factorials. For example, if one letter is repeated three times and a second letter is repeated twice, we divide by $3! \times 2! = 6 \times 2 + 12$. More generally, if one letter is repeated p times, a second r times, and a third t times, divide by $p!r!t!$, and so on. For each repeated letter, divide by the factorial of the number of repetitions.

Example 7. How many different looking arrangements can be made with the letters of MISSISSIPPI?
Solution: If all the letters were different, there would be 11! different-looking arrangements. However, there are four Is, four Ss and two Ps, so we must divide by 4!4!2!, which yields

$$\frac{11!}{4! \times 4! \times 2!} = \frac{11 \times 10 \times 9 \times 8 \times 7 \times 6 \times 5 \times 4 \times 3 \times 2 \times 1}{(4 \times 3 \times 2 \times 1)(4 \times 3 \times 2 \times 1)(2 \times 1)}$$

. This fraction looks truly scary, but the entire denominator cancels. You can see that the first factor in parentheses in the denominator cancels the last four factors in the numerator, and the rest of the denominator, which is $(4 \times 3 \times 2 \times 1)(2 \times 1)$, cancels against the 8 and the 6 in the numerator, the entire fraction to $11 \times 10 \times 9 \times 7 \times 5$. This is a very large number, 34,650, but after all the canceling the arithmetic is manageable.

Exercises 16.2

In problems 1–12, find how many different looking orders can be made with the letters of the given words. Answer the later parts of each problem by using the results of the earlier parts.

1. a. AN b. MAN c. MANY
2. a. ON b. ONE c. HONE d. PHONE
3. a. ME b. ELM c. LAME d. BLAME
   e. RAMBLE f. RAMBLES g. SCRAMBLE
4. a. TOES b. SOOT c. TOOT
5. a. SANE b. SEEN c. SEES
6. a. WAITERS b. SWEATER c. SWEETER
   d. TEETERS
7. ALABAMA
8. OKEEFENOKEE
9. NGORONGORO
10. ZOONOOZ (publication of the San Diego Zoo)
11. ABRACADABRA
12. FIDDLEFADDLE

13. How many batting orders are possible if a baseball coach must first choose nine of his fifteen players to be in the game?

14. In how many orders can fifty people line up to get on a bus? Do not multiply out your answer.

15. The American League East has 5 teams. In how many orders can the teams finish:
    a. With the Boston Red Sox in first place.
    b. With the Red Sox in first and the Baltimore Orioles in last.
    c. With the Tampa Bay Devil Rays in second.

16. A club with twenty-five members wants to choose a president, vice president, secretary, and treasurer. In how many ways can this be done if nobody can hold more than one office?

16.3 Combinations

Permutations count not only the objects selected but also the order in which the selections are made. We now consider questions in which the order does not matter. A number arrived at in this way is called a **combination**.

Example 1. A committee of ten people is to send three representatives to a conference. How many possible groups of three are there?
Solution 1: As in example 3 of section 2, we are to select three people from ten candidates, but the order in which the three are selected does not matter, whereas in the earlier example it was critical (the first person was president, the second vice president, and the third secretary). Dividing the answer to the earlier problem, $10\times9\times8 = 720$, by 3!, the number of orders in which each trio can be selected, yields 120 as the number of possible groups of three.
Solution 2: Imagine the ten candidates lined up, each holding a card. Three of the cards have the letter R (for "representative"), while the rest have Os (for "other"). Now the various choices of three representatives correspond to the different looking arrangements of the letters in the "word" OROOROOROO. As in section 2, we count these by first pretending that all ten letters are different, which would lead to 10! arrangements, then dividing out by 3! (for the 3 Rs) and 7! (for the seven Os) to get $\dfrac{10!}{3!7!} = 120$.

Example 1 counts the number of three-member subsets that can be formed using members of a ten-member set. This is said to be the number of **combinations of ten objects, taken three at a time**. More generally, if n objects are available, any choice of k of them (where $0 \le k \le n$) is said to be a combination of n objects taken k at a time.

Example 1 shows a general pattern. For example, to count the number of 4-member subsets that can be formed from a 15 member set, one way would be to choose the four members in order (15 ways to choose the first, 14 ways to choose the second, 13 ways to choose the third, and 12 ways to choose the fourth, making 15×14×13×12 ways, counting order) and then divide by the number of orders that each selection of four could have been chosen, which is 4! or 24. A second way would be to first pretend that give R cards to 4 of the members and O cards to the others (to 15-4 or 11). Then different selections of 4-member subsets correspond to different looking arrangements of four Os and 11 Rs. There are $\frac{11!}{4!\times(11-4)!}=\frac{11!}{4!\times 7!}$ such arrangements. In general,

> **The number of different combinations of *n* objects taken *k* at a time is $\frac{n!}{k!(n-k)!}$. It is abbreviated $\binom{n}{k}$.**

We want to use the symbol $\binom{n}{k}$ and the formula $\frac{n!}{k!(n-k)!}$ even when all n objects are chosen. But there is only one way to choose all of the n objects available, so for any value of n, $\binom{n}{n}=1$. But when $k = n$, the formula above states that $\frac{n!}{n!(n-n)!}=\binom{n}{n}=1$. In order to make that true, we define 0! as 1.

Example 2. The top four teams in a six-team hockey league will be in the playoffs. How many different sets of four teams could be in the playoffs?

Solution 1: There are $\binom{6}{4}$ ways to select four teams out of six. This is

$$\frac{6!}{4!(6-4)!}=\frac{6\cdot 5\cdot 4\cdot 3\cdot 2\cdot 1}{(4\cdot 3\cdot 2\cdot 1)(2\cdot 1)}=15.$$

Solution 2: Selecting four teams to be in the playoffs is the same as selecting two teams not to be in the playoffs, which can be done in $\binom{6}{2}$ ways, and $\binom{6}{2}=\frac{6!}{2!(6-2)!}=\frac{6\cdot 5\cdot 4\cdot 3\cdot 2\cdot 1}{(2\cdot 1)(4\cdot 3\cdot 2\cdot 1)}=15.$

The reasoning in solution 2 holds generally; every selection of k objects from n available objects determines a corresponding group of objects which are not selected. Colloquially, each in-group determines a corresponding outgroup. You can see from our formula that in general

$$\binom{n}{k}=\binom{n}{n-k}.$$

The next few examples show how the techniques introduced here make it possible to treat some counting problems which would otherwise be very difficult.

Example 3. Four-card hands are dealt from an ordinary deck of 52 cards. How many of these hands do not have four aces?
Solution: It is easy to count the hands which have four aces, since there is only one of them. Of the $\frac{52}{4}$ different four-card hands, one that has four aces, and the other $\frac{52}{4} - 1$, or $\frac{52!}{4!48!} - 1 = 541,450 - 1 = 541,449$ do not have four aces.

Sometimes we have to combine the fundamental counting principle with our knowledge of combinations, as in the following example.

Example 4. How many five-card poker hands have exactly three kings?
Solution: Four of the 52 cards in the deck are kings. Each hand with three kings may be considered as resulting from two successive choices:

1. Choose the kings.
2. Choose two of the 48 cards that are not kings.

There are $\binom{4}{3} = 4$ ways to choose the kings and

$\binom{48}{2} = \frac{48!}{2!46!} = \frac{48 \cdot 47}{2} = 24 \cdot 47 = 1{,}128$ ways to choose the other cards. Then, by the fundamental counting principle, there are $4 \cdot 1{,}128 = 4{,}512$ different hands with three kings.

Example 5. A bridge hand consists of 13 cards, dealt from an ordinary 52-card deck. How many bridge hands have eight spades, four hearts, and one diamond?

Solution: There are 13 cards of each suit, so there are $\binom{13}{8}$ ways to select eight spades, $\binom{13}{4}$ ways to select four hearts, and $\binom{13}{1}$ ways to select the diamond. Then there are

$$
\begin{aligned}
\binom{13}{8}\binom{13}{4}\binom{13}{1} &= \frac{13!}{8!5!} \cdot \frac{13!}{4!9!} \cdot \frac{13!}{1!12!} \\
&= \frac{13 \cdot 12 \cdot 11 \cdot 10 \cdot 9}{5 \cdot 4 \cdot 3 \cdot 2 \cdot 1} \cdot \frac{13 \cdot 12 \cdot 11 \cdot 10}{4 \cdot 3 \cdot 2 \cdot 1} \cdot 13 \\
&= 13 \cdot 11 \cdot 9 \cdot 13 \cdot 11 \cdot 5 \cdot 13 \\
&= 11{,}962{,}665
\end{aligned}
$$

hands of the type specified.

Exercises 16.3

1. Seven people hope to get seats on a flight, but there is room for just five of them. How many different sets of five of them could go?

2. One hundred twenty passengers hold tickets for a flight on Overbook Airlines but there is room for only 118. How many pairs of ticket-holders may be "bumped" off the flight?

3. A pizzeria offers ten optional toppings on its pizzas. How many different kinds of pizzas have just three of the optional toppings?

4. There are eight courses, all at different times, that Ali could take this year, but she can only take three of them. How many choices of three courses are available to her?

5. A soccer coach must select 11 of 15 players to start a game. How many different elevens can be chosen?

6. Lucy knows 14 pieces, but she will get to play just four of them in her audition. How many different sets of four pieces could she choose?

7. How many different five-card poker hands have:
   a. Two jacks and three queens
   b. Four aces and a king
   c. Two tens, two queens, and a king
   d. All diamonds

8. There are thirteen men and fifteen women available to serve on a twelve-person jury.
   a. How many different all-male juries could be formed?
   b. How many different all-female juries could be formed?
   c. How many juries of six men and six women could be formed?
   d. How many juries of 8 women and 4 men could be formed?

9. How many bridge hands have:
   a. All clubs
   b. Twelve clubs
   c. Ten clubs and three hearts
   d. No clubs
   e. Only spades and hearts

10. A Chinese restaurant has a dinner special of three dishes from the eight listed in group A and two from the eight in group B. How many different dinner specials if all five selections are different?

Exploration: A pattern in Pascal's Triangle (problems 11-14)

11. In the notation for combinations, $\binom{n}{k} = \dfrac{n!}{k!(n-k)!}$, Pascal's triangle is

$$\binom{0}{0}$$
$$\binom{1}{0} \quad \binom{1}{1}$$
$$\binom{2}{0} \quad \binom{2}{1} \quad \binom{2}{2}$$
$$\binom{3}{0} \quad \binom{3}{1} \quad \binom{3}{2} \quad \binom{3}{3}$$
$$\binom{4}{0} \quad \binom{4}{1} \quad \binom{4}{2} \quad \binom{4}{3} \quad \binom{4}{4}$$

Remember: By definition, $0! = 1$.
Compute:

a. $\binom{1}{0}\binom{1}{1}+\binom{1}{1}\binom{1}{0}$

b. $\binom{2}{0}\binom{2}{2}+\binom{2}{1}\binom{2}{1}+\binom{2}{2}\binom{2}{0}$

c. $\binom{3}{0}\binom{3}{3}+\binom{3}{1}\binom{3}{2}+\binom{3}{2}\binom{3}{1}+\binom{3}{3}\binom{3}{0}$

d. $\binom{4}{0}\binom{4}{4}+\binom{4}{1}\binom{4}{3}+\binom{4}{2}\binom{4}{2}+\binom{4}{3}\binom{4}{1}+\binom{4}{4}\binom{4}{0}$

12. Here are rows 0 to 4 of Pascal's triangle:

$$\begin{array}{ccccccccc} & & & & 1 & & & & \\ & & & 1 & & 1 & & & \\ & & 1 & & 2 & & 1 & & \\ & 1 & & 3 & & 3 & & 1 & \\ 1 & & 4 & & 6 & & 4 & & 1 \end{array}$$

a. Extend the triangle down to row ten, which begins with 1, 10, and 45.

b. Locate your answers to problem 11 in Pascal's triangle.

c. Using your answer to part a, make a prediction for the value of

$$\binom{5}{0}\binom{5}{5}+\binom{5}{1}\binom{5}{4}+\binom{5}{2}\binom{5}{3}+\binom{5}{3}\binom{5}{2}+\binom{5}{4}\binom{5}{1}+\binom{5}{5}\binom{5}{0}$$

d. Evaluate the expression in part c by doing the arithmetic.

e. Find the answer to part d in the tenth row of Pascal's triangle.

14. **This problem explains the discovery made in problems 11 and 12.** There are 5 men and 5 women available to be on a 5-person committee. Write expressions in terms of combinations $\binom{n}{k}$ for the number of ways to form the committee:

a. With no restrictions whatever

b. No men and five women

c. One man and four women

d. Two men and three women

e. Three men and two women

f. Four men and one woman

g. Five men and no women

h. Explain why the sum of the answers in parts b-g must equal the answer to part a.

16.4 Probability Spaces

How do you measure uncertainty? Gamblers, insurance companies, and others have wrestled with that issue for years. Here we introduce a type of mathematical model called a probability space, which helps us understand and quantify uncertainty.

We begin with an example, tossing a coin, which we think of as an experiment with two possible outcomes, heads or tails. Events with just one outcome are called **simple events,** while events with more than one outcome are called **compound events.** To each possible outcome, we assign a number, called the **probability** of that outcome, to indicate how likely it is. If we view the coin as fair—as likely to land one way as the other—we assign to each outcome the probability $\frac{1}{2}$, but if we think the coin is twice as likely to land heads as tails, we assign the probabilities $\frac{2}{3}$ to heads and $\frac{1}{3}$ to tails.

More generally, consider an experiment that has a finite number of possible outcomes The set of all such outcomes, S, is called the **possibility set** for the experiment, and subsets of S are called **events**. Each possible outcome of the experiment (each member of S) is called a **simple event**. To each simple event x, assign a number $P(x)$, called the **probability of x**, which is a measure of how likely the outcome x is to occur. Assign the probabilities in such a way that:

1. The sum of the probabilities of all outcomes in S is 1.
2. For each simple event x, $P(x) \geq 0$.

If A is an event (subset of S) containing more than one outcome, $P(A)$, the probability of A, is defined as the sum of the probabilities of the outcomes (simple events) in A. Therefore, assigning probabilities to the simple events implicitly determines the probabilities of all other events as well.

The possibility set S, together with the probabilities assigned to the various events, is called a **probability space.**

Example 1. An experiment consists of noting the number that turns up when a die is rolled. Identify the possibility set and assign probabilities that are appropriate if the die is fair (all faces are equally likely). Then find the probability of rolling an odd number.
Solution: The possibility set $S = \{1,2,3,4,5,6\}$. If each outcome is as likely as any other, $P(1) = P(2) = P(3) = P(4) = P(5) = P(6)$. Also, since the sum of the probabilities of the outcomes is 1,

$$P(1) + P(2) + P(3) + P(4) + P(5) + P(6) = 1.$$

To satisfy these conditions, assign 1/6 as the probability of each outcome. Now the probability of rolling an odd number is the probability of the subset $A = \{1,3,5\}$. Since the probability of any event is the sum of the probabilities of its simple events,

$$P(A) = P(\{1,3,5\}) = P(1) + P(3) + P(5) = \frac{3}{6} = \frac{1}{2}.$$

Example 2. Modify example 1 to reflect the fact that the die is loaded to turn up 6 one-third of the time, while the other five outcomes are all equally likely.
Solution: As in example 1, $S = \{1,2,3,4,5,6\}$, and

$$P(1) + P(2) + P(3) + P(4) + P(5) + P(6) = 1.$$

This time, however, $P(6) = \frac{1}{3}$, so $P(1) + P(2) + P(3) + P(4) + P(5) = \frac{2}{3}$.

Then $P(1) = P(2) = P(3) = P(4) = P(5) = \left(\frac{1}{5}\right)\frac{2}{3} = \frac{2}{15}$.

Here $P(A) = P(1) + P(3) + P(5) = \frac{2}{15} + \frac{2}{15} = \frac{2}{15} = \frac{2}{5}$.

Examples 1 and 2 show clearly that a probability space is a model. The choice of the possibility set and the assignment of probabilities are at the discretion of the model maker. The model will be useful to the extent that it sheds light on the situation under consideration.

We have considerable freedom in creating probability spaces, but there are a few basic properties that hold in every probability space.

1. If $P(A)$ is the probability that an event A will occur, then

the probability that A will *not* occur is $1 - P(A)$.

2. If A is an event which is certain to occur, $P(A) = 1$, but if A is certain *not* to occur, then $P(A) = 0$.
3. If every possible outcome in an event A is also in an event B (so that A is a subset of B), then $P(A) \leq P(B)$.

Example 3. The outcome of an experiment is the total shown when a pair of fair dice is rolled. Describe the probability space for this experiment.
Solution: Since each die will show a number from 1 to 6, the possibility set is $S = \{2,3,4,5,6,7,8,9,10,11,12\}$. If all of these eleven possibilities were equally likely, we would assign them each the probability $\frac{1}{11}$, but experience shows that all the possibilities are not equally likely. For example, 7 is a more common total than 12. Figure 16.4.1 shows why.

1st die						
6	7	8	9	10	11	12
5	6	7	8	9	10	11
4	5	6	7	8	9	10
3	4	5	6	7	8	9
2	3	4	5	6	7	8
1	2	3	4	5	6	7
2nd die	1	2	3	4	5	6

Figure 16.4.1

If the dice are fair, all 36 table entries in figure 16.4.1 are equally likely. Model that by assigning each case the probability $\frac{1}{36}$. Only one case has a total of 2, so $P(2) = \frac{1}{36}$. Similarly, $P(3) = \frac{2}{36} = \frac{1}{18}$, $P(4) = \frac{3}{36} = \frac{1}{12}$, $P(5) = \frac{4}{36} = \frac{1}{9}$, etc., so we have the following assignment of probabilities:

Outcome	2	3	4	5	6	7	8	9	10	11	12
Probability	$\frac{1}{36}$	$\frac{2}{36}$	$\frac{3}{36}$	$\frac{4}{36}$	$\frac{5}{36}$	$\frac{6}{36}$	$\frac{5}{36}$	$\frac{4}{36}$	$\frac{3}{36}$	$\frac{2}{36}$	$\frac{1}{36}$

We will see other cases in which the question of designing a probability space that accurately models a situation is interesting and subtle. That issue is particularly important in questions of **conditional probability**, which are considered in section 16.6.

Example 4. A pair of fair dice is rolled. What is the probability that the total shown is no greater than 10?

Solution: The probability space for this example is the one from Example 4. The event A, that the total shown is no greater than 10, is $A = \{2,3,4,5,6,7,8,9,10\}$. Then

$$P(A) = P(2) + P(3) + P(4) + P(5) + P(6) + P(7) + P(8) + P(9) + P(10)$$

$$= \frac{1}{36} + \frac{2}{36} + \frac{3}{36} + \frac{4}{36} + \frac{5}{36} + \frac{6}{36} + \frac{5}{36} + \frac{4}{36} + \frac{3}{36} = \frac{33}{36} = \frac{11}{12}$$

Another solution: Let $\overline{A}$ denote the **complement** of A—the set of outcomes **not** in A.

Then $\overline{A} = \{11,12\}$, and $P(\overline{A}) = P(11) + P(12) = \frac{2}{36} + \frac{1}{36} = \frac{3}{36} = \frac{1}{12}$. Then

$$P(A) = 1 - P(\overline{A}) = 1 - (P(11) + P(12)) = 1 - \left(\frac{2}{36} + \frac{1}{36}\right) = \frac{36}{36} - \frac{3}{36} = \frac{33}{36} = \frac{11}{12}.$$

Example 5. What is the probability that a five-card poker hand contains exactly three aces?

Solution: There are $\binom{52}{5}$ different poker hands in the possibility set S, and they are all equally likely if the deck is well shuffled, so we assign to each hand the probability $\frac{1}{\binom{52}{5}}$. To find how many of these hands have exactly three aces, note that there are $\binom{4}{3}$ ways to select the aces and $\binom{48}{2}$ ways to select two cards that are not aces, making $\binom{4}{3}\binom{48}{2}$ such hands.

Then the probability of the hand having just three aces is $\frac{\binom{4}{3}\binom{48}{2}}{\binom{52}{5}}$. This is evaluated in problem 5 below.

Example 6. King Arthur, in a burst of egalitarian spirit unusual for his time, arranged for himself and 39 others to be seated randomly about the round table. What was the probability of Arthur sitting next to his favorite, the fair Guinevere?

Solution 1: All that matters here is where Arthur and Guinevere sit. The other 38 people are essentially irrelevant. The $\binom{40}{2}$ ways to select a pair of seats for the royal couple are equally likely. Of these, 40 pairs have Arthur and Guinevere next to each other, since each such pair can be thought of as one of the 40 seats and the one to its right.

Thus the probability is $\frac{40}{\binom{40}{2}} = \frac{40}{\frac{40!}{2!\cdot 38!}} = \frac{40}{\frac{40\cdot 39}{2}} = \frac{40\cdot 2}{40\cdot 39} = \frac{2}{39}$.

Solution 2: Wherever Arthur sits, there are 39 equally likely places left for Guinevere. Two of these are next to him, so the probability is $\frac{2}{39}$.

Example 7 shows how to treat probability experiments that involve two or more stages.

Example 7. A fair die is rolled. If the number 1 turns up, a ball will be selected at random from urn R containing two white balls and three black ones. If any other number turns up on the die, a ball will be chosen at random from urn T, which contains two black balls and one white one. Find the probability that the ball obtained by this procedure is black.
Solution: The ball drawn will be either black or white, so it is easy to list the possibility set S = {black, white}. To assign probabilities to the two outcomes, use the tree diagram in figure 16.4.2.

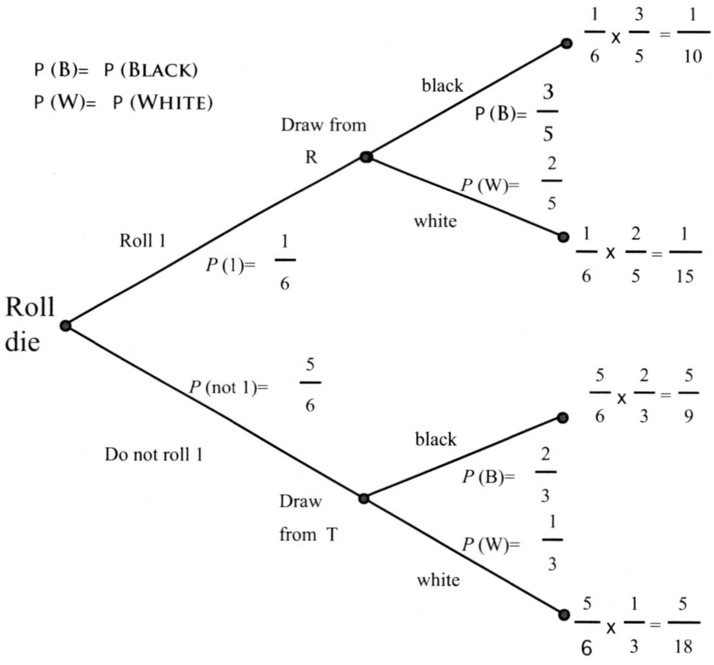

Figure 16.4.2

Each path through the tree is a way to obtain one of the possible outcomes. We assign to each path the product of the probabilities of the various legs of that path, as shown in figure 16.4.2. Now the event of drawing a black ball can happen in two ways—either by rolling a 1 on the die and drawing a black ball from the first urn or by rolling a number other than 1 on the die and drawing a black ball from the second urn. The probabilities of these happening are, respectively, $\frac{1}{6} \times \frac{3}{5} = \frac{1}{10}$ and $\frac{5}{6} \times \frac{2}{3} = \frac{5}{9}$, so the probability

of drawing a black ball is $\frac{1}{10} + \frac{5}{9} = \frac{59}{90}$.

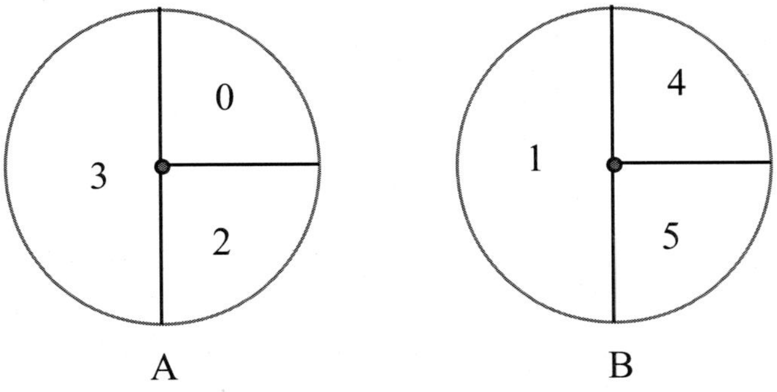

Example 8. For spinners A and B shown above, the probabilities of the various numbers are proportional to the sizes of the associated areas. Thus for spinner A, $P(3) = \frac{1}{2}$, $P(0) = P(2) = \frac{1}{4}$. If spinners A and B are spun simultaneously, what is the probability that spinner A will show a larger number than spinner B?
Solution: Make a tree diagram to show all possible cases and their probabilities (Figure 16.4.3).

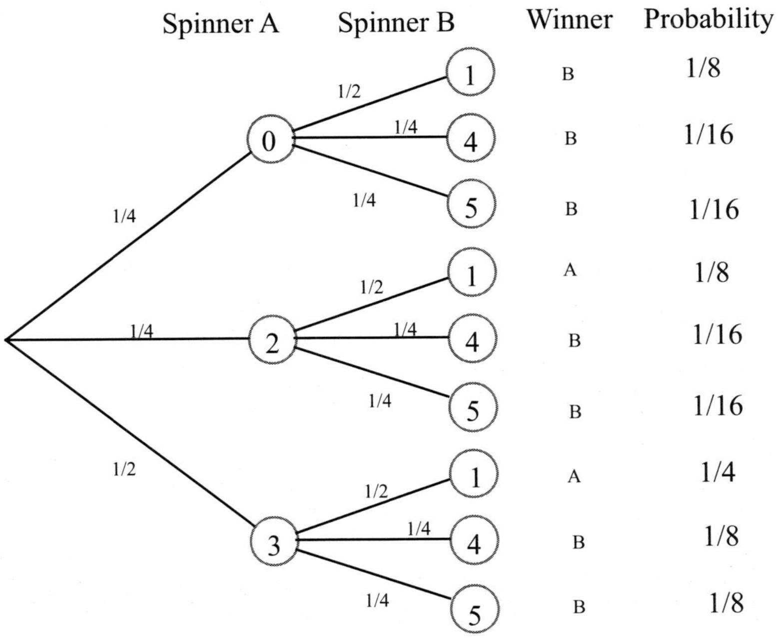

Figure 16.4.3

Add the probabilities in the cases where A is the "winner" (is larger than B) to get $\frac{1}{8}+\frac{1}{4}=\frac{3}{8}$ as the probability of A showing a larger number than B.

Odds

When you see a gent
Payin' all kinds of rent
For a flat that would flatten the Taj Mahal,
Call it sad, call it funny
But it's better than even money
That the guy's only doin' it for some doll!
Guys and Dolls

Many bettors, for example at horse races, describe relative likelihood in terms of odds rather than probabilities. The odds on (or for) an event A is the ratio of the probability that A will happen to the probability that A will not happen. This terminology arose from bookmakers (bookies). If you are given 10 to 1 odds when you bet on a certain horse, that means if the horse wins you will win \$10 for each dollar you bet.

Definition: If A is an event with probability $P(A)$, the odds on (or in favor of) A are defined as $\frac{P(A)}{1-P(A)}$.

Example 9.

a. If the probability of an event A is $\frac{1}{4}$, what are the odds in favor of A?

b. If the odds on an event B are 3 to 2, what is $P(B)$?

Solution: a. If $P(A) = \frac{1}{4}$, then $1 - P(A) = \frac{3}{4}$, and the odds on A are 1 to 3, which is the ratio of $\frac{1}{4}$ to $\frac{3}{4}$.

b. If $\frac{P(B)}{1-P(B)}=\frac{3}{2}$, solve this equation for $P(B)$.

$$\frac{P(B)}{1-P(B)}=\frac{3}{2}$$
$$2P(B)=3[1-P(B)]$$
$$2P(B)=3-3P(B)$$
$$5P(B)=3$$
$$P(B)=\frac{3}{5}$$

Exercises 16.4

In problems 1–4, describe the possibility set for each experiment.

1. The first letter of a person's family name is noted.

2. A pair of dice is rolled and the numbers that turn up are multiplied.

3. A card is drawn from a deck and its suit noted.

4. A voter ranks three candidates for office in order of preference.

5. The expression $\dfrac{\binom{4}{3}\binom{48}{2}}{\binom{52}{5}}$ from example 5 above is typical of the complicated arithmetic that arises in probability questions. The trick to evaluating those expressions is to keep sight of the usual rules for arithmetic in spite of the added complexity introduced by the new expressions.

   i. Recall the meaning of the "binomial coefficient" $\binom{n}{r}$ as $\dfrac{n!}{r!(n-r)!}$ to rewrite each factor of the given expression in terms of factorials.

   a. $\binom{4}{3} =$ b. $\binom{48}{2} =$ c. $\binom{52}{5} =$

   ii. Simplify each of the expressions found in parts a-c.

   For part b, for example, $\binom{48}{2} = \dfrac{48!}{2!(48-2)!}$. The parentheses indicate that 48-2 is to be found before the factorial sign is applied to the factor (48 – 2)!, so (48 – 2)! = (46)! and $\binom{48}{2} = \dfrac{48!}{2!(46)!}$.

   Write corresponding expressions for:

   d. $\binom{4}{3}$ e. $\binom{52}{5}$

   iii. Use the definition of $n!$ as the product of all the whole numbers from 1 to n to rewrite each expression found in part 2. For example,

$$\binom{48}{2} = \frac{48!}{2!(46)!} = \frac{48\times47\times46\times45\times44\times\cdots\times2\times1}{2\times1(46\times45\times44\times\cdots\times2\times1)}$$

f. Do the same for the expressions obtained in parts d and e above.

iv. Use the associative property of multiplication to rewrite the numerator with a factor equal to the largest factorial in the denominator. For example, rewrite $\frac{48!}{2!(46)!}$ as $\frac{48\times47\times(46\times45\times44\times\cdots\times2\times1)}{2\times1(46\times45\times44\times\cdots\times2\times1)}$.

g. Do the same for your results for step 3 above.

v. Recall from chapter 7 that any fraction with equal numerator and denominator is a form of 1 and may thus be canceled. Thus, for example,

$$\frac{48\times47\times(46\times45\times44\times\cdots\times2\times1)}{2\times1(46\times45\times44\times\cdots\times2\times1)} = \frac{48\times47}{2\times1}$$

h. Do the same for the expressions you calculated in part g.

vi. Cancel as much as possible in the fraction that remains. For example,

$$\frac{48\times47}{2\times1} = 24\times47$$

i. Do the same for the other two expressions.

vii. Substitute the results of step j into your original expression. You should have

$$\frac{24\times47}{\frac{52\times51\times50\times49\times48}{5\times4\times3\times2\times1}}$$

viii. Simplify by inverting the divisor and multiplying. In our example, we now have

$$24\times47\left(\frac{5\times4\times3\times2\times1}{52\times51\times50\times49\times48}\right)$$

ix. Reduce the remaining fraction as much as possible. In our example, you should have

$$\frac{47}{26\times17\times49}$$

Even after all that simplification, the fraction that remains is awkward and for most practical purposes is best rewritten as a decimal approximation, using a calculator.

6. A number from one to ten is chosen at random. What is the probability that it is:

a. Greater than 3

b. A prime number

7. The die in example 1 and the die in example 2 are rolled simultaneously. What is the probability that the total is:

   a. 2 b. 3

   c. 8 d. Greater than 5

   e. An odd number f. Eleven or less

8. A famous football coach, when asked why his teams rarely threw passes, answered, "When you throw the ball, there are three things that can happen, and two of them are bad." Comment.

9. A deck of cards is shuffled and a card is drawn. It is the two of diamonds. This card is laid aside and a second card is drawn. What is the probability that the second card is:

   a. A diamond b. A spade

   c. A two d. A three

   e. Not the jack of diamonds

10. Ten people line up single file in random order. What is the probability that two of them, Alice and Phil, are next to each other?

11. Cards numbered 1 through 5 are distributed in random order to Al, Bob, Cal, Don and Ed.

    a. Describe the possibility set for this experiment.

    b. What is the probability that Cal gets a 5?

    c. What is the probability that Don gets a smaller number than Ed?

    d. What is the probability that Al gets the 1, Bob the 2, and Cal the 3?

12. Find the probability of each of the following kinds of poker hands:

    a. A flush (five cards of the same suit)

    b. A full house (three cards of one kind and two of another kind)

    c. Four of a kind.

13. What is the probability that a bridge hand dealt from a well-shuffled deck will have exactly 12 cards of the same suit?

14. Two balls are selected at random from an urn containing 3 white balls and 6 black ones. What is the probability that both the selected balls:

    a. Are white

    b. Are black.

    c. Have different colors

15. In baseball, the conventional wisdom is for a pitcher who can throw a fastball and a curve to use the fastball with three balls and two strikes, because most pitchers can aim their fastball more accurately than their curve, thus minimizing the chance of walking the hitter. Suppose you can get your fastball over the plate 90% of the time and your curve 75% of the time, but, since the hitter expects a fastball, he can get a hit with it 40% of the time when it is in the strike zone, while he will get a hit on only 10% of the curves which come into the strike zone. Assuming the batter swings only at pitches in the strike zone, what is the probability you will get the batter out if you throw:

    a. Your fastball

    b. Your curve

16. A fair die is rolled. If a 1 or 2 is shown, a ball is selected at random from an urn containing 3 black balls and one white one, but if a 3, 4, 5, or 6 is shown on the die, a ball is drawn at random from a second urn, which contains 2 black balls and 5 white ones. Illustrate this two-step experiment with a tree diagram like figure 15.4.2, and compute the probability that the ball drawn by this process is black.

17. Bradley Efron of Stanford University designed the set of non-standard dice shown in figure 16.4.4. (Opened flat so you can see the numbers on all the faces, some of which repeat.)

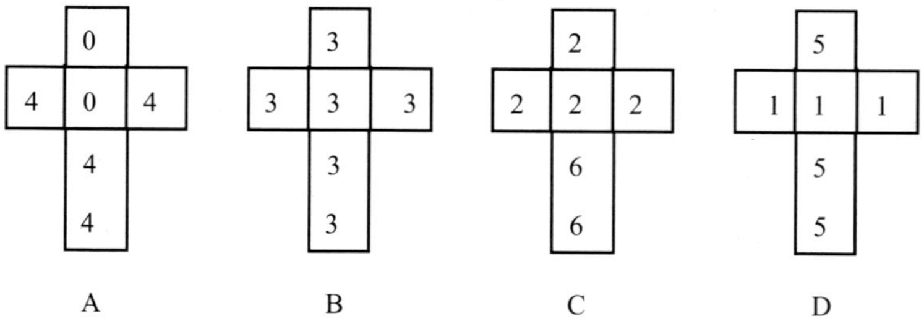

Figure 16.4.4

In contests between people each rolling one of these dice to see who rolls the highest number, find the probabilities that:

a. A beats B. b. B beats C. c. C beats D.

d. D beats A.

e. Comment on the meaning of the results from parts a through d.

16.5 Binomial Probabilities

A particularly important type of probability question involves repetitions of a simple experiment having two possible outcomes. These repetitious experiments are known as Bernoulli trials[1]. Both the possible outcomes and the probabilities of the experiment are assumed to be the same for each repetition. Tree diagrams and the binomial theorem are the keys to understanding Bernoulli trials, as our examples will show.

Example 1. A fair coin is tossed three times. Find the probability that it will land heads twice and tails once.
Solution: If we use the possibility set S = {3H, 2H1T, 1H2T, 3T}, the issue is how to assign probabilities to the members of S. For this we turn to the tree diagram in figure 16.5.1.

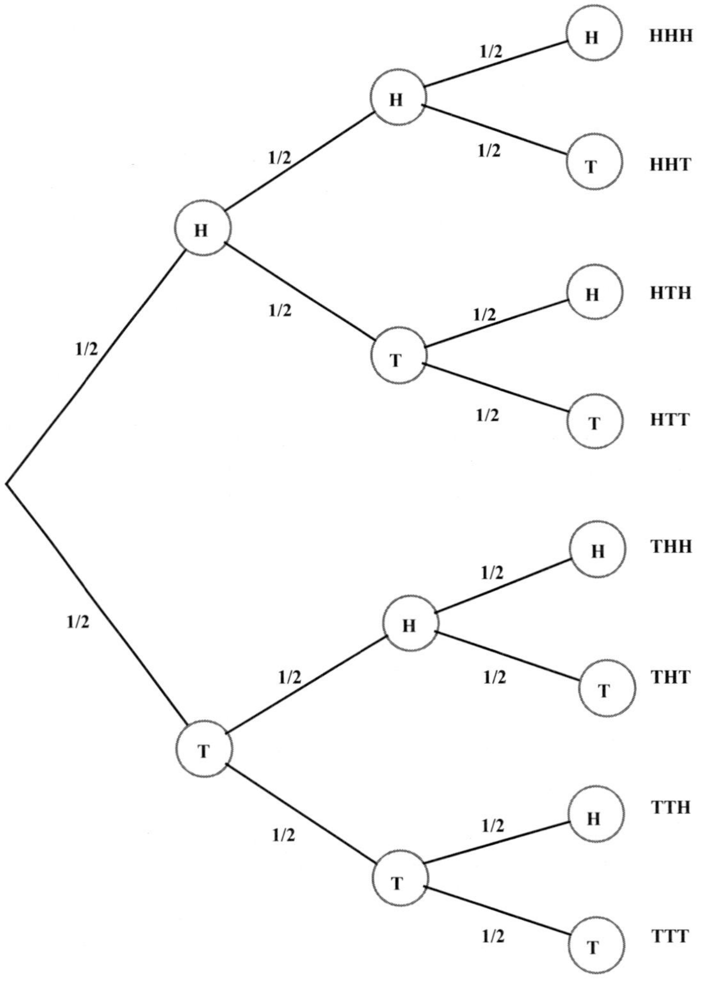

Figure 16.5.1

Here, as in example 7 of section 16.4, the probability of any path is found by multiplying the probabilities along each of its edges. If the coin is fair,

1 Jacob Bernoulli (1654–1705), See note, section 13.2

these probabilities are each $\frac{1}{2}$, and the probability of each path through the tree is then $\frac{1}{8}$. Three of the paths result in two heads and one tail, so the probability of getting two heads and one tail is $\frac{3}{8}$.

Had we used the possibility set

$$S' = \{HHH, HHT, HTH, HTT, THH, THT, TTH, TTT\}$$

corresponding to the paths in the tree diagram, then we would have assigned each outcome the probability $\frac{1}{8}$.

The probability of getting two tails and one head is the sum of probabilities of the simple events HHT, HTH, and THH, which is $\frac{3}{8}$.

We can represent the distribution of the various probabilities graphically as in figure 16.5.2. Figure 16.5.2 resembles a histogram, but the heights are theoretical probabilities instead of numbers of observed cases that are tallied in histograms. The total shaded area is 1, which reflects the fact that the sum of the probabilities of the four possible outcomes (numbers of heads and tails) is 1.

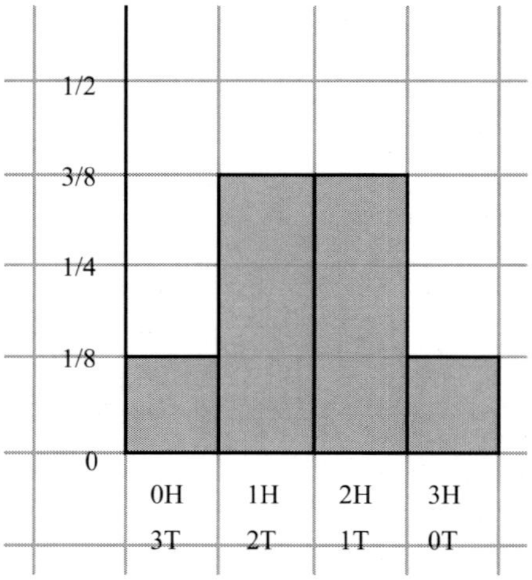

Figure 16.5.2

What is graphed here is called a **probability density function**.

In example 1, as with an earlier question about rolling two dice (example 3 of section 16.4), it was possible to analyze the problem in terms of a more detailed set of possibilities in which all outcomes are equally probable. The next example shows how to deal with cases where even with an expanded possibility set the outcomes are not equally probable.

Example 2. Modify example 1 for a coin that is weighted to make its prob ability of landing heads on each toss $\frac{2}{3}$.

Solution: As in example 1, S = {3H, 2H1T, 1H2T, 3T} is the possibility set, but again S' = {HHH, HHT, HTH, HTT, THH, THT, TTH, TTT} can also be used. Here, however, all eight outcomes in S' are not equally likely . To see how to assign probabilities, modify the tree diagram (figure 15.5.1) by putting $\frac{2}{3}$ on each segment leading to H and $\frac{1}{3}$ on each segment leading to T, as in figure 16.5.3.

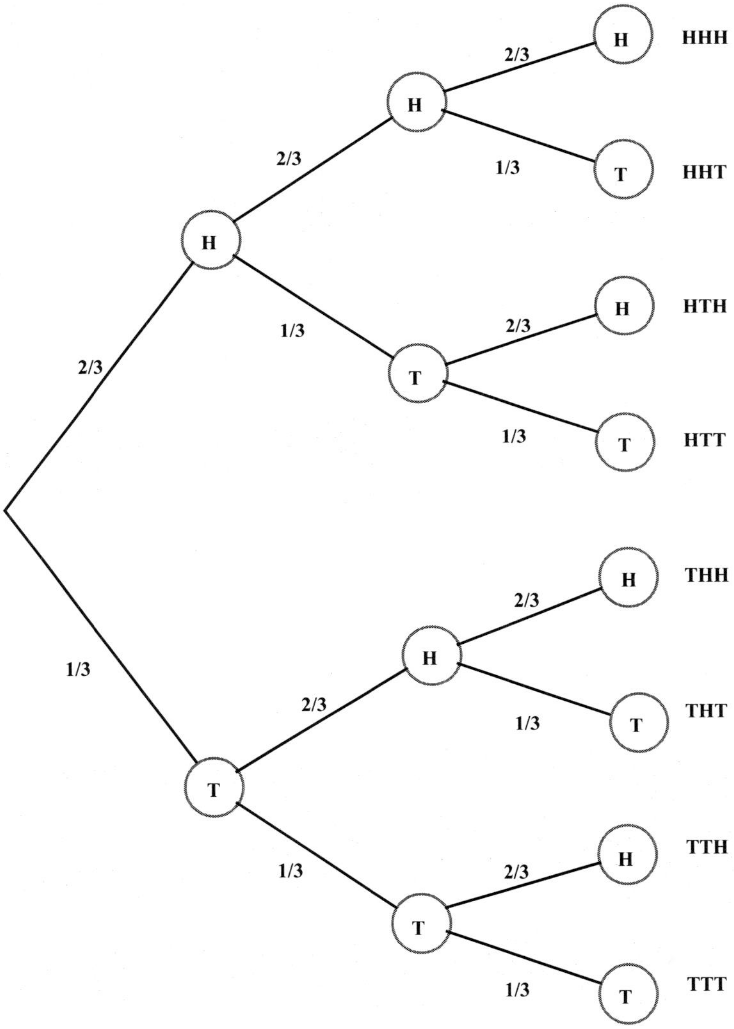

Figure 16.5.3

Again, multiplying the probabilities along each part of a path gives the probability for that path. There are $\binom{3}{2}$ distinguishable arrangements of two Hs and one T. Since $P(\text{HHT}) = P(\text{HTH}) = P(\text{THH})$,
$P(\text{2H1T}) = P(\text{HHT}) + P(\text{HTH}) + P(\text{THH}) = 3\left(\frac{2}{3}\right)\left(\frac{2}{3}\right)\left(\frac{1}{3}\right) = \frac{4}{9}$.

Similar analysis applies to the event 1H2T. There are $\binom{3}{1} = 3$ paths in figure 16.5.3 which result in a head and two tails. Each of these paths has the probability $\left(\frac{1}{3}\right)\left(\frac{1}{3}\right)\left(\frac{2}{3}\right) = \frac{2}{27}$. Then $P(\text{1H2T}) = 3\left(\frac{2}{27}\right) = \frac{2}{9}$.

Figure 16.5.4 is the graph of the probability density function in this case.

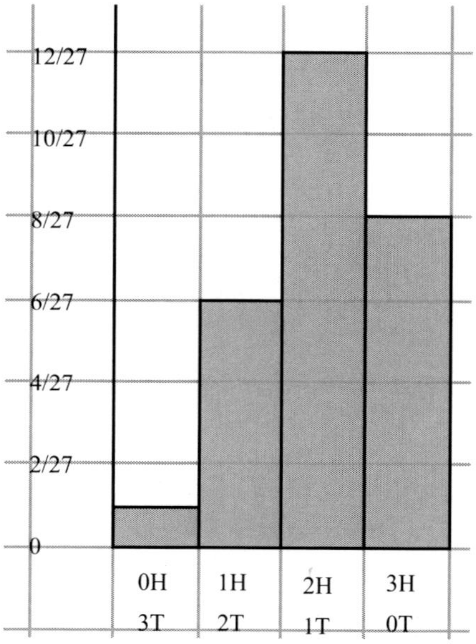

Figure 16.5.4

Again the total shaded area is 1, but the graph is not symmetrical. Here the probability of H is greater than the probability of T, which puts more of the shaded area to the right.

In both previous examples, the coins were tossed three times, but the reasoning would apply to any number of tosses. In general, consider n repetitions of an experiment (such as tossing a coin) with two possible outcomes A and B. Since A and B are the only possibilities, $P(A) + P(B) = 1$.
If $P(A) = p$, then $P(B) = 1 - P(A) = 1 - p$. Again we may represent the experiment with a tree diagram, but now each path has n edges. If there are k As and $n - k$ Bs on a path, the probability of that path is $p^k(1 - p)^{n-k}$. The number

of such paths is $\binom{n}{k}$. That is the number of combinations of k things chosen from n. (It is also the number of distinguishable arrangements of an n-letter "word" made of k As and $n - k$ Bs.). This leads to the basic result about Bernoulli trials.

> ***Bernoulli Trials***
> If an experiment with two outcomes, A and B, whose probabilities are p and 1-p, is repeated n times, the probability that A will occur exactly k times is $\binom{n}{k}p^k(1-p)^{n-k}$.

Example 3. On a multiple-choice quiz with ten questions, each question has four wrong answers and one correct answer. What is the probability of answering eight questions correctly by sheer guessing?

Solution: The probability of a correct guess on any one question is $\frac{1}{5}$, so the probability of guessing right on eight and wrong on two is

$$\binom{10}{8}\left(\frac{1}{5}\right)^8\left(\frac{4}{5}\right)^2 = \frac{10!}{8!(10-8)!}\left(\frac{1}{5^8}\right)\left(\frac{4^2}{5^2}\right) = \frac{10\cdot 9}{1\cdot 2}\cdot\frac{1}{390{,}625}\cdot\frac{16}{25}.$$

This is approximately 0.000 07. The probability of answering eight questions correctly by chance alone is less than one in 10,000.

Example 4. In the situation described in example 3, what is the probability of guessing right on at least one question?

Solution: "At least one" means one or more, which in this case is any number from 1 to 10. We could calculate the separate probabilities of getting one right, two right, three right, and so on up to ten right, but the indirect approach is best: First find the probability of the complementary event of guessing 0 right, then subtract that number from one to get the probability of getting at least one right. The probability of not getting at least one answer right is the probability of getting all ten wrong, which is

$\binom{10}{0}\left(\frac{1}{5}\right)^0\left(\frac{4}{5}\right)^{10} = \left(\frac{4}{5}\right)^{10}$, or about 0.107. Then the probability of guessing at least one question right is $1-\left(\frac{4}{5}\right)^{10} = 1-0.107 = 0.893$.

Exercises 16.5

1. Find the probability of guessing the answer to every question correctly on a true-false test with:
   a. 3 questions
   b. 4 questions
   c. 5 questions
   d. n questions

2. Find the probability that five tosses of a fair coin yield:
   a. 0 heads
   b. 1 head
   c. 2 heads
   d. 3 heads
   e. 4 heads
   f. 5 heads
   g. Add your answers to parts a-f.
   h. At least 3 heads

3. A coin is weighted so that the probability it lands tails is 0.6.
   a. Make a tree diagram of the possibilities for how it could land on four consecutive tosses.
   b. Use your tree from part a to find the probability of getting exactly 3 heads in four tosses with this weighted coin.

4. If a pair of fair dice is rolled twice,
   a. What is the probability that both rolls will yield "box-cars" (double sixes)?
   b. What is the probability that at least one roll yields box-cars?

5. A fair die is rolled three times.
   a. What is the probability it lands one or two all three times?
   b. What is the probability that the number 5 never turns up?

6. A Monopoly player who is in jail gets up to three chances to roll the dice. If any of these rolls yields a double (both dice land the same) the player gets out of jail.
   a. What is the probability that the player will get out of jail on the first roll?
   b. What is the probability that the player will not get out on the first roll but will get out on the second?
   c. What is the probability that the player will not get out on the first two rolls but will get out on the third?
   d. What is the probability that the player will fail to get out of jail in three rolls of the dice?

7. A multiple-choice test of twenty questions has five possible answers to each question. If a student knows the answers to 15 of the questions and answers the remaining five by random choice of options, what is the probability that this student will

   a. Answer all twenty questions correctly?

   b. Answer exactly 19 questions correctly?

   c. Answer exactly 18 questions correctly?

   d. Answer at least 17 questions correctly?

8. Ten percent of the items produced in a certain factory are defective.

   a. A box contains ten of the items produced, selected at random. What is the probability that the items selected include more than one defective item?

   b. Five boxes, each containing ten items produced at the factory, are selected. What is the probability that three of the boxes have more than one defective item?

9. A game consists of tossing a fair coin until two successive heads appear. What is the probability that this game will end in exactly:

   a. Two tosses b. Three tosses

   c. Four tosses d. Five tosses

10. Modify problem 9 as follows: A game consists of tossing a coin until two successive heads appear. If the probability of the coin landing heads on a toss is 3/4, what is the probability the game will end in exactly:

    a. Two tosses b. Three tosses

    c. Four tosses d. Five tosses

11. If four coins are tossed, the possible outcomes are
    S = {0H4T, 1H3T, 2H2T, 3H1T, 4H0T}

    a. Make a tree diagram analogous to figure 16.5.1 to find the probabilities of each of these five outcomes if the coins are fair and tossed independently.

    b. Use the formula for Bernoulli trials to check your answers from part a.

    c. Graph the probability density function for this experiment. Your graph should be analogous to figure 16.5.2.

12. Suppose that in problem 11 each coin, instead of being fair, has $P(\mathrm{H}) = \frac{2}{3}$.

    a. Make a tree diagram analogous to figure 16.5.3 to find the probabilities of each of the five possible outcomes if the coins are tossed independently.

    b. Use the formula for Bernoulli trials to check your answers from part a.

    c. Graph the probability density function for this experiment. Your graph should be analogous to figure 16.5.4.

16.6 Conditional Probability

Example 1. A drawer contains 14 long white socks, 5 long black socks, 4 long gray socks. 2 short white socks, and 8 short red socks.

a. If one sock is chosen at random from the drawer, what is the probability that it is white?

b. If a sock is chosen at random from the short socks in the drawer, what is the probability that it is white?

c. If a sock is chosen at random from the white socks in the drawer, what is the probability that it is short?

Solution: Part a: The phrase "at random" means that all socks are equally likely to be chosen. Since there are 33 socks in all, each sock has a probability of 1/33. Then $P(\text{white}) = \frac{\#\text{ white socks}}{\#\text{ socks}} = \frac{16}{33}$.

Part b: Since there are ten short socks, equally likely to be drawn, each has probability 1/10. Two of the ten short socks are white, so the probability of drawing a white sock at random is now 2/10.

Part c: If we know that the chosen sock is white, it is one of the 16 white socks, 14 long and 2 short. Again, these socks are equally likely to be chosen, so we assign each one the probability 1/16, and the probability that the chosen sock is short is 2/16 or 1/8.

The above example shows how additional information, by ruling out some possibilities, changes the probability space under consideration. It is an example of what is called a **conditional probability** problem-one in which an additional condition (in this case, that the chosen sock is short or that it is white) changes the probability space. The next example shows that conditional probability problems sometimes lead to results that defy intuition.

Example 2. A family is selected at random from a roster of families that have just two children but no twins.

a. What is the probability that the family has two boys?

b. If it is known that the family has one boy, what is the probability that it has two boys?

c. If it is known that the family's oldest child is a boy, what is the probability that the family has two boys?

Assume for this problem that the probability a child is a boy is 1/2.

Solutions: The possibility set for two children is {BB. BG. GB. GG}, and if boys and girls are equally likely to be born, each of these four possibilities has probability 1/4. That is the answer to part a. For part b, we are given the additional information that the family has at least one boy. That rules out the case GG and leaves a smaller set of possibilities, {BB, BG, GB}. The three possibilities are equally likely, so each is assigned the probability 1/3. Finally, for part c, the set of possibilities is reduced even further, because if we know the oldest child is a boy, then the only possible cases are {BB and BG}. These two cases are equally likely, so if the family's oldest child is a boy, the probability that the second child is a boy is 1/2.

Example 2 is remarkable for showing the value of information, even when it might seem at first to be irrelevant. It may seem surprising, for example, to learn that knowing that the first child is a boy raises the chance of the other child being a boy more than knowing that at least one child (either the first or the second) is a boy. The next example is an even stronger illustration of the power of additional information.

Example 3. Three cards are marked as shown and placed in a hat.

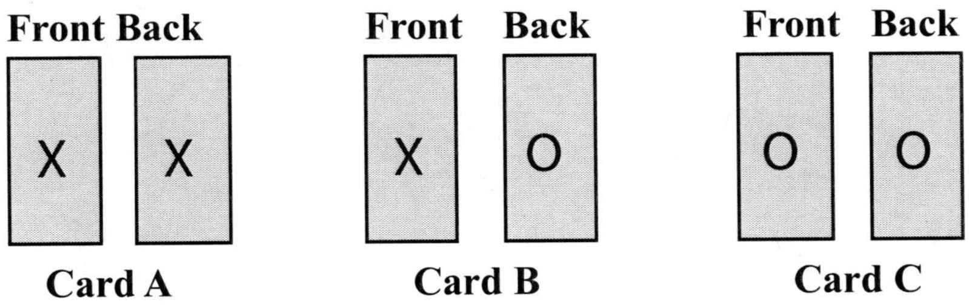

You select a card at random and are allowed to see one side of it, which has an **X**. What is the probability that the other side of the card also has an **X**?[2]

Take a few minutes to ponder this before reading on.

Solution: Most people reason, correctly, that the card cannot be card C, so it must be one of the other two. Therefore they conclude that the probability the other side of the drawn card has an **X** is 1/2. However, that conclusion is not correct! Consider the initial experiment as drawing a card,

[2] This problem is a modification of a probability puzzle known as Bertrand's Box Paradox, after Joseph Bertrand, who mentioned it in 1889. The Monty Hall problem, named after the host of the TV game show, "Let's Make a Deal," aroused immense curiosity after it was written up in Parade Magazine

selecting a side, and noting what is on that side. That experiment can be represented by a tree diagram.

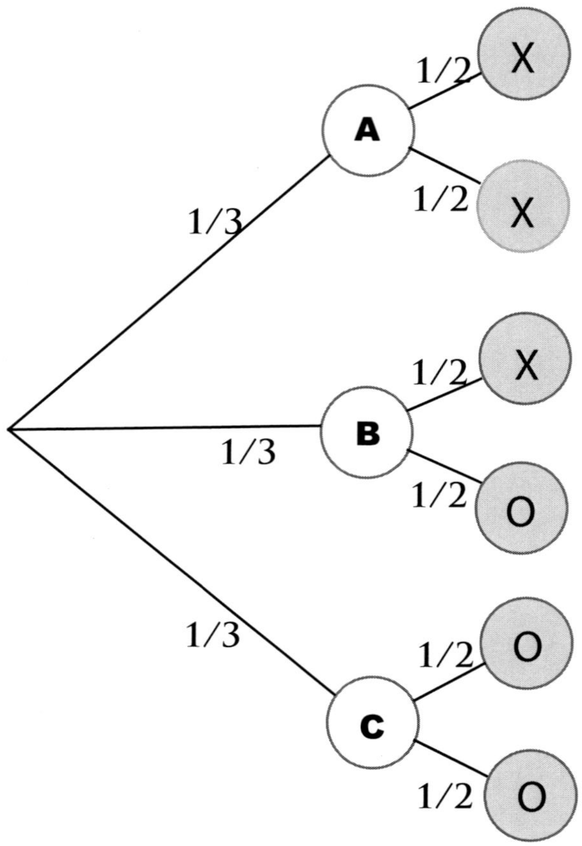

Eliminating card C "prunes" the bottom branch off tree, leaving the diagram at left below.

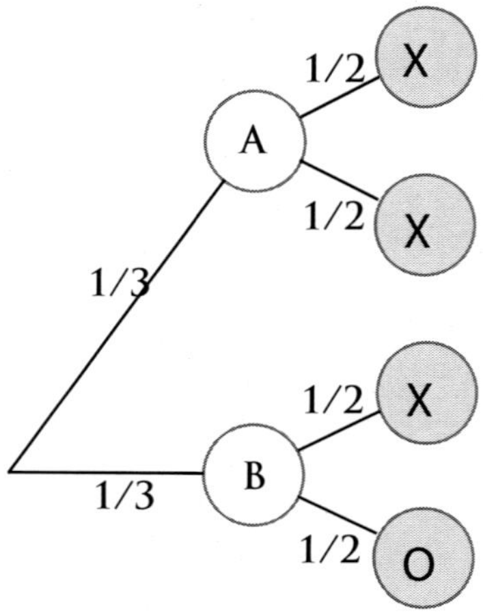

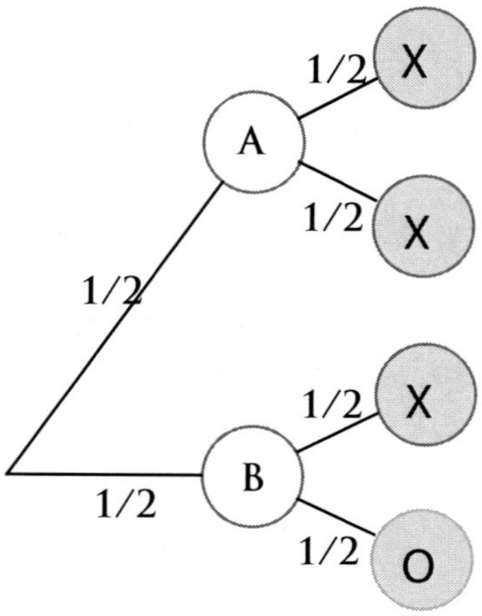

It is no longer the tree diagram of a probability experiment, because the probabilities at the initial vertex add up to 2/3 instead of 1, and therefore the probabilities of the paths through the tree add up to 2/3 instead of 1. In the tree diagram at right, the initial probabilities are changed to 1/2, making this tree diagram represent the probability space that applies when card C is eliminated. The four paths through that tree are equally likely (each has probability 1/4), and three of them end with an **X**. Two of those three involve card A and have an **X** on the other side. Therefore, if drawing a card and looking at one face shows an **X**, the probability that the other side of the card also shows an **X** is 2/3. Problem 1 below outlines an experimental check on the reasoning presented above.

A close cousin of example 2 has become famous as the "Monty Hall problem," named after the host of a televised game show, *Let's Make a Deal*. If you are the guest on the show, you are shown three doors and told that behind one of them there is a car, while behind the other two are goats. If you guess which door hides the car, you win the car. At this point, it seems, you have a probability of 1/3 of winning the car, and you choose a door. After you choose a door, but before it is open, the host, who knows where the car is, opens a different door and reveals a goat behind it. The host then asks you if you would like to change your choice of door. Should you change? Most people assume it makes no difference, but changing improves your chances! Surprised? If you are interested, look up the Monty Hall problem on the internet.

Exercises 16.6

1. You may find the conclusion of example 2 amazing, and in your heart you may not even believe it is correct. Here is a way to test that conclusion experimentally.
   a. Make a set of three cards like those in example 2, put them in a box, and mix them up.
   b. Ask someone to select a card at random and tell you only whether it shows an **X** or an **O**.
   c. Have the person select, at random, a face of one of the remaining cards, and note whether it has an **X** or an **O**.
   d. Repeat this experiment sixty times. If, as claimed in example 2, the probability of a match is 2/3, you would expect that to happen approximately 40 times in your 60 repetitions, but if the probability of a match is only 1/2, as some would think, you would expect only about 30 matches in 60 repetitions.

2. A bowl has 18 mini chocolate bars. Ten of them are milk chocolate. Five of the milk chocolate bars have nuts and five have no nuts. The eight bars which are not milk chocolate are bittersweet. Six of them with nuts and two without.

   a. If a chocolate bar is drawn at random from the bowl, what is the probability that it is bittersweet?:

   b. If a chocolate bar is drawn at random from the bowl, what is the probability that it does not have nuts?

   c. If a bar drawn at random from the bowl has nuts, what is the probability that it is milk chocolate?

   d. If a bar drawn at random from the bowl and is bittersweet chocolate, what is the probability that it has nuts?

3. (Adapted from *Innumeracy*,[3] by Paulos). Imagine a 98% accurate test for cancer. That is, the test correctly indicates the presence or absence of cancer 98% of the time. Also suppose that the incidence of cancer in a certain population people is 0.5%. If a person chosen at random from that population tests positive for cancer by this test, what is the probability that the person has cancer?

Vocabulary for Chapter 16

Bernoulli trials (16.5)
Combinations (16.3)
Complement of an event (16.4)
Compound event (16.4)
Conditional probability (16.6)
Event (16.4)
Odds (16.4)
Permutation (16.2)
Possibility set (16.4)
Probability (16.4)
Probability density function (16.5)
Probability space (16.4)
Simple event (16.4)
Tree diagram (16.1)

Self-Quiz for Chapter 16

1. A restaurant offers two kinds of salads, three kinds of soup, nine entrees, four beverages, and six desserts. How many different dinners are possible if each dinner includes:

   A soup or salad (but not both), an entrée, a beverage, and a dessert.

2. A parade has 12 floats. In how many possible orders could they appear?

[3] John Allen Paulos. *Innumeracy-Mathematical Illiteracy and its Consequences*. Farrar, Straus, and Giroux, 1989 (reissued 2001)

3. In a scholarship competition the top ten finalists are ranked from first to tenth. How many such rankings are possible if 20 candidates are in the finals?

4. How many different looking arrangements of the letters of the word CINCINNATI are possible?

5. A basketball coach must select five of the twelve players to start a game. In how many ways can that be done if it is unimportant who plays what position?

6. How many five-card poker hands have two kings and three red cards that are not kings?

7. In a troop of 14 scouts, three must cook dinner, four must wash dishes, and seven must gather wood. In how many ways can the crews be chosen if nobody is assigned to more than one task?

8. A bowl contains candies of four colors: three red, seven purple, five yellow, and four green. A candy is drawn at random from the bowl, and the color is noted.
   a. List the set of possibilities for this experiment.
   b. Assign probabilities to the possible outcomes in a reasonable way.
   c. Find the probability that the candy drawn is not red.

9. A fair coin is tossed. If it lands heads, a fair die is rolled, but if it lands tails, a biased die is rolled. The biased die is equally likely to land 2, 3, or 4, but it never lands 1 or 5, and it lands 6 half the time.
   a. Draw a tree diagram illustrating this two-stage experiment.
   b. Find the probability that the outcome of the experiment is an even number on the die.

10. If a fair coin is tossed twelve times, what is the probability it will land heads just three times?

11. If a pair of fair dice is rolled six times, find the probability that:
    a. Two rolls yield doubles.
    b. At least two rolls yield doubles.

12. a. If a pair of fair dice are thrown, what is the probability that the sum shown on the dice is ten?
    b. If you know that one of the dice lands 5, what is the probability that the sum shown on the dice is ten?
    c. If you know that the first die lands 5, what is the probability that the sum shown on the two dice is ten?

Chapter Seventeen

An Exploratory Approach to Exponents and Exponential Functions

Exponents were first introduced as shorthand notation for repeated multiplication, but the notation soon took on a life of its own. The concept of exponent was expanded to include negative and fraction exponents, and exponential functions were recognized as fundamental, both for computation and for modeling processes of growth and decay. This chapter explores those developments, beginning with homemade multiplication devices that are an excellent hands-on activity for middle school students.

17.1 A First Multiplier

To understand this material best, build the multipliers as you read the first few sections. Use graph paper, because accurate spacing is extremely important. The most convenient graph paper for these activities has lines 1/5 inch apart, but 1/4 will do almost as well. With a class, you can give out the dot pattern already made and have the students fill in the numbers, but there seems to be some value to having the students make the dot pattern for themselves on graph paper. Many students do this inaccurately at first, which shows that they have not really taken in the pattern. If they make the pattern for themselves, you can be more confident that they have grasped it.

If the students are not familiar with the adder, which you may recall from exercises 2.2, have them make a small one, as in figure 17.1.1, and explore it until they understand how to use it and how it works.

To use the adder, locate one of the two addends in each outer column of numbered dots. Then find the dot in the center column that lines up with the two dots selected in the outer columns. The sum is the number of that center dot.

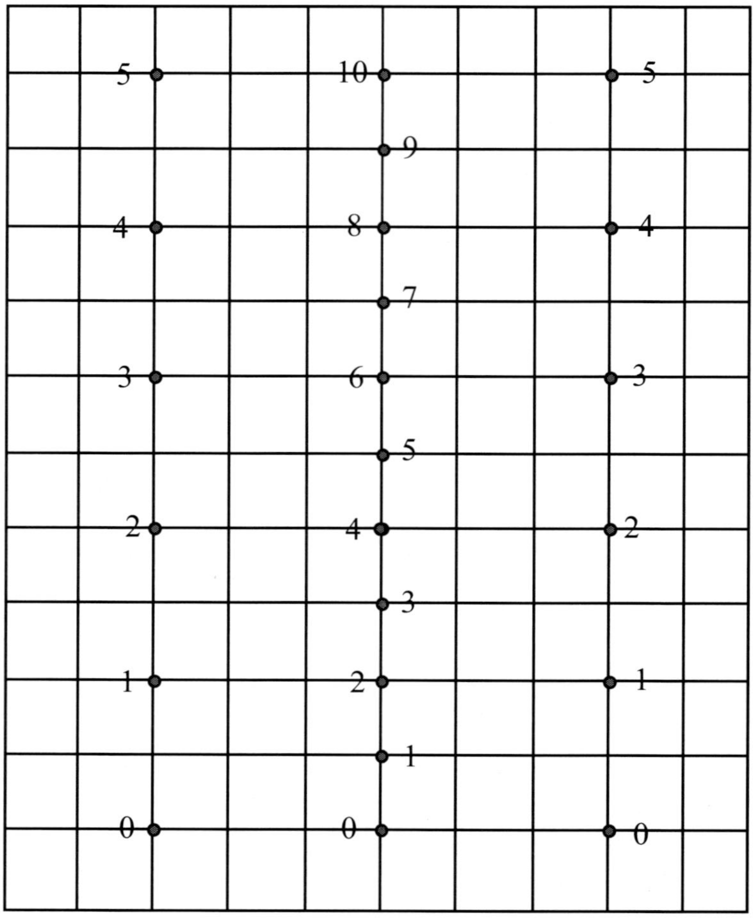

Figure 17.1.1

The multiplier will work in much the same way as the adder, using the same dot pattern but numbered differently. The idea is to number the dots in such a way that if you line up a number in each outer column with a dot in the middle column, the dot in the middle is the product of those in the outer columns.

The big question, of course, is how to number the dots. For the adder we began with a row of 0s, because 0 is the additive identity; adding zero to any number leaves that number alone. For multiplication, 1 plays an analogous role. It is the multiplicative identity; multiplying any number by 1 does not change it. With that in mind, start with a row of 1s, as in figure 17.1.2. Now, what number belongs at *A*? The natural response is 2, so let's try it in figure 17.1.3.

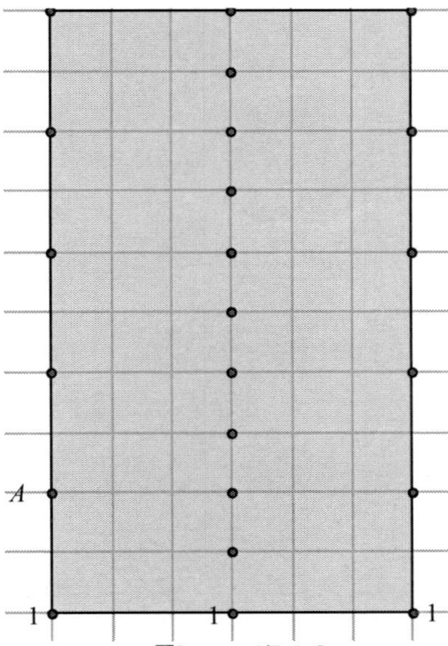

Figure 17.1.2

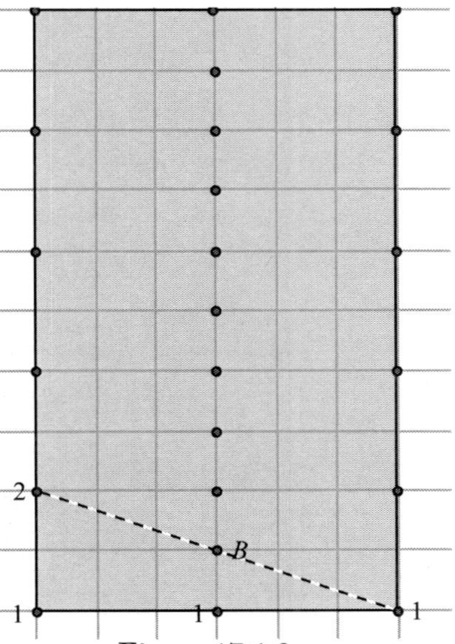

Figure 17.1.3

Now, what number belongs at *B*? There is really no choice, since if this device is to multiply as intended, $B = 1 \times 2 = 2$, and we turn to *C* and *D* (figure 17.1.4).

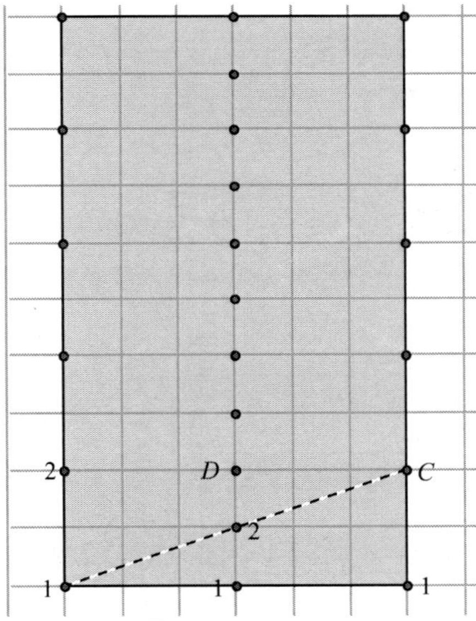

Figure 17.1.4

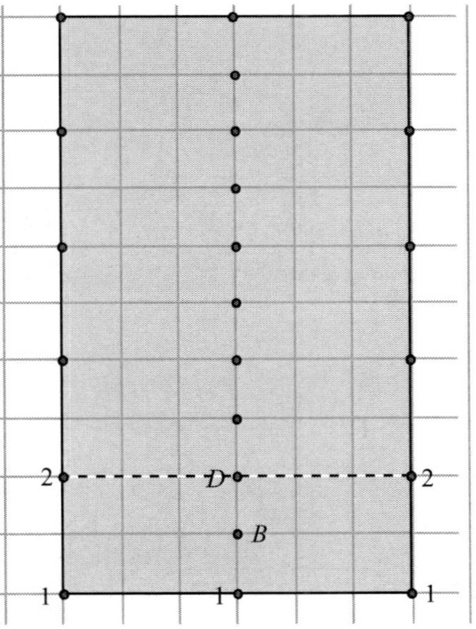

Figure 17.1.5

Multiplying as indicated on the slanted line shows $1 \times C = 2$, so evidently $C = 2$. Then, multiplying along the horizontal line in figure 17.1.5 leads to $D = 2 \times 2 = 4$. Continuing this way, the multiplication shown in figure 17.1.6 dictates that $E = 4$, and similarly $F = 4$. Then from figure 17.1.7 you can see that $G = 8$ and $H = 16$.

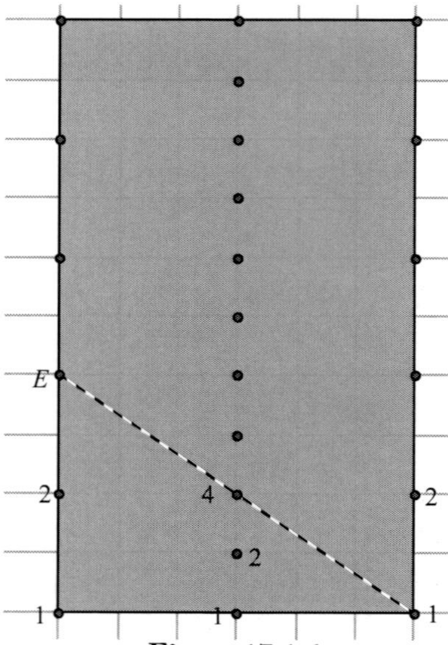

Figure 17.1.6

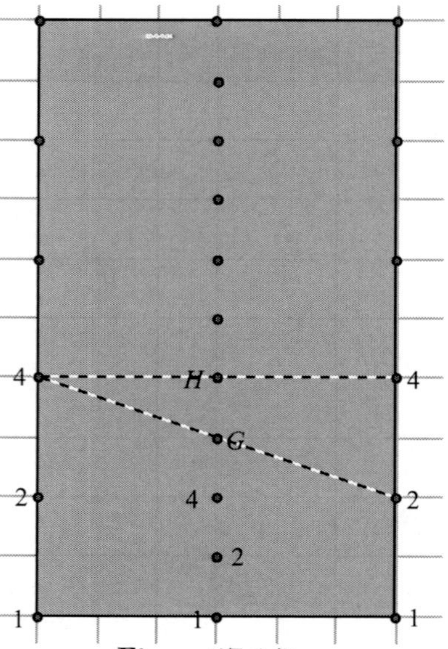

Figure 17.1.7

Continuing this process produces figure 17.1.8.

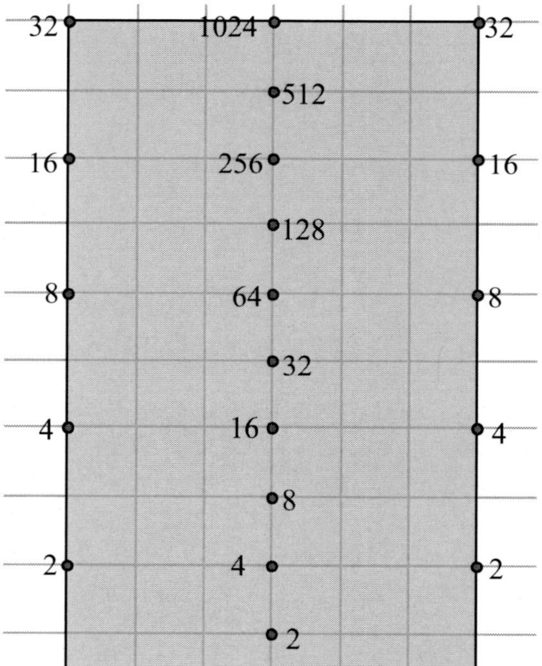

Figure 17.1.8

Check a few multiplications, such as 4×32, for yourself to be sure that this device really does multiply. Still, you could not market it as a practical device. Your customers would want their money back as soon as they found out they could not use this device to multiply by 3, 5, or many other numbers! You might hope to put these numbers in, but that is not easy to do accurately. For example, if you try to put 3 halfway between 2 and 4 as in figure 17.1.9, then the multiplier would indicate that $3 \times 3 = 8$.

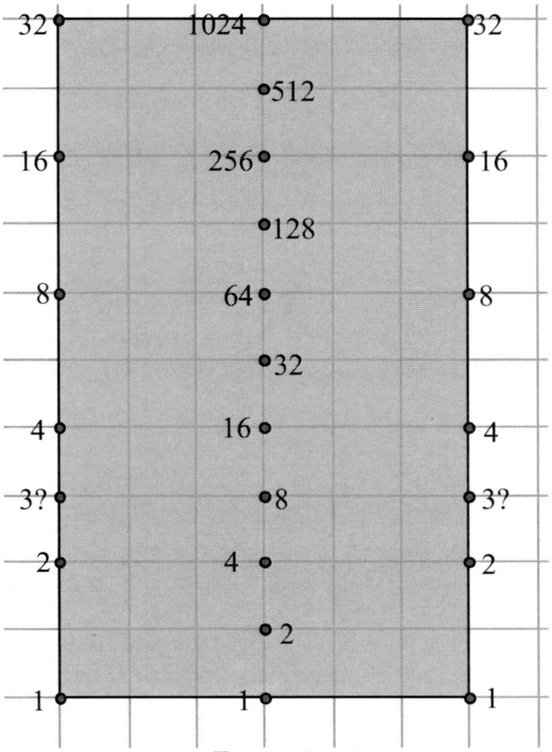

Figure 17.1.9

Evidently, 3 belongs closer to 4 than to 2, but exactly where is hard to determine. Other numbers are also difficult to place accurately. Impractical though it is, however, this first try at a multiplier is not a total failure. A closer look at it will show how to build better ones.

Instead of focusing on numbers that are missing, look at those which appear in figure 17.1.8. What family of numbers do they belong to? (If you answered "even numbers," check again. Some even numbers, like 6 and 10, are not there, and 1, which is odd, appears.)

You no doubt recognized 1, 2, 4, 8, 16, . . . as powers of 2, and in those terms, the multiplier in figure 17.1.8 can be rewritten as in figure 17.1.10. Look carefully at this new figure. What do you see?

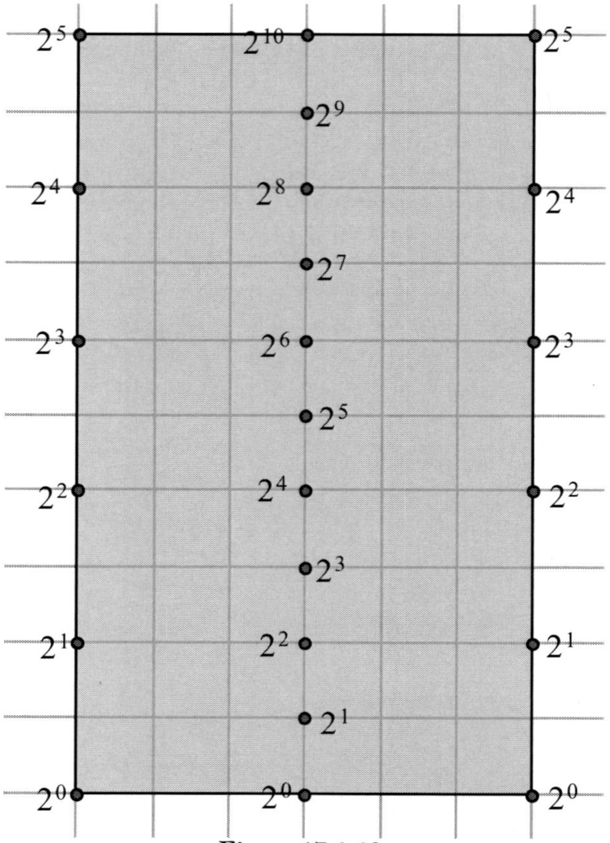

Figure 17.1.10

Figure 17.1.10 is, in effect, an adder for exponents! This is easy to understand, if you think what the exponents mean. If you multiply 2^3 by 2^5, for example, think of 2^3 and 2^5 as trains of 2s, one with three cars, the other with five. Multiplying these numbers is like hooking the trains together to form a single train, which of course has $3 + 5 = 8$ cars, as in figure 17.1.11.

$$(2 \times 2 \times 2) \times (2 \times 2 \times 2 \times 2 \times 2) = 2 \times 2 \times 2 \times 2 \times 2 \times 2 \times 2 \times 2$$

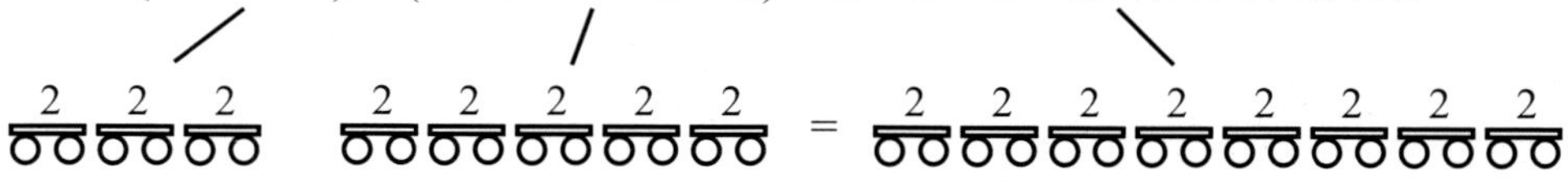

Two two trains (make one big two train)
Figure 17.1.11

Exercises 17.1

These exercises are exploratory in nature. The answers will emerge later in this chapter. Thinking about these issues in advance will both prepare you for the material to come and, if you are successful, boost your confidence in your ability to do mathematics on your own.

1. The adder in exercises 2.2 could be used to subtract. Explain how to use the multiplier in figure 17.1.8 to divide.

2. (This topic is explored further in section 17.3.) The adder in exercises 2.2 could be extended to negative numbers.
   a. Extend figure 17.1.10 to negative powers of 2, even though we have not yet defined those powers.
   b. Make an analogous extension of the multiplier in figure 17.1.8 to numbers below 1, numbering the dots below 1 so that the device continues to multiply.
   c. Use your work from parts a and b to propose a meaning for negative powers. In particular, what would be a sensible way to define 2^{-1} and 2^{-2}?

3. (This topic is explored further in section 17.4.) The adder in exercises 2.2 works for fractions as well as for whole numbers, even though only whole numbers are marked.
   a. Augment figure 17.1.10 by putting in numbered dots for $2^{1/2}$, $2^{3/2}$, $2^{5/2}$, $2^{7/2}$ and $2^{9/2}$ into all three columns and for $2^{11/2}$, $2^{13/2}$, $2^{15/2}$, and $2^{17/2}$ in the center column, even though we have not yet defined those powers.
   b. Augment the multiplier in figure 17.1.8 by putting numbers at dots whose positions correspond to the fractional powers marked on your augmented version of figure 17.1.10 from part a.
   c. Use your work from parts a and b to propose a meaning for fractional powers. In particular, what would be a sensible way to define $2^{1/2}$ and $2^{3/2}$?

17.2 More and Better Multipliers

How can the multiplier be improved? Can we make one that has 3 on it? Our first try had 4s where we might have wanted 3s, but there was really no choice if the multiplier was to work at all. Looking back on it, the choice of a row of 1s seems natural. Then, once the first 2 was put in, there was no more choice in numbering the other dots. What if the initial 2 is replaced with some other number? If you try 3 instead, as in figure 17.2.1, you will be led (try it and see) to figure 17.2.2.

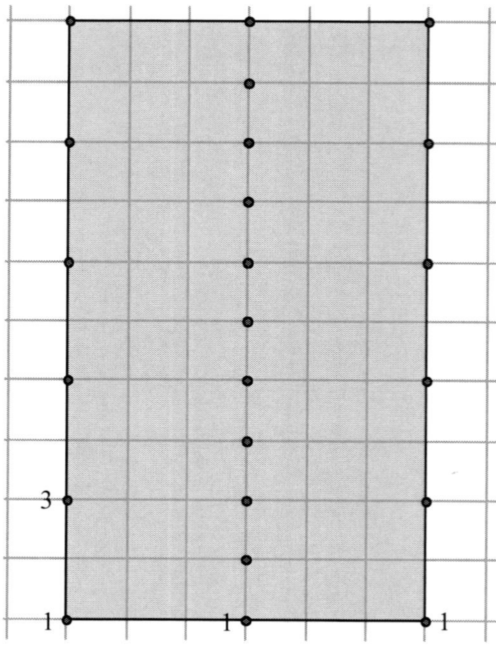

Figure 17.2.1

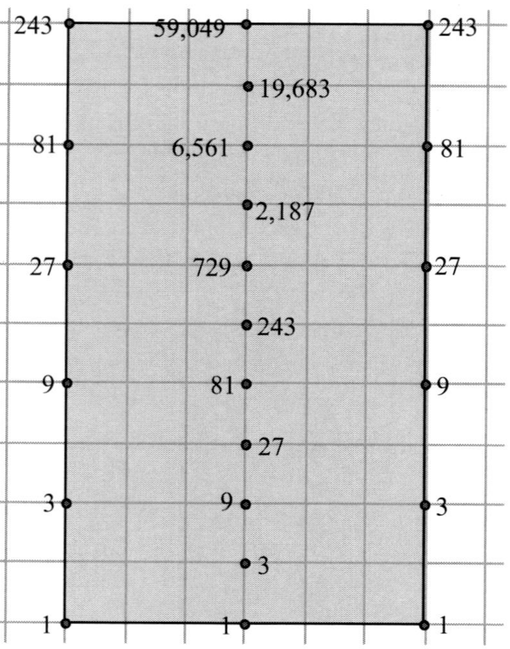

Figure 17.2.2

The good news is that this device really does multiply, and that 3 appears on it. The bad news is that 2 no longer appears. This new multiplier has powers of 3 and nothing else. It is even less practical than the first try, for it will be very hard to place 2, 4, 5, 6, 7, 8 accurately. Indeed, only powers of 3 can be placed precisely on this multiplier.

After putting in a row of 1s, a single choice of an additional number seems to determine the entire multiplier. If we call this number b, the multiplier has the form shown in figure 17.2.3.

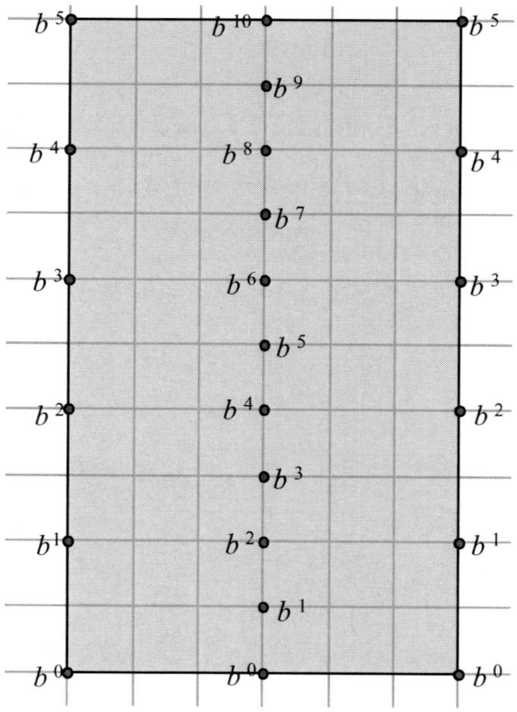

Figure 17.2.3

Our first multiplier fit this form with $b = 2$, and it skipped many numbers. Our second attempt, with $b = 3$, skipped even more numbers. Evidently, powers of two increase too rapidly, and powers of three increase even faster. On the other hand, powers of 1 do not increase at all. The only hope is to base a multiplier on powers of a number larger than 1 but smaller than 2. We will use $b = 1.1$, a value for which it is not hard to make a relatively practical device. The first step is to calculate the needed powers of 1.1. This is not hard to do, either by hand or with a calculator. Then round those powers to the nearest hundredth, and enter the results on a grid. The result is the multiplier in figure 17.2.4 below. In figure 17.2.4, the dots in the grid pattern serve double duty as decimal points.

10.83	117.39	10.83
	106.72	
9.85	97.02	9.85
	88.20	
8.95	80.18	8.95
	72.89	
8.14	66.26	8.14
	60.24	
7.40	54.76	7.40
	49.79	
6.73	45.26	6.73
	41.14	
6.12	37.40	6.12
	34.00	
5.56	30.91	5.56
	28.10	
5.05	25.55	5.05
	23.23	
4.59	21.11	4.59
	19.19	
4.18	17.45	4.18
	15.86	
3.80	14.42	3.80
	13.11	
3.45	11.92	3.45
	10.83	
3.14	9.85	3.14
	8.95	
2.85	8.14	2.85
	7.40	
2.59	6.73	2.59
	6.12	
2.36	5.56	2.36
	5.05	
2.14	4.59	2.14
	4.18	
1.95	3.80	1.95
	3.45	
1.77	3.14	1.77
	2.85	
1.61	2.59	1.61
	2.36	
1.46	2.14	1.46
	1.95	
1.33	1.77	1.33
	1.61	
1.21	1.46	1.21
	1.33	
1.10	1.21	1.10
	1.10	
1.00	1.00	1.00

Figure 17.2.4

Practical matters

Check this multiplier for accuracy as follows. Select a number in each outer column and line up their decimal points with the decimal point of a number in the center column. For example, if you line up 1.46 and 2.36, you will find that the line between the decimal points of these numbers passes through the decimal point of 3.45 in the center column. According to the multiplier, then, $1.46 \times 2.36 = 3.45$. This is obviously not perfectly accurate, since the correct product must have four digits to the right of the decimal point. If you use long multiplication or a calculator, you will see that $1.46 \times 2.36 = 3.4456$. If you round this answer to the nearest hundredth, you get 3.45, which you found with the multiplier. The error in the multiplier is caused by the fact the true powers of 1.1 have too many digits to write conveniently on a device of this type. Therefore, numbers on the multiplier in figure 17.2.4 are rounded, and the rounding introduces inaccuracies. The size of the error introduced by rounding in this way is the absolute value of the difference between the true answer and the answer obtained from the multiplier. That is $|3.4456 - 3.45| = 0.0044$, or slightly more than one tenth of 1% of the correct answer. Not bad for a homemade device! Problems 1-4 in the exercises below guide you through similar checks.

Another practical question is the one that plagued us initially: Can this device multiply by simple numbers, such as 2, 3, 4, or 5? Those numbers do not appear on the multiplier, but we could place them on it with much more accuracy than we could on the first attempts. In our first device, if we put 3 midway between the nearest entries on the sides, 2 and 4, the device would have indicated that $3 \times 3 = 8$. If we tried the same thing on the multiplier in figure 17.2.4, midway between 2.85 and 3.14, it would indicate that $3 \times 3 = 8.95$. This is still not correct, but it is much closer than 8 to the correct answer. In subsequent sections, you will see how the idea behind this multiplier can be made to produce a truly practical device.

One way to improve the accuracy of our device would be to keep more decimal places, but numbers with many digits clutter a multiplier. Therefore, it is more practical to tabulate the entries, as in figure 17.2.5. Now, for example, to multiply 4.5950 (entry 16) by 13.1100 (entry 27), just add the exponents 16 and 27, and look up entry number 43, which is 60.2401. This is, at least approximately, 4.5950×13.1100. The idea is that $4.5950 \times 13.1100 = 1.1^{16} \times 1.1^{27} = 1.1^{(16+27)} = 1.1^{43} = 60.2401$.

n	Approximate value of 1.1^n	n	Approximate value of 1.1^n	n	Approximate value of 1.1^n
1	1.1	18	5.5599	35	28.1024
2	1.21	19	6.1159	36	30.9127
3	1.331	20	6.7275	37	34.0039
4	1.4641	21	7.4002	38	37.4043
5	1.6105	22	8.1403	39	41.1448
6	1.7716	23	8.9543	40	45.2593
7	1.9487	24	9.8497	41	49.7852
8	2.1436	25	10.8347	42	54.7637
9	2.3579	26	11.9182	43	60.2401
10	2.5937	27	13.1100	44	66.2641
11	2.8531	28	14.4210	45	72.8905
12	3.1384	29	15.8631	46	80.1795
13	3.4523	30	17.4494	47	88.1975
14	3.7975	31	19.1943	48	97.0172
15	4.1772	32	21.1138	49	106.7190
16	4.5950	33	23.2252	50	117.3909
17	5.0545	34	25.5477		

Figure 17.2.5

Exercises 17.2

Questions 1-4 help you get an idea of how accurate (or inaccurate) the multiplier in figure 17.2.4 really is.

1. Use the multiplier in figure 17.2.4 to estimate:
   a. 2.36×8.14
   b. 8.95×1.95
   c. 6.12×4.18

2. Compute the three products from question 1 either with a calculator or by pencil and paper. These are the true answers. The results you found in question 1 are approximations, because of the rounding that went into making figure 17.2.4.

3. For each of the three multiplications from questions 1 and 2, find the error in the multiplier by subtracting the smaller of the true answer and the multiplier answer from the larger.

4. In each case, from questions 1-3, find the error as a percentage of the true answer.

5. Estimate where to put 2, 3, and 4 in figure 17.2.4. Although you cannot place these numbers with perfect accuracy, you can do so far better than would be possible with the crude multipliers in section 17.1.

6. In parts a and b, use the table in figure 17.2.5 to estimate the products:
   a. 1.9487×4.1772
   b. 4.5950×7.4002
   c. Find the correct answers to the multiplications estimated in parts a and b.
   d. Find the errors in your estimates from parts a and b as percentages of the correct products.

7. a. Explain how to use the table in figure 17.2.5 to divide.
   b. Use the table in figure 17.2.5 to estimate $45.2593 \div 2.1436$.
   c. Calculate $45.2593 \div 2.1436$.
   d. Find the error in your estimate from part b as a percentage of the true answer obtained in part c.

17.3 Negative Exponents and Scientific Notation

Multipliers are adders for exponents. The adder from Section 2.2 works for negative numbers as well as positive. Can the adder for exponents be extended to negative exponents, as in the right-hand part of figure 17.3.1?

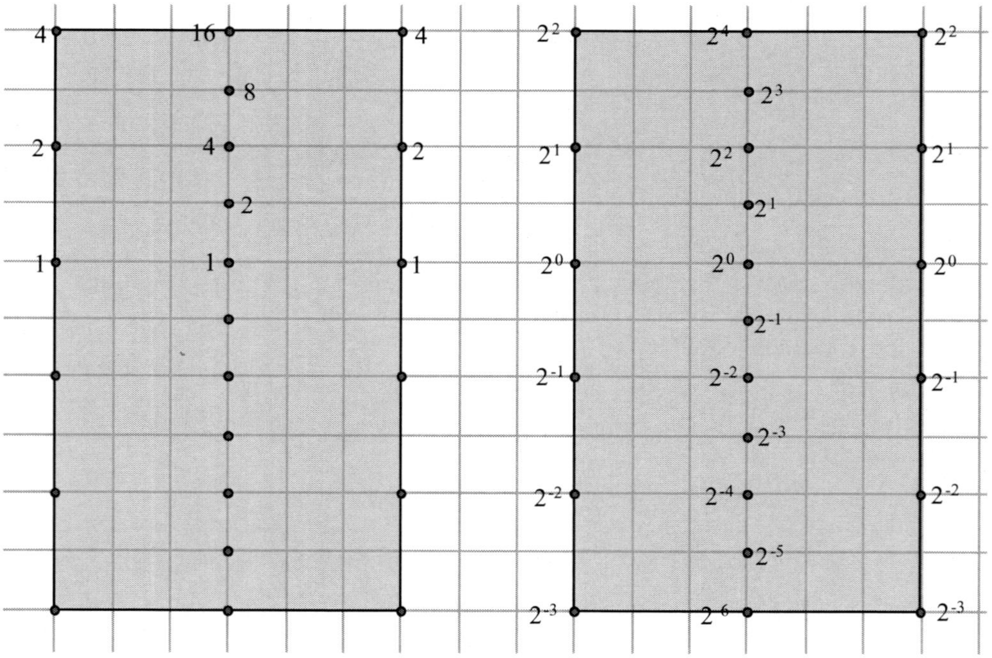

Figure 17.3.1

Numbering the remaining dots on the left will help us to define negative powers of 2. The clue to doing this in a useful way is the patterns themselves. On the top part of the multiplier, moving up one dot doubles the number, and moving down one dot cuts the number in half. Extending this to the dots below 1 yields figure 17.3.2.

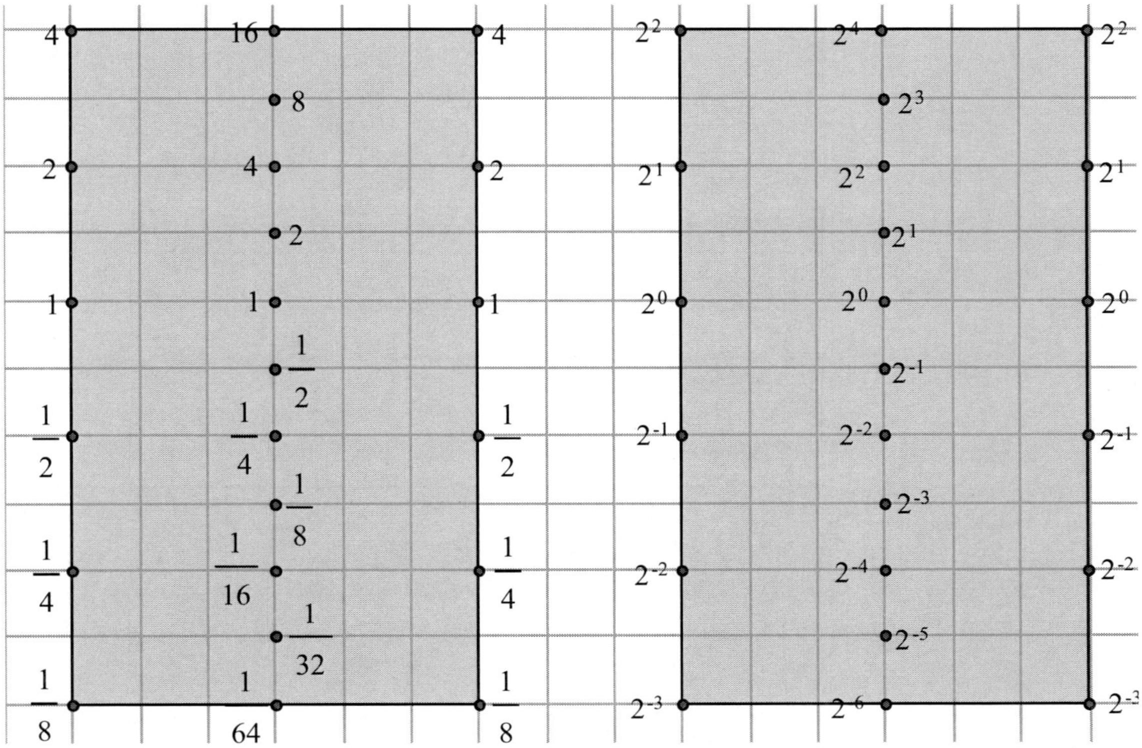

Figure 17.3.2

Try a few multiplications, such as $\frac{1}{4} \times \frac{1}{8}$ or $\frac{1}{8} \times 2$, to see that the extended multiplier does indeed multiply. Such checks will convince you that it is reasonable to define $2^{-1} = \frac{1}{2}$, $2^{-2} = \frac{1}{4}$, $2^{-3} = \frac{1}{8}$, and in general $2^{-n} = \frac{1}{2^n}$ for any positive integer n. Then the statement $2^m \times 2^n = 2^{m+n}$, which was found to hold for positive integers m and n, holds even if one or both of m and n is negative.

The discussion of powers of two applies as well to powers of other numbers. In chapter 8, for example, we already defined $10^{-1} = \frac{1}{10}$, $10^{-2} = \frac{1}{10^2} = \frac{1}{100}$, etc. More generally, extending the multiplier in figure 17.2.3 to negative exponents leads us to define $b^{-1} = \frac{1}{b}$, $b^{-2} = \frac{1}{b^2}$, $b^{-3} = \frac{1}{b^3}$ and $b^{-n} = \frac{1}{b^n}$, for any positive integer n and any number $b \neq 0$.

Definition of negative powers

> **For any number $b \neq 0$ and any positive integer n, $b^{-n} = \dfrac{1}{b^n}$.**

This definition may take time to get used to, but it fits all the patterns. For positive exponents, the key is that if m factors of a number b are multiplied by n factors of the same number, there are $m + n$ factors in all, so that

$b^m b^n = b^{m+n}$. Defining $b^{-n} = \dfrac{1}{b^n}$ extends this rule to all integer exponents.

Basic laws of exponents

> **For any number $b \neq 0$ and any integers m and n, $b^m b^n = b^{m+n}$.**

A second law of exponents, which is explored in the exercises, will turn out to be very useful in calculating powers and roots.

> **For any number $b \neq 0$ and any integers m and n, $(b^m)^n = b^{mn}$.**

To illustrate this law, consider the fourth power of 2^2, or $(2^2)^4$.
This is $2^2 \times 2^2 \times 2^2 \times 2^2 = (2 \times 2)(2 \times 2)(2 \times 2)(2 \times 2) = 2^8 = 2^{2 \times 4}$.

Extending the Multiplier: Scientific Notation

One of the problems with the multiplier in figure 17.2.4 is that it multiplies numbers only between 1 and 10 (to be precise, 10.83). Now we can remedy that defect. Example 1 will give you the idea.

Example 1. Estimate 5,050 × 133 with the multiplier in figure 17.2.4.
Solution: Write 5,050 as 5.05×10^3 and 133 as 1.33×10^2.
Then $5{,}050 \times 133 = 5.05 \times 10^3 \times 1.33 \times 10^2$
$= 5.05 \times 1.33 \times 10^3 \times 10^2$
Multiply the first two factors by the multiplier, getting 6.73, and apply the basic law of exponents to the powers of 10 to get $6.73 \times 10^5 = 673{,}000$. Of course, the result is only approximate, but the method extends the multiplier to numbers of any size whatever.

This extends the multiplier to very small numbers as well as to very large ones, as example 2 demonstrates.

Example 2. Estimate 0.000195 × 0.00235 with the multiplier in figure 17.2.4.
Solution:

$$\begin{aligned} 0.000195 \times 0.00235 &= 1.95 \times 10^{-4} \times 2.35 \times 10^{-3} \\ &= 1.95 \times 2.35 \times 10^{-4} \times 10^{-3} \\ &= 4.54 \times 10^{-7} \\ &= 0.000000454 \end{aligned}$$

Expressing a number as the product of a number between 1 and 10 and a power of 10 is more than a trick for extending the multiplier. It is uniquely suited for writing very large and very small numbers that would otherwise require large numbers of zeros. It is called **scientific notation**, since such large and small numbers are traditionally encountered in science[1]. For example, Avogadro's number, a fundamental constant in chemistry, is approximately 602,000,000,000,000,000,000,000. Written this way, in ordinary base ten notation, you have to count the zeros to even read the number. In this case, scientific notation, 6.02×10^{23}, is more convenient.

To express a number in scientific notation:

> **Write the number as a product of two numbers, where:**
> **1. The first factor is a number between 1 and 10**
>
> **2. The second factor is a power of ten.**

Scientific notation can indicate just how precise a number is. In 1986 the US Postal Service moved 44,700,000,000 pieces of mail. You might assume that this number is approximate, since it is unlikely that all the mail for a year was counted perfectly. If the number was actually rounded to the nearest million, you would write it 4.4700×10^{10}. This indicates that the first two zeros following the seven in 44,700,000,000 are intended to be taken as accurate, but that the remaining six zeros are included only to occupy the units, tens, hundreds, etc. places in the decimal system and thus move the 44,700 to its proper position. The six right-most zeros are thus called **place holders**. In contrast, the digits in 44,700, including the two zeros, are called **significant digits**. Confusion over whether or not a digit is significant can arise only when that digit is zero. Any non-zero digit is automatically assumed to be significant.

To indicate that the number 44,700,000,000 is to be taken as absolutely correct, down to the last zero, write it as $4.4700000000 \times 10^{10}$. Writing "unnecessary" zeros indicates that they are significant.

Calculators use scientific notation to display numbers that would otherwise have too many digits. For example, if you try to compute 15^{15}, you will see something like 4.3789389E17 or 4.3789389   17, depending on the particular model. In scientific notation, these are 4.3789389×10^{17}. In such cases scientific notation is often referred to as **floating point** notation.

[1] Scientific notation arose in 19th century Europe, but details of its invention and first use are difficult to ascertain.

Example 3. Express 0.0000017 (the percentage of ozone in air, by weight) in scientific notation.
Solution: To get a factor between 1 and 10, put a decimal point between the one and the seven. This decimal point is six places to the right of where it appears in 0.0000017. Therefore, in scientific notation, $0.0000017 = 1.7 \times 10^{-6}$.

Example 4. The surface of the earth is about 510,070,000 square kilometers. Express this number in scientific notation:

a. If it is perfectly accurate.

b. If it is accurate to the nearest 1,000.

Answers:

a. 5.10070000×10^8.

b. 5.10070×10^8.

Exercises 17.3

1. Why was b^{-n} not defined for $b = 0$?

2. Compute:

   a. 3^{-2}

   b. 10^{-4}

   c. 5^{-3}

   d. 2^{-4}

   e. 7^{-1}

   f. $\left(\frac{1}{2}\right)^{-1}$

   g. $\left(\frac{2}{3}\right)^{-2}$

   h. $\left(-\frac{2}{3}\right)^{-2}$

   i. $\left(\frac{3}{5}\right)^{-2}$

   j. $\left(-\frac{3}{4}\right)^{-3}$

   k. $(0.03)^{-1}$

3. Extend the multiplier in figure 17.2.3 to negative powers of *b*.

4. Use the rule $b^m \times b^n = b^{m+n}$ to compute:
   a. $10^{-5} \times 10^5$
   b. $8^{-3} \times 8^4$
   c. $2^7 \times 2^{-4}$
   d. $3^5 \times 3^{-6}$
   e. $5^{-17} \times 5^{14}$
   f. $10^{-21} \times 10^{50}$
   g. $10^{-6} \times 10^{-7}$

5. Fill in the blanks:
   a. Division is the inverse of ____________.
   b. Subtraction is the inverse of ____________.
   c. To multiply powers of the same number, ______ their exponents.
   d. To divide powers of the same number, ________ their exponents. Specifically, $b^m \div b^n = b^{—}$. (Fill in the exponent on b).

6. Use the rule from question 5 to find:
   a. $10^4 \div 10^7$
   b. $10^{-3} \div 10^{-5}$
   c. $10^{-8} \div 10^{-6}$
   d. $10^{13} \div 10^{18}$
   e. $10^{-5} \div 10^4$

7. In parts a-c, calculate the given expressions.
   a. $(3^2)^4$
   b. $(2^{-4})^3$
   c. $(4^{-2})^{-3}$

   In parts d-f, compute the given powers.
   d. 3^8
   e. 2^{-12}
   f. 4^6

   Compare the results of parts a and d, b and e, c and f respectively. They should be consistent with the law $(b^m)^n = b^{mn}$.

8. In parts a-c, use the table in figure 17.2.5, together with the law of exponents $(b^m)^n = b^{mn}$, to estimate:
   a. 2.1436^3
   b. 1.6105^7
   c. 2.8531^4
   d. Check your answers to parts a, b, and c by means of a calculator.
   e. Find the percent error in each of your answers from parts a, b, and c.

9. Express in scientific notation:
   a. 299,792,500 (the speed of light in meters per second)
   b. 635,000,000,000 (approximately the number of possible bridge hands)
   c. 0.0000000000000000000000017 (the approximate mass, in grams, of a proton).

10. In favorable conditions, *Escherichia coli*, a bacterium which inhabits the human gut, splits in two every 15 minutes to reproduce. At that rate, in 24 hours one bacterium can become roughly 40,000,000,000,000,000,000,000,000,000. Express this number in scientific notation, indicating that only one figure is significant.

11. Avogadro's number, given in the text as $6.02 = 10^{23}$, has been calculated with increasing accuracy over the years.

    In the 1970s the best estimate was 602,204,500,000,000,000,000,000.

    By the 1980s this had improved to 602,213,670,000,000,000,000,000.

    At this writing, in 2009, the most accurate experimental determination of Avogadro's number that is recognized on the internet by the Committee on Data for Science and Technology is 602,214,141,790,000,000,000,000.

    Express each of these estimates in scientific notation, assuming that the terminal zeros are all place holders.

12. At an average of 70 beats per minute, how many times will a heart beat in an 80-year lifetime? Write your answer in scientific notation in a way that indicates it is accurate to two significant figures.

13. Use the multiplier in figure 17.2.4 to estimate these products and quotients. Leave your answers in scientific notation.

    a. 259×345

    b. 2850×45900

    c. 4.18×0.00000556

    d. $25.55 \div 0.0673$

    e. $0.002111 \div 0.00038$

14. If each square centimeter of skin has about 4×10^6 microbes and you have about 2×10^4 square centimeters of skin, about how many microbes do you have on your skin?

15. In 1881, von Jolly estimated the mass of the Earth to be 6,150,000,000,000,000,000,000,000,000 grams. Express this number in scientific notation, indicating that four digits are significant.

16. A bidder on a federal contract (for the C5-A cargo plane) submitted 1,466,366 pages of documentation with the bid. Express that number of pages in scientific notation:

    a. If it is accurate to the nearest thousand

    b. If it is accurate to the nearest hundred

17. In 1985 certified air carriers in the USA flew 277,182,916,000 revenue passenger miles. Express that number in scientific notation:

    a. If it is accurate to the nearest thousand.

    b. If it is accurate to the nearest hundred.

17.4 Rational Exponents

To square a number x with the multiplier in figure 17.2.4 is particularly easy. Find x on an outside column, and there in the center column, at the same level, is x^2. You can see this from figure 17.4.1, which is part of figure 17.2.4. For example, to square 4.18, locate it in an outer column of numbers. On the same level in the center column is 17.45, which is approximately, the square of 4.18.

You can also use the multiplier to estimate square roots. For example, to estimate $\sqrt{21.11}$, just locate 21.11 in the center column, and read $\sqrt{21.11} = 4.59$ at the same level in the outer column.

5.56	30.91	5.56
	28.10	
5.05	25.55	5.05
	23.23	
4.59	21.11	4.59
	19.19	
4.18	17.45	4.18
	15.86	
3.80	14.42	3.80
	13.11	
3.45	11.92	3.45
	10.83	
3.14	9.85	3.14
	8.95	
2.85	8.14	2.85

Figure 17.4.1

A similar calculation can be done directly from our table of powers of 1.1. Figure 17.2.5 (reproduced as figure 17.4.3) shows that $1.1^{32} = 21.1138$. Cut the exponent 32 in half to get $\sqrt{21.1138} = 1.1^{16} = 4.5950$.

This procedure is far less tedious than calculating roots by hand, so it is natural to try to extend it to other cases. Can we find $\sqrt{15.86}$ that way? Figure 17.4.1 shows $\sqrt{15.86}$ is about midway between 3.80 and 4.18, so we could estimate it at 3.99, which is midway from 3.80 to 4.18. Such estimates based on midpoints are not accurate on multipliers, as you saw when we tried to put 3 midway between 2 and 4 in our original multiplier. Still, in this case the error introduced is relatively small. You can check with a calculator that $\sqrt{15.86}$ is about 3.9825.

If you try to do this with the table in figure 17.4.3, you first find that $15.8631 = 1.1^{29}$. If you cut the exponent in half, you get $1.1^{14.5}$, which raises a new question. What could fraction powers mean?

We first used exponents for repeated multiplication; later, we generalized to negative exponents by following the numerical patterns, assuming that if adding exponents goes with multiplication, then subtracting exponents, even below zero, goes with division. By analogy, if multiplying exponents goes with raising to a power, then dividing exponents should go with taking roots. Figure 17.4.2 shows an adder for exponents, with $2^{1/2}$ placed where it must go to make the adder work.

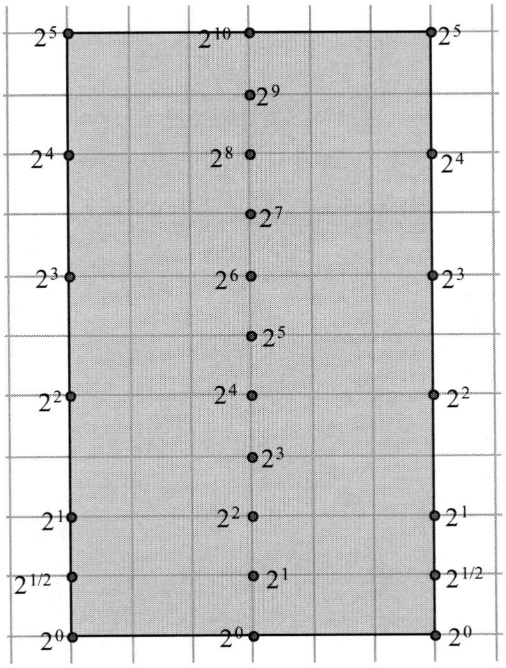

Figure 17.4.2

Even if you agree this is the logical place for $2^{1/2}$ in figure 17.4.2, you might still wonder how to define $2^{1/2}$. But remember, an adder for exponents is a multiplier. Whatever $2^{1/2}$ might be, its square is 2^1 or 2. Therefore, we define $2^{1/2}$ as $\sqrt{2}$.

Following this lead, you can see that $2^{3/2}$ fits midway between 2^1 and 2^2 in figure 17.4.2, so squaring $2^{3/2}$ yields 2^3. Therefore we define $2^{3/2}$ as $\sqrt{2^3}$ and, more generally, $2^{n/2}$ as $\sqrt{2^n}$. This is also reasonable when n is an even number. For example, $2^{4/2}$ is $\sqrt{2^4} = \sqrt{16} = 4$, which is what you get if you first reduce 4/2 and compute $2^{4/2}$ as 2^2.

The same thinking applies just as well to powers of other numbers. Therefore, we adopt the following definition of half powers.

For any positive number b, $b^{\frac{n}{2}} = \sqrt{b^n}$.

The above definition is remarkably practical, when used with a table like figure 17.4.3 (which is a reprint of figure 17.2.5).

n	Approximate value of 1.1^n	n	Approximate value of 1.1^n	n	Approximate value of 1.1^n
1	1.1	18	5.5599	35	28.1024
2	1.21	19	6.1159	36	30.9127
3	1.331	20	6.7275	37	34.0039
4	1.4641	21	7.4002	38	37.4043
5	1.6105	22	8.1403	39	41.1448
6	1.7716	23	8.9543	40	45.2593
7	1.9487	24	9.8497	41	49.7852
8	2.1436	25	10.8347	42	54.7637
9	2.3579	26	11.9182	43	60.2401
10	2.5937	27	13.1100	44	66.2641
11	2.8531	28	14.4210	45	72.8905
12	3.1384	29	15.8631	46	80.1795
13	3.4523	30	17.4494	47	88.1975
14	3.7975	31	19.1943	48	97.0172
15	4.1772	32	21.1138	49	106.7190
16	4.5950	33	23.2252	50	117.3909
17	5.0545	34	25.5477		

Figure 17.4.3

Example 1. Estimate the square root of 19.1943 using figure 17.4.3.
Solution: $19.1943 = 1.1^{31}$, so $\sqrt{19.1943} = 1.1^{31/2} = 1.1^{15.5}$. Estimate this as midway between 1.1^{15}, which is 4.1772, and 1.1^{16}, which is 4.5950, so

$\sqrt{19.1943}$ is about $\dfrac{4.1772 + 4.5950}{2} = 4.3861$. You can check with a

calculator that this estimate is about 0.005 too high, an error of about a tenth of one percent of the correct answer.

Example 2. Use figure 17.4.3 to estimate $\sqrt{12.5}$.
Solution: Since 12.5 is about midway between 11.9 and 13.1 (which are approximately the two nearest entries in the table), 12.5 is roughly $1.1^{26.5}$. Find $\sqrt{12.5}$ by halving the exponent, which is $1.1^{13.25}$. So far, you have not met fraction exponents other than halves, but that need not deter you. Whatever number $1.1^{13.25}$ is, it should be about a fourth of the way from 1.1^{13} to 1.1^{14}. By the table, that is a fourth of the way from 3.4523 to 3.7975. Figure 17.4.4 shows how to estimate this number, which is roughly

3.54, found as $3.45 + \dfrac{3.79 - 3.45}{4}$.

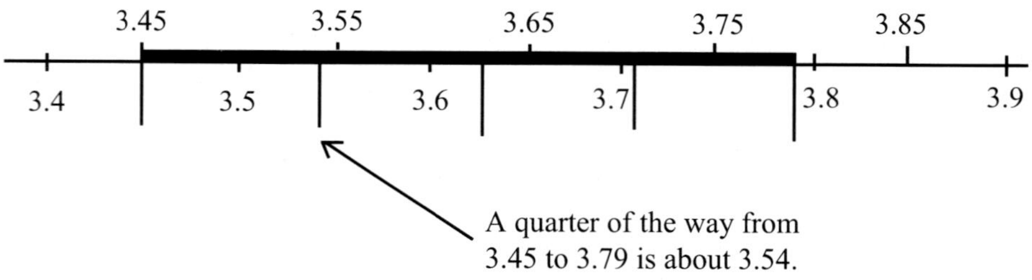

Figure 17.4.4

Again, you can check with calculator that $\sqrt{12.5}$ is about 3.5355, which, rounded to the nearest hundredth, is indeed 3.54.

Other Fraction Exponents
These examples raise new questions. How should we define the $\frac{1}{3}$ power of a number? What about the $\frac{1}{4}$ and $\frac{1}{5}$ powers and more generally the $\frac{m}{n}$ power? Can these be defined in useful ways? Do the definitions make sense even for negative fraction exponents? One of the first to take on these issues was John Wallis, who, in his *Arithmetica infinitorum (Infinite Arithmetic)* of 1655, implicitly gave the definition that is still used today.

Definition of nth roots and rational powers.

1. **For any number $b > 0$, the positive n^{th} root of b, written $\sqrt[n]{b}$, is a positive number x for which $x^n = b$.**
2. **For any number $b > 0$ and positive integers m and n, $b^{m/n}$ is the positive n^{th} root of b^m.**
   In symbols, if $b > 0$ and m and n are positive integers, $b^{\frac{m}{n}} = \sqrt[n]{b^m}$.
3. **We further define $b^{\frac{-m}{n}} = b^{\frac{m}{-n}} = b^{-\frac{m}{n}} = \frac{1}{\sqrt[n]{b^m}}$**

For example, $8^{2/3}$ is defined to be the cube root of 8^2–i.e, the cube root of 64, which is 4. Again, $8^{-2/3}$ is defined to be $\frac{1}{8^{2/3}} = \frac{1}{4}$.

Example 3. Use figure 17.4.3 to approximate $\sqrt[3]{5.5599}$.
Solution: In the table, 5.5599 is step 18, which means that $1.1^{18} = 5.5599$. Then, since the cube of 1.1^6 is 1.1^{18}, the cube root of 1.1^{18} is 1.1^6. The table shows this number as 1.7716. Then $\sqrt[3]{5.5599} = 1.7716$. Like any computation based on the table, this answer is approximate, but a check with a calculator shows it to be accurate to four decimal places.

The next examples show how to extend the reasoning from Example 3.

Example 4. Use figure 17.4.3 to approximate $23.2252^{2/3}$.
Solution: Because $23.2252 = 1.1^{33}$, $23.2252^{2/3} = (1.1^{33})^{2/3} = 1.1^{33\times\left(\frac{2}{3}\right)} = 1.1^{22}$. From figure 17.4.3, this is 8.1403.

Example 5. Use figure 17.4.3 to approximate $\sqrt{\frac{34.0039}{7.4002}}$.
Solution: From figure 17.4.3, $34.0039 = 1.1^{37}$ and $7.4002 = 1.1^{21}$. Then $\frac{34.0039}{7.4002} = 1.1^{37-21} = 1.1^{16}$, since dividing powers of any number corresponds to subtracting the exponents. No need to calculate $\frac{34.0039}{7.4002}$ yet. The square root of $\frac{34.0039}{7.4002}$ is its $\frac{1}{2}$ power. Since $\frac{34.0039}{7.4002} = 1.1^{16}$, its $\frac{1}{2}$ power is $(1.1^{16})^{1/2} = 1.1^8$, which the table shows as 2.3579.

By now you can see that exponents are more than a short way to write repeated multiplication. They are fundamental in mathematics and should be taught thoroughly and carefully in schools.

Exercises 17.4

Part 1. Evaluation with rational exponents

1. Evaluate: a. $4^{1/2}$ b. $9^{1/2}$ c. $49^{1/2}$

   d. $0.01^{1/2}$ e. $0.25^{1/2}$ f. $6.25^{1/2}$

   g. $\left(\frac{4}{9}\right)^{1/2}$ h. $\left(\frac{169}{16}\right)^{1/2}$

2. The 1/3 power of a number is called the **cube root** of that number. As a rule, cube roots are even more tedious than square roots to compute by trial and error, but the following examples yield readily to that technique. Calculate them.

   a. $8^{1/3}$ b. $125^{1/3}$ c. $1{,}000^{1/3}$ d. $0.008^{1/3}$

   e. $(-27)^{1/3}$ f. $\left(\frac{1}{64}\right)^{1/3}$ g. $\left(\frac{8}{125}\right)^{1/3}$ h. $\left(\frac{-1}{216}\right)^{1/3}$

3. In parts a-c, evaluate and compare:

   a. $\left(\frac{1}{4}\right)^{1/2}$ and $\frac{1^{1/2}}{4^{1/2}}$ b. $\left(\frac{1}{9}\right)^{1/2}$ and $\frac{1^{1/2}}{9^{1/2}}$ c. $\frac{1^{1/2}}{100^{1/2}}$ and $\left(\frac{1}{100}\right)^{1/2}$

   d. What can you say in general about $\frac{1^{1/2}}{a^{1/2}}$ and $\left(\frac{1}{a}\right)^{1/2}$?

4. Evaluate and compare:

   a. $\frac{4^{1/2}}{9^{1/2}}$ and $\left(\frac{4}{9}\right)^{1/2}$ b. $\frac{36^{1/2}}{49^{1/2}}$ and $\left(\frac{36}{49}\right)^{1/2}$ c. $\frac{81^{1/2}}{25^{1/2}}$ and $\left(\frac{81}{25}\right)^{1/2}$

   d. In general, what can you say about $\frac{c^{1/2}}{d^{1/2}}$ and $\left(\frac{c}{d}\right)^{1/2}$?

5. Evaluate: a. $(100^{1/2} + 36^{1/2})^{1/2}$ b. $(625^{1/2} - 576^{1/2})^{1/2}$

6. If our definition agrees with the basic laws of exponents, then the various ways to look at the 3/2 power of a number explored below should all yield the same result. Check that this is so.

   a. First cube 4, then take the square root of that number: Find $4^{3/2}$ as $\sqrt{4^3}$.

   b. Take the square root of 4 and cube that. Find $4^{3/2}$ as $(\sqrt{4})^3$

   c. $4^{3/2} = 4^{1/2+1} = (4^{1/2})4^1 =$

   d. $4^{3/2} = 4^{2-1/2} = (4^2)(4^{-1/2}) =$

7. Use your definition of the 3/2 power from question 8 to evaluate:
   a. $9^{3/2}$ b. $100^{3/2}$ c. $81^{3/2}$ d. $(9+16)^{3/2}$

   e. $(100-64)^{3/2}$ f. $\left(\frac{1}{4}\right)^{3/2}$ g. $\left(\frac{16}{25}\right)^{3/2}$ h. $0.09^{3/2}$

8. As a check on the concept of fractional exponents that is emerging, evaluate the following to see if they are the same as 4^2.
   a. $(4^4)^{1/2}$ b. $(4^{1/2})^4$

9. Evaluate the following rational powers:
   a. $4^{5/2}$ b. $9^{5/2}$ c. $100^{5/2}$ d. $\left(\frac{1}{4}\right)^{5/2}$ e. $\left(\frac{4}{9}\right)^{5/2}$

10. a. Calculate $8^{2/3}$, both as $(8^{1/3})^2$ and as $(8^2)^{1/3}$.
    b. Calculate $27^{2/3}$, both as $(27^{1/3})^2$ and as $(27^2)^{1/3}$.
    c. Calculate $125^{2/3}$, both as $(125^{1/3})^2$ and as $(125^2)^{1/3}$.

11. If our approach so far makes sense, then the second power (square) of any number should be the same as the sixth power of its cube root. In particular, we should have
    $$(8^{1/3})^6 = 8^{6/3} = 8^2 \text{ and } (27^{1/3})^6 = 27^{6/3} = 27^2.$$
    Check that these are indeed true.

12. We have seen that negating an exponent corresponds to taking a reciprocal. For example, $2^{-1} = \frac{1}{2}$ and $10^{-1} = \frac{1}{10}$. Now extend that reasoning from whole number exponents to fraction exponents to evaluate:

    a. $9^{-1/2}$ b. $25^{-1/2}$ c. $\left(\frac{1}{4}\right)^{-1/2}$ d. $\left(\frac{49}{16}\right)^{-1/2}$

    e. $16^{-3/2}$ f. $\left(\frac{9}{4}\right)^{-3/2}$ g. $8^{-2/3}$ h. $27^{-2/3}$

13. What could these lists possibly be about?

$\frac{1}{4}$	$\frac{1}{2}$	$\frac{3}{4}$	1	$\frac{5}{4}$	$\frac{3}{2}$
2	4	8	16	32	64

    They are found (in old notation, of course) on a clay tablet known today as MCL 2078 from ancient Mesopotamia. The tablet is over 3,000 years old. How would you interpret it?

14. Evaluate:

a. $9^{1/2} + 16^{1/2}$ b. $25^{1/2} + 144^{1/2}$ c. $49^{1/2} + 225^{1/2}$

d. $(9+ 16)^{1/2}$ e. $(25+ 144)^{1/2}$ f. $(49+ 225)^{1/2}$

g. Compare your answers to parts a and d, b and e, c and f. In general, is it true that $(a^2+b^2)^{1/2} = a+b$?

Part 2. Calculating with rational exponents

15. Use the table in figure 17.4.3 to estimate:

a. $\sqrt{97.0172}$ b. $\sqrt{54.7637}$ c. $\sqrt[3]{9.8497}$

d. $\sqrt[3]{41.1448}$ e. $17.4494^{1/3}$ f. $2.5937^{3/2}$

g. $4.5950^{3/2}$ h. $28.1024^{2/5}$

16. a. Use a calculator or a computer to compute each of the numbers from problem 15 to four decimal places.

b. Each estimate found in part a differs from the true answer by an amount we shall call the error. In each of the eight cases, subtract the smaller of the true answer and the estimate from the larger of the two to find the error.

c. In all eight cases, find what percent of the true answer the error is.

17. Use the table in figure 17.4.3 to estimate:

a. $\sqrt{3.4523 \times 13.110}$ b. $\sqrt{\dfrac{60.2401}{8.9543}}$

c. $(9.8497 \times 1.7716)^{1/5}$

18. Use a calculator to check your results from parts a, b, and c of problem 17.

17.5 An Introduction to Exponential Functions

The ideas underlying the multipliers in this chapter were first developed by Jobst Bürgi (1552-1632), possibly as early as 1588.[2] In 1620, he published a table of "red numbers" and "black numbers" much like our table of powers of 1.1. Bürgi's table, equivalent to powers of 1.0001, was much more accurate than ours but also more difficult to create. It took him over 23,000 steps to get to 10.

Independently of Bürgi, starting around 1594, John Napier (1550–1617) of Merchiston, Scotland, worked on the same concept.[3] The first result of his effort, the *Mirifi logarithmorum canonis descriptio (A Description of the Wonder-working Law of Logarithms)*, appeared in 1614. It consisted of 90 pages of tables and 57 pages of text. His table, equivalent in accuracy to a table of powers of 1.000001, was an immense effort. Modern readers find Napier's original work difficult to follow, since exponents were not invented until decades later, and his work is complicated in several ways that are ultimately irrelevant. Still, his intent is clear from this passage:

> Seeing there is nothing (right well-beloved student of mathematics) that is so troublesome to mathematical practice, nor doth more molest and hinder calculators, than the multiplications, divisions, square and cubical extractions of great numbers, which beside the tedious

---

2 Jobst Bürgi (1552-1632) was from St. Gall Canton, Switzerland and trained as a clockmaker. He had no formal higher education, but he became an expert at constructing astronomical instruments and performing mathematical calculations. One employer described him as a second Archimedes. In 1603 he came to Prague to work for the Emperor Rudolph, where Kepler (see note, Exercises 9.2) was working out his famous laws of planetary motion. The two men worked well together and admired each other deeply, but Kepler, who needed Bürgi's tables for his calculations, fumed at him for delaying their publication, writing, "this man, a procrastinator and guardian of his secrets, abandoned his child at birth and did not rear it for publicity." Unable to wait for Bürgi and unwilling to use Napier's tables (because he could not understand how they had been constructed) Kepler eventually calculated his own tables of logarithms.

3 In his time, Napier was best known for his non-mathematical work. After the Spanish Armada failed to conquer England in 1588, most English were smugly confident that their island was invulnerable, but Napier insisted that it had been sheer luck that a storm had wrecked the invading fleet. Therefore, he devised new armaments, including a rapid-fire gun, which he claimed could kill 30,000 Turks without the loss of a single Christian (i.e., Protestant). He is said to have tried this weapon successfully on some animals in a field. He also wrote prophetically of an artillery weapon that would "clear a field of four miles circumference of all living creatures exceeding a foot of height," an "enclosed military chariot, double musket proof" with a "living mouth of metal," and of "devices for sayling under water." Napier's most celebrated work in his lifetime was a vehemently anti-Catholic tract called *A Plaine Discovery of the Whole Revelation of Saint John* (1593). This was organized like Euclid's *Elements*, and it built up to the conclusion that the Pope was anti-Christ and the world would end by a certain year in the near future. This book sold extremely well, went through 21 editions and was translated into several languages. Today, Napier's theology is a mere footnote, but his contributions to mathematics endure. An outstanding article on Napier and his work is N. T. Gridgeman, "*John Napier and the History of Logarithms*" in *Scripta Mathematica*, volume 29 numbers 1 and 2 (Spring-Summer, 1973).

expense of time are for the most part subject to slippery error, I began therefore to consider in my mind by what certain and ready art I might remove those hindrances.

Napier wrote a second book, *Mirifi logarithmorum canonis constructio (Construction of the wonder-working law of logarithms)* (1619), to explain in detail the construction of his logarithm table. His tables pioneered the use of the decimal point and did much to popularize decimal fractions.

Henry Briggs, a young mathematics professor, upon seeing Napier's logarithms, visited him in Scotland and, with Napier's approval, decided to make an improved table of logarithms, based on ten. The brilliant and tireless Briggs eventually produced tables of logarithms of all whole numbers from 1 to 20,000 and from 90,000 to 100,000, accurate to 14 decimal places! Before he could fill in the gap from 20,000 to 90,000, as he planned, Adriaan Vlacq, a bookseller from Gouda, in the Netherlands, issued a table of logarithms of all whole numbers from 1 to 100,000. Vlacq's table, compiled by using Brigg's existing tables and filling in the gaps approximately, was good to seven place accuracy, but it was successful commercially.

Exponential functions

The tables mentioned above were created as computational devices. They offered the only easy way to carry out the increasingly difficult computations that powered the scientific and technological triumphs of the 17th and 18th centuries. By the mid 18th century, the relationship between exponents and logarithms was clearly understood, and interest in exponential functions began to increase. The electronic age has rendered calculation with logarithms obsolete, but exponential and logarithmic functions are perhaps more important today than ever before.

Definition: For a given positive number b, the exponential function with base b is the function b^x.

Earlier, we calculated 2^x for particular values of x. Now we will consider 2^x as a function. In particular, we consider the graph of this function.

Example 1. Graph the function 2^x.
Solution: When you are not familiar with a function, it is good to begin by calculating and plotting some points on its graph.

x	-4	-3	-2	-1	0	$\frac{1}{2}$	1	$1\frac{1}{2}$	2	$2\frac{1}{2}$	3	$3\frac{1}{2}$	4
2^x	$\frac{1}{16}$	$\frac{1}{8}$	$\frac{1}{4}$	$\frac{1}{2}$	1	$\sqrt{2}$	2	$2\sqrt{2}$	4	$4\sqrt{2}$	8	$8\sqrt{2}$	16

The graph is obtained by plotting the points calculated in the table and then drawing a smooth curve through them.

Figure 17.5.1 shows the graph 2^x, together with the graphs of three more exponential functions.

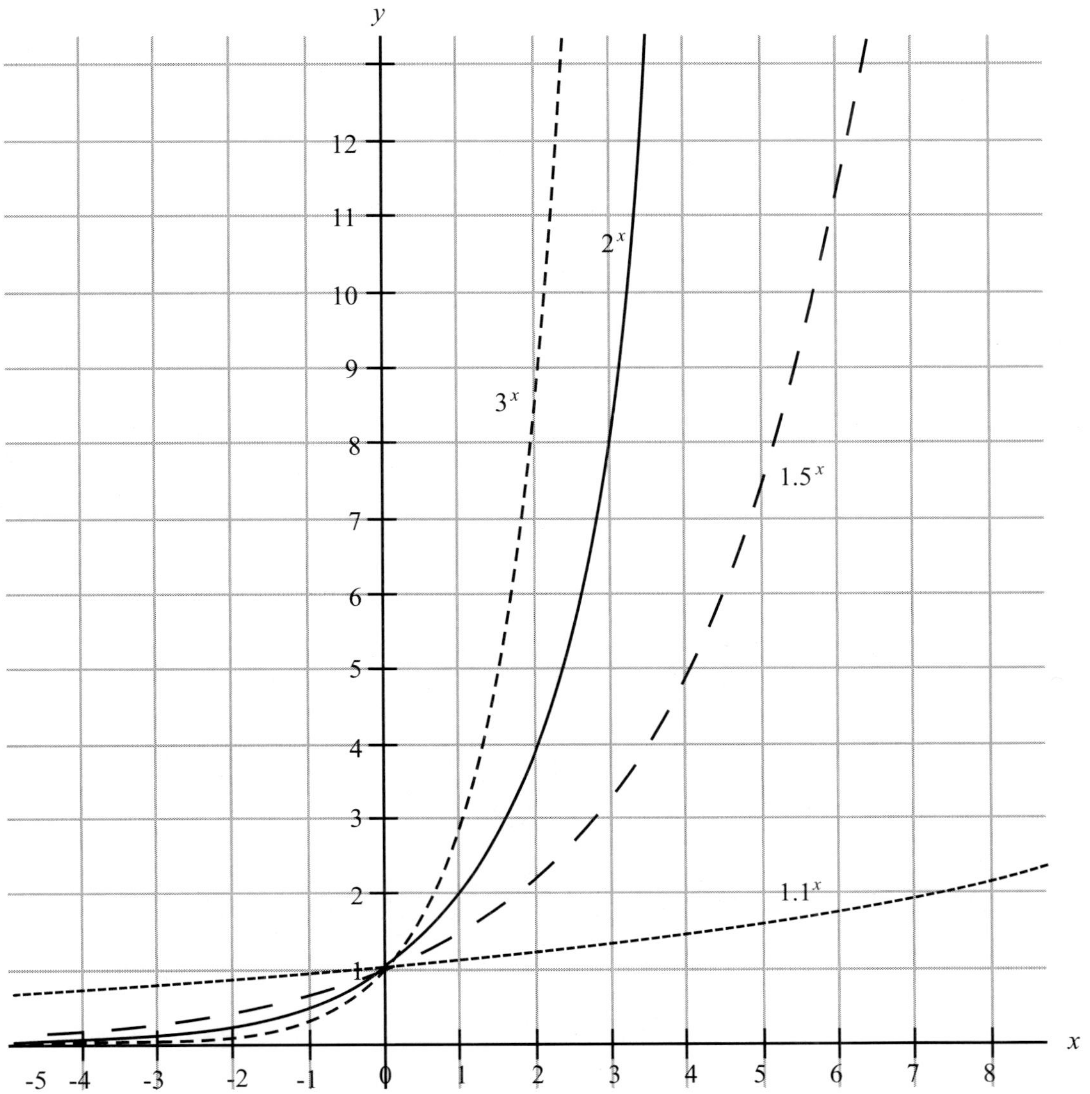

Figure 17.5.1

The graph of every exponential function passes through the point (0,1), because for any number $b \neq 0$, $b^0 = 1$. If $b > 1$, the curve flattens out along the negative x-axis, eventually becoming indistinguishable from the axis itself. This is described mathematically by saying that the curves are **asymptotic** to the negative x-axis. The larger b, the more quickly b^x increases to the right, and the more quickly it approaches the negative x-axis. We based our multiplier on 1.1, because it climbs relatively slowly. Bürgi based his table on powers of 1.0001, because the graph of 1.0001^x is almost a straight line for a long time. Like any exponential function, 1.0001^x eventually climbs very steeply, but even at $x = 23{,}000$ it is still not quite ten units up.

Figure 17.5.2 shows the graphs of two exponential functions b^x for $0 < b < 1$.

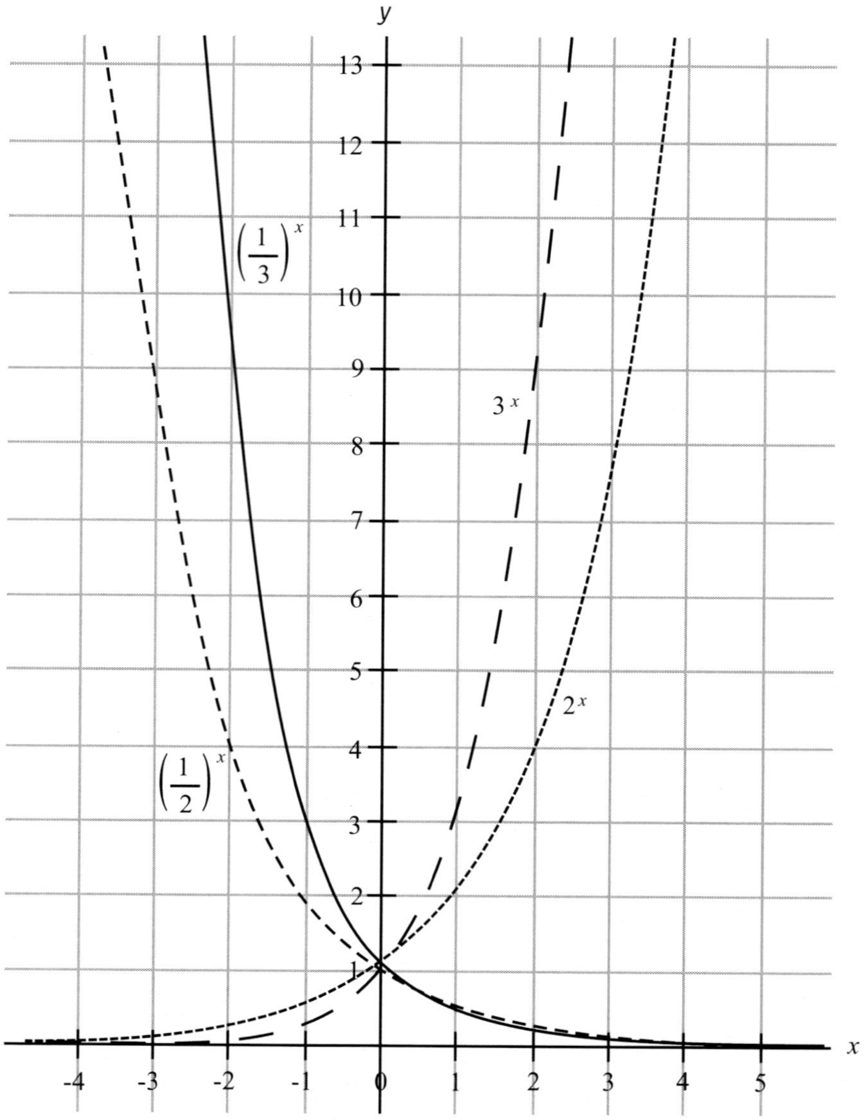

Figure 17.5.2

These curves fall from the left, pass through the point (0,1), and are asymptotic to the positive x-axis. Replacing x with $-x$ in a graph has the effect of running the x-values backward, which reflects the graph about the y-axis. The graph of $\left(\frac{1}{2}\right)^x$ is the reflection of the graph of 2^x about the y-axis, because $\left(\frac{1}{2}\right)^x = (2^{-1})^x = 2^{-x}$. Similarly, the graphs of $\left(\frac{1}{3}\right)^x$ and 3^x are mirror images with respect to the y-axis.

Features of the graphs of exponential functions b^x ($b > 0$)

Since $b^0 = 1$, the graph of b^x passes through (0,1), regardless of b.

For $b > 1$ the graph of b^x is asymptotic to the negative x-axis and rises steeper and steeper as x increases. The larger b, the more rapidly the curve steepens.

For $0 < b < 1$, the graph of b^x is asymptotic to the positive x-axis and falls steeply from the left. The smaller b, the more rapidly the curve descends from the left.

For $b = 1$, b^x is $1^x = 1$, and the graph of b^x is the horizontal line $y = 1$.

The reason for the importance of exponential functions is that many quantities increase or decrease at rates proportional to the amount present. Increases or decreases of this kind are called, respectively, **exponential growth** and **exponential decay**. The following examples illustrate this growth and decay.

Example 2. The water hyacinth, native to South America, became a major pest when introduced into Africa for its pretty blossoms. It was first observed in the Congo River in 1952. Despite intensive eradication efforts, the plant spread quickly. By 1957 an estimated 36 tons per hour of water hyacinths were observed floating downstream past Leopoldville (now Kinshasa), hindering navigation and upsetting the ecology.

If a population of water hyacinths doubles its area every day and begins with a single square meter, estimate how long it would take the population to cover a reservoir with a surface area of 200 km^2? This is just what happened to an African reservoir in 1965.[4]

Solution: Tabulate the area covered by water hyacinths after each day of growth:

After day number	1	2	3	4	5	6
Area (square meters)	2	4	8	16	32	64

By now you can guess the rule that after x days the area covered is 2^x. Since 200 km^2 $= 200 \times (1{,}000)^2 = 2 \times 10^2 \times 10^3 \times 10^3$ m^2, the problem is to find for what number x the value of 2^x first exceeds 2×10^8. To solve this approximately, note that $2^{10} = 1{,}024$, which is just slightly more than 10^3. Therefore, $2^{20} = (2^{10})(2^{10})$ is approximately $(10^3)^2 = 10^6$, and $(2^{10})^{2/3}$ is approximately 10^2. Then 200 km^2 is approximately $2 \times 2^{20/3} \times 2^{10} \times 2^{10}$ m^2 or $2^{27^2/_3}$ m^2.

We can then estimate the time needed to cover the reservoir under these circumstances as between 27 and 28 days.

4 See L. G. Holm, L. W. Weldon, and R. D. Blackburn, "*Aquatic Weeds*," *Science*, November 7, 1969, pp 699-709.

Example 3. If you invest $10,000 on January 1 at 10% interest compounded annually, how much would your investment be worth on each of the next five January 1 dates?

Solution:

Initial number of dollars	10,000
Interest for year 1 (10% of 10,000)	1,000
Value of investment after 1 year	11,000
Interest for year 2 (10% of 11,000)	1,100
Value of investment after 2 years	12,100
Interest for year 3 (10% of 12,100)	1,210
Value of investment after 3 years	13,310
Interest for year 4 (10% of 13,310)	1,331
Value of investment after 4 years	14,641
Interest for year 5 (10% of 14,641)	1,464.10
Value of investment after 5 years	16,105.10

Do these numbers look familiar? They are just 10,000 times the numbers computed in section 17.2 as powers of 1.1. Indeed, the value of this investment after n years will be $10{,}000 \times 1.1^n$. This example, however, is not strictly a case of exponential growth, because the interest is paid not continuously but in lump sums at the end of each year.

Example 4. A nuclear explosion released the radioactive isotope iodine 131 (written I-131) into the atmosphere. This iodine fell to earth, where it was absorbed into grass that was eaten by milk cows. Humans who drink the milk will subsequently store much of the iodine from that milk in their thyroid glands. Commenting on the hazard to those people, an official pointed out that while live grass absorbs I-131, cut grass does not. Furthermore, I-131 decays with a half-life of about eight days, meaning that after any 8-day period half the I-131 present at the start of that period will have disappeared. The official pointed out that feeding cows on grass which had been stored as hay would lower the radioactivity in their milk, and that storing the milk before drinking it would further reduce the I-131 intake of humans by allowing more time for the radioactivity to decay. What fraction of the original I-131 remains 8, 16, 24, 32, and 40, days after the grass is cut?

Solution:
After 8 days, half the original I-131 remains.
After 16 days, one fourth of the original I-131 remains.
After 24 days, one eighth of the original I-131 remains.
After 32 days, one sixteenth of the original I-131 remains.
After 40 days, one thirty-second of the original I-131 remains.
After 48 days, one sixty-fourth of the original I-131 remains.
In general, after x days, the fraction of the original I-131 remaining is 2^{-x}.

Logarithms

Logarithms may be thought of as what you get if you look at exponential functions backward, so to speak. If you begin with the value of the exponential function b^x, then the logarithm is the value of x involved. For example, if you work with the exponential function 2^x and begin with 8, then x is 3, since $2^3 = 8$. We say that 3 is the **base 2 logarithm** of 8 and write $\log_2 8 = 3$. (The word "base" as used here has nothing to do with the base of a system of numeration.) Again, if you are working with base ten logarithms, so that the exponential function involved is 10^x, you would say the base ten logarithm of 100 is 2, since $10^2 = 100$, and you would write $\log_{10} 100 = 2$.

In short, $\log_b y = x$ means that $b^x = y$. The logarithm itself is the exponent, while the base, which is written here as a subscript, is the number which is raised to a power.

Example 5. What is the base ten logarithm of 0.001?
Solution: Since $10^{-3} = 0.001$, the base ten logarithm of 0.001 is −3.

Example 6. Express $7^4 = 2{,}401$ in logarithmic notation.
Solution: $4 = \log_7 2{,}401$.

Example 7. Express $\log_{1/2} 8 = -3$ in exponential notation.

Solution: $\left(\frac{1}{2}\right)^{-3} = 8$

The rules for working with logarithms are all restatements of corresponding rules for exponents. However, we will not pursue the study of logarithms here.

Exercises 17.5

1. The radioactive isotope radium 228 decays at about 10% per year.
   a. If you begin with a kilogram of this isotope, how many kilos do you have after 1, 2, 3, 4, and 5 years?
   b. Express the amount which remains after t years as a function of t.
   c. Sketch the graph of the function from part b.

2. C. Northcote Parkinson has stated that in any public administrative department not actually at war, the staff grows about 6% per year, regardless of the amount of work to be done.[5] Parkinson cited as an example the British Colonial Office, which had a staff of 1,139 in 1947. If the staff did grow 6% per year, how many employees would it have in 1954? (According to Parkinson, the actual figure was 1,661, even though the Colonial Office had ceased to administer India and numerous colonies in Africa during that period.)

[5] C. Northcote Parkinson, *Parkinson's Law: The Pursuit of Progress*, London, John Murray (1958).

3. Cosmic rays striking nitrogen in the air create carbon 14 (C-14), which decays with a half-life of 5,568 years. In the air C-14 disappears through decay as fast as it is created, so the ratio of C-14 to C-12 (ordinary carbon) is fairly constant. In a living organism the C-14:C-12 ratio is initially the same as in air. After the organism dies, the C-14 decays and is not replenished, but the C-12 does not decay. Therefore the C-14:C-12 ratio decreases.

   a. Carbon in wood from an archaeological dig has a C-14:C-12 ratio equal to half that found in the air. How long ago did the tree die?

   b. Charcoal from the ashes of a fire in a cave has a C-14:C-12 ratio equal to one fourth of that found in the air. How long ago did the fire burn?

   c. In 1901 a frozen mammoth was discovered in the eroding banks of a Siberian river. It was remarkably well preserved, with freshly eaten buttercups in its stomach. Its C-14:C-12 ratio was about one eighth of the C-14:C-12 ratio in air. About how old is this mammoth? (The stuffed remains are now on display at the Zoological Museum in St. Petersburg, along with a more recent and even better preserved find, the Duma mammoth.)

4. The Richter scale measures earthquake intensity. Each increase of one unit on the Richter scale corresponds to a tenfold increase in the strength of an earthquake, so an increase of 0.5 on the Richter scale corresponds to an increase in strength of $\sqrt{10}$ or about 3. The great Alaskan quake of 1964 was rated at 8.3 on the Richter scale, and the Sumatran quake of 2004 (which caused a devastating tsunami) was rated at 9.3.

   a. How many times as strong as the Alaskan quake was the Sumatran quake?.

   b. What increase in earthquake strength corresponds with an increase of 2.5 on the Richter scale?

5. Doubling the frequency of any note produces a note one octave higher. There are twelve semitones in an octave. In the well-tempered scale, the ratio of frequencies between pairs of notes which are one semi-tone apart is constant.

   a. What is this constant?

   b. If the A below Middle C is 440 cycles per second (cps), what is the frequency of Middle C? (Middle C is three semi-tones above the A.)

   c. If a child can hear tones with frequencies up to 20,000 cps, and an adult can hear tones with frequencies up to 14,000 cps, how many notes on a scale can be heard by the child but not by the adult?

d. Bats can hear tones with frequencies up to 200,000 cps. How many semitones can be heard by bats but not by human children?[6]

6. Express each statement in logarithmic notation.
   a. $5^2 = 25$
   b. $10^7 = 10{,}000{,}000$
   c. $10^{-5} = 0.00001$
   d. $27^{1/3} = 3$
   e. $(0.01)^2 = 0.0001$
   f. $(0.2)^{-3} = 125$

7. Express each statement in exponential notation.
   a. $9 = \log_2 512$
   b. $5 = \log_3 243$
   c. $3 = \log_{10} 1{,}000$
   d. $3 = \log_7 343$
   e. $-2 = \log_5 0.04$
   f. $0.25 = \log_{81} 3$

8. Use the values $\log_{10} 2 = 0.3010$, $\log_{10} 3 = 0.4771$, and $\log_{10} 5 = 0.6990$ to calculate the base ten logarithms of these numbers:
   a. 4
   b. 8
   c. 6
   d. 9
   e. 12
   f. 15
   g. 16
   g. 18
   h. 20
   i. 24
   j. 30
   k. 32

[6] See Bibby, Neil: "*Tuning and temperament: closing the spiral*" in *Music and Mathematics: From Pythagoras to Fractals*, John Fauvel, Raymond Flood, and Robin Wilson, editors. Oxford University Press, Oxford, 2003.

Vocabulary for Chapter 17

Asymptotic (17.5)
Base of logarithms (17.5)
Cube root (17.4)
Exponential function (17.5)
Exponential growth and decay (17.5)
Logarithm (17.5)
Place holder (17.3)
Scientific notation (17.3)
Significant digit (17.3)

Self-Quiz for Chapter 17

1. If 1.1^8 is about 2.1436, what power of 1.1 is approximately equal to:
   a. 2.1436^5
   b. 2.1436^{10}
   c. $\sqrt{2.1436}$

2. If 1.1^{21} is about 7.40 and 1.1^{22} is about 8.14, what power of 1.1 is about 8.00?

3. The number of German marks in circulation at the height of the post World War I hyperinflation has been estimated at 496,585,346,000,000,000,000. Write this number in scientific notation in a way that indicates that this number is accurate only to the nearest trillion.

4. Evaluate:
   a. $64^{\frac{2}{3}}$
   b. $64^{\frac{3}{2}}$
   c. $64^{-\frac{1}{3}}$

5. A certain model automobile, through normal wear and tear, loses 20% of its value each year. If it was worth \$18,000 brand new, what is it worth after 6 years?

6. Express in logarithmic form: $16^{1.75} = 128$

7. Express in exponential form: $\log_8 \sqrt{2} = \frac{1}{6}$

Chapter Eighteen
Lines, Circles and Angles

The geometry in this chapter dates back to ancient Greece. Euclid[1], whose *Elements* served for two millennia as the model for presenting mathematics, built his geometry on five **postulates** (basic statements, assumed to be true). In modern terms they are:

1. It is possible to draw a line between any two points.
2. It is possible to extend any line segment (finite portion of a line).
3. It is possible to draw a circle with any given center and radius.
4. All right angles are equal.
5. For any given line and any point not on the line, there is just one line through the point parallel to the given line.

These postulates are closely linked to actually constructing geometric objects with traditional construction instruments: the straightedge for drawing lines, the compass for drawing circles, and protractor for drawing and measuring angles. Here we will use the hands-on approach, from Euclid, enhanced by computer programs like *The Geometer's Sketchpad*, which can simulate traditional construction methods in ways that foster creative exploration.

18.1 Lines

Euclid's first postulate notwithstanding, many people have trouble drawing a line through two given points. Therefore, we begin with some exercises that involve drawing lines with a pencil and **straightedge**, which is a just ruler on which marks for measurement are ignored. First, however, some general hints for making good-looking geometric drawings:

1. Use a sharp pencil.
2. Make lines as thin as possible without their being invisible. Do not retrace a line to make it darker.
3. Indicate a point on paper by means of a tiny dot, or perhaps as the intersection of two thin lines or curves, never by a fat dot.

The rationale for these hints is that, ideally, points have no size, and lines and curves have no width. No drawing can live up to this ideal, but denot-

[1] See note on Euclid, page 94.

ing a point by a fat dot is obviously imprecise, since it is not clear where in the fat dot the point is supposed to be. Similarly, lines drawn with wide, heavy pencil marks seem to intersect in a whole region rather than in a single point. You will at times have to compromise on these issues. For example, when you draw a geometric figure on a chalkboard, students more than a few feet away will not be able to see your work unless you make the lines and dots heavy. Also, students with vision or coordination problems may need extra time or extra help in making geometric sketches, and some students may simply not be able to make them very neatly.

Line designs

Here is an enjoyable way to practice using a straightedge. Connect dots in the remaining two-thirds of the picture, using the same kind of pattern as used in the first third. Then color the resulting design in a checkerboard style. Take care to work precisely. When you are done, the result will look like a curved surface. As you watch students in this activity, you will notice that some put their straightedge directly on the points to be connected, causing the pencil point to miss the dot when they actually draw the line. One way to see how much space to allow for this is to first put the pencil directly on one of the points, then slide the straightedge up against the pencil.

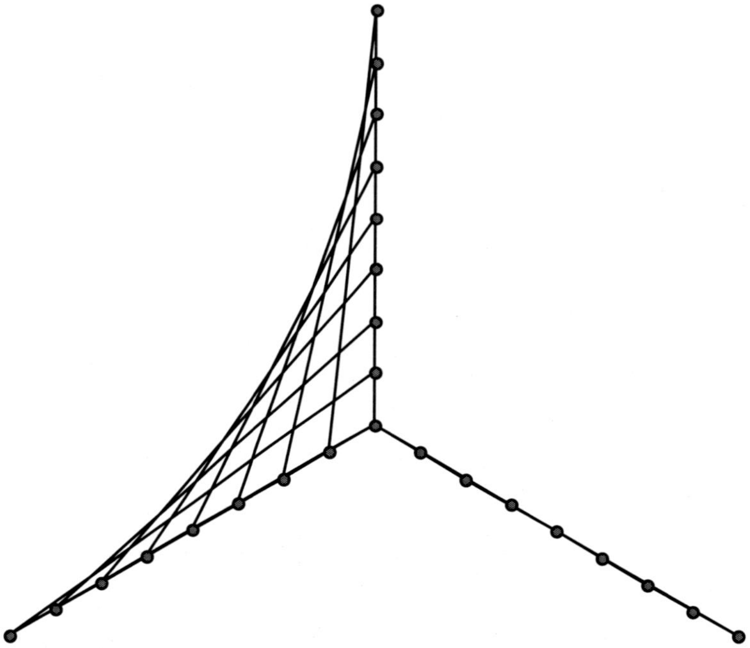

Figure 18.1.1

Templates for similar designs are available commercially, but you and your students can make up your own. To do so, mark regularly spaced points on two or more line segments. Connecting points in some regular order will make a line design. Make up more of these line designs, varying the positions of the lines and the numbers of points connected. This low-key activity enjoyably combines art and mathematics. In case students want to redo their designs, keep extra templates like figure 18.1.1 available.

Perspective Drawing

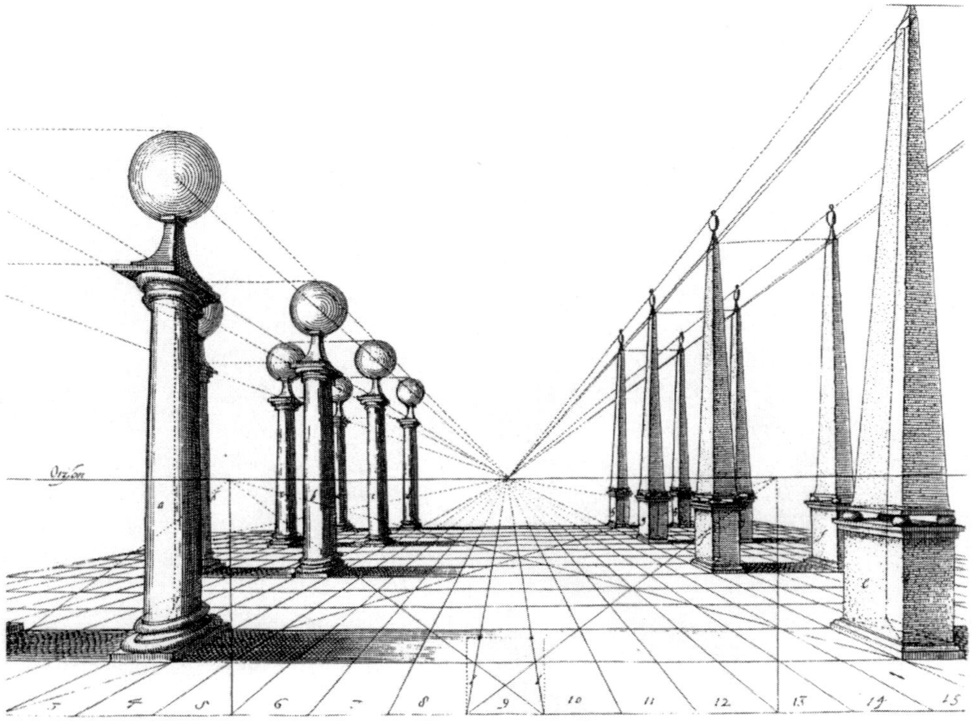

Figure 18.1.2

Figure 18.1.2 is a sketch for a perspective drawing typical of those done in the Renaissance[2]. Perspective drawing was not used at all in medieval Europe, though ancient Roman art from Pompeii used perspective. The idea of perspective drawing is summed up in figure 18.1.3. The painting is done on a plane, which corresponds to the glass shown here. In effect, the artist draws on the glass, as pictured by Durer below.

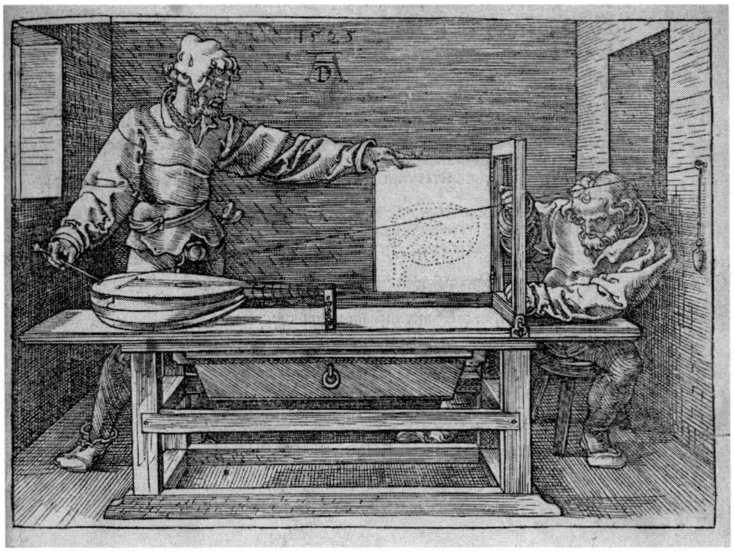

Figure 18.1.3

2 This picture, by Dutch renaissance artist Jan Vredman de Vries, was included in a 1996 history of mathematics calendar by JF Ptak, then a Washington, DC book dealer.

Three important theorems of perspective

> 1. **A perspective view of a line is a line.**
> 2. **Parallel lines, as they appear in a perspective drawing, are either parallel or meet at a point, called the <u>vanishing point</u> for lines with that direction.**
> 3. **If a painting is in perspective, there is a unique point from which it looks correct.**

We will exploit these ideas in making perspective drawings. Work through each of the following activities to create a drawing of your own, following the directions. Depending on interest, need, and time, you can do them with your students in much the same manner. In many cases you need to draw a line parallel or perpendicular to a given line. We will discuss how to do that later. For now, careful freehand approximations will do.

Rectangular box, one-point perspective The edges of a rectangular solid are parallel in fours, but drawn that way (figure 18.1.4), it doesn't quite look realistic.

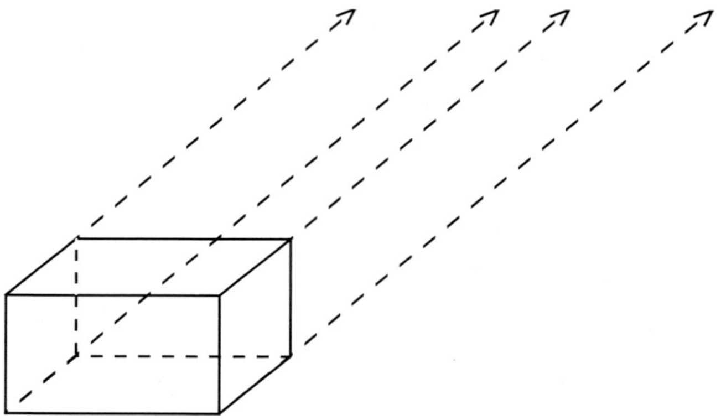

Figure 18.1.4

To create a "one point" perspective drawing, start with a rectangle *ABCD* for the front of the box, and select a **vanishing point** *V* at which the four parallel edges of the box which lead away from us would appear to meet, if they were extended to the horizon. Draw line segments *AV, BV, CV, DV* (figure 18.1.5).

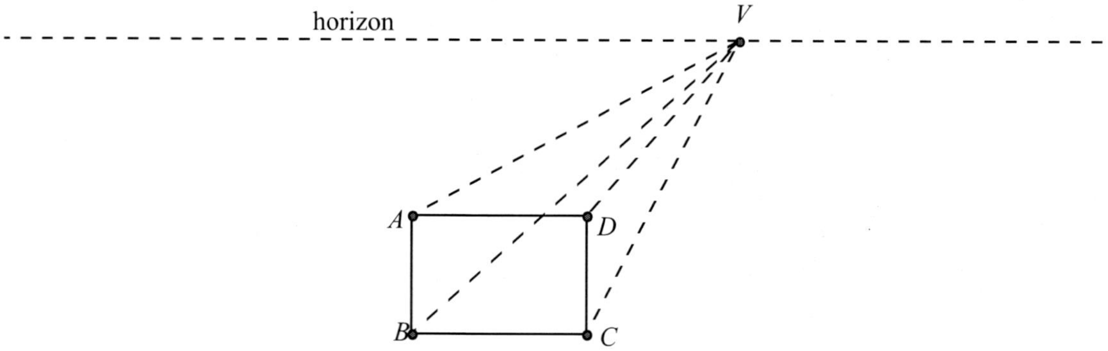

Figure 18.1.5

Now select a point E for a rear corner of the solid, on AV, and draw lines EF and EH parallel to DA and AB (figure 18.1.6).

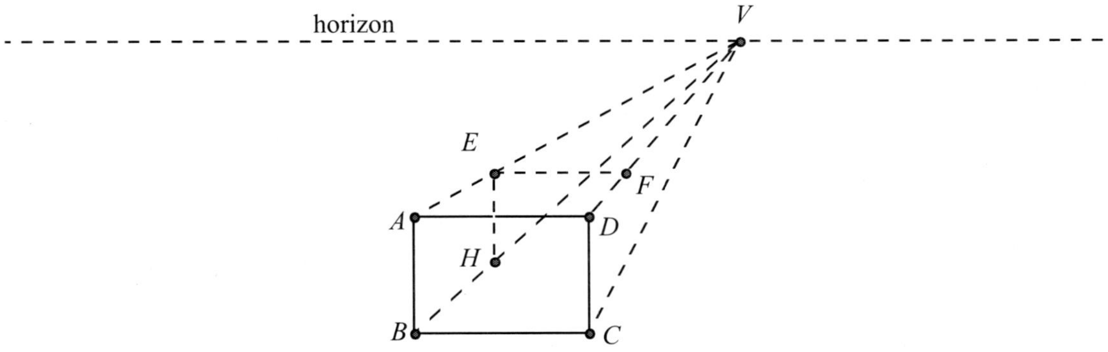

Figure 18.1.6

To complete the rear face, draw lines from H parallel to BC and from F parallel to AB, as in figure 18.1.7.

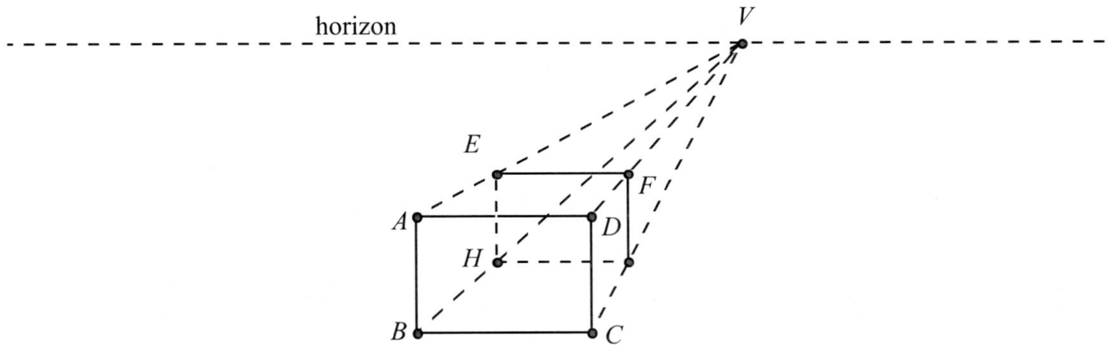

Figure 18.1.7

Finally, erase all unnecessary lines, and make those lines solid which are not seen through the box, as in figure 18.1.8.

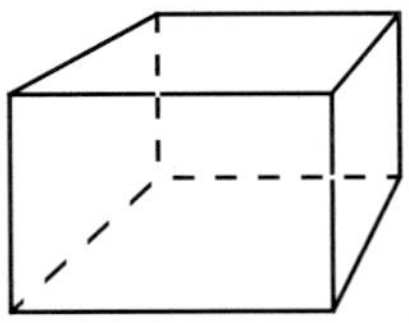

Figure 18.1.8

This is called one-point perspective, because it uses only one vanishing point. Now we will make a two-point perspective drawing of a rectangular solid.

Two-point perspective drawing of a rectangular solid

To draw a rectangular box in two-point perspective, follow these steps.

1. Draw a horizontal line, to serve as the horizon, and draw a vertical line segment AB, which will be the vertical edge closest you. (figure 18.1.9)

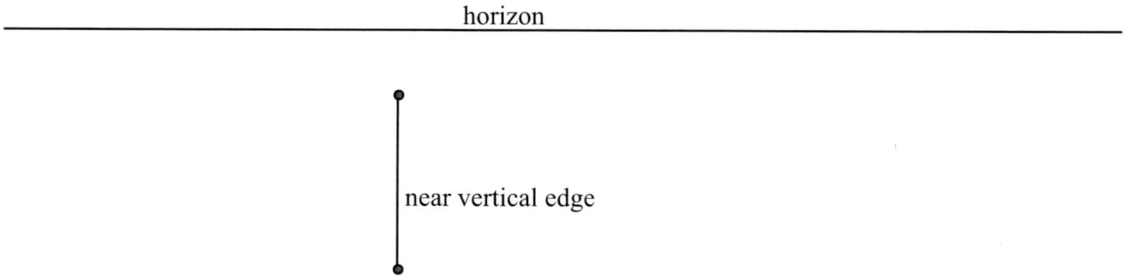

Figure 18.1.9

2. Select two points V and W on the horizon to be vanishing points, and draw lines AV, AW, BV, and BW (figure 18.1.10).

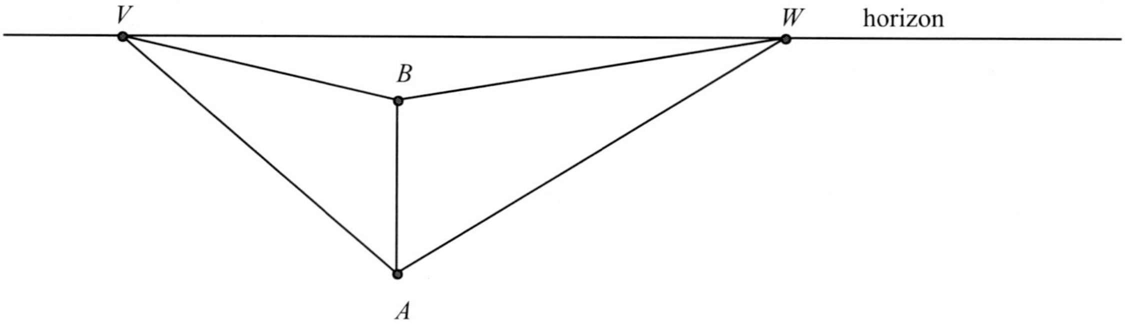

Figure 18.1.10

3. Select a third corner of the box C wherever you like on line AV, and a fourth corner D on line AW, again, wherever you want. Then draw lines CW and DV (figure 18.1.11).

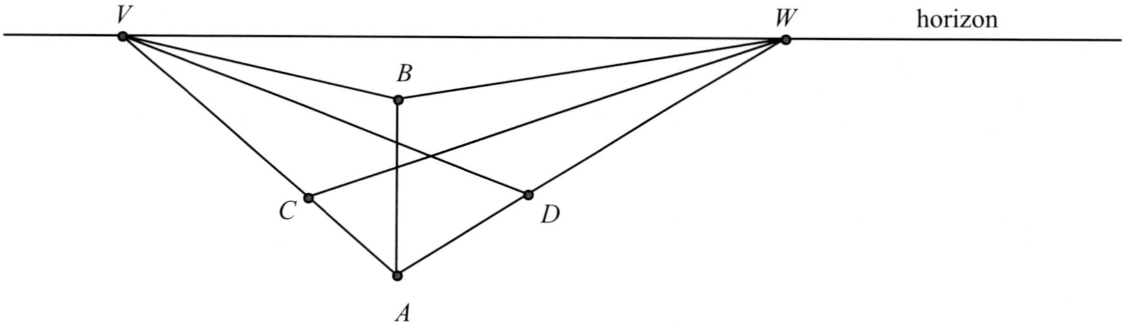

Figure 18.1.11

4. Draw vertical lines from D to line BW, naming the point of intersection E, and from C to line BV, naming the point of intersection F (figure 18.1.12).

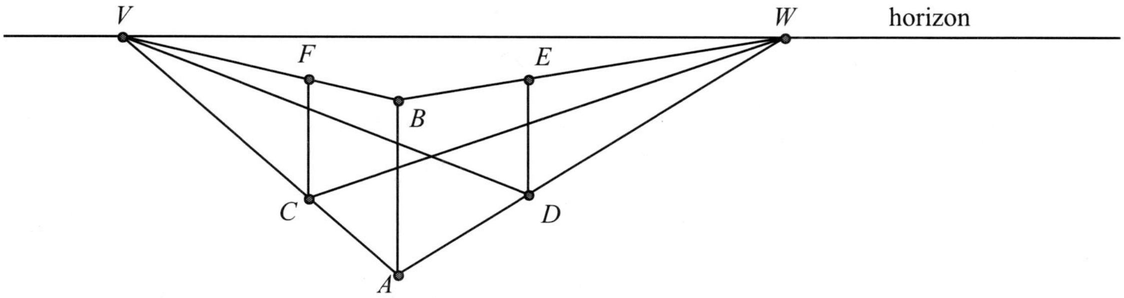

Figure 18.1.12

5. Draw lines *EV* and *FW*, and name their intersection *G*. Then draw a line from *G* straight down to the intersection of lines *DV* and *WC*. Call that intersection *H* (figure 18.1.13)

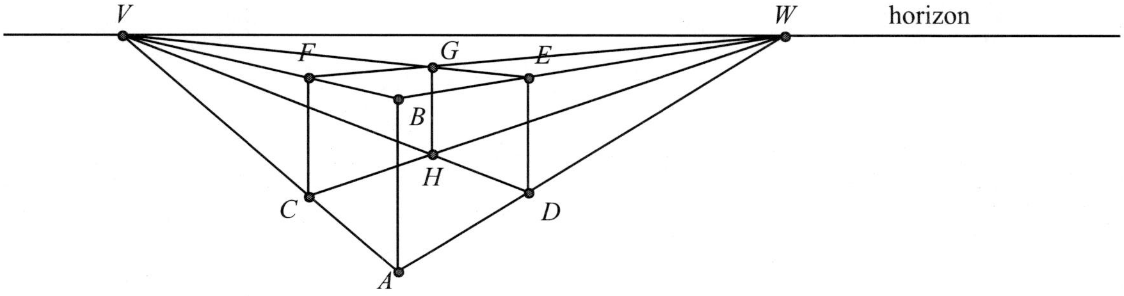

Figure 18.1.13

6. Erase unnecessary lines and labels, and dot those lines which are seen only by looking through the box, as in figure 18.1.14.

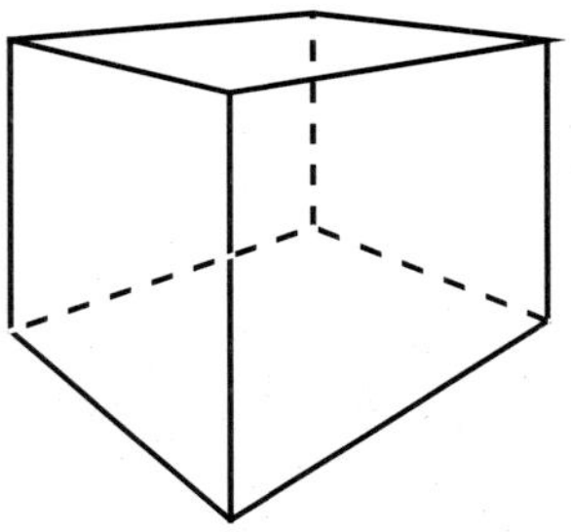

Figure 18.1.14

The drawing will look realistic only when seen from the one right viewing point, which corresponds to the eye in figure 18.1.3. You will need to experiment to find that point.

Checkerboard, two-point perspective

1. Draw a horizontal line to serve as the horizon, and select two points on the horizon to be the vanishing points for lines parallel to sides of the checkerboard.

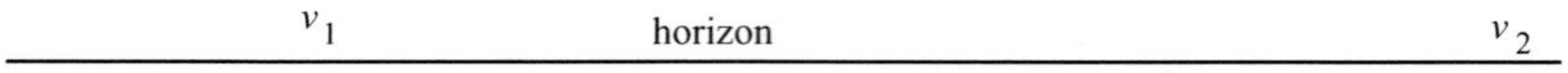

2. Draw pairs of lines from v_1 and v_2 to represent the sides of the board.

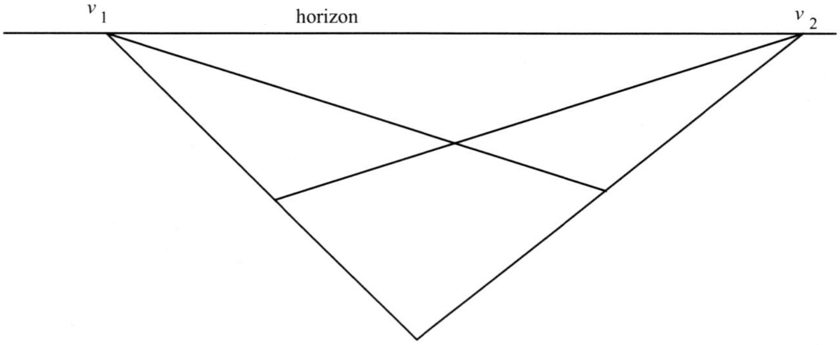

3. Draw the diagonals of the board, thus determining its center.

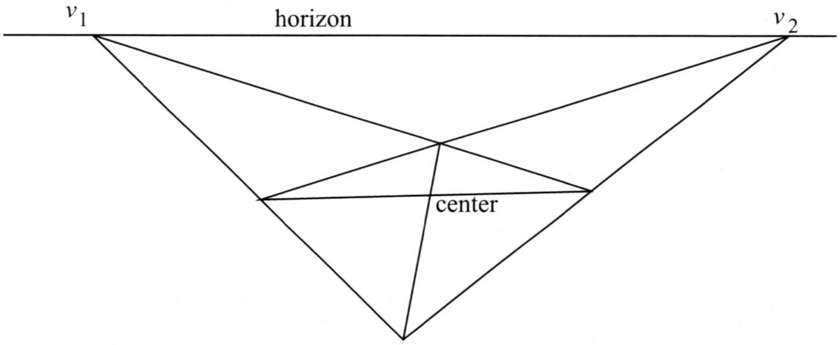

4. Draw lines from v_1 and v_2 through the center of the board to the opposite sides.

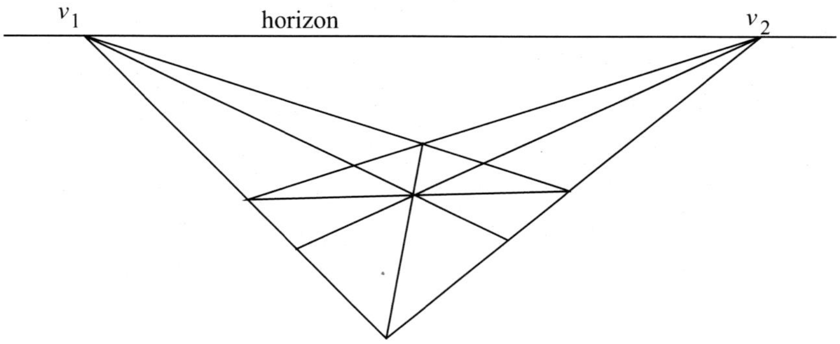

5. Draw missing diagonals of the four quarters of the board.

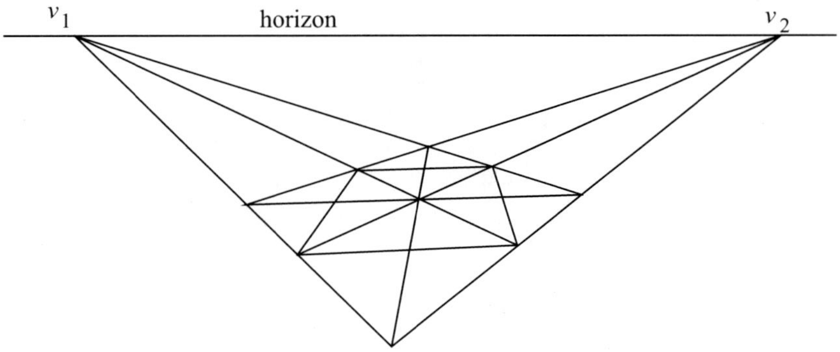

6. Draw lines from v_1 and v_2 through midpoints of quarters.

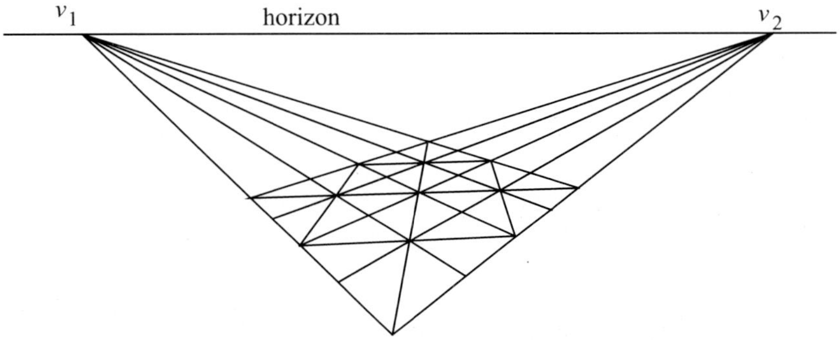

7. Repeat steps 5 and 6 on each quarter of the board.

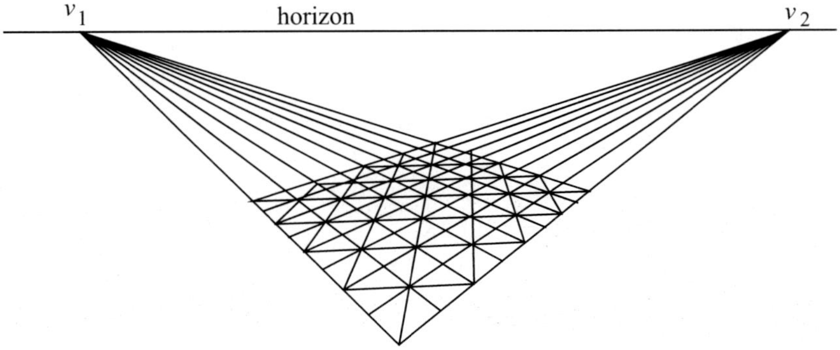

8. Erase unnecessary lines and put in shading.

As with any perspective drawing, there is one particular point from which it looks right. Experiment until you find that point, at least approximately.

18.2 Constructing with Circles

Euclid's third postulate, that one can draw a circle of any specified center and radius, gives circles special prominence among curves. The traditional instrument for drawing circles is a **compass**, usually made out of metal, with one foot to be held fixed at the center of the circle while the other foot, with a pencil or pen at its tip, is rotated about the center.

The compasses used in schools can be sources of difficulty. Many do not hold the pencil rigidly, are stiff and hard to open to the right radius, or are loose and do not maintain a constant radius well. Also, the sharp points on some compasses are dangerous. Today a variety of relatively safe, effective, and inexpensive compasses are available, as are computer programs like *The Geometer's Sketchpad*, with which these pictures were made. Whatever the tool, however, students still need and enjoy practice with construction.

1. Hexagonal build-out

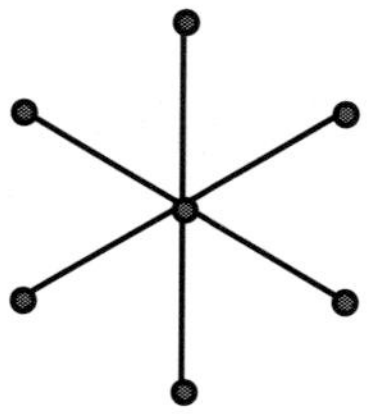

Begin with the pattern of points and lines shown above, and draw circles at each of the outer points so that they pass through the center point, like this.

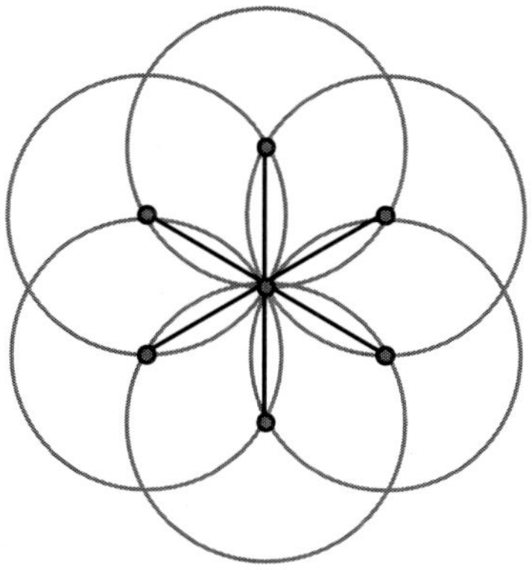

Mark the new points of intersection of those circles, and draw line segments from them to the six points at the edge of the original pattern, like this:

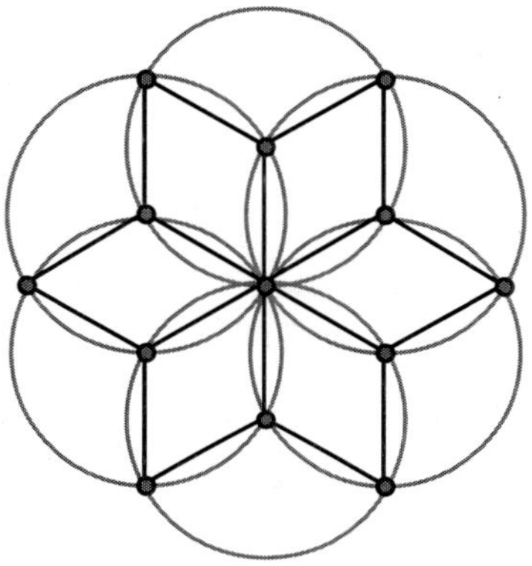

Erase the circles to get this:

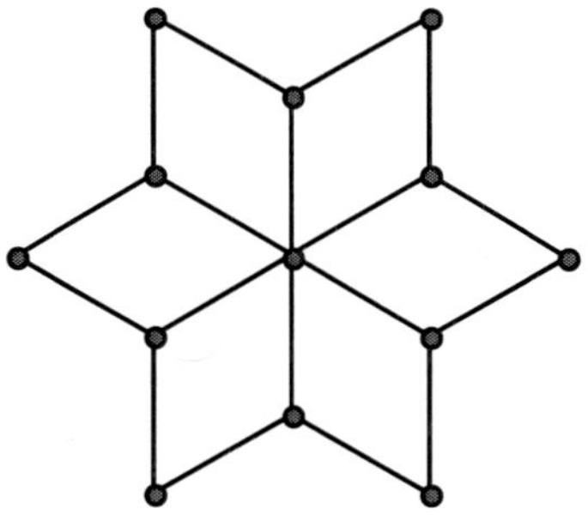

Already the drawing is attractive, and some students might want to leave it there and color it in. However, you can go further, drawing a circle (still the same size) at each point on the tip, like this:

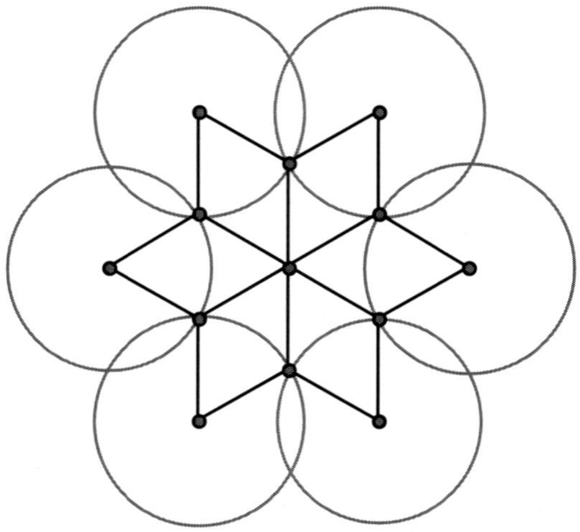

Now mark the new points of intersection of these circles, and connect them to the centers of the circles like this:

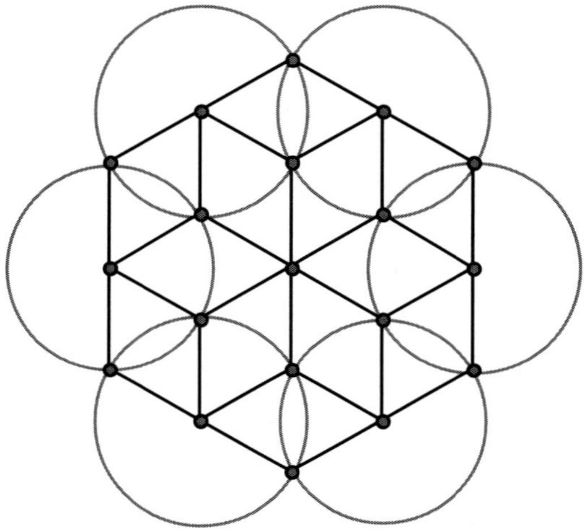

Now, erase the circles, like this:

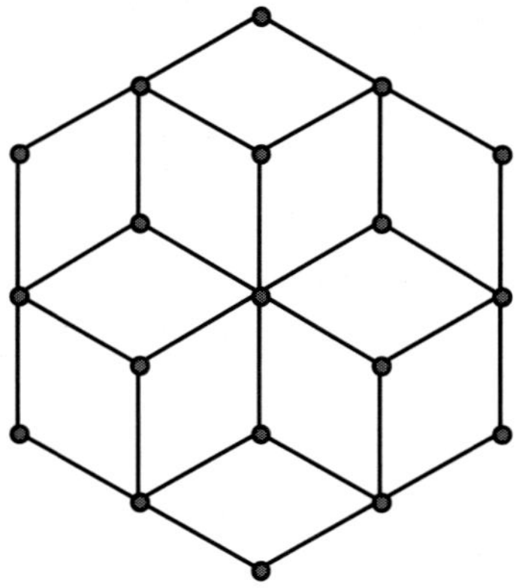

Finally, color or shade the picture artistically.

This activity helps to develop skill at using a compass. Most beginners will not find it easy, but, given time and some guidance, they can make the sketch well enough to be pleased by it. It is best to begin with a pattern of points enlarged to about twice the size shown here, being sure it is well centered on the page.

2. Octagonal build-out

Try the same activity beginning with eight points around a central point to make an octagonal build-out:

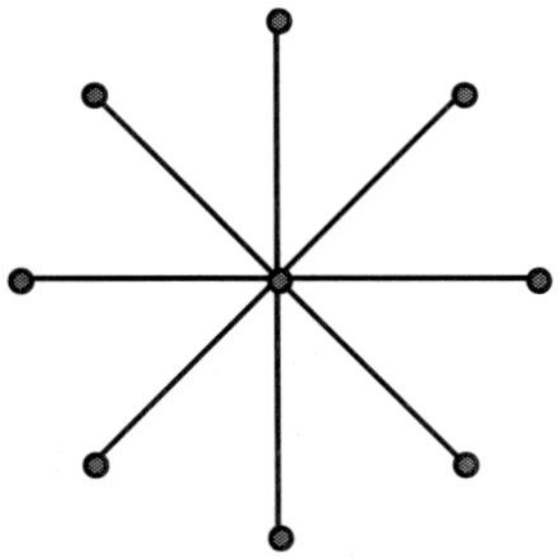

Can you make this design that way?

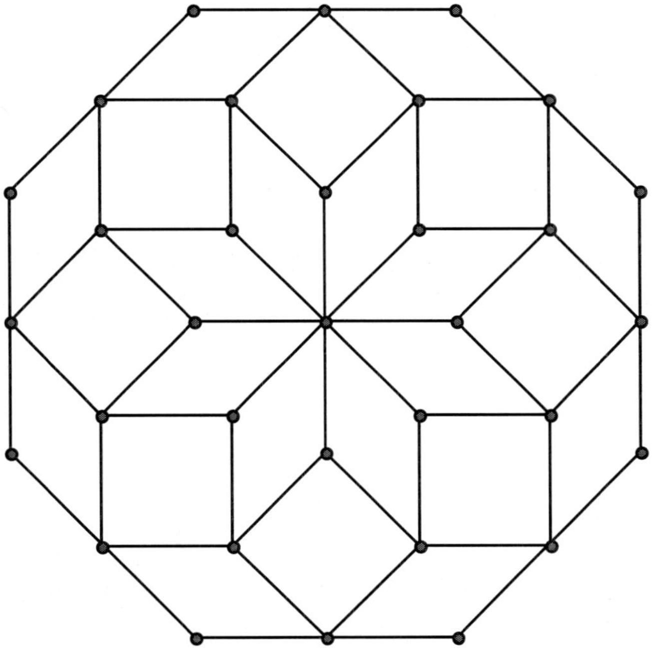

You can make even fancier designs of this type by starting with larger numbers (always an **even** number) of points around a center point. However, these fancier designs take a great deal of patience and skill, and they take a lot of room, so you will want to work on oversized paper if you decide to make them.

Not only do these designs offer excellent practice with the basic tools of geometry, but they offer a chance to incorporate art into the mathematics classroom. Since the visual arts are often under-funded and overlooked in schools today, this is a precious teaching opportunity.

18.3 Angles

If you ask a class what an angle is, you usually get vague replies like, "An angle is where two lines meet." To bring out more precision, draw two crossed lines and ask if the point of intersection is the angle. (It is, after all, "where two lines meet.") Students may then change the picture to something like figure 18.3.1. This is how many people picture angles, and it is close to the definition of an angle as two rays with a common endpoint (A **ray** is a half-line. It extends without end, but only in one direction).

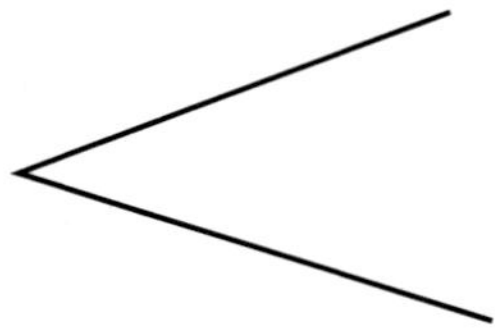

Figure 18.3.1

This definition is found in many books, but there is more to the concept of angle. To pursue it, ask if some angles are larger than others. Specifically, ask which angle in figure 18.3.2 is larger.

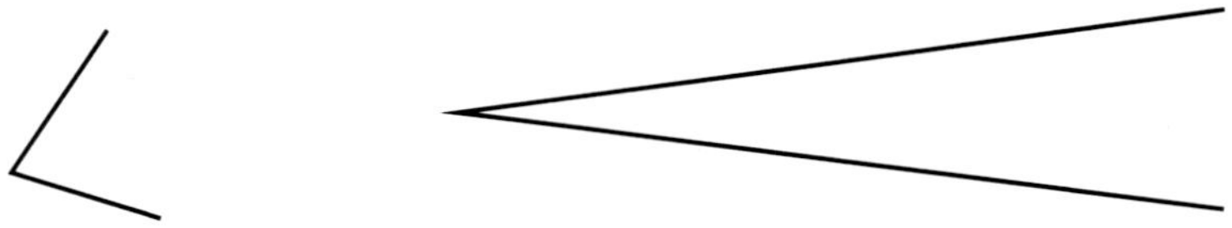

Figure 18.3.2

Most students will say the angle on the left is the larger of the two, even though the picture is smaller. What they are trying to deal with, but have not yet identified, is the rotation needed to match up the lines that are the sides of the angle. This is the heart of the idea, and we will embody it in formal definition:[3]

Definition: An angle is a rotation.

Whenever anything rotates or turns, angles are involved, even if no pictures are drawn. Picture an angle as in figure 18.3.3, where the two line segments meet at the **hub** or **center of rotation**. Many different rotations can carry a given line segment onto another one, so a curved arrow is used to show the rotation explicitly. (A problem with defining an angle as two rays with a common endpoint is that this definition does not distinguish between the three angles shown in figure 18.3.3.)

[3] Since ancient times, some writers, such as Euclid, have viewed angles as geometric figures, while others have considered them to be rotations.

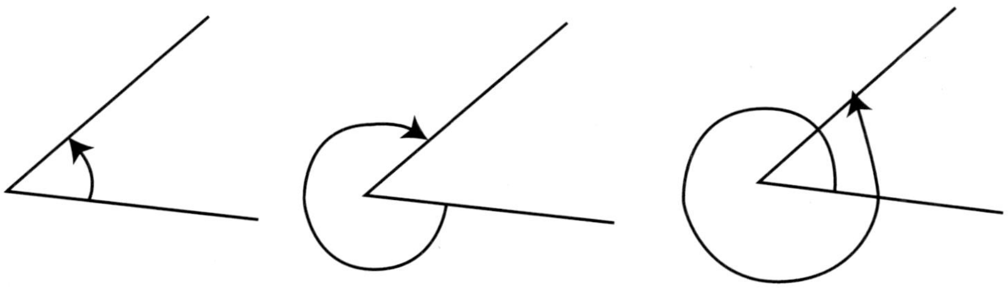

Figure 18.3.3

If the curved arrow is omitted, assume that the rotation is the smallest possible one and that the direction of rotation, if not stated explicitly, is irrelevant. In trigonometry, counterclockwise rotations are considered positive, and clockwise rotations are considered negative.

How are angles measured?
A natural unit of measure for angles, it would seem, is a complete turn, but the traditional unit of angle measure is the **degree**, defined as $\frac{1}{360}$ of a complete turn. Degrees were used in ancient Mesopotamia and possibly earlier. The Mesopotamians did arithmetic in base 60, and their number system may have been related to their calendar. For finer measurements, they divided the degree, like the hour, into 60 **minutes** and the minute into 60 **seconds**. (The word "second" reflects the fact that the first division was into minutes.) This system is still used today; we write, for example, 60°47'13" for an angle of 60 degrees, 47 minutes, and 13 seconds. However, today it is also common to use decimal fractions of a degree, and in theoretical mathematics and physics, degrees are usually replaced by radians. Angles between 0° and 90° are called **acute** angles; those between 90° and 180° are called **obtuse** angles. An angle of 90° is a **right** angle, and an angle of 180° is a **straight** angle. An angle between 180° and 360° is a **reflex** angle.

In navigation, angles are measured clockwise from north, as on the compass rose in figure 18.3.4.

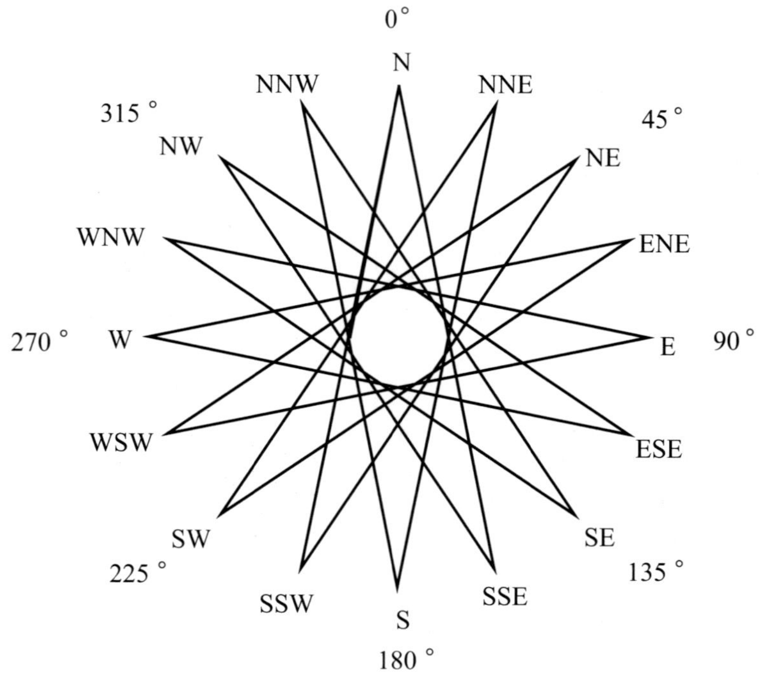

Figure 18.3.4 A compass rose

The basic tool for measuring angles is a **protractor**. Essentially, a protractor is just a circle, marked off into 360 equal parts, though in practice it is common to use only half the circle, as in figure 18.3.5.

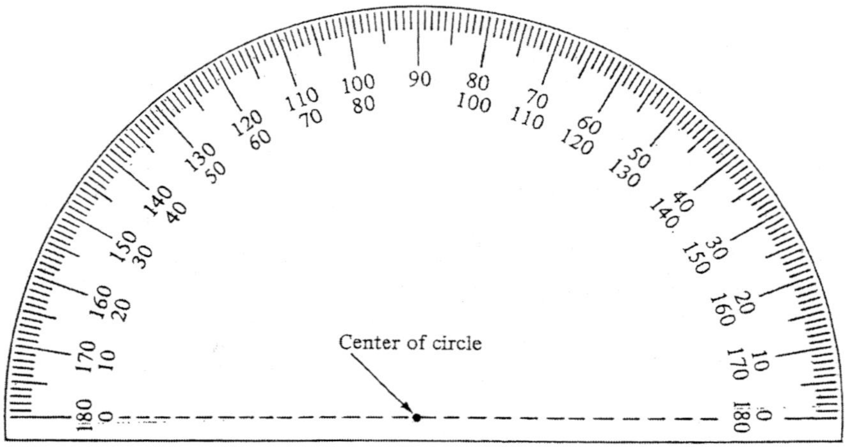

Figure 18.3.5

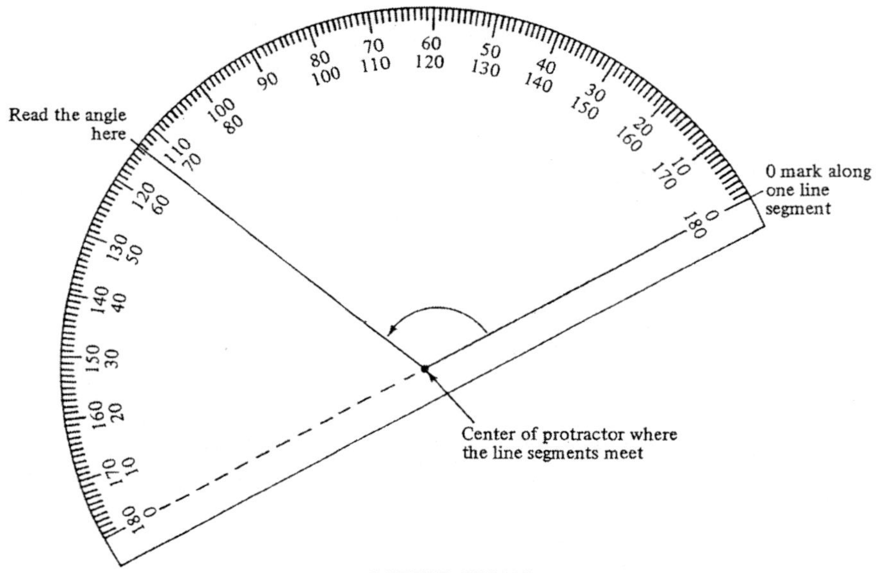

Figure 18.3.6

A good protractor is made of clear plastic, is large enough to read easily, and has its 0 marks set in from the straight edge, so that even if that edge is uneven from misuse or poor manufacturing, accuracy is not affected. Most protractors, like the one in figure 18.3.5, have scales marked for both clockwise and counterclockwise angles.

To measure an angle represented by a picture, place the protractor with its center where the lines meet and one of the 0 marks on one side of the angle. Then, read the number where the other side of the angle crosses the circular edge of the protractor. In figure 18.3.6, the angle is between 113° and 114°. The exercises below show how to use a protractor to measure angles larger than 180°.

Learning to use a protractor well takes practice. Beginners may have trouble placing the protractor properly, and, even when the protractor is correctly placed, students may need help reading the angle. Therefore, provide opportunities for students to measure angles in a setting where they can help each other and you can supervise their work. Here is one such opportunity.

Exploration (to do either individually or as a group)
Begin with a triangle large enough to take up most of the space on a full-sized page (This triangle must be drawn carefully, using a straightedge. A freehand sketch will not do.)

a. Measure all three angles of your triangle as accurately as possible. With most classroom protractors it is possible to measure to the nearest whole degree. Measuring each angle involves only two sides of the triangle. Learning to focus exclusively on those two sides, while ignoring the third side, takes practice.

b. Add up the three angle measurements found in part a, and have students compare and discuss their results.

c. Repeat steps a, b, and c for two or three additional triangles. What do you notice?

If the students measure accurately, they should find that the sum of the three angles of each of their triangles is close to 180°. A deviation of a degree or two in this sum is to be expected, because the thickness of the pencil lines and the lines on the protractor means that even careful measurements can differ by a degree in a classroom setting. Check the work, too, of any students who find that the angles of a triangle add up to a number that is not in the range from 177° to 183°. They may be placing the protractor incorrectly, reading it incorrectly, or even adding incorrectly. Also check the work of students who always get 180° perfectly. They may be fudging their answers to get 180°, because they have heard in advance that this is the number they are supposed to get.

Just as a ruler can be used to "construct" points a given distance apart, a protractor can be used to make line segments whose directions differ by a given angle. This is usually called constructing the angle. For example, to construct a 36° angle, first mark the point where the line segments will meet and place the protractor with its center at that point. Next mark two points at the edge of the protractor with 36° between them; 0 and 36 will do, as will 10 and 46 and many other pairs. Draw line segments from the center to each of the marks on the edge, and complete the job by putting in a curved arrow to show the angle's size and direction.

Exercises 18.3

1. How many complete turns are there in an angle of:

   a. 1,440° b. 5,400° c. 0°
   d. 810° e. 900° f. 45°
   g. 60° h. 120° i. 300°

2. How many degrees are there in:

   a. $\frac{1}{10}$ of a complete turn? b. $\frac{1}{3}$ of a complete turn?
   c. $\frac{3}{4}$ of a complete turn? d. $\frac{1}{12}$ of a complete turn?
   e. $\frac{3}{5}$ of a complete turn? f. $\frac{5}{6}$ of a complete turn?
   g. $\frac{3}{2}$ of a complete turn? h. $11\frac{2}{5}$ of a complete turn?

3. Explain clearly what is meant by "doing a one eighty."

4. a. Through how many degrees does the small (hour) hand of a clock turn in an hour?

   b. Through how many degrees does the small (hour) hand of a clock turn in $\frac{3}{4}$ hour?

   c. Through how many degrees does the small (hour) hand of a clock turn in $1\frac{1}{2}$ hours?

   d. Through how many degrees does the big (minute) hand of a clock turn in 25 minutes?

   e. Through how many degrees does the big (minute) hand of a clock turn in 2 hours 15 minutes?

   f. Through how many degrees does the second hand of a clock turn in 24 hours?

5. To an observer standing on the earth, the sun appears to circle the earth at a steady rate, completing the circle every 24 hours. How many degrees does the sun appear to move through the sky in one hour?

6. An ignorant monarch is said to have refused a shipment of English machinery because it was marked "1,900 revolutions per minute." Perhaps the shipment would have been accepted if it had been marked in degrees per minute (at least the students in that country would have liked it.) How many degrees per minute should it have been labeled?

7. Use a protractor to measure these angles. In parts b and c, first measure the angles in the dotted arcs, then subtract the results from 360°.

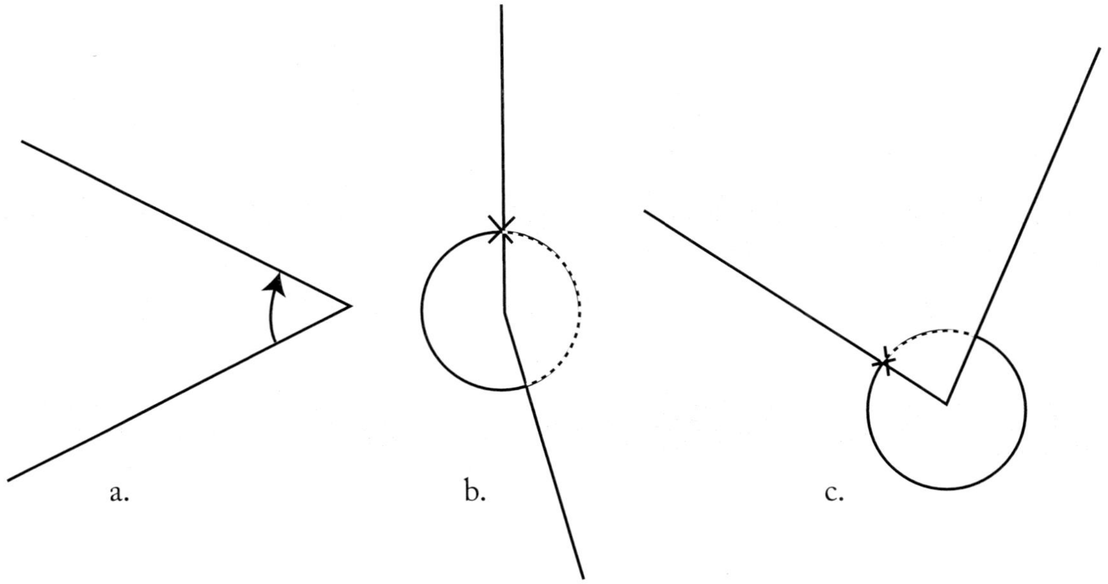

8. Use a protractor to construct angles of:

   a. 45° b. 60° c. 37° d. 90° e. 135° f. 230°

   g. 270° h. 400° i. 5047°

9. On the compass rose in figure 18.3.4, what headings in degrees correspond to these directions?

   a. ENE b. ESE c. SSE

   d. WSW e. WNW

10. If you are on a course of 20°, what will be your heading after a turn of:

    a. 15° to the right b. 45° to the right

    c. 90° to the right d. 280° to the right

    e. 05° to the left f. 45° to the left

11. A course of 60° takes you from Banning Pass to Lake Havasu City. What course takes you from Lake Havasu City to Banning Pass?

12. After heading due north, you turn 17° to the right, then turn 39° to the left. What is your heading now?

13. If you begin facing due north, what will your heading be after five successive turns of 75° to the left?

Most airplanes have radios to receive signals from automatic "omni" stations, such as those shown on this chart at Bellingham, Victoria, and Vancouver. When tuned to 113.0, the radio indicates the "radial" or direction from Bellingham to the plane. When tuned to 113.7 it gives the radial from Victoria. The chart also shows similar navigation aids broadcasting from Vancouver at a frequency of 115.9 and the Whidbey Island Naval Air Station, frequency 279.0. By using the radials from two or more stations, a pilot can locate his or her position, even in poor weather. (On this chart the 0° radials are not vertical, because they point to magnetic north, while true north is straight up on this chart.) The actual pattern of runways is shown for each airport.

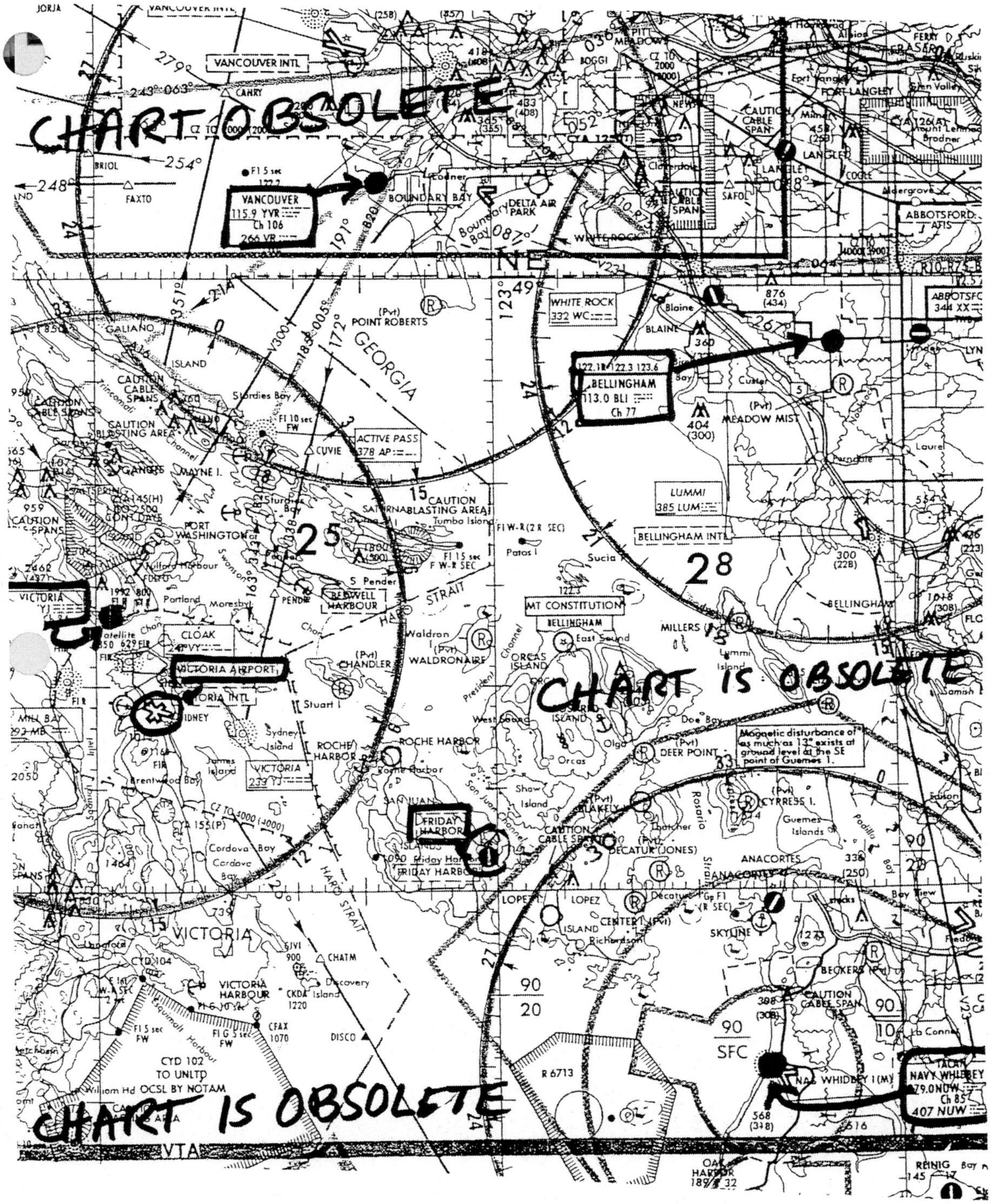

14. At Friday Harbor, what radial are you on from:
    a. Victoria   b. Bellingham   c. Whidbey

15. At Victoria Airport, what radial are you on from:
    a. Bellingham   b. Victoria   c. Vancouver

16. You are on the 210° radial from Bellingham and the 180° radial from Vancouver.
    a. What is the nearest airport shown on the chart?
    b. What magnetic direction should you fly to reach that airport?
    c. If each inch on the chart corresponds to 8 miles, how far is that airport?
    d. At 150 miles per hour, how long will it take to reach that airport?

17. Intersecting lines determine the four angles pictured below.

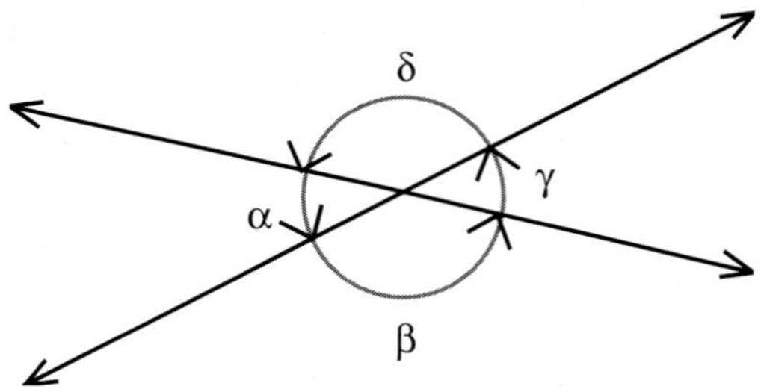

    a. Show that $\alpha+\beta=\beta+\gamma=\gamma+\delta=\delta+\alpha=180°$ (Two angles that add up to 180° are said to be **supplementary angles**. The conclusion of this exercise is sometimes stated as: Adjacent angles are supplementary.)
    b. Opposite angles are equal (i.e., $\alpha=\gamma$ and $\beta=\delta$.) Geometry books state this by saying "Vertical angles are equal."

18. The picture below shows a triangle *ABC* with side *BC* extended beyond *C* to form an exterior angle called δ. Show that $\alpha+\beta=\delta$.

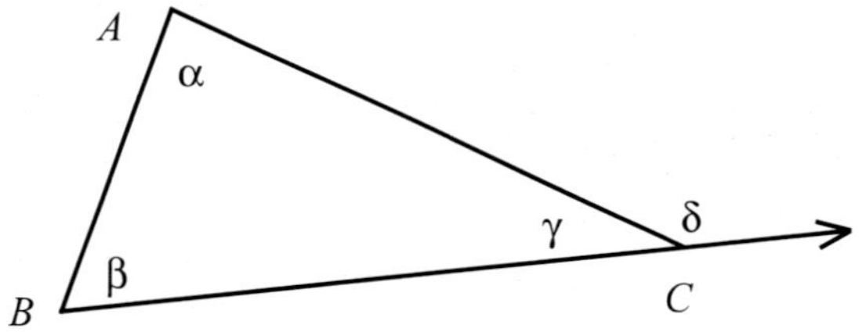

19. Eratosthenes[4] calculated the circumference of the Earth about 250 B.C.E. By shadows, he determined that when the sun was directly overhead in Seyene, Egypt (near modern Aswan), it was $7\frac{1}{4}°$ south of the vertical in Alexandria, 5,000 *stadia* to the north. Then (figure 18.3.7)

[4] See note on Eratosthenes, section 4.2.

the radii to the two cities form an angle of $7\frac{1}{4}°$ at the center of the Earth. Since $7\frac{1}{4}°$ is about $\frac{1}{50}$ of 360°, Eratosthenes concluded that the distance between Alexandria and Seyene is about $\frac{1}{50}$ of the circumference of the Earth.

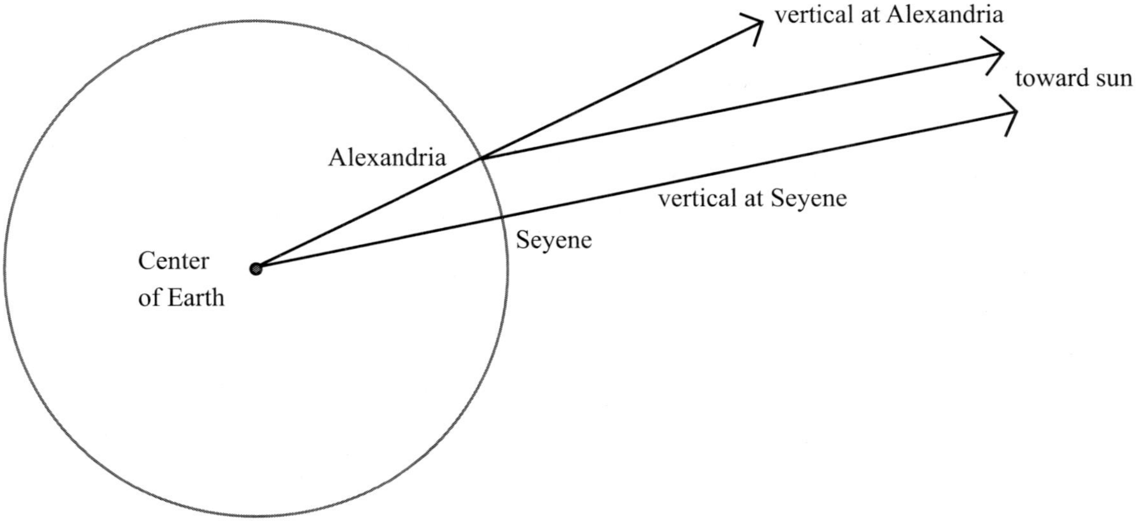

Figure 18.3.7

a. According to Eratosthenes, what is the Earth's circumference in *stadia*?

b. There is some doubt about the length of the *stadium*, Eratosthenes' unit of measure. One estimate is 517 feet. Another is 559 feet. Use each of these figures with Eratosthenes' result to estimate the circumference of the Earth. (The actual value is about 24,480 miles.)

18.4 Euclidean or Non-Euclidean?

Measuring the angles of triangles combines practice with discovery. From personal experience (perhaps from the exploration in section 18.3) you probably know, at least for each triangle you tried, that the sum of the angles is about 180°.

Do the angles of a triangle add up to precisely 180°? This question does not have a simple answer. Whenever the measurements are carefully made, the angle sum differs from 180° by less than can be accounted for by errors of measurement (any measurement has some error), so for all practical purposes we may assume that the angles of a triangle add to 180°. The theoretical question is not so easily disposed of, however. To see why, consider a situation that might have occurred when it was believed that the earth was flat. Suppose at that time you could measure angles quite accurately. You measure the angles of a few triangles in the classroom, and in each case, the angles total 180°, at least within the limits of measurement, but now you measure the angles of a much larger triangle, formed by ships several miles apart at sea. To your surprise, despite great care in measurement, you now find the angles add to more than 180°! Furthermore, the larger the triangle, the more the angle sum exceeds 180°.

This is no mystery if you realize the earth is round. It impossible to draw truly straight lines on a spherical surface; the sides of what looks like a triangle on a sphere actually bow outward.

Similar issues can arise in subtle ways for triangles that are not on a spherical surface. What if we tried to measure the angles of a triangle, with vertices on the Moon, Mars, and Earth? Imagine an observer at each vertex shining precisely aimed beams of laser light that are picked up at the other two vertices. Each observer has a device that precisely measures the angle between the beams of light received at that vertex. Will the three angles add up to 180°? The truth is, we do not know, but it is easy to find reasons why they will not. First off, the Moon, Earth, and Mars are in constant motion relative to each other, and it takes light several minutes to travel between Mars and the Earth. That makes it hard to say what is meant by measuring the angles simultaneously. Second, we are measuring the angles between two light rays we send out, but we know that light does not travel in straight lines. Light rays are bent by passing through media such as air or water, as well as by gravity. On a very large scale, then, the idea of accurately measuring the angles of a triangle is impossible in principle, because we cannot know what is meant by a straight line.

For over 2,000 years it was generally believed that the angles of any triangle add up to exactly 180°, because that is stated and proved in Euclid's *Elements*, a work so highly esteemed that its conclusions were long regarded as absolute. The *Elements* begins by assuming the five postulates mentioned at the start of this chapter and then uses them to prove other statements, called **theorems**. The truth of the theorems depends on that of the postulates. The first four postulates did not arouse much controversy, but in the centuries since Euclid, many people were dissatisfied with the fifth postulate. Unlike the other postulates, the parallel postulate did not seem self-evident. People believed it was true, but they thought it could be proved from the other four axioms. Over the centuries, many mathematicians sought in vain to prove the parallel postulate. Girolamo Saccheri (1667-1733) tried an indirect approach, assuming the postulate to be false and trying to show that this assumption leads to a contradiction. He reached many strange conclusions, among them that the sum of the angles of a triangle is not 180°, and that the discrepancy increases with the size of the triangle, but he did not produce a contradiction. He remarked that although he had not reached a contradiction, his conclusions were absurd, and therefore the fifth postulate must be true. Had he faced the fact that he had not found a contradiction, he would be known as the discoverer of non-

Euclidean geometry. Instead, that honor waited until the next century, when, independently, Nikolai Ivanovich Lobachevsky (1793-1856) in Russia in 1826 and Janos Bolyai (1802-1860) in Hungary in 1829, reached the revolutionary conclusion that an alternate to the parallel postulate leads to an alternate geometry. When Bolyai's father, who had spent many frustrating years trying to prove the parallel postulate, wrote to his friend Gauss about his son's revolutionary idea, Gauss replied that he, too, had long thought the same but feared the controversy that would come with stating such unorthodox thoughts in public. In time, it was shown that some non-Euclidean geometries are just as free of contradiction as Euclidean geometry, and the area became one of intense study. However, in the public mind and in schools, Euclidean geometry was still regarded as absolute truth. Interestingly, Euclid himself seems to have had reservations about the parallel postulate, which he avoided as long as possible in the *Elements*. The parallel postulate and alternatives to it were actively debated in ancient Greece, even before Euclid wrote the *Elements*. In our work we will use Euclidean geometry, even as we remain aware of other possibilites.

18.5 Angles and Chords: An Exploration

1. On a large circle with center O, select two points and call them A and B. These points divide the circle into two

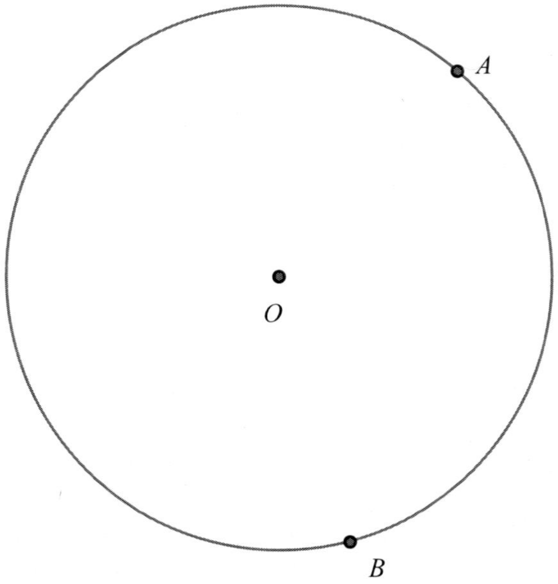

parts called **arcs**. Unless A and B happen to be diametrically opposite, one of the two arcs will go more than half way around the circle (it will be a **major arc**), and the other will be a **minor arc**. Mark a point P_1 on the major arc, and draw **chords** from P_1 to A and to B as in figure 18.5.1. (A **chord** of a circle is a line segment with endpoints on the circle.) An angle formed by two chords of a circle with a common endpoint called an **inscribed angle**.

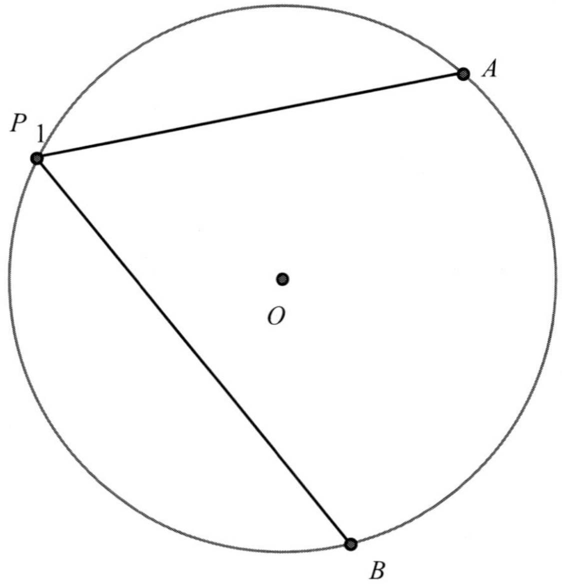

Figure 18.5.1

a. Measure inscribed angle AP_1B as accurately as you can, and record the answer.

b. Select a second point P_2 on the same major arc, as in figure 18.5.1, draw chords AP_2 and BP_2, and measure inscribed angle AP_2B.

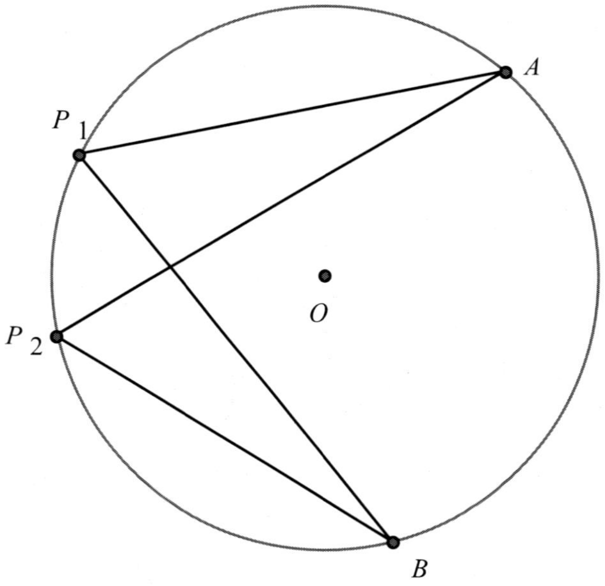

Figure 18.5.2

c. Mark three more points P_3, P_4, and P_5 on major arc $\overset{\frown}{AB}$ and repeat the process, measuring inscribed angles AP_3B, AP_4B, and AP_5B.

d. What do you notice about the angles you have measured?

e. Repeat the experiment with two points P_6 and P_7 on the minor arc $\widehat{AB}$. What do you notice about the angles you have measured?

f. How do the angles measured in part e relate to the angles measured in parts b, c, and d?

g. Repeat this experiment with a different circle, drawn by you, and points A and B selected on that circle in positions somewhat different from the positions of A and B on the first circle. What do you notice?

h. Find the center of each circle used in your experiments above, and draw radii (the plural of radius) to A and B. In each case, measure angle AOB, which is called the **central angle subtended by arc $\widehat{AB}$**.

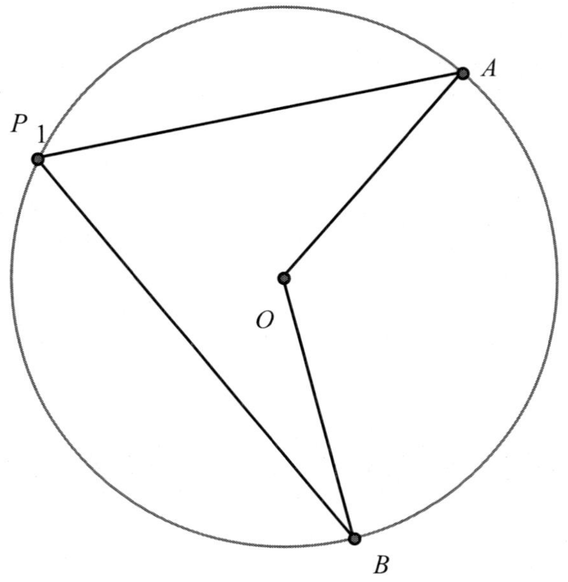

i. How is the central angle related to the inscribed angles? Answer in a sentence or two.

j. If you have access to a computer with the program "*The Geometer's Sketchpad*," repeat the above investigation. First select the points A and B on your circle, then connect them to the center O and a point P on major arc , and measure the angles APB and AOB. Then drag point P along major arc, watching what happens to the angles being measured. Finally, vary the point B, then repeat the experiment with a new point B.

The result of problem 1 is usually summed up in geometry books as follows: Problem 2 will help you see why that statement is true.

An angle inscribed in a circle is half the central angle to the same arc.

2. The above result is easy to prove in the case where the center O is on line PA or PB, as in figure 18.5.3.

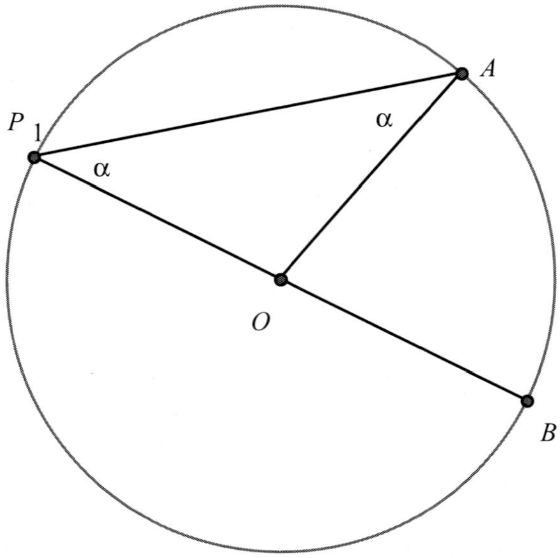

Figure 18.5.3

In that case, since OP and OA are two radii of the same circle and are therefore of equal length, triangle OPA is isosceles. Then its base angles, OPA and OAP, are equal (call them α), and the third angle of the triangle, POA, must be $180^\circ - 2\alpha$. That leaves 2α for angle AOB.

If one side of angle APB does not contain O, then, as in figure 18.5.4, either O is inside angle APB (case 1) or O is outside angle APB (case 2).

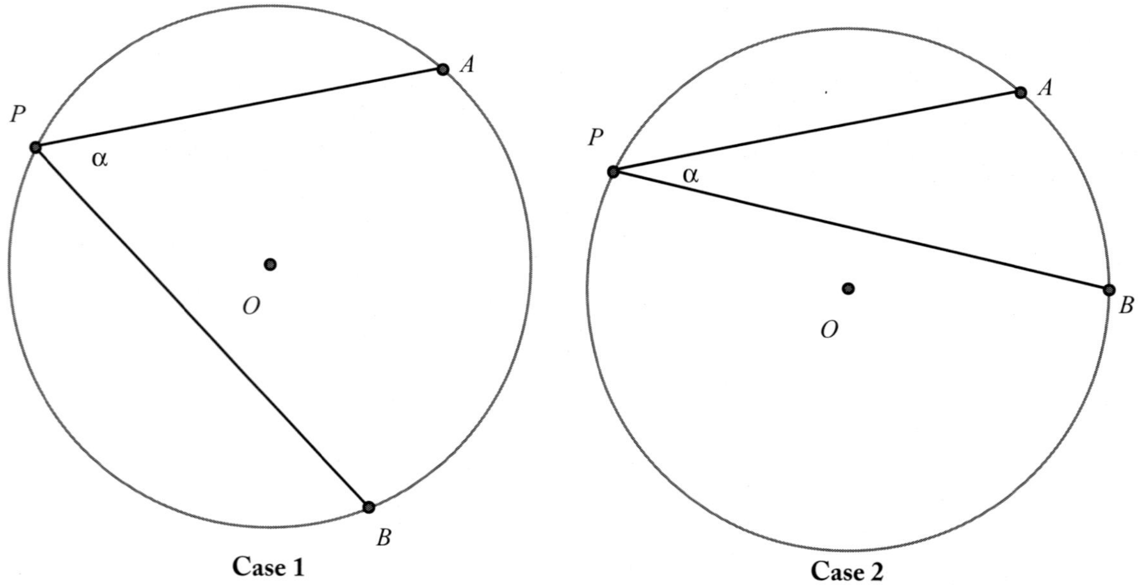

Figure 18.5.4

In the first case, draw the diameter from P and apply the argument above to each half of the picture, as in figure 18.5.5.

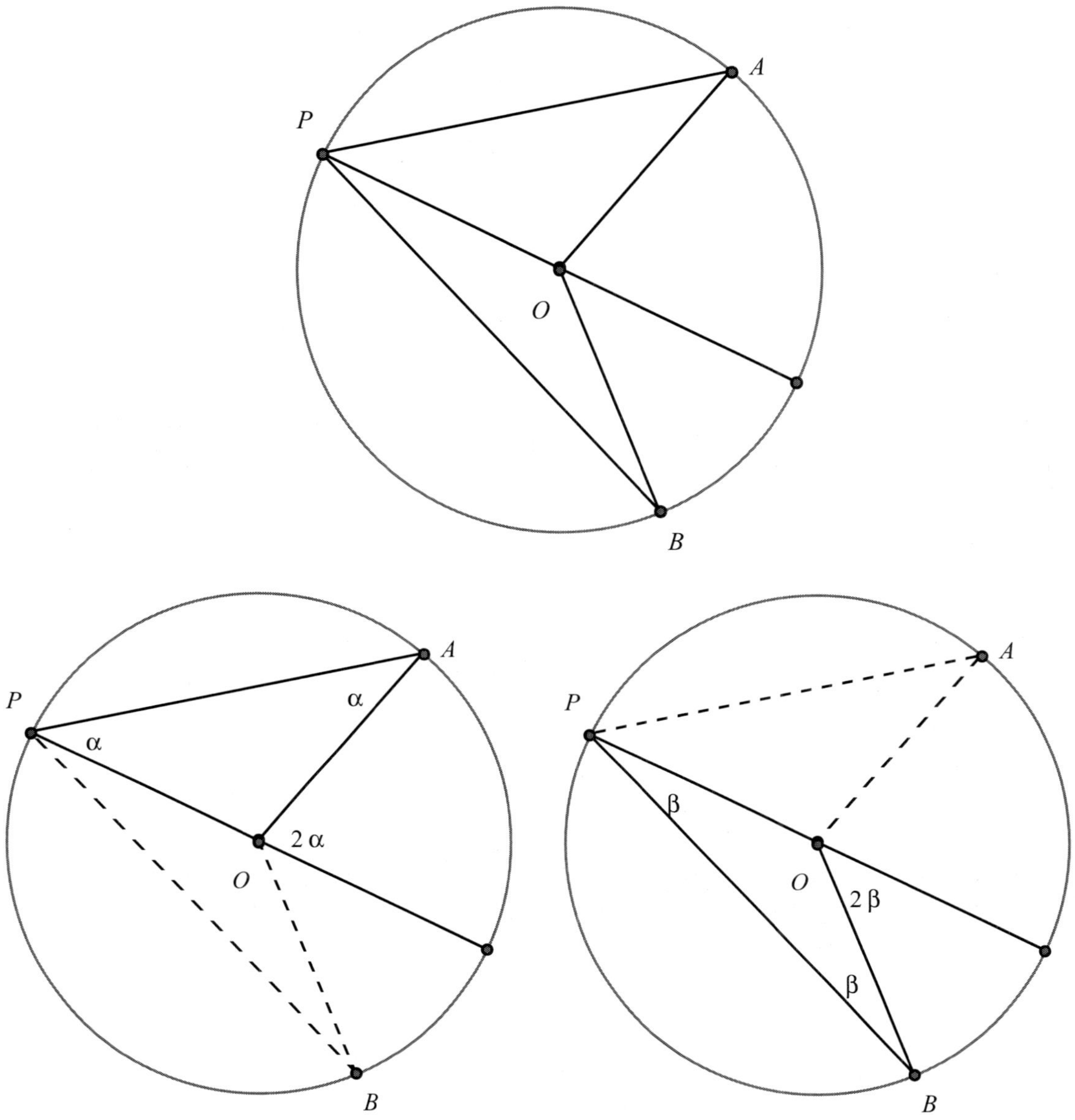

Figure 18.5.5

Now put the two parts of this picture together:

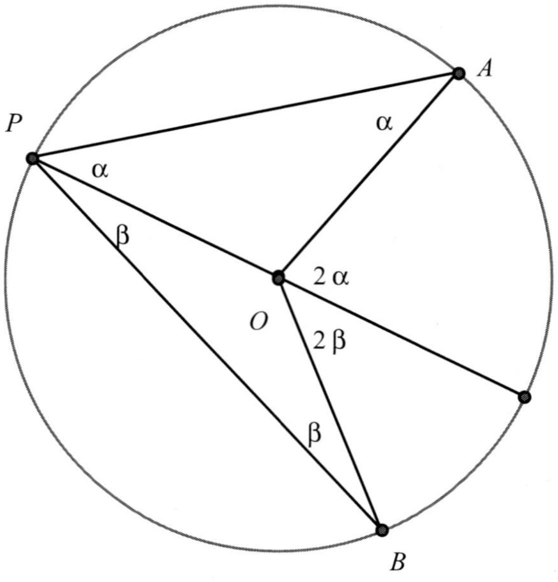

Evidently angle *APB* is α+β and angle *AOB* is 2α+2β or twice (α+β). Modify the above reasoning by using subtraction instead of addition, to explain why the result on inscribed angles holds in case 2.

3. Two chords of a circle intersect within the circle as pictured. Show that the angle α between them is the average of the arcs $\widehat{AC}$ and $\widehat{BD}$. Hint: Draw a chord *BE* parallel to *DC*.

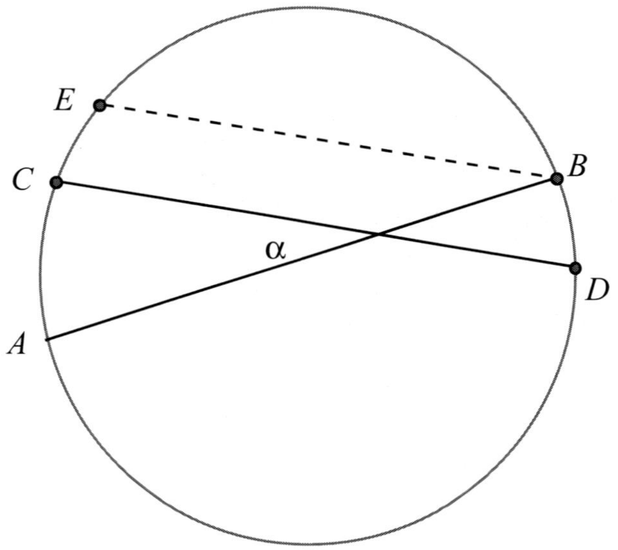

4. Chords *AB* and *CD*, extended, meet at a point *F* outside the circle, as pictured, forming an angle α. Show that $\alpha = \dfrac{\widehat{AC} - \widehat{BD}}{2}$. Hint: Draw chord *BE* parallel to *CD*.

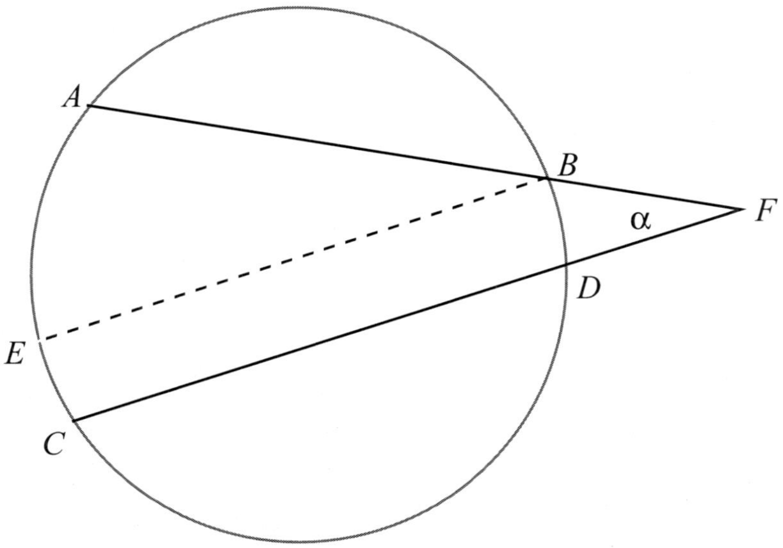

18.6 Congruent and Similar Triangles

Recall that two geometric objects are congruent if they have the same shape and the same size. For polygons, that means that one of them can be laid over the other in such a way that the corners and sides match. Then, of course, the angles also match. Of all the polygons, triangles play a special role in geometry. Sometimes you may want to know whether or not a pair of triangles are congruent, but you don't know all the angles and side lengths. Here are the criteria:

Congruence Criteria for Triangles

> **Triangles are congruent if they agree in any of the following ways:**
> **<u>All three sides</u> (Side, Side, Side, or SSS)**
> **<u>Two sides and the angle between them</u> (Side, Angle, Side or SAS)**
> **<u>Two angles and the side between them</u> (Angle, Side, Angle or ASA)**

Example 1. In the figure pictured, show that distances *DE* and *AB* are equal.

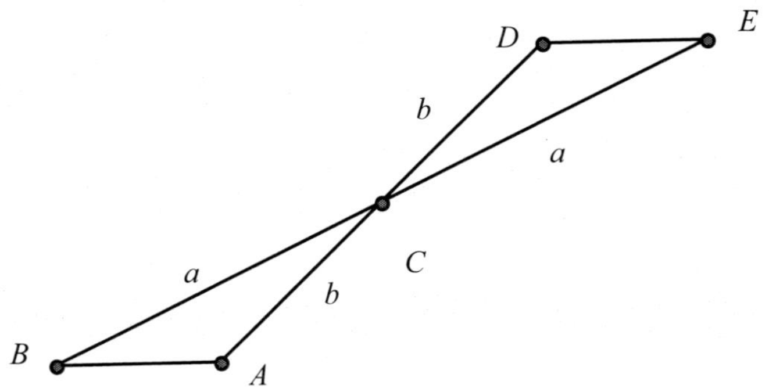

Solution: We know that sides *AC* and *DC* are equal, as are *BC* and *EC*. Also, $\angle ACB = \angle ECD$, as shown in problem 17 of Exercises 18.3. Then triangle ABC is congruent to triangle DEC by SAS. Therefore, lengths DE and AB are equal, since they are corresponding sides of congruent triangles.

Similar triangles are at least as useful as congruent triangles. If triangles $A_1B_1C_1$ and $A_2B_2C_2$ are similar (figure 18.6.1), then

$$\frac{a_1}{a_2} = \frac{b_1}{b_2} = \frac{c_1}{c_2}$$

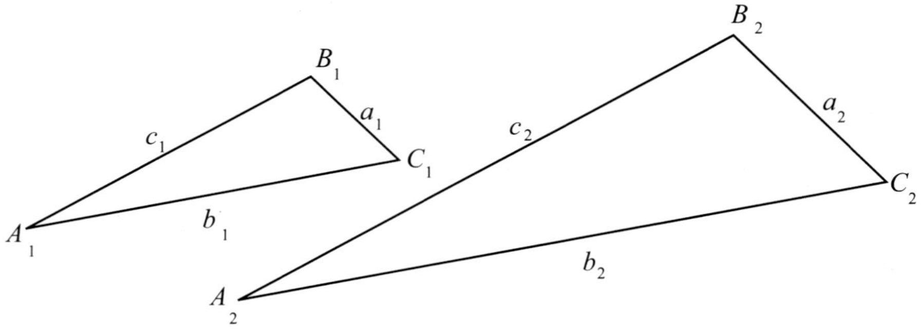

Figure 18.6.1

Example 2. Indirect measurement with similar triangles To measure the height of a building standing on level ground, put your eye as close as possible to the ground, and have a helper move a vertical meter stick until the top of the meter stick lines up with the top of the building and your eye. (Figure 18.6.2) Then measure the distance g from your eye to the foot of the building and to c, the foot of the meter stick.. The ratio of the distances g/c is the height of the building in meters.

This depends on the fact that the right triangles AFG and ABC are similar to each other. To see why, note that C is on AG, and B is on AF, so that $\angle BAC = \angle FAG$. But triangles AFG and ABC both have right angles, and since they have two pairs of equal angles and each of their angles add to 180°, their third angles are also equal. The scale ratio between these triangles is the ratio of the lengths of any pair of corresponding sides. In particular, it is the ratio of the bases, which we know is g/c. The vertical sides are in the same proportion, so $\frac{h}{1} = \frac{g}{c}$, or, equivalently, $h = \frac{g}{c}$. If, for example, g is 12.5 times as long as c, then the building is 12.5 m tall.

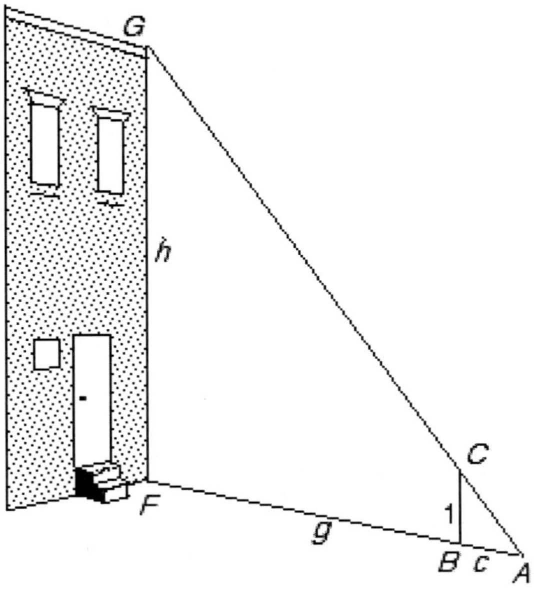

Figure 18.6.2

Example 3 (Another indirect measurement with similar triangles.) A dead pine stands on level ground near a flag pole which is 9 meters tall. The pole casts a shadow 13 m long, while the pine casts a shadow 27 m long. How tall is the pine?

Solution: The top of each object, its base, and the tip of its shadow are the vertices of a triangle (figure 18.6.3). If the tree and the pole are vertical, triangles *TUV* and *PQR* are similar (You should be able to explain why). If h is the height of the tree, then $\frac{h}{9} = \frac{27}{13}$, so h =18.7 m.

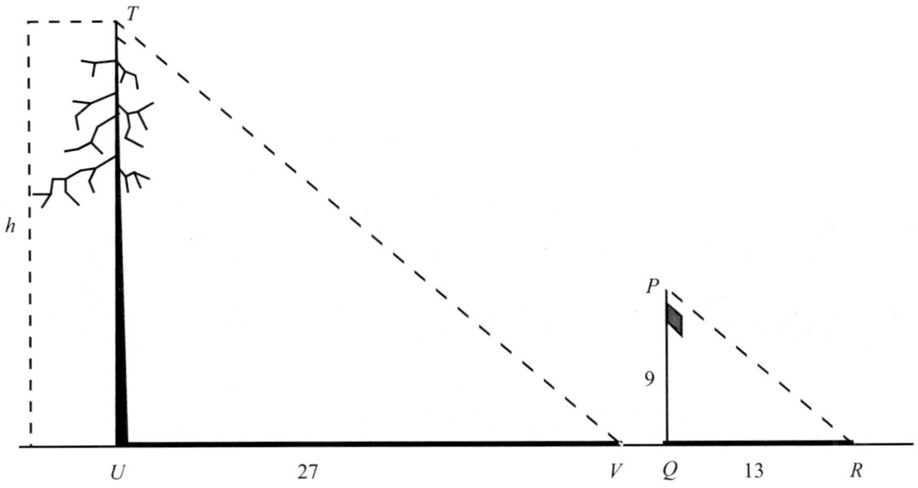

Figure 18.6.3

Indirect measurement with similar triangles is an excellent class activity, combining measurement, calculation with fractions, and the ideas of proportion and similarity. This activity takes careful planning, however. Be sure that it is easy to reach base of the object to be measured (landscaping, other buildings, fences, and streets can easily spoil the activity). If you plan to use shadows, check in advance that the weather and position of the sun will create the necessary shadows where and when you need them.

Example 4. In the figure below, *A, B, C,* and *D* lie on a circle, and chords *AE* and *BD* intersect at *C*. If the length of *AB* is 5/9 that of *DE*. and triangle *ABC* has an area of 50, what is the area of triangle *CDE?*

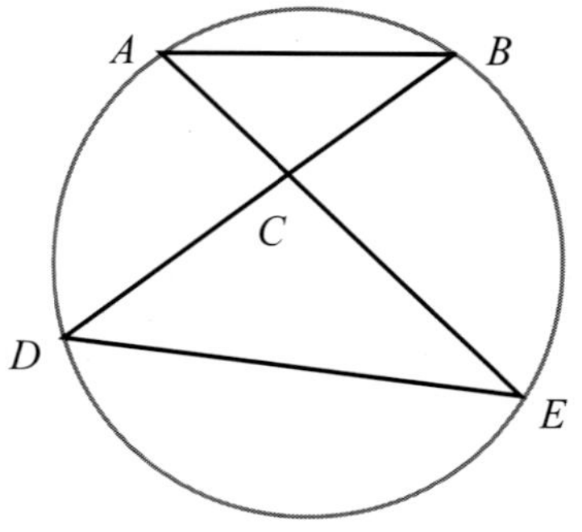

Solution: $\angle BAE = \angle BDE$, because both these angles are inscribed in the circle and subtend arc $\overset{\frown}{BE}$. Likewise, $\angle ABD = \angle AED$, because both angles are inscribed in the circle and subtend arc $\overset{\frown}{AD}$. Since two angles of triangle *ABC* are equal to angles of triangle *DEC*, the third angles of those triangles are also equal (since the angles of any triangle add up to 180°). Then triangles *ABC* and *DEC* are similar to each other. Sides *AB* and *DE*, each opposite equal angles at *C*, correspond to each other, so the ratio of their lengths is the scale of the similarity, namely 9/5. The ratio of the areas of similar figures is the square of the scale, or 81/25. Since triangle *ABC* has area 50, triangle *DEC* has area $50\times\left(\frac{9}{5}\right)^2 = 50\times\frac{81}{25} = 162$.

Exercises 18.6

1. On a sunny day a flagpole casts a 9 meter shadow on level ground, while a vertical meter stick casts a shadow 0.4 meters long on the same ground.
   a. How tall is the flagpole?
   b. Could this same method be used on a uniformly sloping hillside?

      Explain your reasoning in a sentence or two, possibly with a picture.

2. A sea captain sailing due north sights another ship at a bearing of 40°. Ten minutes later, the other ship is still at a bearing of 40° but is noticeably closer than before. What is the practical meaning of this, and what does it have to do with similar triangles?

3. a. Draw a triangle, and find the midpoint of each of the three sides. (You can do this easily by folding one

vertex of your triangle onto another, especially if you draw it on patty paper, which is partially transparent.)

b. Connect the three midpoints of the sides found in part a, thus forming a new triangle. How is this new triangle related to the original triangle?

4. In the figure below, angles *ABC*, *ADE*, and *AFG* are equal, as are angles *ACD* and *AEF*.

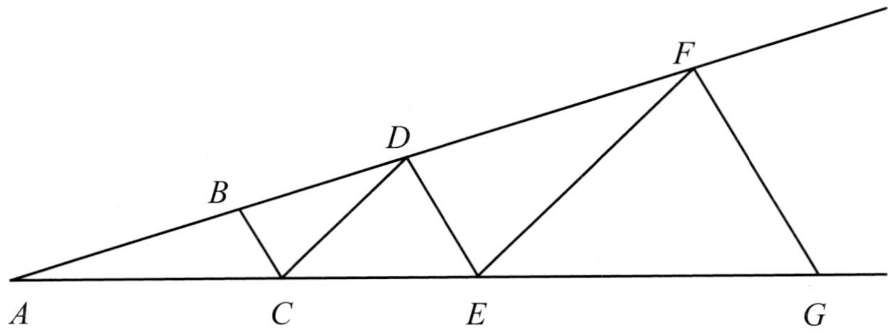

a. Show that triangles *ABC*, *ADE*, and *AFG* are similar.

b. Show that the ratio of lengths *BC* to *FG* is the square of the ratio of lengths *DE* to *FG*.

5. In the figure below, the lengths *BC*, *BE*, and *AD* are labeled. How long is *AE*?

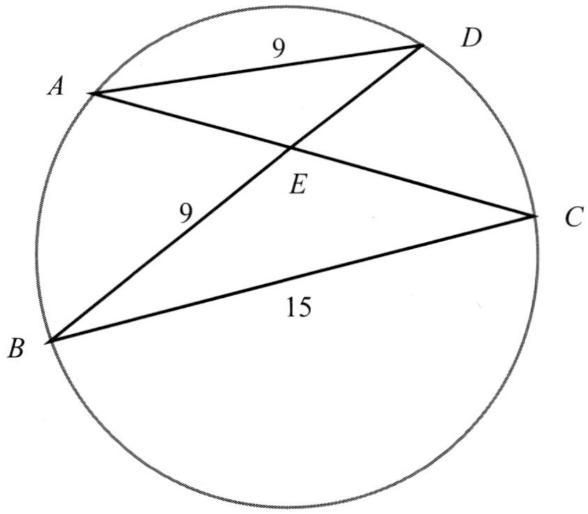

6. Cut out a triangle. If the triangle has an obtuse angle, put it at vertex *C* as pictured below. Otherwise, any vertex can be *C*.

a. Fold the triangle along a line through *C* so that one of the other vertices lies along side *AB*, then crease the fold sharply. If *D* is where the fold line meets side *AB*, then angles *ADC* and *BDC* are right angles. Unfold the triangle.

b. Fold again, so that A and D coincide, and a third time so that B and D coincide. (These fold lines are pictured below as GH and FE.)

c. Fold a fourth time, so that C coincides with D. (This fold line is pictured below as GF.)

d. Angles A, B, and C are all folded together at point D. What do you notice about them?

e. Use rectangle $EFGH$ to show that the area of triangle ABC is half its base times its height.

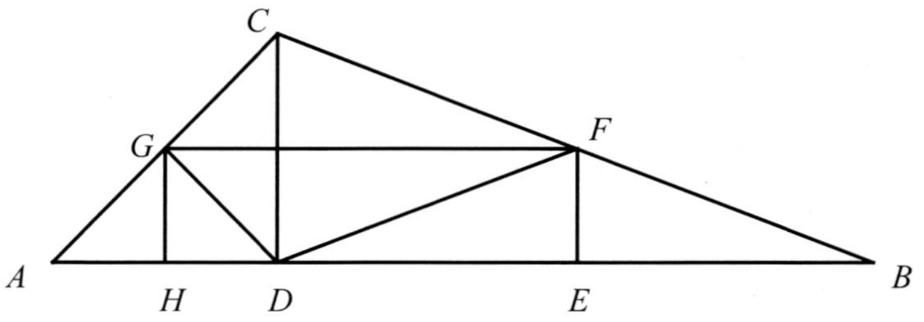

7. In the figure below, CD is the altitude to the hypotenuse of right triangle ABC and so is perpendicular to AB at D.

a. Express each of the angles β, γ, and δ in terms of α.

b. Show that the three triangles ABC, ACD, and CBD are all similar to each other.

c. Show that $h = \sqrt{pq}$. (h is called the **geometric mean** or the **mean proportional** of p and q.)

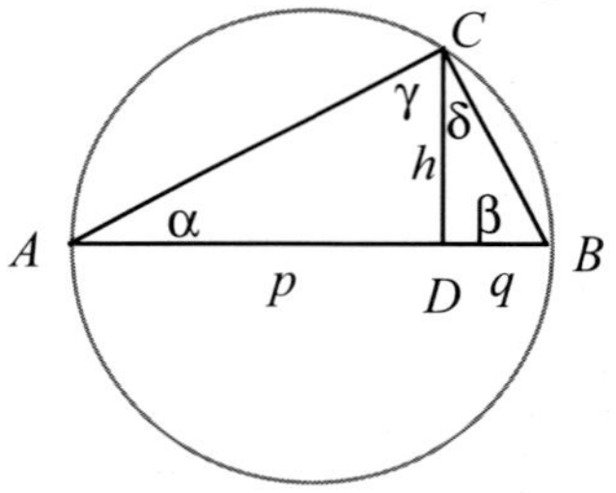

d. Show algebraically that for any two numbers $p>0$ and $q>0$ the arithmetic mean is greater than or equal to the geometric mean, $\frac{p+q}{2} \geq \sqrt{pq}$.

e. Use the picture below to show the result from part d geometrically.

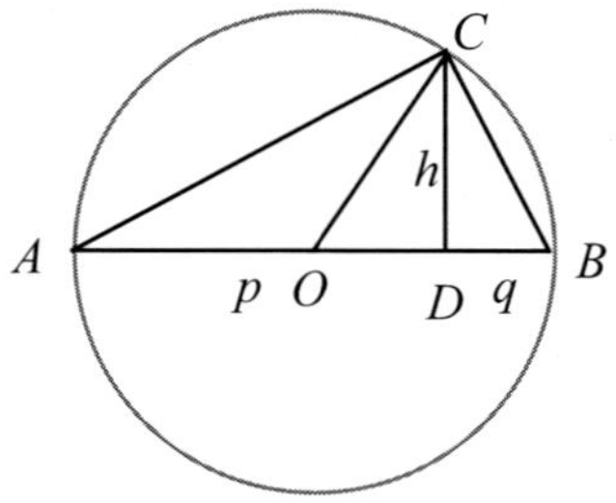

Exploration: Ptolemy's theorem (questions 8-14)[4]
For any quadrilateral inscribed in a circle, the product of the diagonals equals the sum of the products of the opposite sides.

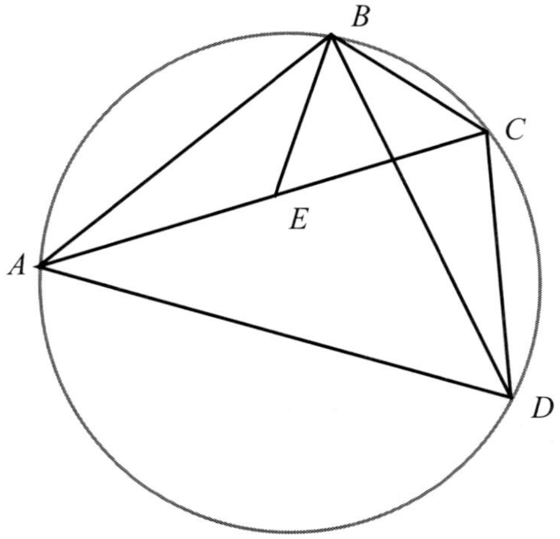

Let $ABCD$ be the given quadrilateral. Construct line BE so that $\angle ABE = \angle DBC$. (i.e., copy $\angle DBC$ and let E be the point of intersection of the newly constructed side of the angle with segment AC).

8. Show that triangles ABD and EBC are similar to each other. (Hint: Two inscribed angles in a circle subtending the same arc are equal-see Exercises 18.5 #1, 2.)

9. Show that $AD \times BC = BD \times EC$.

10. Show that triangles ABE and DBC are similar to each other.

11. Show that $AB \times CD = BD \times AE$.

12. Use problems 9 and 11 to show that $AD \times BC + AB \times CD = BD \times EC + BD \times AE$.

13. Factor BD on the right, and substitute for $EC+AE$ to complete the proof of Ptolemy's theorem.

[4] Little is known of the life of Ptolemy. He wrote his *Mathematical Synthesis* in Alexandria, where he lived and made astronomical observations between 127 and 151. He was reported as still alive after 161. The *Mathematical Synthesis* contains, among other things, Ptolemy's theorem, cited above, which was key to his calculation of a table of chords (much like a modern table of sines) for angles from 0° to 180° in steps of 1/2 °. Ptolemy wrote an atlas that gave latitude and longitude (not, of course, centered on Greenwich, England) for many important towns. Ptolemy created a system to describe and predict the motions of the planets and their appearances in the sky. This system was widely accepted and was unchallenged for over 13 centuries, until displaced by the ideas of Copernicus, Galileo, Kepler, and Newton. Arab writers referred to Ptolemy's *Synthesis* as *Almagest ("The Greatest")*, a name which is still used to day for that work.

14. Show that if the inscribed quadrilateral in Ptolemy's theorem is a rectangle, then Ptolemy's theorem reduces to the Pythagorean theorem.

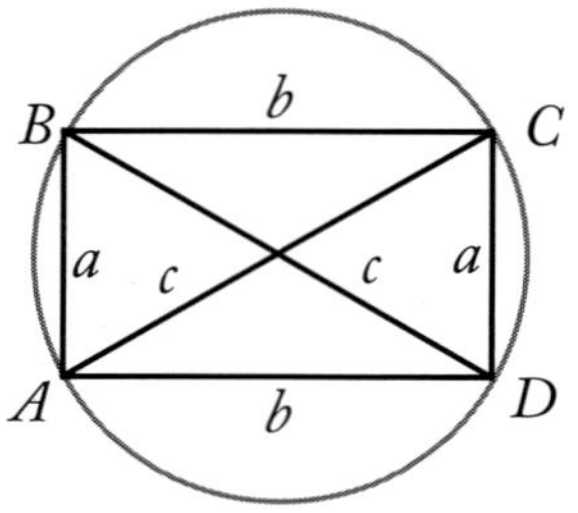

18.7 Classical Constructions

Geometric constructions lie at the heart of Euclidean geometry, both intellectually and in spirit. Constructions form the basis for many of the results of geometry and their Euclidean proofs. Also, certain construction problems, which baffled ancient Greek mathematicians, served over the ensuing centuries to stimulate new and unexpected developments in mathematics. We begin by describing some basic constructions. Work through each of them, illustrating the steps with careful diagrams.

The traditional tools of construction are:

I. **A straightedge**. A straightedge can be used to draw a line through any two points, but it is not marked and therefore cannot be used to measure distances like a ruler.

II. **A compass**. A compass can be used to construct a circle at any specified point, of any radius that can be specified as the distance between two points, which have already been constructed. (We assume the compass may be moved from place to place without collapsing, so that the distance between two known points may be used as the radius of a circle centered elsewhere.)

Also, in describing a construction, you may use any construction which has previously been established. For example, the first construction below shows how to construct a line perpendicular to a given line through a given point not on that line. In later constructions, you can simply appeal to this construction if you need it rather than repeat all the steps.

1. **Perpendicular to a line through a point not on the line.** Given a line l and a point P not on l.
   1. Draw a circle centered at P, large enough to intersect l at two points. Call those points A and B.
   2. Draw circles with radius AP centered at A and at B. They meet at P and at another point, to be called Q.
   3. Draw PQ, the desired line through P perpendicular to l.

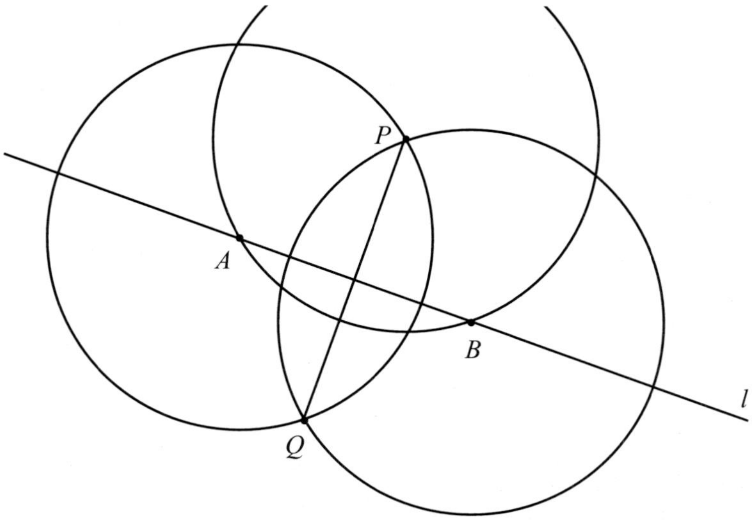

2. **Perpendicular to a line through a point on the line.** Given a line *l* and a point *P* on *l*.
    1. Draw a circle centered at *P* of any convenient radius.
    2. The circle from step 1 meets *l* in two points. Call them *C* and *D*.
    3. Draw circles with radius *CD* centered at *C* and *D*.
    4. The circles from step 3 meet in two points. Call them *E* and *F*.
    5. Draw *EF*, the desired line through *P* perpendicular to *l*.

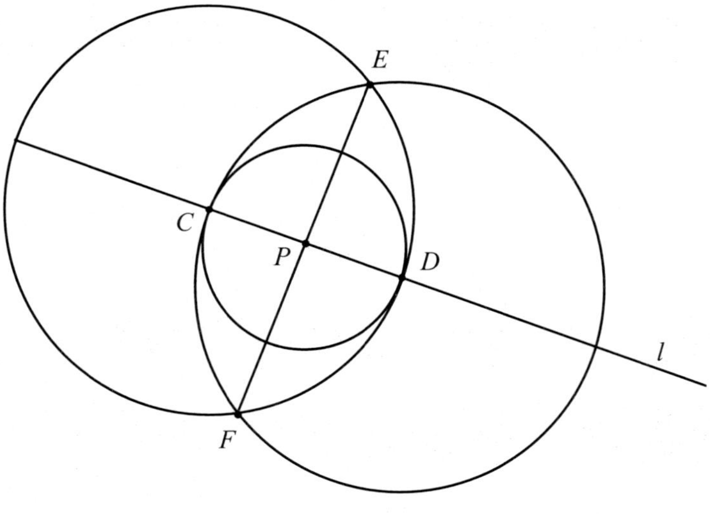

3. **Line parallel to a given line through a point not on the given line.** Given line *l* and point *P* not on *l*.
    1. Construct a line through *P* perpendicular to *l* (construction 1 above). Call this line *m*.
    2. Construct a line through *P* perpendicular to *m* (construction 2 above). This is the desired line through *P* parallel to *l*.

Note: Constructions 1 and 2 are cited and used, but their steps are not restated.

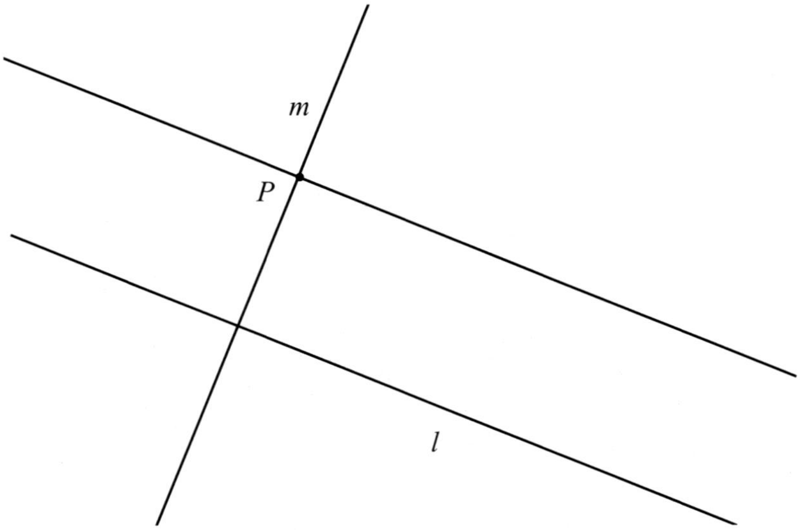

4. Equilateral triangle

1. Draw a line, and mark points *A* and *B* on it.
2. Draw circles with radius *AB* centered at *A* and at *B*.
3. The circles from step 2 meet at two points, *C* and *D*. Then *A*, *B* and *C* are the vertices of an equilateral triangle, as are *A*, *B*, and *D*.

   This is an example of a construction with more than one solution.

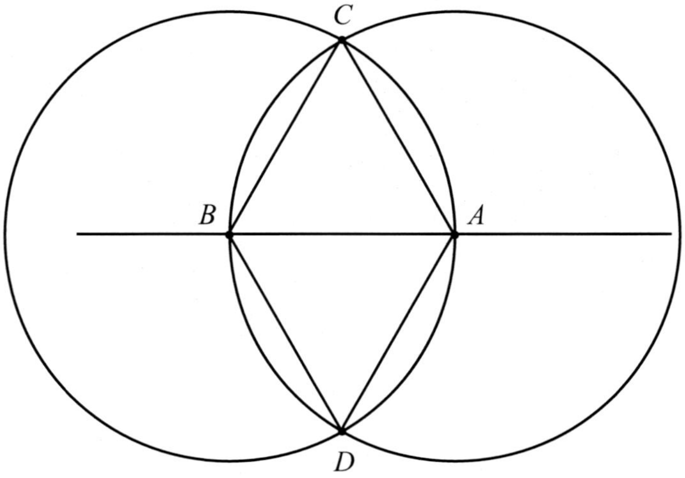

5. Square

1. Draw a line l, and mark points A and B on it.
2. Construct lines m and n perpendicular to l at A and B respectively (construction 2 above).
3. Draw a circle, radius AB, centered at A. This circle meets line m at two points. Call one of these points D.
4. Draw a line r parallel to l through D.
5. If C is the point of intersection of lines r and n, then A, B, C, and D are vertices of a square.

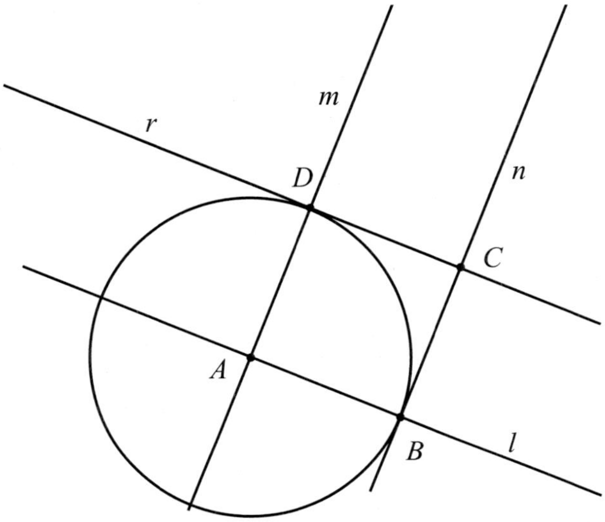

6. Regular hexagon

1. Select a point O, and draw a circle of any convenient radius centered at O.
2. Select a point P_1 on the circle, and draw a circle centered at P_1 with radius P_1O.
3. The circles from steps 1 and 2 meet at two points. Call them P_2 and P_6.
4. Draw a circle from P_2 with radius OP_1. This circle meets the circle centered at O in two points, P_1 and a new point, to be called P_3.
5. Draw a circle from P_3 with radius OP_1. This circle meets the circle centered at O in two points, P_2 and a new point, to be called P_4.
6. Draw a circle from P_4 with radius OP_1. This circle meets the circle centered at O in two points, P_3 and a new point, to be called P_5.
7. The points P_1, P_2, P_3, P_4, P_5, and P_6 are the vertices of a regular hexagon.

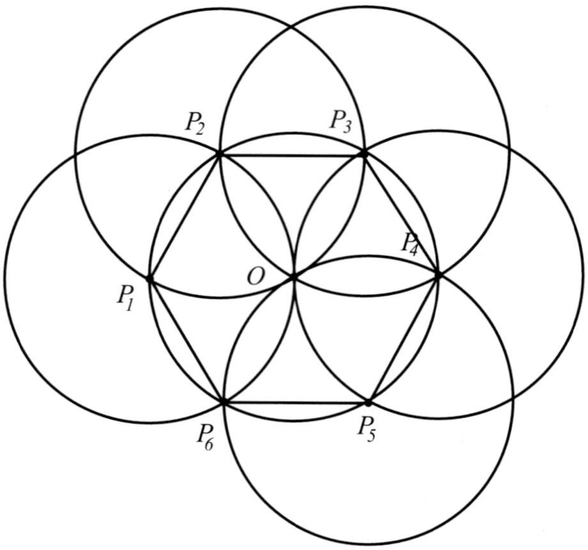

7. **The perpendicular bisector of a line segment.**
    1. Given a line segment *AB*, draw circles with radius *AB* centered at *A* and *B*.
    2. The circles from step 1 meet at two points. Call them *C* and *D*.
    3. Draw line *CD*. Line *CD* passes through the midpoint of *AB* and thus bisects it. Also, line *CD* is perpendicular to *AB*. Therefore, *CD* is called the **perpendicular bisector** of line segment *AB*.

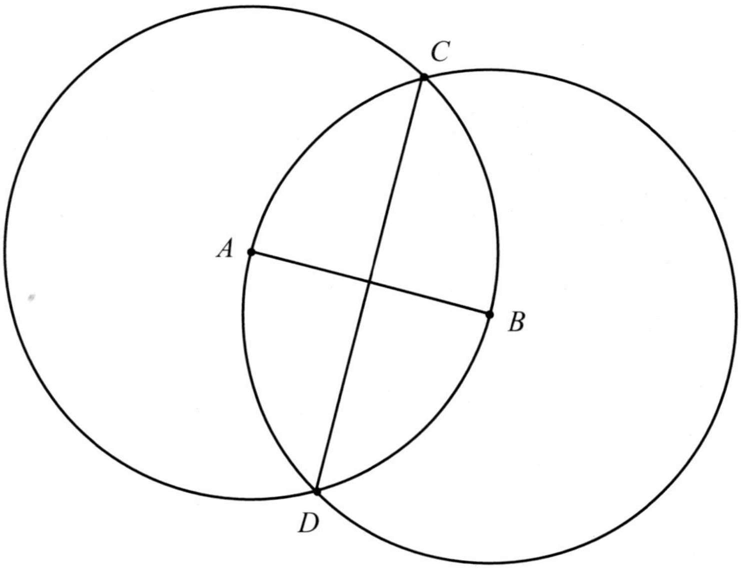

8. To bisect an angle

1. Given angle *AOB*, draw a circle at *O* with arbitrary radius.
2. The circle from step 1 intersects rays *OA* and *OB* at *P* and *Q* respectively.
3. Draw circles of radius *OP* centered at *P* and *Q*.
4. The circles from step 3 meet at *O* and a new point, which we call *R*.
5. Draw *OR*. Line *OR* is called the **angle bisector** of angle *AOB*.

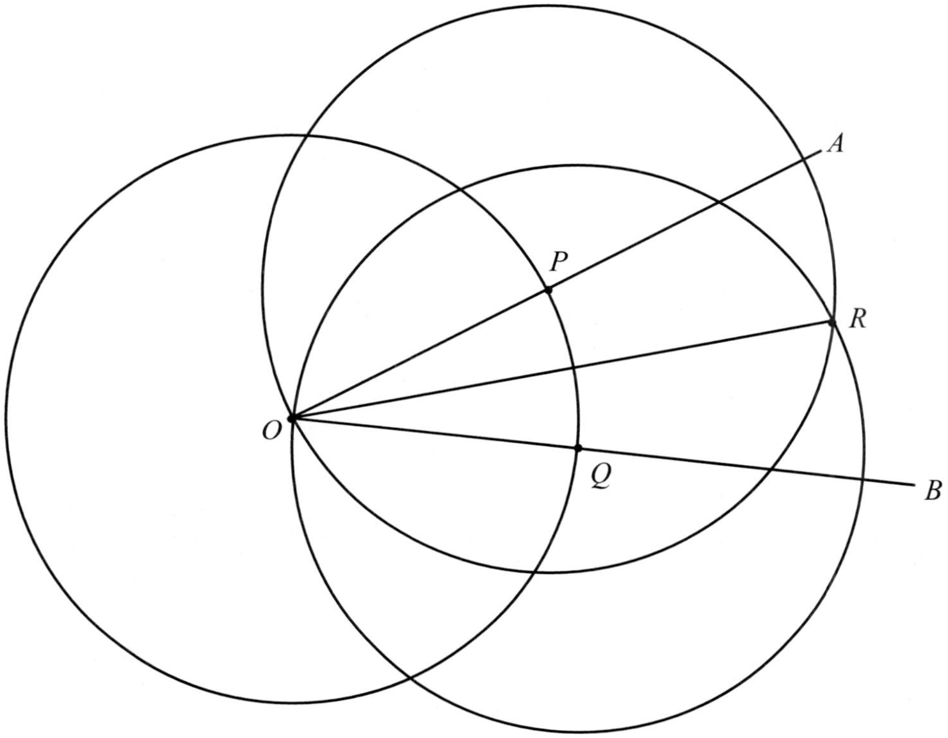

9. To copy an angle

(What this means: Given two rays *m* and *n* from a given point *O*, and a point *P* on some line *l*, construct a second ray from *P* that meets *l* at the same angle as formed by the two original rays.)

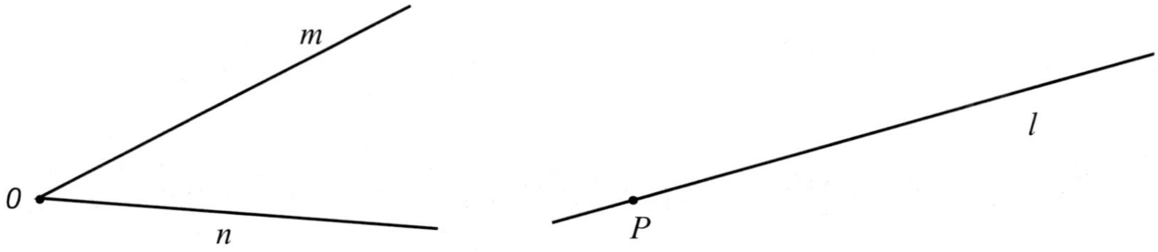

1. Construct circles of equal radii centered at O and at P. Let A and B be the intersections of the circle centered at O with m and n respectively, and let C be the intersection of the circle centered at P with l.
2. Draw a circle of radius AB centered at C.
3. Let D be a point of intersection of the circles drawn with centers P and C. Draw ray PD. Ray PD forms an angle with l at P equal to angle mOn.

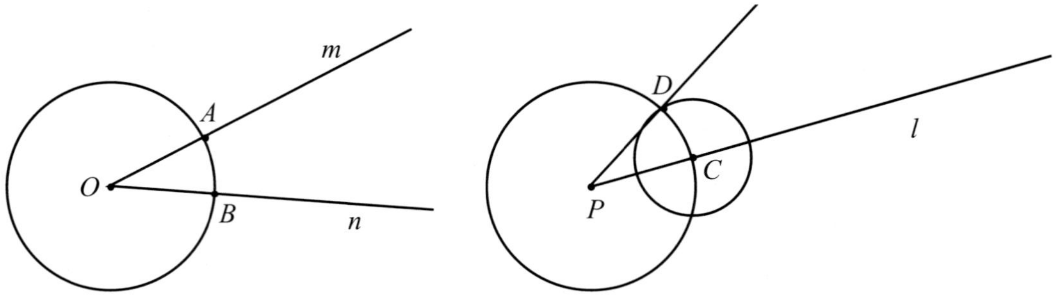

10. To cut a line segment into *n* equal pieces. (Written here for $n = 3$.)
Given a line segment AB.

1. Select a point C that is not on line AB.
2. Draw ray AC.
3. With a compass, mark off three equal lengths AD, DE, EF on AC.
4. Draw line FB.
5. Draw lines parallel to FB through points D and E (construction 3). These lines meet AB in points P and Q, which cut AB into three segments of equal length.

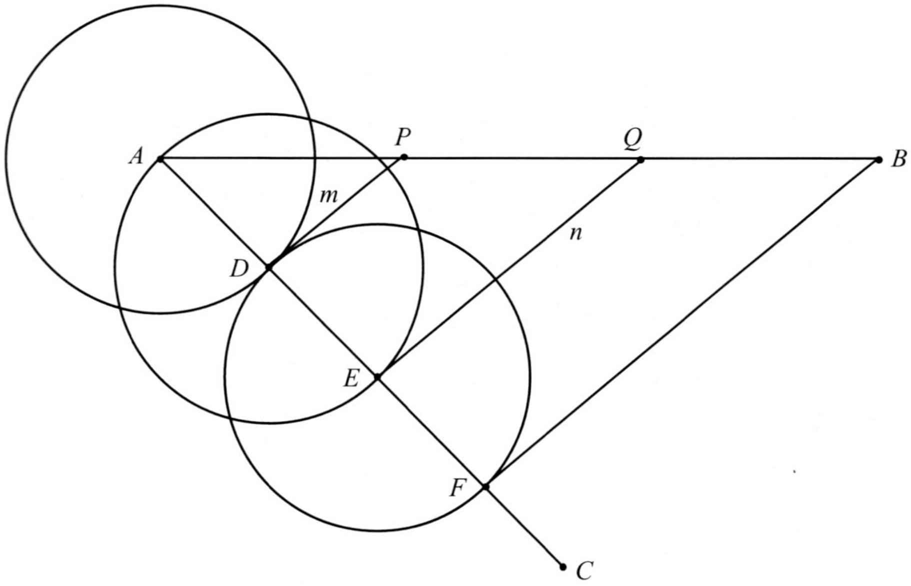

The list of basic constructions can be expanded, of course, but the above list is a good foundation to begin with. Notice that in describing a construction, the only operations allowed are:

1. To draw a line through two given points
2. To draw a circle with a specified center and radius
3. To give names to objects, such as points, lines, circles, or angles that are determined as a result of previous steps
4. To use a construction already established.

Sometimes, when there is no need to be specific, it is permissible to be a bit relaxed and allow, for example, a circle of arbitrary radius.

The exercises below explore constructions farther.

Exercises 18.7

Patty paper, the semi-transparent, coated squares used to separate hamburger patties, makes a handy and inexpensive classroom tool for certain geometric constructions. Questions 1-4 explore this tool.

1. a. Draw a line on a piece of patty paper, and mark a point P not on the line. Then fold the paper carefully, so that the fold goes through point P and the line matches up with itself. Crease your fold sharply to make a line. This line is a perpendicular to the given line through the given point P, as called for in construction 1.
   b. On a new piece of patty paper, draw a line and name a point on the line. Explain how to fold the paper to make a line perpendicular to the given line through the given point (construction 2).
   c. Explain how to use patty paper to construct a line parallel to a given line through a point not on that line (construction 3).

2. In the diagram for construction 1, imagine lines AP, BP, AQ and BQ drawn in.
   a. Explain why triangle APQ is congruent to triangle BPQ.
   b. Explain why angles APQ and BPQ are equal.
   c. Let M be the point of intersection of lines PQ and AB. Show that triangles APM and BPM are congruent.
   d. Show that PQ is perpendicular to AB.

3. Let AB be a chord of a circle with center O, and let M be the midpoint of chord AB.
   a. Use congruent triangles to show that OM is perpendicular to AB.
   b. Use congruent triangles to show that any line perpendicular to AB through O passes through M.

4. Explain how to use patty paper to make two lines that meet at an angle of:
   a. 45°
   b. 22.5°

5. Explain how to use patty paper to bisect a line segment (construction 7).

6. Explain how to use patty paper to construct the bisector of an angle (construction 8).

7. a. Refer to the diagram illustrating construction 10. Explain why one can be sure that $AP = PQ = QB$.
   b. Draw a line segment several inches long, then use construction 10 to divide your line segment into five pieces, all the same length.

8. a. Use the construction for a regular hexagon presented in the text as the basis for a construction of a regular dodecagon (12-sided polygon).
   b. Use the construction for a square presented in the text as the basis for a construction of a regular octagon (8-sided polygon).
   c. Show how, in general, if you are given a regular polygon with n sides, you can construct a regular polygon with $2n$ sides.

9. In step 2 of construction 1 circles are drawn with radius AP. Could other radii be used in place of AP? If other radii may be used, what conditions, if any, must they satisfy?

Vocabulary for Chapter 18

Acute angle (18.3)
Angle (18.3)
Arc (18.5)
Center of rotation (18.3)
Chord (18.5)
Compass (18.2)
Degree (18.3)
Geometric mean (18.6, exercises)
Hub (18.3)
Inscribed angle (18.5)
Major arc (18.5)
Mean proportional (18.6, exercises)
Minor arc (18.5)
Minute (18.3)
Obtuse angle (18.3)
Perspective (18.1)
Postulate (18.1)
Protractor (18.3)
Ray (18.3)
Reflex angle (18.3
Right angle (18.3)
Second (18.3)
Straight angle (18.3)
Straightedge (18.1)
Theorem (18.4)
Vanishing point (18.1)

Self-Quiz for Chapter 18

1. Use a protractor to construct angles of:

   a. 37° b. 130° c. 217°

2. Joey and his dad are standing on level ground at night. From a bright streetlight, Dad's shadow is 20 feet long, while Joey's is 9 feet long. If Dad's height is 74 inches, how tall is Joey?

3. In the picture below, *DE* is 10 units long and is parallel to *BC*. What fraction of the area of triangle *ABC* is inside triangle *FGH*? Hints:

   a. Show that triangles *ADE* and *ABC* are similar.

   b. Use the ratio of similarity to find length *BC*.

   c. Use the length of *BC* to find the length of *FH*.

   d. What fraction of the height of *ABC* is the height of *GFH*?

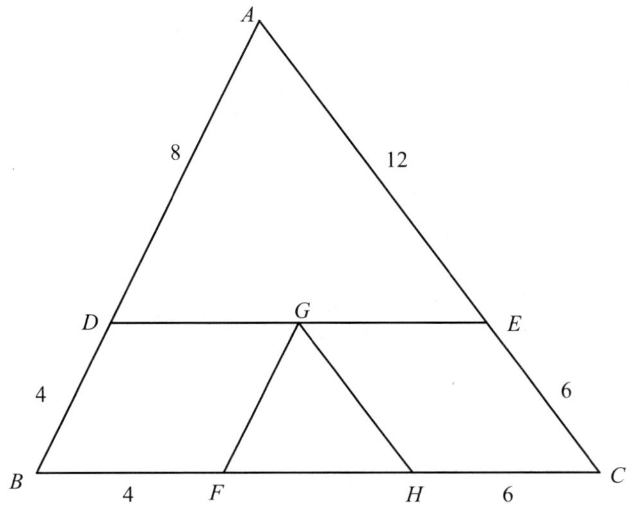

4. In the figure below, points B, C, and E lie on a circle with center O. Point D lies on the extended diameter BE, and angle ABD is a right angle. Furthermore, the $\measuredangle BAC = \measuredangle ACB$. Show that $\measuredangle BAC = 2\measuredangle BDA$.

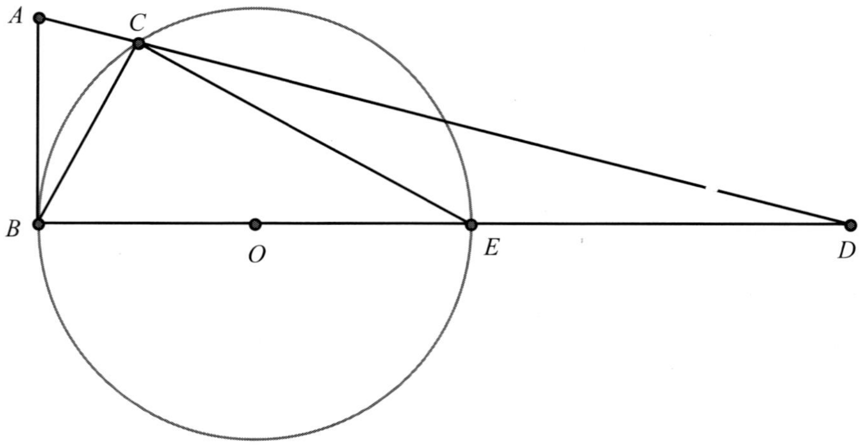

5. Points A, B, C, and D lie on a circle. If the ratios of the arcs determined by these points are: $\overset{\frown}{AB}:\overset{\frown}{BC} = 2:4$, $\overset{\frown}{BC}:\overset{\frown}{CD} = 4:3$, $\overset{\frown}{CD}:\overset{\frown}{DA} = 1:2$, then how many degrees are in each of those four arcs?

6. Describe a straight-edge and compass construction for a regular octagon.

Chapter Nineteen
Polygons, Tessellations, and Polyhedra

Over the centuries polygons and polyhedra have been a source of interest and inspiration in many cultures. This chapter uses these beautiful shapes as an enjoyable way to explore basic geometry.

19.1 Polygons

Of the polygons in Figure 19.1.1, some, such as *B* and *D*, look the same no

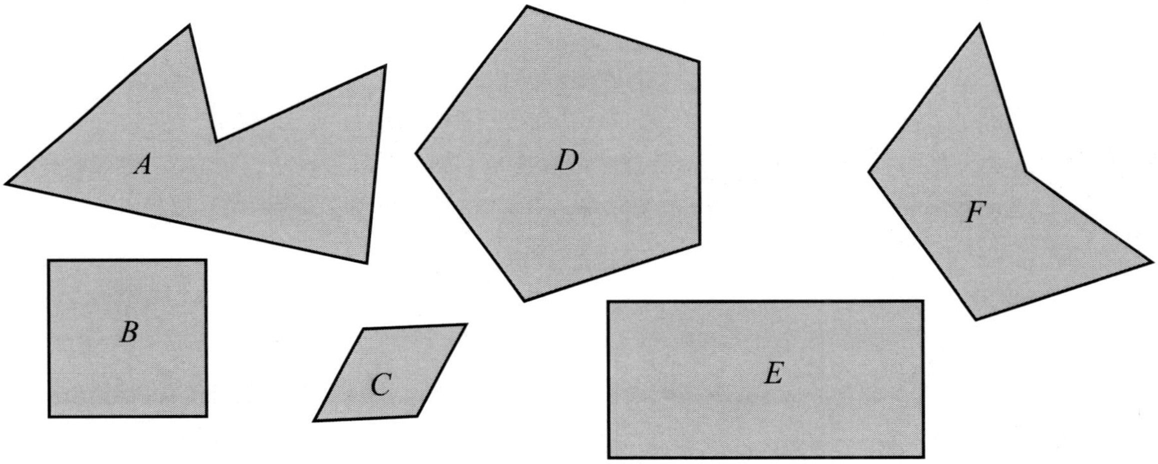

Figure 19.1.1

matter which side they rest on. These are regular polygons. Polygons that do not have that property are not regular.

Can a polygon have all its edges of equal length, yet not be regular? Yes. In figure 19.1.1, both *C* and *F* exemplify that. Can a polygon have all its angles equal and yet not be regular? Yes again. *E* is an example of that kind.

A regular polygon has all angles equal and all sides the same length.

Some terminology: The sides of a polygon are called **edges**, the corners are called are called **vertices**. Singular: **vertex**. (Think of one Kleenex, two Kleenices!) A regular triangle is called an **equilateral triangle**. A polygon with *n* sides is called an ***n*-gon**. For small numbers of sides, we have special names, as follows:

number of sides	**name of polygon**
3	triangle
4	quadrilateral
5	pentagon
6	hexagon
7	heptagon
8	octagon
9	nonagon
10	decagon
12	dodecagon

The fact that the interior angles of any triangle add up to 180° is part of a larger pattern. The interior angles of a quadrilateral add up to 360°, those of a pentagon add up to 540°, and so on. Here is why. Any quadrilateral can be cut into a pair of triangles by a diagonal, as in figure 19.1.2.

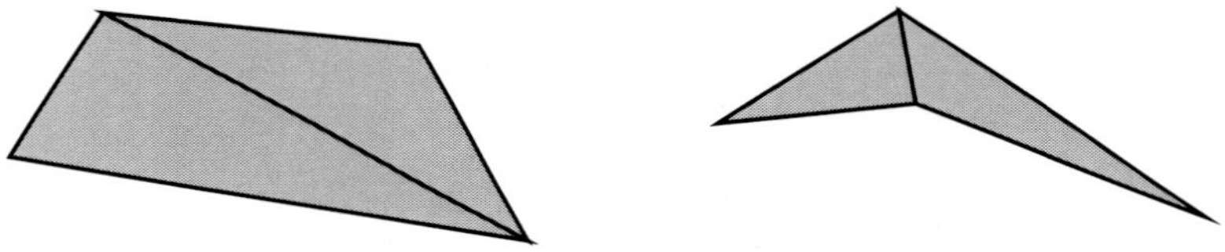

Figure 19.1.2

The angles of these triangles add up to those of the quadrilateral. If the angles of every triangle add up to 180°, then those of the two triangles together add up to 360°.

Similar considerations apply to polygons with more sides, as the following exercises show.

Exercises 19.1

1. a. Draw diagonals to cut each of these pentagons into three non-overlapping triangles.

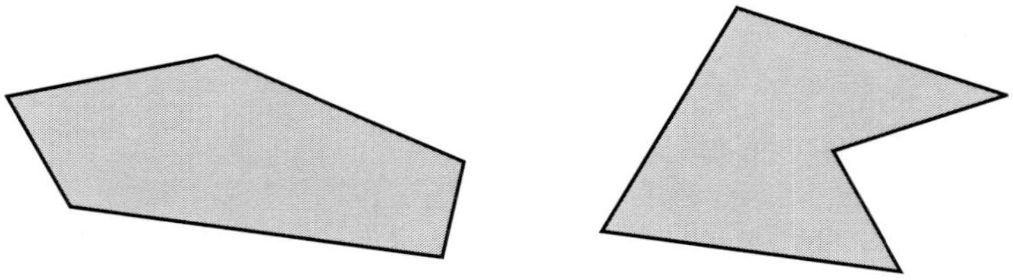

   b. What is the sum of the angles of the three triangles in each case?

   c. What is the sum of the angles of a pentagon?

2. a. Draw diagonals to cut each of these hexagons into non-overlapping triangles.

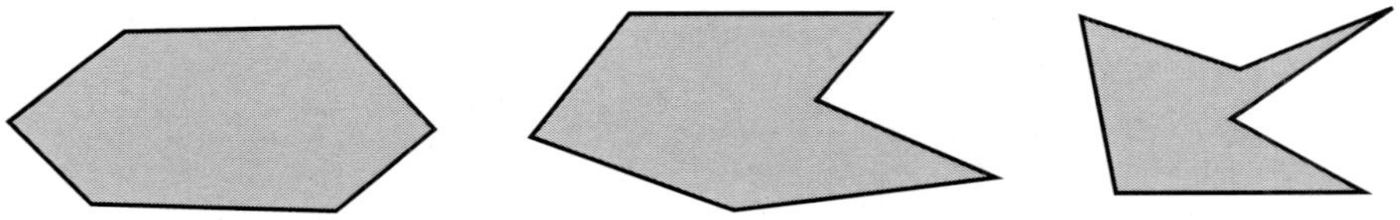

   b. What is the sum of the angles of the four triangles in each case?

   c. What is the sum of the angles of a hexagon?

3. a. Draw diagonals to cut each of these heptagons (7 sides) into non-overlapping triangles.

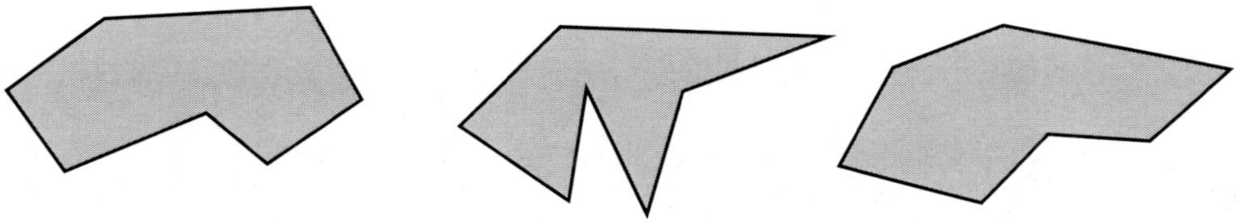

   b. What is the sum of the angles of the triangles in each case?

   c. What is the sum of the angles of a heptagon?

4. Fill in the table on the next page, based on what you know so far. The last case, for *n*-gons (with *n* unspecified) is a generalization which you should try to guess, using the patterns from the cases already filled in. The case n=3 is included just to help bring out the patterns.

Number of sides	3	4	5	6	7	n
Number of triangles	1					
Sum of angles in triangles = sum of angles in polygon	180°					

5. If all the angles of a triangle are equal, they must each be a third of 180°, since 180° is the sum of angles in a triangle. Likewise, the angles of a quadrilateral with all angles equal can be found by dividing the sum of the angles by 4.
   a. Use the idea described to find the size of each angle in a pentagon having all angles equal.
   b. Use the method from part a to find the size of each angle in a hexagon having all angles equal.

6. a. Where have you seen regular hexagons used? Try to name at least three common uses for them.
   b. Why are stop signs, and no other traffic signs, made in the form of regular octagons?

7. Heads of bolts are usually regular hexagons or squares.
   a. What advantage does a hexagonal bolt head have over a square head?
   b. What advantage does a square bolt head have over a hexagonal head?
   c. Why are regular polygons with odd numbers of sides rarely used as heads of bolts?

8. Construct a square, using a protractor and ruler.
   a. Decide how long you want each side to be, and mark off that distance on a line. These points are marked *A* and *B* in the picture.

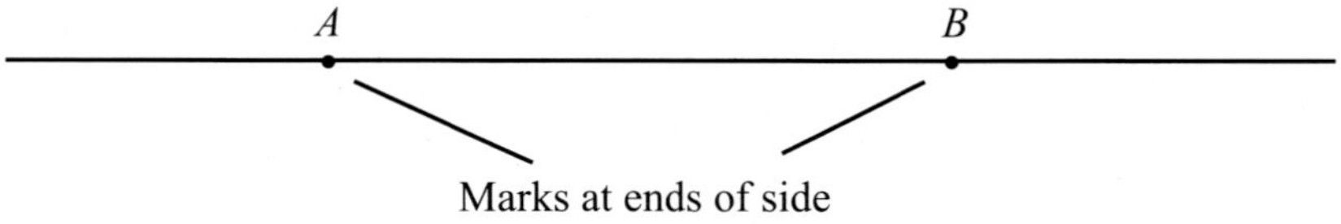

   b. With a protractor, construct 90° angles at *A* and *B*.

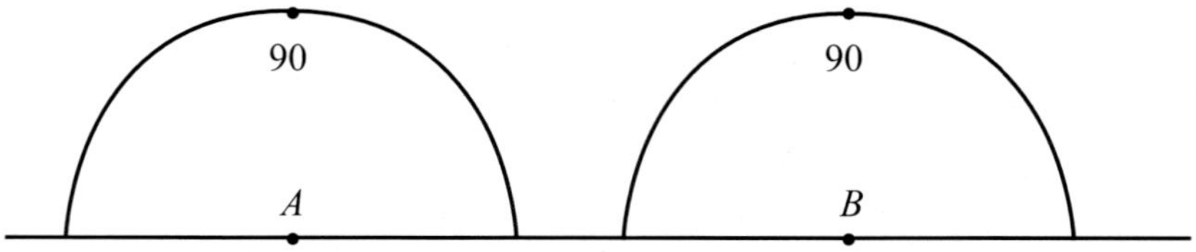

   c. Mark off distance *AB* on each perpendicular, thus

determining two new points, C and D. Connect C and D to finish the square.

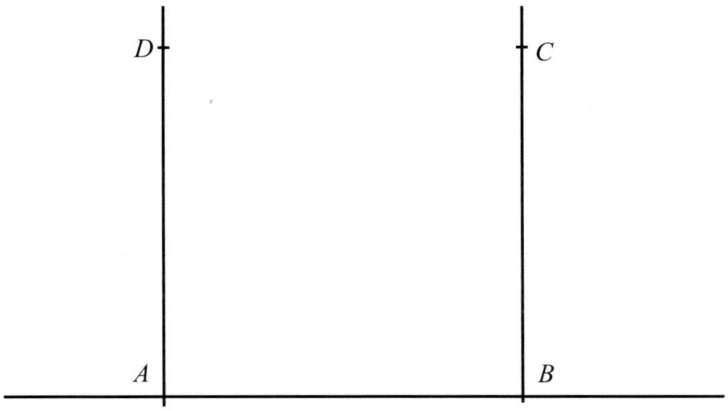

d. Check your work by looking at it from every side. The square should look the same no matter which side it rests on.

9. Make a regular pentagon. PRECISION HERE IS CRITICAL. ANY ERROR CARRIES OVER TO LATER STEPS IN THE PROCESS.

   a. First determine the size of the angles. Imagine walking around the edges of a regular pentagon (figure 19.1.3), starting off on an edge that runs from west to east. After making five equal turns of x° each, you are once again facing east, so x° must be $\frac{1}{5}$ of 360°, which is 72°.

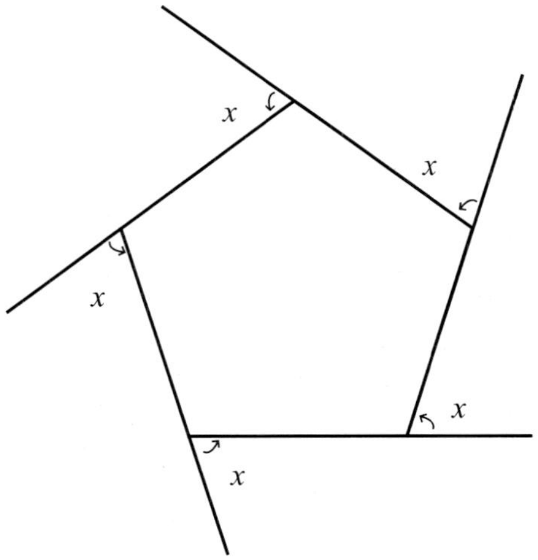

Figure 19.1.3

(These are **exterior** angles, in contrast to the interior angles that we use most of the time.)

b. As with the square, begin by marking two points A and B on a line. Choose A and B a convenient distance apart. For classroom purposes, 2, 3, or 4 inches is often good.

c. At point B make the appropriate angle, as shown.

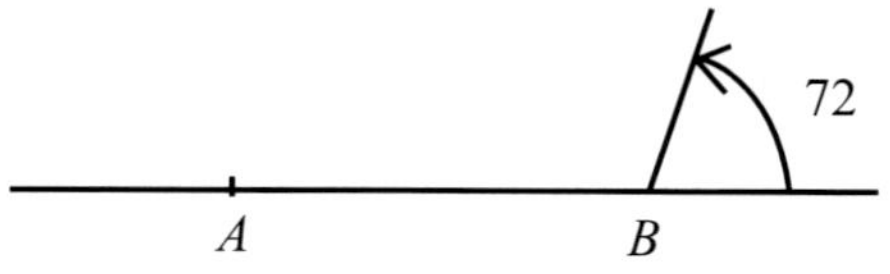

d. Draw a line from B in the direction of the new side, as found in step c.

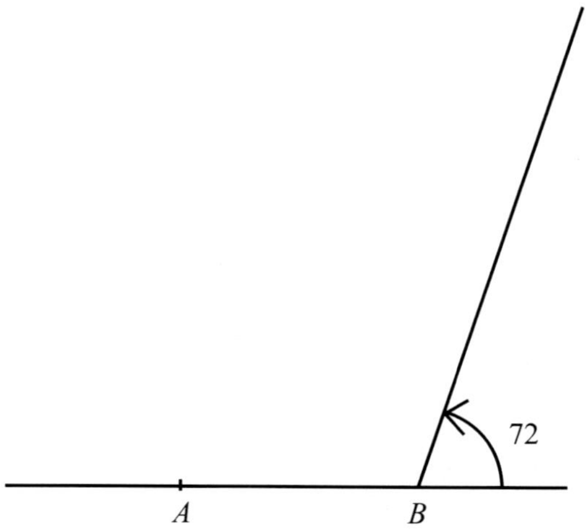

e. Mark point C on this new line so that distance AB = distance BC.

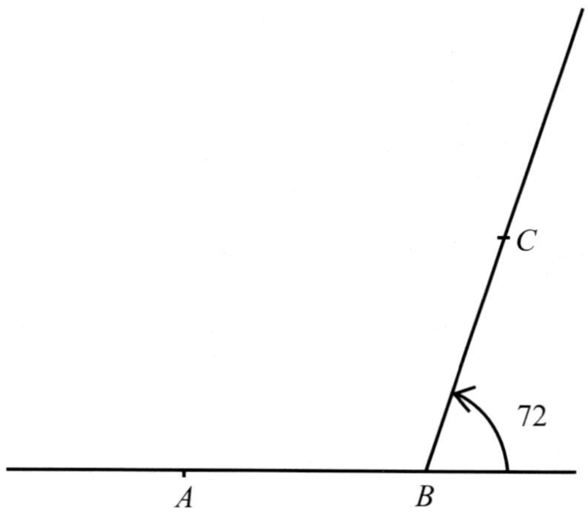

f. Construct an angle at C equal to the one at B, essentially by turning the paper so that BC is horizontal and repeating steps c and d.

g. Mark point D on this new side so that distance AB = distance BC = distance CD.

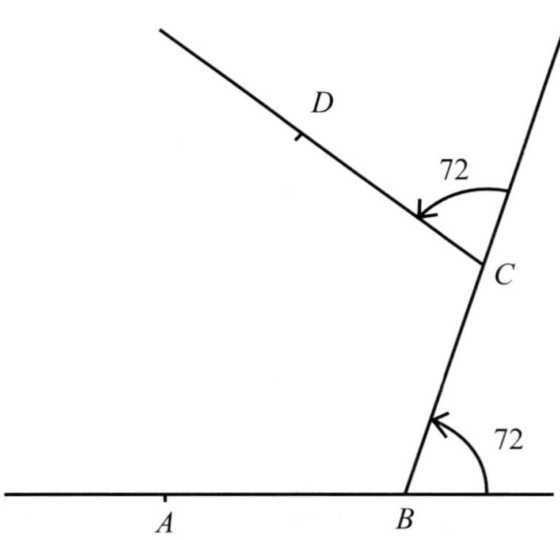

h. Repeat the above process, constructing an angle at D and marking off a new point E. Then draw side AE to complete the pentagon.

10. a. Construct a regular hexagon, using the same method as in question 9.
    b. Check your work by looking at the hexagon from all sides.

11. Use the method from questions 9 and 10 to construct:
    a. A regular octagon.
    b. A regular heptagon (7-gon) (The crucial angle involves a fraction of a degree; approximate it as accurately as you can.)

Exploration: Regular star polygons (questions 12-18)

A **regular star polygon** is formed by connecting the vertices of a regular polygon in a pattern, as in these examples:

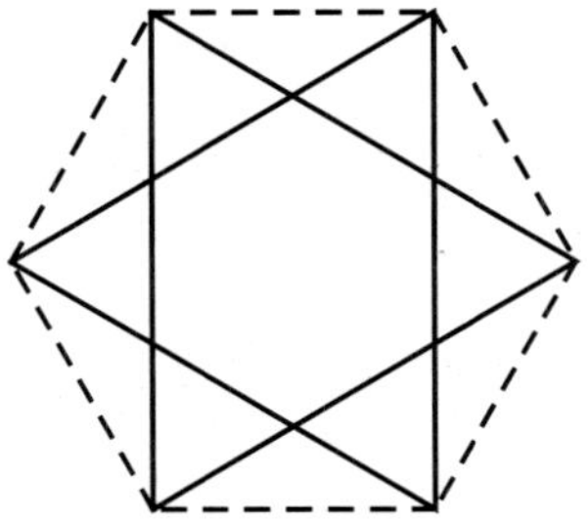

6/2

Connect every second vertex

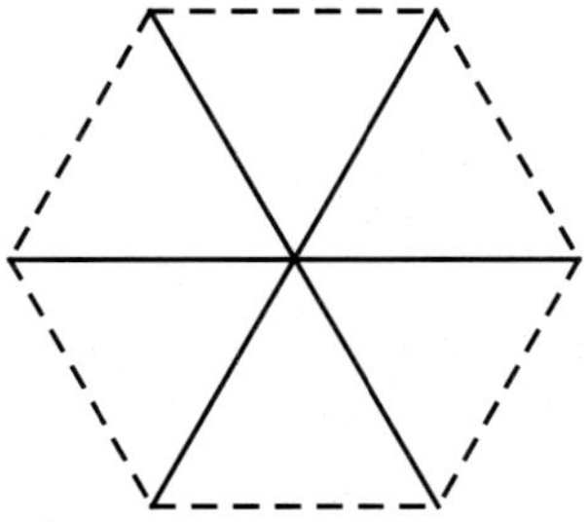

6/3

Connect every third vertex

Do you understand the designations 6/2 and 6/3? In that notation, the ordinary hexagon would be 6/1, since it is formed by connecting "every first" vertex.

12. Draw in the star octagons indicated in the octagons below. (Be sure to draw lines from every vertex.)

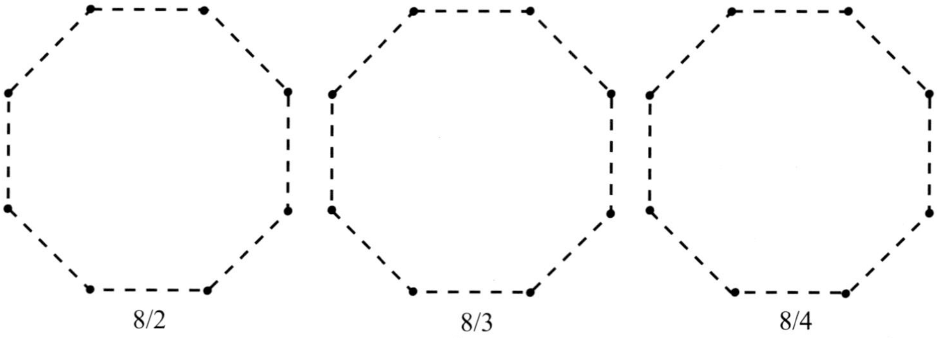

13. Why, in question 12, were you not asked to draw the star octagons 8/5 and 8/6? If you do not see how to answer this question right away, draw those two star polygons, and see what you get.

14. Count the number of sides (line segments) drawn in
    a. 8/2
    b. 8/3
    c. 8/4

15. Add your answers from the three parts of question 14 and compare the sum with the number of diagonals of a regular octagon. (The number of diagonals of a regular polygon with n sides was found in question 8 of exercises 9.2).

16. Sketch all the different kinds of star polygons that can be formed by connecting the vertices of a regular polygon with:
    a. 4 sides
    b. 5 sides
    c. 6 sides (see above)
    d. 7 sides
    e. 8 sides (see questions 12 and 13)
    f. 9 sides
    g. 10 sides
    h. 11 sides
    i. 12 sides

17. Use information from question 16 to find a formula for the number of different types of star polygons with n vertices.

18. Some star polygons can be drawn by going from vertex to vertex until you return to where you began, but other star polygons cannot be drawn that way, because they break into two or more separate pieces. How must the number of vertices in the n-gon be related to the number of vertices skipped by each diagonal of the star polygon for the star to be a single piece?

19. The figure below is known as Kürschák's Tile.[1] It includes a regular dodecagon with four vertices at midpoints of the sides of the square tile. The triangles on the inside of the dodecagon's edges are equilateral.
    a. Show that all of the thin obtuse triangles are congruent to each other.
    b. Show that each of the four corner triangles is equilateral.
    c. Use parts a and b to find what fraction of the area of the square is contained in the dodecagon.

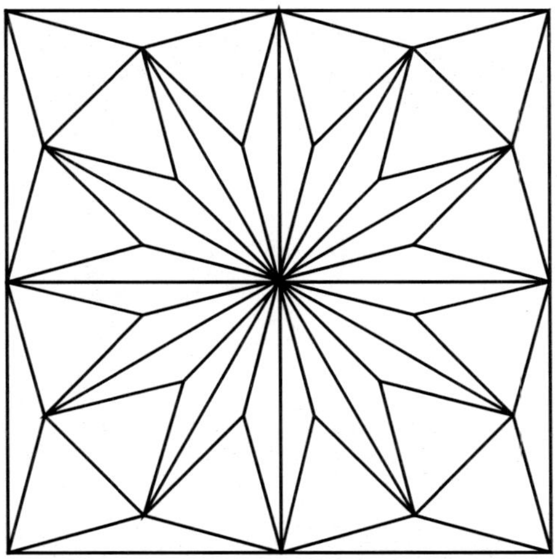

1 Jösef Kürschák (1864-1933) was a mathematics professor at the Technical University in Budapest.

19.2 Tessellations

Floors are often tiled with squares (figure 19.2.1a), but you may also have seen triangular and hexagonal tilings, or **tessellations**, as they are called, like those in figures 19.2.1b and c.

a b c

Figure 19.2.1 Regular tessellations

The tessellations in figure 19.2.1 are called **regular tessellations**, meaning that each uses congruent copies of just one regular polygon, meeting edge to edge and vertex to vertex, and can be extended without limit to cover the plane.

There are many other tiling patterns that cover the plane, some with just one shape of tile, others with several shapes, some with tiles shaped as regular polygons, others with completely different shaped tiles, possibly not polygons at all. Tessellations are intrinsically fascinating, and designs made with them are useful in dozens of ways, from floor and ceiling tiles to designs incorporated in baskets, wire fences, and fabrics.

The three regular tessellations shown in figure 19.2.1 are, interestingly, the only regular tessellations, a fact which is established in the exercises below. However, if we loosen the rules a bit, we can come up with additional tessellations which are in many ways cousins of regular tessellations and are quite pretty. These are called **semi-regular tessellations**. They, too, are made up entirely of regular polygons, meeting edge to edge and vertex to vertex with all vertices alike. (The same kinds of polygons, in the same order, meet at each vertex.)

One way to create a semi-regular tessellation is to **truncate** (cut the corners off) the polygons in a regular tessellation. Figure 19.2.2 shows truncation of triangles and squares.

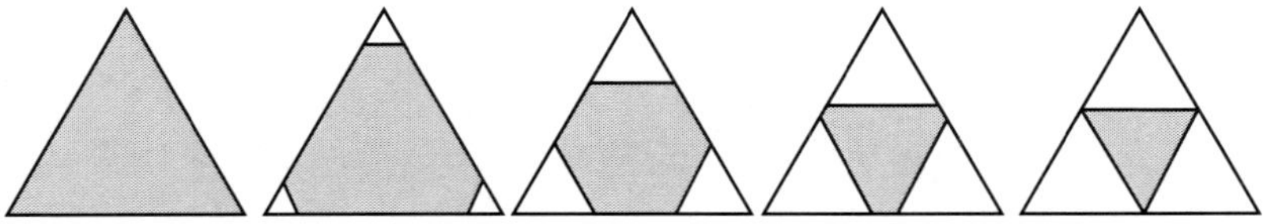

a b c d e

Truncating an equilateral triangle.
Stages c and e leave regular polygons (shaded).

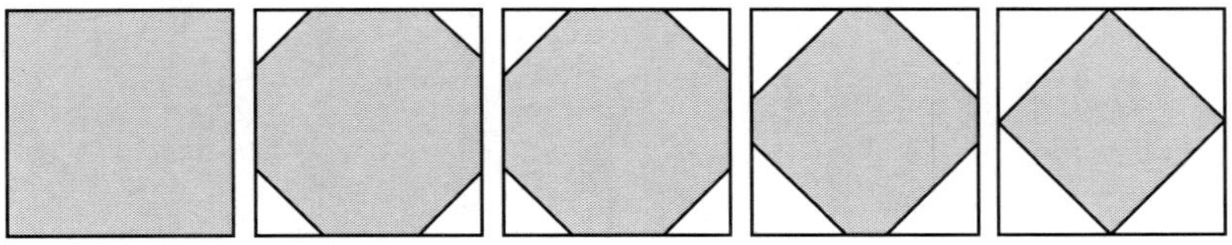

a b c d e

Truncating a square.
Stages c and e leave regular polygons (shaded).

Figure 19.2.2

Figure 19.2.3 gives you some idea of what can happen when the triangles in a regular tessellation are truncated. You can get another regular tessellation, consisting of hexagons (which is nearly the case in the lower left illustration), and you can also come up with the semiregular tessellation at the lower right.

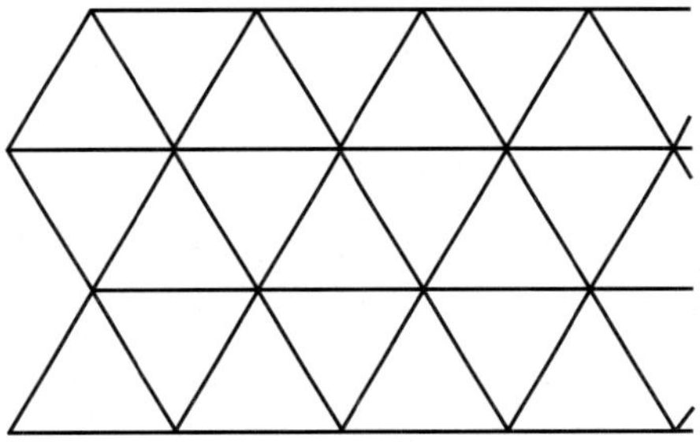

Start with equilateral triangles.

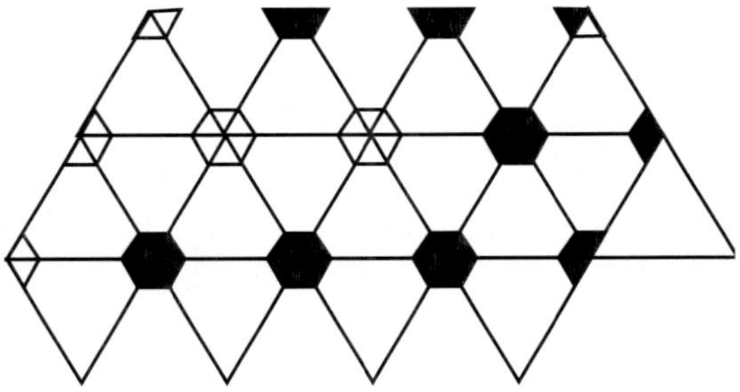

Truncate a little at each vertex.
Corner hexagons are regular, but remaining white hexagons are not.

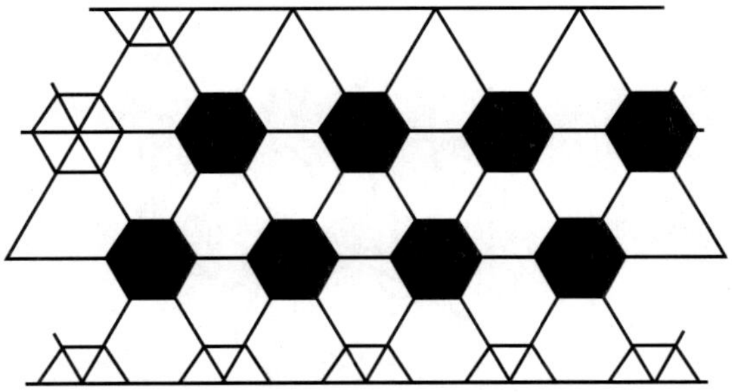

Truncate a bit more at each vertex.
Now white hexagons are nearly regular.
(If they were, we would have another regular tessellation.)

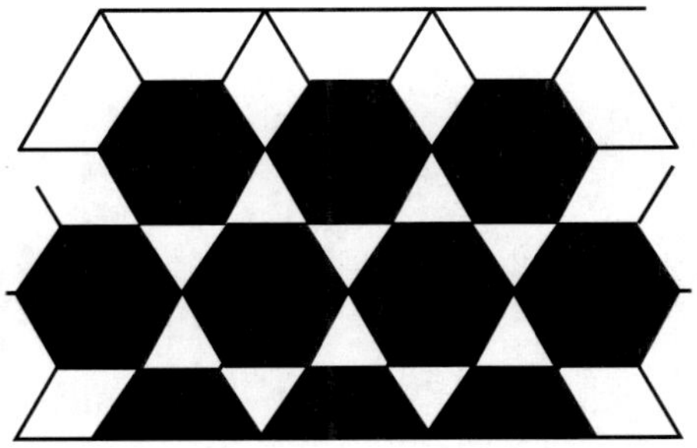

Truncate so much that all that remains of the original hexagons is little equilateral triangles.

Figure 19.2.3

Figures 19.2.4 and 19.2.5 show similar treatments of the other two regular tessellations by truncation. When the cuts are just the right size, the polygons which remain after truncation are regular, and the result is a **semi-regular tessellation.**

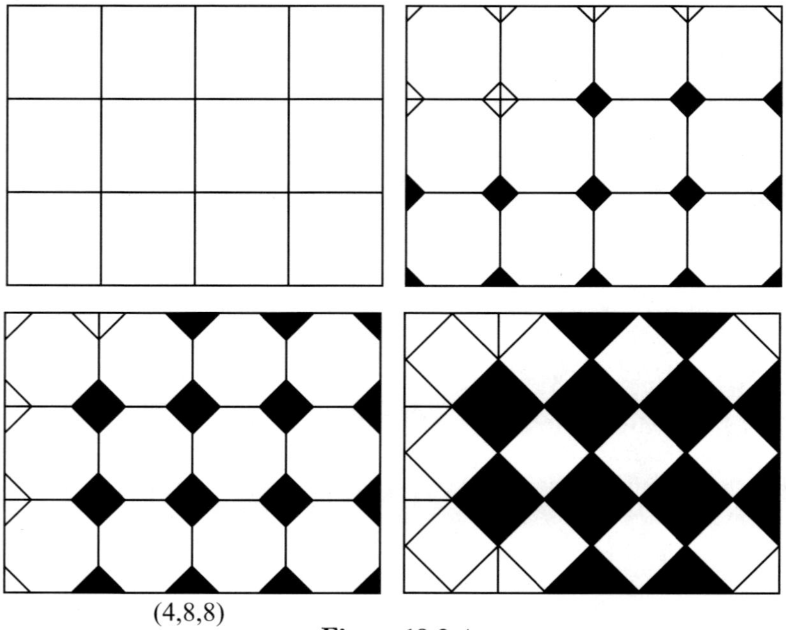

Figure 19.2.4

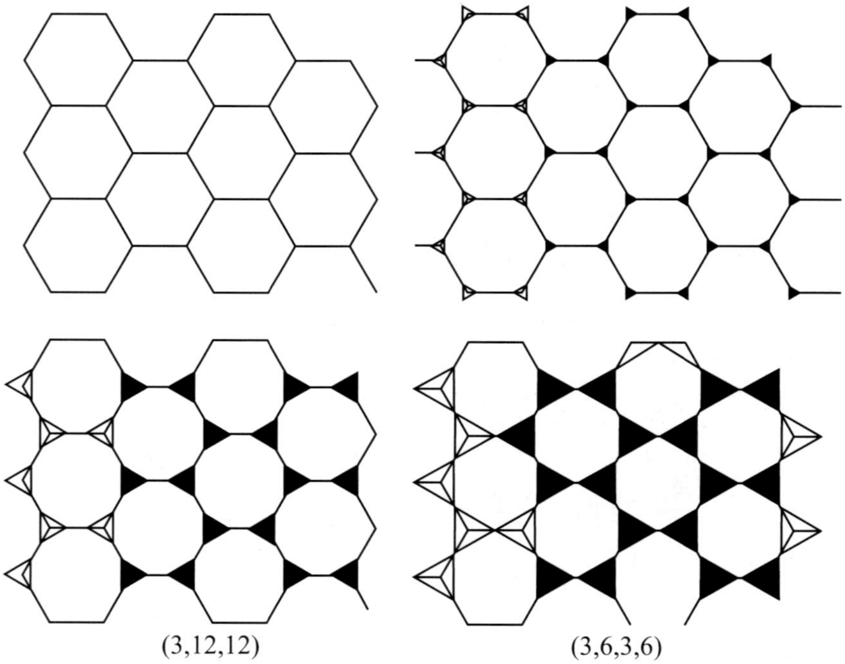

Figure 19.2.5

The polygons which meet at a vertex are specified in what is called a **vertex symbol**. It consists of the number of sides of each polygon at that vertex, in order. For example, the regular tessellation in figure 19.2.1b has the vertex symbol (3, 3, 3, 3, 3, 3) or (3^6) for short, meaning an ant walking around any vertex would cross six triangles before coming back to its starting point, as shown in figure 19.2.6. Similarly, the vertex symbols for the tessellations in figures 19.2.1a and 19.2.1c are, respectively, (4, 4, 4, 4) or (4^4) and (6, 6, 6) or (6^3). In the same spirit, the vertex designation for the tessellation in the lower left part of figure 19.2.4 is (4, 8, 8) or $(4, 8^2)$, at least if the ant starts on a square and then crosses two octagons, as in figure 19.2.7.

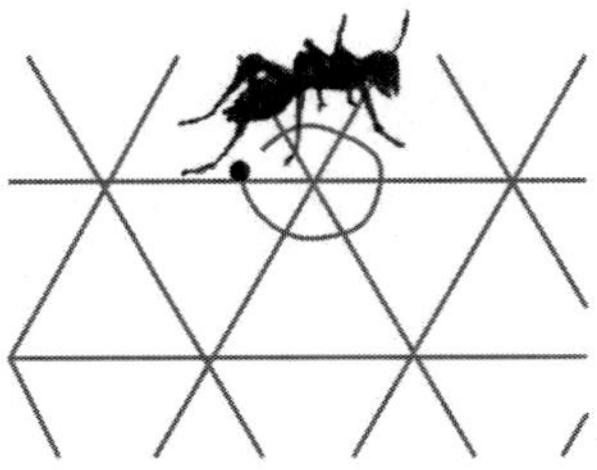

A path an ant might take around a vertex.

Illustrations by Joey Stein.

Figure 19.2.6

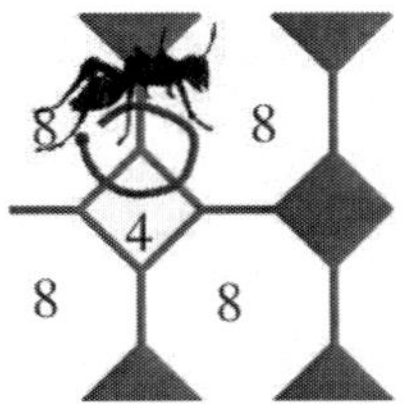

Figure 19.2.7

Had the ant started on an octagon, the same vertex would be given the symbol (8, 8, 4) or (8, 4, 8). These symbols are considered equivalent, since they are assumed to read cyclically, as in figure 19.2.8.

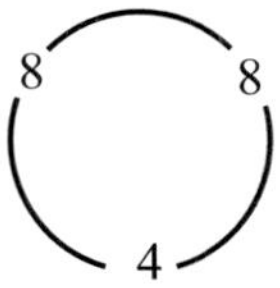

Figure 19.2.8

As it turns out, there are only eight semiregular tessellations, and you can find them by experimentally, by trying to fit regular polygons together. To do this you need a supply of various regular polygons, with edges all the same length. A key observation in such experiments is that at any vertex the angles in any tessellation add up to 360°. Figure 19.2.9 shows this for the tessellation (4, 8, 8).

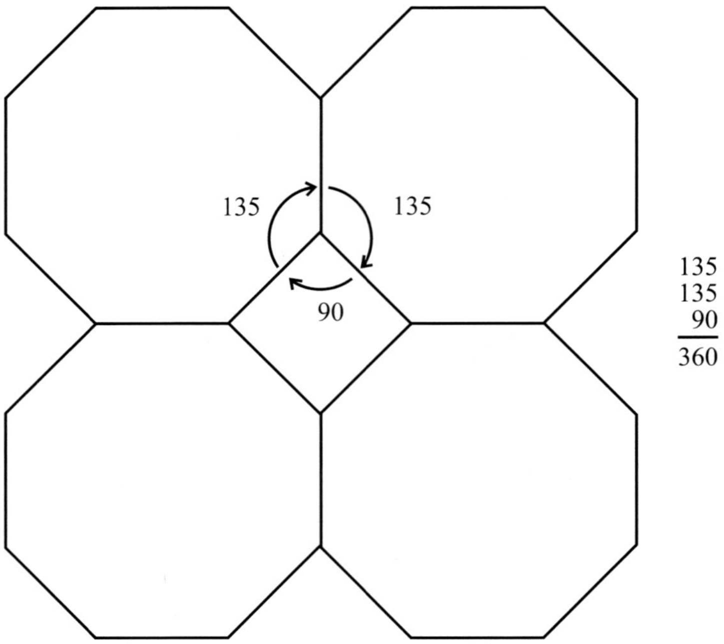

Figure 19.2.9

If you have no physical models, you can experiment anyway, using the interior angles of regular polygons, trying to find assortments of three or more which add up to exactly 360°. This will allow you to discover all of the semiregular tessellations. You will also discover some combinations of regular polygons whose interior angles allow them to fit together neatly around a point and yet still do not yield regular tessellations. Question 3 below explores this further.

Exercises 19.2

Exploration: Semi-regular tessellations (questions 1-7)

1. The angles at any vertex of a tessellation total exactly 360°. Use the table below to show that there are only three ways to fit congruent regular polygons together to make a regular tessellation.

Number of sides	Number degrees in each interior angle
3	60
4	90
5	108
6	120
7	$128\frac{4}{7}$
8	135
9	140
10	144
11	$147\frac{3}{11}$
12	150
13	$152\frac{4}{13}$

2. There are just eight semi-regular tessellations. Three of them are pictured in figures 19.2.3-5. Another is started below.
   a. In each of the semi-regular tessellations pictured, show that the interior angles of the polygons which meet at each vertex add up to exactly 360°.
   b. Extend the sketch below to include another row of squares and triangles.

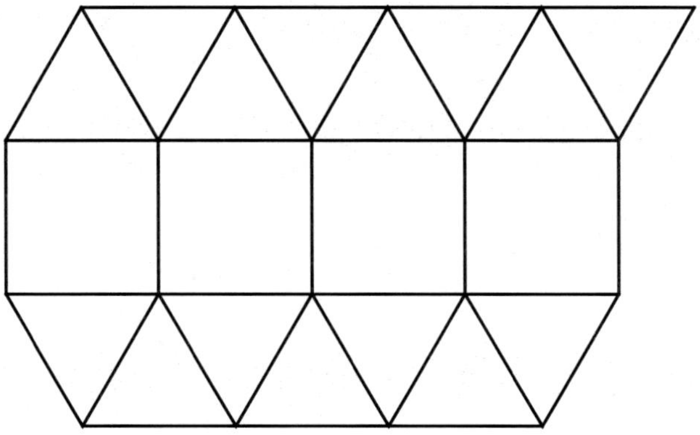

3. Sketch or put polygons together to make the semi-regular tessellations with these vertex symbols:
   a. (4, 6, 12)
   b. (3, 3, 3, 3, 6)
   c. (3, 4, 6, 4)
   d. (3, 3, 4, 3, 4)

4. a. If you tried to form the vertex of a semi-regular tessellation with two pentagons and a decagon, what would their interior angles add up to?
   b. Can you make a semi-regular tessellation with the vertex symbol (5, 5, 10)? Explain your answer.

5. a. If you tried to form the vertex of a semi-regular tessellation with two triangles, a square, and a dodecagon, what would their interior angles add up to?
   b. Can you make a semi-regular tessellation with vertex symbol (3, 3, 4, 12)? Explain your answer.

6. Could you make a semi-regular tessellation with vertex symbol (3, 4, 3, 12)? Explain your answer.

7. (A challenge) There are 17 ways to arrange regular polygons around a vertex, with 360° as the sum of the angles at that vertex. Three of these ways correspond to regular tessellations, and eight to semiregular. Three of the others are mentioned in questions 3-5 above. Find the other three, and write their vertex symbols.

8. Pictured on the next page is a tiled wall from the Alhambra, a Moorish palace in Granada, Spain. Study it carefully.
   a. Find three sets of rectangles in the wall.
   b. Find black regular hexagons of two sizes.
   c. Trace the path of each strip of light colored tiles to find a beautiful pattern.

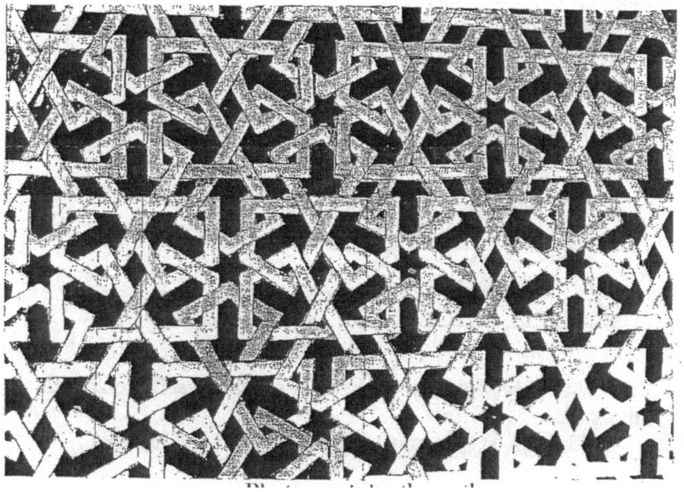

Photograph by the author

19.3 Polyhedra

The three-dimensional analogs of polygons are called **polyhedra** (singular: **polyhedron**). A polyhedron is bounded by polygons called **faces**. The lines where adjoining faces meet are the **edges** of the polyhedron, and the points where adjoining edges meet are the **vertices** (singular: **vertex**). Polyhedra are an excellent introduction to three-dimensional geometry. A particularly good way to study them is to actually make them from paper or cardboard. You can buy these shapes already made, but actually constructing the polygons and taping them together is much more instructive.

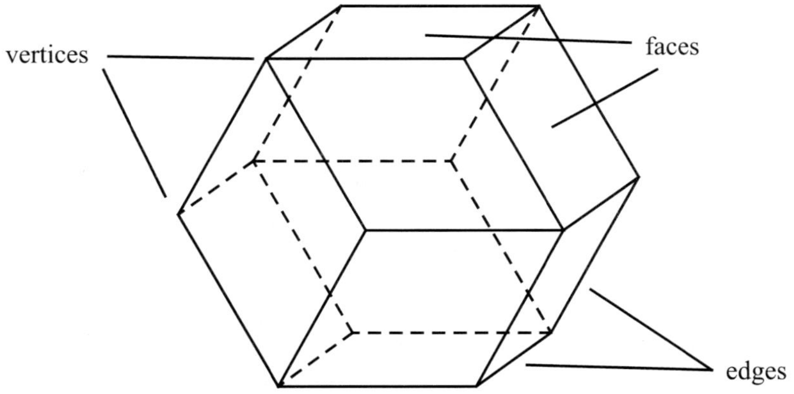

Figure 19.3.1 Parts of a polyhedron
(Garnet crystals often occur in this shape, called a **rhombic dodecahedron**)

A polyhedron is said to be **regular** if:

1. All its faces are congruent regular polyhedra meeting edge to edge.
2. The vertex symbols, defined just as for tessellations, are the same at every vertex.
3. All angles between adjoining faces are equal.

These conditions are so restrictive that the kinds of regular polyhedra are, in the words of Lewis Carroll, "distressingly few."[1] The five regular polyhedra shown in figure 19.3.2 are often called the **Platonic polyhedra** in honor of Plato, who wrote about them in his dialogue *Timaeus*.[2] Ancient Greeks associated regular polyhedra with what they took to be the four basic elements, the tetrahedron with fire, the octahedron with air, the cube with earth, and the icosahedron with water. They associated the dodecaheron with the whole universe.

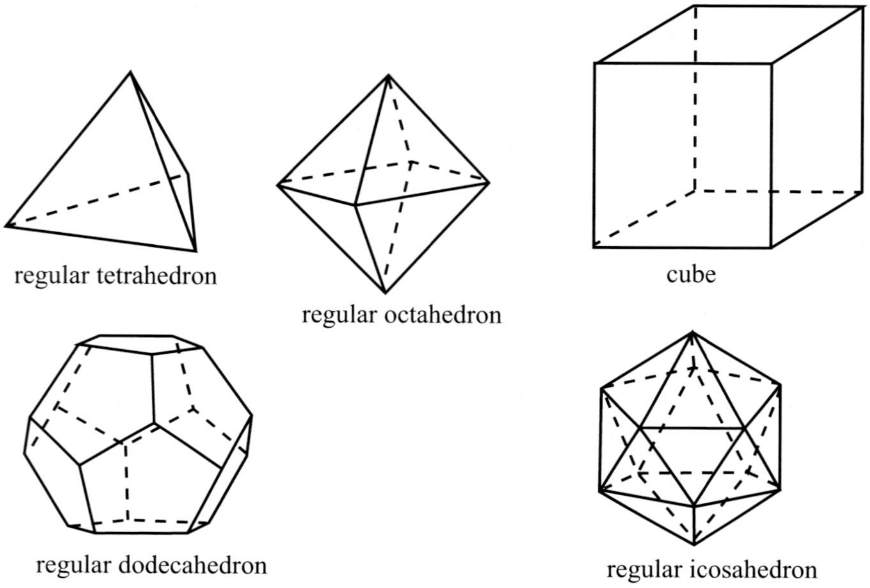

Figure 19.3.2 The five Platonic polyhedra

Except for the cube, the regular polyhedra all are named for the number of their faces. The Greek prefix "tetra" means 4, "octa" means 8, "dodeca" means 12 (literally 2 and 10), and "icosa" means 20. Of the five Platonic polyhedra, the regular tetrahedron is the easiest to make. In an equilateral triangle, connect the midpoints of the sides with lines. Then cut out the figure, fold on these lines (figure 19.3.3), and tape the corresponding edges together. The regular tetrahedron is sometimes used to package liquids, such as cream. It is also the crystal shape of the mineral called, appropriately, tetrahedrite.

1 See note, section 14.2.

2 The regular polyhedra seem to have been discovered independently in more than one ancient society. Hundreds of carved stone balls, estimated to be some 4,000 years old, have been found in Scotland, and many of them have edges which correspond to Platonic polyhedra. See Marshall, Dorothy N. *"Carved Stone Balls,"* ***Proceedings of the Society of Antiquaries of Scotland*** 180, p40-72, 1976-77.

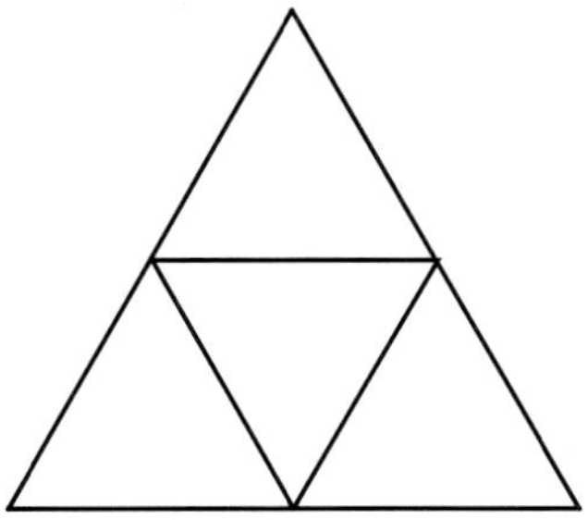

Figure 19.3.3 A net pattern for a regular tetrahedron[3]

The cube, most familiar of the regular polyhedra, is also easy to make. Begin with a pattern of squares like that shown in figure 19.3.4. Fold them along the lines to make your cube, and tape the corresponding edges together. Among the many minerals which typically form cubic crystals are table salt and iron pyrite (fool's gold.)

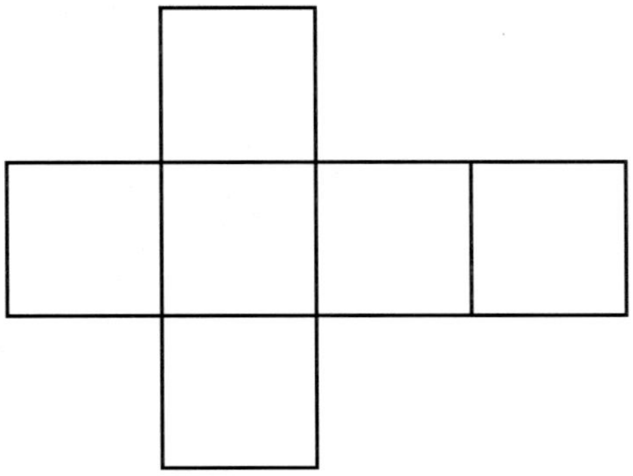

Figure 19.3.4 A net pattern for a cube

Figures 19.3.5 and 19.3.6 show net patterns for the regular octahedron and icosahedron. Diamond and magnetite (a magnetic iron ore used in early compasses) are among the many minerals which crystallize as a regular octahedron. There are no minerals at all which crystallize as regular icosahedra or dodecahedra, but pyrite crystals sometimes take the form of a slightly irregular dodecahedron called a pyritohedron.

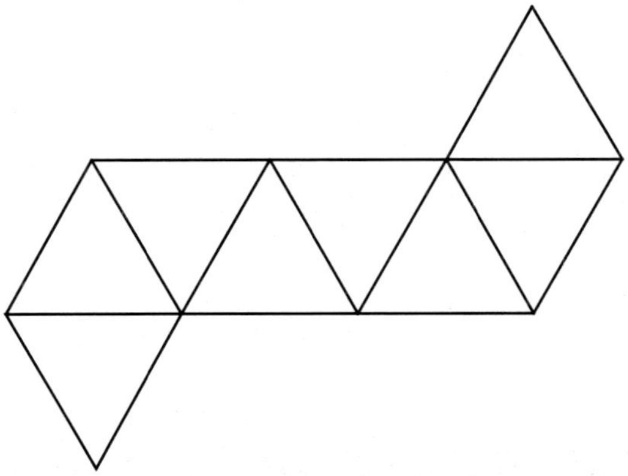

Figure 19.3.5 A net pattern for a regular octahedron

[3] The net for polyhedra was invented by Albrecht Dürer (1471-1528), who is best known today for his graphic art.

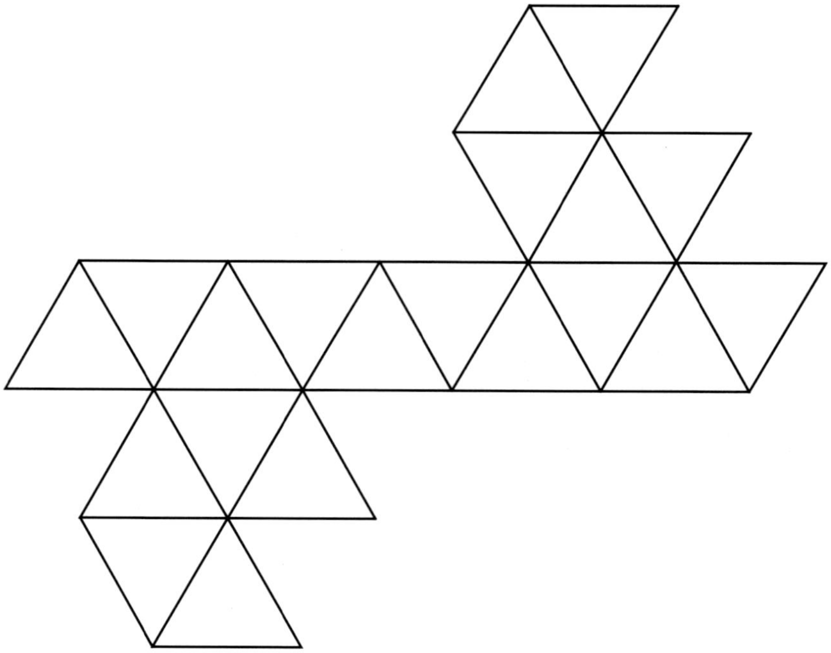

Figure 19.3.6 A net pattern for a regular icosahedron

To make a regular dodecahedron, you can use a pattern much like those shown for the other four regular polyhedra. However, here is a construction which is more fun and will prove helpful in making other polyhedra.

1. Make two identical, large pentagons, perhaps 4 inches on a side, from cardboard which is stiff enough not to bend too easily, yet not too thick. (The cardboard backs of writing tablets are often good for this purpose.)
2. Draw the diagonals of your pentagons, thus forming a 5-point star or "pentagram," as in figure 19.3.7, in each.

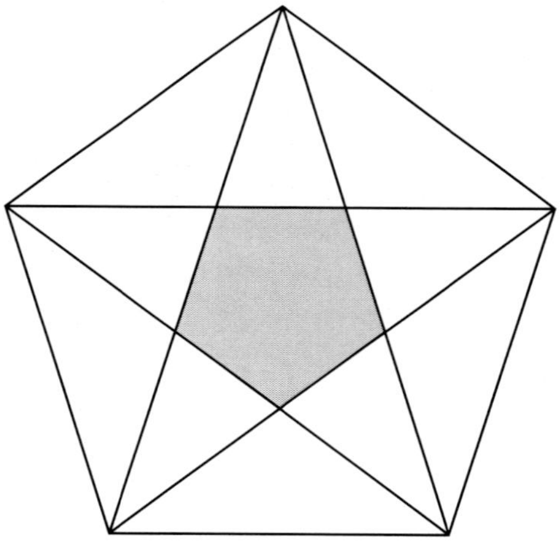

Figure 19.3.7

3. Notice the small regular pentagon formed in the middle of figure 19.3.7 (shaded). Draw its diagonals, and extend them all the way to the edges of the large original pentagon, as in figure 19.3.8.

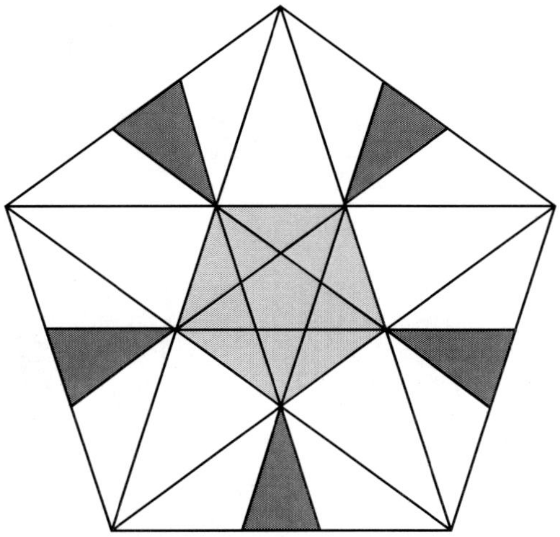

Figure 19.3.8

4. Cut out the small triangles, which are shaded dark in figure 19.3.8. The result is a shape like 19.3.9. Remember, you need **two** of these, identical in size and precisely made.

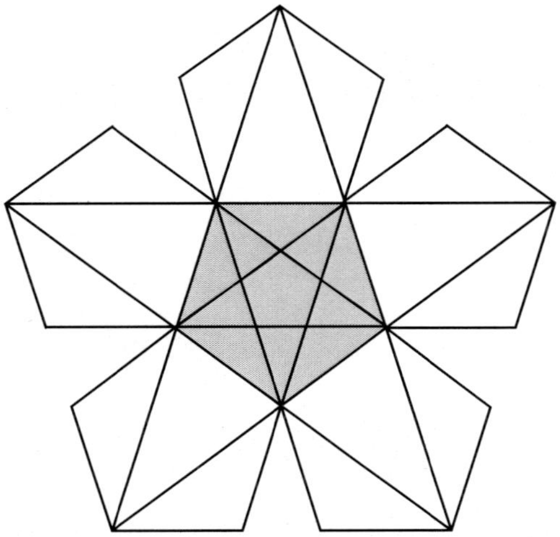

Figure 19.3.9

5. Turn one of the two shapes upside down as in figure 19.3.10.

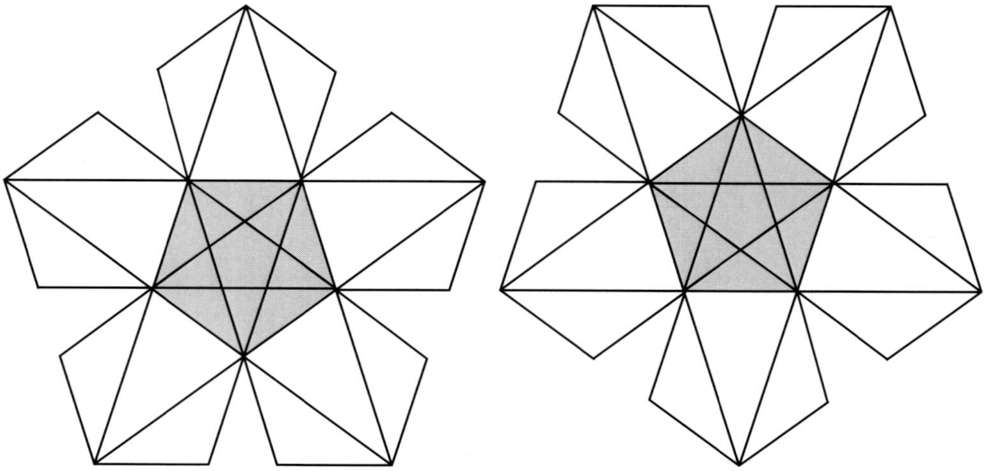

Figure 19.3.10

6. With a knife held against a straight edge, such as a metal ruler, score (cut part way through) the outline of each shaded inner pentagon, so that the cardboard folds neatly along those lines.

7. Slide one piece over the other, to produce figure 19.3.11.

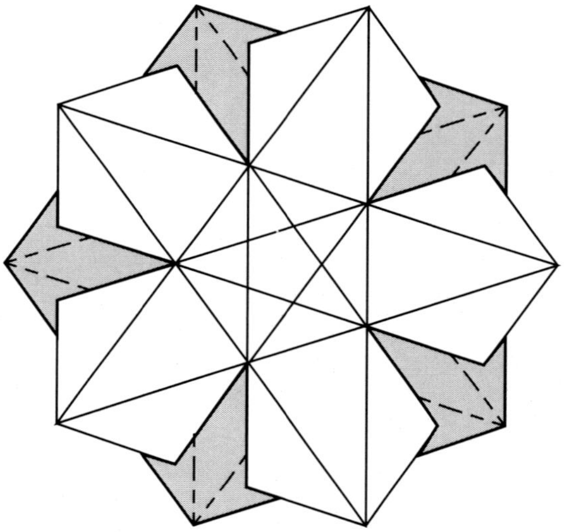

Figure 19.3.11

8. Weave a rubber band as shown in heavy black in figure 19.3.12.

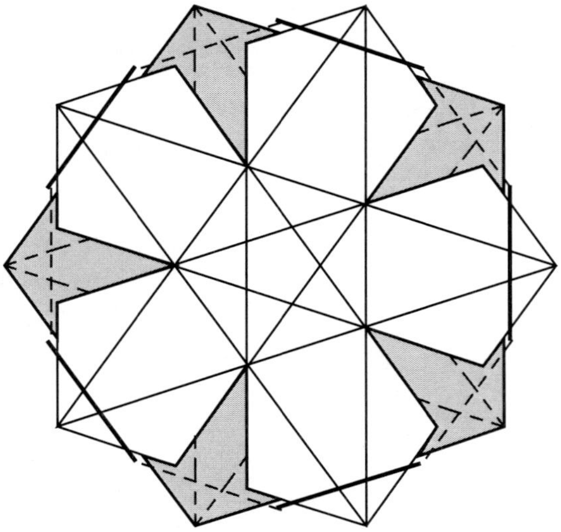

Figure 19.3.12 (Rubber band dotted where it goes behind)

9. Let go, and the figure will spring into shape as a regular dodecahedron (figure 19.3.13).

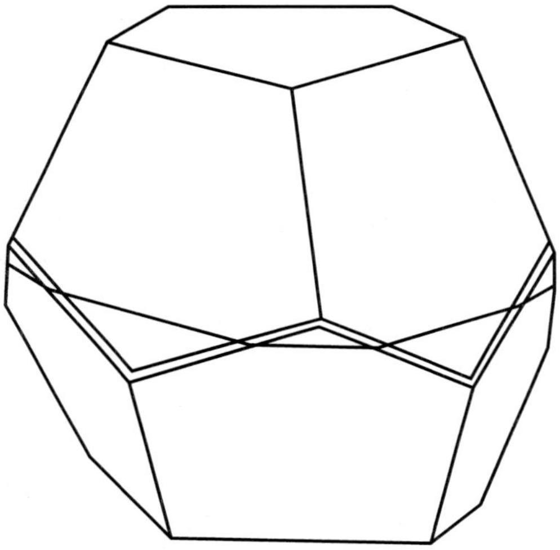

Figure 19.3.13 A "popup" dodecahedron

A word of caution. The popup dodecahedron works well only if it is built precisely, and if the rubber band is neither so strong that it constricts the figure nor so loose that it fails to spring the figure up into position. You may have to experiment with various types of cardboard, rubber bands, and sizes of models until you get a combination that works well.

Making polyhedron models, and popup dodecahedra in particular, is a nice activity for grades 5 and above, when the children are developing the necessary eye-hand coordination to make models that really look nice. (There is considerable variability at all ages in such coordination. Therefore, it is best to praise and display student work in this area rather than assign it a grade. Often students who are having trouble making their shapes look good are all too aware of how their work looks, and having a teacher say the obvious and unflattering truth contributes nothing but embarrassment.) Building polyhedron models is one of the few opportunities for craftsmanship in school mathematics.

A good use of the popup dodecahedron is as a greeting card, especially in the December holiday season. The students can put designs and greetings on the dodecahedra, then send them, possibly with a rubber band already woven and ready to pop up when a tack or pin is removed. After making a few of them, the students get quite good at it, and they can sell them to raise funds for class activities.

Exercises 19.3

For these exercises, you may find it convenient to refer to actual models of the shapes involved.

1. In the cube shown below, the vertices are numbered. If you connect the odd-numbered vertices with line segments, what figure is formed?

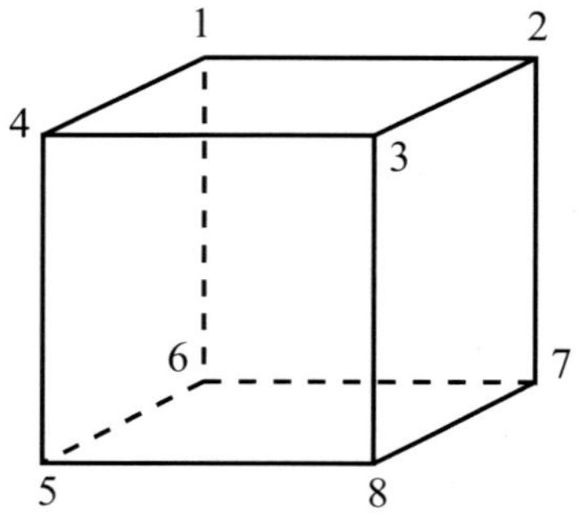

2. In the dodecahedron shown below, eight of the vertices are marked with heavy dots. Of what figure are these dots the vertices? To see this, it may help to use a dodecahedron instead of relying on the picture.

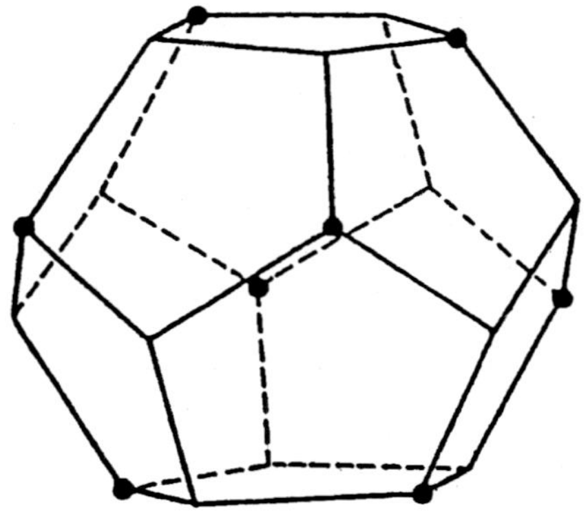

3. Show how to select four vertices of a regular dodecahedron so that these four vertices are also vertices of a regular tetrahedron. (Hint: Combine the results of questions 1 and 2).

4. This question concerns the centers of the faces of regular polyhedra. The following pictures show the centers of the faces involved.

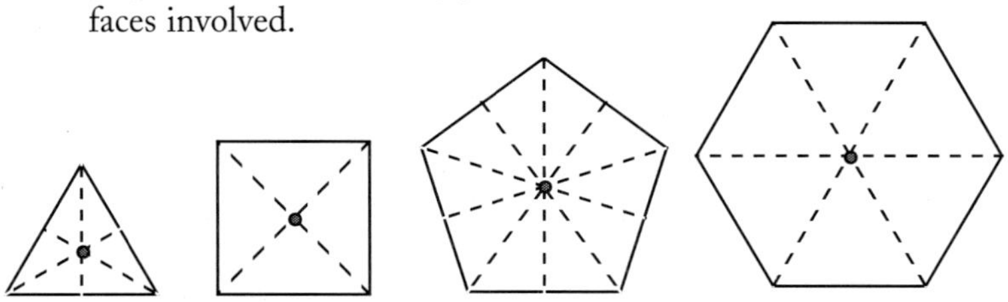

Describe the figure whose vertices are the midpoints of the faces of a:

   a. Regular tetrahedron
   b. Cube
   c. Regular octahedron
   d. Regular dodecahdron
   e. Regular icosahedron

5. Describe how to cut a cube so as to form a regular hexagon cross section.

6. What figure is formed by connecting the midpoints of the edges of a regular tetrahedron?

7. **Why there are only five regular polyhedra** In a tessellation, the angles at each vertex add up to 360°, but in a polyhedron the angles at each vertex add up to **less than** 360°. This fact, together with the fact that each vertex of a polyhedron is the junction of at least three faces, will show why there are only five regular polyhedra.

   a. The tetrahedron, octahedron, and icosahedron have, respectively, three, four, and five equilateral triangles meeting at each vertex. Why can't there be a regular polyhedron with six or more equilateral triangles at each vertex?
   b. The cube has three squares meeting at each vertex. Why can there be no regular polyhedron which has four or more squares meeting at each vertex?
   c. In a regular dodecahedron, three pentagons meet at each vertex. Why can there be no regular polyhedron which has four or more pentagons meeting at each vertex?

d. Why is it impossible to build a regular polyhedron from polygons with six or more sides?

e. Combine parts a-d to explain why there are only five regular polyhedra.

Exploration. Finding the volume of a regular tetrahedron and octahedron with edges of length *e*. (questions 8-12)

The most straightforward approach would be to use the formula for the volume of a pyramid, $V = Bh/3$, where B is the area of the base and h is the height. The base area B is not difficult to find, since it is just the area of an equilateral triangle, but the height is a bit harder to calculate, since you first need to find where the foot of the altitude meets the base.

In this problem, a somewhat indirect approach works better. Begin by putting the tetrahedron into a cube like this:

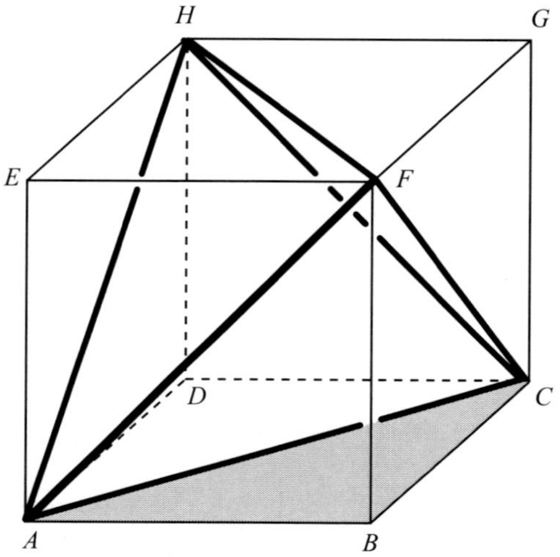

8. a. Find the volume of the cube, if its edge is a.

   b. Explain why the four tetrahedra *ABCF*, *CGHF*, *AEFH*, and *ADCH* are congruent. (These tetrahedra are not regular, but because they are congruent, they have equal volumes.)

   c. Express the volume of tetrahedron *ABCF* in terms of the edge of the cube.

   The formula $V = \frac{1}{3}Bh$ is convenient here.

   d. Express the volume of tetrahedron *ACFH* in terms of a.

9. Express a, the length of each edge of the cube, in terms of e, the length of an edge of tetrahedron *ACFH*.

10. Use your answers to questions 8 and 9 to express the volume of tetrahedron *ABCF* in terms of the length of its edge, *e*.

11. Find the volume of a regular octahedron by viewing it as two square pyramids attached at their bases. The height of each such pyramid is equal to half the diagonal of its square base.

12. (The idea for this question is from Philip Mallinson) Another way to find the volume of a regular octahedron is suggested by the picture below, which shows a regular tetrahedron, with a smaller regular tetrahedron at each corner and a regular octahedron inside. (It may help to make a physical model.) Since all regular tetrahedra are similar to each other, the large tetrahedron is similar to the four small ones at the corners.

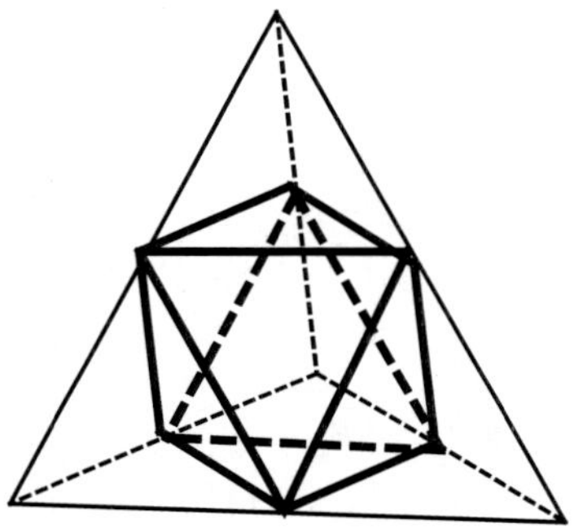

    a. What is the ratio of the volume of the large tetrahedron to the volume of each small tetrahedron?
    b. Use the result of part a and the fact that the volumes of the octahedron and four small tetrahedra together equal that of the large tetrahedron, to find the ratio of the volume of the octahedron to one small tetrahedron.
    c. Check the result of part b against that of question 11.

13. If you broaden your definitions of polygon and polyhedron slightly, four more beautiful shapes count as regular polyhedra. Specifically, if you allow the sides of a polygon to cross over each other, then star pentagon shown below counts as a regular polygon.

If you also allow the faces of a polyhedron to cut through each other, then the four polyhedra shown below qualify as regular polyhedra.[4]

Great dodecahedron   Great icosahedron   Great stellated dodecahedron   Small stellated dodecahedron

Figure 19.3.14

The following steps will help you make the pieces for the great dodecahedron and either the great stellated dodecahedron or the small stellated dodecahedron.

a. Make 12 identical copies of figure 19.3.7, two in each of 6 colors. (For this problem, heavy-weight paper works even better than cardboard.)

b. From the results of part a, cut out the obtuse-angled triangles at each edge. (Save the pentagrams to use in part c.) You now have 60 congruent obtuse-angled triangles. Tape these together to make a great dodecahedron. Special challenge: Arrange the colors so that each pentagonal face is a single color, as pictured above.

c. Cut the triangular corners off at each pentagram left from part b. Tape these 60 triangles together to make either a small stellated dodecahedron or a great stellated dodecahedron. (You have enough pieces to make either one but not both.) Special challenge: Arrange the colors so that each pentagram shows as a single color, as shown in figure 19.3.14.

[4] These shapes were studied extensively by Kepler (see note, exercises 9.2) in his book, *Harmonices Mundi*, (Linz, 1619). Nobody knows when these shapes were first made or discovered. A marble picture of a small stellated dodecahedron, dating to 1425-1427, is in the Basilica of San Marco, Venice.

19.4 Semi-regular Polyhedra

A particularly interesting and appealing class of polyhedra are those which are called **semi-regular**. Like semi-regular tessellations, a semi-regular polyhedron is built with two or more kinds of regular polygons, all with edges of the same length, and with identical sets of polygons meeting in the same order at every vertex. (Every vertex has the same vertex symbol.) Thirteen beautiful semi-regular polyhedra were studied by Archimedes and are still called **Archimedean polyhedra**.

An enjoyable way to explore the semi-regular polyhedra is to actually build models of them, and the main goal of this section is to help you do that. Building polyhedra teaches geometry "through the eye and the hand," with relatively little conscious effort. Each semi-regular polyhedron pictured below is listed with the information needed to build it. Specifically, for each polyhedron the vertex symbol is given, as well as the name and the number of each kind of polygon needed. The names, incidentally, are of secondary importance, but they strike many beginners as so long and whimsical as to be fun in their own right.

Practical tips for building polyhedra.

1. **Build the simplest polyhedra (those with fewest faces) first. Then work your way up to more difficult ones.**
2. **Make all the polygonal faces before you begin to put them together.**
3. **Take advantage of "mass production" techniques. For example, make your equilateral triangles six at a time by drawing the diagonals connecting opposite vertices of a regular hexagon (figure 19.4.1)**

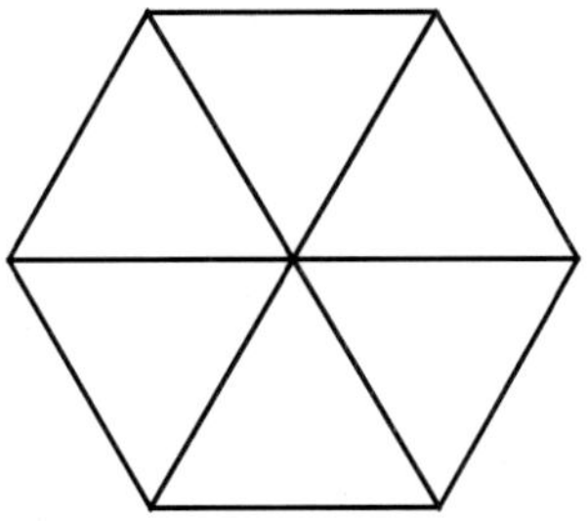

Figure 19.4.1 A regular hexagon makes 6 equilateral triangles.

Again, regular pentagons may be made six at a time as in figure 19.3.9. This is particularly helpful, because, as you will see, all semi-regular polyhedra which have pentagonal sides (many do not) have 12 of them. If you make your pentagons six at a time this way, save the small triangles, shaded in figure 19.3.8, for possible later use in making the small stellated dodecahedron and great stellated dodecahedron (figure 19.3.14). These spectacular figures call for patience and skill in construction.

Once you have made the pieces, assemble a single vertex, using the picture in the text and the vertex symbol. Then add new faces so as to make each new vertex have the proper vertex symbol.

To illustrate, suppose you decide to start off making a truncated tetrahedron, (figure 19.4.3). This is a good first choice, because it requires only four hexagons and four triangles.

Begin by making five identical regular hexagons, one of which will be cut up for triangles. Making each edge 1" long usually results in a convenient size. Ideally, use one color paper for four of your hexagons and a second color for the fifth, which you then mark off into six triangles as in figure 19.4.1. Next, cut out four of the triangles and the four other hexagons.

Now put the pieces together, using figure 19.4.3 and the vertex symbol (3, 6, 6) as your guide. Specifically, make the pattern in figure 19.4.2 by taping together two hexagons and a triangle. Now tape edge *AC* to edge *BC,* and you will form, at *C*, a three-dimensional vertex with symbol (3, 6, 6). From here on you will have no trouble seeing where to put the remaining polygon faces.

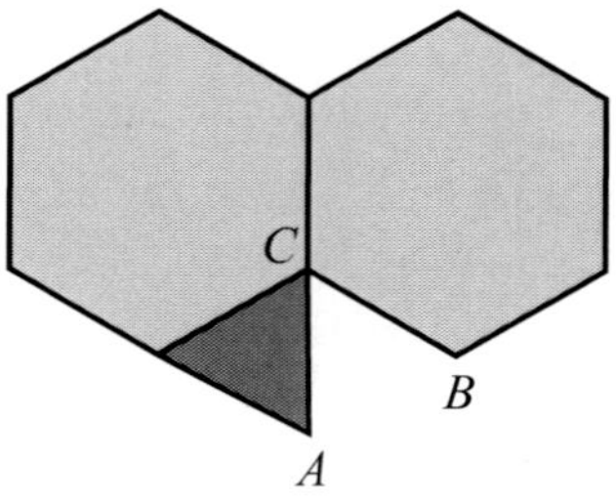

Figure 19.4.2

After you have built the model, look it over carefully. You will understand the name "truncated tetrahedron"—this is a tetrahedron with the corners cut off.

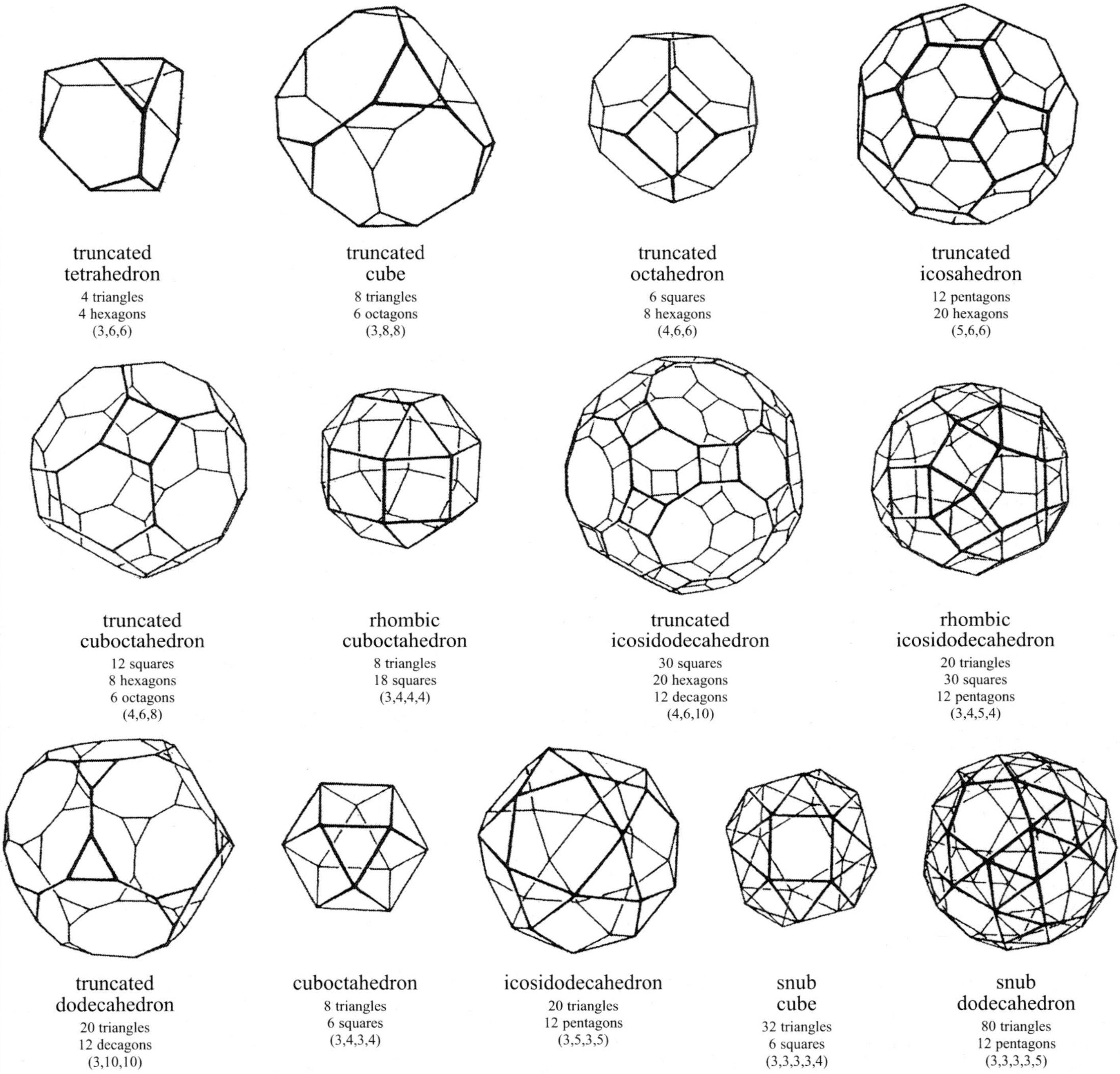

Figure 19.4.3 The Archimedean Polyhedra

Having once made the above polyhedra, you will begin to notice certain relationships among them. For example, if you increase the truncations on either the truncated cube or the truncated octahedron, in the spirit of figures 19.2.4 and 19.2.5, you get, in either case, the cuboctahedron. It gets its name from somehow being midway between the cube and the octahedron. Similarly, increasing the truncations on either the truncated dodecahedron or the truncated icosahedron produces the icosidodecahedron, which is midway between the icosahedron and the dodecahedron.

If you truncate a polyhedron or a tessellation which is not regular, you produce faces which are not regular polygons. For example, figure 19.4.4 shows that if you truncate the semi-regular tessellation (4, 8, 8) you will form, in addition to regular octagons like A and squares like B, triangles like those shaded, which are not equilateral.

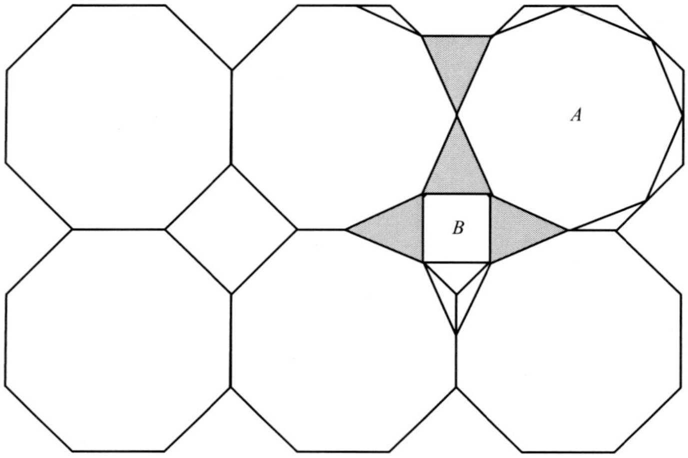

Figure 19.4.4 Truncating (4, 8, 8)

Two other families of semi-regular polyhedra are worth noting. These are the **prisms** (figure 19.4.5) and the **alternating prisms** (figure 19.4.6).

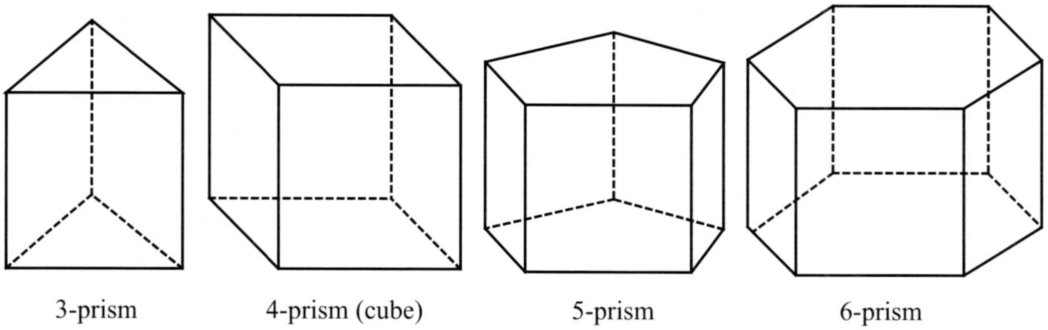

Figure 19.4.5

A regular octahedron is an alternating 3-prism. Figure 19.4.6 shows an alternating 5-prism and an alternating 6-prism. (Alternating prisms are sometimes called **antiprisms**.)

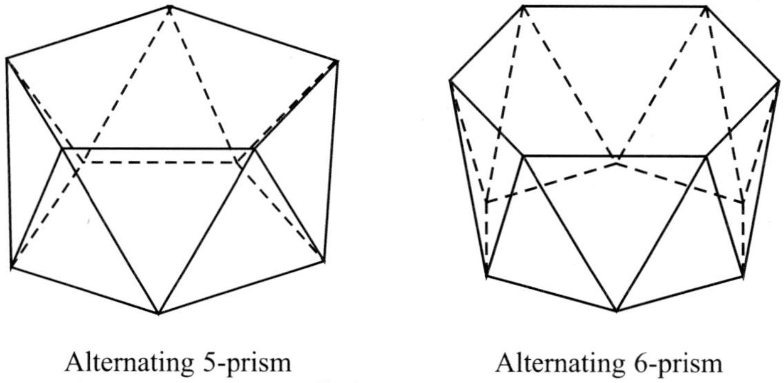

Figure 19.4.6

It was long believed that the Archimedean polyhedra, prisms and alternating prisms were the only semi-regular polyhedra, but another, based on rotating one "cap" (consisting of 5 squares and 4 triangles) of the rhombic cuboctahedron by 45° was noticed about a century ago.[1]

Making advanced polyhedra takes special care, both in making the pieces and in putting them together. More information about these and other polyhedron models can be found in many books. Of these, are Cundy and Rollett, *Mathematical Models*, 2nd edition, Oxford U. Press, 1961 and Wenninger, *Polyhedron Models*, Cambridge U. Press, 1971, are particularly good sources.

[1] The earliest reference seems to be Sommerville, D.M.Y., "*Semi-regular Networks of the Plane in Absolute Geometry*", ***Transactions of the Royal Society*** (Edinburgh), 41 (1905) 725-47 and plates I-XII.

Vocabulary for Chapter 19

Alternating prism (19.4)
Archimedean polyhedra (19.4)
Dodecahedron (19.3)
Edge (19.1)
Equilateral triangle (19.1)
Face (19.3)
Icosahedron (19.3)
Net for a polyhedron (19.3)
Platonic polyhedra (19.3)
Polyhedron (19.3)
Prism (19.4)
Regular polygon (19.1)
Regular polyhedron (19.3)
Regular tessellation (19.2)
Semi-regular polyhedron (19.4)
Semi-regular tessellation (19.2)
Tessellation (19.2)
Truncate (19.2)
Vertex (19.1)
Vertex symbol (19.2)

Self-Quiz for Chapter 19

1. Without measuring, find the sum of the interior angles of this polygon.

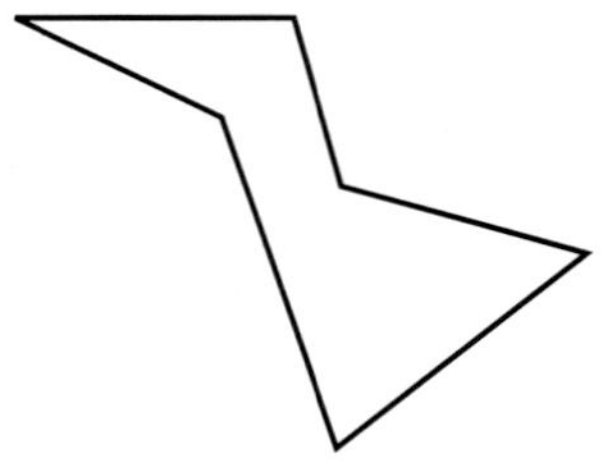

2. What is the interior angle of a regular dodecagon (12-gon)?

3. With a ruler and protractor, construct a regular octagon. Make each side 2 inches long.

4. Sketch a semi-regular tessellation with vertex symbol (3, 3, 3, 4, 4).

5. What is the vertex symbol of:
   a. A semi-regular 8-prism?
   b. An alternating 7-prism?

6. Find the volume of a regular tetrahedron inscribed in a cylinder whose base is a circle with radius 10.

Solid Rock (Polyhedron Song)

by Bob Stein
To the tune of *Get You a Copper Kettle*

Get you some colored cardboard, get you a role of tape
Put 'em together and what have you got, a polyhedral shape!

Chorus: You just sit there in the classroom, work at home at night
Watch them solids shapin' up in cardboard colors bright

Get you some regular pentagons, equilateral triangles, too
Don't use no bad construction jobs, they take up too much glue.

Chorus

Our class, we made solids. The others, they did, too
We ain't done no 'rithmetic, since 1962.

Chorus

Get You a Copper Kettle

Get You a Copper Kettle, a song about making moonshine, may be found on the web.

Answers and Hints to Selected Questions

Exercises 1.1

1. a. commutative b. associative

3. Making up your own questions and evaluating your own work is preferable to doing exercises someone else has made up. It enhances your control of your own education and your responsibility for it.

5. If you sense a drop in interest on the part of the students, it may well be time to move to another topic. Move on before a topic goes stale. Students will enjoy revisiting it at a later date.

Exercises 1.2

1. To prove a general statement such as the commutative or associative property, you would need to show it is true in all cases. Showing it true in a particular case does not guarantee it to be true in all cases.

6. Hint: What digit can M be? The M in the bottom row comes from carrying, but if you add two digits, you never carry a very large number. What are the possibilities? Once M is determined, make S large enough so that you have to carry when you add M and S. Then what can O be? (O cannot be the same as M).

Exercises 1.3

1. a. 2×4 (4×2 is also correct) d. 12×4 (or 4×12)

5. Only very rarely does a big league baseball team go for an entire season without scoring at least one run in every game. If they fail to score in even one game, what will Gullible Gulliver's number be for the season?

10. b. Partial solution: You could add $100 + 200 + 300 + \ldots + 36{,}500$ directly, using Gauss' idea. Another way: Add $1 + 2 + 3 + \ldots + 365$, then multiply the sum by 100 to get 6,679,500.

13. You could add all the odd numbers from 1 through 999 by Gauss' method, then separately add all the numbers from 5 through 995 which end in 5 and subtract the second total from the first. Alternatively, you could just use Gauss' method directly as in question 11b.

Exercises 1.4

1. b. Hint: Reverse the order of the factors. Multiply 351×82.

3. Pierre's cookies cost $3 \times 40¢$, and Alicia's cost $5 \times 40¢$ for a total of \$1.20 + \$2.00. The aunt sees it as buying 8 cookies at 40¢ each or $(3 + 5)40¢$.

5. a. $6 \times 18 = 6 \times 10 + 6 \times 8$.

9. The multiplication is correct: $65 \times 23 = 5 \times 23 + 60 \times 23 = 115 + 1{,}380 = 1{,}495$. This way of writing long multiplication is widely used in the tropical Americas.

Self-Quiz for Chapter 1

1.

27	93	120
49	74	123
76	167	243

2.

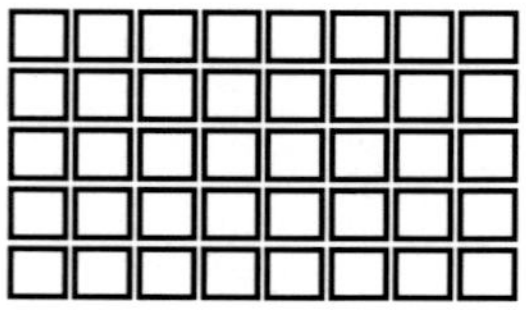

3.

1	2	3	4
500	499	498	497
501	501	501	501

498	499	500	250×501
3	2	1	250×501
501	501	501	500×501

$1 + 2 + 3 + 4 + \ldots + 499 + 500 = 250 \times 501 = 125{,}250$

4a.

	3	7	10
8	24	56	80
6	18	42	60
14	42	98	140

b.

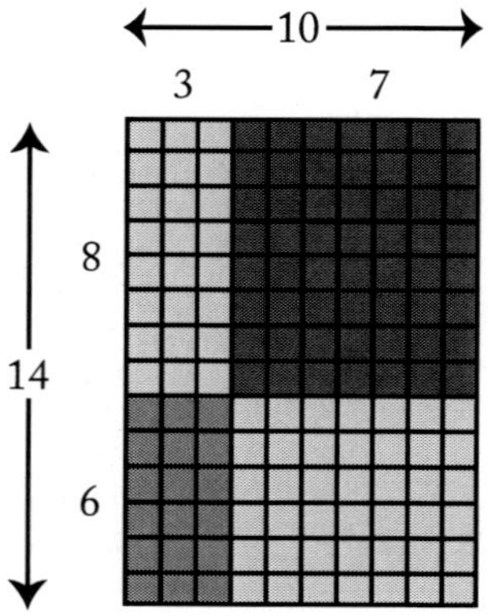

5.

	30	6	36
4	120	24	144
80	2,400	480	2,880
84	2,520	504	3,024

6. Lattice Multiplication

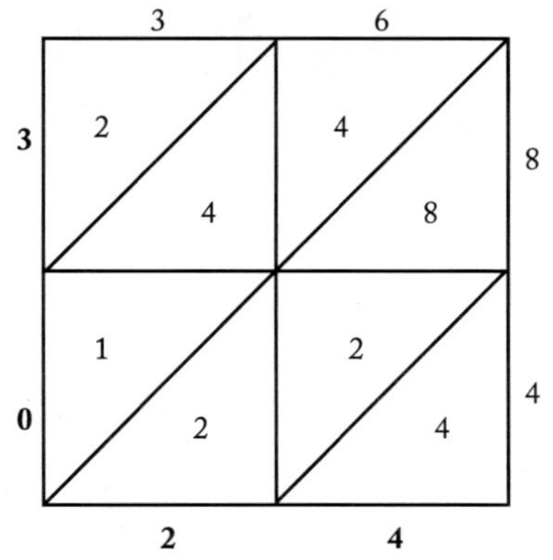

Repeated Doubling

$1 \times 84 = 84$

$2 \times 84 = 168$

$4 \times 84 = 336$

$8 \times 84 = 672$

$16 \times 84 = 1{,}344$

$32 \times 84 = 2{,}688$

$$36 \times 84 = 4 \times 84 + 32 \times 84$$
$$= 336 + 2{,}688$$
$$= 3{,}024$$

Exercises 2.1

1. a. We lost money last month.
c. The airplane is descending 500 feet per minute.
e. The rocket is leaving Earth at 3 miles per second.
g. The quarterback lost 7 yards.

Exercises 2.2

1. a. $6 + (-8) = -2$ c. $-6 + (-2) = -8$

2. a.

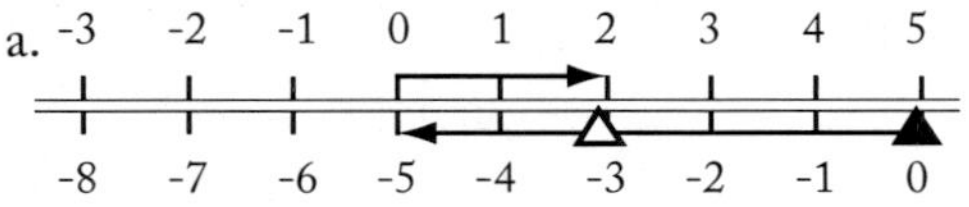

3. a. The picture below shows $-4 + 8 = 4$.

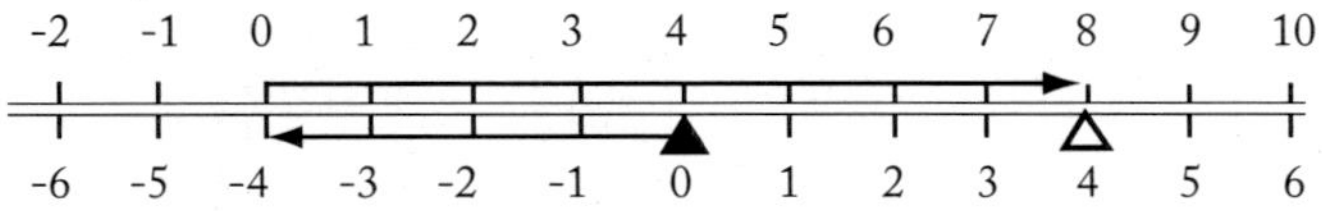

The picture below shows $8 + (-4) = 4$.

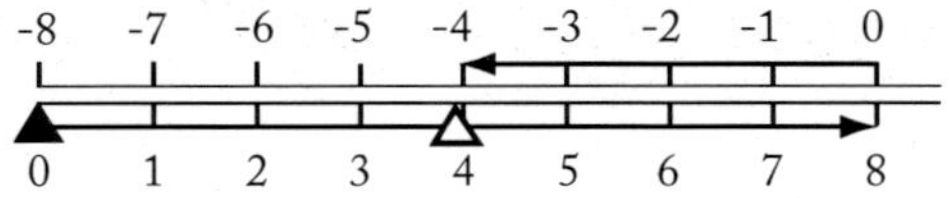

b. This illustrates the commutative property of addition.

4. b.

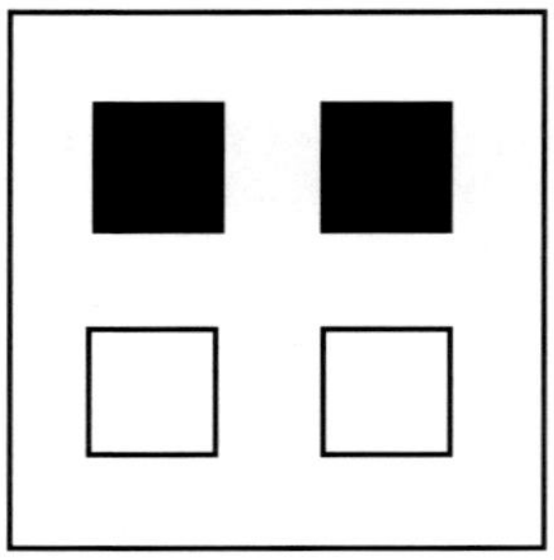

6 positive (black), 2 negative

Zero pairs in square

$$\begin{aligned} 6 + (-2) &= (4 + 2) + (-2) \\ &= 4 + [2 + (-2)] \\ &= 4 + 0 \\ &= 4 \end{aligned}$$

8. Hint: Draw a line from dot 5 on the left to dot 2 on the right. Where does this line cross the middle column? Experiment further; connect various dots on the outside with straight lines and observe where these lines cross the middle column.

10. A few questions one class generated are:
- Does the adder work for larger numbers?
- Does it work with fractions?
- Does it work with negative numbers?
- Why does it work (assuming it does work)?

11. a.

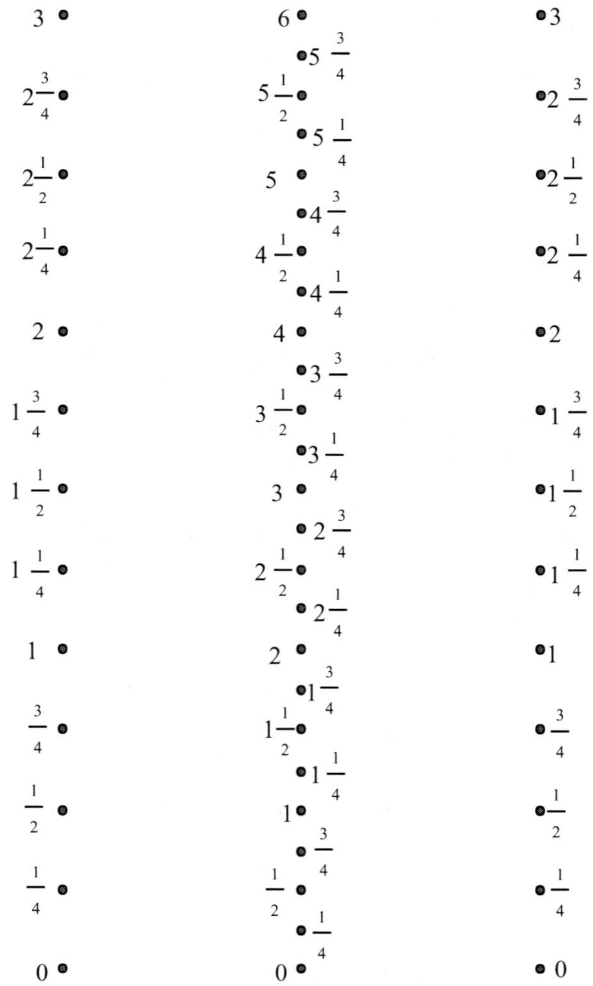

12.

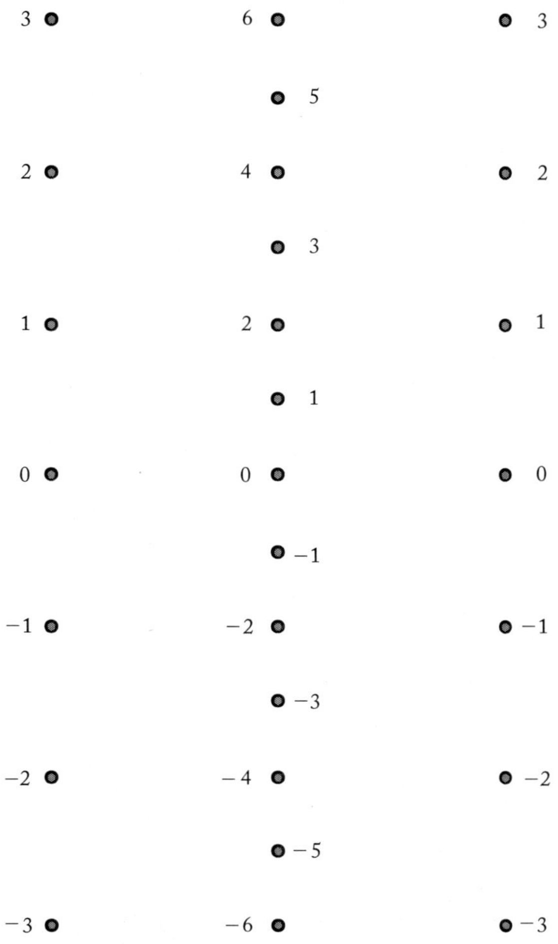

14. a. -39 c. −47 e. 1 g. -931 i. -58 k. -101

15. a.

-9	13	4
7	-6	1
-2	7	5

b.

20	-11	9
-31	18	-13
-11	7	-4

17. Hint: Write the numbers from -10 to 10 in the top row of a box puzzle, and then write the same numbers backwards in the second row.

Exercises 2.3

1. The product in part c is negative, the others positive.

2. c.

	-11	5	-6
-8	88	-40	48
-2	22	-10	12
-10	110	-50	60

4. d. $-(s + t) = -s + (-t)$

Self-Quiz for Chapter 2.

1. a.

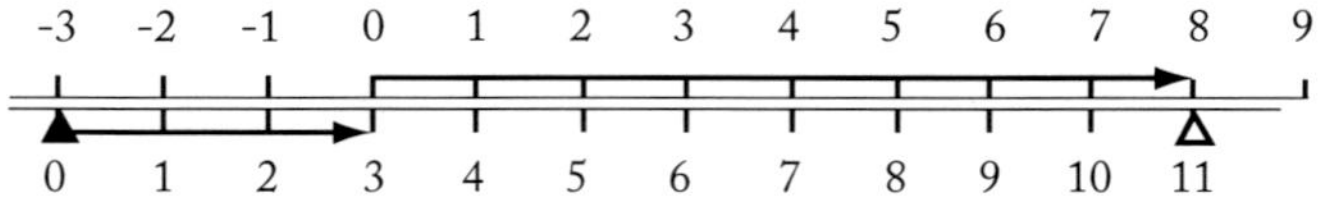

b.

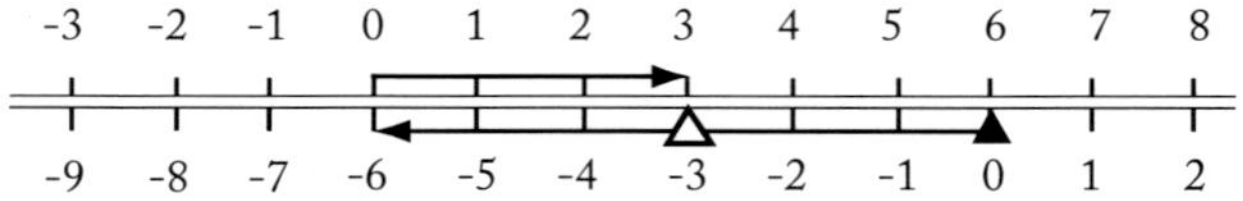

2. a. 23. To picture this, consider an election with 32 negative votes and 55 positive votes, the positives win by 23 votes.

b. -29. To picture this, consider an election with 121 positive votes and 150 negative votes. The negatives win by 29 votes.

c. -115. There is no contest. The negative votes all reinforce each other instead of offsetting positive votes.

3.

-27	93	66
49	-74	-25
22	19	41

4. a. $-7(4) = (-7) + (-7) + (-7) + (-7) = -28$

b. One way to see this: $-3 \times 0 = 0$, since anything times 0 is 0. Now rewrite 0 as $-2 + 2$ to see that $-3 \times (-2 + 2) = 0$. Now apply the distributive property, and you have $-3 \times (-2) + (-3) \times (2) = 0$. But if you already know that $-3 \times (2) = -6$, then $-3 \times (-2) + (-6) = 0$. Evidently, $-3 \times (-2)$ is a number which can be added to -6 to get 0, and that number is (positive) 6.

5.

	-4	-9	-13
12	-48	-108	-156
-7	28	63	91
5	-20	-45	-65

Exercises 3.1

1. b. Hint: What family of numbers has just been introduced?
c. Hint: Count each triangular number by horizontal rows.

2. b. Hint: This question is related to the other questions in this section.

3. a.
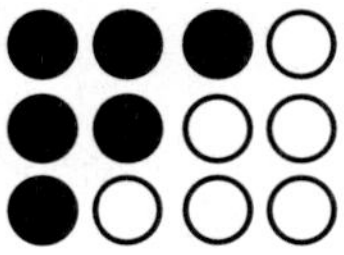

d. Yes. It can be pictured as half of an $n \times (n + 1)$ array.

4. b. Hint: This is analogous to question 1. Count the dots of each shade.

5. Square numbers.

10. b. Hint: Try squaring 11, 21, and 31 by long multiplication. Do you see why the last digits of their squares are all 1s? Repeat this experiment with 12, 22, 32, then with 13, 23, and 33. Soon you will see what is going on.

11. b. Hint: Draw lines like those given in question 10 connecting the units digits of the first and last entries in your list, and so on.

12. a. 1
c. 121
e. Hint: These are some of the odd square numbers. Just which odd square numbers you get is a pattern hunt all its own.

13. a. 21
c. 78
f. Hint: The white dots in figure 3.1.7 are three times a triangular number. Six times another triangular number are the black dots.

14. a. 21

15. b. 3
d. 10

16. Partial answer: For 6, 10, 15, multiply 6×15 and add 10, to get 100, which is the square of 10.

Exercises 3.2

2. a. Hint: See section 3.1.

c. Hint: How many gumdrops are on each level of a tetrahedral number?

4. Hint: Use the list of the first twenty cubes generated in question 3.

7. b. Hint: The sums are a family of three-dimensional figurate numbers, but they do not include the smallest numbers in that family.

10. b. The floors or levels of a tetrahedral number are triangular numbers (see question 2 above). But in section 3.1, question 7 you saw that adding two consecutive triangular numbers always yields a square. In this picture, then, the floors are the square numbers, starting with 1 at the top.

d. Now concentrate on the colors; the white balls are a tetrahedral number, as are the black balls. These are consecutive tetrahedral numbers.

11. b. The numbers being added are squares, alternately the even ones or the odd ones. If you consider a tetrahedral number and count the floors in pairs, you will understand the pattern geometrically.

12. b. Tetrahedral numbers

c. In the list of tetrahedral numbers, circle 1, skip 4, circle 10, skip 20, etc. The alternate tetrahedral numbers circled are those found in part a.

13. b. Two consecutive tetrahedral numbers

14. a. Again you get alternate tetrahedral numbers, but not beginning at 1.

15. b. Partial answer: $1 \times 6 + 2 \times 5 + 3 \times 4 + 4 \times 3 + 5 \times 2 + 6 \times 1 = 56$

Self-Quiz for Chapter 3

1.

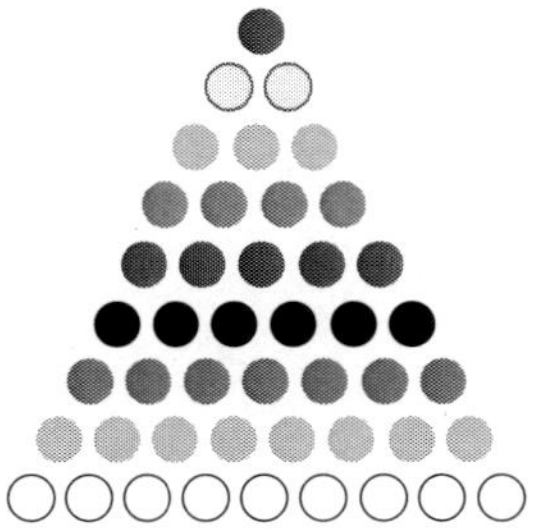

2.

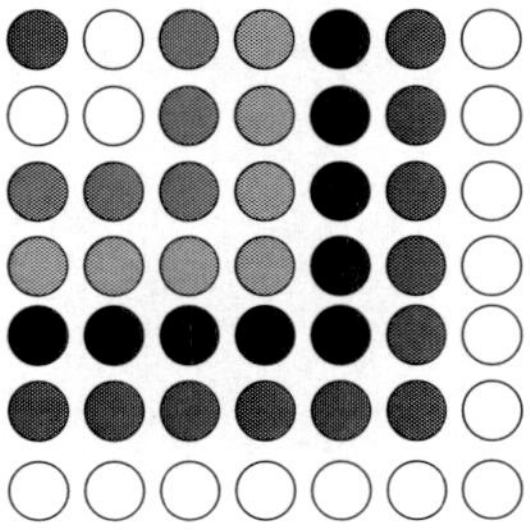

3., 4. Solutions are models and are not included here.

5. (Illustration is for the triangular numbers 6 and 10, but a similar picture can be drawn for any pair of consecutive triangular numbers.)

Begin with consecutive triangular numbers.

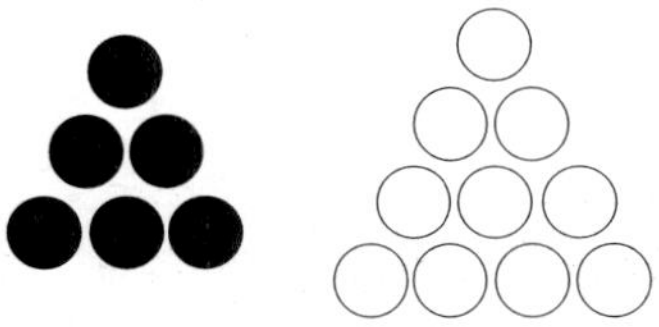

Rearrange the balls to fit in a square corner.

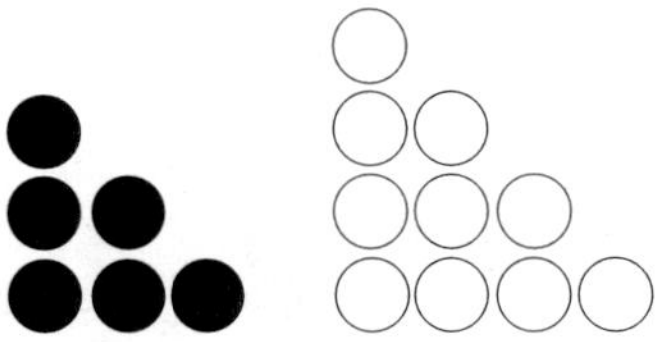

Put one of them upside down next to the other to form a square.

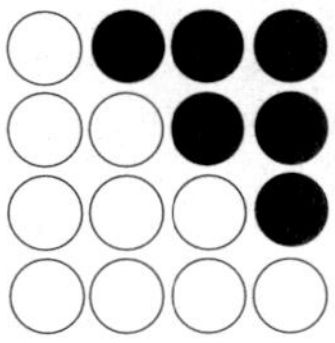

6. A square cannot end in 2, 3, 7, or 8 (see the diagram for Exercises 3.1, #10).

Exercises 4.1

2. b. The heavy lines show the pattern.

1	2	3	4	5	6	7	8	9	10
11	12	13	14	15	16	17	18	19	20
21	22	23	24	25	26	27	28	29	30
31	32	33	34	35	36	37	38	39	40
41	42	43	44	45	46	47	48	49	50
51	52	53	54	55	56	57	58	59	60
61	62	63	64	65	66	67	68	69	70
71	72	73	74	75	76	77	78	79	80
81	82	83	84	85	86	87	88	89	90
91	92	93	94	95	96	97	98	99	100

3. Multiples of 4 are shown

1	2	3	4	5	6	7	8	9	10
11	12	13	14	15	16	17	18	19	20
21	22	23	24	25	26	27	28	29	30
31	32	33	34	35	36	37	38	39	40
41	42	43	44	45	46	47	48	49	50
51	52	53	54	55	56	57	58	59	60
61	62	63	64	65	66	67	68	69	70
71	72	73	74	75	76	77	78	79	80
81	82	83	84	85	86	87	88	89	90
91	92	93	94	95	96	97	98	99	100

5. c. These numbers are all even-i.e. they are themselves multiples of 2.

6. a. They are multiples of 6
c. Multiples of 15
e. Multiples of 20
g. Multiples of 12

9. a. The square numbers are on the diagonal through 1, shown in bold type.

7	14	21	28	35	42	**49**
6	12	18	24	30	**36**	42
5	10	15	20	**25**	30	35
4	8	12	**16**	20	24	28
3	6	**9**	12	15	18	21
2	**4**	6	9	10	12	14
1	2	3	4	5	6	7

b. Starting with a square on the main diagonal (say $a \cdot a$), the number one to the right and one down is one less, or $a \cdot a - 1$.

d. $(a - 1)(a + 1) = a \cdot a - 1$

10. d. Analogous to question 9 d.

e. $68 \times 72 = (70-2)(70+2) = 70^2 - 2^2 = 4{,}900 - 4 = 4{,}896$

11. a. Starting with a square on the main diagonal (say $a \cdot a$), the number three to the right and three down is nine less, or $a \cdot a - 9$.

c. $17 \cdot 23 = (20 - 3)(20 + 3) = 20 \cdot 20 - 9 = 400 - 9 = 391$

e. $16 \cdot 22 = 361 - 9 = 352$

13. a. $(a - b)(a + b) = a \cdot a - b \cdot b$

18. c. These are squares of triangular numbers.

Exercises 4.2

1. If x is a factor of y, then y is a multiple of x, which means there is a whole number w for which $xw = y$. This not at all the same as saying that x is a multiple of y, which means there is a whole number z for which $x = yz$.

3. b. Hint: Note the link between the number of rectangles and whether or not n is prime.

4. a. $1 \cdot 60$ (The trivial factorization), $2 \cdot 30$, $3 \cdot 20$, $4 \cdot 15$, $5 \cdot 12$, $6 \cdot 10$

b. $2 \cdot 30 = 2(2 \cdot 3 \cdot 5)$

$3 \cdot 20 = 3(2 \cdot 2 \cdot 5)$

$4 \cdot 15 = (2 \cdot 2)(3 \cdot 5)$

$5 \cdot 12 = 5(2 \cdot 2 \cdot 3)$

$6 \cdot 10 = (2 \cdot 3)(2 \cdot 5)$

c. In every case there are two factors of 2, one of 3, and one of 5.

6. a. Prime

c. $3 \cdot 11$

e. $3 \cdot 17$

g. $3 \cdot 29$

i. $7 \cdot 13$

k. Prime

n. $7 \cdot 17$

r. Hint: Use divisibility tests

7. a. 4 times

c. 6 times

e. Hint: They are squares (e.g. 9 appears as $9 \cdot 1$, $3 \cdot 3$, and $1 \cdot 9$), but not all squares fit this pattern. (e.g. 16 appears as $1 \cdot 16$, $2 \cdot 8$, $4 \cdot 4$, $8 \cdot 2$, and $16 \cdot 1$, or five times). Which squares are they?

8. a. Yes. Last digit is 0

c. No. Last 2 digits, 78, are not a multiple of 4

e. Yes. $1 + 2 + 3 + 0 + 0 + 0 - (0 + 0 + 1 + 3 + 2) = 0$

9. e. The 1st number crossed out by 11 but by no smaller number is $11 \times 11 = 121$.

f. Any number with a prime factor smaller than 11 will be crossed out by that factor, so it will have already been crossed out by the time you cross out multiples of 11.

10. b. $2{,}491 = (50 - 3)(50 + 3) = 47 \times 53$

11. This student factored the first two digits of 149 as $14 = 2 \times 7$, then factored the last digit, 9, as 3×3.

14. b. $1{,}591 = 1{,}600 - 9 = (40 - 3)(40 + 3) = 37 \times 43$

Exercises 4.3

1. Hint: Is 237 a multiple of both 2 and 3?

3. a. 1, 2, 4, 8, 16, 32

5. a. 1, 2, 3, 4, 6, 8, 9, 12, 18, 24, 36, 72

6. a. Yes (1 is always a factor) c. Yes

f. No (36 is a multiple of 2, but 369 has no factors of 2)

7. a. Yes

c. No (168 has only one factor of 3)

e. No

g. No

i. No

8. a. 1, 2, 3, 6, 9, 18

9. a. No. $15 = 3 \times 5$, and 135,725 is a multiple of 5 but not of 3.

d. Hint: Use divisibility tests for 5 and 11.

10. 3. Each 0 corresponds to a factor of 10, which, in turn contains a factor of 2 and a factor of 5. Factoring this large product completely yields plenty of factors of 2 but only three factors of 5 (from 5, 10, and 15).

Exercises 4.4

3. a. Hint: A good way to attack this problem is to list the squares of primes that are less than 83. Then look for three numbers in your list of squares that add up to 83.

5. Hint: You can write an even number as two plus a smaller even number. You can write an odd number as three plus an even number.

Self-Quiz for Chapter 4

1. 13, 26, 39, 52, 65, 78, 91, . . .(The list extends without end. Any five numbers from the list will do.)

2. 60, 120, 180. Any multiple of 60 is a common multiple of 12 and 15.

3. Circle 97 and 107. ($77 = 7 \times 11$, $87 = 3 \times 29$, $117 = 3 \times 3 \times 13$)

4. a. $1{,}000 = 10 \times 10 \times 10 = 2 \times 2 \times 2 \times 5 \times 5 \times 5$

b. $2 \times 2 \times 7 \times 7$

c. 11×13

5. a. Yes. Ends in 5

b. Yes. Digits add to 45, which is a multiple of 9.

c. Yes. Last two digits, 24, are a multiple of 4

d. Yes. Sum of digits, 36, is a multiple of 9

e. Yes. $2 - 2 + 3 - 3 + 4 - 4 + 5 - 5 = 0$

6. a. Yes. $15 = 3 \times 5$, and 12,345 is a multiple of both 3 and 5

b. No. 45 is a multiple of 9, but 12,345 is not.

c. No. Digits add to 19, so 3 is not a factor. Then 36, which is a multiple of 3, is not a factor either.

d. Yes. Passes divisibility tests by 5 and 11, so it is a multiple of $5 \times 11 = 55$.

7. $437 = 441 - 4 = (21 - 2)(21 + 2) = 19 \times 23$

8. a. 1, 3, 9, 27, 81

b. 1, 2, 3, 5, 6, 9, 10, 15, 18, 30, 45, 90

c. 1, 37

Exercises 5.1

1. a. $3^4 = 81$, $4^3 = 64$

c. p^q and q^p are not generally equal.

2. Black dots show 3^2 and 5^2, all dots show $(3 + 5)^2$.

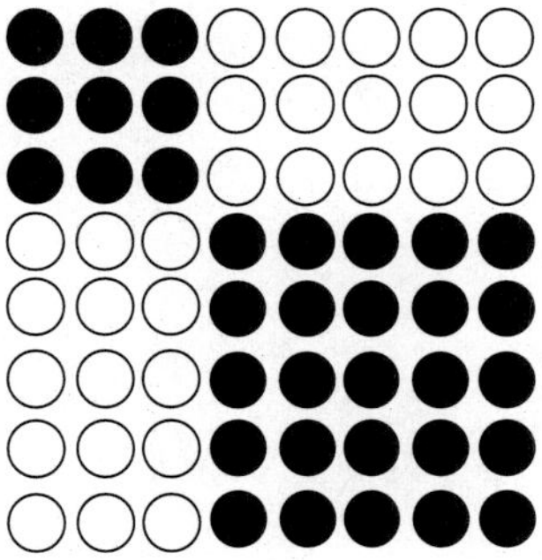

3. a. 4

b. 8

c. 2^5=32 sheets thick

d. Hint: How many sheets thick would the paper be after nine folds? What about the length, and width of the paper after it was folded nine times?

5. b. Hint: List the triangular numbers and compare them to the numbers 3, 10, 21, ...

6. Hint: Find the cost of the first two nails, the first three nails, etc., and compare each sum to the cost of the next nail.
Answer: $2^{24} - 1 = 16{,}777{,}215$

7. $1 + 7 + 7^2 + 7^3 + 7^4 = 2{,}801$. Now add the speaker, for a grand total of 2,802.

8. b. Partial answer:

$6 = 4 + 1 + 1$
$8 = 4 + 4$
$10 = 9 + 1 = 4 + 4 + 1 + 1$
$12 = 4 + 4 + 4 = 9 + 1 + 1 + 1$
$13 = 9 + 4 = 4 + 4 + 4 + 1$
$15 = 9 + 4 + 1 + 1$
$18 = 9 + 9 = 16 + 1 + 1$
$19 = 9 + 9 + 1 = 16 + 1 + 1 + 1$
$20 = 16 + 4 = 9 + 9 + 1 + 1$
$21 = 16 + 4 + 1$
$22 = 9 + 9 + 4 = 16 + 4 + 1 + 1$
$23 = 9 + 9 + 4 + 1$

9. For example, $3^3 - 3 = 27 - 3 = 24 = 6 \times 4$. Do this kind of calculation in each case, then list the factors which multiply 6. Except for 0, which occurs in the first case, these are figurate numbers from section 3.2.

10. Odd numbers up to 21 are given here; the other numbers are left for you to do.

1	**3** = 1 + 2
5 = 2 + 3	**7** = 3 + 4
9 = 4 + 5 = 2 + 3 + 4	**11** = 5 + 6
13 = 6 + 7	**15** = 7 + 8 = 4 + 5 + 6 = 1 + 2 + 3 + 4 + 5
17 = 8 + 9	**19** = 9 + 10

21 = 10 + 11 = 6 + 7 + 8 = 1 + 2 + 3 + 4 + 5 + 6

11. b. Starting with 3, every second...
Starting with 6, every third...
Starting with 10, every fourth...
Starting with 15, every fifth...

12. b. Hint: The prime factorizations of these numbers are curiously alike.

13. Hint: The rule is related to the factors. For example, 15 can be done 3 ways (see answer for question 10), and its factors are 1, 3, 5, and 15. On the other hand, 8 can be done 0 ways, and its factors are 1, 2, 4, and 8. Count the number of odd factors of the number to see the rule.

14. a. You get 324 either way.

15. c. 3,600 – 9 = 3,591

18. 968 or $2^3 \times 11^2$

19. a. The rows sum to 1, 2, 4, 8, etc. (powers of 2)

c. Hint: They lie on lines parallel to the sloping edges of the triangle.

f. To see the pattern, write the number of ways to reach each letter directly on the letter itself

20-23. Partial answers

Number of ways to make	Answers to #20	Answers to #21	Answers to #22
1	2	3	4
2	3	6	
3	4	10	
4	5		
n	$n + 1$		$\frac{(n+1)(n+2)(n+3)}{1 \times 2 \times 3}$

24. a. 2 b. 5 c. 13

g. Hint: The answers to parts a-f are some but not all of the Fibonacci numbers. Describe the pattern.

25. a. (first line) $1^2 + 1^2 + 2^2 + 3^2 + 5^2 + 8^2 = 8 \times 13$

26. b. Partial answer: The next two lines: $13 \times 34 = 21^2 + 1$ and $21 \times 55 = 34^2 - 1$

28. a. 8. (1 + 1 + 1 + 1, 1 + 1 + 2, 1 + 2 + 1, 2 + 1 + 1, 2 + 2, 1 + 3, 3 + 1, 4)

29. Hint: These are all Fibonacci numbers

Exercises 5.2

1. a.

Factors of $5^2 \cdot 7$		(5^2)	7
Factors of $3 \cdot 7^2$	(3)		(7^2)
Factors of LCM	3	5^2	7^2

4.

Factors of $2^5 \cdot 7^2 \cdot 11$	(2^5)		7^2	11	
Factors of $2^4 \cdot 5 \cdot 7^3 \cdot 11^2$	2^4	5	7^3	(11^2)	
Factors of $5^7 \cdot 7^4 \cdot 31$		(5^7)	(7^4)		(31)
Factors of LCM	2^5	5^7	7^4	11^2	31

7. c. 720 e. $60p^2q^3$

8. Five packs of buns and six packs of hot dogs.

10. a. $3.78

Exercises 5.3

1. Partial answer (parts a, b, c) for the case $n = 5$ only.

a. $2^{(5-1)}(2^5-1) = 2^4(2^5-1) = 16(32-1) = 16 \times 31 = 496$

b. Factors of 496: 1, 2, 4, 8, 16, 31, 2×31, 4×31, 8×31, 16×31 = 1, 2, 4, 8, 16, 31, 62, 124, 248, 496

c. $1 + 2 + 4 + 8 + 16 + 31 + 62 + 124 + 248 = 496$. Since 496 is the sum of all its factors except for 496 itself, 496 is a perfect number.

4. a. Partial answer: Factor 1,210 into primes as $2 \times 5 \times 11 \times 11$. Then factors of 1,210 are: 1, 2, 5, 10, 11, 22, 55, 110, 121, 242, 605, 1,210. Add all these factors except for 1,210 to get 1,184. Now do likewise for 1,184 and show that you get 1,210.

5. Hint: Add up all the partitioned areas and factor out $1 + 2 + 4 + 8 + 16$.

Exercises 5.4

1. a. 2,100 b. 32,000

2.

Tally	Base five	Spoken in base five	Base ten
𝍸 𝍸 𝍸 |	31	three hands one	16
𝍸 𝍸 𝍸 ||	32	three hands two	17
𝍸 𝍸 𝍸 |||	33	three hands three	18
𝍸 𝍸 𝍸 ||||	34	three hands four	19
𝍸 𝍸 𝍸 𝍸	40	four hands	20
𝍸 𝍸 𝍸 𝍸 |	41	four hands one	21
𝍸 𝍸 𝍸 𝍸 ||	42	four hands two	22
𝍸 𝍸 𝍸 𝍸 |||	43	four hands three	23
𝍸 𝍸 𝍸 𝍸 ||||	44	four hands four	24
(𝍸 𝍸 𝍸 𝍸 𝍸)	100	hand of hands	25
(𝍸 𝍸 𝍸 𝍸 𝍸) |	101	hand of hands one	26
(𝍸 𝍸 𝍸 𝍸 𝍸) ||	102	hands of hands two	27

3. a. 321 (which would be written as 86 in our usual base ten notation).

4. a. 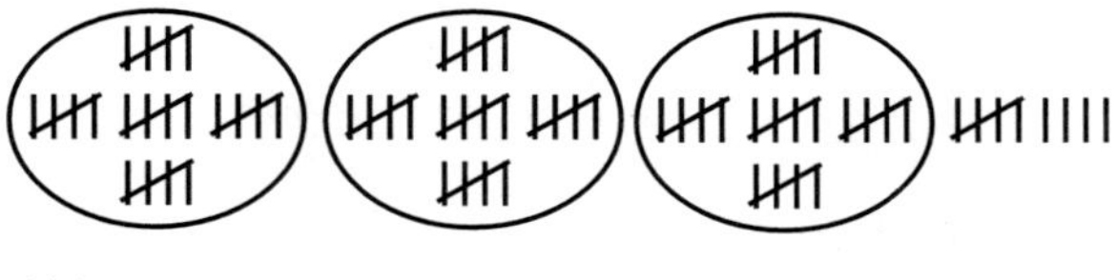

5. a. 444

b. We would write this in base ten as 124.

7. a. 23 + 22 = 100

10.

	2	4	11
3	11	22	33
1	2	4	11
4	13	31	44

12. a. 212 (base-ten equivalent: 23)

c. 112,111 (base-ten equivalent: 391)

14. a. 208

c. 509

15. a. 2,231

b. 173

16. a. 

Exercises 5.5

1. Partial answer: The column at the right has all odd numbers, no evens. One way to describe this is: Skip 0, write 1, skip 2, write 3, skip 4, . . . Similarly, in the next column over, skip 0 and 1, write 2 and 3, skip 4 and 5, write 6 and 7, . . You can extend this to the other columns. Another way to see the pattern is to write each number in base 2.

4. Multiplication is commutative, so the product is independent of the order in which the numbers are written. However, if the numbers are nowhere near equal in size, it is usually less work to double the larger and halve the smaller.

6. Partial answer:

a. 2, 5, 8 is one AP in the list. Can you find the other three?

b. 2, 6, 10

c. AP free

d. Partial answer: 6, 10, 14 is one AP here. Another is 5, 12, 19. Can you find a two more APs hidden here?

7. The fourth number is 4.
The fifth number is 9 (skip 5 because of 1, 3, 5; skip 6 because of 0, 3, 6; skip 7 because of 1, 4, 7; skip 8 because of 0, 4, 8).
The sixth number is 10.
The seventh number is 12 (skip 11 because of 9, 10, 11).

Self-Quiz for Chapter 5

1. a. 128

b. 125

c. 58

d. 100

2. 120

3. 1 9 36 84 126 126 84 36 9 1

4. A number is an idea, regardless of how or whether it is expressed in writing. A numeral is a written mark or sign which is intended to stand for a number.

5. 414

6. 1,133

7. 2,111

8. Thirty seven

9. a. 2,023 (In base-ten, $69 + 194 = 263$)

b. 22,013 (In base-ten, $116 \times 13 = 1{,}508$)

Exercises 6.1

1. b. (Partial answer)

2. a. (partial answer)

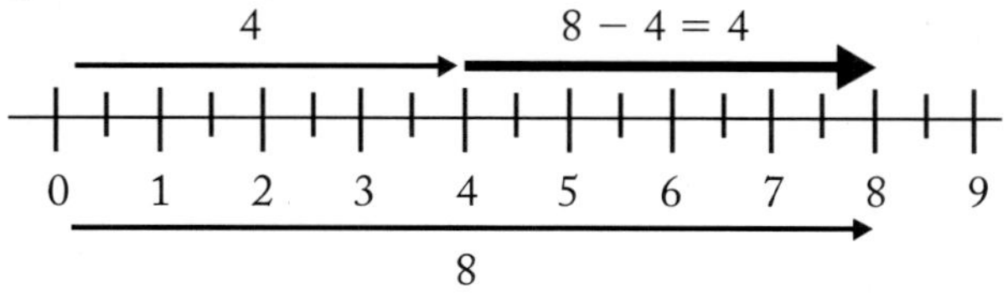

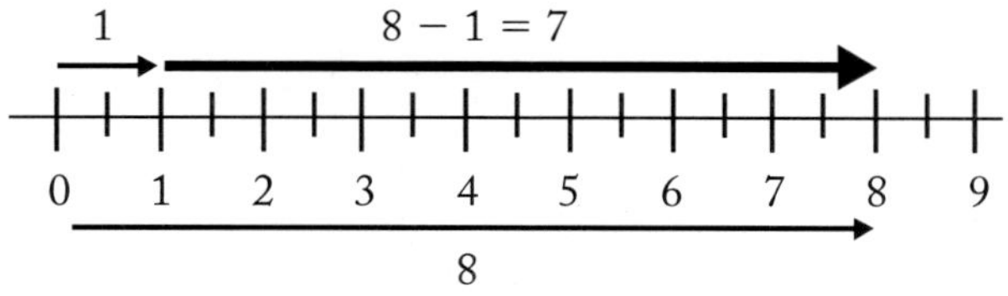

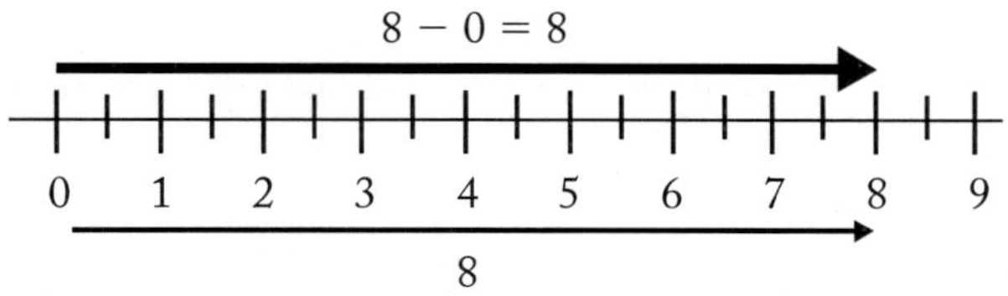

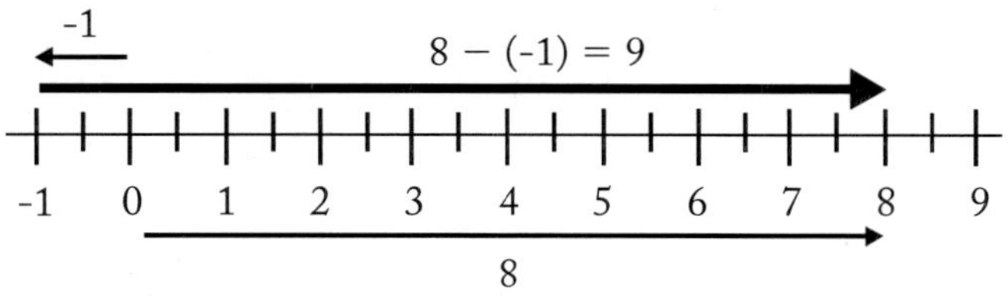

b. (One case) Think of −1-(-3) as the number which must be added to −3 to get −1.

Subtraction question $a - b$	Answer to subtraction	Equivalent addition $a + (-b)$	Result of adding the opposite
2 − 9	-7	2 + (-9)	-7
7 − (-4)	11	7 + (+4)	11
-1 − (-3)	2	-1 + (+3)	2

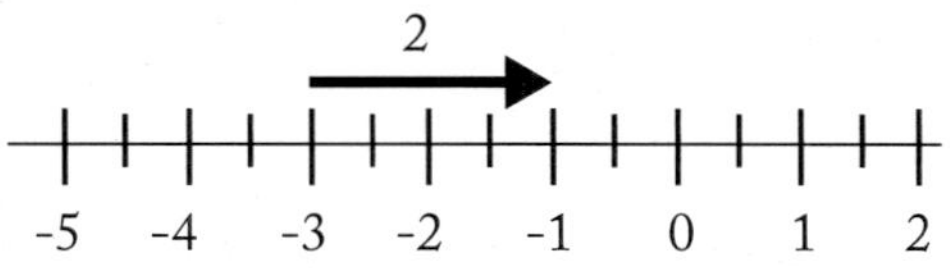

-1+(+3) is a vector starting at −1, pointing right, 3 units long. It ends at 2.

3. a. Hint: The middle row asks what, when added to 12, makes 25. Once you get that, the middle column asks what plus 13 is 22.

b. Hint: Start either with the bottom row or with the right hand column.

d.

35	22	57
-7	-9	-16
28	13	41

7. a. Find the number you are subtracting from in the center column and the number to be subtracted in one of the side columns. Connect these dots with a straight line and extend the line to the other side column. The number where the extended line meets the third columns is the difference.

8. a. 14,777 feet. b. 14,495 − (-282) = 14,777

10. a. 22 yards. b. -5 − 17=-22

11. a. 20° F b. -27 − (-47)=20

13. Subtraction is not associative. If it were, 9 − (3 − 2), which is really 8, would equal (9 − 3) − 2, which instead is 6 − 2 or 4.

14. For any numbers x, y, and z, $x - (y - z) = x - y + z$

Exercises 6.2

6. Activities that involve making change are all valuable, as are questions like, "If Nellie has read 137 pages of a 419 page book, how many pages does she have left to read?" Keeping track of money in an account is another good activity of this type.

8. d. The patterns continue to hold for negative integers. E.g., for squares

$(-3)^2$		$(-2)^2$		$(-1)^2$		0^2		1^2		2^2
9		4		1		0		1		4
	-5		-3		-1		1		3	
		2		2		2		2		2

e. Hint: Organize the information like this to bring out the patterns:

Power	2	3	4
Number of rows down to get constant difference	2	3	4
Constant differences	2	6	24

Note that $2 = 1 \times 2$, $6 = 1 \times 2 \times 3$, and $24 = 1 \times 2 \times 3 \times 4$.

For first powers, we would expect a consistant difference of 1 on the first row down.

Exercises 6.3

1. a.

	3	2	5
2	6	4	10
6	18	12	30
8	24	16	40

c.

	4	-16	-12
2	8	-32	-24
3	12	-48	-36
5	20	-80	-60

e.

	-21	4	-17
5	-105	20	-85
-8	168	-32	136
-3	63	-12	51

These examples emphasize the missing factor view of division.

3. a. 3 Check: (-8)3=-24.

4. If x and y have the same sign, $x \div y$ is positive. If x and y have different signs, $x \div y$ is negative.

5. a. $7 - 2(-5) - 5(6) = 7 - (-10)-30 = 7 + 10 - 30 = -13$

d. $7(25) = 175$ e. $(-11)^2 = 121$ h. $(40 \div 10)\times 2= 4\times 2 = 2$

6. a.

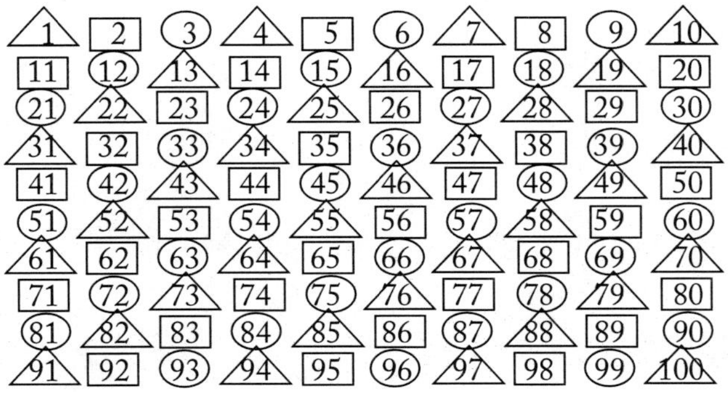

7. a. The sum of two numbers with triangles is always a number with a square.

8. a. Partial answer. For $1^2 \div 4$, the remainder is 1 (the quotient is 0). For $3^2 \div 4$ the remainder is 1 (the quotient is 2).

c. Hint: If a number k is even, it is a multiple of 2 and can be written $k = 2h$, where h is an integer. Show then that k^2 is a multiple of 4. On the other hand, if a number j is odd, it can be written $j = 2m + 1$ for some whole number m. Show that if you square an odd number $2m + 1$, you get one more than a multiple of 4. You may use a box puzzle with borders or refer to figure 3.1.6 and the problem that goes with it.

9. d. The evidence from parts a–c suggests that:

i. Every prime which is 1 more than a multiple of 4 can be expressed as a sum of two squares.

ii. No prime which is 3 more than a multiple of 4 can be expressed as a sum of two squares.

Statement ii can be shown as follows:
If an odd prime can be written as a sum of two squares, one of those must be odd and the other even. Now every even square is a multiple of 4, and every odd square is 1 more than a multiple of 4, so their sum is 1 more than a multiple of 4.

Statement i is also true, but it cannot be explained without more advanced mathematics.

10. a. 3 c. 4

11. a. Cast out the 9 and add the other digits: $1 + 2 + 6 = 9$. Cast out this 9 to get 0 as the digital root.

c. 2

12. a. $3 \times 99 + 8 \times 9 + 3 + 8 + 5 = 3 \times 9 \times 11 + 8 \times 9 + 3 + 8 + 5$
$= 9(3 \times 11 + 8) + 3 + 8 + 5.$

13. b. A common copying error is to interchange two digits. What happens then?

14. a. The digital root of 39,812 is 5.
Digital root of 17,734 is 4
Digital root of 39,812 − 17,734 is $5 - 4 = 1$. Since 1 is not the digital root of 22,178, the subtraction is incorrect.

15. a. Digital root of 237 is 3
Digital root of 529 is 7
Digital root of 237×529 is digital root of $3 \times 7 = 21$ or 3.
Is 3 the digital root of 125,273?

17. a. 9,990

Exercises 6.4

2. The school should use four buses. (In reality, many schools will try to get by with only three buses.)

3. 49 m.

5. e. Hint: You have divided by 7, 11, and 13. How much is $7 \times 11 \times 13$? Try multiplying a three digit number by this product? What do you notice?

6. e. All the remainders are 0.

Hint for parts l,m,n: This is much like the trick in question 5. What number plays the role here that 1,001 plays in question 5?

9. Hints: After you bring down a digit in long division, you subtract a multiple of the divisor, unless you need to bring down a second digit, as happens a couple of times here. That happens when, even after bringing down a digit, the result is less than the divisor. In that case, there is a 0 in the quotient. Also, note that if the divisor is multiplied by 8, the product has only two digits, but in some cases a multiple of the divisor has three digits.

Exercises 6.5

1. a. $468 = 3 \times 150 + 18$. $150 = 8 \times 18 + 6$. $18 = 3 \times 6 + 0$.
The last non-0 remainder, 6, is the GCF of 468 and 150.

2. a. Partial solution: The first line in the table is

a, b	GCF	LCM	product ab
4, 6	2	12	24

b. Hint: Multiply the GCF by the LCM in each line. What do you notice?

3. b. Odd numbers.
c. Prime numbers
d. Hint: If the table is extended, what happens with 16? With 25? With 36? With 49? It depends on what kind of numbers are squared. Find the relationship and state it explicitly.

Self-Quiz for Chapter 6

1. a. 57 b. -123 c. -50

2.

-13	5	-8
12	9	21
-1	14	13

3. i. Take away: $a - b$ is the result of removing b objects from a objects.

ii. Add the opposite: $a - b$ is the result of adding $-b$ to a.

iii. Missing addend: $a - b$ is the number which, if added to b, yields a.

4. a. -47 − (-51) asks what number must be added to -51 to get -47. Since -51 + 4 = -47, the answer to that question is 4, and -47 − (-51) = 4.

b. -47 + (+51) = 4

c. Begin with 47 negatives. We would like to take away 51 negatives, but we have only 47 available. Therefore, add four zero pairs. Then we have 51 negatives and four positives. Take away the 51 negatives, and four positives remain.

5.

	11	-4	7
3	33	-12	21
-5	-55	20	-35
-2	-22	8	-14

6. i. Repeated subtraction: 11 ÷ 4 asks how many 4s can be subtracted from 11.

ii. Missing factor: 11 ÷ 4 asks what 4 must be multiplied by to get 11.

iii. Multiplication by the reciprocal: 11 ÷ 4 asks for the product $11 \times \frac{1}{4}$.

7. a. 38 − 9 = 29

29 − 9 = 20

20 − 9 = 11

11 − 9 = 2

We have taken away 9 four times but cannot take away any more 9s. Therefore the quotient is 4 and the remainder is 2.

b. With 38 tiles, you can make four rows of 9, and you have two tiles left as remainder.

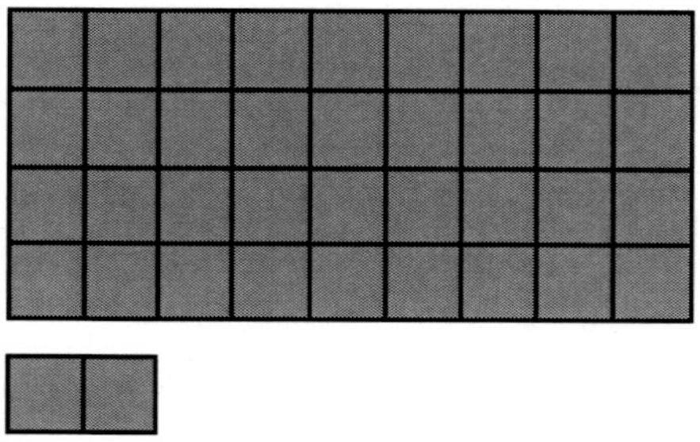

8. a. $$\begin{array}{r} 12 \\ 35\overline{)427} \\ \underline{35} \\ 77 \\ \underline{70} \\ 7 \end{array}$$ remainder 7

b. 35

12

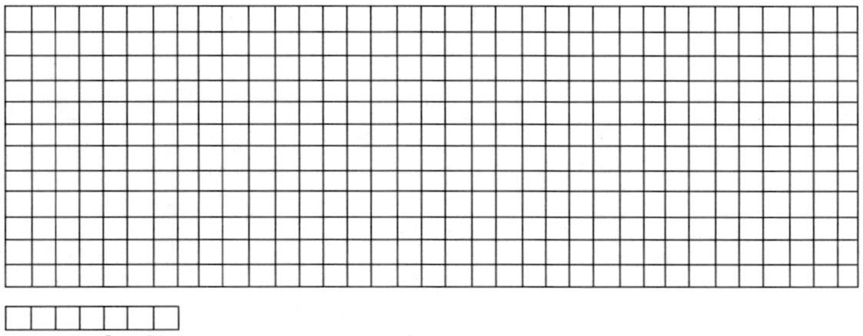

remainder 7

9. 2

Exercises 7.1

3. a. A foot is a third of a yard.
b. A day is a seventh of a week.

4. a. 8 c. $\frac{3}{8}$

5. a. $\frac{1}{36}$ c. $\frac{1}{144}$

6. Parts a and b.

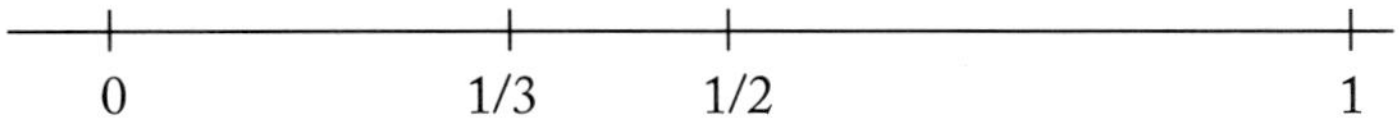

8. a. $\frac{1}{2}$ f. $\frac{3}{16}$

10. One fourth of a dollar is a quarter, so $\frac{3}{4}$ of a dollar is 3 quarters or 75¢.

Four friends can change $3 for 12 quarters, then divide the quarters equally, so that each of the four gets 3 quarters or 75¢.

11. $A = 1\frac{7}{8}$

12. c. If two fractions $\frac{a}{b}$ and $\frac{a}{c}$ have the same numerator but different denominators, the one with the smaller denominator is the larger of the two. (This assumes that the numbers involved are positive.)

Exercises 7.2

1. (Partial answer)

	area of each small rectangle	area of shaded rectangle	length and width of shaded rectangle	multiplication illustrated
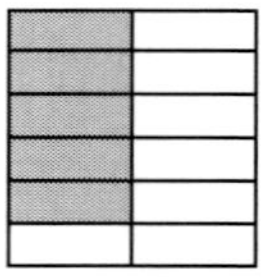	$\frac{1}{12}$	$\frac{5}{12}$	length $\frac{5}{6}$ width $\frac{1}{2}$	$\frac{5}{6} \times \frac{1}{2} = \frac{5}{12}$
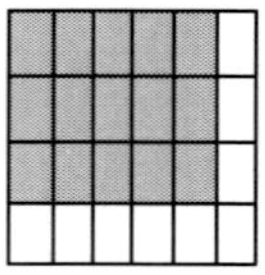	$\frac{1}{24}$	$\frac{15}{24}$	length $\frac{5}{6}$ width $\frac{3}{4}$	$\frac{5}{6} \times \frac{3}{4} = \frac{15}{24}$

2. a.

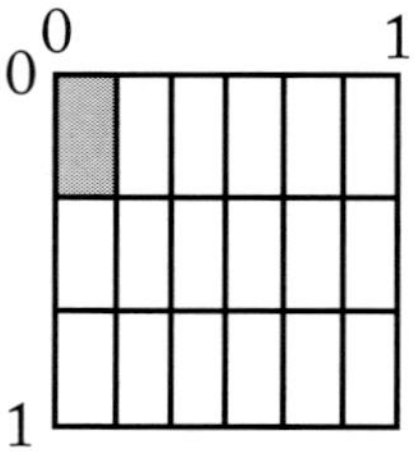

3. Multiplications illustrated: a. $\frac{5}{6} \times \frac{3}{2} = \frac{15}{12}$

4. c.

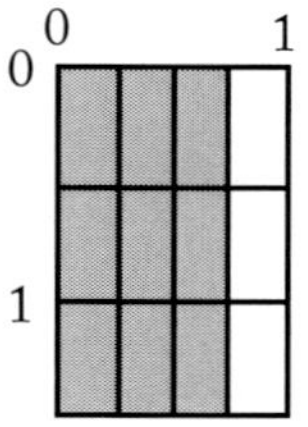

$\frac{3}{2} \times \frac{3}{4} = \frac{9}{8}$

6. a. $\frac{3prs^2t}{20uv}$

Exercises 7.3

1. Hint: Reduce as much as possible before multiplying.

2. a. 2

4. a. $\frac{1}{4}$ d. $\frac{9}{10}$

5. a. $\frac{1}{2}$ c. $\frac{3}{4}$

7. a-e. Hint: Reduced to lowest terms, all these answers are the same.

f. One fourth of the balls are in the upper left hand corner of the picture, and they are, by the colors, 1 + 3. Three times as many (the other three fourths of the balls) are in the other three corners, and their number, by the colors, is 5 + 7. Thus $\frac{1+3}{5+7} = \frac{1}{3}$.

8. a. 2 c. 4

11. Hint: The dots for each set of equal fractions lie on a line through (0,0).

Exercises 7.4

2. a. $\frac{8}{8} = 1$ c. $\frac{6}{8} = \frac{3}{4}$

4. c. $\frac{7}{12}$

5. a. 20 b. 75

6. d. $\frac{34}{60} = \frac{17}{30}$

8. b.

$\frac{1}{3}$	$\frac{1}{4}$	$\frac{7}{12}$
$\frac{5}{6}$	$\frac{7}{12}$	$\frac{17}{12}$
$\frac{7}{6}$	$\frac{10}{12}$	$\frac{24}{12}$

9. a. $8\frac{7}{8}$ gallons b. $9\frac{9}{20}$ miles

12. a. $\frac{3}{4} = \frac{2^2-1}{2^2}$ c. $\frac{15}{16} = \frac{2^4-1}{2^4}$

13. a. $\frac{3+1}{9} = \frac{4}{9} = \frac{8}{2\cdot 9} = \frac{9-1}{2\cdot 9} = \frac{3^2-1}{2\cdot 3^2}$ c. $\frac{40}{81} = \frac{3^4-1}{2\cdot 3^4}$

e. Hint: The solutions given for parts a and c are written to help you see the pattern. Check to see that your answers for parts b and c fit this pattern as well, then extend that pattern.

g. $\frac{1}{3} + \frac{1}{9} + \cdots + \frac{1}{3^n} = \frac{\frac{3^n-1}{2}}{3^n} = \frac{3^n-1}{2\cdot 3^n}$

14. a. $\frac{4+1}{16} = \frac{5}{16} = \frac{\frac{16-1}{3}}{16} = \frac{4^2-1}{3 \cdot 4^2}$

15. a. $\frac{5}{6}$ b. $\frac{7}{12}$ c. $\frac{9}{20}$

d. The answers to parts a-c fit a pattern. First find the pattern in the numerators. To see the pattern in the denominators it may help to factor them.

16. a. $\frac{1}{2}$ b. $\frac{1}{3}$

17. a. $\frac{2}{8}$ b. $\frac{2}{15}$

Hint: With numerators of 2, factor the denominators to bring out the pattern.

19. a. 2 b. 2

Exercises 7.5

1. a. Start with 11 fifths, take away 3 fifths; you have 8 fifths left.

b. $\frac{11}{5} - \frac{3}{5} = \frac{11}{5} + \frac{-3}{5} = \frac{8}{5}$

c. $\frac{11}{5} - \frac{3}{5}$ is the number which may be added to $\frac{3}{5}$ to get $\frac{11}{5}$. The picture below shows that this number is $\frac{8}{5}$.

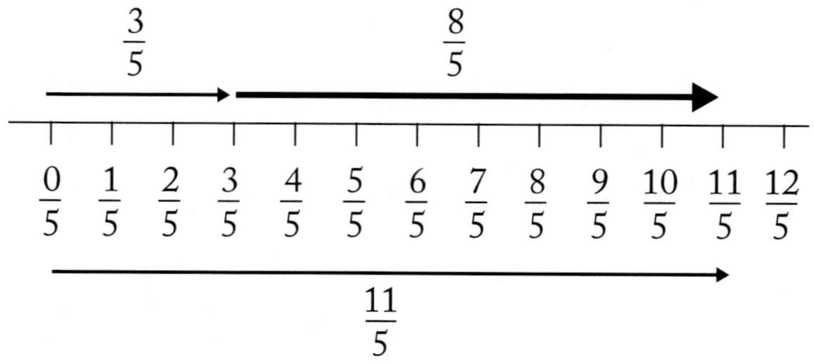

4. a.

$\frac{3}{8}$	$\frac{6}{8}$	$\frac{9}{8}$
$\frac{5}{8}$	$\frac{1}{8}$	$\frac{6}{8}$
$\frac{8}{8}$	$\frac{7}{8}$	$\frac{15}{8}$

5. Hint: $35 - \left(2\frac{2}{3} + 3\frac{1}{3} + 5\frac{2}{3}\right)$

6. $\frac{33}{40}$ gallon

Exercises 7.6

2. a. $\frac{3}{7} \div \frac{4}{5} = \frac{\frac{3}{7}}{\frac{4}{5}} = \frac{\frac{3}{7}}{\frac{4}{5}} \times \frac{\frac{5}{4}}{\frac{5}{4}} = \frac{\frac{3}{7} \times \frac{5}{4}}{\frac{20}{20}} = \frac{\frac{15}{28}}{1} = \frac{15}{28}$

3. a. $\frac{5}{8} \div \frac{1}{4} = \frac{5}{8} \times \frac{4}{1} = \frac{20}{8} = \frac{5}{2}$ Check: $\frac{5}{2} \times \frac{1}{4} = \frac{5}{8}$

g. In general, $\frac{p}{q} \div \frac{s}{t}$ is the reciprocal of $\frac{s}{t} \div \frac{p}{q}$.

4. b. $11 \div 2\frac{1}{4}$

5. $\frac{2}{5}$ mile

Exercises 7.7

1. a. $\frac{5}{3}$ b. $\frac{5}{4}$

2. Hint: Use the answer to each part to save work on the next one, canceling before multiplying as much as possible.

3. Hint: Use the answer to each part to save work on the next one.

Another interesting approach to this question is to take advantage of the rule for factoring a difference of squares (see questions 9–13 of Exercises 4.1) by writing each factor with a common denominator, so that, for example, $1-\frac{1}{3^2}=\frac{3^2-1}{3^2}=\frac{9-1}{9}=\frac{8}{9}=\frac{2\times 4}{3\times 3}$.

To see a pattern, in your answers, first put each answer in lowest terms, then multiply those with odd denominators by 2/2.

4. a. $\frac{4+4+4}{4}$ c. $\frac{4\times 4+4}{4}$ e. $4+4-\frac{4}{4}$ g. $4+4+\frac{4}{4}$

In many cases you may find several ways.

5. a. $\frac{3}{2}$ b. $\frac{5}{3}$ d. $\frac{13}{8}$

Self-Quiz for Chapter 7

1. a.

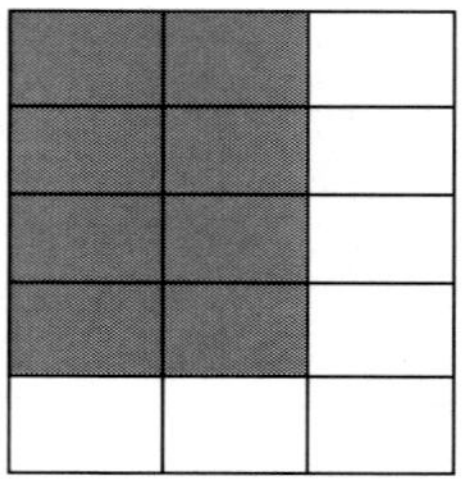

b. $\frac{8}{15}$

2. a. $\frac{15}{35}$ b. $\frac{14}{17}$

3. $8\frac{9}{20}$

4. a. 0 b. $4\frac{17}{24}$

5. a. You have $\frac{7}{5}$, take away $\frac{3}{5}$, and $\frac{4}{5}$ remains

b. $\frac{7}{5}+\frac{-3}{5}=\frac{7+(-3)}{5}=\frac{4}{5}$

c. $\frac{3}{5}+x=\frac{7}{5}$ and $x=\frac{4}{5}$

6. $\frac{35}{24}$

7. a. To divide a number n by 2 is to find a number h for which $2h=n$. To divide a number n by 1/2 is to find a number k for which $\frac{1}{2}k=n$.

b. (Answers will vary)

8. $\frac{5}{6} - (-\frac{1}{4}) = \frac{5}{6} + \frac{1}{4} = \frac{10}{12} + \frac{3}{12} = \frac{13}{12}$

Exercises 8.1

1. a.

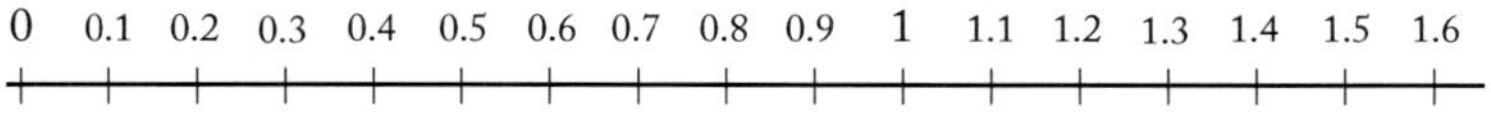

b.

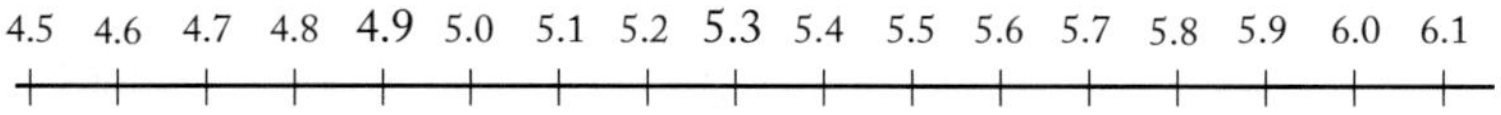

e.

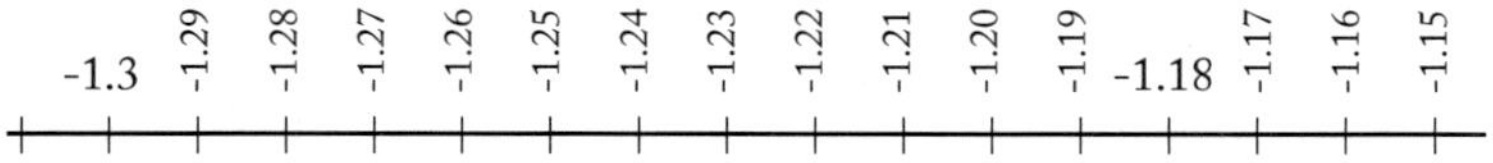

4. a.

3.1	2.8	5.9
4.7	9.3	14.0
7.8	12.1	19.9

b.

-0.4	8.01	7.61
12.5	-7.6	4.9
12.1	0.41	12.51

d.

	1.1	0.4	1.5
0.5	0.55	0.2	0.75
3.1	3.41	1.24	4.65
3.6	3.96	1.44	5.4

5. a. $11,000,000,000,000

8. a. $6\frac{2}{100}$

9. h. The product is in the neighborhood of 120. Misplacing the decimal point a digit in either direction would change the result to a number around 12 or around 1,200, which are obviously too small and too large, respectively.

10 a. 3.24 is about $3\frac{1}{4} \times 12 = 39$. Then $3\frac{1}{4} \times 13$ is $39 + 3\frac{1}{4}$ or $42\frac{1}{4}$. Then $3.24 \times 13\frac{1}{4}$ is slightly larger yet, close to 43.

Exercises 8.2

1. a. $1.036 \div 25 = 1.036 \times 1/25 = 1.036 \times 4/100 = 4.144/100 = 0.04144$

4. b. $8 \times 0.375 = 3$

5. a. $\frac{1}{7} = 0.142857142857\ldots$ (the six digit cycle 142857 repeats endlessly).

b. Hint: Add the first three digits to the last three digits.

6. f. All these infinite decimals have six digit cycles, and the cycles are all obtained from the cycle for $\frac{1}{7}$ by beginning at different digits.

7. a. Yes.

b. You get $0.\overline{9}$ (9 repeats endlessly), which is another decimal representation of 1. (Note: Beginners often think that $0.\overline{9}$ is less than 1, though they can't say by how much.)

9. The width, in km, is $0.35 \div 0.67$. Dividing yields .05223880597. . ., but no calculation can yield an answer more accurate than the numbers used in that calculation. Therefore an answer of 0.52 km is appropriate here.

10. $185,000,000

Exercises 8.3

1.

$\frac{1}{2} = 0.5$	$\frac{1}{11} = 0.\overline{09}$	$\frac{1}{20} = 0.05$
$\frac{1}{3} = 0.\overline{3}$	$\frac{1}{12} = 0.08\overline{3}$	$\frac{1}{21} = 0.\overline{047619}$
$\frac{1}{4} = 0.25$	$\frac{1}{13} = 0.\overline{076923}$	$\frac{1}{22} = 0.0\overline{45}$
$\frac{1}{5} = 0.2$	$\frac{1}{14} = 0.0\overline{714285}$	$\frac{1}{23} = 0.\overline{0434782608695652173913}$
$\frac{1}{6} = 0.1\overline{6}$	$\frac{1}{15} = 0.0\overline{6}$	$\frac{1}{24} = 0.041\overline{6}$
$\frac{1}{7} = 0.\overline{142857}$	$\frac{1}{16} = 0.0625$	$\frac{1}{25} = 0.04$
$\frac{1}{8} = 0.125$	$\frac{1}{17} = 0.\overline{0588235294117647}$	$\frac{1}{26} = 0.0\overline{384615}$
$\frac{1}{9} = 0.\overline{1}$	$\frac{1}{18} = 0.0\overline{5}$	$\frac{1}{27} = 0.\overline{037}$
$\frac{1}{10} = 0.1$	$\frac{1}{19} = 0.\overline{052631578947368421}$	$\frac{1}{28} = 0.03\overline{571428}$

2. a. In problem 1, the fractions with finite decimals are:

$$\frac{1}{2}, \frac{1}{4}, \frac{1}{5}, \frac{1}{8}, \frac{1}{10}, \frac{1}{16}, \frac{1}{20}, \frac{1}{25}$$

c. Their denominators have prime factors of 2 and 5 only.

d. A finite decimal can be written as a fraction with a power of 10 as denominator. Reducing such a fraction cannot introduce any new prime factors in the denominator. Therefore, if a fraction in lowest terms has a finite decimal equivalent, its denominator can have no prime factors other than the prime factors of 10, namely 2 and 5.

3. a. Finite: $\frac{1}{32}, \frac{1}{40}, \frac{1}{80}, \frac{1}{125}$

4. a.

	$\frac{1}{2} = \frac{1}{2^1 5^0} = 0.5$	$\frac{1}{4} = \frac{1}{2^2 5^0} = 0.25$	$\frac{1}{8} = \frac{1}{2^3 5^0} = 0.125$
$\frac{1}{5} = \frac{1}{2^0 5^1} = 0.2$	$\frac{1}{10} = \frac{1}{2^1 5^1} = 0.1$	$\frac{1}{20} = \frac{1}{2^2 5^1} = 0.05$	$\frac{1}{40} = \frac{1}{2^3 5^1} = 0.025$
$\frac{1}{25} = \frac{1}{2^0 5^2} = 0.04$	$\frac{1}{50} = \frac{1}{2^1 5^2} = 0.02$	$\frac{1}{100} = \frac{1}{2^2 5^2} = 0.01$	$\frac{1}{200} = \frac{1}{2^3 5^2} = 0.005$
$\frac{1}{125} = \frac{1}{2^0 5^3} = 0.008$	$\frac{1}{250} = \frac{1}{2^1 5^3} = 0.004$	$\frac{1}{500} = \frac{1}{2^2 5^3} = 0.002$	$\frac{1}{1{,}000} = \frac{1}{2^3 5^3} = 0.001$

b. The number of digits after the decimal point in the finite decimals is the larger of the exponents of 2 and 5 in the factorization of the denominator.

5. a. 2 c. 28 e. $n - 1$

6. a.

n	3	6	7	9	11	12	13	14	15	17	18	19	21	22
maximum length of period of $1/n$ by problem 5	2	5	6	8	10	11	12	13	14	16	17	18	20	21
Actual length of period of $1/n$	1	1	6	1	2	1	6	6	1	16	1	18	6	2

8. b. The number of nonrepeating digits before the repeating digits begin is the larger of the exponents on 2 and 5 in the prime factorization of the denominator.

12. b. $\frac{451}{999}$ d. $\frac{77.56}{99} = \frac{7,756}{9,900}$

Exercises 8.4

3. 18

6. $21,000

7. 34%

8. a. $2.40 b. $4.50 c. $9.50

9. a. about 22% b. about 28%

10. a. 23.5% b. 23.5%
c. The commutative property of multiplication

11. A net gain of 5%

12. 12.5%

16. $38,220

20. 13/17 liters or about 0.76 liters

21. At 40%, the solution has 0.4 × 11=4.4 liters of antifreeze. At 25% it has 0.25 × 11=2.75 liters of antifreeze, so 1.65 liters of antifreeze must be added. Therefore, the mixture drained must contain 1.65 liters of water. The mixture is 75% water, so to remove 1.65 liters of water, remove 1.65 × 4/3 liters of the mixture, or 2.2 liters.

22. a. 58 1/3% b. 110

Self-Quiz for Chapter 8

1.

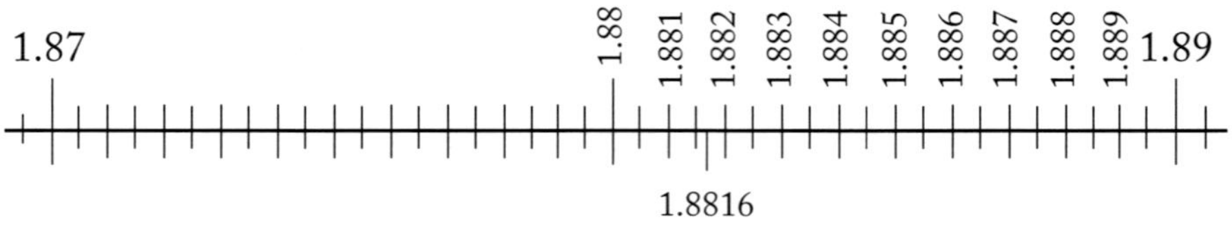

2. $\frac{723}{10,000}$

3. 0.523809 . . .(the six digits, 523809 repeat endlessly)

4.

	-0.2	30	29.8
0.03	-0.006	0.9	0.894
2	-0.4	60	59.6
2.03	-0.406	60.9	60.494

5. \$24.44

6. About 28%

Exercises 9.1

1. Hint: If you have trouble seeing the message, plot your points with extra heavy dots, and look at the paper from a distance. The dots will form letters.

3. You should get a simple picture.

4. g. The *x*-coordinate of the midpoint is the average of the *x*-coordinates of the endpoints. The *y*-coordinate of the midpoint is the average of the *y*-coordinates of the endpoints.

Exercises 9.2

1. a. Clark's rule: 30 mg   Cowling's rule, $29\frac{1}{6}$ mg

Dilling's rule, 30 mg   Young's rule, $33\frac{1}{3}$ mg

3. b. $2n + 9$

4. b. $2(n + 9)$

5. a. There is no input for which the two rules produce the same output. From the tables given for these rules, you can see that for any input n, the output from the rule $2(n + 9)$, is 9 greater than the output from the rule $2n + 9$.

6. The number of miles from the lightning to you is one fifth the number of seconds between the lightning flash and the thunder.

8. b. A regular polygon with n sides has $\frac{n(n-3)}{2}$ diagonals.

9. b. Hint: The answers are a family of numbers from Chapter 3.

11. b. On the basis of observations for 1, 2, 3, 4, and 5 points it would seem that the number of regions formed is 2^{n-1}, where n is the number of points.

c. For the case with 6 points, predict $2^{6-1} = 2^5 = 32$ regions.

12. The cube of the mean distance from the sun equals the square of the length of the year. (The data support this observation to several decimal places, which is all that could be expected here, since there were certainly limits to the accuracy of the observations.)

Exercises 9.3

1. a.

Inputs n	6	5	3	1	0	-1	-2
Outputs $3n-7$	11	8	2	-4	-7	-10	-13

c. The graph of $3n - 7$

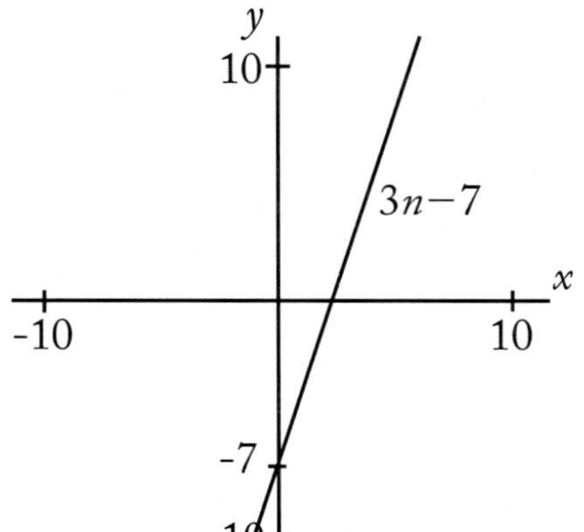

3. a.

Inputs n	-3	-2	-1	0	1	2	3
Outputs n^2	9	4	1	0	1	4	9

d. The graph of $y = n^2$

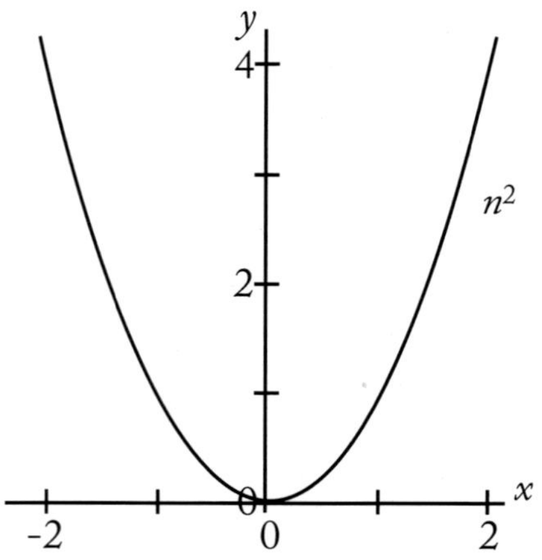

4. i. Hint: What point do all of these lines go through?

Exercises 9.4

1. a. −2 c. 4/3 e. $(4 - b)/4$

2. (parts a and c graphed below)

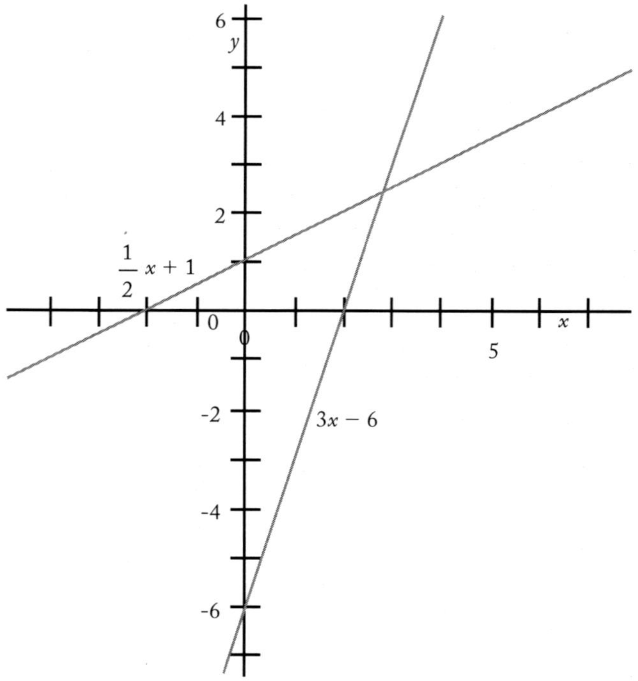

5. −2

6. d.

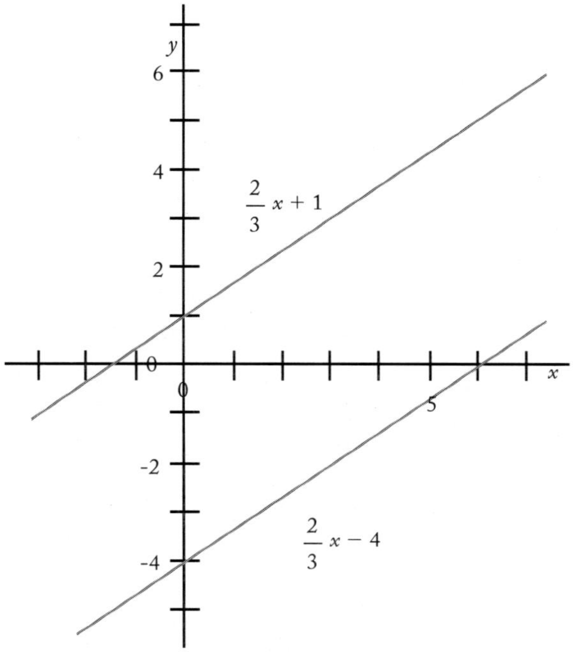

7. a.

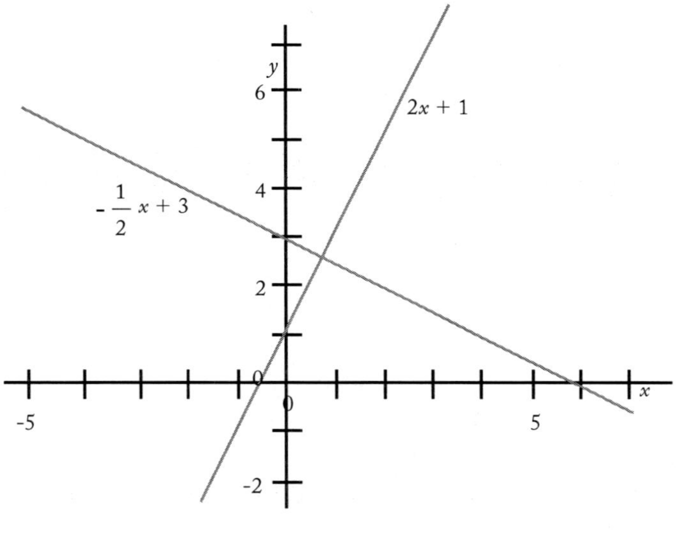

e. If the product of the slopes of two lines is −1, then the lines are perpendicular.

8. a. Slope $\frac{2}{3}$, y-intercept $\frac{1}{3}$

b. Slope 0, y-intercept −3

c. Slope 4, y-intercept −1

Self-Quiz for Chapter 9

1. (Included in answer to question 5)

2. a. Double the number, then subtract four.

b. Subtract four from the number, then double what is left.

c. If you subtract 4, then double, you double a smaller number and get less than if you first doubled and then subtracted 4.

3. $3n + 5$

4. Yes. $3\left(5-\frac{2}{3}\right) = 3\left(4\frac{1}{3}\right) = 13$

5.

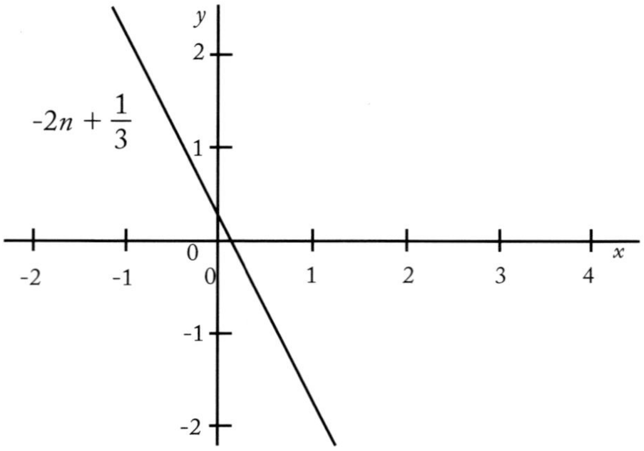

6. A rule or function is said to be linear if its graph is a line.

7. A line with slope −3 has the property that between any two points on it, the change in y-coordinates is −3 times the change in the x-coordinates.

8. $-\frac{1}{7}$

Exercises 10.1

1. Take a number

   Add 3

   Multiply by 2

   Add your original number

   Divide by 3

   Subtract your original number

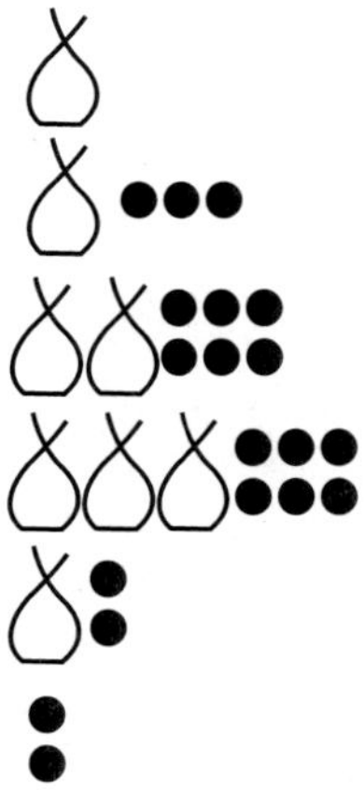

3. Take a number

   Add 7

   Multiply by 3

   Subtract 1

   Subtract your original number

   Divide by 2

   Subtract your original number

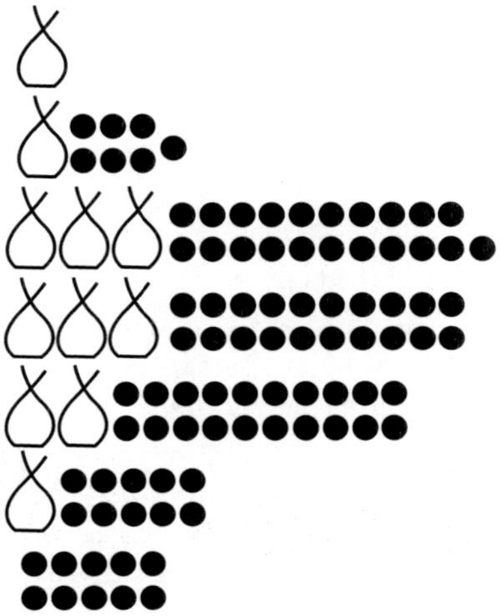

5.

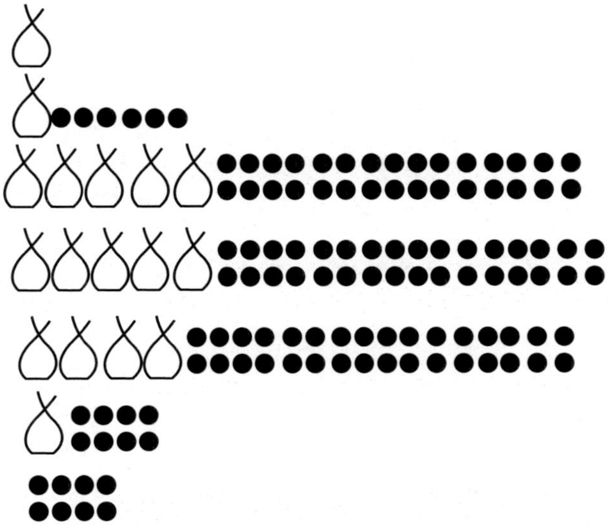

7. Take a number, add 4, multiply by 5, add your original number, add 1, divide by 3, add 1, divide by 2, subtract your original number.

9. $x, x + 4, 5x + 20, 6x + 20, 6x + 21, 2x + 7, 2x + 8, x + 4, 4$

Exercises 10.2

1. a.

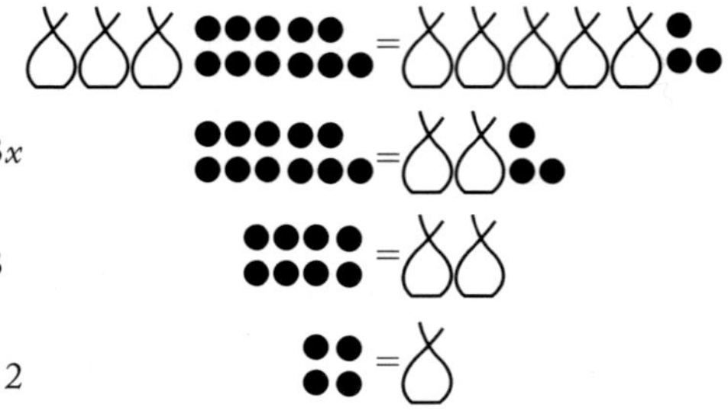

	Bags and Stones	Algebra
		$3x + 11 = 5x + 3$
Subtract $3x$		$11 = 2x + 3$
Subtract 3		$8 = 2x$
Divide by 2		$4 = x$

d.

	$3x + 11 = 5x + 3$
Subtract 5x	$-2x + 11 = 3$
Subtract 11	$-2x = -8$
Multiply by -1	$2x = 8$
Divide by 2	$x = 4$

3. a.

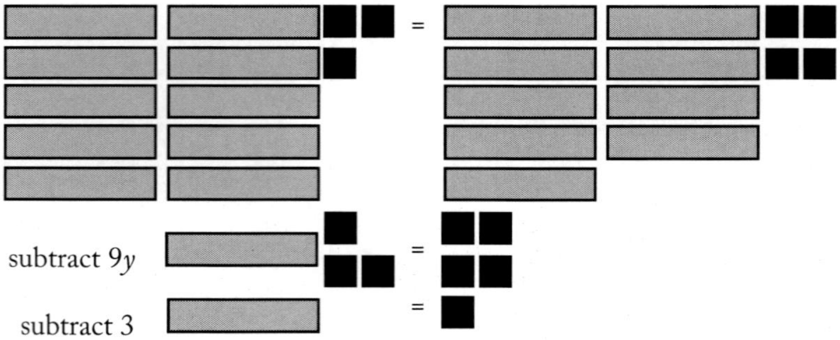

	Algebra Tiles	Algebra
		$10y + 3 = 9y + 4$
subtract $9y$		$y + 3 = 4$
subtract 3		$y = 1$

Second way: (without bags and stones, because of negative numbers.)

Subtract 4	$10y-1 = 9y$
Subtract $10y$	$-1 = -y$
Multiply (or divide by) -1	$1 = y$

Third way: (without bags and stones, because of fractions.)

Divide by 10	$y + \frac{3}{10} = \frac{9}{10}y + \frac{4}{10}$
Subtract $\frac{9}{10}y$	$\frac{1}{10}y + \frac{3}{10} = \frac{4}{10}$
Subtract $\frac{3}{10}$	$\frac{1}{10}y = \frac{1}{10}$

Multiply by 10 $y = 1$

Fourth way:

Subtract $3 + 9y$ (or $9y + 3$) $y = 1$

5. $5t + 1 = 2t + 9$ Subtract $2t + 1$ to get $3t = 8$,

divide by 3: $t = \frac{8}{3}$

6. a. 1 c. 3 e. 3 g. 3/4

Exercises 10.3

1. a. T b. T c. F d. T e. F f. F g. F h. T

3. a. T b. F c. F d. F

4. a. T c. F e. F

8. Hint: Read the very fine print at the bottom.

9. Hint: Look at the part of the "Certificate of Disbursement" that has the words, "THIS HEREBY. . .. " Where does that sentence really begin?

10. Hint: Do you assume you get to choose between the Atari and $1995? How much do you assume the Atari is worth?

Exercises 10.4

1. a. $9c$ ¢ b. $9c + 3$ ¢

d. $9c + 3 = 12c - 15$ Then $c = 6$; pencils cost 6¢ each, and I have $9c + 3$ ¢ when $c = 6$, or 57¢. (This is also $12 \times 6 - 15$¢.)

2. Hint: If you try trial and error, guess numbers which are common multiples of 4 and 3, so you can easily find two thirds and one fourth of your guess. If you prefer algebra and let n be the number, how do you express two thirds of the number and one fourth of the number?

5. Hint: As a start, if s = the son's age in years, express the wife's age in terms of s. Next, express the son's and wife's ages 4 years from now.

6. 12

7. Hint: The methods from example 1 work here. The answer is $18.

8. a. 4

9. Hint: $5n - 3 = n + 11$.

10. Hint: If the woman is now w years old, fill in a table like the one in example 3, with all ages expressed in terms of w.

11. If p is the parrot's present age in years, what will be its age in fourteen years? How can you represent two years older than three times its present age? Equate these two expressions to solve the problem.

12. 14 kids. From this you can easily find the amount of money.

13. Hint: Work backward. The \$42 left at the end was 2/3 of what I had initially, which therefore must have been \$63. Continue that way. Another way: In a guess and check table, try various initial amounts until you get one that ends up with \$42.

14. \$70.

16. Team A won game 1 by 12 to 6; team B won game 2 by 24 to 18.

17. Hint: This can be solved, like question 14, either by trial and error (use a guess and check table) or by working backwards. But be careful: When I bet half my money, I kept the rest of it, which I did not bet. Answer: I started with \$20.

18. Hint: One twelfth of the cargo is lost, so each merchant loses one twelfth of his cargo. The answers involve fractions of guilders.

20. The horses cost \$140 and \$220.

21. Hint: If n is one of the numbers, the other number is $999 - n$.

23. Hint: If the son did c problems correctly, the remaining $26 - c$ were incorrect. Figure out how much he earns from the correct problems, how much he owes from the others, equate those expressions and solve. Trial and error works well in this case, since you can see in advance that the answer must be a whole number less than 26.

25. \$92

27. Treat this much like question 23.

31. I bought 6 pears, 12 peaches, and 27 plums.

32. Hint: You can let n be the number of nickels, and figure out how much n nickels are worth. What does $56 - n$ stand for? Or, just use trial and error.

33. Hint: If x is the number of years Diophantus lived, then $\frac{1}{6}x+\frac{1}{12}x+\frac{1}{7}x+5$ was his age when his son was born.

34. Hint: Make a table to keep track of how many dollars the travelers, the hotel, and the bellboy have at each stage.

Exercises 10.5

1. $y = 8$

3. Inconsistent

5. $w = 16$, $z = 0$.

7. $z = 5$, $v = 1/2$

9. $v = -1/25$, $u = 8/5$

10. Inconsistent

15. Hint: $s = p + 1$, and $3(s - 7) = p$

16. Hint: Make a table of how many acres each has before and after the sale.

18. Let s and w be the ages, in years, of son and wife. Then $7s = w$. Represent their ages next year in terms of s and w, and write an equation stating that the son's age next year will be one sixth the wife's age then.

23. Each wood apple costs 5, and each citron costs 8. (No units of money are specified in this ancient question.)

Exercises 10.6

1. a. live births per thousand per year   b. feet per minute

c. parts per million   d. cents per pound

e. cents per dollar of sales   f. cents per \$1,000 assessed value

3. Slope is vertical change (rise or fall) per unit horizontal change.

4. $\dfrac{8.1 \text{ grams}}{\text{cubic centimeter}} \times 12 \text{ cubic centimeters} = 97.2 \text{ grams}$

5. (partial solution) The man's rate of drinking beer, in barrels per day, is $\frac{1}{20}$. If the woman would drink the barrel dry in w days, her rate of beer consumption, in barrels per day, is $\frac{1}{w}$. Each day, at those rates, the man and woman together drink $\frac{1}{20} + \frac{1}{w}$ barrels of beer. Since they will drain the barrel in 14 days, they drink together at a rate of $\frac{1}{14}$ barrels per day. Equate the expressions for their drinking rate and solve for w.

6. \$0.61 per quart

7. $\dfrac{5.8 \text{ pounds}}{\text{kilogram}} \times \dfrac{1.81 \text{ U.S. dollars}}{\text{British pound}} \times \dfrac{1 \text{ kilogram}}{2.2 \text{ pounds}} = \$4.77/\text{pound}$

8. \$1.80 per gallon

9. Hint: The hands meet at 12, and the large hand moves 12 times as fast as the small hand.

10. 600 km

11. Hint: Each hour the filling system adds $\frac{1}{4}$ the volume of the pool, and the drain system removes $\frac{1}{6}$ of the volume of the pool. What is the net change? Answer: 12 hours

12. If Maya has m mangos, then $37m + 105 = 45(m - 3)$, and $m = 30$.

13. a. $2 \times 58^\circ + 30^\circ = 146^\circ\text{F}$, but the correct value is $1.8 \times 58^\circ + 32^\circ = 136.4^\circ\text{F}$.

c. They are equal at 10°C.

14. a. The man runs $\frac{2}{5}$ of the bridge length to the near end, $\frac{3}{5}$ of it to the far end. In the extra time the man needs to reach the far end rather than the near end, the train travels the length of the bridge. Thus, in the time the man takes to run $\frac{1}{5}$ the length of the bridge, the train travels the entire length of the bridge. Then the train moves five times as fast as the man, or 65km/h.

Algebraic solution:
If s is number of meters train travels to near end of bridge and v is trains' velocity in km/h,

$\frac{s}{v} = \frac{12}{13}$ and $\frac{s + 30}{v} = \frac{18}{13}$

15. Hint: When 1 ounce of milk-coffee mixture is poured from the milk glass into the coffee glass, how much of it is milk and how much coffee? Another way: Each container ends up with as much fluid as it had at first.

16. They meet $109\frac{3}{8}$ miles from Venice, after $3\frac{15}{16}$ days of travel.

Exercises 10.7

1. $r \geq s$

3. $x \geq y + 3$

5. $c + d < \frac{c^2}{3}$

7. $a > 4$

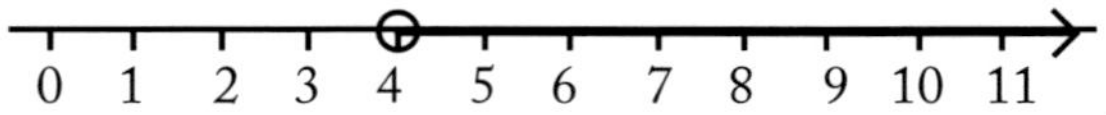

9. $c \geq 12$

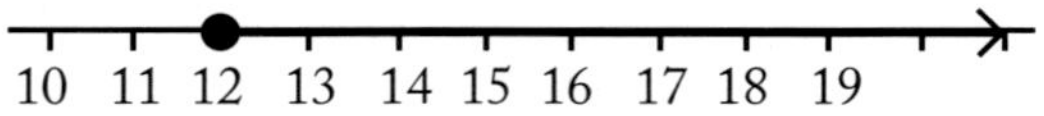

11. $e < -10$

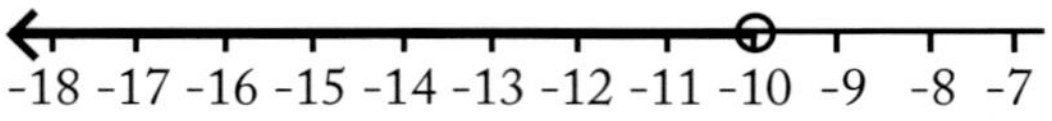

15. $k < \frac{2}{75}$ (graph omitted)

17. $y < \frac{52}{17}$

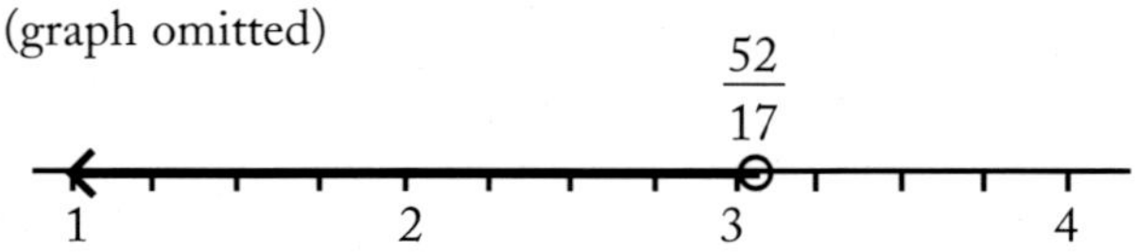

19. $90 < m < 130$

Self-Quiz for Chapter 10

1.

a. Take a number

b. Add three

c. Multiply by 5

d. Add three

e. Add the number you began with

f. Divide by six

g. Subtract your original number

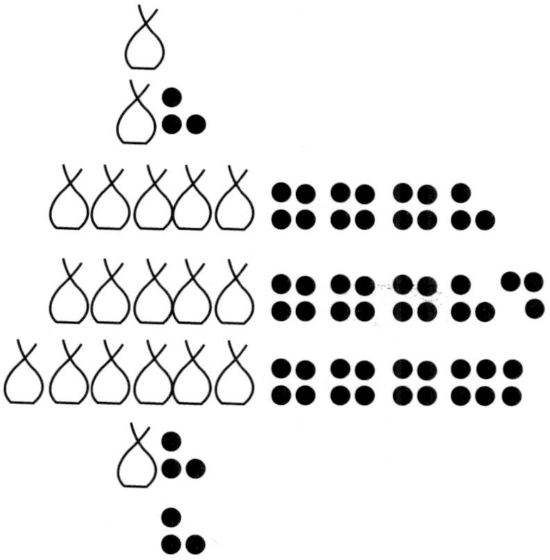

2. $\frac{3}{4}\left(\frac{-2}{3}-1\right)=\frac{1}{2\left(\frac{-2}{3}\right)}+\frac{1}{4}$.

The left hand side is $= \frac{3}{4}\left(\frac{-5}{3}\right)=\frac{-15}{12}=-\frac{5}{4}$

The right hand side is $= \frac{1}{\left(\frac{-4}{3}\right)}+\frac{1}{4} = \frac{-3}{4}+\frac{1}{4}=\frac{-2}{4}=\frac{-1}{2}$

Since $-\frac{5}{4}\neq\frac{-1}{2}$, $\frac{-2}{3}$ is not a root of this equation.

3. $t = -5$

4. The velvet cost 5 and 5/21 pesos per vara.

5. She walked 8 km in 2 hours for an average rate of 4 km per hour.

6. $3 < t$

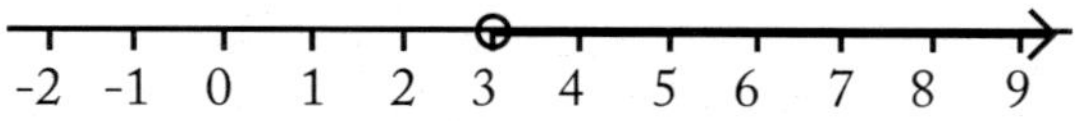

7. $x = 3, y = -2$

Exercises 11.1

1. a. Partial answer

parallelogram	K	L	M	R
base		2	4	
height		1	2	
area		2	8	

b. For parallelograms, Area = base × height

2. a. Partial answer

triangle	A	B	C	D	E	F	G	O	P
base	1	2		3		1			
height	1	1		1		1			
area	$\frac{1}{2}$	1		$1\frac{1}{2}$		$\frac{1}{2}$			

b. For triangles, Area = $\frac{1}{2}$ base × height

3. g. Partial answer

		Number of dots on boundary				
		3	4	5	6	7
Number of dots in interior	**0**	$\frac{1}{2}$	1		2	
	1	$1\frac{1}{2}$				
	2		3			
	3					$5\frac{1}{2}$
	4					

h. Each new boundary dot increases the area by 1/2.

i. Each new interior dot increases the area by 1.

j. Pick's rule for the areas of polygons with vertices at square lattice points: The area is the number of interior points plus half the number of boundary points minus 1. If there are i lattice points in the interior and b lattice points on the boundary,

$$\text{Area} = i + \frac{b}{2} - 1$$

4. a. The areas are all equal. The triangles all have base AB and height equal to the distance between the parallel lines.

8. Partial answer

Can Make	1	2		4	5			8	9	10			13	
Can't Make			3			6	7				11	12		14

9. a. Partial answer

Area of lattice square	1	2	3	4	5	6	7	8	9	10	11	12	13	14	15	16	17
Key move	(1,0)	(1,1)		(2,0)	(2,1)			(2,2)	(3,0)	(3,1)			(3,2)			(4,0)	(4,1)

b. Hint: Square each of the numbers in the key move.

10. A prime number that is the area of a lattice square is a number that can be expressed as the sum of two square numbers. In question 9 of exercises 6.3, those prime numbers were seen to be 2 and those which are one more than a multiple of 4. (For example, 17 is one more than a multiple of 4, but 15 is not.)

Exercises 11.2

2. b.

L	W	P	A
4	4	16	16
5	3	16	15
6	2	16	12
7	1	16	7

5. a. Hint: Number of possible rectangles is a simple fraction of the perimeter.

7. Hint: There are really two rules, one if the length of the longest side is odd, the other if it is even. In one case the answer is a triangular number added to itself, and in the other it is the sum of consecutive triangle numbers—i.e., a square number.

8. a.

perimeter	number of triangles	triangles, given by lengths of sides
3	1	1, 1, 1
4	0	
5	1	2, 2, 1
6	1	2, 2, 2
7	2	3, 2, 2 3, 3, 1
8	1	3, 3, 2
9	3	4, 4, 1 4, 3, 2 3, 3, 3
10	2	4, 4, 2 4, 3, 3
11	4	5, 5, 1 5, 4, 2 5, 3, 3 4, 4, 3
12	3	5, 5, 2 5, 4, 3 4, 4, 4

b. Students enjoy finding patterns in this problem, but a complete solution is difficult. *Focus*, vol 22, #5 (May-June 2002) has an article about how one high school class solved this problem.

Exercises 11.3

1. 4 m

2. a. It means that any objects shown on the map are 500,000 times as far apart in reality as their images on the map.

b. 115 km

c. About 44 miles

3. $4{,}000{,}000 \times 4\frac{3}{4}$ inches = 300 miles

5. a. $\left(\frac{50}{20}\right)^2 = \frac{25}{4} = 6\frac{1}{4}$.

Exercises 11.4

1. 1.331

3. $\frac{216}{343}$

5. $\frac{1}{6}$

6. b. A $1 \times 1 \times 6$ block has a surface area of 26 square units, A $1 \times 2 \times 3$ block has a surface area of 22 square units.

7. b. The surface area of a $1 \times 1 \times 12$ block is 50 square units.

The surface area of a $1 \times 2 \times 6$ block is 40 square units.

The surface area of a $1 \times 3 \times 4$ block is 38 square units.

The surface area of a $2 \times 2 \times 3$ block is 32 square units.

8. a. $1 \times 1 \times 36$, $1 \times 2 \times 18$, $1 \times 3 \times 12$, $1 \times 4 \times 9$, $1 \times 6 \times 6$, $2 \times 2 \times 9$, $2 \times 3 \times 6$, $3 \times 3 \times 4$.

c. The more nearly the length, width, and height of the block are equal, the smaller the surface area for a given volume.

Exercises 11.5

1. a. $8{,}000\pi$ or about 25,133 miles

b. Hint: The circumference of the new circle is 8000pi miles + 100 feet.

3. About 6 inches in diameter.

5. The number of letters in the words gives you digits of π.

Exercises 11.6

1. a. (Partial answer) $V = \pi\left(\frac{25}{2\pi}\right)^2 \cdot 8$

2. a. Hint: The height is 3(69) = 207 mm.

4. (Partial answer) The base area is $3\times70^2 + (35)^2$ square mm and the height is 210 mm.

Self-Quiz for Chapter 11

1.

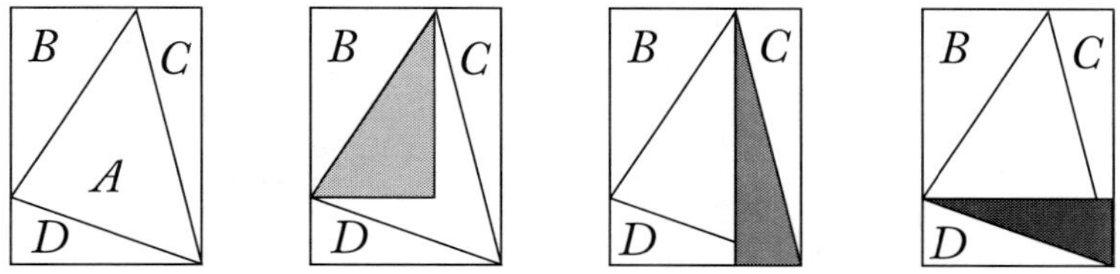

i ii iii iv

Enclose the triangle in a rectangle, and label the parts of the rectangle as shown in i. Part ii shows B is half of a rectangle 2 squares wide and 3 long, so B has area 3. Similarly, part iii shows C is half of a rectangle 1 square wide and 4 high, so C has area 2. Finally, D is half of a 1×3 rectangle, so D has area $1\frac{1}{2}$. Taking areas B, C, and D away from the 3×4 rectangle which encloses triangle A shows the area of A is $12 - 3 - 2 - 1\frac{1}{2} = 5\frac{1}{2}$.

A second solution: Triangle A has 5 interior dots and three on the boundary, so by Pick's rule its area is $5 + \frac{3}{2} - 1 = 5\frac{1}{2}$.

2. Enclose the tilted rectangle in a 5×5 square as shown in i.

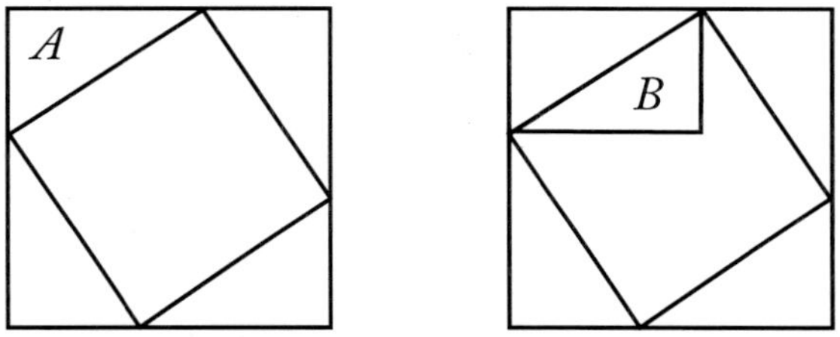

i ii

Triangle A, with triangle B (in figure ii) forms a 2×3 rectangle, so triangle A has area 3. Take the areas of 4 copies of triangle A from the 25 unit area of the enclosing square, leaving $25 - 12 = 13$ as the area of the tilted square.

3. Only 5, 10, 37, and 65 are sums of two squares.

4. You can always cut a parallelogram into two parts and reassemble those parts to make a rectangle with length and width equal to the base and height of the parallelogram. Cutting and reassembling does not change the area.

5. You can put two copies of any triangle together to make a parallelogram with the same base and height. Therefore the area of the triangle is half the area of the parallelogram, or half the base times the height.

6. 2,160 cubic feet

7. $\frac{1}{10}$ pound

8. $\frac{17}{\pi}$ feet or about 5.4 feet

9. $\frac{756^2 \times 481}{3}$ cubic feet or 91,636,272 cubic feet

10. $\frac{2}{3}$

11. $43 \times 6^3 = 9{,}288$ pounds

12. The steel drilled out of each hole is $(\frac{1}{4})^2$ the volume of the entire plate, or $\frac{1}{16}$ the volume. Drilling three such holes removes $\frac{3}{16}$ of the volume, so what remains weighs $30 \times \frac{13}{16} = 24\frac{3}{8}$ pounds.

Exercises 12.1

1. a. 49 b. 24 (6 each) c. $49 - 24 = 25$ d. $\sqrt{25} = 5$

e.

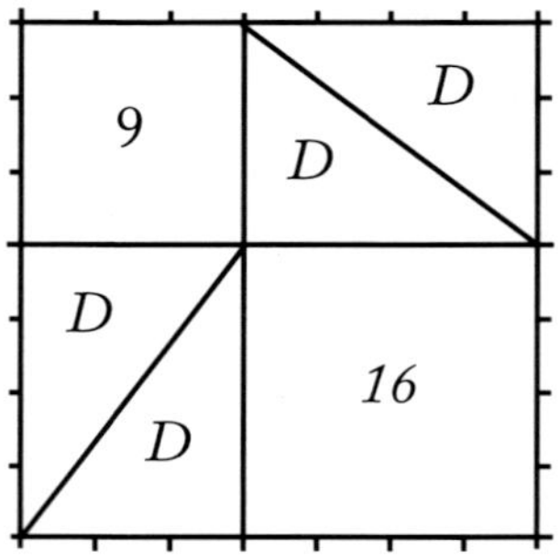

2. c. 289 d. $\sqrt{289} = 17$

4. a. $\sqrt{16^2 + 30^2} = \sqrt{256 + 900} = \sqrt{1{,}156} = 34$ c. 51

6. $\sqrt{96^2 + 40^2} = 104$ feet

7. $\sqrt{27^2 + 120^2} = \sqrt{729 + 14{,}400} = \sqrt{15{,}129} = 123$ km

9. In the equation $a^2 + b^2 = c^2$, if the length of the hypotenuse, c, and the length of one leg of the right triangle, a, are known, then the length of the second leg b may be calculated as $b = \sqrt{c^2 - a^2}$.

10. a. $\sqrt{169^2 - 119^2} = \sqrt{28{,}561 - 14{,}161} = \sqrt{14{,}400} = 120$

11. 50 feet

12. 4 feet

Exercises 12.2

1. a. 93 c. 302 e. 286

5. b. Hint: $\sqrt{8}$ is a simple multiple of $\sqrt{2}$.

6. $90\sqrt{2}$ or about 127'3"

8. b. 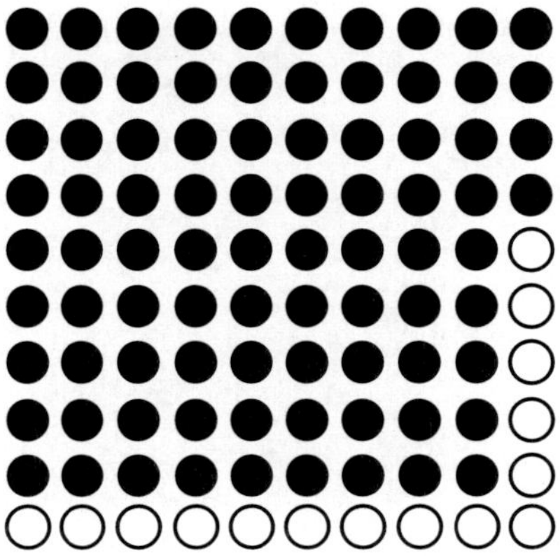 $9\frac{4}{19}$ or about 9.21

9. a. $2\sqrt{3}$ c. $3\sqrt{11}$

10. $\sqrt{5{,}281^2 - 5{,}280^2} = 102.76$

Exercises 12.3

1. a. 5 c. 10

2. a. 17

3. a. $2\sqrt{2}$

4. $x^2 + y^2 = 81$

7. $x^2 + (y - 7)^2 = 11$

11. $(x - 4)^2 + (y - 7)^2 = 25$

13. $(t - 25)^2 + (u + 23)^2 > 70^2$

17. (x, y) is further from (3, 4) than from (1, 1)

21. $(x - 2)^2 + (y - 5)^2 > 12^2$

23. Hint: You may be able to write the express these distances with the distance forumula and show that they are all the same without calculating all of them.

25. $|y - 4| = 2|y - 1|$

27. $|t - 7| = \frac{1}{2}|t - 13|$

30. c. $3wt^2\sqrt{3t}$

Exercises 12.5

1. a. $1 + \frac{1}{2} = \frac{3}{2}$ b. $1 + \cfrac{1}{2 + \cfrac{1}{2}} = \frac{7}{5}$ c. $1 + \cfrac{1}{2 + \cfrac{1}{2 + \cfrac{1}{2}}} = \frac{17}{12}$

d. $1 + \cfrac{1}{2 + \cfrac{1}{2 + \cfrac{1}{2 + \cfrac{1}{2}}}} = \frac{41}{29}$

2. a. If $\frac{a}{b}$ is a fraction in the sequence, the next is $\frac{a+2b}{a+b}$. This leads to the prediction $\frac{41+2\times 29}{41+29}=\frac{99}{70}$ for the next in the sequence. Indeed,

$$1+\cfrac{1}{2+\cfrac{1}{2+\cfrac{1}{2+\cfrac{1}{2+\frac{1}{2}}}}}=1+\cfrac{1}{2+\cfrac{1}{2+\cfrac{1}{2+\cfrac{1}{\frac{5}{2}}}}}=1+\cfrac{1}{2+\cfrac{1}{2+\cfrac{1}{2+\frac{2}{5}}}}=1+\cfrac{1}{2+\cfrac{1}{2+\cfrac{1}{\frac{12}{5}}}}$$

$$=1+\cfrac{1}{2+\cfrac{1}{2+\frac{5}{12}}}=1+\cfrac{1}{2+\cfrac{1}{\frac{29}{12}}}=1+\cfrac{1}{2+\frac{12}{29}}=1+\cfrac{1}{\frac{70}{29}}=1+\frac{29}{70}=\frac{99}{70}$$

3. a. $\left(\frac{3}{2}\right)^2=\frac{9}{4}$ $\quad\left(\frac{7}{5}\right)^2=\frac{49}{25}$ $\quad\left(\frac{17}{12}\right)^2=\frac{289}{144}$ $\quad\left(\frac{41}{29}\right)^2=\frac{1{,}681}{841}$

b. $\frac{9}{4}=2+\frac{1}{4}$ $\quad\frac{49}{25}=2-\frac{1}{25}$ $\quad\frac{289}{144}=2+\frac{1}{144}$ $\quad\frac{1{,}681}{841}=2-\frac{1}{841}$

$\frac{9{,}801}{4{,}900}=2+\frac{1}{4{,}900}$

c. The squares are alternately above and below 2 and are closing in on 2, so the fractions before squaring are closing in on $\sqrt{2}$.

4. a. $\frac{4}{3}$ b. $\frac{13}{10}$ c. $\frac{43}{33}$ d. $\frac{142}{109}$

e. Squaring, $\left(\frac{4}{3}\right)^2=\frac{16}{9}$, $\left(\frac{13}{10}\right)^2=\frac{169}{100}$, $\left(\frac{43}{33}\right)^2=\frac{1{,}849}{1{,}089}$, $\left(\frac{142}{109}\right)^2=\frac{20{,}164}{11{,}881}$
The squares are not approaching 3, since they are all less than 2, so the continued fractions themselves are not approaching $\sqrt{3}$.

5. a. $\frac{9}{4}$ b. $\frac{38}{17}$ c. $\frac{161}{72}$ d. $\frac{682}{305}$

e. If $\frac{a}{b}$ is a fraction in the sequence, the next is $\frac{2a+5b}{a+2b}$.

f. Prediction: $\frac{2\times 682+5\times 305}{682+2\times 305}=\frac{2{,}889}{1{,}292}$ for the next in the sequence.

g. $2+\cfrac{1}{4+\cfrac{1}{4+\cfrac{1}{4+\cfrac{1}{4+\cfrac{1}{4}}}}}=2+\cfrac{1}{4+\cfrac{72}{305}}=2+\frac{305}{1{,}292}=\frac{2{,}889}{1{,}292}$

h and i.

$$\left(\frac{9}{4}\right)^2=\frac{81}{16}=5+\frac{1}{16},\quad \left(\frac{38}{17}\right)^2=\frac{1{,}444}{289}=5-\frac{1}{289},$$

$$\left(\frac{161}{72}\right)^2=\frac{25{,}921}{5{,}184}=5+\frac{1}{5{,}184},\quad \left(\frac{682}{305}\right)^2=\frac{465{,}124}{93{,}025}=5-\frac{1}{93{,}025},$$

$$\left(\frac{2{,}889}{1{,}292}\right)^2=\frac{8{,}346{,}321}{1{,}669{,}264}=5+\frac{1}{1{,}669{,}264}$$

j. The squares approach 5, so the continued fractions approach $\sqrt{5}$.

6. a. $\frac{19}{6}, \frac{117}{37}, \frac{721}{228}, \frac{4{,}443}{1{,}405}$

If $\frac{a}{b}$ is a fraction in the sequence, the next is $\frac{3a+10b}{a+3b}$.

Prediction: $\frac{3\times 4{,}443+10\times 1{,}405}{4{,}443+3\times 1{,}405}=\frac{27{,}379}{8{,}658}$

Check:

$$3+\cfrac{1}{6+\cfrac{1}{6+\cfrac{1}{6+\cfrac{1}{6+\cfrac{1}{6}}}}}=3+\cfrac{1}{6+\cfrac{1}{6+\cfrac{37}{228}}}=3+\cfrac{1}{6+\cfrac{228}{1{,}405}}=\frac{27{,}379}{8{,}658}$$

$$\left(\frac{19}{6}\right)^2=\frac{361}{36}=10+\frac{1}{36},\quad \left(\frac{117}{37}\right)^2=\frac{13{,}689}{1{,}369}=10-\frac{1}{1{,}369},$$

$$\left(\frac{721}{228}\right)^2=\frac{519{,}841}{51{,}984}=10+\frac{1}{51{,}984},\quad \left(\frac{4{,}443}{1{,}405}\right)^2=\frac{19{,}740{,}249}{1{,}974{,}025}=10-\frac{1}{1{,}974{,}025}$$

$$\left(\frac{27{,}379}{8{,}658}\right)^2=\frac{749{,}609{,}641}{74{,}960{,}964}=10+\frac{1}{74{,}960{,}964}$$

The squares approach 10, and the continued fractions approach $\sqrt{10}$.

7. a. $\sqrt{101}$

8. The sequence $n + \frac{1}{2n}$, $\quad n + \cfrac{1}{2n + \cfrac{1}{2n}}$, $\quad n + \cfrac{1}{2n + \cfrac{1}{2n + \cfrac{1}{2n}}} \ldots$

approaches $\sqrt{n^2 + 1}$.

Self-Quiz for Chapter 12

1. $\sqrt{85^2 - 84^2} = \sqrt{169} = 13$

2. Newton's method: Try 3, which divides into 7.5 with a quotient of 2.5. Average 2.5 and 3 to get next guess 2.75. Divide 7.5 by 2.75, get 2.73 as quotient. Average 2.73 and 2.75 to get next guess, 2.74. Divide 7.5 by 2.74 to get 2.737. Now $2.737 < \sqrt{7.5} < 2.74$, so to two decimal places $\sqrt{7.5} = 2.74$.

3. To square a number, multiply it by itself. To find the square root of a number, find a number whose square is the given number. For example, the square of 9 is 81, and the square root of 9 is 3.

4. 13

5. $(x + 5)^2 + (y - 2)^2 = 49$

6. a. $\dfrac{4\sqrt{6}}{5\sqrt{5}}$ b. $8|x|\sqrt{3xy}$

7. $\dfrac{151}{115}$

8. $\sqrt{4 + 1 + 1} = \sqrt{6}$ or about 2.45

9. $|x - 7| > 5$. Then $x > 12$ or $x < 2$. (picture omitted)

Exercises 13.1

1. a.

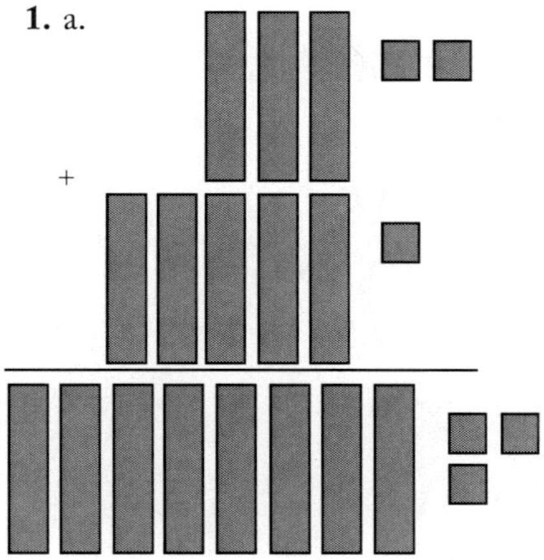

$3x + 2$

$+$

$5x + 1$

$8x + 3$

c.

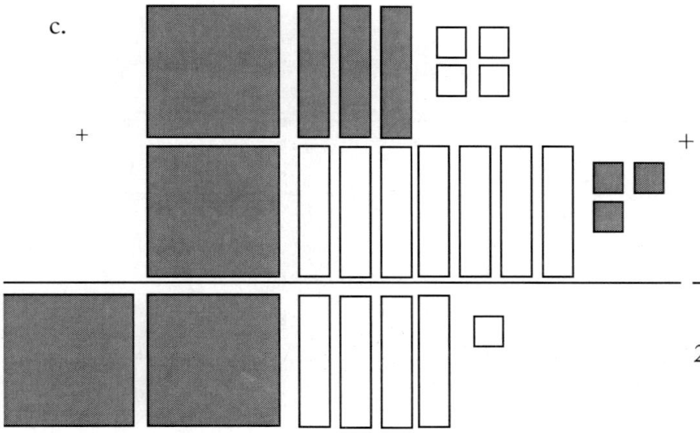

$x^2 + 3x - 4$

$+$

$x^2 - 7x + 3$

$2x^2 - 4x - 1$

3. $2a^2 + 3ab - b^2$

5. $10y^3 + 4y^2 - y + 9$

7. $-x^3y + x^2y^2 + 4xy^3 + y^4$

9. $-2.4d^2 + 4.07d - 5.9$

11. $-u + y$

13. $-y^3 + 6y^2 - 3y - 8$

15. $-5y^3 - 2y^2 - 2y + 6$

Exercises 13.2

1. $(x + 2)6$

3. $x(x\text{-}2)$

7. $63t^8s$

9. $-276\, p^5r^5q^8$

12.

$$\begin{array}{r} -2x + 1 \\ \times \quad 3x - 4 \\ \hline 8x - 4 \\ -6x^2 + 3x \\ \hline -6x^2 + 11x - 4 \end{array}$$

15.

	$5w^2$	w	$5w^2 + w$
$4w^3$	$20w^5$	$4w^4$	$20w^5 + 4w^4$
-2	$-10w^2$	$-2w$	$-10w^2 - 2w$
$4w^3 - 2$	$20w^5 - 10w^2$	$4w^4 - 2w$	$20w^5 + 4w^4 - 10w^2 - 2w$

20. $d^2e^2 - 25$

22. $b^4 - a^2$

24. $k^3 - g^3$

26. $n^4 + n^2p^2 + p^4$

28. $4r^2 + 4r + 1$

31. b. Use Pascal's triangle. The coefficients are 1, 6, 15, 20, 15, 6, 1. Now fill in the appropriate powers of a and x to get

$$(a+x)^6 = a^6 + 6a^5x + 15a^4x^2 + 20a^3x^3 + 15a^2x^4 + 6ax^5 + x^6$$

32. a. When $a = x = 1$, $(a + x)^n = 2^n$, and the sum of powers in the binomial formula reduces to the sum of the coefficients. For example, the third line of Pascal's triangle
$(a + x)^3 = a^3 + 3a^2x + 3ax^2 + x^3$ becomes
$(1 + 1)^3 = 1^3 + 3 \cdot 1^2 \cdot 1 + 3 \cdot 1 \cdot 1^2 + 1^3$, or $2^3 = 1 + 3 + 3 + 1$
Similar explanations show that all rows of Pascal's triangle (below the lone 1 at the top) add up to powers of 2.

33. b. $$\frac{\left(1+\sqrt{5}\right)^2 - \left(1-\sqrt{5}\right)^2}{2^2\sqrt{5}} = \frac{\left(1+2\sqrt{5}+5\right) - \left(1-2\sqrt{5}+5\right)}{4\sqrt{5}} = \frac{4\sqrt{5}}{4\sqrt{5}} = 1$$

c. By the binomial theorem,

$$\left(1+\sqrt{5}\right)^3 = 1^3 + 3\cdot 1^2\sqrt{5} + 3\cdot 1\left(\sqrt{5}\right)^2 + \left(\sqrt{5}\right)^3$$

$$= 1 + 3\sqrt{5} + 3\cdot 5 + 5\sqrt{5} = 16 + 8\sqrt{5}$$

Similarly, $\left(1-\sqrt{5}\right)^3 = 16 - 8\sqrt{5}$

Then $$\frac{\left(1+\sqrt{5}\right)^3 - \left(1-\sqrt{5}\right)^3}{2^3\sqrt{5}} = \frac{\left(16+8\sqrt{5}\right) - \left(16-8\sqrt{5}\right)}{8\sqrt{5}} = \frac{16\sqrt{5}}{8\sqrt{5}} = 2$$

e. 5

34. a. $a^2 - x^2$ c. $a^4 - x^4$

Exercises 13.3

1.

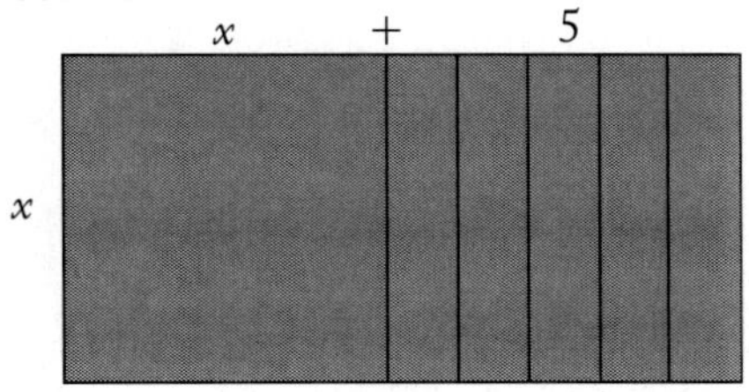

$(x^2 + 5x) \div x = x + 5$

5.

$$\begin{array}{r|l} & x - 4 \\ \hline x + 5 & x^2 + x - 30 \\ & \underline{x^2 + 5x} \\ & -4x - 30 \\ & \underline{-4x - 20} \\ & -10 \end{array}$$

Quotient x-4, remainder -10

7. $a + 4$

9. $c + 8$

11. $e - 5$ remainder 2

13. $g^2 - 7g - 2$ remainder 6

15. $j^2 - 18j$ remainder 3

17. $8z^4 + 24z^2 + 72$ remainder 189

19. (partial answer)

Remainder when $x^2 - x - 20$ is divided by		Value of $x^2 - x - 20$ when	
$x - 1$	-20	$x = 1$	-20
$x - 2$	-18	$x = 2$	-18
$x - 3$	-14	$x = 3$	-14
$x - 4$		$x = 4$	
$x - 5$		$x = 5$	
x		$x = 0$	
$x + 1$		$x = -1$	
$x + 2$		$x = -2$	
$x + 3$		$x = -3$	

20.

$$\begin{array}{r|l} & x + (a - 1) \\ \hline x - a & x^2 - x - 20 \\ & \underline{x^2 - ax} \\ & (a - 1)x - 20 \\ & \underline{(a - 1)x - a(a - 1)} \\ & a(a - 1) - 20 \\ & = a^2 - a - 20 \end{array}$$

21. a. (partial answer).

Remainder when $x^3 - 4$ is divided by		Value of $x^3 - 4$ when	
$x - 1$	-3	$x = 1$	-3
$x - 2$	4	$x = 2$	4
$x - 3$	23	$x = 3$	23
$x - 4$		$x = 4$	
$x - 5$		$x = 5$	
x		$x = 0$	
$x + 1$		$x = -1$	
$x + 2$		$x = -2$	
$x + 3$		$x = -3$	

c.

$$\begin{array}{r}
x^2 + bx + b^2 \\
x - b \overline{\big)\, x^3 \qquad\qquad\qquad - 4} \\
\underline{x^3 - bx^2} \qquad\qquad\quad \\
bx^2 \qquad\qquad\quad \\
\underline{bx^2 - b^2x} \qquad\quad \\
b^2x \qquad - 4 \\
\underline{b^2x - b^3} \quad\; \\
b^3 - 4
\end{array}$$

22. If a polynomial in x is divided by $(x - c)$, the remainder is the value of the polynomial when $x = c$. The result discovered here is called the **remainder theorem.** In question 20, when $x^2 - x - 20$ is divided by $x - a$, the remainder is the value of $x^2 - x - 20$ when $x = a$. In question 21, when $x^3 - 4$ is divided by $x - b$, the remainder is the value of $x^3 - 4$ when $x = b$.

A corollary of the remainder theorem, known as the **factor theorem,** states that the value of a polynomial is 0 for $x = c$ if and only if $(x - c)$ is a factor of the polynomial. The factor theorem is one reason why factoring is important, not only for quadratic equations but for other equations as well.

Exercises 13.4

1. a. The first factorization is $6x(x + 2)$, while the second is $6(2+x)x$,

2. a. $7x$ b. 1 c. $-3y$ d. $7x(1-3y)$

5. a. $10k$ b. 3 c. $5t$ d. $10k(3+5t)$

6. a. $2x$ b. $9y$ c. $(2x+9y)(2x-9y)$

9. a. $12xy^3$ b. 1 c. $(12xy^3+1)(12xy^3-1)$

12. $a = x, b = -4$ $(x-4)^2$

15. $a = 2y, b = -1$ $(2y-1)^2$

18. $a = 3s, b = 2$ $(3s-2)(9s^2+6s+4)$

23. $a = -2, b = 6s^3$ $(-2+6s^3)(4-12s^3+36s^6)$

24. (Partial answer) Difference of cubes

29. (Partial answer) Difference of squares

31. $y(y + 3)$

33. $w(5w - 1)$

35. $10s(s - 4)$

37. $x(x^2 + 3x + 1)$

39. $x^5(1 + x)$

41. $(2a - 5b)(2a + 5b)$

43. $(12r^3 - 9)(12r^3 + 9)$

45. $(6 - v)^2$

47. $(4s + 7t)^2$

49. $(1 + m)(m^2 - m + 1)$

51. $(4y - 10)(16y^2 + 40y + 100)$

53. $(3j + 2)(j + 2)$

55. $2w(5w^2 - 6w + 2)$

Exercises 13.5

1. $3(a - 2)(a + 2)$

3. $2(6 - c)(6 + c)$

5. $2e^2(3e + 1)$

7. $7(g - 2)(g - 3)$

9. $k(y - 11)(y + 11)$

11. $3(m - 4n)^2$

13. $3(r - 7s)(r + 2s)$

15. $5y^3(y^2 + 4y + 5)$

17. $(2 - x)(2 + x)(4 + x^2)$

19. $(c^2 + d^2)(x + 3)$

21. $4(j + 3)(j^2 - 3j + 9)$

23. $(s + t - u)(s + t + u)$

25. This is a difference of squares. $(7 - y - A)(7 + y + A)$

27. Hint: This is a difference of squares. One of the squares is 1.

29. $12ab^3c$

31. $(k - 1)(k + 1)(k - 4)$

33. $30(m - 1)(m + 1)$

35. $10(r - z)(r + z)(r^2 + z^2)$

37. $(u^2 + 1)(u - 1)(u + 1)(u^2 + u + 1)$

39. $(z - 5)(z + 4)(z - 3)$

41. $(X + 2)(X + 3)(X + 4)$

Exercises 13.6

1. $\frac{8a^2}{15d^2}$

3. 4

5. $\frac{p^2}{4aq}$

7. 1

9. $\frac{3f^2 - 3f}{6f^2 - 10f - 4}$

11. $q - p$

13. $2a(a - b)^2(X - B)$

15. $\frac{2a}{b}$

17. $\frac{8x}{7ay}$

19. $\frac{x-1}{x-2}$

21. $\frac{9a + x}{20}$

23. $\frac{2j - 8y}{15}$

25. $\frac{2}{1 - v^2}$

27. $\frac{5t + 24}{12}$

29. a.

Inputs n	0	1	2	3	4	5	-2	12	-9
Outputs $\frac{3n+6}{2n-3}$	-2	-9	12	5	$\frac{18}{5}$	3	0	2	1

b. The inputs and outputs may be interchanged. If the input A yields the output B, then the input B yields the output A.

c. Since the input of 4 yields $\frac{18}{5}$, the input of $\frac{18}{5}$ should yield 4.

d. $$\frac{3\left(\frac{18}{5}\right)+6}{2\left(\frac{18}{5}\right)-3}=\frac{\frac{54}{5}+\frac{30}{5}}{\frac{36}{5}-\frac{15}{5}}=\frac{\frac{84}{5}}{\frac{21}{5}}=\frac{84}{5}\times\frac{5}{21}=\frac{84}{21}=4$$

e. $$\frac{3\left(\frac{3n+6}{2n-3}\right)+6}{2\left(\frac{3n+6}{2n-3}\right)-3}=\frac{\frac{9n+18}{2n-3}+6\frac{2n-3}{2n-3}}{\frac{6n+12}{2n-3}-3\frac{2n-3}{2n-3}}$$

$$=\frac{\frac{9n+18+12n-18}{2n-3}}{\frac{6n+12-6n+9}{2n-3}}$$

$$=\frac{21n}{2n-3}\times\frac{2n-3}{21}=\frac{21n}{21}=n$$

30. a.

Inputs n	0	1	12	3	4	-8	-3	7
Outputs $\frac{2n+6}{n-2}$	-3	-8	3	12	7	1	0	4

31. $$\frac{2\left(\frac{2n+6}{5n-2}\right)+6}{5\left(\frac{2n+6}{5n-2}\right)-2}=\frac{\frac{4n+12+30n-12}{5n-2}}{\frac{10n+30-10n+4}{5n-2}}=\frac{34n}{34}=n$$

32. You get $\frac{25n+4}{4n+20}$. This is not n, so this rule does not have the special property of the other rules studied here.

33. Yes for a, b, g, h, j.

35. Hint: The relationship involves only two of the letters A, B, C, and D.

Self-Quiz for Chapter 13

1. $\frac{-5}{27}$

2. $-7x^5 + 4x^3 + 12x^2 - 21x - 18$

3. $10y^4 + y^3 - 3y^2 - y - 7$

4. $5z^6 - 4z^5 + 2z^4 - 25z^3 + 25z^2 - 14z + 2$

5. $4w^2 + 2w + 1$, remainder 3

6. a. $(s - 2t)(s + 2t)(s^2 + 4t^2)$ b. $3(2x + 7)(x + 1)$

7. $12(x + 1)^2(2x + 1)$

8. $\frac{w^2-2}{2+6w}$

9. $16x^2 + 16x - 21$

10. $\frac{y^2z+2y^4+3z^2}{y^3z^2}$

11. $\frac{-2b}{(a+b)^2(a-b)}$

12. $\frac{1}{1-x}$

Exercises 14.1

1. a. Factor: $(t - 3)(t + 3) = 0$. If $t - 3 = 0$, then $t = 3$. If $t + 3 = 0$, then $t = -3$.

   b. Square roots: $t = \pm 3$

2. a. $t - 1 = \pm 5$. If $t - 1 = -5$, then $t = -4$. If $t - 1 = 5$ then $t = 6$

4. c. Subtract 6. $z^2 + 4z + 4 = 169$ or $(z + 2)^2 = 169$ and $z + 2 = \pm 13$.

   If $z + 2 = -13$, then $z = -15$. If $z + 2 = 13$, then $z = 11$.

5. a. $(x - 17)(x + 9) = 0$. If $x - 17 = 0$, then $x = 17$. If $x + 9 = 0$, then $x = -9$.

   d. $t = 4$ or $t = 8$

7. b. Although we saw that $y^2 + 4y - 9$ is a prime polynomial in the arena of polynomials with integer coefficients, $y^2 + 4y - 9$ is not a prime polynomial in the larger arena of polynomials with real coefficients, because $y^2+4y-9=\left(y+2-\sqrt{13}\right)\left(y+2+\sqrt{13}\right)$.

9. b.

$$\begin{aligned} x^2+3x+1&=2\\ x^2+3x+1+\frac{5}{4}&=2+\frac{5}{4}\\ x^2+3x+\frac{4}{4}+\frac{5}{4}&=\frac{8}{4}+\frac{5}{4}\\ x^2+3x+\frac{9}{4}&=\frac{13}{4}\\ \left(x+\frac{3}{2}\right)^2&=\frac{13}{4}\\ x+\frac{3}{2}&=\pm\sqrt{\frac{13}{4}}\\ x&=-\frac{3}{2}\pm\frac{\sqrt{13}}{2} \end{aligned}$$

Exercises 14.2

1. $z = -13$, $z = 3$

2. $t = -2$, $t = 1$

5. $x=\frac{3}{2}$, $x=-\frac{2}{5}$

7. a.
$$\begin{aligned}\left(2+\sqrt{-1}\right)^3&=2^3+3\left(2^2\right)\sqrt{-1}+3\left(2^1\right)\left(\sqrt{-1}\right)^2+\left(\sqrt{-1}\right)^3\\ &=8+12\sqrt{-1}-6-1\sqrt{-1}\\ &=2+11\sqrt{-1}\\ &=2+\sqrt{-121}\end{aligned}$$

d. Hint: Add $2+\sqrt{-1}+2-\sqrt{-1}$. Explain why that is relevant.

9. Center $\left(\frac{-3}{2},\frac{7}{2}\right)$, radius $\frac{\sqrt{58}}{2}$

12. a.

$$\begin{aligned} x &= 1+\frac{1}{1+x} \\ x-1 &= \frac{1}{1+x} \\ (x-1)(1+x) &= 1 \\ x^2-1 &= 1 \\ x^2 &= 2 \\ x &= \pm\sqrt{2} \end{aligned}$$

13. $t=\dfrac{15\pm\sqrt{(-15)^2-4(1)(9)}}{2(1)}=\dfrac{15\pm\sqrt{225-36}}{2}=\dfrac{15\pm\sqrt{189}}{2}=\dfrac{15\pm3\sqrt{21}}{2}$

Therefore, the polynomial factors as $\left(t-\dfrac{15+3\sqrt{21}}{2}\right)\left(t-\dfrac{15-3\sqrt{21}}{2}\right)$.

15. $v=\dfrac{13\pm\sqrt{(-13)^2-4(1)(11)}}{2(1)}=\dfrac{13\pm\sqrt{169-44}}{2}=\dfrac{13\pm\sqrt{125}}{2}=\dfrac{13\pm5\sqrt{5}}{2}$

Therefore, the polynomial factors as $\left(v-\dfrac{13+5\sqrt{5}}{2}\right)\left(v-\dfrac{13-5\sqrt{5}}{2}\right)$.

20. Begin with $ax^2+bx+c=0$

Subtract c $\quad ax^2+bx=-c$

Divide by a $\quad x^2+\dfrac{b}{a}x=-\dfrac{c}{a}$

Complete the square by adding $\left(\dfrac{b}{2a}\right)^2$ $\quad x^2+\dfrac{b}{a}x+\left(\dfrac{b}{2a}\right)^2=\left(\dfrac{b}{2a}\right)^2-\dfrac{c}{a}$

Rewrite both sides of the equation $\quad \left(x+\dfrac{b}{2a}\right)^2=\dfrac{b^2}{4a^2}-\dfrac{c}{a}$

$$=\frac{b^2-4ac}{4a^2}$$

Take square roots $\quad x+\dfrac{b}{2a}=\pm\dfrac{\sqrt{b^2-4ac}}{2a}$

Subtract $\dfrac{b}{2a}$ from both sides $\quad x=-\dfrac{b}{2a}\pm\dfrac{\sqrt{b^2-4ac}}{2a}$

Add the fractions over the common denominator $\quad x=\dfrac{-b\pm\sqrt{b^2-4ac}}{2a}$

Exercises 14.3

1. Let x be the number. Then $x(x + 6) = 891$, and $x^2 + 6x - 891 = 0$. Factoring, $(x - 27)(x + 33) = 0$, and $x = 27$ or -33. Then the number is either -33 or 27.

3. 36 and 17

5. Let w be the width of the smaller rectangle. Then $\frac{4}{3}w$ is the length of the smaller rectangle, and $\frac{4}{3}w^2$ is its area. The larger rectangle has width $w + 2$ and length $\frac{4}{3}w + 4$, so its area is $(w+2)\left(\frac{4}{3}w+4\right)$.

Then $(w+2)\left(\frac{4}{3}w+4\right)=2\left(\frac{4}{3}w^2\right)$. The roots of this equation are -1 and 6, but only the positive root makes sense in this context.

Therefore, the smaller rectangle is 6 by 8, the larger 8 by 12.

7. Hints: Let n be the number of lemons bought. Then they cost $\frac{400}{n}$ ¢ each. If this price were lowered by 4¢, the cost per lemon would be $\frac{400}{n} - 4$.

9. Let the width of the original sheet be w cm. Then the length is $2w$, and the area of the base of the box is $(w - 4)(2w - 4)$. The volume of the box is $4(w - 4)(2w - 4) = 60$. The roots of this equation are 7 and -1, but only a positive number can be a width. Then the original sheet was 7cm by 14 cm.

11. 70 sheep

12. Hint: Use the Pythagorean theorem.

13. 90 hours

15. 420 km/hour

17. 1.2 liters of liquid were drained each time.

19. 10 hours for the larger pipe, 15 hours for the smaller

21. The width is 5, the length 12.

Self-Quiz for Chapter 14

1. $x = -8$ or 5

2. $(2w + 3)(w + 11) = 0$. Then $w = -3/2$ or -11.

3. 20.25

4. $\left(x+\frac{5-2\sqrt{7}}{3}\right)\left(x+\frac{5+2\sqrt{7}}{3}\right)$

5. 9 cm by 18 cm

Exercises 15.1

1. a. Ratio b. Categorical c. Ratio d. Ratio
e. Categorical f. Ratio g. Ratio
h. Ordinal i. Ordinal

2. a. Al wins with 6 first-place voted. b. 9
c. 9. Note that Al wins, but most voters would prefer either of the other candidates to Al!

3. a. Al gets 12 points, Bo gets 19, and Cy gets 14.

Exercises 15.2

1. a. $\varnothing$, $\{a\}$, $\{b\}$, $\{c\}$, $\{a, b\}$, $\{a, c\}$, $\{b, c\}$, $\{a, b, c\}$

b. $\varnothing$, $\{a\}$, $\{b\}$, $\{c\}$, $\{d\}$, $\{a, b\}$, $\{a, c\}$, $\{a, d\}$, $\{b, c\}$, $\{b, d\}$, $\{c, d\}$
$\{a, b, c\}$, $\{a, b, d\}$, $\{a, c, d\}$, $\{b, c, d\}$, $\{a, b, c, d\}$

c. The text shows a set with 2 elements has 4 subsets.
Part a shows a set with 3 elements has 8 subsets.
Part b shows a set with 4 elements has 16 subsets.

d. On the basis of the above, we predict that a set with 5 elements has $2^5 = 32$ subsets.

2. a. Hint: Might some people not be in the phone book?

3. Hint: Who was most likely to be absent when the interviewers first visited the neighborhood?

4. Few of the university students taking freshman calculus would have had the capability to take that course while they were still in high school. Those who had reached that level a full year before leaving high school are probably more sophisticated for their age.

5. People are sometimes stimulated to improvement by the mere awareness that they are being studied.

Exercises 15.3

1. A line graph is misleading in this case, because the data is categorical and has no natural order that would correspond to position on the horizontal axis.

2. Hint: If you want to emphasize the change between last year and this, which of the two graphs would you choose?

5. a.

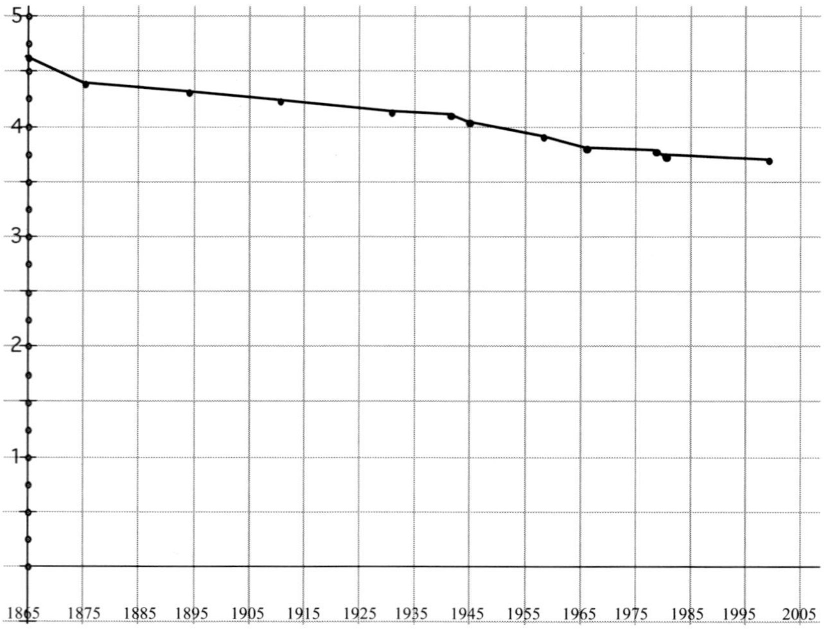

b. The graph appears to be dropping steadily to the right. Continuing it that way allows us to estimate the year when it will get down to three and a half minutes.

6.

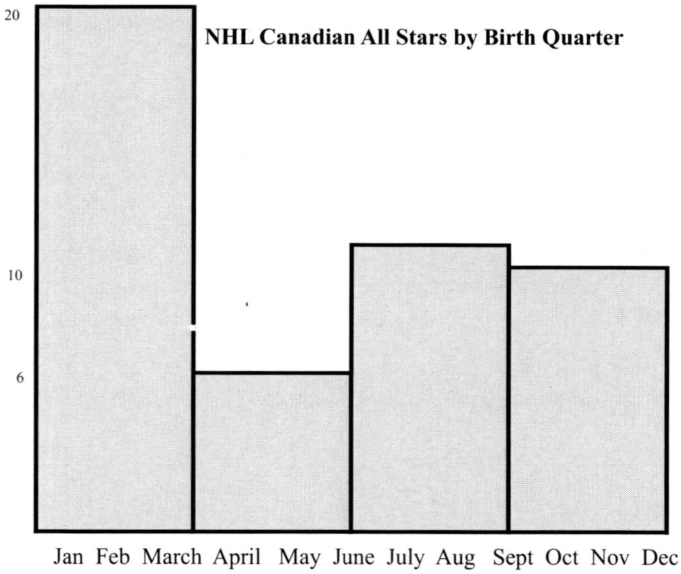

7. a.

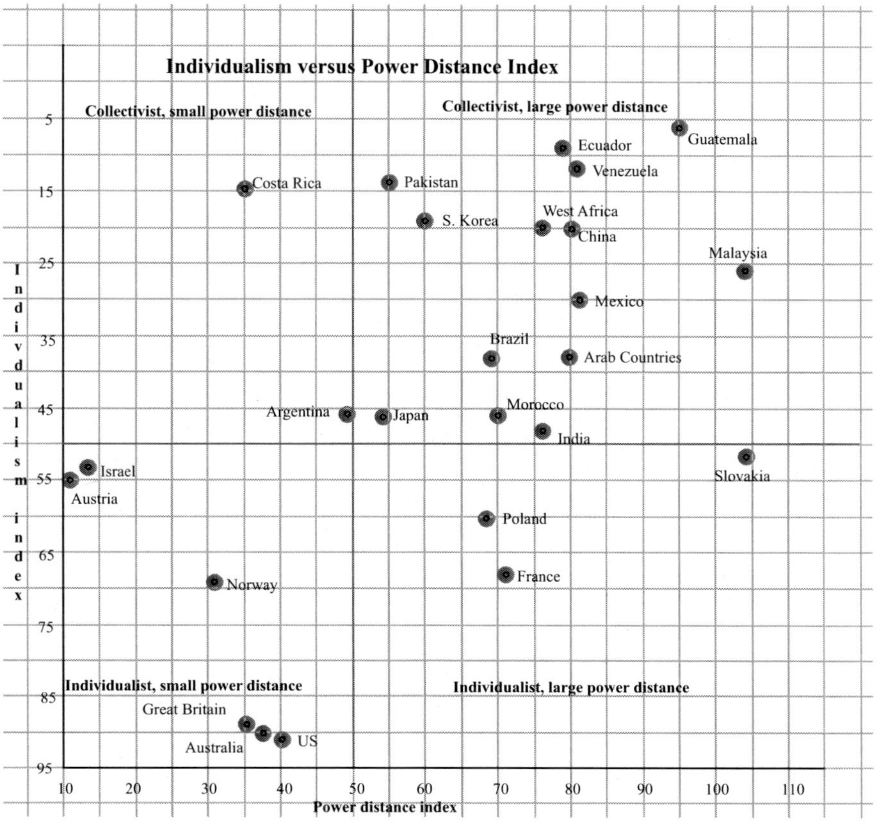

b. Hofstede's data is ordinal.

Self-Quiz for Chapter 15

3.

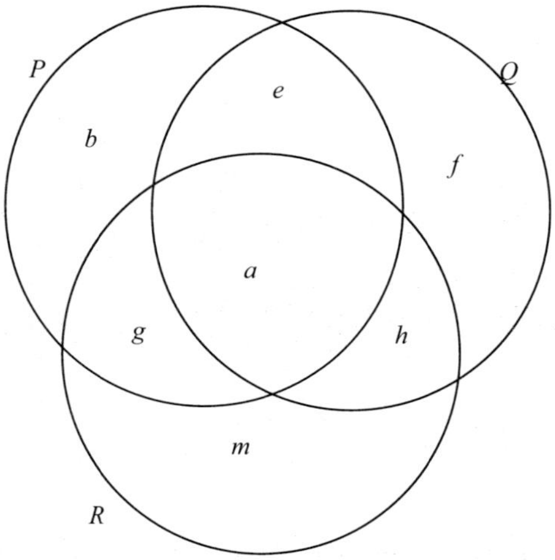

4. a. Range is from 48 to 100, a range of 42.

b. The median is 84.5

c.

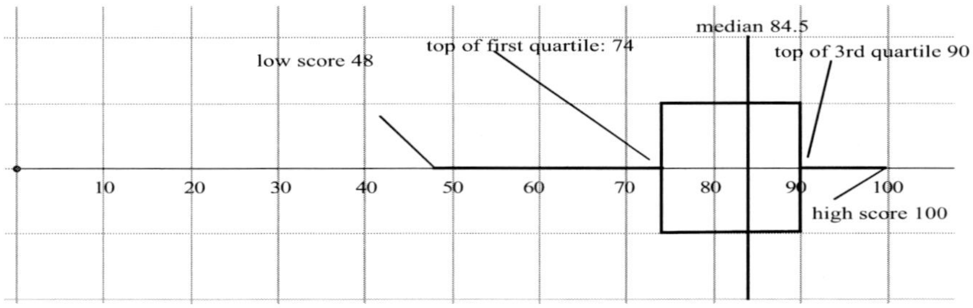

d. i

5.

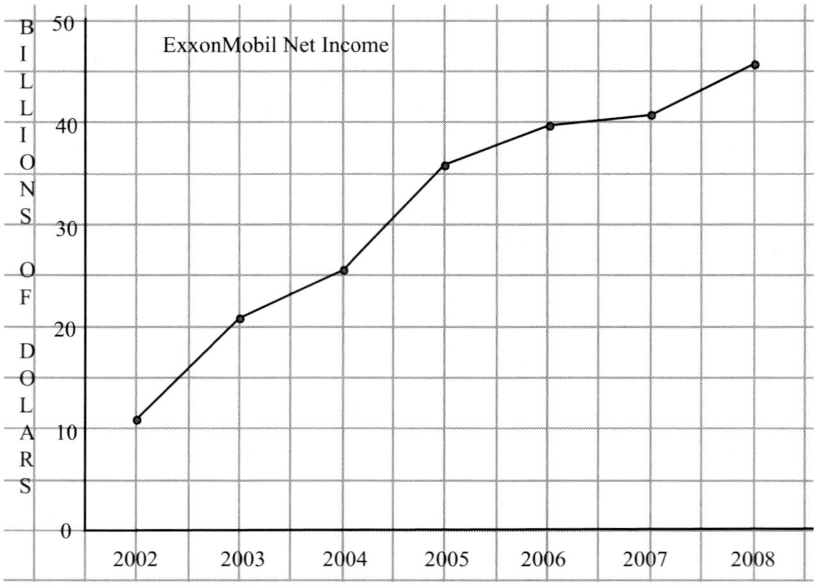

Exercises 16.1

1. 160

3. 720

5. 162

7. $26^2 \times 10^4 = 6{,}760{,}000$

9. $26^2 \times 10^4 + 26^3 \times 10^3 + 10^5 = 24{,}436{,}000$

11. 1,048,576

13. 24

Exercises 16.2

1. a. 2	b. 6.	c. 24
3. a. 2	d. 120	f. 5,040
5. a. 24	b. 12	c. 6

7. $\frac{7!}{4!} = 210$

9. $\dfrac{10!}{2!2!4!2!}=18{,}900$

11. $\dfrac{11!}{5!2!2!}=83{,}160$

13. $\dfrac{15!}{(15-9)!}=\dfrac{15!}{6!}=15\cdot14\cdot13\cdot12\cdot11\cdot10\cdot9\cdot8\cdot7$

15. a. 4!=24 b. 3!=6 c. 4!=24

Exercises 16.3

1. $\binom{7}{5}=\dfrac{7!}{5!(7-5)!}=\dfrac{7!}{5!2!}=21$

3. $\binom{10}{3}=\dfrac{10!}{3!(10-3)!}=\dfrac{10!}{3!7!}=120$

5. $\binom{15}{11}=\dfrac{15!}{11!(15-11)!}=\dfrac{15!}{11!4!}=1{,}365$

7. a. There are $\binom{4}{2}$ ways to choose the jacks and $\binom{4}{3}$ ways to choose the queens, making $\binom{4}{2}\binom{4}{3}$ hands with 2 jacks and 3 queens.

9. a. $\binom{13}{13}=1$ c. $\binom{13}{10}\binom{13}{3}=\dfrac{13\cdot12\cdot11}{1\cdot2\cdot3}\times\dfrac{13\cdot12\cdot11}{1\cdot2\cdot3}=286^2=81{,}796$

11. a. $1\cdot1+1\cdot1=2$ b. $1\cdot1+2\cdot2+1\cdot1=6$

13. a. $\binom{10}{5}$ c. $\binom{5}{1}\binom{5}{4}$ e. $\binom{5}{3}\binom{5}{2}$

Exercises 16.4

1. $S=\{a,b,c,d,\ldots,x,y,z\}$

2. $S=\{1,2,3,4,5,6,8,9,10,12,15,16,18,20,24,30,36\}$

5. e. $\dfrac{52!}{5!47!}$

g. (2nd part) $\binom{48}{2}=\dfrac{48!}{2!(48-2)!}=\dfrac{55\cdot54\cdot53\cdot52\ldots2\cdot1}{5\cdot4\cdot3\cdot2\cdot1\cdot50\cdot49\cdot48\ldots2\cdot1}$

h. (2nd part) $\dfrac{55\cdot54\cdot53\cdot52\cdot51}{5\cdot4\cdot3\cdot2\cdot1}$

7. Let P_1 refer to probability of die 1 and P_2 refer to probability of die 2.

a. $P(2)=P_1(1)P_2(1)=\dfrac{1}{6}\cdot\dfrac{2}{15}=\dfrac{2}{90}$

c. $P(8)=P_1(2)P_2(6)+P_1(3)P_2(5)+P_1(4)P_2(4)+P_1(5)P_2(3)+P_1(6)P_2(2)$

$$=\frac{1}{6}\cdot\frac{1}{3}+\frac{1}{6}\cdot\frac{2}{15}+\frac{1}{6}\cdot\frac{2}{15}+\frac{1}{6}\cdot\frac{2}{15}+\frac{1}{6}\cdot\frac{2}{15}=\frac{1}{18}+\frac{4}{45}=\frac{13}{90}$$

e. $P(3) = P_1(1)P_2(2) + P_1(2)P_2(1)$
$$= \frac{1}{6}\cdot\frac{2}{15} + \frac{1}{6}\cdot\frac{2}{15} = \frac{4}{90}$$
$P(5) = P_1(1)P_2(4) + P_1(2)P_2(3) + P_1(3)P_2(2) + P_1(4)P_2(1)$
$$= \frac{1}{6}\cdot\frac{2}{15} + \frac{1}{6}\cdot\frac{2}{15} + \frac{1}{6}\cdot\frac{2}{15} + \frac{1}{6}\cdot\frac{2}{15} = \frac{8}{90}$$
$P(7) = P_1(1)P_2(6) + P_1(2)P_2(5) + P_1(3)P_2(4) + P_1(4)P_2(3) + P_1(5)P_2(2) + P_1(6)P_2(1)$
$$= \frac{1}{6}\cdot\frac{1}{3} + \frac{1}{6}\cdot\frac{2}{15} + \frac{1}{6}\cdot\frac{2}{15} + \frac{1}{6}\cdot\frac{2}{15} + \frac{1}{6}\cdot\frac{2}{15} + \frac{1}{6}\cdot\frac{2}{15} = \frac{1}{18} + \frac{5}{45} = \frac{15}{90}$$
$P(9) = P_1(3)P_2(6) + P_1(4)P_2(5) + P_1(5)P_2(4) + P_1(6)P_2(3)$
$$= \frac{1}{6}\cdot\frac{1}{3} + \frac{1}{6}\cdot\frac{2}{15} + \frac{1}{6}\cdot\frac{2}{15} + \frac{1}{6}\cdot\frac{2}{15} = \frac{1}{18} + \frac{3}{45} = \frac{11}{90}$$
$$P(11) = P_1(5)P_2(6) + P_1(6)P_2(5) = \frac{1}{6}\cdot\frac{1}{3} + \frac{1}{6}\cdot\frac{2}{15} = \frac{7}{90}$$
$$P(\text{odd}) = P(3) + P(5) + P(7) + P(9) + P(11) = \frac{4}{90} + \frac{8}{90} + \frac{15}{90} + \frac{11}{90} + \frac{7}{90} = \frac{1}{2}$$

9. a. $\frac{12}{51}$ c. $\frac{3}{51}$ e. $\frac{50}{51}$

10. Hint: In how many ways can a pair of places in the line of ten people be chosen? How many of those pairs are adjacent?

12. a. $\dfrac{4 \times \binom{13}{5}}{\binom{52}{5}}$ (You must explain why it is correct).

c. Hints: How many ways are there to choose four of a kind? For each choice of four of a kind, how many ways are there to choose the fifth card of the hand? Use the fundamental counting principle to find out the number of hands that have four of a kind. How many five-card hands are possible with no restrictions?

14. a. $\dfrac{\binom{3}{2}}{\binom{9}{2}}$

Exercises 16.5

1. a. $\frac{1}{8}$ c. $\frac{1}{32}$

2. b. $\binom{5}{1}\frac{1}{32} = \frac{5}{32}$ h. Hint: "At least three heads" means 3, 4, or 5 heads)

4. a. The probability of 12 on one roll of fair dice is $\frac{1}{36}$. The probability of 12s on two consecutive rolls is $\left(\frac{1}{36}\right)^2 = \frac{1}{1{,}296}$.

b. Hint: Consider the probability of the complementary event.

5. a. $\left(\frac{1}{3}\right)^3 = \frac{1}{27}$

6. a. $\frac{1}{6}$ b. $\frac{5}{36}$

7. a. $\frac{1}{3{,}125}$ d. $\binom{5}{2}\left(\frac{1}{5}\right)^2\left(\frac{4}{5}\right)^3 = \frac{640}{3125} = \frac{128}{625}$

9. a. $\frac{1}{4}$

b. Hint: For the game to end on the fourth toss, the third and fourth tosses must be heads and the second must be tails.

10. a. $\frac{9}{16}$ b. $\frac{9}{64}$

11.c

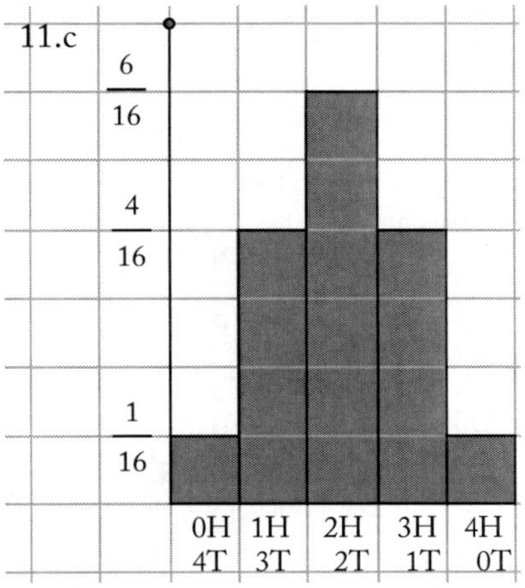

Exercises 16.6

2. a. 8/18 c. 5/11

Self-Quiz for chapter 16

1. $5 \cdot 9 \cdot 4 \cdot 6 = 1{,}080$

2. 12!

3. $20 \cdot 19 \cdot 18 \cdot 17 \cdot 16 \ldots 12 \cdot 11$

4. $\frac{10!}{2!3!3!}$

5. $\frac{12!}{5!7!}$

6. $\binom{4}{2}\binom{24}{3} = 24 \cdot 23 \cdot 22 = 12{,}144$

7. 120,120

8. a. {red, purple, yellow, green}

b. $P(\text{red}) = \frac{3}{19}$, $P(\text{purple}) = \frac{7}{19}$, $P(\text{yellow}) = \frac{5}{19}$, $P(\text{green}) = \frac{4}{19}$

c. $P(\text{not red}) = 1 - P(\text{red}) = 1 - \frac{3}{19} = \frac{16}{19}$

9. a

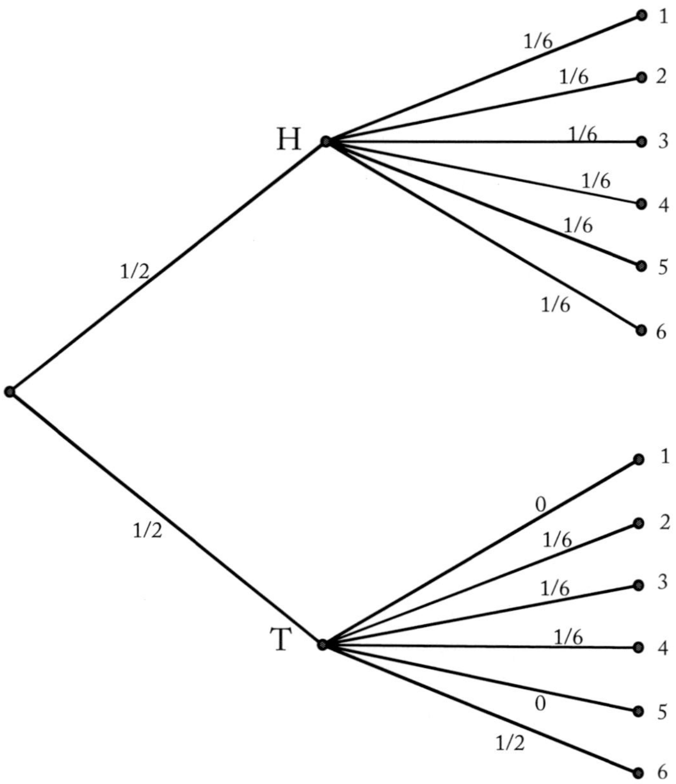

b. $5\times\frac{1}{2}\times\frac{1}{6}+\frac{1}{2}\times\frac{1}{2}=\frac{1}{2}$

10. $\binom{12}{3}\left(\frac{1}{2}\right)^{12}=\frac{55}{210}$

11. a. $\binom{6}{2}\left(\frac{1}{6}\right)^{2}\left(\frac{5}{6}\right)^{4}$

b. $1-\left[P(0\text{ doubles})+P(1\text{ double})\right]=1-\binom{6}{0}\left(\frac{1}{6}\right)^{0}\left(\frac{5}{6}\right)^{6}-\binom{6}{1}\left(\frac{1}{6}\right)^{1}\left(\frac{5}{6}\right)^{5}$

$=1-\frac{15{,}625}{46{,}656}-\frac{6\cdot 3125}{46{,}656}=1-\frac{34{,}375}{46{,}656}=\frac{12{,}281}{46{,}656}$. This is about 0.26.

12. a. $\frac{1}{12}$ b. $\frac{1}{11}$ c. $\frac{1}{6}$

Exercises 17.1

1. Hint: You can use the adder to subtract by using the missing addend notion of addition, lining up the sum (in the middle column) with one of the addends in an outer column.

In a similar way, line up the product in the center column with the divisor in an outer column.

Exercises 17.2

1. a. 19.19 c. 25.55

2. a. 19.2104 c. 25.5816

3. a. 0.0204 c. 0.0316

4. a. The error, as a percentage of the correct answer,

is $\frac{0.0204}{19.2104} = 0.0010619\ldots$ or about 0.1% of the true answer.

6. a. 8.1403

c. (for part a only) 8.14010964

d. (for part a only) The error is $8.1403 - 8.14010964 = 0.00019036$, or less than 0.0025% of the product.

7. a. To divide one approximate value of 1.1^n by another, subtract their exponents. For example, 17.4494 corresponds to the exponent 30, while 3.4523 corresponds to the exponent 13. Then $17.4494 \div 3.4523$ corresponds to the exponent $30 - 13 = 17$, so it is approximately 5.0545.

Exercises 17.3

1. Hint: For what value of b is the fraction $\frac{a}{b}$ not defined?

2. a. $\frac{1}{9}$ c. $\frac{1}{125}$ e. $\frac{1}{7}$ g. $\frac{9}{4}$ i. $\frac{25}{9}$ k. $\frac{100}{3}$

4. a. 10^0 c. 2^3 e. 5^{-3} g. 10^{-13}

5. a. multiplication b. addition c. add d. subtract

6. a. 10^{-3} c. 10^{-2} e. 10^{-9}

7. a. 6,561 b. $\frac{1}{4,096}$

8. a. $2.1436^3 = (1.1^8)^3 = 1.1^{24} = 9.8497$. This is exact.

d. Answer for part a only.

e. The error in using the approximation is less than 0.002% of the correct answer.

11. Answer for the 1980s: 6.0221367×10^{23}

13. a. 8.95×10^4

14. 8×10^{10}

15. 6.150×10^{24}

Exercises 17.4

2. a. 2 e. −3

3. a. $\frac{1}{2}$ and $\frac{1}{2}$

4. b. $\frac{6}{7}$ and $\frac{6}{7}$

7. a. 27 d. 125

9. a. 32

12. a. $\frac{1}{3}$ d. $\frac{4}{7}$

14. a. 7 d. 5

15. a. 9.8497 d. 3.4523

Exercises 17.5

1. a. 0.9 kg after 1 year, $(0.9)^2 = 0.81$ kg after 2 years, $(0.9)^3 = 0.729$ kg after 3 years, $(0.9)^4 = 0.6561$ kg after 4 years, $(0.9)^5 = 0.59049$ after 5 years

b. Amount, in kg, is $(0.9)^t$

c.

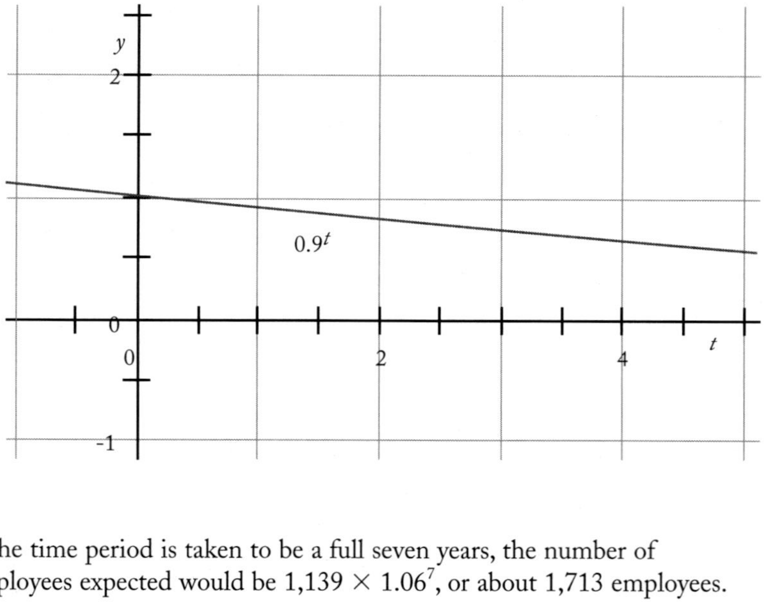

2. If the time period is taken to be a full seven years, the number of employees expected would be $1{,}139 \times 1.06^7$, or about 1,713 employees.

3. a. 5,568 years ago b. 11,136 years ago c. 16,704 years old.

4. b. $100\sqrt{10}$ or about 316

5. a. $\sqrt[12]{2}$ b. $440(\sqrt[12]{2})^3 = 440\sqrt[4]{2}$ cps, or about 523 cps.

c. 6 semi-tones

6. a. $\log_5 25 = 2$ c. $\log_{10} 0.00001 = -5$

7. a. $2^9 = 512$ c. $10^3 = 1{,}000$

8. a. $\log_{10} 4 = 2\log_{10} 2 = 0.6020$ b. $\log_{10} 8 = 3\log_{10} 2 = 0.9030$

e. Partial answer: $\log_{10} 12 = 2\log_{10} 2 + \log_{10} 3 = 1.0791$

Self-Quiz for Chapter 17

1. a. 40 b. 80 c. 4

2. The interval from 7.40 to 8.14 is 0.74 units long, and 8.00 lies 60/74 of the way along, or about 80% of the way. Therefore, $1.1^{21.8}$ is about 8.00.

3. $4.96585346 \times 10^{20}$.

4. a. 16 b. 512 c. 1/4

5. $\$18{,}000 \times (0.8)^6 = \$4{,}719$

6. $\log_{16} 128 = 1.75$

7. $8^{1/6} = \sqrt{2}$

Exercises 18.3

1. a. 4 c. 0 e. $2\frac{1}{2}$ g. $\frac{1}{6}$ i. $\frac{5}{6}$

2. a. 36 c. 270 e. 216 g. 540

4. a. 30 c. 45 e. 810

5. 15

7. Hint: Estimating in advance can prevent gross errors. You can see at a glance that the angle in b is a bit more than 180°, so that a measurement in the neighborhood of 190° or 200° would be reasonable. Similarly, in part c you would expect an angle a bit more than 270°.

9. a. $67\frac{1}{2}°$ c. $157\frac{1}{2}°$ e. $292\frac{1}{2}°$

10. a. 35° c. 110° e. 5° f. 335°

12. 338°

13. Hint: A turn of 5 × 75° or 375° to the left is 15° more than a complete turn.

14. a. 100° c. Approximately 295°

16. a. Victoria International

b. Approximately 15°

c. 14 miles

d. About 6 minutes

19. Hint: The three interior angles of the triangle add up to 180°, and so do the internal and exterior angles at *C*.

20. Approximately 255,000 stadia

Exercises 18.6

1. a. 22.5 m

2. The triangles formed by the ships themselves and the point of intersection of their courses are similar. If both ships maintain their speed and course, they will collide.

3. b. Similar on a scale of 1/2.

5. Hint: Use inscribed angles to show that the triangles are similar and which parts of the two triangles correspond to each other in the similarity relationship. Then find the scale of similarity, and use it to find length *AE*.

8. *BE* is constructed so that $\angle ABE = \angle DBC$. Then $\angle ABE + \angle EBD = \angle CBD + \angle DBE$, and $\angle ABD = \angle CBE$. Also $\angle BDA = \angle BCA$ (they both intercept arc *AB*). But $\angle BCA = \angle BCE$. Since two angles of triangle *BCE* equal two angles of triangle *BDA*, the triangles are similar.

9. The ratios of corresponding sides of similar triangles are in proportion. Thus $AD/BD = EC/BC$. Multiply by $BC \times BD$ to get the desired identity.

10. As in problem 9

11. As in problem 10

12. Add the equations obtained in problems 9 and 11.

Exercises 18.7

2. a. Sides *AP* and *BP* are radii of the same circle and so have equal lengths. Similarly, $AQ=BQ$. Side *PQ* is common to both triangles, so they are congruent by SSS.

b. Triangles *APQ* and *BPQ* are both isosceles, so $\angle APQ = \angle AQP$ and $\angle BPQ = \angle BQP$. Since the two isosceles triangles are congruent, all four base angles are equal.

c. $\angle APQ = \angle BPQ$ as established above. Since $\angle APQ = \angle APM$ and $\angle BPQ = \angle BPM$, $\angle APM = \angle BPM$. Since side $AP = BP$ and side *PM* is common to both triangles, that triangles *APM* and *BPM* are congruent by the criterion *SAS*.

d. $\angle PMA = \angle PMB$, because those angles are corresponding parts of congruent triangles. Since $\angle PMA = \angle PMB$, and since also $\angle PMA + \angle PMB = \angle AMB = 180°$, $\angle PMA = \angle PMB = 90°$, which is to say *PQ* is perpendicular to *AB*.

3. a. Hint: You can use congruence criterion *SSS.*

b. Hint: Use the Pythagorean theorem.

4. a. Hint: Make a right angle and fold it in half.

7. a. Outline: Show that triangles *ADP, AEQ,* and *AFB* are all similar. Then use the scales.

8. a. Construct the perpendicular bisector of each side of the regular hexagon. Where the bisectors meet the circle which passes through the vertices of the hexagon are the six new vertices, which, together with the vertices of the hexagon, are the vertices of a regular dodecagon. You can simplify this construction by determining one vertex as described above and then using it as one vertex in a new regular hexagon. The vertices of the two hexagons are those of the dodecahedron.

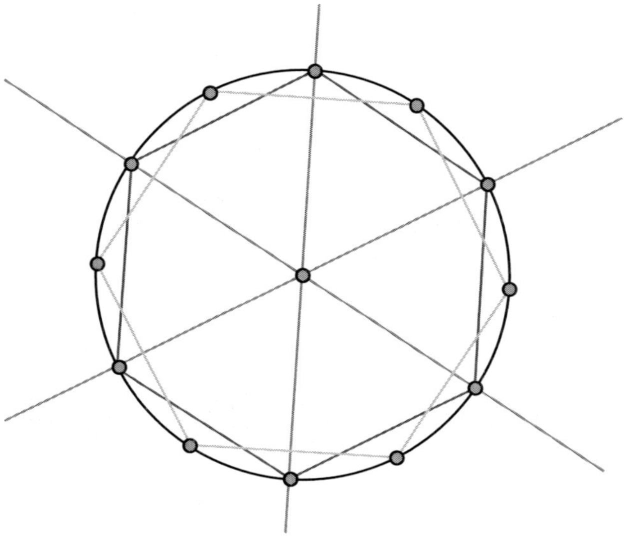

Self-Quiz for Chapter 18

1. (no solution given here)

2. Joey is 74(9/20) inches tall, or 74×0.45 inches. (About 33.3 inches)

4. Hint: Show first that $\measuredangle EDC = \measuredangle ECD$

5. $\overset{\frown}{AB}=48°$, $\overset{\frown}{BC}=96°$, $\overset{\frown}{CD}=72°$, $\overset{\frown}{DA}=144°$.

6. 1. Draw a square (construction 5) and label the vertices *A, B, C,* and *D.*

2. Draw the diagonals of square *ABCD* and label their point of inter section *O.*

3. Draw circle, center *O,* radius *OA.*

4. Bisect angle *AOB* and extend the angle bisector to meet the circle at a point *P.*

5. Extend line segment *PO* to meet the circle at a new point *R.*

6. Construct perpendicular to *PR* at *O* and extend this perpendicular to meet the circle at points *S* and *T.*

Then *A,B, C, D, P, R, S,* and *T* are vertices of a regular octagon.

Exercises 19.1

1. b. 540°

3. b. 900°

4. (Partial answer)

number of sides	3	4	5	6	7	n
number of triangles	1	2	3	4	5	$n - 2$
sum of angles in triangles = sum of angles in polygon	180°	360°		720°		

5. a. As seen in question 4, the angles of a pentagon add up to 540°. If all five angles are the same size, each is a fifth of 540° or 108°.

6. a. Regular hexagons often appear in tile patterns on floors and sidewalks, old fashioned chicken wire, honeycombs, and nuts and bolts.

b. Hint: What if someone could not read the writing?

7. c. Bolts with odd number of sides cannot be turned with most adjustable wrenches. Fire hydrants in some localities have pentagonal bolts to make it hard for unauthorized users to turn on the water.

12.

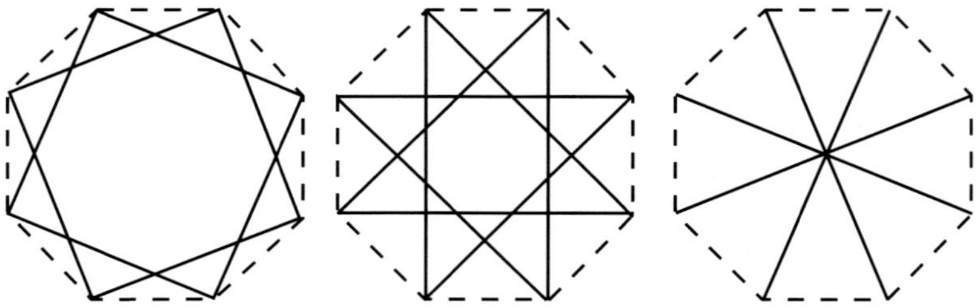

8/2 8/3 8/4

13. An 8/5 is an 8/3 traced in reverse order (try it and see!).

17. Hint: Distinguish between polygons with even numbers of sides and polygons with odd numbers of sides.

18. Hint: For the star polygon n/k (n-sided polygon, connecting every k^{th} vertex), how is the number of pieces related to factors of n and k?

19. c. 3/4

Exercises 19.2

2. a. (partial solution)

Vertex symbol	Sum of interior angles
(3,6,3,6)	$60° + 120° + 60° + 120° = 360°$
(4,8,8)	$90° + 135° + 135° = 360°$
(3,12,12)	$60° + 150° + 150° = 360°$

4. b. Hint: Two pentagons and a dodecagon fit together perfectly for a single vertex. Indeed, you can extend the pattern to surround the dodecagon with a pentagon adjoining each edge. If you try to extend the pattern to a tessellation of the entire plane, however, you will run into trouble. Describe this problem.

5. a. 360°

b. Hint: Try sketching equilateral triangles and squares around three consecutive vertices of the dodecagon pictured below in such a way that every vertex symbol is (3,3,4,12).

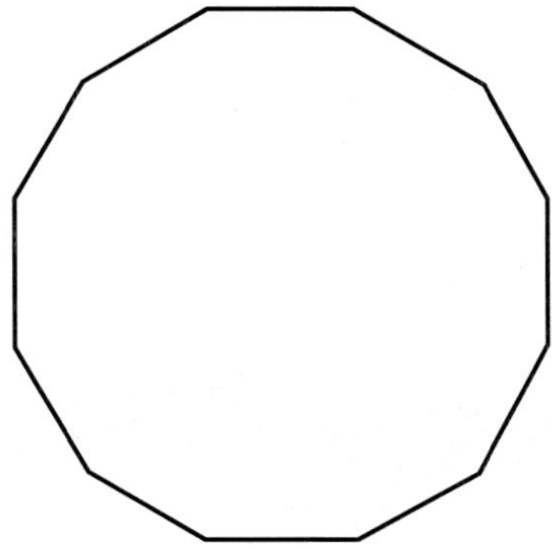

6. Use the same dodecagon and the same approach as in question 5b.

Exercises 19.3

1. Hint: It is a regular polyhedron.

2. Hint: It is a regular polyhedron.

4. a. Regular tetrahedron c. Cube e. Regular dodecahedron

5.

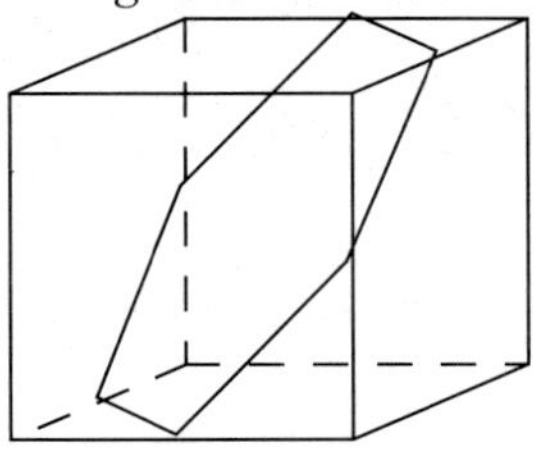

6. Look at the picture for question 12 of exercises 19.3.

8. c. The area of base ABC is $\frac{a^2}{2}$. Height BF is a. Then

$$V = \frac{1}{3}Bh = \frac{1}{3}\left(\frac{a^2}{2}\right)a = \frac{a^3}{6}$$

d. The volume of the cube is a^3. Subtract the volumes of the four tetrahedra at the corners to get $\frac{1}{3}a^3$ as the volume of tetrahedron *ACFH.*

9. By the Pythagorean theorem, $e = a\sqrt{2}$. Then $a = \frac{e}{\sqrt{2}}$.

10. $V = \frac{e^3}{6\sqrt{2}} = \frac{e^3\sqrt{2}}{12}$

12. a. Hint: The ratio of the volumes of two similar figures is the cube of the scale.

b. The inscribed octahedron has half the volume of the big tetrahedron and twice that of each of the small tetrahedra.

Self-Quiz for Chapter 19

1. 720°

2. 150°

3. (not shown)

4.

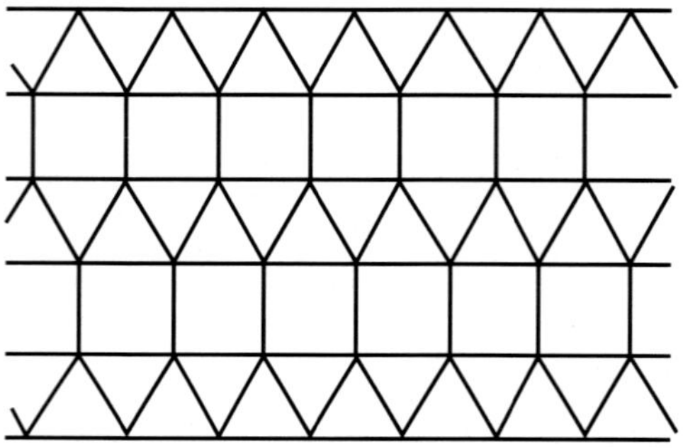

5. a. (4,4,8) b. (3,3,3,7)

6. $250\sqrt{6}$

Selected References
with Annotations

Adams, James L. *Conceptual Blockbusting. A Guide to Better Ideas.* W. H. Freeman and Co., San Francisco, 1974. A readable and enjoyable book that really can lead to better ideas.

Apostol, Tom M., producer. *Project MATHEMATICS!* Nine classic videos on key mathematical topics, e.g. The Pythagorean theorem, similarity, and polynomials with award winning animation by James Blinn. Available now on DVDs in Spanish or English. Check the web for the best way to get these.

Association of Teachers of Mathematics. *Some Lessons in Mathematics.* The University Press, England, 1964.

Beiler, Albert H. *Recreations in the Theory of Numbers: The Queen of Mathematics Entertains, 2nd ed.,* Dover, New York, 1966. A wonderful if idiosyncratic tour of the fascinating world of number theory.

Bourgoin, J. *Islamic Patterns.* Dover Publications, New York, 1977. A coloring book of tile patterns taken from a late-19th century work of a noted French art historian.

Burkhardt, Hugh. *The Real World and Mathematics.* Blackie and Son, Ltd, Glasgow, 1981. A readable introduction to the practical side of mathematics.

Chabert, Jean Luc, editor. *A History of Algorithms, from the Pebble to the Microchip* English translation by Chris Weeks, Springer Verlag, Berlin, 1999.

Connolly, Paul and Vilardi, Teresa, editors. *Writing to Learn Mathematics and Science.* Teachers College Press, New York, 1989. A collection of 23 essays on the connections between language, writing, and mathematics.

Conway, John H and Richard K. Guy. *The Book of Numbers.* Springer Verlag, New York, 1996. Profuse illustrations, an informal writing style, and more than you ever dreamed about figurate numbers.

Cromwell, Peter. *The Search for Quasi-Periodicity in Islamic 5-fold Ornament. The Mathematical Intelligencer,* Volume 31 #1 (Winter, 2009). The gorgeous pictures of Islamic tilings alone make this article worth looking at.

Cundy, H. Martyn, and A.P. Rollett. *Mathematical Models.* Oxford, New York, 1961. A classic by two British high school teachers on mathematical models and how to make them.

De Villiers, Michael D. *Rethinking Proof with the Geometer's Sketchpad.* Key Curriculum Press, Emeryville, CA, 1999. Contains an excellent discussion of the nature and function of proof and its proper role in the school curriculum.

Dunham, William. *Journey through Genius.* John Wiley and Sons, 1990. Reprinted in paperback by Penguin Books, NY, 1991. Lively and readable accounts of selected historical topics in mathematics.

Edwards, A.W.F. *Pascal's Arithmetical Triangle, the Story of a Mathematical Idea.* Johns Hopkins Paperbacks edition, Baltimore, 2002. A clear account of how this triangle, or in some cases only parts of it, were discovered in various contexts and cultures, and how they were assimilated and extended by Pascal, Wallis, and others.

Gardner, Martin. *Gardner's Mathematical Games.* Mathematical Association of America, Washington, D.C. 2005. This CD has all of Gardner's "Mathematical Games" columns from the *Scientific American* from 1965 to 1986. Gardner's non-technical style brought a whole generation of fans to mathematics and introduced several key mathematical ideas to the general public. They are as good today as ever. Collections of selected Gardner columns have long been in print, but this is the first time the entire set has been made available in a single, convenient, searchable format.

Gerdes, Paulus. *Geometry from Africa. Mathematical and Educational Explorations.* The Mathematical Association of America, Washington, D.C. 1999. Perhaps the first book to appear in North America that really does justice to the topic, Gerdes' book is fun to browse in and draws the reader in.

Gray, Shirley. *On the Equals Sign Our "Twins": A Tour through Original Sources. The Mathematical Intelligencer,* volume 30, number 2. Spring, 2008.

Haldane, J. B. S., "*On Being the Right Size,*" reproduced in Nagel and Newman, *World of Mathematics.* Simon and Schuster, New York, 1956. You can read this essay in five minutes, but it will set you thinking for much longer than that.

Holt, Jim. *"Numbers Guy. Are Our Brains Wired for Math?" The New Yorker*, March 3, 2008 pages 42-47. Stanislas Dehaene's research into how the brain treats numbers.

Jacobs, Harold. *Mathematics, A Human Endeavor, A Book for Those Who Think They Don't Like the Subject. 3rd edition.* W. H. Freeman and Co. San Francisco, 1994. Jacobs based this high school textbook on Martin Gardner's publications, and it soon became a favorite for undergraduate liberal arts math courses. It remains a delightful classic.

Jenks, Stanley M. and Donald M. Peck. *Building Mental Imagery In Mathematics.* Holt, New York, 1968. An informal treatment, especially good on fractions, areas, and functions or rules.

Katz, Victor, and Michalowicz, Karen Dee, Editors. *Historical Modules for the Teaching and Learning of Mathematics.* Mathematical Association of America, Washington, D.C., 2005. This CD is a gold mine of ideas and teaching activities, designed for classroom use. Portions may be copied for classroom use.

Lenchner, George. *Creative Problem Solving in School Mathematics.* Houghton Mifflin, Boston, 1983. Problem solving for elementary and middle schools, with many examples from the Mathematical Olympiads for Elementary Schools. A masterpiece.

Lenchner, George. *Math Olympiad Contest Problems for Elementary and Middle Schools.* Glenwood Publications, East Meadow, NY, 1997. A collection of all the problems from the Mathematical Olympiads for Elementary and Middle Schools for the period 1979-1997, including solutions.

Kim, Scott. *Ten Ways to Use Puzzles in Mathematics Education.* http://scottkim.com/education.This brief, down-to-Earth article by Kim, known as a puzzle maker, dancer, artist, and general master of symmetry, is practical and offers links to wonderful classroom resources.

Kraitchik, Maurice. *Mathematical Recreations.* W. W. Norton, New York, 1942. Reprinted by Dover Publications, New York, 1953.

Ma, Liping. *Knowing and Teaching Elementary Mathematics.* Lawrence Erlbaum and Associates, Inc. Mahwah, N.J., 1999. Casts issues of mathematics education in a clear light and offers new insights on them.

Marcy, Steve and Janis. *Punchine Problem Solving, Punchline Bridge to Algebra, Punchline Algebra, Mathimagination.* Collections of practice problems keyed to coded messages that make the work enjoyable and build in a self-correcting mechanism. Marcy Mathworks, Los Angeles, CA.

Miss Mason's School. *"Math for Five Year Olds"* (videotape) New Jersey, 196_? Shows ordinary five-year olds happily learning arithmetic with positive and negative numbers, solving equations, and generally doing mathematics more sophisticated than they are usually thought capable of learning. Unfortunately, this videotape is out of print. It may still be possible to get a copy from Sharon Tomalin at the Newgrange School, 52 Lafayette Ave., Trenton, NJ. (609 394 2255). Cost unknown.

Morrison, Philip, and Morrison, Phylis. *Powers of Ten: A Book About the Relative Size of Things in the Universe and the Effect of Adding Another Zero.* W. H. Freeman & Co., San Francisco, CA, 1985. Dramatic pictures of the effects of repeatedly zooming in or out. Also available as a CD.

Polya, George. *Mathematical Discovery. On Understanding, Learning, and Teaching Problem Solving.* John Wiley & Sons, New York, 1965. The bible of mathematics education for high school teachers, but to really benefit, you need to work some of the problems, and they are not easy.

Preston, Richard. "*The Mountains of Pi,*" *The New Yorker,* March 2, 1992. New York. How the Chudnovsky brothers, with mail order parts, turned their Manhattan apartment into a supercomputer to find more and more digits of π.

Project *MATHEMATICS! "The Theorem of Pythagoras."* (25-minute videotape and workbook) California Institute of Technology, 1988. Beautiful narration, award winning graphics, and careful organization make this outstanding video worth watching over and over. Widely available from school districts and county offices of education, may be copied for noncommercial purposes.

Sawyer, W. W. *Vision in Elementary Mathematics*. Penguin Books, Baltimore, 1964. Sawyer's informal, chatty style makes mathematical ideas simple and understandable.

Schattschneider, Doris. *Visions of Symmetry. Notebooks, Periodic Drawings, and Related Work of M. C. Escher.* W. H. Freeman and Co, New York, 1990. A beautiful "coffee table" book of Escher's art, with explanations of his mathematical investigations.

Senechal, Marjorie, and Fleck, George, editors. *Shaping Space: A Polyhedral Approach.* Birkhäuser, Boston, 1988. Delightful and informative introduction to polyhedra, their construction, classification, and occurence in nature and art.

Steinhaus, Hugo. *Mathematical Snapshots, 3rd edition.* Oxford University Press, New York, 1969. Photographs and drawings, with brief text, illustrating mathematical ideas. This book has been reissued in paperback by Dover.

Takenouchi, Osamu et al, translators. *Jinkoki.* Wasan Institute, Tokyo, 2000. A translation of a collection of practical Japanese mathematics works from the early stages of the Edo Period (1603-1867).

Van Delft, Pieter and Jack Botermans. *Creative Puzzles of the World.* Harry N. Abrams, Inc. New York, 1978. Gorgeous full-color illustrations and explanations of some of the world's best puzzles (but not Rubik's cube).

Wenninger, Magnus. *Polyhedron Models.* Cambridge U. Press, New York, 1971. Photographs and descriptions of gorgeous polyhedra, some extremely difficult to make.

Wentworth, Daniel F., J. K. Couchman, John MacBean, and Adam Stecher. *Mapping Small Places.* Winston, Minneapolis, 1972. Hands-on activities for students in the upper elementary or junior high grades.

Williams, Bill. *A Sampler on Sampling.* John Wiley & Sons, New York, 1978.

Wilson, Robin J. *Lewis Carroll in Numberland: his fantastical mathematical logical life. An agony in eight fits.* Alan Lane (part of Penguin Books), 2008. A readable and authoritative look at the life and work of Charles L. Dodgeson (Lewis Carroll).

Wilson, Robin J. *Stamping through Mathematics* Springer Verlag, New York, 2001. A beautifully illustrated book about mathematical postage stamps.

Wirtz, Robert, Morton Botel, Max Beberman, and W.W. Sawyer. *Math Workshop for Children, revised ed.* Encyclopedia Britannica Press, Chicago, 1967. Though never widely adopted, these K-6 texts pioneered many of the approaches used in this book, including box puzzles, and bags and stones. This outstanding series may be reissued in a revised form sometime around 2006.

Index

NOTE: Page references in **bold** refer to definitions.